Characterization of Semiconductor Heterostructures and Nanostructures

Characterization of Semiconductor Heterostructures and Nanostructures

Edited by

Carlo Lamberti

Department of Inorganic, Physical and Materials Chemistry and
NIS Center of Excellence, University of Torino, and INSTM
Centro Di Riferimento, Torino, Italy

ELSEVIER

Amsterdam • Boston • Heidelberg • London • New York • Oxford • Paris
San Diego • San Francisco • Singapore • Sydney • Tokyo

Elsevier
Radarweg 29, PO Box 211, 1000 AE Amsterdam, The Netherlands
The Boulevard, Langford Lane, Kidlington, Oxford OX5 1GB, UK

First edition 2008

Copyright © 2008 Elsevier B.V. All rights reserved

No part of this publication may be reproduced, stored in a retrieval system
or transmitted in any form or by any means electronic, mechanical, photocopying,
recording or otherwise without the prior written permission of the publisher

Permissions may be sought directly from Elsevier's Science & Technology Rights
Department in Oxford, UK: phone (+44) (0) 1865 843830; fax (+44) (0) 1865 853333;
email: permissions@elsevier.com. Alternatively you can submit your request online by
visiting the Elsevier web site at http://elsevier.com/locate/permissions, and selecting
Obtaining permission to use Elsevier material

Notice
No responsibility is assumed by the publisher for any injury and/or damage to persons
or property as a matter of products liability, negligence or otherwise, or from any use
or operation of any methods, products, instructions or ideas contained in the material
herein. Because of rapid advances in the medical sciences, in particular, independent
verification of diagnoses and drug dosages should be made

British Library Cataloguing in Publication Data
A catalogue record for this book is available from the British Library

Library of Congress Cataloging-in-Publication Data
A catalog record for this book is available from the Library of Congress

ISBN: 978-0-444-53099-8

For information on all Elsevier publications
visit our website at books.elsevier.com

Printed and bound in Hungary

08 09 10 11 12   10 9 8 7 6 5 4 3 2 1

Working together to grow
libraries in developing countries

www.elsevier.com | www.bookaid.org | www.sabre.org

ELSEVIER   BOOK AID International   Sabre Foundation

To the older trees, because that's from them that we came from but most importantly, to the young trees, because to them belongs the future.

In my naïve mind I like to think of knowledge dissemination as an old tree, with its roots strongly and deeply set in the ground, and with its seeds gently taken by a lovely wind that spreads them all around. Let the ground stick close to the old tree's roots and let the wind caress its branches, as long as possible ... till new young trees grow up from the ground and become stronger and stronger and can finally replace the old one.

Table of Contents

Preface .... *ix*

Chapter 1: Introduction: the interdisciplinary nature of nanotechnology
and its need to exploit frontier characterization techniques
Carlo Lamberti 1

Chapter 2: Ab initio studies of structural and electronic properties
Maria Peressi, Alfonso Baldereschi, and Stefano Baroni 17

Chapter 3: Electrical characterization of nanostructures
Anna Cavallini and Laura Polenta 55

Chapter 4: Strain and composition determination in semiconducting
heterostructures by high-resolution X-ray diffraction
Claudio Ferrari and Claudio Bocchi 93

Chapter 5: Transmission electron microscopy techniques for imaging
and compositional evaluation in semiconductor heterostructures
Laura Lazzarini, Lucia Nasi, and Vincenzo Grillo 133

Chapter 6: Accessing structural and electronic properties of semiconductor
nanostructures via photoluminescence
Stefano Sanguinetti, Mario Guzzi, and Massimo Gurioli 175

Chapter 7: Power-dependent cathodoluminescence in III-nitrides
heterostructures: from internal field screening to controlled band-gap modulation
*Giancarlo Salviati, Francesca Rossi, Nicola Armani, Vincenzo Grillo
and Laura Lazzarini* 209

Chapter 8: Raman spectroscopy
Daniel Wolverson 249

Chapter 9: X-ray absorption fine structure in the study of semiconductor
heterostructures and nanostructures
Federico Boscherini 289

Chapter 10: Nanostructures in the light of synchrotron radiation:
surface-sensitive X-ray techniques and anomalous scattering
*Till Metzger, Vincent Favre-Nicolin, Gilles Renaud, Hubert Renevier,
and Tobias Schülli* 331

Chapter 11: Grazing incidence diffraction anomalous fine structure
in the study of structural properties of nanostructures
Maria Grazia Proietti, Johann Coraux, and Hubert Renevier .371

Chapter 12: The role of photoemission spectroscopies in heterojunction research
Giorgio Margaritondo 407

Chapter 13: ESR of interfaces and nanolayers in semiconductor heterostructures
Andre Stesmans and Valery V. Afanas'ev 435

Subject Index *483*

Preface

After the publication of the review "C. Lamberti, The use of synchrotron radiation techniques in the characterization of strained semiconductor heterostructures and thin films, Surf. Sci. Rep., **53** (2004) 1–197", Elsevier contacted me for the coordination of a new book collecting the most used characterization techniques for the investigation of semiconductor heterostructures and nanostructures. After an initial inertia from my part, I accepted the task. The immediate enthusiastic replay of most of the contacted chapter coordinators has made my work easier than expected. I would like to express my gratitude to all these colleagues, as well as to all colleagues who have kindly acted as competent reviewers, as their hard work has importantly helped the chapter coordinators and myself to improve the quality of the final product. At the end of the task, looking to the final result, I am proud to count within the chapter coordinators dear friends and eminent scientists that are in the top positions in the ranking lists of the corresponding disciplines. I am even more proud to underline that, among such list of eminent scientists, women play an important role, leading four chapters.

"Characterization of Semiconductor Heterostructures and Nanostructures" is structured in chapters, each one devoted to a specific characterization technique used in the understanding of the properties (structural, physical, chemical, electrical, etc.) of semiconductor quantum wells and superlattices. A chapter is devoted to the ab initio modeling. The book has basically a double aim. The first one lies on the educational ground. The book provides the basic concept of each of the selected techniques with an approach understandable by master and Ph.D. students in Physics, Chemistry, Material Science, Engineering, and Nanotechnology. The second aim is to provide a selected set of examples from the recent literature of the TOP results obtained with the specific technique in understanding the properties of semiconductor heterostructures and nanostructures. Each chapter has consequently this double structure: a first part devoted to explain the basic concepts, serving the largest possible audience, and a second one to the discussion of the most peculiar and innovative examples, allowing the book to have the longest possible shelf life. So students should not get frustrated if they find more difficulties in the understanding of the second part of the chapters. My advice is to focus on the first parts; they can always came back to the second parts in the ensuing years, when their experience will be improved. Of course, the book is devoted also to the specialized community of scientists working in the fields of design, growth, characterization, and testing of heterostructures-based devices in both academic and industrial laboratories. Such readers should skip the first parts of the chapters, focusing on the final ones.

On top of this, the book has a further and somewhat even more ambitious goal. In this regard, the topic of quantum wells, wires, and dots should be seen as a pretext of applying top level characterization techniques in understanding the structural, electronic, etc. properties of matter at the nanometer (even sub-nanometer) scale. In this way, it is aimed to become a reference book in the much broader, and extremely hot, field of Nanotechnology.

1

Introduction: the interdisciplinary nature of nanotechnology and its need to exploit frontier characterization techniques

Carlo Lamberti

Department of Inorganic, Physical, and Materials Chemistry and NIS Centre of Excellence, Università di Torino, Via P. Giuria 7, 10125 Torino, Italy and INSTM Centro di Riferimento, Torino, Italy

> **Abstract** High-performance electronic and optoelectronic devices based on semiconductor heterostructures are required to obtain increasingly strict and well-defined performances, needing a detailed control, at the atomic level, of the structural composition of the buried interfaces. This goal has been achieved by an improvement in the epitaxial growth techniques and by the parallel use of increasingly sophisticated characterization techniques and of increasingly accurate ab initio calculations. This chapter, introducing the book *Characterization of Semi-conductor Heterostructures and Nanostructures*, is divided into five sections. In Section 1, the impact of nanoscience and nanotechnology in our society is described, using the point of view of the articles, the citations and the journals devoted to the field. The multidisciplinary nature of nanotechnology is reported in Section 2, while the dynamic interplay among growth/synthesis techniques, theoretical modeling, and characterization techniques in the design and improvement of semiconductor heterostructure-based devices is discussed in Section 3. Section 4 reports the purposes of the book and the layout of the chapters. Finally, in Section 5, the strength of combined experimental and theoretical investigation of a selected nanomaterial is underlined by an example.
>
> **Keywords** nanotechnology, nanoscience, heterostructures, nanostructures, characterization techniques, ab initio calculations

1. The scientific and editorial blow up of nanotechnology in the new millennium

The term nanotechnology refers to a branch of applied science and technology whose unifying theme is the control of matter at the atomic and molecular scale, normally 1–100 nm (1 nm = 10^{-9} m), and the fabrication of devices within that size range. The appeal in such approach lies in the fact that the structural, physical, chemical, electronic, optical, etc. properties of nanometer-dimensioned materials differ markedly from those of the corresponding bulk (unconfined) materials. Nanotechnology, among the most advanced frontiers of Science, is certainly showing the higher degree of multidisciplinarity, generated from the well-accorded interplay among different fields such as materials science, applied physics, interface and colloid science, device physics, supramolecular chemistry, surface science, and engineering. Nanotechnology results from a combined extension of such sciences into the nanoscale.

Scientists, politicians, media, and industries have much expectation concerning what new science, technology, and application may result from these lines of research. Such expectations have strongly stimulated the effort made in the previous years by university and industrial laboratories in the field of nanoscience and nanotechnology. A simple way to evaluate such effort is to look to the number of papers published per year that are found using *nanochemistry* OR *nanophysics* OR *nanotechnology* OR *nanoscience* as search keys (Fig. 1(a)), and to the number of citations that such papers have received (Fig. 1(b)). It is evident that such simple and superficial statistical study is far to be comprehensive, as most of the papers that actually report results in this field do not necessarily used one of those four keywords. It is evident from the last row of Table 1 that the 32 journals belonging to the *Subject Category*[1] named "Nanoscience & Nanotechnology" have published 8939 papers in 2006, while only 1125 have been found using those four keywords (Fig. 1(a)). So, the data reported in Fig. 1 are not important as absolute values, which are underestimated by a factor of about 10, but for the trend they are showing. It is evident that this trend shows an impressive acceleration starting from the new millennium.

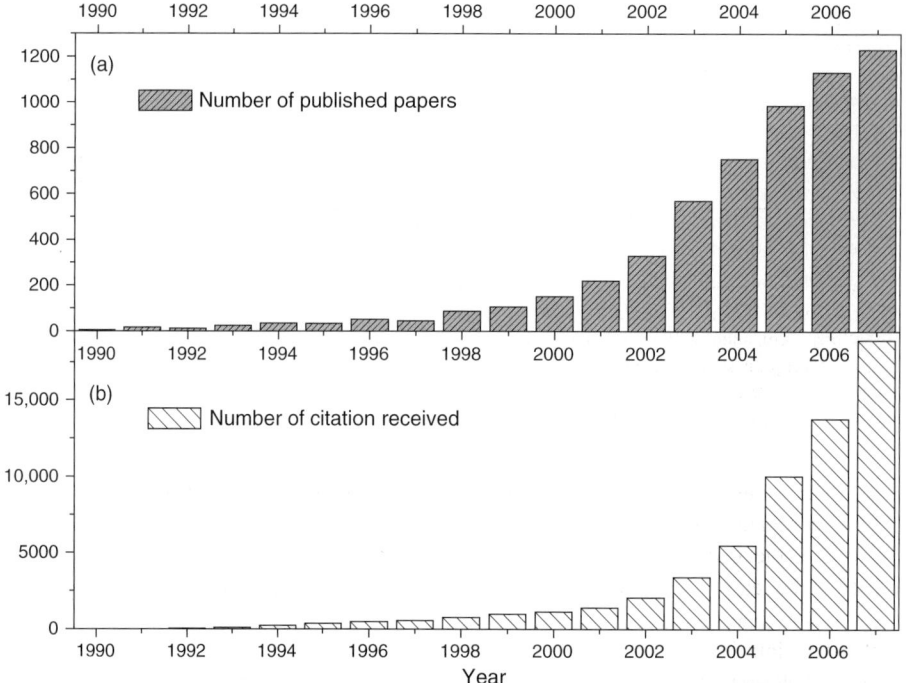

Fig. 1. (a) Number of paper published per year found using nanochemistry OR nanophysics OR nanotechnology OR nanoscience as search keys. Spanned period 1990 to 2007; total number of papers 5788 source ISI web of science. (b) Number of citations received per year by the papers reviewed in (a): total number of citations 59,747.

---

[1] *Subject Categories* are classes where journals are sorted according to the scientific topic of the published papers.

Table 1

List of a selection of scientific journals belonging to the Subject Category named "Nanoscience & Nanotechnology" sorted by publishing age. The last column refer to the aggregated values obtained by the 32 journals together. Data reported in this line allow to evaluate the impact of the Subject Category and to locate each single journal belonging to the category in a pondered ranking scale

Journal name	Publisher	Published since	Papers published in 2006	Impact factor 2006[a]	Immediacy index 2006[b]
Nanotechnology	IoP	1994	1042	3.037	0.534
J. Nanopart. Res.	Kluwer first, now Springer	1999	101	2.156	0.119
Nano Lett.	ACS	2001	555	9.960	1.485
Lab Chip	RSC	2001	190	5.821	1.111
Physica E	Elsevier	2001	503	1.084	0.163
J. Nanosci. Nanotechnol.	American Scientific Publishers	2001	586	2.194	0.309
IEEE Trans. Nanotechnol.	IEEE	2002	111	1.909	0.207
Small	Wiley-VCH	2005	210	6.024	1.152
Curr. Nanosci.	Bentham Science Publ. Ltd	2005	33	2.080	0.242
Nano Today	Elsevier	2006	14	–	1.357
Nature Nanotechnol.	Nature Publishing Group	2006	24	–	0.833
ACS Nano	ACS	2007	0	–	–
Subject Category: 32 journals	–	2005	8939	2.459	0.400

[a] The journal impact factor of year 2006 is the average number of times articles from the journal published in 2005 and 2004 have been cited in 2006. Journals published first in 2006 will have their first impact factor evaluation for year 2007, usually divulgated in spring 2008.

[b] The journal immediacy index of year 2006 is the average number of times an article published in 2006 is cited in 2006. The journal immediacy index indicates how quickly articles in a journal are cited.

This scientific and technological phenomenon has been so significant that the most important publishing companies have decided to create new journals specifically devoted to this topic (see Table 1 for a selection). The Institute of Physics (IoP) has been the first publisher in this field, founded in the early 1990s the journal *Nanotechnology* (impact factor 3.037), which has published more than 4000 papers so far. This number makes *Nanotechnology* the most important journal in the field in terms of published papers. In 1999 Kluwer Academic Publishers founded the journal *J. Nanopart. Res.* Then, in 2001 appeared *Nano Lett.*, from the American Chemical Society (ACS), which probably represents today the most prestigious journal in the field, with more than 2800 published papers and an impressive impact factor that, starting form 5.033 in 2002, progressively increased to almost 10 in 2006. Among the ACS journals, excluding the review journals (*Chem. Rev.* and *Accounts Chem. Res.*), *Nano Lett.* has become the highest impact factor journal, even exceeding the prestigious *J. Am. Chem. Soc.*, first published in 1879. In the same year 2001, the Elsevier journal *Physica E* changed its name into *Physica E-Low-Dimensional Systems & Nanostructures*, while the Royal Society of Chemistry (RSC) launched the title *Lab Chip* (impact factor as high as 5.821) and also *J. Nanosci. Nanotechnol.* (impact factor 2.194) appeared. An important impact factor (6.024) has been achieved by the title *Small*, published by Wiley-VCH since 2005 (see Table 1). In 2006, Elsevier launched the title *Nano Today*, while the prestigious Nature Publishing Group entered the business with the journal *Nat Nanotechnol*. The success of *Nano Lett.* encouraged ACS to found a second title in the field: *ACSNANO*, whose first articles appeared at the end of 2007.

The American Institute of Physics (AIP) and the American Physical Society (APS) adopted a different strategy. Starting from 2000, they selected the most pertinent papers published on the regular AIP and APS journals (*Phys. Rev. Lett., Phys. Rev. A, Phys Rev. B, Appl. Phys. Lett, J. Appl. Phys.*, etc.) in a web collection named *Virtual Journal of Nanoscale Science & Technology* (http://www.vjnano.org/). Currently, more than 50 journals (see http://www.virtualjournals.org/vjs/partpub.jsp for a complete list), including *Science* and *Nature*, joined this virtual journal.

In 2005 the ISI web of Science (http://scientific.thomson.com/products/wos/) has introduced a new subject category (see footnote 1) named "Nanoscience & Nanotechnology." Twenty-seven journals joined that subject category that year, and they are 32 in 2006. Among them, we count journals that already had a long history, like *J. Vac. Sci. Technol. B* (impact factor 1.597) started in 1983, or *Scripta Materialia* (impact factor 2.161), or *Mater. Sci. Eng. A* (impact factor 1.490), or *Microporous Mesoporous Mater.* (impact factor 2.796), and devoted to zeolites and zeotypes that are crystalline porous materials with regular empty cavities and cages in the nanometer and sub-nanometer regime. The 32 journals belonging to the *Subject Category* "Nanoscience & Nanotechnology" in 2006 have 8939 edited papers, exhibit an aggregate impact factor of 2.459 and an aggregate immediacy index of 0.400 (see last row in Table 1). It is strange to note that the Academic Press/Elsevier journal *Superlattices Microstruct.* did not join this new *Subject Category*.

From the data reported in Fig. 1 and Table 1, the tremendous editorial blowup of nanoscience and nanotechnology in the past few years is evident. The editorial blowup mirrors the efforts produced worldwide in the laboratories. This research is found with public and private money and represents the expectation that the society has in respect to this new branch of science.

2. Heterostructures and nanostructures: definition and applications, from optoelectronics to catalysis

On a historical ground, the future born of nanotechnology was probably first foreseen by physicist Richard Feynman at the American Physical Society meeting at Caltech on December 29, 1959, with his famous talk "There's Plenty of Room at the Bottom."

The first applications of nanotechnology were probably semiconductor heterostructures and nanostructures, also defined as low-dimensional systems, i.e., systems that are confined in one, two, or three spatial dimensions, resulting in 2D, 1D, and 0D systems, respectively [1–8]. In the fields of solid-state physics and optoelectronics, 2D, 1D, and 0D systems are usually labeled as quantum wells, quantum wires, and quantum dots (or boxes), respectively.

However, the interest in low-dimensional systems is not restricted to the fields of solid-state physics [1,7,9–12], interface physics [13–17], and optoelectronics [7,18–21], but examples of application of confined systems can be found also in several fields such as metallurgy [22,23], chemistry [24–54], catalysis [55–65], photocatalysis [66–73], and drug delivery [74–81]. All these systems are generally labeled as nanostructures as the confinement usually lies in the nanometer scale. The ultimate frontier in nanotechnology, the single molecule device, has been reached recently [82–88].

Introduction

Emblematic among the fields mentioned so far is the case of catalysis. Catalysis is the science that is aimed to increase the speed and the selectivity of a given chemical reaction. A catalyst is a species that allows to reduce the activation barrier of a chemical reaction, and remains unchanged at the end of the process. Two examples of nanostructured catalysts will now be given. The first one concerns gold. Gold in bulk is known to be chemically inert and gold jewels have been passed on through millenaries, still keeping a perfect conservation. However, when gold is small enough, with particle diameters below 10 nm, it turns out to be surprisingly active for many chemical reactions [89–92], for instance, CO oxidation and propylene epoxidation. As a second example, we will mention zeolites and zeotypes: a new class of materials that, starting from the late 1970s, entered aggressively in the market of catalysts used in industrial plants. Zeolites are nanoporous crystalline alumosilicates constituted by corner-sharing [TO$_4$] tetrahedra, where [TO$_4$] represents a silicon or an aluminum atom. The introduction of a trivalent Al(III) atom in a [TO$_4$] unit (substituting the tetravalent Si(IV) atom) induces a net negative charge to zeolitic framework, which must be compensated by the presence of charge-balancing extra-framework cations. Such cations act as Lewis acid centers, being electron acceptors, but when they are protons, the zeolite becomes a Brønsted solid acid (i.e., a proton donor). Starting from the basic constituents, the framework of any zeolite will be realized by progressively connecting two adjacent [TO$_4$] units by sharing an oxygen atom, which thus becomes "bridged" between the two T atoms (T–O–T).

The remarkably great flexibility of the T–O–T angle (from ≈100° up to 180°) allows to realize, using the unique [TO$_4$] unit as the sole building block, an impressive large number of different zeolites that results in crystalline aluminosilicates with ordered empty cavities in the 0.5–1.5 nm (Fig. 2) range hosting catalytic metal centers or proton donor centers [24,25,27,93–96]. Such materials act as nanoreactors as the chemical reactions occur inside their cavities. The pore opening and the dimension of the internal cavities discriminate the

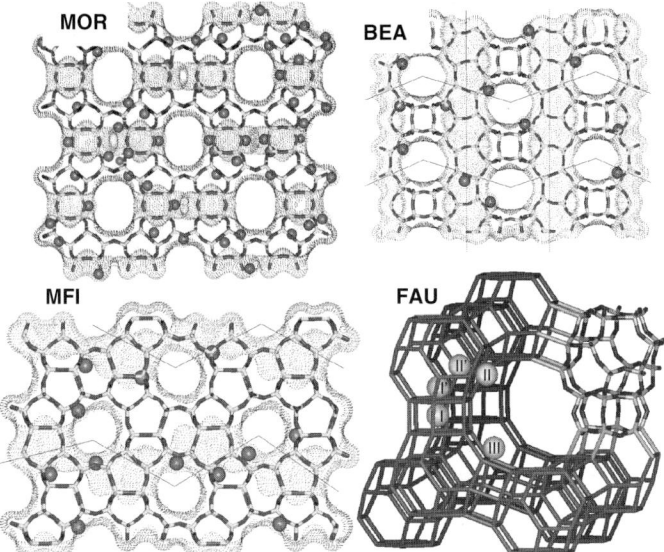

Fig. 2. Sticks representation of four different nanostructurated zeolitic frameworks: mordenite (MOR), ZSM-5 (MFI), β (BEA) and Y (FAU). Light gray the tetrahedral coordinated Si or Al atoms; dark gray O atoms. Extra-framework charge-balancing counterions, playing the role of active catalytic centers, are represented as spheres.

reactant molecules that reach the catalytic centers and the product molecules formed. Therefore, we are dealing with shape-selective catalysts, which represent the best artificial attempt to simulate the enzymes created in the nature. Only recently, the scientists involved in the field of catalysis have explicitly used the word "nano" in their articles and books [65], but it is evident that people working in the field of catalysis studied nanoscience since decades.

3. Dynamic interplay among growth/synthesis techniques, theoretical modeling, and characterization techniques in the design and improvement of semiconductor heterostructure-based devices

Great efforts have been made in improving the preparation methods (epitaxial growth [97–113], etching/regrowing [5,114,115], self-organized growth [116–118], Langmuir–Blodgett films [26–27,119,120], nanolithography [50,121–123], scanning tunneling microscopy (STM) tip-assisted deposition [113], surfactant assemblies as supramolecular templates [124], bottom-up self-assembly approaches [50,113,125–131], intrazeolite encapsulation [24,25,27,28,50,132–134],

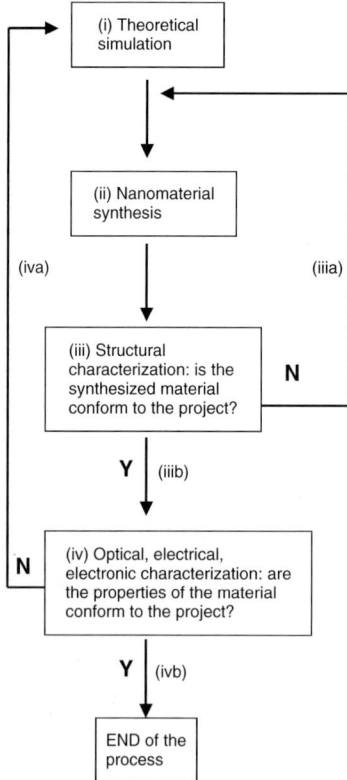

Fig. 3. Flow chart schematizing the strictly interconnected interplay among theoretical calculations (i), synthesis/growth techniques (ii) and characterization techniques (iii and iv) in the improvements in the realization of nanostructures with desired performances.

etc.) of low-dimensional systems to optimize the size control of the confinement regions and of the confined systems, to guarantee a spatial homogeneity, and to optimize the preparation reproducibility. Parallelly, important theoretical efforts have been made to predict the properties of low-dimensional systems, and relevant progresses have been achieved in their characterization [21,123,135–143].

Improvements in the realization of nanostructures can be realized by a strict interplay among the progress achieved on these three grounds, as basically schematized in the flow chart in Fig. 3 and described in the following. (i) theoretical calculations predict the physical properties of a given nanostructure; (ii) the preparation techniques try to realize it; (iii) structural characterization techniques check whether the actually realized nanostructure corresponds to the desired one or not; (iiia) if not, the preparation conditions have to be optimized and step (ii) has to be repeated; (iiib) if yes, then optical, electrical, electronic, chemical reactivity, etc. properties are checked to verify whether the desired nanostructure has actually the foreseen properties (iv); (iva) if not, then the level of theory used in step (i) has to be improved and the game has to restart again from the beginning; (ivb) if yes, then end of the process. Point (ivb) represents the final point of the scientific work and the future of the device lies now on an engineering/economical level where the production rate, the realization costs, and the demand of the device are the main driving forces. Of course, the interplay often moves in the opposite direction, i.e., theoretical models help in the interpretation of the previously not understood (or wrongly interpreted) experimental results.

4. Purposes of the book and chapters layout

The book is structured in chapters, each one devoted to a specific characterization technique used in the understanding of the properties (structural, physical, chemical, electrical, etc.) of semiconductor quantum wells, superlattices, and nanostructures in general. A chapter is devoted to the ab initio modeling. Basically, the book has two aims. The first one deals with the educational point of view. The first part of each chapter provides the basic concept of each of the selected techniques with an approach understandable by Master and PhD students in Physics, Chemistry, Material Science, Engineering, and Nanotechnology. The second aim is to provide a selected set of examples from the recent literature of the latest results obtained with the specific technique in understanding the properties of semiconductor heterostructures and nanostructures. Each chapter has this dual structure: a first part devoted to explain the basic concepts, providing the larger possible audience, and a second one to the discussion of the most peculiar and innovative examples, allowing the book to have the longer possible shelf life. Of course, the book is devoted to the specialized subset of scientists working in the field design, growth, characterization, and testing of heterostructure-based devices based in both academic and industrial laboratories. But the final goal is somewhat more ambitious, and in this regard, the topic of quantum wells, wires, and dots should be seen as a pretext of applying top level characterization techniques in understanding the structural, electronic, etc. properties of matter at the nanometer (even sub-nanometer) scale. In this way, it is aimed to become a reference book in the much broader, and extremely hot, field of nanotechnology.

Besides growth and synthesis techniques, step (ii) for which we refer the reader to specified books and reviews articles [5,24–28,50,97,110–112,116–120, 122,124,125,127,131,144–149], the remaining three fundamental steps of the flow chart reported in Fig. 3 have been considered in the book. Chapter 2, coordinated by Maria Peressi (University of Trieste, Italy), is devoted to the study of structural and electronic properties of semiconductor heterostructures and nanostructures by ab initio calculations (step (i) in Fig. 3).

The experimental investigation of structural properties (step (iii) in Fig. 3) is provided in the following chapters: Chapter 4 discussing X-ray diffraction (XRD) (coordinated by Claudio Ferrari, IMEM-CNR, Parma, Italy); Chapter 5 presenting transmission electron microscopy (TEM) (coordinated by Laura Lazzarini IMEM-CNR, Parma, Italy); Chapter 9 introducing extended X-ray absorption fine structure (EXAFS) (authored by Federico Boscherini, University of Bologna, Italy); Chapter 10 dealing with surface-sensitive X-ray techniques and anomalous scattering (coordinated by Till Metzger, ESRF, Grenoble, France); and Chapter 11 treating the coupled crystallographic and spectroscopic diffraction anomalous fine structure (DAFS) technique (coordinated by Maria Grazia Proietti, CSIC-Universidad de Zaragoza, Spain and by Hubert Renevier, CEA Grenoble, France).

The characterization of the nanostructures properties (step (iv) in Fig. 3), is explained in the following chapters: Chapter 3 for the electrical properties of nanostructures (coordinated by Anna Cavallini, University of Bologna, Italy); the methods for investigating the radiative recombination channels of semiconductor nanostructures are discussed in Chapter 6 (coordinated by Stefano Sanguinetti, Università Milano II, Italy) and in Chapter 7 (coordinated by Giancarlo Salviati, IMEM-CNR, Parma, Italy) that are devoted to photoluminescence and cathodoluminescence spectroscopies, respectively; phonons in nanostructures are described in Chapter 10 (authored by Daniel Wolverson, University of Bath, UK), describing Raman spectroscopy; the band discontinuities in semiconductor heterojunctions are explained in Chapter 12 (authored by Giorgio Margaritondo, EPFL, Lausanne, Switzerland), where photoelectron spectroscopies are introduced; and last but not the least, Chapter 13 (coordinated by Andre Stesmans, University of Leuven, Belgium) deals with the characterization of interfaces and nanolayers in semiconductor heterostructures by means of electron spin spectroscopy.

By moving from bulk materials down to 2D-, 1D-, and 0D-confined heterostructures, the volume (and thus the number of atoms) forming the active regions in nanostructured devices is reduced by order of magnitudes. This means that the understanding of the structural properties of such nanostructures often requires high photon flux techniques like those exploiting synchrotron radiation sources [150–152]. In fact, synchrotron radiation has distinct advantages as a photon source, notably high brilliance and continuous energy spectrum; by using the latter characteristic atomic selectivity can be obtained and this is of fundamental help to investigate the structural environment of atoms present only in a few angstroms (Å) thick interface layers of heterostructures. The third-generation synchrotron radiation sources have allowed to reach the limit of measuring a monolayer of material, corresponding to about 10^{14} atoms/cm^2. Chapters 9–12 describe synchrotron radiation techniques.

5. A key example of multidisciplinarity in nanotechnology: the atomically defined –O–Ti–O–T–O– quantum wires guested in ETS-10 titanosilicate

The last section of this chapter is devoted do briefly describe an example of a nanoporous titanosilicate material that, as zeotype, belongs to the class of materials for catalysis. It has however become rather famous for its nanostructured nature.

Titanium dioxide colloids have attracted tremendous interest because of their potential utility in photocatalytic [153–157], photovoltaic [158,159], and battery [160] applications; moreover, they exhibit nonlinear optical properties [161]. Also heteroepitaxial technique has been used to grow TiO$_2$ films (see e.g., Refs [162,163]). These interests have strongly encouraged the study of TiO$_2$ nanoparticles [154,156,157,164–174]. Among all these studies, we mention the studies of Anpo et al., who combined UV–Vis diffuse reflectance spectroscopy, photoluminescence, X-ray absorption near-edge structure (XANES), and EXAFS to investigate TiO$_2$ nanoclusters hosted inside zeolite Y [156,157]. A blue shift of the energy gap of the TiO$_2$ particle hosted inside the zeolite, with respect to bulk titania, has been observed. These studies represent examples of semiconductor quantum dots hosted inside the ordered nanovoids of zeotype materials, acting as inverse template agents. We focus now the attention on the different case of semiconductor nanowires already present inside the zeolitic framework. The two great advantages in obtaining the semiconductor quantum wire directly during the synthesis as part of the framework, with respect to postsynthesis encapsulation inside the voids, are the elimination of the risks of (i) the formation of a bulk semiconductor segregated phase outside the zeolitic crystals and (ii) an incomplete filling of the voids, which could result in inhomogeneity of the size confinement.

Engelhard titanosilicate ETS-10 is a microporous crystalline material belonging to the family of Ti-substituted silicates [175]. Owing to the high degree of disorder, and also due to the presence of two polymorphs, a straightforward structural determination from XRD was not possible; and Anderson et al. have solved the structure using combined high-resolution TEM, XRD, solid-state NMR, and molecular modeling techniques [176–178]. They proposed a model where the ETS-10 framework is composed of corner-sharing [SiO$_4$] tetrahedra and [TiO$_6$] octahedra (Fig. 4(a)) linked through bridging oxygen atoms, forming 12-membered rings, which give rise to two sets of perpendicular channels having an elliptical cross-section 7.6 × 4.9 Å (Fig. 4(c)). One of the most important characteristics of ETS-10 is that the [TiO$_6$] octahedra are linked together with formation of linear ...–O–Ti–O–Ti–O–Ti–O–... chains (see Fig. 4(b)). Each Ti atom is also linked, in the perpendicular plane, to four Si atoms through oxygen bridges [176–178] resulting in the 3D structure shown in Fig. 4(c).

As bulk TiO$_2$ is a wide band gap semiconductor [179], $E_g = 3.02$ eV for rutile (see gray curve in Fig. 4(d)) and 3.18 eV for anatase, the presence of an atomically defined ...–O–Ti–O–Ti–O–Ti–O–... chains, embedded inside a highly insulating siliceous matrix, $E_g(SiO_2) \sim 9$ eV [180], implies that ETS-10 can be considered as a 1D quantum wire of atomic definition. This observation was first in 1997 [33], proving that the confinement of electrons and holes inside the ...–O–Ti–O–Ti–O–Ti–O–... wires resulted in a blue shift of the energy gap ($E_g = 4.03$ eV, evaluated at the inflection point of the black spectrum in Fig. 4(d)) of ($E_g = 0.85$ and 1.01 eV, when computed from anatase or from rutile, respectively. The experimental blue shift was comparable with that predicted by the simple model of a particle confined along the two directions (xy, in Fig. 4(b)) inside an infinite potential barrier:

$$\Delta E_g = \frac{h^2}{4\mu d^2} = 0.84 \text{ eV} \qquad (1)$$

where h is the Planck constant (6.6256 × 10^{-34} J s), $\mu \sim 2m_e$ is the reduced effective mass of the electron–hole pairs along the wire direction, and $d \sim 6.7$ Å is the wire diameter. Equation (1) has

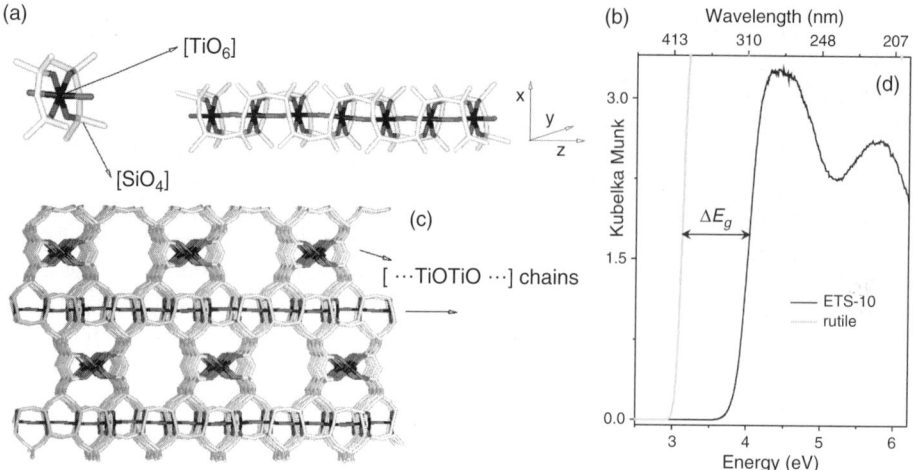

Fig. 4. The framework structure of ETS-10 showing chains of corner-sharing [TiO$_6$] octahedra which run along two perpendicular directions, and which are isolated by corner-sharing [SiO$_4$] tetrahedra: (a) single element of the chain; (b) single chain; (c) three-dimensional view. For clarity, extraframework (charge-balancing) cations are omitted. Ti (black), O (dark gray), Si (light gray). Part (d) reports the DRS UV-VIS spectrum of ETS-10 (black curve, reported from [33]) compared with that of rutile bulk (gray curve). The shape of this spectrum reflect the DOS of the unoccupied valence states. The blue shift of the band gap (ΔE_g), with respect to bulk TiO$_2$, is also evidenced.

been written by considering the potential outside the wire infinite and neglecting the exciton binding energy. Both approximations are well acceptable as the E_g of the host SiO$_2$ matrix (≈ 9 eV) is much greater than that of TiO$_2$ (≈ 3 eV), and because of the high dielectric constant of TiO$_2$ ($\epsilon \approx 180$ [179]): this implies that the exciton binding energy is in the meV range [181]. The oversimplified model that is behind Eq. (1) already gives a qualitative agreement between the predicted energy shift ($\Delta E_g = 0.84$ eV) and the experimental ones ($\Delta E_g = 0.85$ or 1.01 eV). Subsequently, the band structure of the ...–O–Ti–O–Ti–O–Ti–O–... quantum wires embedded inside ETS-10 has been computed by ab initio periodic models [37,45].

Recently, Yilmaz et al. [182] reported original I–V experiment on a similar titanosilicate (ETS-4), also characterized by ...–O–Ti–O–Ti–O–Ti–O–... linear chins, reporting a non-linear behavior at higher bias voltages. The nonohmic I–V behavior has been explained in terms of space–charge-limited effect due to the fact that the conduction occurs inside an atomically defined ...–O–Ti–O–Ti–O–Ti–O–... quantum wire embedded in an highly insulating SiO$_2$ crystalline hosting matrix.

Till now, one of the most promising application of ETS-10 and ETS-4 materials is in the field of photocatalysis [69,71,183]. The combination of a wide band gap semiconducting oxide with a 3D 12-membered ring microporous framework offers many potential advantages in photocatalysis such as excellent diffusion of reactant molecules, trapping and, in particular, shape selectivity. In the perfect structure, these quantum wires are surrounded by an envelope of SiO$_4$ groups. These wires are characterized by full (O-based: 2p) valence and empty (Ti-based: 3d$_{xy}$ and 3d$_{x^2}$) conduction bands [37,45], separated by a band gap of about 4.03 eV, which is slightly higher than that of TiO$_2$ (3.2 eV) due to confinement effects. They can act as an antenna-like system able to collect the light and to form electron–hole pairs in the –O–Ti–O–Ti– chain. Like in TiO_2, these pairs can recombine or diffuse toward the wire, to its end, originating a photocatalytic activity. In a perfect crystal, this happens where the wires emerge on the external surface, but in the real material this can happen also at internal defects (Ti vacancies). In the cited works [69,183], it has been demonstrated that there are three main features of the degradation process: (i) the photocatalytic active sites are titanols located on the external surfaces where the –O–Ti–O–Ti– chain emerge, exposing a surface Ti–OH titanol group; (ii) the fundamental role in determining the shape selectivity is played by the channels and channel mouth shape; (iii) both activity and selectivity can be improved by controlled defect production [69]. As a high selectivity toward the photodegradation of large aromatic molecules has been observed, it has been concluded that molecules entering the pore system were protected from photodegradation, which occurred on the external surface. The ETS-10 thus displays a peculiar inverse shape selectivity. The interpretation that the active photocatalytic centers are Ti–OH species at defective sites agrees well with the parallel studies by the Howe's group [184,185], who have shown how a highly defective proton exchanged sample of ETS-10 was much more active for the gas-phase photooxidation of ethylene than a well-crystalline as-synthesized sample. The local structure of the actually involved Ti–OH sites and the photochemical mechanism are however still a matter of debate.

ETS-10 has presented a severe problem for the characterization techniques. It has already been shown how its structure has been solved by only combining high-resolution TEM, XRD, solid-state NMR, and molecular modeling techniques [176–178]. Also the EXAFS characterization of this material was far from being straightforward. EXAFS spectroscopy (see Chapter 9) is an important tool for the characterization of new materials, particularly when an important degree of disorder in the structure prevents the straightforward use of diffraction techniques, as was the case for ETS-10 [177]. In the first EXAFS results on ETS-10, Davis et al. [186] performed a single scattering study treating all the first shell oxygen as equivalent and finding an averaged Ti–O distance of 2.00 ± 0.01 Å. Subsequently, Sankar et al. [187] presented a local model for the Ti environment able to reproduce the whole EXAFS signal, i.e., up to the second Ti and Si shells, whose contributions appear up to about 3.6 Å in R space. This model is based on the presence of four equivalent Ti–O–Si contributions, perpendicular to the –Ti–O–Ti–O– chain (characterized by an unique Ti–O distance of 2.02 Å), and two unequivalent Ti–O contributions along the chain, characterized by significantly different first shell Ti–O distances of 1.71 and 2.11 Å [187]. The presence of a short and a long Ti–O distance along the –Ti–O–Ti–O– chain is not present in the model proposed by Wang and Jacobson [188] based on single-crystal diffraction, or in the computational results of Sankar himself, based on a periodic atomistic approach [187], or in those of Damin et al. [45] based on a periodic ab initio method. In all the cases, two equivalent axial Ti–O distances are found at 1.872 Å (by Wang and Jacobson [188]), at 1.88 Å (by Damin et al. [45]) and at 1.90 Å (atomistic approach by Sankar et al. [187]). Prestipino et al. [189] have been able to analyze EXAFS and XANES spectra of ETS-10 starting from the periodic ab initio investigation by Damin et al. [45], and with the single-crystal data (collected on a single polymorph) by Wang and Jacobson [188].

The example of ETS-10 titanosilicate, chosen to conclude this chapter, is emblematic and tells us that when the structure and the properties of a nanostructured materials become highly complex, the synergic use of different structural and nonstructural characterization techniques properly supported by ad hoc, ab initio simulations becomes a must to fully understand and exploit the synthesized nanomaterial.

References

[1] L.L. Chang, L. Esaki and R. Tsu, Appl. Phys. Lett., 24 (1974) 593.
[2] H. Sakaki, L.L. Chang, R. Ludeke et al., Appl. Phys. Lett., 31 (1977) 211.
[3] L.L. Chang and L. Esaki, Surf. Sci., 98 (1980) 70.
[4] L. Esaki, IEEE J. Quantum Electron., 22 (1986) 1611.
[5] R. Notzel and K.H. Ploog, Adv. Mater., 5 (1993) 22.
[6] J. Faist, F. Capasso, D.L. Sivco et al., Science, 264 (1994) 553.
[7] P. Bhattacharya, III–V Quantum Wells and Superlattices, INSPEC: London, 1996.
[8] G. Scamarcio, F. Capasso, C. Sirtori et al., Science, 276 (1997) 773.
[9] L. Esaki and L.L. Chang, Phys. Rev. Lett., 33 (1994) 495.
[10] L. Esaki, Nanostructured Mater., 12 (1999) 1.
[11] A. Wacker, Phys. Rep., 357 (2002) 1.
[12] T. Chakraborty and V.M. Apalkov, Adv. Phys., 52 (2003) 455.
[13] R.S. Bauer and G. Margaritondo, Phys. Today, 40 (1987) 27.
[14] F. Capasso and G. Margaritondo, Heterojunction Band Discontinuities: Physics and Device Applications, North-Holland: Amsterdam, 1987.
[15] G. Margaritondo, Electronic Structure of Semiconductor Heterojunctions, Kluwer: Dordrecht, 1988.
[16] G. Margaritondo, Prog. Surf. Sci., 46 (1994) 275.
[17] G. Margaritondo, Prog. Surf. Sci., 56 (1997) 311.
[18] R. Cingolani and K. Ploog, Adv. Phys., 40 (1991) 535.
[19] G. Khitrova, H.M. Gibbs, F. Jahnke et al., Rev. Mod. Phys., 71 (1999) 1591.
[20] F. Rossi and T. Kuhn, Rev. Mod. Phys., 74 (2002) 895.
[21] C. Lamberti, Surf. Sci. Rep., 53 (2004) 1.
[22] A. Pundt, Adv. Eng. Mater., 6 (2004) 11.
[23] D.V. Louzguine-Lugin and A. Inoue, J. Nanosci. Nanotechnol., 5 (2005) 999.
[24] G.A. Ozin and C. Gil, Chem. Rev., 89 (1989) 1749.
[25] G.A. Ozin, A. Kuperman and A. Stein, Angew. Chem. Int. Ed. Engl., 28 (1989) 359.
[26] J.M. Lehn, Angew. Chem. Int. Ed. Engl., 29 (1990) 1304.
[27] G.A. Ozin, Adv. Mater., 4 (1992) 612.
[28] G.A. Ozin and S. Ozkar, Chem. Mater., 4 (1992) 511.
[29] N. Sabbatini, M. Guardigli and J.M. Lehn, Coord. Chem. Rev., 123 (1993) 201.
[30] J.P. Sauvage, J.P. Collin, J.C. Chambron et al., Chem. Rev., 94 (1994) 993.
[31] V. Balzani, A. Juris, M. Venturi et al., Chem. Rev., 96 (1996) 759.
[32] B. Hasenknopf, J.M. Lehn, B.O. Kneisel et al., Angew. Chem. Int. Ed. Engl., 35 (1996) 1838.
[33] E. Borello, C. Lamberti, S. Bordiga et al., Appl. Phys. Lett., 71 (1997) 2319.
[34] V. Balzani, S. Campagna, G. Denti et al., Accounts Chem. Res., 31 (1998) 26.
[35] P.R. Ashton, V. Balzani, J. Becher et al., J. Am. Chem. Soc., 121 (1999) 3951.
[36] A.K. Cheetham, G. Ferey and T. Loiseau, Angew. Chem. Int. Ed. Engl., 38 (1999) 3268.
[37] S. Bordiga, G.T. Palomino, A. Zecchina et al., J. Chem. Phys., 112 (2000) 3859.
[38] V. Balzani, A. Credi, F.M. Raymo and J.F. Stoddart, Angew. Chem. Int. Ed. Engl., 39 (2000) 3349.
[39] A. Caneschi, D. Gatteschi, N. Lalioti et al., Angew. Chem. Int. Ed. Engl., 40 (2001) 1760.
[40] G. Ferey, Chem. Mater., 13 (2001) 3084.
[41] B.L. Chen, M. Eddaoudi, S.T. Hyde et al., Science, 291 (2001) 1021.
[42] M. Eddaoudi, J. Kim, N. Rosi, et al., Science, 295 (2002) 469.
[43] J.M. Lehn, Science, 295 (2002) 2400.
[44] N.L. Rosi, J. Eckert, M. Eddaoudi et al., Science, 300 (2003) 1127.
[45] A. Damin, F.X. Llabrés i Xamena, C. Lamberti et al., J. Phys. Chem. B, 108 (2004) 1328.
[46] S. Bordiga, C. Lamberti, G. Ricchiardi et al., Chem. Commun. (2004) 2300.
[47] J.D. Badjic, V. Balzani, A. Credi et al., Science, 303 (2004) 1845.
[48] C. Altavilla, E. Ciliberto, D. Gatteschi and C. Sangregorio, Adv. Mater., 17 (2005) 1084.

[49] V. Balzani, Small, 1 (2005) 278.
[50] G.A. Ozin and A. Arsenault, Nanochemistry: A Chemical Approach to Nanomaterials, Royal Society of Chemistry: Cambridge, 2005.
[51] A. Zecchina, S. Bordiga, J.G. Vitillo et al., J. Am. Chem. Soc., 127 (2005) 6361.
[52] C. Prestipino, L. Regli, J.G. Vitillo et al., Chem. Mater., 18 (2006) 1337.
[53] J. Hafizovic, M. Bjorgen, U. Olsbye et al., J. Am. Chem. Soc., 129 (2007) 3612.
[54] V. Balzani, A. Credi and M. Venturi, Nano Today, 2 (2007) 18.
[55] M. Sasaki, M. Osada, N. Sugimoto et al., Microporous Mesoporous Mater., 21 (1998) 597.
[56] P. Van Der Voort, P.I. Ravikovitch, K.P. De Jong et al., Chem. Commun. (2002) 1010.
[57] P. Schwerdtfeger, Angew. Chem. Int. Ed. Engl., 42 (2003) 1892.
[58] A. Zampieri, P. Colombo, G.T.P. Mabande et al., Adv. Mater., 16 (2004) 819.
[59] T.A.R. Nijhuis, T. Visser and B.M. Weckhuysen, Angew. Chem. Int. Ed. Engl., 44 (2005) 1115.
[60] J.A. van Bokhoven, C. Louis, J.T. Miller et al., Angew. Chem. Int. Ed. Engl., 45 (2006) 4651.
[61] J.G. Mesu, T. Visser, A.M. Beale et al., Chem. Eur. J., 12 (2006) 7167.
[62] K.C. Szeto, K.P. Lillerud, M. Tilset et al., J. Phys. Chem. B, 110 (2006) 21509.
[63] K.C. Szeto, C. Prestipino, C. Lamberti et al., Chem. Mater., 19 (2007) 211.
[64] N. Weiher, A.M. Beesley, N. Tsapatsaris et al., J. Am. Chem. Soc., 129 (2007) 2240.
[65] A. Zecchina, E. Groppo and S. Bordiga, Chem.-Eur. J., 13 (2007) 2440.
[66] J.A. Byrne, A. Davidson, P.S.M. Dunlop and B.R. Eggins, J. Photochem. Photobiol. A-Chem., 148 (2002) 365.
[67] Z.G. Zou, J.H. Ye and H. Arakawa, J. Phys. Chem. B, 106 (2002) 13098.
[68] T.T.Y. Tan, M. Zaw, D. Beydoun and R. Amal, J. Nanopart. Res., 4 (2002) 541.
[69] F.X. Llabrés i Xamena, P. Calza, C. Lamberti et al., J. Am. Chem. Soc., 125 (2003) 2264.
[70] S. Takabayashi, R. Nakamura and Y. Nakato, J. Photochem. Photobiol. A-Chem., 166 (2004) 107.
[71] S. Usseglio, P. Calza, A. Damin et al., Chem. Mater., 18 (2006) 3412.
[72] S. Usseglio, A. Damin, D. Scarano et al., J. Am. Chem. Soc., 129 (2007) 2822.
[73] B. Muktha, G. Madras, T.N.G. Row et al., J. Phys. Chem. B, 111 (2007) 7994.
[74] C. Allen, D. Maysinger and A. Eisenberg, Colloid Surf. B-Biointerfaces, 16 (1999) 3.
[75] H. Kawaguchi, Prog. Polym. Sci., 25 (2000) 1171.
[76] A. Rosler, G.W.M. Vandermeulen and H.A. Klok, Adv. Drug Deliv. Rev., 53 (2001) 95.
[77] I.G. Loscertales, A. Barrero, I. Guerrero et al., Science, 295 (2002) 1695.
[78] R. Gasparac, P. Kohli, M.O. Mota et al., Nano Lett., 4 (2004) 513.
[79] D. Moinard-Checot, Y. Chevalier, S. Briancon et al., J. Nanosci. Nanotechnol., 6 (2006) 2664.
[80] S. Moritake, S. Taira, Y. Ichiyanagi et al., J. Nanosci. Nanotechnol., 7 (2007) 937.
[81] B. Xiao, P.S. Wheatley, X.B. Zhao et al., J. Am. Chem. Soc., 129 (2007) 1203.
[82] S.J. Tans, A.R.M. Verschueren and C. Dekker, Nature, 393 (1998) 49.
[83] J.K. Gimzewski and C. Joachim, Science, 283 (1999) 1683.
[84] M.C. Hersam, N.P. Guisinger and J.W. Lyding, Nanotechnology, 11 (2000) 70.
[85] J.H. Schon, H. Meng and Z. Bao, Nature, 413 (2001) 713.
[86] M. Cavallini, F. Biscarini, J. Gomez-Segura et al., Nano Lett., 3 (2003) 1527.
[87] S.W. Hla and K.H. Rieder, Annu. Rev. Phys. Chem., 54 (2003) 307.
[88] A. Bruckbauer, D.J. Zhou, L.M. Ying et al., J. Am. Chem. Soc., 125 (2003) 9834.
[89] J.D. Aiken and R.G. Finke, J. Mol. Catal. A-Chem., 145 (1999) 1.
[90] M. Haruta and M. Date, Appl. Catal. A-Gen., 222 (2001) 427.
[91] R.M. Crooks, M.Q. Zhao, L. Sun et al., Accounts Chem. Res., 34 (2001) 181.
[92] M.C. Daniel and D. Astruc, Chem. Rev., 104 (2004) 293.
[93] M.E. Davis and R.F. Lobo, Chem. Mater., 4 (1992) 756.
[94] B. Notari, Adv. Catal., 41 (1996) 253.
[95] D. Barthomeuf, Catal. Rev. Sci. Eng., 38 (1996) 521.
[96] S. Kitagawa and M. Kondo, Bull. Chem. Soc. Jpn., 71 (1998) 1739.
[97] K. Ploog, Angew. Chem. Int. Ed. Engl., 27 (1988) 593.

[98] T. Wang, E.H. Reihlen, H.R. Jen and G.B. Stringfellow, J. Appl. Phys., 66 (1989) 5376.
[99] M. Ozeki, N. Ohtsuka, Y. Sakuma and K. Kodama, J. Cryst. Growth, 107 (1991) 102.
[100] J. Hergeth, D. Grützmacher, F. Reinhardt and P. Balk, J. Cryst. Growth, 107 (1991) 537.
[101] A. Antolini, P.J. Bradley, C. Cacciatore et al., J. Electron. Mater., 21 (1992) 233.
[102] F. Genova, A. Antolini, L. Francesio et al., J. Cryst. Growth, 120 (1992) 333.
[103] A. Antolini, L. Francesio, L. Gastaldi et al., J. Cryst. Growth, 127 (1993) 189.
[104] K. Streubel, J. Wallin, G. Landgren et al., J. Cryst. Growth, 143 (1994) 7.
[105] M. Heuken, J. Cryst. Growth, 146 (1995) 570.
[106] C.T. Hsu, Thin Solid Films, 335 (1998) 284.
[107] Y. Luo, D. Slater, M. Han et al., Langmuir, 14 (1998) 1493.
[108] N. Georgiev and T. Mozume, Appl. Phys. Lett., 75 (1999) 2371.
[109] P.B. Joyce, T.J. Krzyzewski, G.R. Bell et al., Phys. Rev. B, 62 (2000) 10891.
[110] P. Masri, Surf. Sci. Rep., 48 (2002) 1.
[111] T. Kingetsu and M. Yamamoto, Surf. Sci. Rep., 45 (2002) 79.
[112] J.R. Arthur, Surf. Sci., 500 (2002) 189.
[113] J. Shen and J. Kirschenr, Surf. Sci., 500 (2002) 300.
[114] R. Notzel, N.N. Ledentsov, L. Daweritz et al., Phys. Rev. B, 45 (1992) 3507.
[115] E. Kapon, Proc. IEEE, 80 (1992) 398.
[116] S. Mann and G.A. Ozin, Nature, 382 (1996) 313.
[117] V.A. Shchukin and D. Bimberg, Rev. Mod. Phys., 71 (1999) 1125.
[118] K. Acobi, Prog. Surf. Sci., 71 (2003) 185.
[119] J. Roncali, Chem. Rev., 92 (1992) 711.
[120] A. Ulman, Chem. Rev., 96 (1996) 1533.
[121] T.H. Fang, C.I. Weng and J.G. Chang, Nanotechnology, 11 (2000) 181.
[122] D. Wouters and U.S. Schubert, Angew. Chem.-Int. Ed, 43 (2004) 2480.
[123] J. Loos, Adv. Mater., 17 (2005) 1821.
[124] H. Yang, N. Coombs and G.A. Ozin, Nature, 386 (1997) 692.
[125] R.P. Andres, T. Bein, M. Dorogi et al., Science, 272 (1996) 1323.
[126] O. Dag, G.A. Ozin, H. Yang et al., Adv. Mater., 11 (1999) 474.
[127] V.A. Shchukin and D. Bimberg, Rev. Mod. Phys., 71 (1999) 1125.
[128] P. Politi, G. Grenet, A. Marty et al., Phys. Rep.-Rev. Sec. Phys. Lett., 324 (2000) 271.
[129] N.N. Ledentsov, M. Grundmann, F. Heinrichsdorff et al., IEEE J. Sel. Top. Quantum Electron., 6 (2000) 439.
[130] D. Bimberg and N. Ledentsov, J. Phys.: Condens. Matter, 15 (2003) R1063.
[131] P. Bhattacharya, S. Ghosh and A.D. Stiff-Roberts, Annu. Rev. Mater. Res., 34 (2004) 1.
[132] T. Bein, ACS Symp. Ser., 499 (1992) 274.
[133] O. Dag, A. Kuperman and G.A. Ozin, Adv. Mater., 6 (1994) 147.
[134] K. Moller and T. Bein, Chem. Mater., 10 (1998) 2950.
[135] D. Grützmacher, R. Mayer, M. Zachau et al., J. Cryst. Growth, 93 (1988) 382.
[136] D. Grützmacher, J. Cryst. Growth, 107 (1991) 520.
[137] R.G. Vansilfhout, J.F. Vanderveen, S. Ferrer and C. Norris, Surf. Sci., 264 (1992) 281.
[138] S.H. Tolbert and A.P. Alivisatos, Annu. Rev. Phys. Chem., 46 (1995) 595.
[139] V.H. Etgens, R.M. RibeiroTeixeira, P.M. Mors et al., Europhys. Lett., 36 (1996) 271.
[140] D.P. Woodruff, Surf. Sci., 500 (2002) 147.
[141] S. Ferrer and Y. Petroff, Surf. Sci., 500 (2002) 605.
[142] C.T. Williams and D.A. Beattie, Surf. Sci., 500 (2002) 545.
[143] C.J. Hirschmugl, Surf. Sci., 500 (2002) 577.
[144] P. Jensen, Rev. Mod. Phys., 71 (1999) 1695.
[145] A.N. Shipway, E. Katz and I. Willner, Chem. Phys. Chem., 1 (2000) 18.
[146] Y.N. Xia, P.D. Yang, Y.G. Sun et al., Adv. Mater., 15 (2003) 353.
[147] M. Law, J. Goldberger and P.D. Yang, Annu. Rev. Mater. Res., 34 (2004) 83.

[148] J. Stangl, V. Holy and G. Bauer, Rev. Mod. Phys., 76 (2004) 725.
[149] J.V. Barth, G. Costantini and K. Kern, Nature, 437 (2005) 671.
[150] G. Margaritondo, Introduction to Synchrotron Radiation, Oxford: New York, 1988.
[151] G. Margaritondo, Rivista Del Nuovo Cimento, 18 (1995) 1.
[152] G. Margaritondo, Surf. Rev. Lett., 7 (2000) 379.
[153] A.L. Linsebigler, G.Q. Lu and J.T. Yates, Chem. Rev., 95 (1995) 735.
[154] M. Anpo, N. Aikawa, Y. Kubokawa et al., J. Phys. Chem., 89 (1985) 5017.
[155] H. Yamashita, Y. Ichihashi, M. Harada et al., J. Catal., 158 (1996) 97.
[156] H. Yamashita, Y. Ichihashi, M. Anpo et al., J. Phys. Chem., 100 (1996) 16041.
[157] M. Anpo, H. Yamashita, Y. Ichihashi et al., J. Phys. Chem. B, 101 (1997) 2632.
[158] Y.I. Kim, S.W. Keller, J.S. Krueger et al., J. Phys. Chem. B, 101 (1997) 2491.
[159] P.V. Kamat, Chem. Rev., 93 (1993) 267.
[160] S.Y. Huang, L. Kavan, I. Exnar and M. Gratzel, J. Electrochem. Soc., 142 (1995) L142.
[161] R.F. Khairutdinov, Uspekhi Khimii, 67 (1998) 125.
[162] D.R. Burgess, P.A.M. Hotsenpiller, T.J. Anderson and J.L. Hohman, J. Cryst. Growth, 166 (1996) 763.
[163] P.A. Morris Hotsenpiller, G.A. Wilson, A. Roshko et al., J. Cryst. Growth, 166 (1996) 779.
[164] D. Duonghong, J. Ramsden and M. Graetzel, J. Am. Chem. Soc., 104 (1982) 2977.
[165] R.F. Howe and M. Gratzel, J. Phys. Chem., 89 (1985) 4495.
[166] I. Bedja and P.V. Kamat, J. Phys. Chem., 99 (1995) 9182.
[167] B.I. Lemon and J.T. Hupp, J. Phys. Chem., 100 (1996) 14578.
[168] N. Serpone, D. Lawless, R. Khairutdinov and E. Pelizzetti, J. Phys. Chem., 99 (1995) 16655.
[169] E. Joselevich and I. Willner, J. Phys. Chem., 98 (1994) 7628.
[170] C. Kormann, D.W. Bahnemann and M.R. Hoffmann, J. Phys. Chem., 92 (1988) 5196.
[171] L. Kavan, T. Stoto, M. Gratzel et al., J. Phys. Chem., 97 (1993) 9493.
[172] W.Y. Choi, A. Termin and M.R. Hoffmann, J. Phys. Chem., 98 (1994) 13669.
[173] M. Anpo, T. Shima, S. Kodama and Y. Kubokawa, J. Phys. Chem., 91 (1987) 4305.
[174] N. Serpone, D. Lawless and R. Khairutdinov, J. Phys. Chem., 99 (1995) 16646.
[175] M. Kuznicki US patent, 4853202, 1989.
[176] M.W. Anderson, O. Terasaki, T. Ohsuna et al., Nature, 367 (1994) 347.
[177] M.W. Anderson, O. Terasaki, T. Ohsuna et al., Philos. Mag. B, 71 (1995) 813.
[178] T. Ohsuna, O. Terasaki, D. Watanabe et al., Stud. Surf. Sci. Catal., 84 (1994) 413.
[179] F.A. Grant, Rev. Mod. Phys., 31 (1959) 646.
[180] S.M. Sze, Physics of Semiconductor Devices, John Wiley and Sons: New York, 1981.
[181] J. Pascual, J. Camassel and H. Mathieu, Phys. Rev. B, 18 (1978) 5606.
[182] B. Yilmaz, A. Sacco and J.D. Deng, Appl. Phys. Lett., 90 (2007) Art. No. 152101.
[183] P. Calza, C. Paze, E. Pelizzetti and A. Zecchina, Chem. Commun., (2001) 2130.
[184] R.F. Howe and Y.K. Krisnandi, Chem. Commun. (2001) 1588.
[185] Y.K. Krisnandi, P.D. Southon, A.A. Adesina and R.F. Howe, Int. J. Photoenergy, 5 (2003) 131.
[186] R.J. Davis, Z. Liu, J.E. Tabora and W.S. Wieland, Catal. Lett., 34 (1995) 101.
[187] G. Sankar, R.G. Bell, J.M. Thomas et al., J. Phys. Chem., 100 (1996) 449.
[188] X.Q. Wang and A.J. Jacobson, Chem. Commun. (1999) 973.
[189] C. Prestipino, P.L. Solari and C. Lamberti, J. Phys. Chem. B, 109 (2005) 13132.

2

Ab initio studies of structural and electronic properties

Maria Peressi[1,2], Alfonso Baldereschi[1,2], and Stefano Baroni[2,3]

[1]*Dipartimento di Fisica Teorica, University of Trieste, Trieste, Italy*
[2]*CNR-INFM DEMOCRITOS Theory@Elettra Group, Trieste, Italy*
[3]*Scuola Internazionale Superiore di Studi Avanzati, Trieste, Italy*

Abstract Ab initio calculations are not only an important tool for reliable predictions of structural and electronic properties of individual semiconductor heterojunctions, but are also essential for understanding the underlying mechanisms, establishing a rationale for the peculiarities of entire classes of heterojunctions, and gaining the necessary insight to engineer their properties for specific applications. In this chapter, we present the basic ingredients of an ab initio approach to the problem and discuss their application to prototypical systems. Among the electronic properties, particular emphasis is placed on the band offsets, which are the key parameters governing the transport properties of a junction, but other features such as localized electron states at interfaces and cross-sectional scanning tunneling microscopy images are also discussed. We focus mostly on ab initio computations and theoretical models derived from first-principles numerical experiments.

Keywords ab initio calculations, density functional theory, supercell, band offset, linear response theory, models

1. Introduction

In the last decades computer simulations have emerged as a fundamental and powerful new tool for condensed matter research. Simulations not only make possible the detailed interpretation of experimental results, particularly at the atomic scale, but in most (actually the most interesting) cases they provide predictions of properties and processes not yet observed in the laboratory. They are also extremely useful in the accurate assessment of existing theories and models and often they provide the inspiration for new ones.

In the case of semiconductor heterostructures, numerical simulations have been essential in establishing general trends in the electronic properties, and in particular the band structure discontinuities at interfaces. An important issue that has been addressed and rationalized is whether the band discontinuities are intrinsic or not, i.e., whether they are determined solely by properties of the constituent bulk materials and hence display commutativity and transitivity, or they also depend on interface-specific features, such as crystallographic orientation and abruptness, and are therefore tunable to some extent. Although simple models and semi-empirical methods have been particularly helpful in understanding the broad features of experimental results, the solution of such a difficult issue has required simulations based on advanced techniques and capable of dealing with atomic-scale processes.

It is possible nowadays to study the structural and electronic properties as well as the energetics of many-electron systems by performing fully ab initio computations, i.e., by solving the quantum-mechanical equations for the system under consideration without any use of empirical parameters or experimental inputs. For further details we refer the reader to recent books providing an extensive exposition of theory and methods for electronic structure calculations [1,2]. This chapter will focus on one of the most efficient and widely used among such advanced methods, i.e., the ab initio pseudopotential method based on density functional theory (DFT). After a brief presentation of some basic models (Section 2), we give a general description of the computational approach and specific tools devoted to the problem of heterostructures in Section 3. We will then present some selected applications to the problem of band alignments (Section 4) starting from the simple and paradigmatic case of ideal GaAs/AlAs and considering also more complex cases such as strained heterojunctions, defected interfaces, and alloy-based heterostructures. We will discuss also band-offset engineering (Section 5), which has been the subject of an intense theoretical effort in order to both understand underlying phenomena and predict offset modifications. Section 6 is devoted to the problem of localized interface states, whose presence is often invoked to interpret anomalous features in the results of photoemission spectroscopy, optical, and transport measurements. Section 7 will show how scanning tunneling microscopy (STM), originally used for surface analysis, can also be employed to characterize semiconductor interfaces when used in cross-sectional geometries. Finally, in Section 8, we will address the properties of heterostructures that have a growing technological interest for magneto-opto-electronic devices, i.e., for spintronic applications.

2. Basic models

Theoretical and numerical investigations based on different approaches have contributed to predict band-offset values and to identify the basic mechanisms responsible for band structure discontinuities. Successful investigations can be divided into two main classes:

(i) models and approaches making simplifying and sometimes even drastic approximations in the description of the interface, but retaining the most relevant physical concepts, and
(ii) accurate computational studies based on self-consistent ab initio approaches, in which interface features such as orientation, stoichiometry, abruptness, and defects are taken into full account.

The focus of this chapter is on the latter approach as it has played a major role in understanding the physical mechanisms governing band discontinuities and in providing accurate predictions. However we mention here some simple but basic models that have the merit of having addressed the problem, helped in understanding relevant factors, and provided a first predictive tool. We limit ourselves to a short overview of the most relevant models and their historical development without attempting an exhaustive and detailed discussion. We refer the interested reader to books, review papers, and original papers for a wider and deeper presentation. For instance, the book by Capasso and Margaritondo [3] contains a review of both experimental and theoretical early works. Fundamental papers on the theoretical approaches are collected in the book by Margaritondo [4]. More recent reviews are those by Yu et al. [5], Franciosi and Van de Walle [6], and Lamberti [7].

The root of the band-offset problem is that the average of the electrostatic potential in an infinite solid is an ill-defined quantity [8] and therefore no universal energy scale exists to which the band structures of different semiconductors can be easily and uniquely referred. The lineup of the average electrostatic potential across the interface between two semi-infinite solids is, on the contrary, well defined and band-edge offsets are obtained by adding to it the bulk band-edge difference resulting when the arbitrary values of the electrostatic potentials in the two materials forming the interface are aligned. The latter quantity is easily obtained from band structure calculations of the two bulk materials and the difficulty resides in accurately predicting the lineup of the average electrostatic potential across the interface that contains all the effects of interface-specific phenomena. Band-offset models differ from each other in the method they propose for approximating the average electrostatic potential lineup without performing any complicated analysis of interface processes. Most models are based on *intrinsic reference levels* which are defined for each material individually.

This is the case of the historical electron affinity rule proposed by Anderson [9] in which the offset of the conduction band minimum is set equal to the difference in electron affinities of the two materials. This rule is the extension to the problem of band discontinuities at semiconductor heterostructures of the famous Schottky–Mott model [10,11] for predicting Schottky barrier heights at metal–semiconductor junctions. The intrinsic limitation of this rule is that the electron affinity is not a bulk quantity, but it is measured at surfaces, and the model assumes that the atomic structure, electron charge density, and electron affinity of the two surfaces forming the interface remain unchanged on forming the contact. This assumption is not at all justified but the rule was generally accepted until the late 1970s. The model was refined by Van Vechten [12] who proposed to extract a "bulk" contribution from the electron affinities by neglecting all terms due to the surface configuration of the material (such as atomic relaxations, surface reconstructions, and electrostatic fields) and use it to predict band alignments. These "bulk" electron affinities were defined in the context of the dielectric theory of electronegativity [13].

Harrison [14,15] proposed a model based on an atomic orbital tight-binding approach and provided a table of the valence band maxima for all tetrahedral semiconductors from which band discontinuities can be obtained directly by subtraction. In this model, the reference level for the band structure energies of the individual semiconductors is set by the use of the term values of neutral atoms in vacuum as diagonal elements of the tight-binding hamiltonian. The effects due to charges on the atoms, which occur in ionic semiconductors, and in particular those on the average electrostatic potential lineup across the interface are neglected in the model. The resulting offsets are only reliable on the scale of a few tenths of 1 eV.

Frensley and Kroemer [16,17] choose the average interstitial crystal potential as the reference level for the band structure of each individual semiconductor. The lineup across the interface between the two average potentials is then estimated with a model distribution of point charges on the atoms and results in a dipole shift that can be expressed in terms of electronegativity differences, i.e., differences between bulk quantities of the two materials forming the interface. This provides a justification for expressing band offsets as differences between intrinsic reference levels that are defined for each material individually.

Other theories focus on the concept of *charge neutrality level* in semiconductors. On forming the contact between two semiconductors, the two charge neutrality levels in general do not coincide, electron charge will flow across the interface thus setting up dipoles that will drive the system toward the exact alignment of the neutrality levels if metallic dielectric screening is assumed for the interface [18]. Due to finite dielectric screening of semiconductors, however, the two charge neutrality levels will have a residual mismatch [19]. Band structure discontinuities can then be obtained in a zeroth-order approximation by exact alignment of the two levels or, in a better approximation, by estimating their mismatch on the basis of the energy difference between the two neutrality levels and the dielectric properties of the interface. Harrison and Tersoff [20], using self-consistent tight-binding theory, have calculated numerically the interface dipoles and discussed their connection with dieletric screening. It should also be mentioned that Cardona and Christensen [21], while studying the dielectric screening of absolute hydrostatic deformation potentials, were lead to introduce a dielectric midgap energy for tetrahedral semiconductors and have argued that this energy reference is related to the charge neutrality point and can be used to evaluate band offsets. One advantage of models based on charge neutrality levels is that they provide quantitative predictions of both heterojunction band offsets and Schottky barrier heights. For ideal, lattice-matched, non polar interfaces such as GaAs/AlAs, the concept of intrinsic charge neutrality levels works well and is supported by good agreement with experimental data and self-consistent calculations. For example, it predicts correctly a negligible dependence of band offsets on the crystallographic orientation of the interface [22]. Its predictions are typically accurate to about 0.1 eV. However, both experimental data and theoretical investigations indicate that in several cases, such as polar interfaces, band offsets depend considerably on interface features like crystallographic orientation. In an attempt to reconcile this evidence with the concept of charge neutrality levels, Flores and co-workers [23] introduced the concept of "extrinsic" charge neutrality levels that depend on interface geometry and include additional dipole contributions with respect to the "intrinsic" levels.

Through all-electron first-principles electronic structure calculations of core levels, Wei and Zunger systematically calculated the "natural" band offsets between several II–VI and separately between III–V semiconductor compounds. They found that the valence band offset in common anion systems is primarily determined by intrinsic bulk effects and that interface charge transfer has a small effect on these quantities, provided that cation d orbitals are correctly taken into account [24–26].

In the late 1970s, self-consistent calculations based on semi-empirical pseudopotentials started to be performed not only for bulk materials but also for supercells modeling heterojunctions, as for instance in the pioneering works by Baraff et al. [27] and Pickett and Cohen [28,29]. The problem of heterojunctions could thus be approached theoretically in a more rigorous way and the way was open to the modern full ab initio supercell calculations that will be discussed in detail in the remaining part of the chapter. At the same time, ab initio calculations suggested the formulation of more refined models based on parameter-free self-consistent calculations of the charge density. This is the case of the *model-solid approach* proposed by Van de Walle and Martin [30,31]. It is based on the fact that the electron charge distribution of both bulk semiconductors and semiconducting interfaces is accurately reproduced by the superposition of the spherical densities calculated for neutral atoms in vacuum and with properly chosen electron configuration. The deviations are due to, generally small, electron density readjustments associated with the formation of covalent bonds and, more importantly, to inter-atomic charge transfers associated with the formation of ionic bonds. These authors, therefore, define as reference energy for each material the average (pseudo)potential in a model solid, in which the charge density is constructed as a superposition of neutral, spherical (pseudo)atomic densities. This reference depends on the density of each type of atom and on the detailed form of the atomic charge density, which must be chosen consistently for the different materials. The bulk band structures of the two semiconductors are then aligned according to these average potential positions. This model has been successfully applied to a large class of lattice-matched heterojunctions. The model-solid approach has been extended to the case of strained heterojunctions: the results were expressed again in terms of an "absolute" energy level for each semiconductor and of deformation potentials describing the effects of strain on the electronic bands [32]. The model-solid approach is at present the most widely used method for quantitative predictions of band offsets at semiconductor interfaces [33–38]. Care should be taken in applying the approach to strongly ionic semiconductors whose electron charge density is less accurately reproduced by a superposition of neutral atomic densities than that of strongly covalent materials.

On lines similar to those of Van de Walle and Martin, Baldereschi, et al. have proposed a model based on an accurate description of the electron charge density in heterojunctions in terms of neutral "building blocks" derived from the bulk charge densities of the two constituents [39]. For lattice-matched common anion (common cation) heterojunctions the building blocks are the charge densities within Wigner–Seitz cells centered on the cations (anions). In the prototypical case of GaAs/AlAs, the model is based on two building blocks, one for GaAs and one for AlAs. These charge distributions are neutral and by symmetry do not have any dipole or quadrupole moments. The interface charge density is obtained by rigid juxtaposition of these blocks, at the price of small discontinuities at their boundaries. Equivalently, one can consider the juxtaposition all over space of blocks containing the *average* charge density, and blocks containing one-half of the charge density *difference* and taken with the plus or minus sign on one side or the other of the interface, respectively. Although the model charge density distribution is very different for differently oriented interfaces because of the different geometry of each juxtaposition, the electrostatic potential lineup across the interface is orientation independent as it depends only on the difference between the charge distributions of the two building blocks. This Wigner–Seitz charge-density model is very accurate but applies to lattice-matched systems only.

The formulation of the model, however, opened the way to further developments, based on the observation that what is relevant, for instance, in a case like GaAs/AlAs, is the difference between the electron charge distributions around the As nuclei surrounded by Ga or Al neighbors. Such differences are quite small with respect to typical density variations in the bulk of a semiconductor, and can be calculated accurately with a low-order perturbation approach. In general, when the two constituent semiconductors are rather similar to each other, their heterojunction can be viewed as a *perturbation* with respect to an appropriate *reference* periodic system: the perturbation being related to the potential difference between different ionic cores on the two sides of the interface. The response, to the lowest order, is a charge density *linear response* of the reference periodic system to the potential perturbation, so that it is possible to derive macroscopic quantities, such as the electrostatic potential lineup across the interface, from the responses to each individual atomic site perturbation. Such reponses can be determined through a direct approach by ab initio calculations. The *linear response theory* approach can be applied to common-ion as well as no-common-ion heterostructures, isovalent and heterovalent systems, polar and nonpolar interfaces, and it can also account for the strain effects which are present in pseudomorphic structures [40–43].

3. Computational approach

3.1. *First-principle self-consistent calculations*

Important progress in the application of ab initio schemes to materials science was made in the 1960s with the DFT proposed by Kohn [44,45]. It was shown that the total energy of a many-electron system is a *functional* of the electron density distribution $n(\mathbf{r})$. The latter therefore, rather than the many-electron wave function, plays a central role in finding approximate solutions of Schroedinger's wave equation for such systems. Within DFT, the many-body problem of interacting electrons is reduced to a system of single-particle Schrödinger equations (Kohn–Sham equations [45]), which must be solved self-consistently (SCF, for self-consistent field) and iteratively. Electron–electron interactions are fully included by adding to the Hartree potential an exchange–correlation term, which is also a functional of the charge density. The application of DFT to real systems has been possible by using approximate formulas for the exchange–correlation functional, among which the local density approximation (LDA) has proven to yield reliable results, at an acceptable computational cost, on the electronic ground-state properties of complex crystalline systems [46–48]. In the LDA, this functional [45] is reduced to a function of the *local* charge density that has been calculated accurately [49] and interpolated using parametrized forms (see, e.g., Ref. [50]). More accurate approximations are possible, such as the generalized gradient approximation, which accounts also for the gradient of the charge density distribution.

In principle the Kohn–Sham single-particle energy eigenvalues cannot be interpreted as removal (or addition) energies (*quasi-particle* energies), and the derived energy gaps from occupied to empty states and in general the excited states are not correct [1,48]. The underestimate of the energy gap for semiconductors is typically from $\approx 20\%$ up to $\approx 50\%$ with respect to the experimental value. The single-particle energies should be corrected for many-body effects [51,52], which are much larger, in general, than the numerical uncertainty of the LDA–SCF values. However, as will be discussed in Section 3.4, this limitation can be solved by calculating quantities related to the ground state and deriving quantities related to excited states using informations from experiments or more refined approaches.

The study of interfaces that we will illustrate in this chapter is based on the pseudopotential method, which is an efficient approach for dealing with semiconductors and metals of practical interest for electronic devices. In the pseudopotential approach, only the valence electrons, which are responsible for the formation of the chemical bonds and determine the relevant low-energy physical properties of the system, are explicitly treated. The pseudopotential – derived from SCF calculations for the isolated atom with an all-electron technique – describes the effects of the nucleus and of the core electrons on the valence electronic states [53–56]. For periodic solids, a plane-wave basis set up to a certain kinetic energy cutoff is generally used to expand the single-particle electronic orbitals.

The choice of a plane-wave basis set allows for a convenient reciprocal space formulation and a straightforward evaluation of the total energy of the system, as well as the forces [57] on the atoms and the macroscopic stresses [58]. The latter quantities can then be used to relax the atomic structure, allowing one to determine the equilibrium structural parameters of a bulk crystal or optimize the interface geometry in a complex superlattice. Integrals over the Brillouin zone (BZ), which are necessary to determine global quantities of the system like the charge density, are performed by a discrete summation over a set of special **k** points [59–62] that are typically representative of a uniform grid covering the BZ.

For metallic systems an electronic level broadening scheme is generally used together with the discrete **k** point summation [63–65].

The choice of all the technical parameters (kinetic energy cutoff, **k** point mesh, broadening) and of a threshold for the iterative solution determines the numerical accuracy of the calculations, which is typically of the order of 10 meV for the individual electron energy eigenvalues.

Major sources of error limiting the global accuracy are the choice of the pseudopotentials and the approximation used for the exchange–correlation functional in the Schroedinger equation. It is essential before simulating a heterostructure to make accurate tests on the constituent bulk materials in order to control the reliability of the results.

Dealing with semiconductor heterostructures, it is common to encounter alloys, which are widely used in semiconductor devices. If one is not interested in the atomic-scale structure of the alloy, simple nonstructural theories can be applied, such as the *virtual crystal approximation* (VCA) [66]. An A_xB_{1-x} alloy (or a pseudobinary $A_xB_{1-x}C$ alloy) is modeled using a single type of atom on the AB lattice (or AB sublattice): the virtual $\langle A_xB_{1-x}\rangle$ atom, whose pseudopotential is the weighted average of those of the "true" A and B atoms.

By its nature, the VCA approach cannot describe correctly the *microscopic* scale structure of the alloys as bond length relaxations are neglected, but it may also fail in describing the *macroscopic* scale structure, such as the average lattice parameter. In most cases, the VCA predicts correctly a lattice constant strictly following Vegard's law (i.e., linear interpolation between the lattice parameters of the constituents), but in other cases, such as for instance in SiGe [67] and GaInP alloys [68], it gives positive deviations from Vegard's law, which are not correct and disappear or even change sign using more refined approaches.

Concerning the average electronic properties of alloys, such as the valence band top edge or the band gap, the VCA often provides correct predictions, as the valence electrons are quite delocalized and feel a potential that is an average of the potentials individually originating from the *real* atoms. This is, for instance, the case of GaPAs [69,70] and other alloys made of rather similar components, where the VCA provides correct predictions for the band-gap bowing. At variance, the VCA fails when the mismatch between the electronic properties of the constituent bulks exceeds some critical value, such as in GaInAs [69].

In conclusion, the VCA has been largely successfully applied to a variety of semiconductor alloys, and even today, in the absence of direct and precise information about the actual morphology of the alloy and its interfaces, its use is justified in a wide class of alloys. For a review, although not recent, of the electronic structure theory of the alloy, we address the reader to Ref. [71].

Several computer codes, based on DFT and designed to perform electronic structure calculations, are freely available to researchers (a list with related links is available, for instance, in Ref. [72]). Among them, the Quantum-ESPRESSO [73] package allows for the numerical simulation of the electronic, structural, and dynamical properties of materials. The package, developed and distributed by the Numerical Simulation Center CNR-INFM DEMOCRITOS in Trieste [74], comprises the set of codes PWscf (plane-wave self-consistent field), FPMD (first-principle molecular dynamics), and CP (Car-Parrinello), which are fully integrated and compatible with each other. The PWscf code in particular is suitable for the study of heterostructures which is outlined in this chapter: it allows for the use of both norm-conserving and ultrasoft pseudopotentials and includes several useful postprocessing tools for extracting relevant physical quantities from the numerical simulations.

In the remainder of this section, we present several aspects of first-principle calculations, which are specific to the study of heterostructures.

3.2. Modeling heterostructures and nanostructures with supercells

Heterostructures can be studied using periodically repeated supercells, which allow for a convenient reciprocal space formulation of the problem, otherwise not possible because of the loss of translational symmetry. Supercells are actually more suitable to describe superlattices rather than isolated interfaces. But as far as the electronic properties are concerned, the relevant modifications produced by a neutral interface are confined to a small region, and the bulk features of the charge density are completely recovered within a few atomic units from the interface. This implies that the relevant interface features can be studied using supercells with a reasonably small number of atoms (i.e., a few atomic planes of each constituent). In Fig. 1 we show the smallest three isovolumic supercells that can be used to describe zinc blende-based ideal (001), (110), and (111) oriented heterojunctions: they all contain 12 atoms, i.e., three double layers of each binary semiconductor constituting the heterojunction. Their dimension parallel to the interface is suitable to describe heterostructures with abrupt interfaces, whereas larger supercells are required to describe atomic mixing at the interface. Longer supercells in the direction perpendicular to the interface are typically needed for heterovalent heterostructures. Typically, the supercells are chosen in such a way that they contain two interfaces that are equivalent in terms of stoichimetry and geometry, to avoid electric fields due to unbalanced charges.

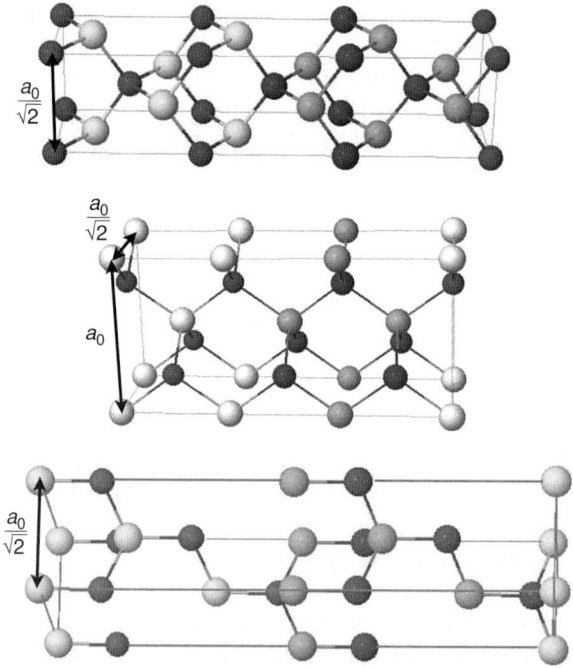

Fig. 1. Typical supercells used in ab initio calculations to describe zinc blende-based (001), (110), and (111) abrupt heterojunctions (from the top to the bottom).

The supercell self-consistent calculations provide the electronic charge density distribution $\rho_{n,\mathbf{k}}(\mathbf{r}) = \Sigma_{n,\mathbf{k} \text{ occ}} |\psi_{n,\mathbf{k}}(\mathbf{r})|^2$ where the sum runs over the occupied bands n and wave vectors $\mathbf{k}$ of the supercell BZ and $\psi_{n,\mathbf{k}}(\mathbf{r})$ is the electronic wave function. Figure 2(a) shows the contour plots of the electronic valence charge density of the GaAs/AlAs (001) interface simulated by a $3+3$ superlattice in three different atomic planes containing the growth axis. Because the atomic structure is periodic in the planes parallel to the interface (the x, y planes), the first obvious simplification is to consider *planar averages* as a function of the z-coordinate only:

$$\bar{f}(z) = \frac{1}{S} \int_S f(x, y, z) dx dy \qquad (1)$$

From the three-dimensional electronic charge density of the GaAs/AlAs(100) heterostructure, we get the one-dimensional plane-averaged charge density $\bar{n}(z)$ and electrostatic potential $\bar{V}(z)$ shown in Fig. 2(b). This shows two distinct – though closely similar – periodic

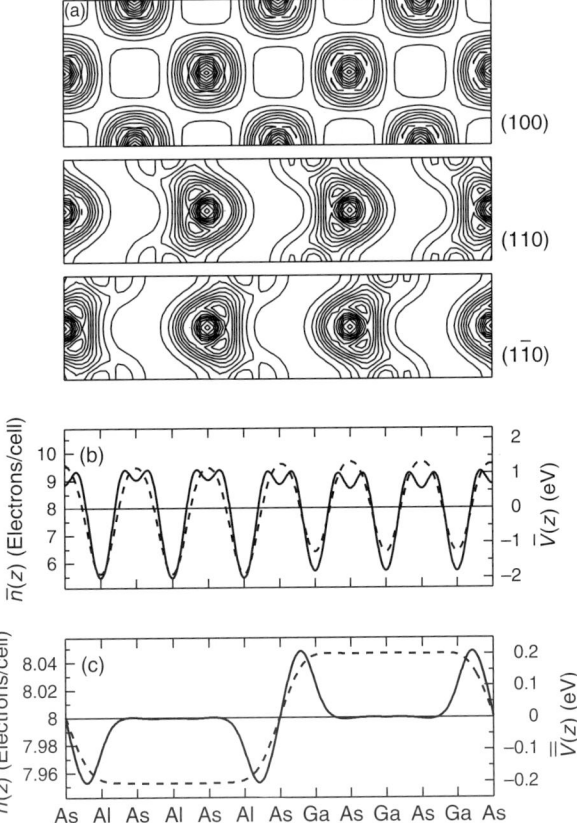

Fig. 2. Contour plots of the self-consistent electron density distribution (a) for GaAs/AlAs(001) heterojunction over different atomic planes containing the growth axis and centered on the interface anion. Planar averages ($\bar{n}(z)$ and $\bar{V}(z)$) (b) and macroscopic averages ($\bar{\bar{n}}(z)$ and $\bar{\bar{V}}(z)$) (c) of the electron density and of the electrostatic potential along the growth direction (data from Ref. [76]).

functions in the two bulk materials, which smoothly join across the interface. Because the system is lattice matched, the period a of $\bar{n}(z)$ and $\bar{V}(z)$ is the same on both sides of the interface and, in this particular case, equal to $a_0/2$, where a_0 is the bulk lattice parameter. The effect of the interface is related to the *difference* between these periodic functions and how they join to each other. Such a difference, which is barely visible in the figure, can be enhanced by getting rid of the bulk-like atomic-scale oscillations using the *macroscopic average* [39], whose concept is borrowed from classical electromagnetism [75]. Starting from the one-dimensional planar average $\bar{f}(z)$, it is particularly convenient to filter it simply with the step function $w(z) = (1/a)\Theta(a/2-|z|)$, commensurated with the period of $\bar{f}(z)$, obtaining:

$$\bar{\bar{f}}(z) = \int w(z')\bar{f}(z')dz' = \frac{1}{a}\int_{z-\frac{a}{2}}^{z+\frac{a}{2}} \bar{f}(z')dz' \qquad (2)$$

The results for the GaAs/AlAs(100) charge and potential are shown in Fig. 2(c). The macroscopically averaged quantities $\bar{\bar{f}}(z)$ show no microscopic oscillations on either side of the interface and recover the constant macroscopic limit in the two bulks. Conversely, deviations from the macroscopic value indicate the interface region and allow one to define unambigously an "interface dipole".

The macroscopic average technique can be extended to interfaces between two materials with different periodicity because of lattice mismatch or even structural differences. This is obtained by filtering the plane-averaged quantities twice, with filter functions appropriate to each material in turn [76]. Different choices of the filter functions will produce profiles with different atomic-scale details, but the same relevant macroscopic parameters.

3.3. Predicting structural properties

The theoretical prediction of the static equilibrium structural properties is obtained by means of total energy minimization. A crucial point is to compute forces on all the atoms. Forces on the nuclei are of course due to the direct nuclear interactions and the interaction with the electrons in the system. In the Born–Oppenheimer approximation, the electrons stay in their instantaneous ground state (gs) as the nuclei move. The force acting on each nucleus I can be calculated from the derivative of the electronic total energy with respect to the position R_I of the nucleus under consideration, $\mathbf{F}_I = -\partial E_{gs}[\{\mathbf{R}_I\}]/\partial \mathbf{R}_I$, using the Hellmann–Feynman theorem (see, for instance, Ref. [1]) which is based on the assumption that the electronic total energy satisfies a variational principle (in the present case, the Hohenberg–Kohn–Sham principle on which DFT is based).

The calculation of the forces allows one to obtain the equilibrium atomic structure: in order for the energy to be variational, one has to solve the SCF problem to obtain $E_{gs}[\{\mathbf{R}_I\}]$ for each individual geometry, then calculate forces, update the geometry, and then repeat the cycle until the forces are smaller than a given threshold. In this kind of "relaxation" calculation, the classical variables of the ions and the quantum variables of the electrons are treated separately.

Optimization of the atomic positions is important in the case of heterostructures. Even for heterostructures with lattice-matched and isovalent constituents such as InAs/GaSb, important lattice distortions can occur at the interface if the two constituents have no common ion, because of the formation of GaAs and InSb bonds which are strongly different with respect to each other (by ≈14%) and with respect to the bonds of the two bulk semiconductors [77]. Local lattice distortions are accompanied by charge displacements (ionic charges partially screened by the electronic charge distribution) that can contribute to the interface dipole.

Besides atomic relaxations at the interface, a different type of distortion of the structure is a scaling of space, i.e., an applied strain tensor. We limit the discussion to pseudomorphically grown heterostructures between lattice-mismatched cubic materials, in which case the substrate determines the parallel lattice constant $a_\|$. The mismatch between the substrate and the epilayer is accommodated by an appropriate strain of the epilayer along the growth direction, corresponding to a lattice constant $a_\perp$ that essentially depends on its elastic properties. From the experimental point of view, the thickness of the pseudomorphic epilayer has to be kept small enough to avoid misfit dislocations. Lattice-mismatched structures include most of the heterostructures between pure semiconductors (at variance with semiconducting solid solutions), such as Si/Ge and many others.

The macroscopic theory of elasticity provides a first hint for the determination of the equilibrium geometry of pseudomorphic heterostructures. Following Refs [32,78], and considering only the (001) growth direction, one has

$$a_\|^{epi} = a_\|^{sub} \equiv a^{sub}$$
$$a_\perp^{epi} = a^{epi}\left[1 - 2\left(\frac{c_{12}}{c_{11}}\right)^{epi} \epsilon_\|^{epi}\right] \quad (3)$$
$$\epsilon_\|^{epi} = \frac{a_\|^{epi}}{a^{epi}} - 1, \quad \epsilon_\perp^{epi} = \frac{a_\perp^{epi}}{a^{epi}} - 1$$

where $\overleftrightarrow{\epsilon}$ is the strain tensor and c_{ij} are the elastic constants of the bulk epilayer; the label sub refers to the cubic substrate and epi to the strained epilayer. More generally, the possibility of inducing strain in both materials constituting the heterojunction must be considered, when the substrate governing the pseudomorphic growth of the heterostructure has an intermediate lattice parameter.

The macroscopic theory of elasticity predicts the interplanar distances accurately enough in the bulk regions, i.e., two to three atomic planes away from the interface, whereas it fails at smaller distances from the interface where the interplanar distances sensitively depend both on the substrate composition and on the interface termination [79], and also, to a lesser extent, on the period of the superlattice [67]. The exact determination of the equilibrium structure can be achieved by looking for those atomic positions and value of the tetragonal deformation c/a of the supercell, which make the forces acting on the atoms and the macroscopic stress tensor components vanish.

The stress tensor can be calculated from the derivative of the total energy with respect to the strain, $\sigma_{\alpha\beta} = -(1/\Omega)(\partial E/\partial \epsilon_{\alpha\beta})$, where Ω is the volume of the supercell, α and β are cartesian indices. A method for calculating the stress tensor has been proposed by Nielsen and Martin [58] based on a generalization of the virial theorem, with explicit reciprocal space expressions that are suitable for ab initio calculations.

3.4. Predicting electronic properties: band structure alignments, density of states, and STM images

The electronic band profiles at the interfaces, and more specifically the valence and conduction *band offsets* (VBO and CBO) control the transport properties in semiconductor heterojunction devices. The band-gap difference $\Delta E_g(A/B) \equiv E_g(A) - E_g(B)$ between the two constituent materials is shared between valence and conduction bands:

$$\Delta E_g(A/B) = \text{VBO}(A/B) + \text{CBO}(A/B) \tag{4}$$

as schematically indicated in Fig. 3.

The band structure of each bulk material can be referred to a given internal reference level, say, the average electrostatic potential $\langle V \rangle$, which is an arbitrary reference. The arbitrariness comes from the long-range character of the Coulomb interaction that makes $\langle V \rangle$ of an *infinite* system ill defined [8,40]. As a consequence, it does not exist an *absolute* energy scale to which bulk band structures can be referred. This makes the problem of band alignment at interfaces a difficult one, which requires an accurate calculation of the interface charge distribution and of the corresponding electrostatic potential.

Although simple models and semi-empirical methods had a fundamental role in understanding most of the broad features of experimental results concerning band alignments, fully self-consistent ab initio calculations, which provide the electronic charge distribution at the interface, were instrumental in demonstrating the importance of interface details on the band alignments, and definitely contributed to explain the intrinsic nature of the band offsets in some classes of heterostructures and the extrinsic character in some others.

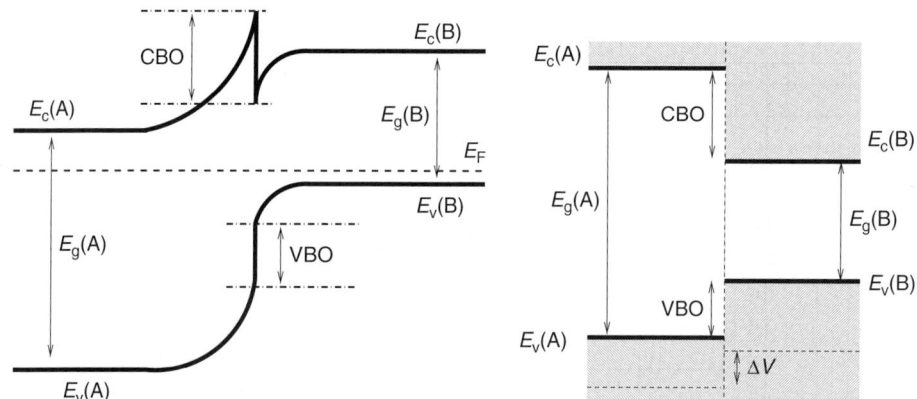

Fig. 3. Schematic spatial profile of the valence and conductions bands along the growth direction for a semiconductor heterojunction, and definition of valence and conduction band offsets, VBO and CBO. Left panel: with band bending due to space-charge effects; the spatial region considered is of the order of 100–1000 atomic units. Right panel: with flat bands, as the focus is on a spatial region of the order of 10 atomic units, where the band bending is negligible. In the latter panel, also the average reference electrostatic potential giving rise to the electrostatic potential lineup ΔV is schematically indicated.

From a theoretical point of view, convenient for such numerical approaches, the band offset can be split into two contributions:

$$\text{VBO}(A/B) = \Delta E_v(A/B) + \Delta V(A/B)$$
$$\text{CBO}(A/B) = \Delta E_c(A/B) + \Delta V(A/B) \quad (5)$$

as schematically shown in the right panel of Fig. 3. Flat bands are represented, as the focus is on a region of the order of 10 atomic units, where the band bending due to space-charge effects is negligible. In general, space-charge effects can modify the band-edge space profile on longer length scales (left panel). The accumulation of opposite charges on the two sides of the interface, and the resulting alternating sign of the charge distribution, causes a curvature of the band edges. The charge redistribution and the band bending are typical of long length scales compared with the screening length of the semiconductors. The formation of a roughly triangular potential well binding the change carriers at interfaces gives the possibility of realizing quasi two-dimensional systems of electrons and holes.

The band structure term ΔE_v (ΔE_c) is the difference between the relevant band edges in the two materials measured with respect to the average electrostatic potential $\langle V \rangle$ in the corresponding bulk crystal, and can therefore be derived from individual bulk calculations. The electrostatic potential lineup ΔV can, in principle, depend on structural and chemical details of the interface, and it is obtained from a self-consistent supercell calculation of the heterostructure as the difference between the *macroscopic averages* of the electrostatic potential in the two bulk regions across the interface, corresponding to the electronic charge distribution $\rho(\mathbf{r})$. The potential lineup is exactly related to the dipole moment of the charge profile:

$$\Delta V = 4\pi e^2 \int z\bar{\bar{\rho}}(z)\, dz \quad (6)$$

where ρ is the *total* (ionic plus electronic) charge density that averages to zero in the bulk-like regions.

At variance with the VBO, however, the electrostatic potential lineup is not a physically relevant quantity, as the partition in Eq. (5) is not unique. ΔV, and the corresponding bulk $\langle V \rangle$ values, must take into account all the *long-range electrostatic* potential contributions coming from the electronic and the ionic charge distributions. All the contributions related to the short-range local components of the potential (exchange–correlation, difference between the local part of the pseudopotential, and an ionic point-charge potential) are bulk quantities, and can be included arbitrarily in anyone of the two terms on the right-hand side of Eq. (5). Typically they are included in the band structure term.

Many-body corrections [51,52] are also embedded in the band structure term ΔE_v (ΔE_c) and do not affect the potential lineup across the interface, which – being a function of the ground-state charge density – can be accurately calculated within a DFT–LDA approach. Therefore, although being a source of uncertainty in the absolute values of calculated band offsets, the many-body effects do not affect the trend of the numerical results concerning the possible *dependence* of the band alignment on interface properties. Many-body corrections are typically of the order of tenths of electron volts for the valence band edge and up to about 1 eV for the conduction band edge [51,52]: it is therefore convenient to calculate the band alignment for the valence band edges, VBO and then obtain the CBO by adding the experimental band-gap difference ΔE_g. As a final remark, we notice that many-body corrections tend to cancel out for the VBO at semiconductor heterojunctions, so that the overall absolute uncertainty on the calculated VBO is typically of the order of 0.1 eV, less than that of the individual valence band edges.

An alternative procedure for obtaining the band offset from supercell calculations is to calculate the local density of states LDOS(ϵ, **r**) defined as

$$\text{LDOS}(\epsilon, \mathbf{r}) = \sum_{k,n} \rho_{k,n}(\mathbf{r}) \delta(\epsilon - \epsilon_{n,k}) \quad (7)$$

In the supercell calculations, the LDOS on the two sides of the interface are automatically referred to the same energy scale, so that the relative positions of the band edges far from the interface, where LDOS(ϵ, **r**) converges to the bulk density of states of the corresponding crystal, give directly the band discontinuities. This approach however is less convenient and less accurate than the potential lineup approach: a larger **k** point mesh and a larger supercell are required, as LDOS has a spatial convergence to bulk features which is slower than that of integrated quantities such as the charge density and the corresponding electrostatic potential.

The LDOS can be used also to simulate cross-sectional scanning tunneling microscopy (XSTM) images of heterostructures. XSTM studies of semiconductor heterostructures have been used to characterize compositional fluctuations, isovalent intermixing, and interfacial roughness in a variety of semiconductor heterostructures, and may also be employed to study interfacial chemical bonding at heterojunctions.

From the experimental point of view, XSTM studies require a cleavage in situ to expose a cross section of the grown heterostructure, which is free from spurious cleavage-induced morphological features. The natural cleavage surface is the {110} surface for most cubic semiconductors. This surface contains the zig-zag bond chains and is usually characterized by the absence of surface states, which would obscure the observation of the properties of interest. Therefore, the XSTM technique conveniently applies to (001) grown heterostructures.

From a theoretical point of view, the study of cross-sectional surfaces is conceptually similar to that of natural surfaces, apart from the unconventional surface geometry that requires the use of large simulation cells. The simplest and most widely used model to simulate STM images is due to Tersoff and Hamann (TH) [80]. The tunneling current is approximately given by

$$I(V) \propto \int_{E_F}^{E_F + eV_b} \text{LDOS}(\epsilon, \mathbf{r}) \, d\epsilon \quad (8)$$

where **r** is the position of the tip, E_F is the Fermi energy, and V_b is the applied tip–sample bias. The TH model is essentially a first-order perturbation theory, which does not take into account the tip–surface interaction and hence, in general, can be applied for fairly large tip–surface separations. In this model the proximity of the tip is assumed not to perturb the electronic structure of the surface. Despite its simplicity and the severe approximations used, the model has been widely applied to metallic and semiconductor surfaces, and has provided, at least qualitatively, a correct description of STM images. Extensions of this model have been proposed to cover also the possibility of p or d character orbitals of the tip [81,82]. Alternative approaches aimed at treating both the system and the tip have also been proposed [83–86].

3.5. Linear response theory

Linear response theory (LRT), as implemented within the LDA–SCF framework, has contributed to understanding the general trends of band offsets and providing a rationale for the results obtained from supercell calculations as described above. It consists in decomposing the hamiltonian of a given, real interface as the sum of the hamiltonian of a reference periodic crystal and a perturbation describing the difference between the real interface and the reference system. With a proper choice of the reference, the strength of the latter term should be sufficiently weak so that its effects can be taken into account by the low order of perturbation theory.

The reference system for studying the interface between two lattice-matched semiconductors could be anyone of the two bulks itself, but to minimize the strength of the perturbation, the optimal choice is the *virtual* periodic crystal, introduced in Section 3.1 in the case of alloys. Considering for definiteness an interface between two binary semi-conductors C_1A_1 and C_2A_2, the virtual crystal consists by virtual anions $\langle A \rangle$ and cations $\langle C \rangle$, which are, in terms of pseudopotentials, $v_{\langle A \rangle, \langle C \rangle} = (1/2)(v_{A_1,C_1} + v_{A_2,C_2})$. The perturbation that builds up the actual system (interface or alloy) amounts to replacing virtual ions with physical ones, in a given pattern, therefore with an appropriate superposition of perturbing potentials $\pm \Delta v_{A,C} = \pm(1/2)(v_{A_1,C_1} - v_{A_2,C_2})$. If the single perturbing potential $\Delta v_{A,C}$ is weak enough to induce a localized charge density electronic response $\Delta \rho_{A,C}$ which is *linear* in the perturbation, the *total* charge density of the real system is the charge density of the virtual reference system plus the superposition of the localized responses $\Delta \rho_{A,C}$, taken with the proper sign.

The relevant quantity for the interface potential lineup problem is the change in the electrostatic potential induced by the perturbations, which is ultimately related to the long-wavelength behavior of the isolated $\tilde{\Delta}\rho_{A,C}(\mathbf{q})$, the Fourier transform of the isolated responses. The electronic – and hence also the total – charge density induced by the isolated substitution has the full point symmetry of the substitutional site, the bare perturbation being spherically symmetric. The response $\Delta \rho_{A,C}$ can be obtained, through a direct approach and with two independent supercell calculations, as the difference between the self-consistent charge density of the perturbed system with an impurity and that of the unperturbed virtual material. In the case of the elemental or binary cubic semiconductors with T_d symmetry, $\Delta \rho_{A,C}$ has no dipole nor quadrupole moment.

Even restricting to the case of *lattice-matched* heterojunctions, differences arise depending on whether the heterojunctions are isovalent or heterovalent, i.e., the constituting bulks are from the same or different group, namely IV, III–V, or II–VI. In the case of isovalent heterojunctions, the pseudopotentials of the ions determining the perturbation are very similar, the perturbation is weak and the induced reponses (charge density and electrostatic potential) have amplitude much smaller than the typical bulk values, both in terms of absolute values and spatial variations. The charge density response is practically confined within a two-atom bulk Wigner–Seitz cell centered at the substitutional site, with an error of $\approx 0.01|e|$ for the integral over the cell; the quadratic term of the charge response is even more localized. In the case of heterovalent perturbations, the response charge densities are more extended: they are only partially contained within a two-atom bulk Wigner–Seitz cell centered at the substitutional site, and the higher order terms are more important. However, in all the cases here discussed, the LRT values of band offsets are reliable to ≈ 0.04 eV; in most cases to ≈ 0.02 eV, which is also the typical numerical accuracy of full SCF calculations. For further details, we address the reader to Refs [40–42]. We will discuss in Section 4 the application of the method to several families of heterostructures through selected examples.

4. Band Offsets

4.1. *Lattice-matched semiconductor interfaces*

The GaAs/AlAs heterojunction is the simplest and most studied among the *isovalent* heterojunctions.

We show in Fig. 4 the electronic charge and potential profiles from LDA–SCF supercell calculations performed for GaAs/AlAs (001), (110), and (111) heterostructures with a sharp interface [39] and, in the (001) case, also for a nonabrupt interface containing a mixed cationic plane, with equal concentrations of Ga and Al atoms [43]. It can be noticed that the charge and

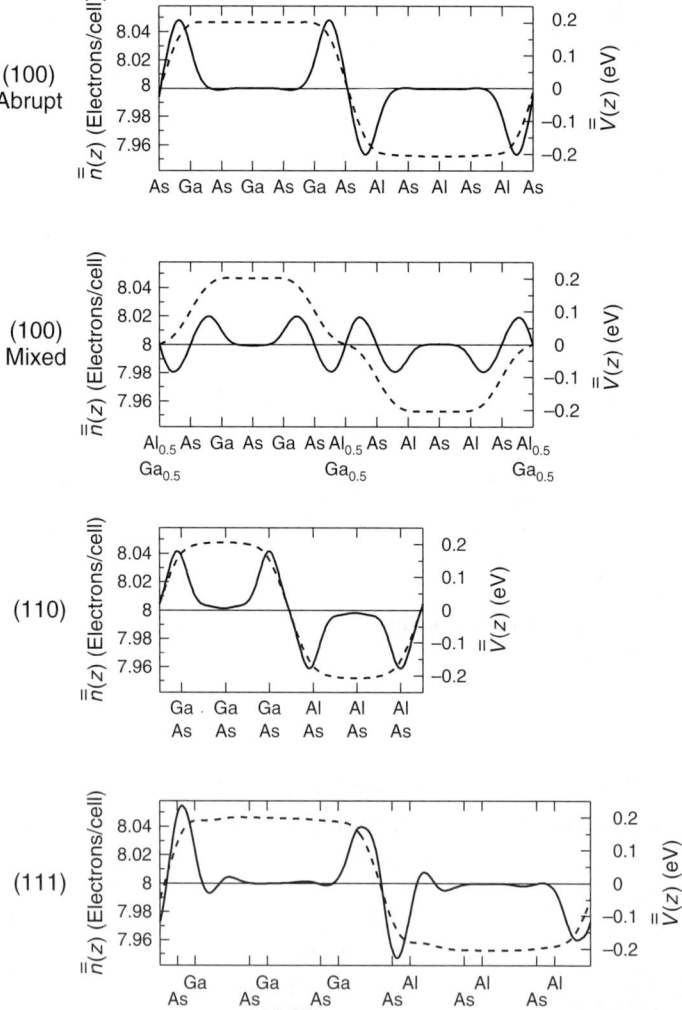

Fig. 4. Macroscopic averages of electron density (solid line) and corresponding electrostatic potential (dashed line) at GaAs/AlAs heterojunction. From the topmost to the lowest panel: (001) abrupt, (001) nonabrupt with a mixed cationic plane, (110) abrupt, (111) abrupt (data from Ref. [76]).

potential profiles, and hence the interface dipoles, have different shape (see Fig. 4, solid lines). Remarkably, the macroscopic averages of the potential give the same electrostatic potential lineup $\Delta V = 0.41 \pm 0.01$ eV in all the cases investigated. Adding the band structure contribution according to Eq. (5), the total LDA–SCF band offset VBO is 0.45 ± 0.02 eV, neglecting many-body and relativistic (spin–orbit) effects, which can be added a posteriori, and which amount to 0.1 ± 0.02 eV [52,87] and 0.03 ± 0.01 eV [88], respectively. We emphasize here that the many-body and relativistic effects enter only in the bulk band structure term, according to the present scheme, and therefore they do not change our main conclusion: the VBO is independent on the crystallographic orientation interface and on the cation disorder at the junction. Adding many-body and relativistic corrections, the resulting final estimate for the VBO at the GaAs/AlAs interface is thus 0.58 ± 0.06 eV, which compares well with the experimental values, ranging from 0.45 to 0.55 eV [5]. GaAs/AlAs shows a so-called type I, or "straddling," band alignment, where the band gap of GaAs is entirely contained within the one of AlAs. Different kinds of alignment are also possible, as schematically indicated in Fig. 5.

The application of LRT provides a rationale of the results obtained with the SCF supercell calculations. The appropriate reference crystal in this case is the virtual crystal $\langle Ga_{1/2}Al_{1/2}\rangle$ As. The isolated atomic substitution, which refers only to the cations as the anions are common, is neutral. The induced potential lineup across the interface is determined by the charge distribution $\Delta \rho_C(\mathbf{r})$ of the linear electronic response (the ionic cores of the reference system are the same) and it can be shown that it is given by

$$\Delta V = \frac{4\pi e^2}{3\Omega} \int r^2 \Delta \rho_c(\mathbf{r})\, d\mathbf{r} \qquad (9)$$

independently of interface orientation and abruptness. We stress that this property derives from the charge neutrality of the perturbation, and is therefore valid also for other isovalent interfaces.

In particular, it has been proven for the more general case of *no-common-ion* heterojunctions C_1A_1/C_2A_2, such as InAs/GaSb [77] or InP/Ga$_{0.47}$In$_{0.53}$As [89]. In the spirit of LRT, the VBO can be obtained by calculating separately the anion and cation contributions to the potential lineup, considering the C_1A/C_2A and CA_1/CA_2 interfaces, with lattice parameter taken equal to the common lattice parameter of the two real constituents, and then summing the two contributions.

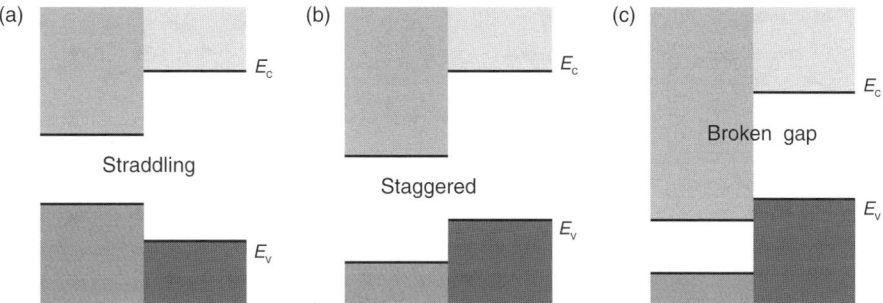

Fig. 5. Schematic possible types of band alignments: (a) straddling, (b) staggered, and (c) broken gap.

The *microscopic local interfacial strain* that can be present in the possible two different abrupt inequivalent interfaces in the polar directions, corresponding to the different terminations C_1–A_2 and C_2–A_1, can also be taken into account in an LRT framework that allows to describe in a physically sound and accurate way the effects of such strain on the VBO in terms of displacements of *effective charges* [77,89]. Remarkably, LRT explains why, despite the interfacial strain can vary with the interface composition, the band offset is almost unchanged. This is in agreement with experimental data, and full LDA–SCF supercell calculations for InP/$Ga_{0.47}In_{0.53}As$(100) [90], which show that although anion intermixing at the junction can reduce by 3% the interface strain, it has virtually no effect on the band offset provided the minimal energy structure is used for a given composition.

The consequences of the bulk-like character of the VBO and CBO are the commutativity and transitivity relationships, which are therefore valid, within the limits of LRT, in the whole class of isovalent semiconductor heterojunctions:

$$\begin{aligned} \text{VBO}(A/B) &= -\text{VBO}(B/A) \\ \text{VBO}(A/B) &= \text{VBO}(A/C) + \text{VBO}(C/B) \end{aligned} \quad (10)$$

Heterovalent heterojunctions are characterized by anions and cations of the constituent bulks, which belong to different groups of the periodic table and hence have different valence. For the sake of clarity, we refer to the case of Ge/GaAs as the simplest prototype of this class of heterostructures but the following considerations apply to other heterojunctions such as those between ZnSe, Ge, and GaAs.

In the (110) direction, each atomic plane is characterized by the same average ionic charge, so that an abrupt junction does not carry any ionic charge contribution and the potential lineup ΔV is due only to the electrons. At variance, ideally abrupt interfaces along a polar direction such as (001) are charged and hence thermodynamically unstable, as already emphasized first by Harrison in 1979 [91] and later discussed also by Martin [92]. The simplest *neutral* and stable interfaces one can envision are terminated by one mixed plane of anions or cations, $As_{0.5}Ge_{0.5}$ or $Ga_{0.5}Ge_{0.5}$ (Fig. 6). These two interfaces

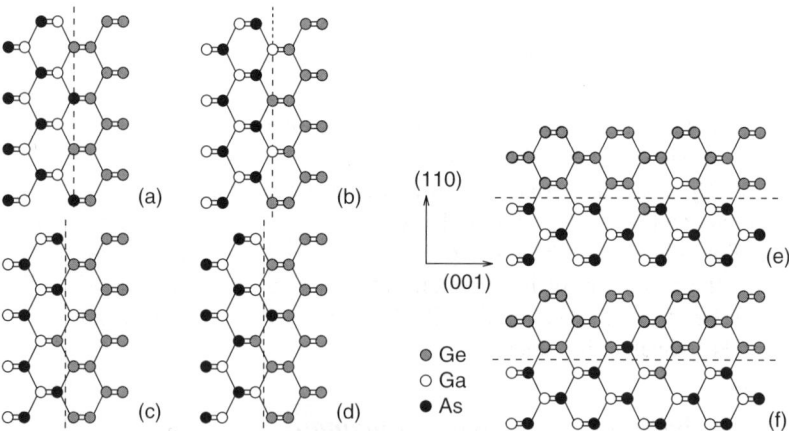

Fig. 6. Atomic configurations for some selected neutral nonabrupt Ge/GaAs interface morphologies in different orientations: (001) with one ((a) and b)) and two ((c) and d)) mixed planes; (110) with 25% of Ga–Ge (e) or As–Ge (f) swaps. The dashed line shows the position of the geometric interface.

are stoichiometrically inequivalent, and, because of ionic point charge contributions due to the different valence of the atoms involved (Ge versus Ga, Ge versus As), they correspond to different interface dipoles and consequently different band offsets. Supercell calculations give the VBO values 0.60 eV (Ge higher), 0.88 eV, and 0.28 eV for the abrupt (110), (001) anion mixed, and (001) cation mixed interface, respectively.

Such differences can be rationalized within LRT. The potential lineup can be split into two contributions:

$$\Delta V_{tot} = \Delta V_{hetero} + \Delta V_{iso} \qquad (11)$$

The latter is purely electronic, and is the only term present in the case of isovalent interfaces. As discussed for GaAs/AlAs, ΔV_{iso} is independent on interface details. ΔV_{hetero} contains both the ionic and the corresponding electronic contribution, and is formally equivalent to the lineup generated by an assembly of *point charges*, corresponding to the bare perturbation, which are screened by the electrons via a proper average $\langle \varepsilon \rangle$ of the dielectric constants of the two constituents. This term depends therefore on the atomic structure of the interface (orientation, abruptness, relaxation, and so on). However, once the structure is known – either experimentally or by independent theoretical calculations – ΔV_{hetero} can be evaluated from simple electrostatics.

In the (110) direction, the virtual crystal is made of atomic planes containing one cation and one anion per unit surface cell. The perturbation transforming the virtual crystal into an ideal *abrupt* interface is therefore neutral in each plane parallel to the interface, so that ΔV_{hetero} vanishes, and $\Delta V(110)_{abrupt} = \Delta V_{iso}$. For the two (001) interfaces, respectively with $As_{0.5}Ge_{0.5}$ or $Ga_{0.5}Ge_{0.5}$ mixed planes, the screened ionic point charge contribution to the offsets is equal in magnitude and opposite in sign: $\Delta V_{hetero} = \pm \pi e^2/2a_0\langle \varepsilon \rangle$ where a_0 is the lattice parameter involved. Assuming $1/\langle \varepsilon \rangle = 1/\varepsilon(Ge) + 1/\varepsilon(GaAs)$, the predicted difference of band offset between these two (001) interfaces is thus $\pi e^2/a_0\langle \varepsilon \rangle \approx 0.8$ eV for Ge/GaAs, in reasonable agreement with the results obtained from SCF supercell calculations that give a difference of about 0.6 eV.

In ZnSe/Ge the valence difference between the constituting elements is twice that of Ge/GaAs, and therefore the bare perturbation charges are also twice as big. Taking into account the dielectric constants of Ge and ZnSe, the predicted difference between the anion and cation mixed (001) interfaces would be $2\pi e^2/a_0\langle \varepsilon \rangle \approx 1.3$ eV.

In general, depending on growth conditions, atomic interdiffusion across the interface may occur over several atomic planes thus reducing the point charge contribution to the offset, and the experimentally observed variations of the offsets are smaller. However, deviations from the *commutativity* and *transitivity* rule as large as ≈ 0.5 eV, i.e., definitely beyond the experimental resolution, have been observed for heterovalent heterostructures, and these are clear fingerprints of the formation of inequivalent interfaces [6,93,94]. Further investigations on this system [95] and others such as Si/GaP, Ge/GaAs [96] have confirmed the tunability of band offset in heterovalent interfaces.

4.2. Strained (lattice-mismatched) semiconductor interfaces

Following the discussion of Section 3.3, we consider here the effects of strain on the band offsets of pseudomorphically grown heterostructures with lattice-mismatched cubic materials.

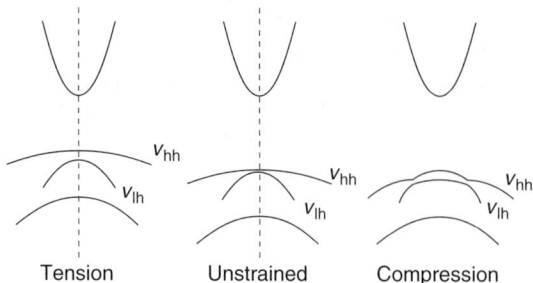

Fig. 7. Schematic representation of the energy band splitting caused by strain.

In a lattice-mismatched A/B heterojunction, one or both materials can be subject to strain, according to the choice of the sustrate and the value of $a_\parallel$.

We start from the strain effect on the electronic structure of an individual semiconductor. Neglecting relativistic effects, the top of the valence band at the Γ point in a bulk unstrained cubic semiconductor is threefold degenerate. A uniaxial (001)-oriented strain lowers the crystal symmetry from T_d to D_{2d}, thus splitting the valence band edge into a singlet and a doublet, as show schematically in Fig. 7. In the case of a tensile (compressive) strain, the singlet is below (above) the doublet. Taking into account the spin degeneracy, the valence band-edge manifold includes six states, which – in the absence of strain – are split by spin–orbit interaction into a quadruplet and a doublet, the split-off band (so). Moving away from the zone center, the quadruplet is split into a pair of doublets: the heavy hole (hh), and the light hole (lh) bands. A (001) uniaxial strain splits the hh and lh levels at Γ. The states split by the combined effect of strain and spin–orbit are found by adding a posteriori the spin–orbit effects to the results of nonrelativistic LDA–SCF calculations. First, the weighted average of the valence band manifold E_v^{ave} and the strain splitting δE_{001} are obtained from the band edges calculated neglecting the effect of spin–orbit:

$$E_v^{\text{hh,lh,so}} = E_v^{\text{ave}} \pm \delta E_{001} \qquad (12)$$

and then adding the effect of spin–orbit [78]:

$$E_v^{\text{hh}} = E_v^{\text{ave}} + \frac{1}{3}\Delta_0 - \frac{1}{2}\delta E_{001}$$
$$E_v^{\text{lh,so}} = E_v^{\text{ave}} - \frac{1}{6}\Delta_0 + \frac{1}{4}\delta E_{001} \pm \frac{1}{2}[\Delta_0^2 + \Delta_0\,\delta E_{001} + \frac{9}{4}(\delta E_{001})^2]^{1/2} \qquad (13)$$

where the $+(-)$ sign refers to lh(so). Δ_0 is the spin–orbit term, δE_{001} is – in absolute value – equal to 2/3 of the total separation between the two topmost states of the valence band manifold calculated *without* the spin–orbit interaction; its sign is negative for an elongation, positive otherwise. This implies that uniaxial tensile strain shifts the hh band above the lh band, whereas uniaxial compressive strain shifts the lh band above the hh band. The average E_v^{ave} is subject only to shifts due to the hydrostatic component of the strain, corresponding to a relative volume change, through the *absolute deformation potential* [78,97]. It can also be calculated as the threefold degenerate top valence state of the material in a cubic configuration with the same volume as the strained material, i.e., with an *effective* cubic lattice constant $\tilde{a} = (a_\parallel^2 a_\perp)^{1/3}$ [98].

The variation of the VBO with strain includes also the variation of the potential lineup ΔV, which however varies very little with $a_\parallel$. The VBO variation with strain is therefore mainly a *bulk* effect. Furthermore, its variation is small when it is calculated from the *averages* of the valence band manifolds (VBO$_{ave}$) and relevant when it is calculated between the *topmost* (VBO$_{top}$) split valence states. Therefore, in general the variation of the VBO with strain is mostly due to the strain-induced *splittings* δE_{001} (Eq. (13)) of the valence band manifold rather than from the *shifts* related to E_v^{ave}.

The simplest case of lattice-mismatched heterojunctions is Si/Ge, which is an example of *isovalent homopolar* interfaces [98–100]. The offset VBO$_{ave}$ is about 0.44 eV (Ge higher) for the configuration corresponding to a substrate made of 50–50% alloy, with a tunability of about 0.06 eV going from a substrate of pure Ge to one of pure Si. For VBO$_{top}$, on the contrary, the corresponding tunability over the whole range is one order of magnitude larger, i.e., about 0.5 eV.

Several common ion lattice-mismatched heterojunctions have been investigated in detail, such as GaAs/InAs [101], GaAs/GaP [102], ZnS/ZnSe [103], GaAs/GaSb [104], InP/InAs, and others [105].

With respect to the isovalent case, *heterovalent lattice-mismatched interfaces* offer an increased flexibility in terms of tunability of the VBO, owing to the peculiar nonbulk character of the band alignment between heterovalent materials. The Si/GaAs heterojunction has been studied [79] in a few selected configurations corresponding to pseudomorphic growth along the (001) orientation to discuss the effects on the VBO of *strain* (i.e., of different substrate concentrations) and of *chemistry* (i.e., of different interface terminations). The simplest terminations which give rise to neutral interfaces and were considered are those with only one mixed atomic plane $\langle Si_{1/2}As_{1/2}\rangle$ (As term.) or $\langle Si_{1/2}Ga_{1/2}\rangle$ (Ga term.). The predicted VBO, with spin–orbit effects included, and measured between the topmost valence states for different interface terminations and substrates, vary in a range of about 1.1 eV due to the combined effects of strain and chemistry. Other heterojunctions characterized by a large tunability of band offset due to the combined effects of strain and chemical composition of interfaces are, for instance, AlN/SiC(001) and GaN/SiC(001) [106,107].

4.3. Ideal and nonideal interfaces (relaxations, disorder, interdiffusion, defects)

Ab initio simulations necessarily deal with schematic and simplified representation of the complex physical systems, i.e., with ideal interfaces, abrupt on the atomic scale (or at most with simple mixed morphologies to satisfy charge neutrality, in case of heterovalent heterostructures), with no defects such as stacking faults, vacancies, antisites, or impurities.

Although heterostructure technology has received a tremendous impetus from the rapid advances in crystal growth techniques such as molecular beam epitaxy (MBE) that allow the synthesis of very high-quality structures, real samples are not always ideal.

Beside native stacking faults, corresponding to deviations of stacking sequence of atomic planes from the ideal sequence, and vacancies, other nonstructural defects are present. In epitaxially grown heterostructures, *atomic interdiffusion* induced during growth by strain inhomogeneities arising from stress relaxation and/or interface roughening, and also post-growth interdiffusion with or without thermal annealing can occur. In particular, in heterostructures making use of alloys, unintentional *composition inhomogeneities* could be present, due to inhomogeneous incorporation of the alloy components during growth. Although such inhomogeneities can be minimized, their residual occurrence can affect the electronic and optical properties. Another source of variations of the electronic properties in alloy-based heterojunctions is the occurrence of spontaneous *ordering* in the constituting alloys. It is important for device design not only to predict the value of band offsets at heterojunctions with given composition, but also to estimate the effects of both composition fluctuations and ordering. The problem has been addressed, for instance, for $In_{0.75}Ga_{0.25}As/In_{0.75}Al_{0.25}As$ heterojunction in Ref. [108] and references therein. Possible composition fluctuations can be accounted for considering that the *real* composition in the region where the band offsets are detected could be slightly different from the nominal one, say within $\approx 5\%$. The corresponding variation in the VBO, estimated interpolating linearly the VBO between the end points, is of the order of the typical numerical accuracy achieved in first-principle calculations, i.e., less than 0.05 eV. A similar variation is due to possible effects of short-range order/disorder that can be accounted for simulating the alloys constituting the heterojunctions with different ordered structures and calculating the corresponding VBO. The final conclusion is that, for alloy-based isovalent interfaces, realistic composition fluctuations and ordering effects are small and not exceeding ≈ 0.1 eV.

Remaining within the class of isovalent interfaces, the case of lattice-matched BeTe/ZnSe (100) interfaces is particularly instructive. Controversial experimental findings have been reported [109]: values ranging from 0.46 to 1.26 eV were measured in different growth conditions, corresponding to the largest VBO variation ever observed at semiconductor heterojunctions, and initially interpreted as due to Se- and Zn-terminated substrates. Such a large dependence of band offset on the interface morphology is unexpected in isovalent interfaces and would imply a clear violation of the LRT. Ab initio calculations reconcile experiments and theoretical picture: Be/Se and Zn/Te abrupt interfaces, as well as mixed interfaces, show equal band offsets in agreement with LRT, whereas offset variations over ≈ 0.8 eV are found in case of formation of strained interfaces between BeSe/ZnSe and ZnTe/ZnSe. Islands with such composition, different from the nominal one (BeTe/ZnSe) may form at the interface, depending on growth conditions (Se- or Zn-rich), as indicated by thermodynamical arguments and total energy calculations. Although additional BeTe/BeSe or BeTe/ZnTe interfaces could in principle be considered, they are unlikely to occur in the case considered where the substrate is ZnSe. However, the corresponding VBO for such interfaces pseudomorphically strained on ZnSe substrate would be in both cases about 0.4 eV (BeTe and ZnTe higher in the two cases respectively), i.e., within the variations discussed in the text.

It is not necessary to invoke the formation of islands of compositions different from the nominal one to explain the variability of the band offset in heterovalent heterostructures. Following previous discussion, sizeable effects on VBO are expected in case of interdiffusion at interfaces. Even in a nonpolar orientation such as (110), the band alignment can depend on the specific interface morphology: atomic disorder and interdiffusion create an additional interface dipole that gives a contribution ΔV_{hetero} to the lineup and consequently modifies the band offset. Focusing on ZnSe/Ge interface as a prototypical case, besides the abrupt (110) configuration (a) with composition profile $\ldots Ge_2-Ge_2-Ge_2-ZnSe-ZnSe-ZnSe-\ldots$, we can consider, as simple cases of atomic intermixing, interfaces containing Zn–Ge (b) and Se–Ge (c) swaps between the two adjacent atomic planes forming the interface. Such swaps maintain the charge neutrality but create an additional dipole. The resulting composition profiles are $\ldots Ge_2-Ge_2-Ge_{2-x}Zn_x-Ge_xZn_{1-x}Se-ZnSe-ZnSe-\ldots$ for case

(b) and ...Ge_2–Ge_2–$Ge_{2-x}Se_x$–$Ge_x ZnSe_{1-x}$–$ZnSe$–$ZnSe$–... for case (c), respectively. According to LRT, the corresponding variation of the VBO with respect to the abrupt case is of $\pm 4\pi e^2 x/a_0 \varepsilon$, where the plus/minus sign holds for case (b)/(c), respectively. Full supercell calculations performed for $x = 0.25$ and 0.75 (higher percentages of swaps would be unrealistic) confirm the LRT predictions, giving a variation of the VBO of $\approx \pm 0.5$ eV [110].

Heterovalent heterostructures could be also affected by the presence of staking fault density, mainly determined by the growth conditions. This problem has been addressed, for instance, in case of ZnSe/GaAs interfaces [111]. From an accurate characterization of the samples, it is evident that the minimum stacking fault is related to the formation of a ternary (Zn, Ga)Se alloy of variable composition with a substantial concentration of cation *vacancies*. Because of vacancies, its average lattice parameter is smaller than the one of GaAs and ZnSe, and therefore this alloy is under tensile biaxial strain when epitaxially grown on GaAs substrates, accumulating a nonnegligible elastic energy. Evidence of formation of ordered binary defected compounds such as Ga_2Se_3 or interface layers with vacancies arranged with other symmetry at ZnSe/GaAs interfaces has also been reported. Ab initio calculations indicate that in some particular thermodynamic conditions (basically Se-rich) the formation of defected interfaces with cation vacancies is even favored over the simpler, undefected, unstrained ones. The band alignments are strongly dependent on the particular interface morphology, in particular in the occurrence of vacancies; on the contrary different morphologies can correspond to the same or to similar band offsets, so that a unique correspondence between morphology, band offset, relative thermodynamic stability cannot be established. Nevertheless, the predicted stability of some defected interfaces and the compatibility of predicted band alignments with measurements support the experimental evidence of (Zn, Ga)Se defected compounds in high-quality ZnSe/GaAs(001) heterojunctions with low native stacking fault density.

We conclude this section mentioning the possibility of heterostructures based on semiconductors with morphologically different phases, the limiting case being the one of crystalline/amorphous semiconductors. Crystalline/amorphous silicon interfaces have received some attention for their use in solar cells and in other optoelectronic devices and have also been studied using the model-solid theory [112]. A direct atomistic simulation of the interface requires the use of a hierarchical combination of computational schemes different with respect to ab initio methods to have a reasonable structural model [113]. The starting point can be a constant pressure molecular dynamics simulation based on the "environment-dependent interatomic potential" (EDIP) to have a structural model of the amorphous part (a tetragonal 160-atom cell is suitable to this aim), followed by the creation of the heterostructure by joining the obtained amorphous sample and a crystalline slab of the same size; the resulting structure of 320 atoms, with cell dimensions $a = 2a_0$ and $c \approx 10a_0$, where a_0 is the lattice parameter of the crystalline phase, is further aged and relaxed through a careful thermal annealing perfomed by constant pressure, constant temperature EDIP molecular dynamics simulations, finally switching in same cases to a tight-binding molecular dynamics simulation.

Ab initio calculations of the electronic structure indicate that a sizeable band lineup can establish even at Si "homojunction", including in this term c-Si/a-Si, strained/unstrained, cubic-diamond/hexagonal-diamond interfaces. The value of the band lineup depends on the specific nature of the phases that form the interface. Remarkably, it is almost the same for all those samples with a nonnegligible concentration of coordination defects in the amorphous region, whereas it is sizeably different for a defect-free sample. This result can be ascribed to the semi-metallic behavior of the defect-rich a-Si, in particular to a Fermi level pinning mechanism acting in the former case, and to the establishment of a semiconductor/semiconductor type of interface in the latter. As expected for isovalent junctions, in the semiconductor/semiconductor case the source of band alignment is not due to interface-specific effects; that the basic mechanism is, instead, a bulk-related effect.

Along the same lines of c-Si/a-Si, also the simulation of c-Si/a-Si:H has been performed, generating models with different realistic H concentrations, up to 11% [114]. Model junctions with a realistic percentage of topological defects are obtained and show either a trend to *amorphization* of the c-Si region, or a trend to *recrystallization* of the a-Si:H region, thus indicating the variety and complexity of the real samples, and also suggesting that numerical samples can be representative of some realistic cases.

5. Designing heterostructures and engineering band offsets
5.1. Bulk strain and composition

We have shown that for lattice-matched isovalent semiconductor heterojunctions, the VBO is mostly determined by the bulk properties of the constituents. Interface details such as orientation and stoichiometry play a very minor role on the VBO, even in the case of no-common-ion heterojunctions, where the atomic interdiffusion can considerably change the composition-induced interfacial strain.

Using only isovalent materials, the only way to *tune* the offset is to act on the *bulk* rather than on the *interface*. This can be done with *bulk strain* (see Section 4.2), or with alloying. It is particularly interesting to study heterojunctions based on GaAs, AlAs, and InAs semiconductors: when combined into alloys, they form $In_xGa_{1-x}As/In_yAl_{1-y}As$ heterojunctions whose electronic properties can be in principle tailored according to the technological needs, acting on composition to control and intentionally modify the band offset. The small lattice mismatch between GaAs and AlAs can be neglected, and the system can be considered lattice matched when $x=y$, without introducing an appreciable error in the calculations. In the heterojunction between lattice-matched alloys, the VBO varies with composition from 0 eV for $x=y=1$ to 0.58 eV for $x=y=0$, i.e., for the case of the GaAs/AlAs interface. The alloy can be described by the VCA or, for selected compositions such as 0.25, 0.5, or 0.75, by fictitiously ordered structures that can be simulated using *real* instead of *virtual* atoms with reasonably small unit cells. Spin–orbit splitting [88] and many-body corrections to the valence band maximum are included in these values. For the many-body correction an interpolation between the values calculated by Zhu and Louie (Table XIV of Ref. [52]) for the end point materials and the 50–50% alloys was used. For $x=0.53$, when the system is lattice matched to InP, we calculated VBO is 0.19 eV, in the range of the experimental measurements [5], and in agreement with other pseudopotential calculations [115]. For $x=0.75$, a composition indeed successfully grown, the calculated VBO is 0.11 eV, obtained by adding to the LDA value of 0.07 eV a self-energy correction of 0.04 eV, taken from the values given in Ref. [52], properly scaled to be adapted to our calculations. The VBO(x) has a significant negative bowing with respect to the linear interpolation between the two end points $x=y=0$ and $x=y=1$ (Fig. 8). The effect on VBO of a possible ordering in the alloys can be estimated comparing the offset calculated for different ordered structure describing the alloy. For the system considered here, such effect does not exceed 0.05 eV, and similar conclusion can be extended to isovalent alloys.

Considering $x \neq y$, the heterojunction is lattice mismatched and in case of pseudomorphic growth one or both materials are under strain. Taking into account interfaces

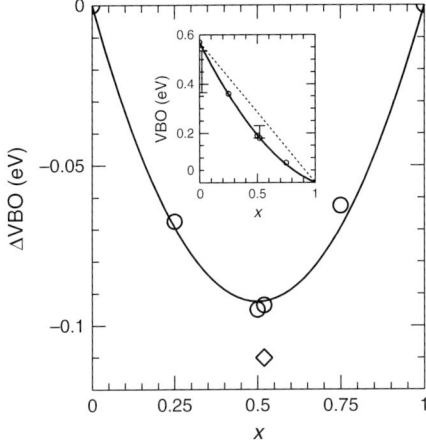

Fig. 8. VBO at In$_{1-x}$Ga$_x$As/In$_{1-x}$Al$_x$As (001) heterojunction as a function of the alloy composition x (inset). The error bars indicate the range of the experimental data. The main panel emphasizes the deviation from the linear interpolation between the two end points. Circles are from supercell LDA–SCF calculations reported in Ref. [76], the solid line is a quadratic fit, given as a guide to the eye; the diamond is from similar pseudopotential calculations from Ref. [116].

between the pure binaries and on different substrates, the LDA VBO goes from 0.83 eV in the case of InAs/AlAs interface with a parallel lattice constant equal to the one of GaAs or AlAs, to 0.44 eV in case of GaAs/AlAs unstrained interface, up to 0.07 eV at InAs/AlAs on InAs substrate and −0.27 eV in the case of GaAs/InAs interface on GaAs substrate. The range of variation of the CBO is even larger, due to the larger effects of strain on the conduction band edges and energy gaps rather than on the valence band edges: the maximum range of variation is almost 1.8 eV, from 0.71 eV occurring at a GaAs/InAs interface with a parallel lattice constant equal to the one of GaAs, up to −1.10 eV at a InAs/AlAs interface with a parallel lattice constant equal to the one of InAs.

Deviations from linearity of VBO with respect to the composition x are compatible with a LRT description. Within the spirit of LRT, one could choose the optimal reference virtual crystal for each composition depending on x; the response function also depends on x, so that the combined effect is a dependence on x of the VBO which is not simply linear. This applies for instance for GaAs/Ga$_{1-x}$Al$_x$As interfaces; however, due to the similarity of the dielectric properties and lattice parameters of GaAs and AlAs, one should expect a very small effect in this case, as it is indeed experimentally observed [5].

5.2. Interlayers

The peculiarity of heterovalent interfaces leads naturally to a practical way for modifying the offset at isovalent heterojunctions such as GaAs/AlAs, or even for creating an offset at a homojunction. It is instructive to examine first the case of homojunctions and consider the following (001) growth sequence: ...As–Ga–As–⟨Ge$_{1/2}$Ga$_{1/2}$⟩–⟨Ge$_{1/2}$As$_{1/2}$⟩–Ga–As–Ga.... Ideally, this sequence of atomic planes can be obtained from bulk GaAs in two steps: creating first a ⟨Ga$_{1/2}$Ge$_{1/2}$⟩-terminated GaAs/Ge interface, and then transforming back the Ge half-space to GaAs with a $\langle Ga_{1/2}Ge_{1/2}\rangle$-terminated GaAs/Ge interface shifted by one interplanar spacing, $a_0/4$. These are precisely the two GaAs/Ge inequivalent interfaces discussed in Section 4.1; therefore, using the difference between the VBO's of these two inequivalent interfaces a net potential drop $\Delta V = \pi e^2/a_0\langle\varepsilon\rangle$ is predicted for the above transformation. This potential drop is the same that would result from a *microscopic capacitor* [116] whose plates are placed at a distance $a_0/4$, carry a surface charge $\sigma = e/a_0^2$, and are filled with a material whose dielectric constant is the same as that of the virtual crystal. The above sequence of atomic planes can also be thought of as due to the transfer of a proton per atomic pair from the As to the Ga planes [91].

Within LRT, this viewpoint is easily generalized to arbitrary concentrations of Ge in a pair of consecutive compensated GaAs planes, $\langle Ge_xGa_{1-x}\rangle\langle Ge_xAs_{1-x}\rangle$ (this ensures local charge neutrality). In this case, one has $\Delta V(x) = 2\pi e^2 x/a_0\langle\varepsilon\rangle(x)$, where $\langle\varepsilon\rangle(x)$ is the effective dielectric constant of the reference system whose optimal choice is a bulk alloy having the same composition as the doped region between the two plates of the microscopic capacitor. The corresponding dielectric constant is $\langle\varepsilon\rangle^{-1}(x) = (1-x)\varepsilon_h^{-1} + x\varepsilon_i^{-1}$ [117], where h indicates the host material, GaAs, and i the interlayer, Ge. Following this reasoning, the behavior of the VBO at small doping ($x \to 0$) depends on the host material, whereas at high doping ($x \to 1$) it is dominated by the electrostatic screening of the dopant. The excellent agreement between the simple predictions of LRT and full supercell LDA–SCF calculations [117] confirms the soundness of the physical picture underlying this LRT approach. Analogous considerations hold for other polar orientations – such as (111) – as well, and also for compensated heterovalent interlayers embedded into a group IV bulk. In fact, the existence of a measurable potential drop across a GaAs interlayer embedded in Ge along (111) has been detected experimentally [118].

The above results can be generalized to the heterojunctions, e.g., doping a GaAs/AlAs interface with ultrathin layers of Si or Ge. In this case, the band offset is the sum of an *intrinsic* term, VBO_I, plus a doping contribution. If the Si dopant atoms are assumed to be uniformly distributed over two consecutive atomic layers, the VBO is simply

$$\mathrm{VBO}(x) = \mathrm{VBO_I} \pm \frac{\pi e^2}{a} x((1-x)\varepsilon_h^{-1} + x\varepsilon_i^{-1}) \qquad (14)$$

where ε_h^{-1} is, in this case, the average of the inverse dielectric constants of GaAs and AlAs. In Fig. 9, the predictions of LRT are compared with full supercell LDA–SCF calculations [117] and experimental data [119] for the GaAs/Si/AlAs VBO, as a function of the Si coverage which is $2x$ (measured in atomic monolayers, so that $x = 1$ corresponds to a full Si bilayer). The supercell calculations were done using the VCA for $x < 1$ to describe the two consecutive doped atomic layers at the interface. The results compare well with experiment [119] up to a coverage $2x \approx 0.5$, whereas a substantial disagreement appears for higher coverages. Such a disagreement between theory and experiments should be ascribed to the simple picture of dopants confined over two atomic planes.

The model of the microscopic capacitor is adequate for thin interlayers, but it cannot be extended to the thick coverage limit. It would predict a dipole monotonically increasing with coverage, and this cannot be reconciled with the energetic stability of the junction; furthermore, in the limit of two isolated interfaces, each of them must be individually neutral, as previously discussed. Experimental measurements [120] in AlAs/Ge/GaAs(001) and GaAs/Ge/AlAs(001) single quantum well structures with thick Ge interlayers (2–16 monolayers),

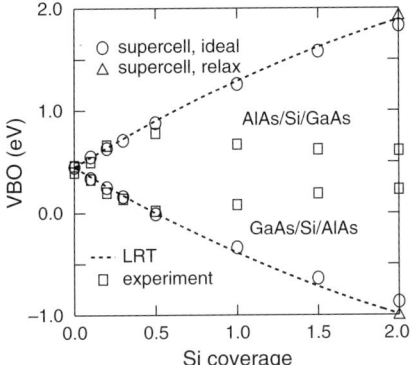

Fig. 9. VBO at GaAs/AlAs heterojunctions as a function of the coverage $2x$ of a Si interlayer. Circles: LDA–SCF supercell calculations with ideal unrelaxed zinc blende atomic positions; squares: experimental results from Ref. [119]; dashed line: predictions of LRT. Triangles indicate results of calculations performed allowing a full microscopic relaxation of the atomic positions, due to the lattice mismatch between Si and GaAs/AlAs (numerical data from Ref. [117]).

show that the band offset between GaAs and AlAs across the interlayer is independent of its thickness in the explored range, and that it is different from that directly measured at GaAs/AlAs interface; this suggests that two neutral and inequivalent interfaces are established already at such thickness. This situation is analogous to the lack of transitivity that is observed in the case of isolated III–V/Ge, ZnSe/III–V, and ZnSe/Ge interfaces [93,121].

6. Localized interface states

The existence of localized electron states at semiconductor heterojunctions is generally claimed whenever anomalous features are observed in photoemission, optical, or transport measurements [6]. Their unambiguous identification, however, is far from being easily performed. Reflectance anisotropy spectroscopy [122] has been recently applied to a few interfaces [123] and seems to be a promising tool for more direct investigations.

Interface states can have different physical origin. In some cases, they are associated with the presence in the neighborhood of the heterojunction of substitutional defects, such as antisites, or vacancies which typically arise at junctions between two morphologically different phases (e.g., cubic GaN grown on wurtzite InGaN/GaN [124]). Localized electron states can also occur at contacts between semiconductors with the same (or similar) structure but different chemical properties, such as lattice-matched semiconductor heterojunctions. The existence of such states and their characteristics depend on the interface morphology. Localized states have been detected at zinc blende/zinc blende heterovalent interfaces, e.g., GaAs/ZnSe [123], at diamond/zinc blende interfaces, e.g., Ge/ZnSe [125], and even at isovalent no-common-ion interfaces, e.g., BeTe/ZnSe [126]. The mechanisms governing their formation as well as the conditions for their existence are still under study. Simple electron-counting schemes indicate that the valence mismatch between neighboring atoms from opposite sites of the junction can result in localized interface states corresponding to donor or acceptor bonds. Therefore, heterovalent heterojunctions, which naturally exhibit such kind of bonds in the interface region, are excellent candidates for the existence of localized interface states.

Occupied and/or empty interface states have been predicted in the lower and/or upper region of the common fundamental gap for GaAs/Ge [127,128] and for ZnSe/Ge [127] for the unreconstructed (001) interface, both charged and in presence of a compensating external electric field [128]. We stress that, in these works, the band alignment is typically taken as an external parameter and is not calculated explicitly for the particular interface morphology.

The general finding for abrupt (110) heterojunctions is that interface states with energy in the common fundamental gap exist only in limited regions of the two-dimensional BZ.

Full supercell ab initio calculations are the ideal tool to study the microscopic origin of the interface states, as in these calculations, one has simultaneously access to band alignments and individual electronic states in a consistent way. True atoms rather than virtual ones are needed to study the atomic origin and the bonding character of the individual electronic states.

The case of ZnSe/Ge(110) interface is a prototypical heterovalent system where interface states induced by acceptor and donor bonds should be particularly evident given the large mismatch between the chemical valences of the constituent atoms. The calculated VBO for this interface is about 1.70 eV (slightly different technical parameters would give some differences) [43,129–131]. For the two (110) configurations with atomic intermixing, the calculated VBO is 2.11 eV (b) and 1.14 eV (c), in reasonable agreement with the predictions of LRT, which is less accurate in this case characterized by a large mismatch of chemical valences and therefore by a strong perturbation.

We show in Fig. 10 the macroscopic averages of the LDOS on the (110) atomic layers of the heterojunctions (a), (b), and (c), respectively, going from the central atomic layer of the Ge slab (upper panels) to the central atomic layer of the ZnSe slab (lower panels). The small, but detectable, differences among the LDOS of the central layers of the three configurations, and the nonvanishing density of states in the energy gap region of Ge, indicate that the supercell is not thick enough to fully recover the bulk features of the electronic spectrum. As it has been already noticed, interface effects on the electronic states become negligible only at a distance of several atomic layers from the interface, at variance with the integrated quantities such as charge density and electrostatic potential. The black areas of the figure represent the excess LDOS with respect to the corresponding constituent bulk DOS. The excess LDOS in an atomic layer, with respect to the corresponding bulk values, indicates the existence of interface-related states which have some probability to be within the layer. Figure 10 shows that localized interface states exist below the valence band for all three configurations. They essentially correspond to interface-modified Se-s states. Panel (a) of the figure, which corresponds to the abrupt case, gives no evidence of the existence of localized interface states in the common fundamental gap, as already observed in Refs [29,132,133]. Conversely, for the interfaces with atomic intermixing, a sizeable excess LDOS is clearly visible in the common gap.

This is the energy region of practical interest because interface states with energy in the gap are easily detectable in experiments as they affect considerably the electronic, transport, and optical properties.

For a precise identification of localized interface states, the spectrum of the heterojunctions must be compared with the projected band structure (PBS) of the two constituent bulks: states which do not overlap any of the PBSs are candidates to be *true* interface states; *resonance* states, on the contrary, are degenerate with the bulk Bloch states of at least one constituent and hence overlap either one or both bulk PBSs. For an easier comparison of the three structures, the results for the (110) abrupt system are represented in the small interface

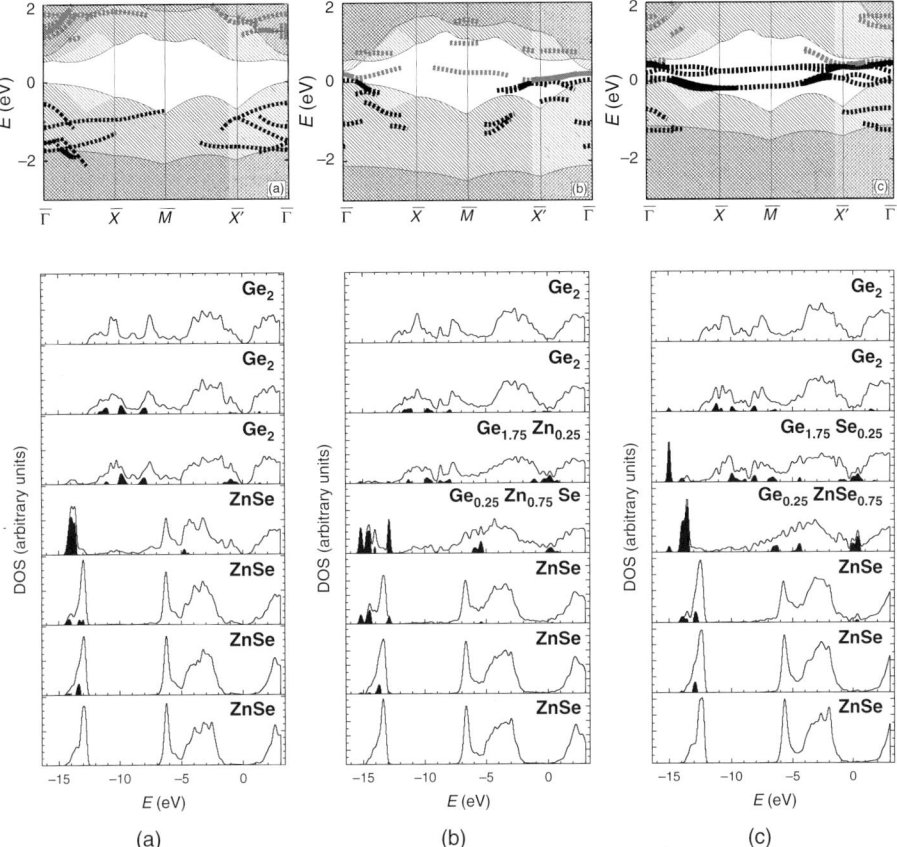

Fig. 10. Upper panels: Local density of states of the ZnSe/Ge(110) interfaces studied in this work: abrupt (a), with 25% Zn–Ge swaps (b), with 25% Se–Ge swaps (c). The profiles correspond to local density of states averaged over layers centered on atomic planes and with thickness equal to one interplanar distance. From top to bottom, the layers go from the one centered on the central plane of the Ge slab to that centered on the central plane of the ZnSe slab. Black areas indicate the excess LDOS with respect to bulk. Lower panels: corresponding electronic structure. For each case, the projected band structures of the two bulk materials (hatched regions), properly aligned with each other according to the calculated VBO, are shown together with localized interface states (solid lines) and resonance states (dashed lines). Occupied (empty) states are indicated with black (grey) lines (band structure data from Ref. [134], LDOS data from Ref. [130]).

BZ corresponding to the (2×2) geometry of the interfaces with atomic intermixing. A complex structure of localized states is found for the two nonabrupt interfaces. A detailed discussion is in Ref. [134].

We conclude that the existence of localized interface states is neither characteristic of given types of heterojunctions (heterovalent rather than isovalent) nor of particular interface orientations, rather it depends on many factors including the detailed atomic-scale structure of the interface. Concerning the nonpolar ZnSe/Ge(110) interface, atomic intermixing is responsible not only for sizeable variations of the band alignment with respect to the abrupt case but also for nonnegligible modifications of the interface density of states. Notably, monitoring simultaneously the existence of localized interface states and the value of band offsets, can be an efficient tool to investigate the crystallographic perfection of an interface on the atomic scale.

7. Simulating cross-sectional scanning tunneling microscopy images for characterization

Successful experimental XSTM studies of semiconductor heterostructures are rather recent achievements [135–139]. A first experimental report of atomic resolution imaging and spectroscopy of GaAs/AlGaAs superlattices dates back to about 15 years ago together with a combined theoretical study [140]. Studies of several clean and defected surfaces of semiconductors, and in particular of several III–V binary compounds, are summarized in Ref. [141].

With STM, anions are visible at negative bias as they contribute to the valence band edge, while cations are seen at positive bias corresponding to empty states at the bottom of the conduction band. The contrast in the images results from a combination of geometric and electronic factors, i.e., from the different electronic structure of individual atoms and also from their different position with respect to the exposed surface. It is therefore essential an accurate optimization of the atomic positions in the numerical simulations of STM images.

A clear example of the importance of geometric and electronic factors is shown, for instance, in a combined experimental and theoretical work on InAs/GaSb superlattices [139]. Besides the difference in the image on the extended scale caused by the different electronic structure of the two materials (for instance, both in the predicted and measured images the InAs surfaces appear lower than GaSb), local variations and peculiar features can be seen in the interface region: for instance, InSb- or GaAs-like interfaces appear different because of different degree of disorder and atomic intermixing, which determine local differences in bond length and geometric distortions.

Not all the structures visible in the experimental images can be clearly identified with experimental tools only. Numerical simulations of XSTM images are therefore essential to help in identifying the atomic-scale configurations compatible with experimentally detected images. This is, for instance, the case of Si-doped GaAs heterostructures, where it is rather difficult to obtain direct information about the position and the environment of Si dopants due to its amphoteric behavior and the possibility that it occupies both As and Ga sites. The systematic study of simulated XSTM images for many configurations has allowed to identify the self-compensated donor–acceptor configurations (reported in Fig. 11) as the simplest ones which are compatible with those experimental observations, and also to exclude other configurations [142]. Other examples of concerted experimental and theoretical efforts for semiconductor surfaces, not only heterostructures, are summarized in Ref. [141].

In spite of the uncertainties due to several factors (tunneling voltage, nature, and shape of the tip in case of experiments, specific technical details of the numerical simulations) not fully under control, the numerical simulation of XSTM images provide a qualitatively correct picture of the microscopic mechanism responsible for the observed images, thus pointing to the value of first-principles simulations as a tool for the characterization of materials and their interfaces.

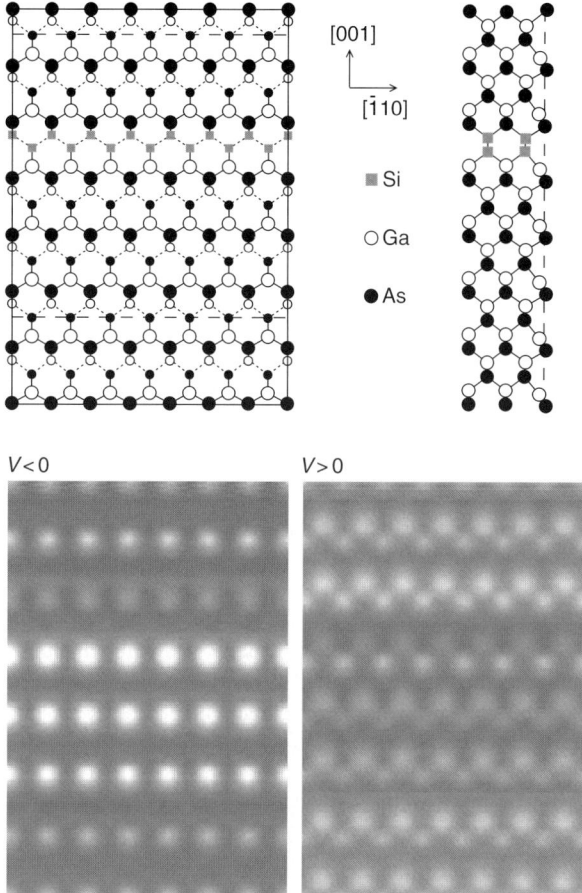

Fig. 11. (001) Si bilayer in GaAs, with a particular configuration of the exposed surface where Si_{Ga} and Si_{As} are sub-surface. Upper panels: ball-and-stick model of the relaxed surface, top and side view (Ga: open circle, As: close circle, Si: square). Dashed lines in the top view indicate the extension in the [001] direction of the supercell used to simulate the system with the interlayer. In the figure, the supercell is repeated in the [110] direction but not in the [001] direction, where it is instead embedded in the perfect clean surface to give a better feel of the image of an isolated interlayer. Lower panels: simulated STM images at bias voltages of -1.5 eV (occupied states, left) and at $+1.2$ eV (empty states, right) respectively. The size of the entire region shown in our simulated STM images corresponds to $\approx$ 2.75 nm; $\times$ 3.89 nm (data from Ref. [142]).

8. Semiconductor heterostructures for spintronic applications

Within the solid-state community the word "spintronic" is rapidly becoming popular to denote electronic-like heterostructures where the relevant physical quantity is the spin of the carriers and its interactions with external magnetic fields rather than the charge of holes and electrons and the associated electronic properties. The main target is the design and the realization of new devices with magnetic layers suitable for spin injection in semiconductors, and whose properties could be tailored in accordance with the needs of electronic industry.

As for the magnetic constituent, considerable attention was directed to the family of the half-metallic so-called Heusler compounds which are metallic in the majority spin channel and semiconducting in the other. NiMnSb is one of the most known among the Heusler alloys and it has been recently successfully grown on GaAs(001) [143]. It has a fcc crystal structure, with atomic basis composed by Ni at $(0,0,0)a_0$, Mn at $(\frac{1}{4},\frac{1}{4},\frac{1}{4})a_0$ and Sb at $(\frac{3}{4},\frac{3}{4},\frac{3}{4})a_0$, and the site at $(\frac{1}{2},\frac{1}{2},\frac{1}{2})a_0$ is empty. For the presence of this empty site, NiMnSb is known as "half-Heusler" alloy, at variance with others such as Co_2MnGe that have also this site occupied and are known as "full-Heusler" alloys. Mn and Sb atoms are surrounded by Ni atoms, with a tetrahedral coordination like in the zinc blende structure. This kind of coordination is considered responsible for the fact that the minority spin band structure is very similar to the one typical of zinc blende semiconductors [144], making the system half-metallic. Furthermore, the good structural matching with conventional semiconductors makes the Heusler alloys excellent candidates for integration with semiconductors.

The same theoretical approach presented above for nonmagnetic semiconductor heterostructures applies also to magnetic/semiconductor interfaces with a few additional complications: (i) *local spin density approximation* instead of LDA to account for the two spin channels; (ii) larger supercells as Heusler alloys have a more complex basis; (iii) Bz integration performed with smearing techniques [64,65] in addition to a larger **k** point set, as done for metals. Extending the concept of band offset, the key parameters are the *spin-resolved band alignments*. For the minority spin channel, we can use the procedure explained above for the nonmagnetic semiconductor/semiconductor interfaces, whereas the majority spin channel resembles a metal/semiconductor contact and therefore the relevant band alignment is the *Schottky barrier* (see Ref. [76] for a review of metal/semiconductor contacts), defined as the difference between the Fermi level of the supercell, which is the same of the Heusler alloys, and the semiconductor valence band top edge.

A few theoretical investigations on magnetic materials/conventional semiconductor interfaces have been performed, e.g., on NiMnSb/CdS [144], NiMnSb/GaAs [145], Co_2MnGe/GaAs [146], and Co_2CrAl/GaAs [147].

For the sake of definiteness, we discuss the case of NiMnSb/GaAs (001) heterojunction. In order to simulate an interface with NiMnSb epilayer epitaxially grown on GaAs substrate, we fix $a = b = a_\parallel/\sqrt{2}$, where $a_\parallel = a_0^{th} GaAs) = 5.54$ Å. ($a_0^{exp}(GaAs) = 5.65$ Å), and we optimize both c (minimizing the stress along its direction) and the internal microscopic degrees of freedom (atomic positions). The GaAs slab remains cubic, unstrained, as expected. Because $a_0^{th}(NiMnSb) = 5.71$ Å ($a_0^{exp}(NiMnSb) = 5.92$ Å), NiMnSb has a lattice mismatch of about 3.1% (experimentally 4.6%) with respect to GaAs and its slab pseudomorphically grown on GaAs is under a compressive bisotropic in-plane stress and elongated along the growth direction: in the region far from the interfaces its perpendicular lattice constant is $a_\perp^{th}(NiMnSb) = 5.91$ Å. The half-metallic character of strained NiMnSb epitaxially grown on GaAs substrate is kept under these pseudomorphic growth conditions.

The (001) interface is simulated by a tetragonal periodically repeated supercell of 40 atoms with a basis rotated by 45° with respect to the cubic axis, containing a slab of GaAs and a slab of NiMnSb (Fig. 12). In the (001) direction, the two constituents show this stacking: ...–Ga–As–Ga–As–... and ...–Ni–(Mn,Sb)–Ni–(Mn,Sb)–.... Therefore, different, in principle nonequivalent, interface configurations can occur according to the termination of the constituent slabs and to the particular way of matching the two structures: for instance, As, Mn, and Sb atoms can occupy the same fcc sublattice, whereas Ni occupies partly the cationic Ga sublattice. In such a case, the simplest configurations are the abrupt interfaces As–Ni and

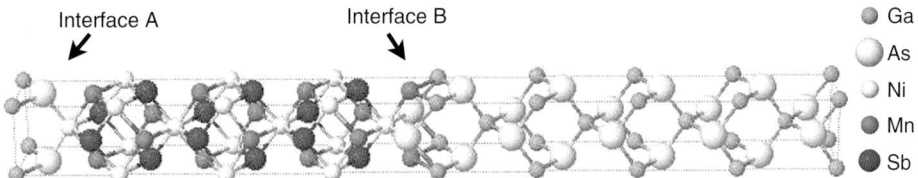

Fig. 12. Supercell used in the calculations of the NiMnSb/GaAs (001) heterostructure of particular interest for spintronics. This supercell has two non equivalent interfaces, labelled (a) and (b), described in the text.

Ga–(Mn,Sb). With an integer number of formula units of GaAs and NiMnSb, we can simulate both interfaces in a single supercell. Another simple configuration is obtained considering a single mixed *anionic* plane (As,(Mn,Sb)) which for instance can be a (Mn,As) plane where As substitutes the Sb (isovalent) atom of the Heusler alloys. This is actually one of the two interfaces (B) (the other is the As–Ni abrupt interface, (A)) described in the simulation cell. From supercell calculations we can extract therefore a *majority Schottky barrier* and a *minority valence band offset*, the latter defined as the difference between the GaAs and the minority NiMnSb valence band top edges. These band alignments amount to 0.46 and -0.09 eV, respectively, taking the average of the two nonequivalent interfaces, which have slightly different values. A schematic picture is shown in Fig. 13.

Typically in magnetic/nonmagnetic interfaces, the half-metallic character peculiar on the magnetic constituent is lost at interface, as it can be noticed from the spin-resolved density of states (Fig. 14). This must be ascribed mainly to the fact that the Mn and Sb atoms (which dominate the minority band gap in the bulk Heusler alloy) show a sizeable density of states at the Fermi energy also in the minority channel when they are in the interface regions. This is

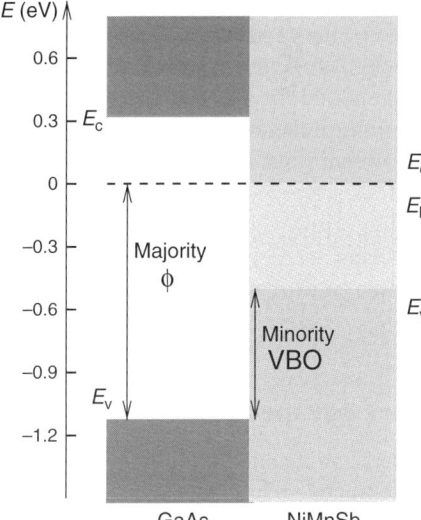

Fig. 13. Schematic picture of the spin-resolved band alignment at NiMnSb/GaAs heterojunction for the particular interface morphology discussed in the text, showing the *majority Schottky barrier* and *minority VBO* (data from Ref. [145]).

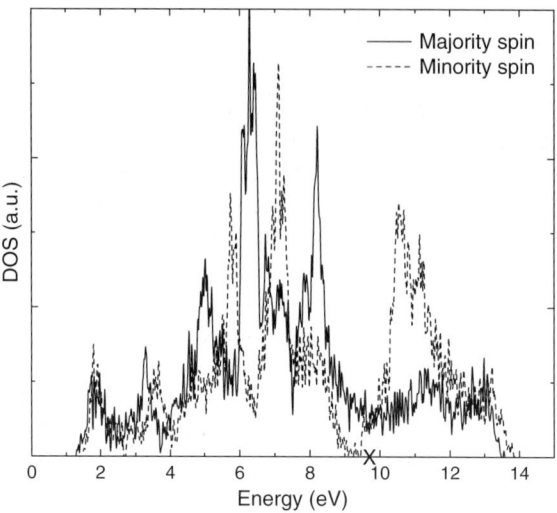

Fig. 14. Majority and minority spin density of states (arbitrary units) of the NiMnSb/GaAs(001) supercell for the particular interface morphology discussed in the text. The Fermi energy is set at zero energy (data from Ref. [145]).

consistent with the theoretical predictions of Refs [144,146] and with many experimental findings, including, for instance, the case of $Co_2MnSi/GaAs(001)$ heterostructure [148] where a spin polarization at the Fermi level of 12% only has been obtained, considerably smaller than expected.

Recent first-principles calculations focused on $Co_2CrAl/GaAs$ interfaces [147] have shown that spin polarization tends to remain relatively high at the (110) interface and reaches almost unity for a specific (110) interfacial structure. Furthermore, the nearly half-metallic interface turns out to be the most stable of the (110) interfacial structures. $Co_2CrSi/GaAs$, $Co_2MnSi/GaAs$, and $Co_2MnGe/GaAs$ heterostructures are also investigated, and similar half-metal-like behavior at (110) interface is reported for all of them.

In general, important variations of the electronic state population at the Fermi level can be induced by the interface, in different ways for different orientation and interface morphology, and in most cases such variations are in the direction of destroying the peculiar half-metallic character of the constituent alloy. However, the existence of possible interface morphologies conserving the half-metallic character is in principle possible by selecting particular ones, although the interface morphology in real samples depends on the growth conditions and cannot be fully controlled.

9. Summary and future perspectives

Considerable progress in understanding and controlling the structural and electronic properties of semiconductor heterostructures has been achieved in recent years by combined experimental and theoretical efforts, and in particular concerning the band alignments which are the key parameters governing the transport properties at interfaces. New developments in computational physics have made possible accurate ab initio calculations of the electronic structure of semiconductor contacts, and the complexity of the systems which can be examined is steadily increasing, from ideal interfaces based on conventional semiconductors such as GaAs/AlAs up to defected interfaces and interfaces with constituents with important structural and electronic mismatch such as the case of Heusler/semiconductor for spintronic applications or multifunctional complex oxides/semiconductor systems. These computations can be used to study the effect of various microscopic features in numerical experiments, and derive models which, within clearly defined limits of applicability, retain the same accuracy as the calculations from which they are derived.

In the present chapter, we have illustrated the basic points of a state-of-the-art theoretical approach, which allows not only to compute band discontinuities for the various classes of heterojunctions, but also to obtain insight into the atomic-scale mechanisms that determine the band lineups, and interpret and predict their trends.

There remain a number of open issues, which still limit the predictive capability of theoretical schemes. The most important concerns the mechanisms responsible for the actual atomic-scale arrangement at the interfaces, and their kinetic versus thermodynamic character. A better understanding of these mechanisms and the possibility of predicting the morphology of the epitaxial structures which are actually established, would clearly improve our ability to engineer interface parameters.

References

[1] R.M. Martin, Electronic Structure: Basic Theory and Practical Methods, Cambridge University Press, Cambridge, 2004.
[2] J. Kohanoff, Electronic Structure Calculations for Solids and Molecules: Theory and Computational Methods, Cambridge University Press, Cambridge, 2006.
[3] F. Capasso and G. Margaritondo (Eds), Heterojunction Band Discontinuities: Physics and Device Application, North-Holland, Amsterdam, 1987.
[4] G. Margaritondo (Ed.), Electronic Structure of Semiconductor Heterojunctions, Kluwer, Dordrecht, 1988.
[5] E.T. Yu, J.O. McCaldin, and T.C. McGill, in: H. Ehrenreich and D. Turnbull (Eds), Solid State Physics, Advances in Research and Applications. Academic Press, Boston, 1992, p. 1.
[6] A. Franciosi and C.G. Van de Walle, Surf. Sci. Rep. 25 (1996) 1.
[7] C. Lamberti, Surf. Sci. Rep. 53 (2004) 1.
[8] L. Kleinman, Phys. Rev. B 7412 24 (1981).
[9] L. Anderson, Solid-State Electron. 5 (1962) 341.
[10] W. Schottky, Naturwissenschaften 26 (1938) 843.
[11] N.F. Mott, Proc. Camb. Phil. Soc. 34 (1938) 568.
[12] J.A. Van Vechten, J. Vac. Sci. Technol. B 3 (1985) 1240.
[13] J.A. Van Vechten, Phys. Rev. 187 (1969) 1007.
[14] W.A. Harrison, J. Vac. Sci. Technol. 14 (1977) 1016.
[15] W.A. Harrison, Electronic Structure and the Properties of Solids, Freeman, San Francisco, 1980, Section 10-F.
[16] W.R. Frensley and H. Kroemer, J. Vac. Sci. Technol. 13 (1976) 810.
[17] W.R. Frensley and H. Kroemer, Phys. Rev. B 16 (1977) 2642.
[18] J. Tersoff, Phys. Rev. B 30 (1984) 4874.
[19] C. Tejedor and F. Flores, J. Phys. C 11 (1978) L19.
[20] W.A. Harrison and J. Tersoff, J. Vac. Sci. Technol. B 4 (1986) 1068.
[21] M. Cardona and N.E. Christensen, Phys. Rev. B 35 (1987) 6182.
[22] A. Muñoz, J. Sánchez-Dehesa, and F. Flores, Phys. Rev. B 35 (1987) 6468.

[23] F. Flores, A. Muñoz, and J.C. Duran, Appl. Surf. Sci. 41 (1989) 144.
[24] S.-H. Wei and A. Zunger, Phys. Rev. Lett. 59 (1987) 144.
[25] S.-H. Wei and A. Zunger, J. Vac. Sci. Technol. B 5 (1987) 1239.
[26] S.-H. Wei and A. Zunger, Appl. Phys. Lett. 72 (1998) 2011.
[27] G.A. Baraff, J.A. Appelbaum, and D.R. Hamann, Phys. Rev. Lett. 38 (1977) 237.
[28] W.E. Pickett, S.G. Louie, and M.L. Cohen, Phys. Rev. Lett. 39 (1977) 109.
[29] W.E. Pickett and M.L. Cohen, Phys. Rev. B 18 (1978) 939.
[30] C.G. Van de Walle and R.M. Martin, Phys. Rev. B 35 (1987) 8154.
[31] C.G. Van de Walle and R.M. Martin, J. Vac. Sci. Technol. B 4 (1986) 1055.
[32] C.G. Van de Walle, Phys. Rev. B 39 (1989) 1871.
[33] M.P.C.M. Krijn, Semicond. Sci. Technol. 6 (1991) 27.
[34] T.Y. Wang and G.B. Stringfellow, J. Appl. Phys. 67 (1990) 344.
[35] D. Rioux and H. Hochst, Phys. Rev. B 47 (1993) 1434.
[36] C. Lamberti, Comput. Phys. Commun. 93 (1996) 53.
[37] C. Lamberti, S. Bordiga, F. Boscherini, et al., J. Appl. Phys. 83 (1998) 1058.
[38] W.Z. Cai and D.L. Miller, J. Vac. Sci. Technol. B 20 (2002) 512.
[39] A. Baldereschi, S. Baroni, and R. Resta, Phys. Rev. Lett. 61 (1988) 734.
[40] S. Baroni, R. Resta, A. Baldereschi, and M. Peressi, in: G. Fasol, A. Fasolino, and P. Lugli (Eds), Proceedings of the NATO Advanced Research Workshop on "Spectroscopy of Semiconductor Microstructures" Plenum, New York, 1989.
[41] A. Baldereschi, M. Peressi, S. Baroni, and R. Resta, in: L. Miglio and A. Stella (Eds), Proceedings of the International School of Physics "Enrico Fermi" (Course CXVII, Varenna, 1991): Semiconductor Superlattices and Interfaces, Academic, New York, 1993, p. 59.
[42] S. Baroni, M. Peressi, R. Resta, and A. Baldereschi, in: P. Jiang and H.-Z. Zheng (Eds), Proceedings of the 21th International Conference on the Physics of Semiconductors, World Scientific, Singapore, 1993, p. 689.
[43] A. Baldereschi, R. Resta, M. Peressi, et al., in: H.W.M. Salemink and M.D. Pashley (Eds), Proceedings of the NATO Advanced Research Workshop on "The Physical Properties of Semiconductor Interfaces at the Sub-Nanometer Scale", NATO ASI ser. E, no 243, Kluwer Academic Publishers, Dordrecht, 1993, p. 89.
[44] P. Hoenberg and W. Kohn, Phys. Rev. 136 (1964) B864.
[45] W. Kohn and L.J. Sham, Phys. Rev. 140 (1965) A1133.
[46] S. Lundqvist and N. H. March (Eds), Theory of Inhomogeneus Electron Gas, Plenum, New York, 1983.
[47] See for instance: R.M. Martin, in: J.T. Devreese and P. Van Camp (Eds), Electronic Structure, Dynamics and Quantum Structural Properties of Condensed Matter, Plenum, New York, 1985, p. 175.
[48] R.O. Jones and O. Gunnarson, Rev. Mod. Phys. 61 (1989) 689.
[49] D.M. Ceperley and B.J. Alder, Phys. Rev. Lett. 45 (1980) 566; J. Perdew and A. Zunger, Phys. Rev. B 23 (1981) 5048.
[50] J. Perdew and A. Zunger, Phys. Rev. B 23 (1981) 5048.
[51] M.S. Hybertsen and S.G. Louie, Phys. Rev. Lett. 55 (1985) 1418; Phys. Rev. B 34 (1986) 5390; ibid. 35 (1987) 5585; ibid. 5602.
[52] X. Zhu and S.G. Louie, Phys. Rev. B 43 (1991) 14142.
[53] W.E. Pickett, Comput. Phys. Rep. 9 (1989) 115, and references therein.
[54] D.M. Bylander and L. Kleinman, Phys. Rev. Lett. 52 (1995) 14566.
[55] N. Troullier and J. L. Martins, Phys. Rev. B 43 (1991) 1993.
[56] D. Vanderbilt, Phys. Rev. Lett. 32 (1985) 8412.
[57] H. Hellmann, in: F. Deuticke (Eds), Eiführung in die Quantaenchemie, Leipzig, 1937 p. 285.
[58] O.H. Nielsen and R.M. Martin, Phys. Rev. Lett. 50 (1983) 697; Phys. Rev. B 32 (1985) 3780; ibid. 3792.
[59] L. Kleinman and J.C. Phillips, Phys. Rev. 116 (1959) 880.
[60] D.J. Chadi and M.L. Cohen, Phys. Rev. B 8 (1973) 4547.

[61] A. Baldereschi, Phys. Rev. B 12 (1973) 5212.
[62] H.J. Monkhorst and J.P. Pack, Phys. Rev. B 13 (1976) 5188.
[63] C.L. Fu and K.M. Ho, Phys. Rev. B 28 (1983) 5480.
[64] N. Marzari, D. Vanderbilt, and M. C. Payne, Phys. Rev. Lett. 79 (1997) 3296.
[65] M. Methfessel and A. T. Paxton, Phys. Rev. B 40 (1998) 3616.
[66] L. Nordheim, Ann. Physik 9 (1931) 607.
[67] M. Peressi and S. Baroni, Phys. Rev. B 49 (1994) 7490.
[68] N. Marzari, S. de Gironcoli, and S. Baroni, Phys. Rev. Lett. 72 (1994) 4001.
[69] R. Magri, S. Froyen, and A. Zunger, Phys. Rev. B 51 (1991) 7947.
[70] L. Bellaiche, S.H. Wei, and A. Zunger, Phys. Rev. B 56 (1997) 10233.
[71] M. Jaros, Rep. Prog. Phys. 48 (1985) 1091.
[72] http://psi-k.dl.ac.uk/data/codes.html.
[73] http://www.quantum-espresso.org and http://www.pwscf.org
[74] http://www.democritos.org
[75] J.D. Jackson, Classical Electrodynamics, Wiley, New York, 1975.
[76] M. Peressi, N. Binggeli, and A. Baldereschi, J. Phys. D: Appl. Phys. 31 (1998) 1273.
[77] B. Montanari, M. Peressi, S. Baroni, and E. Molinari, Appl. Phys. Lett. 69 (1996) 3218.
[78] C.G. Van de Walle and R.M. Martin, Phys. Rev. Lett. 62 (1989) 2028.
[79] M. Peressi, L. Colombo, A. Baldereschi, et al., Phys. Rev. B 48 (1993) 12047.
[80] J. Tersoff and D.R. Hamann, Phys. Rev. Lett. 50 (1983) 1998; Phys. Rev. B 31 (1985) 805.
[81] C.J. Chen, Phys. Rev. Lett. 65 (1990) 448; Phys. Rev. B 42 (1990) 8841.
[82] M. Chen, P.G. Clark, Jr, T. Mueller, et al., Phys. Rev. B 60 (1999) 11783.
[83] P. Sautet, J.C. Dunphy, and M. Salmeron, Surf. Sci. 364 (1996) 335.
[84] M. Di Ventra and S.T. Pantelides, Phys. Rev. B 59 (1999) R5320.
[85] K. Cho and J.D. Joannopoulos, Phys. Rev. Lett. 71 (1993) 1387.
[86] S.H. Ke, T. Uda, R. Pérez, et al., Phys. Rev. B 60 (1999) 11631.
[87] S.B. Zhang, D. Tománek, S.G. Louie, et al., Sol. State Commun. 66 (1988) 585.
[88] Landolt–Börnstein (Ed.), Numerical Data and Functional Relationships in Science and Technology, Group III, Vol. 17a–b, Springer, New York, 1982.
[89] M. Peressi, S. Baroni, A. Baldereschi, and R. Resta, Phys. Rev. B 41 (1990) 12106.
[90] M.S. Hybertsen, Phys. Rev. Lett. 64 (1990) 555.
[91] W.A. Harrison, J. Vac. Sci. Technol. 16 (1979) 1492.
[92] R.M. Martin, J. Vac. Sci. Technol. 17 (1980) 978.
[93] R. Nicolini, L. Vanzetti, Guido Mula, et al., Phys. Rev. Lett. 72 (1994) 294.
[94] M. Funato, S. Aoki, S. Fujita, and S. Fujita, J. Appl. Phys. 82 (1997) 2984.
[95] A. Kley and J. Neugebauer, Phys. Rev. B 50 (1994) 8616.
[96] R.G. Dandrea, S. Froyen, and A. Zunger, Phys. Rev. B 42 (1990) 3213.
[97] R. Resta, L. Colombo, and S. Baroni, Phys. Rev. B 41 (1990) 12358.
[98] L. Colombo, R. Resta, and S. Baroni, Phys. Rev. B 44 (1991) 5572.
[99] C.G. Van de Walle and R. M. Martin, Phys. Rev. B 34 (1986) 5621.
[100] J.E. Bernard and A. Zunger, Phys. Rev. B 44 (1991) 1663.
[101] N. Tit, M. Peressi, and S. Baroni, Phys. Rev. B 48 (1993) 17607.
[102] M. Di Ventra, M. Peressi, and A. Baldereschi, Phys. Rev. B 54 (1996) 5691; J. Vac. Sci. Technol. B 17 (1996) 2936.
[103] M. Di Ventra, M. Peressi, and A. Baldereschi, in: M. Scheffler and R. Zimmermann (Eds), Proceedings of the 23rd International Conference on the Physics of Semiconductors, World Scientific, Singapore, 1996, p. 987.
[104] A. Qteish and R.J. Needs, Phys. Rev. B 43 (1991) 4229.
[105] San-huang Ke, Ren-zhi Wang, and Mei-chun Huang, Phys. Rev. B 49 (1994) 10495.
[106] M. Städele, J.A. Majewski, and P. Vogl, Phys. Rev. B 56 (1997) 6911.
[107] N. Binggeli, P. Ferrara, and A. Baldereschi, Phys. Rev. B 63 (2001) 245306.

[108] A. Stroppa and M. Peressi, Phys. Rev. B 71 (2005) 205303.
[109] F. Bernardini, M. Peressi, and V. Fiorentini, Phys. Rev. B 62 (2000) R16302.
[110] G. Bratina, L. Vanzetti, L. Sorba, et al., Phys. Rev. B 50 (1994) 11723.
[111] A. Stroppa and M. Peressi, Phys. Rev. B 72 (2005) 245304.
[112] C.G. Van de Walle and L. H. Yang, J. Vac. Sci. Technol. 13 (1995) 1635.
[113] M. Peressi, L. Colombo, and S. de Gironcoli, Phys. Rev. B 64 (2001) 193303.
[114] M. Tosolini, L. Colombo, and M. Peressi, Phys. Rev. B 69 (2004) 075301.
[115] M.S. Hybertsen, Appl. Phys. Lett. 58 (1991) 1759.
[116] F. Capasso, A.Y. Cho, K. Mohammed, and P.W. Foy, Appl. Phys. Lett. 46 (1985) 664; F. Capasso, K. Mohammed, and A.Y. Cho, J. Vac. Sci. Technol. B 3 (1985) 1245.
[117] M. Peressi, S. Baroni, R. Resta, and A. Baldereschi, Phys. Rev. B (RC) 43 (1991) 7347.
[118] M. Marsi, S. La Rosa, Y. Hwu, et al., J. Appl. Phys. 71 (1992) 2048.
[119] L. Sorba, G. Bratina, A. Antonini, et al., Phys. Rev. B 43 (1991) 2450.
[120] G. Biasiol, L. Sorba, G. Bratina, et al., Phys. Rev. Lett. 69 (1992) 1283.
[121] G. Bratina, L. Vanzetti, L. Sorba, et al., Phys. Rev. B 50 (1994) 11723.
[122] D.E. Aspnes and A.A. Studna, Phys. Rev. Lett. 54 (1985) 1956.
[123] T. Yasuda, Thin Solid Films 313 (1998) 544; C. Meyne, M. Gensch, S. Peters, et al., Thin Solid Films 364 (2000) 12, and references therein.
[124] L.J. Brillson, T.M. Levin, G.H. Jessen, and F.A. Ponce, Appl. Phys. Lett. 75 (1999) 3835.
[125] G. Margaritondo, F. Cerrina, C. Capasso, et al., Solid State Commun. 52 (1984) 495.
[126] A.V. Platonov, V.P. Kocherechko, E.L. Ivchenko, et al., Phys. Rev. Lett. 83 (1999) 3546.
[127] J. Pollmann and S.T. Pantelides, Phys. Rev. B 21 (1980) 709.
[128] T. Saito and T. Ikoma, Phys. Rev. B 45 (1992) 1762.
[129] M. Peressi, F. Favot, and A. Baldereschi, in: V. Fiorentini and F. Meloni (Eds), Advances in Computational Materials Science I, Conference Proceedings Vol. 55, SIF, Bologna, 1997, p. 13.
[130] A. Baldereschi, M. Peressi, F. Favot, and G. Cangiani, Electrons and Photons in Solids, Quaderni della Scuola Normale Superiore, Classe di Scienze, 2001, p. 85.
[131] G. Bratina, L. Vanzetti, L. Sorba, et al., Phys. Rev. B 50 (1994) 11723.
[132] E.G. Wang, C. Chen, and C.S. Ting, J. Appl. Phys. 78 (1995) 1832.
[133] E.G. Wang, Appl. Surf. Sci. 104 (1996) 626.
[134] M. Peressi, F. Favot, G. Cangiani, and A. Baldereschi, Appl. Phys. Lett. 81 (2002) 5171.
[135] R.M. Feenstra, Physica B 274 (1999) 796.
[136] J. Steinshnider, M. Weimer, R. Kaspi, and G.W. Turner, Phys. Rev. Lett. 85 (2000) 2953.
[137] W.D. Schneider, Surf. Sci. 514(1–3) (2002) 74.
[138] R.S. Goldman, J. Phys. D: (For the presence of this empty site, NiMnSb is known as "half-Heusler" alloy, at variance with others such as Co_2MnGe that have also this site occupied and are known as "full-Heusler" alloys.) Appl. Phys. 37 (2004) R163.
[139] S.G. Kim, S.C. Erwin, B.Z. Nosho, and L.J. Whitman, Phys. Rev. B 67 (2003) 121306.
[140] O. Albrektsen, D.J. Arent, H.P. Meier, and H.W.M. Salemink, Appl. Phys. Lett. 57 (1990) 31.
[141] Ph. Ebert, Surf. Sci. Rep. 33 (1999) 121, and references therein.
[142] X.M. Duan, M. Peressi, and S. Baroni, Phys. Rev. B 72 (2005) 085341; X.M. Duan, S. Baroni, S. Modesti, and M. Peressi, Appl. Phys. Lett. 88 (2006) 022115.
[143] W. Van Roy, G. Borghs, J. De Boeck, J. Cryst. Growth 227–228 (2001) 862.
[144] G.A. de Wijs and R.A. de Groot, Phys. Rev. B 64 (2001) 020402(R).
[145] A. Debernardi, M. Peressi, and A. Baldereschi, Mater. Sci. Eng. C 23 (2003) 743; A. Debernardi, M. Peressi, A. Baldereschi, Compt. Mater. Sci. 33 (1–3) (2005) 263.
[146] S. Picozzi, A. Continenza, and A. J. Freeman, IEEE Trans. Magn. 38 (2002) 2895; J. Appl. Phys. 94 (2003) 4723; J. Phys. Chem. Sol. 64 (2003) 1697.
[147] K. Nagao, Y. Miura, and M. Shirari, Phys. Rev. B 73 (2006) 104447.
[148] W.H. Wang, M. Przybylski, W. Kuch, et al., Phys. Rev. B 71 (2005) 144416.

3

Electrical characterization of nanostructures

Anna Cavallini and Laura Polenta

CNISM and Physics Department, University of Bologna, Bologna, Italy

Abstract Electrical characterization of semiconductors covers a number of different experimental methods giving information on the charge carrier distribution and transport mechanisms. Defects can be intentionally introduced by doping by which it is possible to tailor the electrical properties of the material. However, the unavoidable presence of undesired imperfections, as native or process-induced defects, may affect these properties in an uncontrolled way, with severe consequences on the electrical and optical behavior.

An extensive investigation, therefore, includes the detection and characterization of defect-related energy levels, in view of the technological improvements of material growth and processing.

In this chapter we will describe a selection of electrical characterization methods based on electrical and optical excitation of carriers.

The current–voltage and capacitance–voltage characterization allows one to investigate the mechanisms of charge transport and distribution, but a deep insight about the electrically active defect population and their properties is obtained by spectroscopic methods (thermal and photo-induced spectroscopy).

While the majority of the presented methods are widely used in heterostructure investigation, their successful application to nanostructures is currently at the state-of-the art level, as shown in the case studies on GaN nanowires.

Keywords electrical characterization, deep levels transient spectroscopy, spectral photocurrent, surface photovoltage, GaN nanowires

1. Introduction

The revolutionary introduction of semiconductors in industrial and scientific applications led to a large-scale development of the electrical characterization techniques, which are necessary tools for the optimization of growth procedures and for predicting performance of devices. The technological progress of growth methods, with the synthesis of compound semiconductors, made it necessary the implementation of the investigation techniques to characterize with a high confidence level the electrical and optical behavior of the new materials.

Currently, electrical characterization includes a very large number of techniques, the main part dedicated to the analysis of massive structures. In the past decades, however, the advent of nano- and heterostructures posed the question about the resolution and the improvement of these techniques.

Investigation of samples where surface-or interface-related effects play a major role in the electrical and optical behavior is therefore an interesting challenge for experimental physics. The electrical and optical properties of a semiconductor can be determined only by having a complete knowledge of the chemical composition, of the presence of lattice defects or chemical impurities and of the structure dimensions [1].

Defects can be defined as everything breaking the symmetry of the perfect crystal lattice: point-like imperfections (vacancies, substitutional, interstitials, antisites) or extended imperfections (line defects, planar defects, bulk defects).

Surface itself can be considered as the biggest extended defect: its influence on the optoelectronic properties increases as the surface-to-volume ratio does, hence revealing its key role in nanostructured materials.

Defects can be introduced intentionally by doping or unintentionally by growth, processing or radiation. Structural imperfections may interact with free carriers like scattering centers, traps, or recombination centers, hence largely affecting the material optical and electrical behavior [2]. The knowledge of defects and their characteristics is therefore of crucial importance. Electrically active defects are associated to energy levels within the forbidden gap, which can have a donor- or acceptor-like character: donor levels are defined as neutral when filled by electrons and positive if empty, conversely, an acceptor level is defined as neutral if empty and negative when filled by electrons. The energy levels, classified as "shallow" when their distance from either band is on the order of the thermal energy (few meV) or "deep" otherwise, can be analyzed by spectroscopic techniques, making use of electrical, thermal, or optical excitation of levels.

In this chapter we will present the basic principles of the mostly used techniques for electrical characterization and will focus on the main results in the analysis of nanostructures. Current–voltage, deep level transient spectroscopy (DLTS), spectral photoconductivity, and surface photovoltage (SPV) spectroscopy have revealed to be very powerful methods for nanostructures characterization, as reported in the case studies presented here, concerning GaN nanowires. For the sake of brevity, we do not included in this review the methods for mobility determination for which we refer the reader to specific literature [1].

2. Electrical characterization

In this chapter some of the mostly diffused methods for electrical characterization will be presented, based on the analysis of conductivity or capacitance signals.

The current–voltage and capacitance–voltage methods allow for investigating charge transport and distribution and getting general information on the material. However, dealing with optical and electrical properties of semiconductors, the effects of shallow and deep levels must be considered. DLTS and related methods, based on the analysis of capacitance or conductivity transients, detect and characterize the electrical activity of defects.

Originally developed for the study of bulk silicon, these methods show high potential application also to nanostructures, provided that signal intensity is not below the detection limit, which is not a trivial requirement for this kind of specimens.

The basic requirements for the application of the different techniques has therefore to be tested considering that, due to the specific geometry and the unavoidable surface-related effects, nanoscaled materials can show sensibly different electrical properties with respect to massive structures of the same material.

Table 1
Summary of the main electrical characterization techniques presented in this chapter

Method	Configuration	Signal	Information
I–V Current-voltage	Ohmic	I(V)	Current transport model, resistivity (in ohmic regime)
	Junction	I(V)	Transport model, barrier height, ideality factor
C–V Capacitance-voltage	Junction	C(V)	Barrier height, shallow level density and profiling; deep level density
DLTS Deep level transient spectroscopy	Junction	C(t)	Majority carrier deep traps (energy, capture cross section, density)
PICTS Photo-induced current transient spectroscopy	Ohmic	I(t)	Majority and minority carrier deep traps (energy, capture cross section, density)
	Junction	I(t)	Majority carrier deep traps (energy, capture cross section, density)
P-DLTS Photo-DLTS	Junction	I(t)	Majority carrier deep traps (energy, capture cross section, density)

Table 1 summarizes the properties of the experimental methods presented in this chapter, the configuration requirements (ohmic, i.e., with two ohmic contacts, or junction-like), the analyzed signal, current I or capacitance C as a function of voltage V or time t, and the relevant information achievable.

2.1. Current–voltage characteristics

Semiconductors are commonly classified on the basis of their specific conductivity σ (or resistivity $\rho \approx 1/\sigma$), ranging from semi-insulating regime to quasi-metal one.

The first step for electrical characterization of materials is therefore the investigation of conductivity to determine the main mechanisms controlling the current transport and identify, at the same time, the more suitable excitation for deeper investigation.

The current–voltage characteristic (I–V) gives the knowledge of fundamental parameters for devices. In industrial application the manufacturers of microelectronic devices test their products with this type of characterization monitoring the quality control of semiconductor devices.

The electrical signal of the semiconductor is collected via ohmic or Schottky contacts, depending on the parameters to evaluate [3].

Processing strategies are indeed important to have good performance of electrical contacts: generally speaking, "ohmic conduction" analysis is useful for determining the model of charge carrier transport and in some cases to observe the role of shallow levels [4]. The behavior of the injected carriers in a semiconductor through an ohmic contact is affected by the physical properties of the material in which the carriers flow.

Schottky conduction analysis, differently, allows for extracting the parameters concerning barrier height, series resistance, and ideality factor of the diode. It plays a paramount role due to the large diffusion of rectifying diodes and junctions in electrical and optical devices.

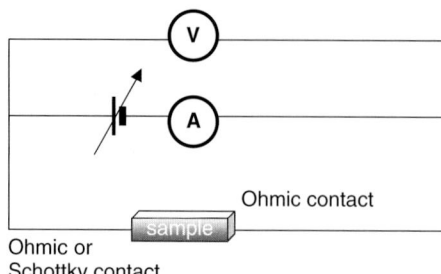

Fig. 1. Sketch of the experimental setup for *I–V* characterization.

In presence of donor- and acceptor-like defects, as well as of trapping centers due to impurities and defects capturing a fraction of the injected carriers, conductivity is dramatically modified. For this reason, *I–V* characteristic is strongly dependent on the concentration and the energy distribution of trapping centers inside the specimen.

The following discussion will take into account the *I–V* characterization from injecting (ohmic) contact or rectifying (Schottky) ones, by assuming a transport by majority carriers only. The connection of the sample to the external circuit has to be via an ohmic contact as shown in Fig. 1.

2.1.1. Current–voltage characteristics with ohmic contacts

The *I–V* characteristics of real semiconductors cannot be explained without taking into account the role of shallow or deep traps.

The theory on current injection in solids [4] can be easily described by considering the simplified model of one-carrier currents, hence neglecting the minority carrier contribution. In particular, from now on, the only case of n-type semiconductors will be considered. The reader is refereed to specific literature for more detailed discussion concerning other configurations [1,4].

The following assumptions can be made to describe the regimes of current–voltage characteristics in the presence of injecting contacts [5]:

1. negligible carrier diffusion,
2. free and trapped carriers (n or p) in quasi-thermal equilibrium,
3. constant mobility μ of carriers.

The phenomenological approach allows for distinguishing four classes of carriers: thermal free electrons, thermal trapped electrons, injected free electrons, and injected trapped electrons. For sake of clarity, the treatment will concern a single trap as shown in Fig. 2.

By increasing the applied voltage, i.e. the injection level, the regimes of:

(a) ohmic conduction
(b) shallow-trap space charge limited conduction (SCLC)
(c) trap-filled limited conduction (TFLC)
(d) trap-free SCLC

are encountered in sequence (Fig. 3).

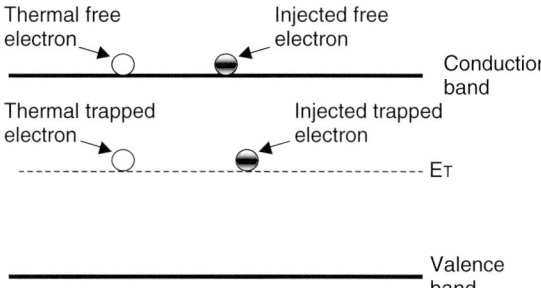

Fig. 2. Band diagram with the four categories of carriers considered in the simplified theory of conduction by injecting contacts.

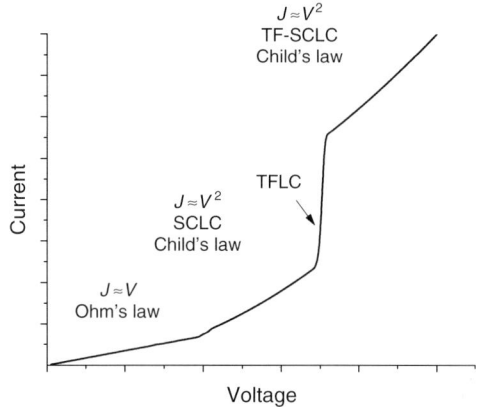

Fig. 3. Current–voltage behavior with injecting contacts.

In each operating regime, one of the previously listed categories of carriers dominates the transport in the entire sample, ignoring the other categories in the Poisson equation.

The current density J is expressed by the continuity equation as

$$J = nq\mu_n E \qquad (1)$$

with different analytical expressions of the electric field E and of the charge density n depending on the regime under consideration. Here q is the carrier charge and μ_n the mobility of the majority carrier (electron).

(a) The regime of ohmic conduction is due to thermal electrons, present in higher concentration with respect to the injected ones: the Ohm's law is therefore valid considering that the applied bias does not modify the free carrier concentration.
Hence there is no space charge Q associated and a spatially constant electric field E takes place ($dE/dx = 0$). Naming n_0 the concentration of thermal free carriers, which follows the Boltzmann statistics [1], the current density is

$$J = \frac{n_0 q \mu_n V}{L} \propto V \quad \text{(Ohm's law)} \qquad (2)$$

L being the layer thickness.

(b) When the injected carriers exceed the number of thermal carriers, they dominate the space charge: so long as most of the traps are still empty, the ratio of density of free (n_0) and trapped (n_t) electrons keeps constant. When this happens, the number of free carriers remains proportional to the space charge (SCL current flow).

The number of the injected trapped electrons is large with respect to both the thermal free and the thermal trapped electrons.

If the trap is shallow enough with respect to the quasi-Fermi level, the ratio θ between the free and the trapped carrier concentrations is approximately independent of bias and J is

$$J = \frac{\chi \theta \varepsilon_s \mu_n V^2}{L^3} \quad (3)$$

where χ is the SCL pre-factor and ε_s the dielectric constant. The quadratic current–voltage relation (Child's law) is due to the fact that the free carrier concentration is bias dependent.

(c) In the TFLC the space charge is dominated by trapped electrons (as in the shallow-trap SCLC) but the current is carried by the injected free electrons whose number is not a constant fraction of trapped electrons. Empty traps become scarce: the injected trapped carriers are still dominating but the number of injected free electrons rapidly rises, and so does the current. This regime is usually characterized by a fixed voltage corresponding to the space charge of the injected trapped electrons, whose value depends on the square of the layer thickness.

(d) By further increasing the bias, hence the injection level, injected free electrons exceed the injected trapped electrons and dominate the space charge. Again the number of free carriers is proportional to the space charge and the SCL current flows. A quadratic dependence similar to that expressed in Eq. (3) is thereby found in the trap-free SCL regime except for a different value of θ which takes into account the injection level.

2.1.2. Current–voltage characteristics with rectifying contact

The analysis of conductivity in presence of Schottky (rectifying) contacts, or junctions, allows for evaluating barrier height, capacitance, and shallow and deep level concentrations [1,2,6].

We will briefly recall the typical effects concerning the formation of a Schottky barrier to understand better the mechanisms activated by the different techniques described in the present section and in the following ones on capacitance methods.

From now on, we will focus on n-type semiconductors, hence the situation concerning p-type one can be easily derived.

The intimate contact between the metal and the n-type semiconductor, neglecting surface states or oxide interfaces, gives rise to a band bending where the ionization of levels is modified. Free electrons move across the interface because Fermi levels must coincide to satisfy the thermodynamic equilibrium. This leads to the creation of an electric field and therefore of a depletion layer, where uncompensated positive donors lie.

The concentration of donors being much less than that of electrons in the metal, the charge balance can be achieved only if an appreciable layer of the semiconductor is depleted, and the depletion layer extends almost only in the semiconductor as shown in Fig. 4, differently from a p–n junction [2].

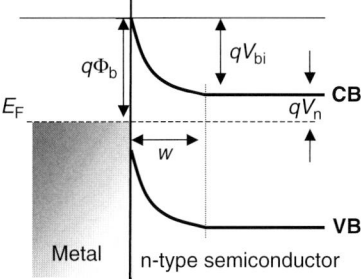

Fig. 4. Band diagram of the intimate contact between metal and semiconductor for a rectifying junction. Semiconductor is n-type and the band bending moves away majority carriers (electrons from the interface). The depletion region is positively charged by the fixed donor ions. CB means conduction band and VB means valence band.

The depletion layer thickness w is described by the equation

$$w = \sqrt{\frac{2\varepsilon_s(V_{bi} - V - kT/q)}{qN_{eff}}} \qquad (4)$$

where V_{bi} is the built-in voltage, V the external bias applied, and N_{eff} the net concentration of donor levels (shallow or deep). To understand the meaning of these parameters, consider the scheme in Fig. 4, describing the physical parameters involved in the formation of the Schottky barrier.

In thermal equilibrium, the band bending originates a potential barrier Φ_b opposed to the flow of electrons from metal to semiconductor. From the semiconductor side, the band bending creates a potential difference between the bottom of conduction band and the vacuum level, of value qV_{bi}. The semiconductor Fermi level is at energy qV_n from the bottom of the conduction band.

To physically understand the charge transport across an ideal diode, it must be considered that the carriers from the metal to the semiconductor always encounter the barrier Φ_b whatever bias is applied. Differently, the current flowing from semiconductor to metal is strictly related to the band bending, which determines the potential barrier to overpass.

When external bias V is applied, the Fermi level of the semiconductor is shifted up (forward bias) or down (reverse bias) changing the charge flux between the two materials: in forward bias the electron barrier for flow from semiconductor to metal is reduced, favoring the passage of electrons from semiconductor to metal. In reverse bias, the barrier increases, reducing electron flow and hence current.

This is schematically depicted in Fig. 5, where the black arrows indicate the current from semiconductor to metal and vice versa in the different regimes. It is trivial to imagine that at thermal equilibrium, where the Fermi levels of metal and semiconductor are aligned, the net currier density is null, as the two flows of carriers are opposite.

The application of forward bias, which reduces the band bending, hence narrowing the depletion region (Eq. (4)), may turn on processes of charge transport such as emission over the barrier, quantum-mechanical tunneling through the barrier, recombination in the space charge region (SCR), i.e., the depletion region, and recombination in the neutral region, where

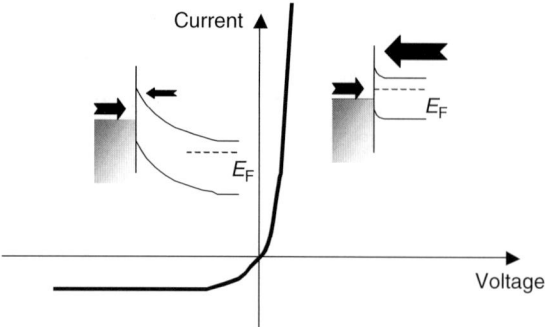

Fig. 5. Schematic representation of the *I–V* curve of an ideal diode, and the relevant band diagrams.

bands are flat (hole injection). The current flowing from the semiconductor to the metal is larger with respect to that flowing from the metal to the semiconductor.

The theory of Crowell and Sze [3,7] well explains the mechanisms of thermoionic emission and diffusion of carriers through a Schottky diode.

In forward bias, the complete equation describing the current is

$$I = I_s \left(e^{(q(V-IR_s)/nkT)} - 1 \right) \quad (5)$$

where V is the applied voltage, R_s is the series resistance, n is the diode ideality factor, and I_s is the saturation current. Saturation current is expressed by the equation

$$I_s = A^{**} T^2 A e^{-(q\phi_b/kT)} \quad (6)$$

where A^{**} is the corrected Richardson constant (taking into account quantum-mechanical effects, metal surface and phonon scattering of electrons), A is the contact area, and Φ_b is the barrier height [2]. In ideal diodes, the saturation current corresponds to the reverse leakage current of the diode.

The analysis of current–voltage characteristics can be simplified if the sample properties allow for neglecting the influence of R_s, as occurs in most semiconductor materials.

When reverse bias is applied, the depletion region widens and conductivity reduces, showing a behavior that can be ascribed to thermoionic emission in an ideal case, but which can depart from ideality due to field dependence of the barrier height, generation in the depletion region, or tunneling effects [2].

The analysis of a current–voltage curve may allow for determining the barrier height, the ideality of the diode, and series resistance due to the material or to contacts.

2.2. Capacitance–voltage characteristics

Capacitance–voltage characterization (*C–V*) is a standard method for the determination of the net doping concentration N_{eff} of semiconductors, by measuring the small-signal depletion capacitance.

In *C–V* measurements, a reverse-biased diode is subject to an AC excitation of small voltage, whose function is to change the trap state of occupation [1,3].

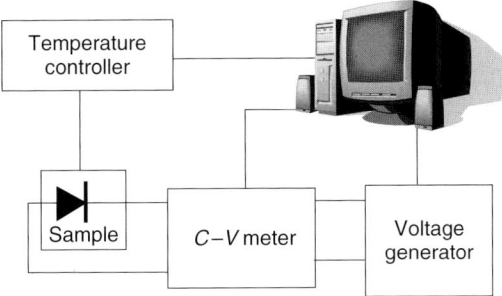

Fig. 6. Schematic capacitance–voltage setup.

A schematic setup is shown in Fig. 6.

Let us consider again a Schottky junction formed by a metal and an n-type semiconductor as shown in Fig. 4. We will neglect for brevity both the effects due to minority carriers and the presence of interfacial layers for which we again refer to specific literature [1,3,6].

Following the Schottky junction theory [3], the two distributions of charges (negative at the metal–semiconductor interface and positive across the depletion region of the junction) can be reasonably approximated to a plane capacitor, also when doping is not uniform [8], taking the surface where bands flatten as second plate.

Hence, by making use of the Poisson equation and the Gauss's law [6], the fixed space charge Q in the depletion region, if A is the contact area and N_{eff} the net concentration of donor levels, is

$$Q = qAwN_{\text{eff}} = A\sqrt{2q\varepsilon_s\left(V_{\text{bi}} - V - \frac{kT}{q}\right)N_{\text{eff}}} \qquad (7)$$

The capacitance C can therefore be expressed by the equation:

$$C = \frac{\partial Q}{\partial V} = A\sqrt{\frac{q\varepsilon_s N_{\text{eff}}}{2\left(V_{\text{bi}} - V - \frac{kT}{q}\right)}} \qquad (8)$$

leading to a quadratic dependence of $1/C$ on voltage V.

The analysis of C–V curves may therefore infer the concentration of effective fixed charge as well as the value of the built-in voltage V_{bi}. Usually the C–V characteristics can be analyzed by plotting $1/C^2$ vs V and deducing N_{eff} from the slope and V_{bi} from the intercept (see Fig. 8).

If deep traps are negligible, the technique allows for measuring directly the dopant distribution, i.e., the concentration and doping profile due to shallow levels, usually completely ionized at room temperature. N_{eff} measured is indeed the equilibrium value of the free electron concentration at the end of the depletion region ($x = w$) if temperature is high enough to avoid any freeze-out effect, occurring when the thermal excitation toward the bands is inhibited by the low temperature.

Therefore, the quantity N_{eff} should be substituted by the quantity n, the electron density in the conduction band at the edge of the depletion region. However, if the doping is uniform and there are no deep levels present, the donor density and the free electron density generally coincide at room temperature.

If the doping concentration is not uniform [8], it is possible to deduce the spatial distribution of carriers by changing the reverse bias voltage, hence the depletion layer w. It can be shown that

$$N_{\text{eff}}(x) = -C^3 \left(q\varepsilon A^2\right)^{-1} \left(\frac{dC}{dV}\right)^{-1} \qquad (9)$$

with x being the depletion region extent at bias V.

2.2.1. Deep trap analysis

A deep trap is a "localized electron state in the bulk of the semiconductor, which is so far removed from the conduction or valence bands that is not ionized at room temperature" [3]. Its occupation level varies in the depletion region where bands are bent, as its energy position with respect to the Fermi level changes. The application of an external bias therefore will modify the band bending, and thus alters the state of charge of the trap and affects the capacitance value.

According to the Shockley–Read–Hall (SRH) statistics [9,10], the probability of occupation of a trap located at the level E_T is related to the emission e_p and e_n and to the capture rates c_p and c_n for both carriers n and p, as well to their mean thermal velocities $<v>$ [1].

A deep level can be defined considering its emissive properties, in particular

- if $e_n \cong e_p$ the defect acts as a recombination center;
- if $e_n \gg e_p$ the defect acts as an electron trap;
- if $e_p \gg e_n$ the defect acts as a hole trap.

To account for the contribution of deep traps, the trap dynamics has to be taken into consideration. Because traps do not fill or empty instantaneously, their contribution to the measured capacitance will be taken into account only if the frequency of the AC signal ω_{AC} is smaller than $2e_n$ [3].

Consider the case $2e_n > \omega_{AC}$ in Fig. 7. On the left-hand side of plane y, all donors (deep and shallow) are ionized because of the band banding, while between y and w only the shallow donors are ionized. The charge density in these two sections is hence different. Increasing reverse bias from V_r to $V_r + \Delta V_r$ the distribution of positive charge increases by the amount $qN_T\Delta y$ and the total capacitance, if donors are uniformly distributed, so that $\Delta y = \Delta w$, is now expressed by

$$C = \frac{\varepsilon(N_D + N_T)}{N_D w + N_T y} \qquad (10)$$

When dealing with semi-insulating materials, where trap density is on the order of a magnitude larger than shallow donors, this procedure can give a good estimation of the density of the deep levels, provided that the AC frequency is small enough and free carriers are increased by thermal or optical excitation [11].

To determine the influence of the AC frequency for detecting all deep traps, a typical C–V relevant to the semi-insulating GaAs is presented in Fig. 8, where curves C–V and, consequently the lines ($1/C^2$ vs V), change as a function of frequency [12]. From the analysis of the C–V characteristics, $N_{\text{eff}} = N_D + N_T$ can be inferred, observing that the lower the frequency, the higher is the N_{eff} calculated, as more deep traps can follow the AC signal variation.

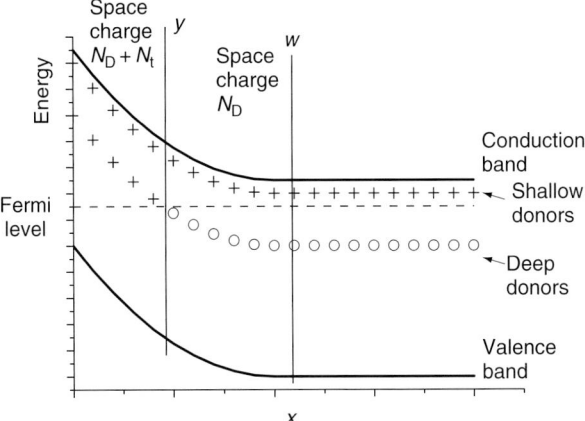

Fig. 7. Band diagram relevant to deep and shallow donors.

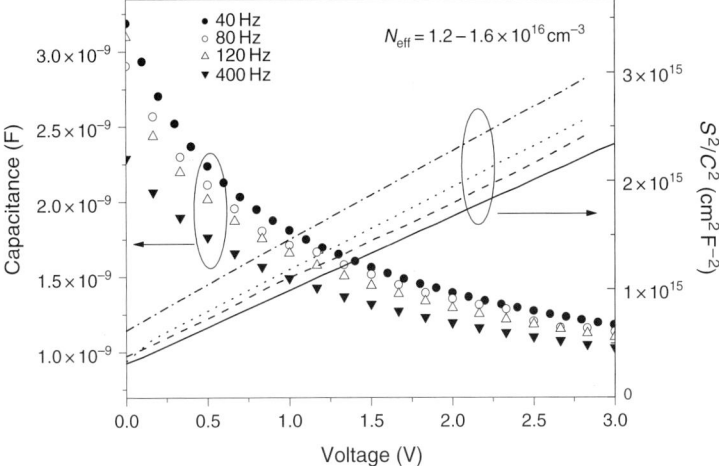

Fig. 8. Typical C–V characteristics relevant to GaAs bulk structures. Temperature is maintained at 420 K to increase the number of free carriers with respect to RT density. The different line slopes (right scale) obtained by changing the AC voltage frequency indicate the presence of deep levels [12].

2.3. Thermal transient spectroscopy

As a direct consequence of C–V measurements, the analysis of capacitance transients as a function of temperature can give detailed information on the deep levels involved. The analysis of current and capacitance transients during constant thermal heating of the sample is the core mechanism of the thermal transient spectroscopy.

Determining the capacitive and conductive behavior of materials by means of the above-described steady-state techniques, one can apply a wide variety of techniques, derived by the DLTS [13,14], suitable for the different properties of the material. The original idea is that the pulsed electrical excitation allows for observing the major role of the level dominating the signal at the given temperature.

As temperature, by changing the equilibrium Fermi level, modifies the ionization level of deep traps, the revolutionary idea was to apply a gradient of temperature by maintaining a pulsed excitation (such as for C–V measurements). By the simultaneous action of temperature and electrical excitations, repetitive signal transients recorded contain information relevant to the level whose ionization degree has been modified by temperature.

The dynamics of the thermal generation–recombination process is governed by the continuity equation expressing the time variation of the occupation of the state. Indicating by $n_t(t)$ the number of captured electrons at time t, it can be demonstrated that [1]

$$n_t(t) = n_t(\infty) + [n_t(0) - n_t(\infty)] \exp\left[-(e_n + c_n + e_p + c_p)\right] \tag{11}$$

where the emission rates are expressed as

$$e_n = \frac{\sigma_n \langle v_n \rangle N_C}{g} \exp\left(-\frac{E_C - E_T}{kT}\right) \tag{12}$$

or

$$e_p = \frac{\sigma_p \langle v_p \rangle N_V}{g} \exp\left(-\frac{E_T - E_V}{kT}\right) \tag{13}$$

where $\langle v \rangle$ is the mean thermal velocity and g the degeneracy factor; σ_n and σ_p are the trap capture cross sections for electrons and holes, respectively.

The capture coefficients for electron and holes, c_n and c_p are defined as

$$c_n = n \langle v_n \rangle \sigma_n \quad \text{and} \quad c_p = p \langle v_p \rangle \sigma_p \tag{14}$$

For $t \to \infty$, the electron concentration at trapping level t is therefore

$$n_t(\infty) = \frac{c_n + e_p}{c_n + c_p + e_n + e_p} N_t \tag{15}$$

Responding to a perturbation which varies their occupation, the electron population evolves exponentially with the following time constant, independent of the concentration of the deep level N_t:

$$\tau = (e_n + c_n + e_p + c_p)^{-1} \tag{16}$$

Significant simplifications can be because only one of the four capture/emission coefficients is dominant under most experimental conditions [15].

Recalling Fig. 7, consider the band bending in the SCR of thickness w associated with the positive charge of the shallow donors and the deep donor represented at energy E_t, crossing the Fermi level at y and yielding different occupation conditions. With E_t representing an electron trap, it can be assumed that $e_n \gg e_p$; moreover, in the region $0 < x < y$, the depletion approximation holds ($n, p \approx 0$), so that emission processes dominate the kinetics of the traps [1]. Equation (15) modifies to

$$n_t(\infty) = \frac{e_p}{e_n + e_p} N_t \approx 0 \tag{17}$$

The transient analysis follows therefore the SRH statistics considering the temperature dependences of the emission rates. Exhaustive treatments of thermal spectroscopy methods are widely reported in many books [1,8,16]; here we briefly describe the basics of DLTS and the analysis principles.

2.3.1. Deep-level transient spectroscopy

DLTS is a widely adopted method for detection and characterization of deep levels contained in materials and devices. It was developed in 1974 by Lang [13,14] and implemented by Miller [17] and Kimerling [18]. The fundamental point is the measurement of transients of differential capacitance of the SCR under electrical pulses of p–n junctions, Schottky diodes or MIS (metal–insulator–semiconductor) structures.

The DLTS technique allows the reliable assessment of deep-level parameters (concentration, activation energy, and capture cross section) in a straightforward manner by the analysis of the capacitance transients $\Delta C(t)$. A negative bias is applied to the sample, so that a certain region is depleted of free carriers. At each period T, a pulse of duration t_p and amplitude V_f is applied to the junction.

The duration of the pulse affects the trap filling, while the amplitude can be related to the region probed by the bias pulse:

(a) $V_f < V$: in this case an interval inside the SCR is selected. This can be useful if one is interested in concentration profiles of traps close to the surface or to the junction region, although more sophisticated techniques, such as DoubleDLTS [1], should be used for that.
(b) $V_f \sim V + V_{bi}$: the whole SCR is investigated with free carriers.

The requirements for applying the DLTS method are that the free carrier concentration (or shallow donors) must be higher than the trap density ($n \gg N_T$). Moreover, it is necessary to use diodes with well-defined characteristics; otherwise, the application of reverse bias may draw an exceedingly large leakage current affecting the capacitance measurement.

In the case of one trap energetically well separated from other traps, the capacitance transient can be expressed by a single exponential [1]:

$$\frac{\Delta C(t)}{C_0} \simeq \frac{N_T}{2N_I(\infty)} e^{-e_n t} \qquad (18)$$

where N_T is the trap concentration, $N_I(\infty)$ is the total space charge in the steady state (N_D), and C_0 is the steady-state capacitance at fixed bias V. The transients are repetitively generated by electrical excitation of the traps located in the SCR of a junction, the thermal scan allows for exploring the energy gap and to detect those traps whose $e_n(T)$ matches with the reciprocal of the time constant τ of the experimental setup. The transients are analyzed by using the rate window concept.

One way to introduce the rate window concept is to consider the DLTS method based on the boxcar correlator (Fig. 9). The capacitance transient subsequent to each bias pulse is sampled at two different times t_1 and t_2, and the difference between the capacitance values at these two instants is considered.

By taking into account Eq. (18), the DLTS signal is therefore

$$S(T) = \Delta C(T) = C(t_1, T) - C(t_2, T) \propto N_T [\exp(-e_n t_1) - \exp(-e_n t_2)] \qquad (19)$$

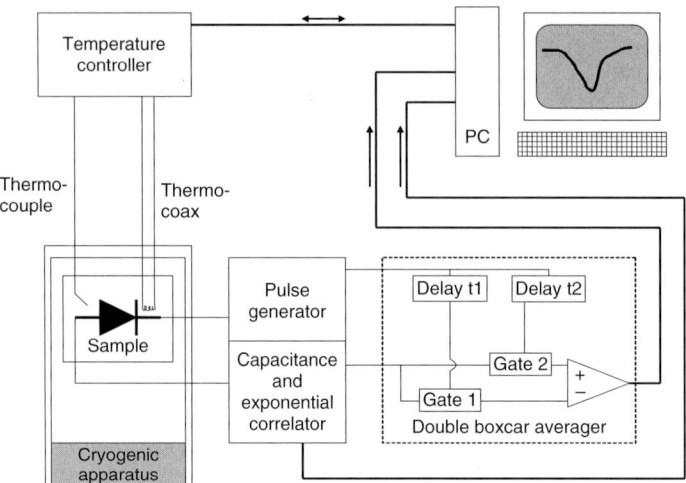

Fig. 9. DLTS experimental setup (courtesy of Dr Lorenzo Rigutti).

The point of maximum signal gives the relationship between e_n and T [16]:

$$\frac{dS(T)}{dT} = \frac{dS(e_n)}{de_n}\frac{de_n}{dT} = 0 \tag{20}$$

This is equivalent to assume $dS(e_n)/de_n = 0$ as the second factor is never zero (see Eq. (14)). The temperature at which the signal has a maximum is thus correlated to the emission rate according to

$$e_n = \frac{\ln(t_2/t_1)}{t_2 - t_1} \tag{21}$$

Hence e_n depends on the time constants fixed at the beginning of the measurement. The time interval $t_2 - t_1$ is called rate window.

The physical parameters of the deep centers can be calculated by making use of the Arrhenius equation or by refining the measurements by fitting the DLTS spectra [14,19]. The former method is faster and easier, although it is not very reliable when dealing with multiple levels with close energy values or related to extended defects [20,21].

Let us consider an electron trap: its emission rate (see Eq. (12)) can be expressed as a function of T as

$$e_n = \lambda_n \sigma_n T^2 e^{-((E_C - E_T)/kT)} \tag{22}$$

where λ_n is a constant depending on the effective mass of the carrier [1].

If σ_n is constant with T, and assuming $A_n = \lambda_n \sigma_n$, Eq. (22) assumes the form of the Arrhenius equation:

$$\frac{e_n}{T^2} = A_n e^{-((E_C - E_T)/kT)} \tag{23}$$

Once the time constants $\tau = e_n^{-1}$ giving the maximum values of ΔC at a given temperature T is measured, the Arrhenius curve is obtained by plotting the points of coordinates $[\ln(e_n T^{-2}); 1/T]$.

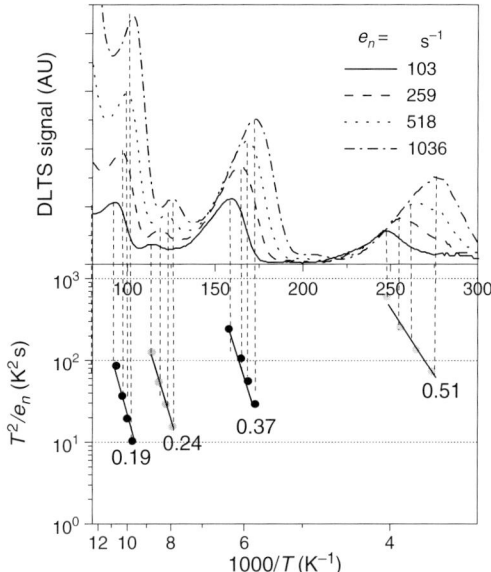

Fig. 10. DLTS spectra at four different emission rates and corresponding Arrhenius plot.

The slope of the interpolating line gives E_A and the intercept with the y-axis gives $A_n = \lambda_n \sigma_n$, from which the capture cross section can be inferred.

Further details on the technique and the effects due to non-ideal conditions (such as non-exponential transients, temperature dependence of the capture cross section, and lambda effects) are reported in the specific literature [1,8,16].

Figure 10 shows a schematic construction of Arrhenius plot from the spectra relevant to four different emission rates.

2.3.2. DLTS-related techniques

The application of DLTS is strictly related to the relative concentration between the free carriers and the traps: if $n \gg N_T$ in n-type material (or $p \gg N_T$ in p-type material) the capacitive signal of DLTS is high enough to give reliable information.

Differently, if the material has high resistivity, the concentration of traps can be even orders of magnitude higher than the free carrier concentration. In this case, the capacitive signal is very small and on the order of electric noise.

Alternative methods are therefore used, applying the rate window concept of DLTS, which analyze the current signal following a pulsed excitation. Even though the data interpretation is often more complex than in DLTS, a wide variety of alternative methods has been developed to study majority or minority carrier traps often making use of light as excitation source. Among these methods, we recall two methods that are widely used for semi-insulating materials: PICTS (photoinduced current transient spectroscopy) and P-DLTS (photo-deep level transient spectroscopy) [1,19].

PICTS is based on the trap filling induced by photocarriers generated by light pulses; the thermally stimulated detrapping induces a temperature-dependent current transient to which

Table 2
Operation modes of thermal transient spectroscopy

Transient spectroscopy	Electrical excitation	Optical excitation	Output signal
DLTS	(V pulses vs t)	none	ΔC transient
PICTS	constant $-V_b$ vs t	hν pulses vs t	ΔI transient
P-DLTS	V pulses vs t	constant hν vs t	ΔI transient

the rate window concept can be applied. PICTS can be used with two ohmic contacts or with a Schottky barrier and an ohmic contact.

Photoinduced generation of free carriers is also used by P-DLTS, where a constant optical excitation is added to a pulsed electrical excitation: because Schottky barrier is used in P-DLTS, only majority traps can be revealed, as occurring for DLTS. For details about these techniques, we refer to specific literature [1]. Here we just propose a table (Table 2) showing the main differences and analogies between the three techniques.

Thermal spectroscopy methods allow for detecting and characterizing deep levels, provided that the excitation is sufficient to induce a significant response of the material without altering its main properties.

Electrical characterization by I–V and C–V on nanometric structures and devices is currently performed [22,23], while DLTS-related techniques have been since then successfully applied to heterostructures and quantum wells [24–26], although a first result on nanowires has been recently published [27].

Thermal spectroscopy is particularly efficient in characterizing defects in semiconductors with energy gap lower than 1.5 eV. For larger gaps, a complete detection of electrically active defects is not possible due to thermal constraints. Once fixed, the range of the emission rate, the range of temperature T where the deep-level transition to the band occurs, is the controlling factor: deeper the trap, higher is the temperature of maximum emission. However, elevated temperatures may cause damage or alter the performance of the electric contacts and, what is more, leakage current exponentially increases with T to detriment of the collected signal.

The use of light excitation can be a very efficient tool to overcome the problem arising from the electronic and/or geometric structure of the material. Some of the above-mentioned thermal spectroscopy methods make use of light to increase the number of mobile carriers (P-DLTS) and to change the occupation of traps (PICTS); however, the exploration of energy gap occurs via temperature.

In contrast, photoinduced spectroscopy methods, as those presented in the following section, make use of the photon energy to probe the forbidden gap.

3. Photoinduced electrical characterization

The physical mechanism on the basis of surface photovoltage spectroscopy (SPS) and spectral photocurrent (SPC) methods is the extrinsic or intrinsic absorption of photons: both the techniques analyze electrical signals arising from the variation of carriers in conduction and valence band subsequent to photon-induced transitions.

In analogy with the photoelectric effect [28], in the process giving rise to these photo-induced transitions, the bound charge absorbing a photon can acquire enough energy to be freed and transit into the main bands.

Intrinsic absorption occurs for photon energy higher or equal to the band-gap value, so that carriers of both type are generated. In contrast, extrinsic absorption is generated by below band-gap energy light, able to excite transitions of carriers from deep levels to either bands (a sketch on the intrinsic and extrinsic transitions is reported in Fig. 11).

SPV and SPC methods might allow for distinguishing surface or bulk states by the comparison of the relevant spectra. Spectral photocurrent induced by sub-band-gap light, as a matter of fact, is less sensitive to surface effects because the photocurrent is collected from the entire bulk of the sample, so that the contribution of the SCR is usually negligible. Therefore, transitions appearing in both types of spectra are usually attributed to bulk transitions, whereas transitions appearing in the SPV spectra alone are assigned to surface states [29]. Table 3 summarizes the main characteristics relevant to the different techniques presented here.

3.1. The surface voltage

If surface has a dominant role, as in nanostructures (nanowires in particular), even in presence of ohmic contacts, the current–voltage analysis cannot neglect the surface contribution. The flow of charge may indeed be mainly controlled by surface space charge layer or barriers due to surface Fermi level pinning [30].

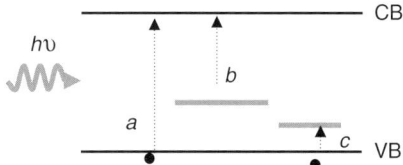

Fig. 11. Transitions giving rise to variations in conductivity or surface charge redistribution: (a) intrinsic transition; (b) deep-level-to-band by an electron; and (c) deep-level-to-band by a hole (extrinsic transitions).

Table 3
Summary of the spectroscopic techniques illustrated in the present paragrap

Method	Configuration	Signal	Information
SPS Surface photovoltage spectroscopy	Non-contact	$V(\lambda)$	Surface properties, lifetimes, minority carrier diffusion length, doping density, energy gap, below and above gap transitions, band tails
SPC Spectral photoconductivity	Planar Junction	$I(\lambda)$ $I(\lambda)$	Energy gap, below and above band-gap transitions, relative density of defects, band tails

Despite the bulk of a homogeneous semiconductor is neutral at equilibrium, near the surface uncompensated charges (surface states) and high electric fields are found, spatially extended for a thin space charge layer under the surface, usually on the order of the bulk Debye length [29].

The surface states carry a charge Q_{SS} per unit area of the surface and, to obey to the neutrality condition, the semiconductor will be characterized by a surface layer whose charge density Q_{SC} is

$$Q_{SS} = -Q_{SC} \tag{24}$$

where the underlying crystal is the sole supplier of the surface charge. Therefore, the carrier density near the surface deviates from its bulk equilibrium value and results in the surface SCR.

This deviation from bulk value can make the surface in

- *accumulation* regime, where the majority carrier concentration at the surface is larger than its bulk value (downward bent bands in n-type, upward in p-type);
- *depletion* regime, where the majority carrier concentration at the surface is smaller than its equilibrium value, but larger than the minority carrier concentration at the surface (upward bent bands in n-type, downward in p-type);
- *inversion* regime, where the majority carrier concentration at the surface is smaller than the minority carrier concentration at the surface (upward bent bands in n-type, downward in p-type) [29,31].

According to the Poisson equation [29], a non-equilibrium carrier density implies a non-zero electric field and potential. So the semiconductor bands are bent at the surface with a band bending defined as the *surface potential* V_S, that is the potential difference. By definition, the energy band is lower the higher the electric potential is, so that a positive V_S corresponds to downward bent bands. V_S is by definition positive for downward-bent bands (p-type material) and thus negative for upward bent bands (n-type material).

For calculating V_S, the dependence of Q_{SS} (net surface charge) and Q_{SC} (the net charge in the SCR) has to be found. For determining an explicit expression for Q_{SC}, the Poisson equation must be integrated from the bulk toward the surface using the boundary conditions.

In the accumulation and inversion regimes,

$$|Q_{sc}| \propto \exp\left(\frac{e|V_S|}{2kT}\right) \tag{25}$$

In the depletion regime, the region where the mobile carriers are negligible ends abruptly at distance w from the surface:

$$|Q_{SC}| \propto \sqrt{|V_S|} \propto q|N_A - N_D|w \qquad (26)$$

3.2. Surface photovoltage spectroscopy

SPS is a non-contact spectroscopic method that also allows for the identification of the semiconductor type conductivity [32], the determination of its energy gap [33], the detection, and the distinction between the surface and the bulk states in the forbidden band gap [29,34].

The SPV method is a well-established technique [35–38] for the characterization of semiconductors, based on the study of changes in surface photovoltage induced by an optical excitation. When a semiconductor surface is illuminated, the photogenerated minority carriers are swept to the surface barrier. Under open-circuit condition (the case of non-contact SPV), the buildup of minority carriers occurs at the surface and the majority carriers are collected in the material beyond the depletion layer edge. This causes a band bending and, in turn, a voltage difference between the surface contact and the bulk neutral semiconductor, equivalent to a *forward bias*. The steady-state condition is reached when the charged carrier flow caused by the band bending is equal in magnitude to the photoinduced carrier flow. This condition defines the *steady-state photovoltage*.

The SPV is the illumination-induced variation of the equilibrium surface potential:

$$\text{SPV} = V_S(\text{illuminated}) - V_S(\text{dark}) \qquad (27)$$

The formation of SPV signal is therefore caused by the charge redistribution subsequent to the charge generation in the surface SCR. It is commonly assumed that no appreciable voltage drop occurs on the quasi-neutral bulk, even under illumination.

In the absence of an external field, the charge neutrality rule must remain valid regardless of illumination. However, both Q_{SS} and Q_{SC} may change, possibly very significantly, on illumination.

The magnitude of V_{SPV} is strongly influenced by the properties of the surface barrier and by the treatment of the semiconductor surface. A depletion (or inversion) type surface barrier assures preferable collection of the excess minority carriers in the surface space charge, while the majority carriers are repelled from the surface and give negligible contribution (Fig. 12).

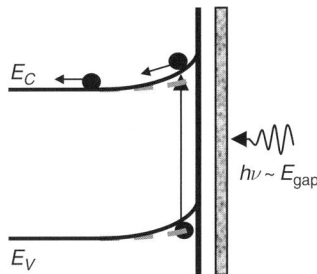

Fig. 12. Band bending and carrier redistribution in n-type semiconductor for band-gap light illumination.

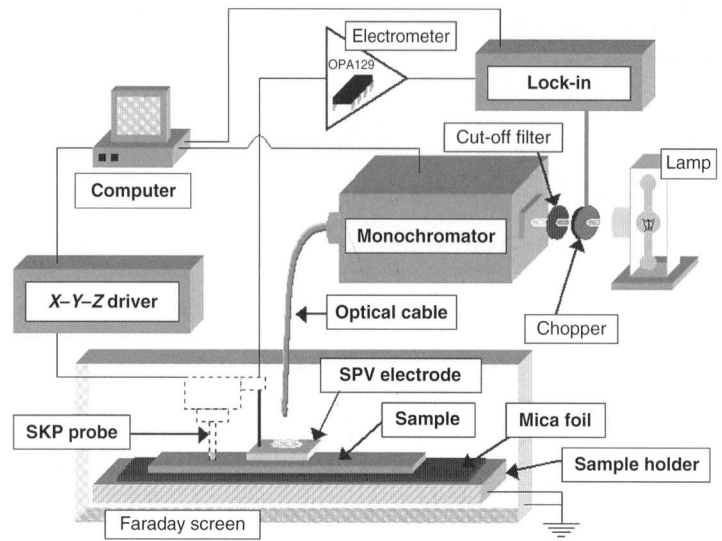

Fig. 13. Experimental setup for SPV measurement (courtesy of Dr Marco Rossi).

A chopped light illuminates a grating monochromator and the out-coming monochromatic beam can be directed to the sample or guided in a fiber optical cable, as shown in Fig. 13. For the SPV measurement the injection conditions must be carefully checked as the Debye length L_D [1] has its usual physical meaning only under low injection conditions [31].

The SPV electrode is a semi-transparent metallic electrode, which is capacitively coupled with the semiconductor surface. It has been obtained by sputtering indium thin oxide (ITO) on quartz. The two capacitors represented are respectively due to the "electrode–sample" (C_{es}) and the "sample–sample holder" (C_{sh}) couplings. This double capacitor scheme allows for maintaining the sample *floating* performing the ground contact only on the sample holder.

The SPV electrode is maintained at a voltage V_{SPV} and the AC signal (induced by the chopped light) is measured by a unit-gain FET operational amplifier. This electrical component has an input resistance larger than $10^{10}\,\Omega$ and it is essential to perform non-contact SPV measurements even when the distance between the electrode and the sample is high; as a matter of fact this "electrometer" is used as an *I/V* converter and it reduces the *infinite* impedance of the air gap between the sample and the electrode.

From the electrometer, the signal goes to a lock-in amplifier performing a phase-sensitive detection of V_{SPV}, with a working frequency referenced to the chopper frequency. A PC collects the data from the lock-in and electronically controls the monochromator and the X–Y–Z driver of the Scanning Kelvin Probe (SKP) system. The sample and the electrode are in a grounded Faraday cage to shield any electromagnetic fields that can be present in the nearby environment. As a matter of fact the *floating* electrode can act like an antenna, and thus, can detect any external field.

As band-to-band transitions always decrease the surface band bending, SPV is positive for n-type material. Therefore, the sign of the SPV signal can be used to detect the conductivity type of the semiconductor.

In semiconductors the light flux decreases with a decay length called "penetration depth," which is the reciprocal of the absorption coefficient α for the incident light. Thus, the light flux at the plane x follows the Beer's law:

$$\Phi(x) = \Phi_0(1-R)\exp(-\alpha x) \qquad (28)$$

where R is the reflection coefficient of illuminated surface and Φ_0 the photon flux impinging the surface.

The generation rate of minority carriers is equal to the rate at which photons are absorbed:

$$G(x)\,dx = \Phi(x) - \Phi(x+dx) = -\left(\frac{d\Phi}{dx}\right)dx = \alpha\Phi_0(1-R)\exp(-\alpha x)\,dx \qquad (29)$$

Under the depletion approximation, the drift current is negligible with respect to the diffusion current in the quasi-neutral region, and vice versa in the depletion region.

Some phenomena, such as trapping of minority carriers by deep levels and surface state, can interfere with SPV measurements. These introduce hysteresis corresponding to long-time-constant components on the light-on/light-off SPV relaxation. By working at frequencies sufficiently higher than the reciprocal of the surface time constant τ, the long-time components contribution can be avoided. Therefore, the upper limit of the working frequency ω is fixed by the decrease of V_{SPV} when $\omega\tau > 1$ [29].

The absorbed photons induce the formation of free carriers by creating electron–hole pairs via band-to-band transitions (above band-gap $h\nu \geq E_{gap}$) and/or release captured carriers via trap-to-band transitions (sub-band-gap $h\nu < E_{gap}$). Thus, a significant amount of charge may be transferred from the surface to the bulk (or vice versa) and/or redistributed within the surface or the bulk.

3.2.1. Above band-gap illumination

Under above band-gap illumination the probability of band-to-band absorption is typically orders of magnitude larger than that of trap-to-band absorption. In most semiconductors, there is a large increase in the absorption coefficient α near the band-gap energy, E_G. Therefore, a significant SPS increase is to be expected at approximately this energy. Therefore the energy gap value can be inferred in analogy to the absorption spectroscopy method, with the advantage that, as opposed to transmission measurements [39], in SPS analyses the sample can be arbitrarily thick and it does not have to be removed from the substrate or grown on a transparent one.

The semiconductor is in excess carrier injection condition (non-equilibrium) and this case can be analyzed with the quasi-Fermi levels formalization [1,40,41].

In the case of steady-state illumination through the free surface, the quasi-Fermi levels of the electron and hole generally depart from the equilibrium Fermi level value, but in-depth the illumination is practically negligible and the quasi-Fermi levels converge to the equilibrium value. In addition, recombination in the surface SCR may be significantly larger than in the bulk, thereby reducing the effective excess carrier density. This may cause the quasi-Fermi levels to deviate less from equilibrium toward the surface than at the edge of the SCR.

Changes in the quasi-Fermi levels indicate a non-zero electron and hole current, even under steady-state conditions. Indeed, under open-circuit conditions only the total current, J, must be zero, whereas J_n and J_p may be non-zero.

Again the Poisson equation allows for obtaining the value of Q_{SC}. It can be demonstrated that if surface charge does not vary, illumination always tends to decrease the band bending.

Moreover, the band bending should approach zero as the excess carrier density increases. Physically, the photoinduced carriers partially screen the fixed surface state charge, thereby reducing the surface band bending. For a large enough excess density, the screening should be complete. Note that if the bands are flat (i.e., $V_s = 0$) in equilibrium, they remain flat under illumination.

3.2.2. Below band-gap illumination

When working with sub-band-gap photons ($h\nu \leq E_{gap}$) SPS reveals intraband transitions. Taking into account an *n-type* semiconductor surface, where Q_{SS} is negative and Q_{SC} is positive, the most common mechanism for sub-band-gap SPV (SPS) involves the direct modification of the surface charge, and hence potential, by excitation of trapped carriers.

Illumination by photons with energy $h\nu \sim E_C - E_t$ (with E_t the trap energy level) may produce electron transitions from a defect (surface or bulk-related) state at an energy E_t into the conduction band, where they are swept quickly to the semiconductor bulk by means of the surface electric field. Hence, Q_{SS} is reduced and the band bending is decreased.

Conversely, illumination by photons with energy $h\nu \sim E_t - E_V$ may produce electron transitions from the valence band into a surface state situated at an energy E_t above the valence band maximum E_V (which are equivalent to hole transitions from the surface state to the valence band). Such transitions increase Q_{SS} and therefore the surface band bending. Obviously, this requires that surface states are not completely filled prior to excitation.

Figure 14 shows a typical example of SPS spectrum obtained in Si nanocrystals [34].

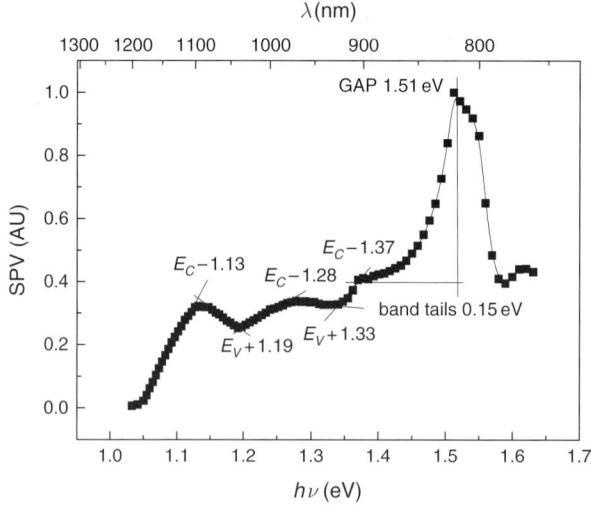

Fig. 14. SPS spectrum referring to Si nanocrystals embedded in the amorphous silicon matrix (reprinted from Ref. [34] with kind permission of Springer Science and Business Media).

3.3. Spectral photoconductivity

Photoconductivity effects in solids were discovered by Smith [42] at the end of the 19th century. In the 1920s, Gudden and Pohl developed the photoconductivity theory [43], demonstrating the dependence of photoconductivity, light absorption and luminescence on the light wavelength and assessing that the interaction occurs between one photon and one electron.

The dark conductivity σ of a semiconductor is given by

$$\sigma = e(n\mu_n + p\mu_p) \tag{30}$$

where μ_n and μ_p are the mobility of the electrons and holes, respectively.

Photoconductivity [44] $\Delta\sigma$ is defined as the increase of conductivity occurring in a semiconductor under optical excitation. When light of near-band-gap energy hits a homogeneous semiconductor, conductivity increases by the amount $\Delta\sigma$ due, in most cases, to the increase of the free carrier densities Δp and Δn:

$$\Delta\sigma = e(\Delta n\mu_n + \Delta p\mu_p) \tag{31}$$

In a basic treatment of photoconductivity in unipolar semiconductors, only the majority carrier transport, electrons for n-type and holes for p-type semiconductors, is usually considered. Moreover, the charge neutrality is assumed to be maintained during illumination, i.e., $\Delta p = \Delta n$. Photogenerated excess carriers are typically orders of magnitude lower than their density in dark ($\Delta p \ll p$ and $\Delta n \ll n$) in semiconductors; conversely, in semi-insulators the excess carriers are much higher than the dark density. Photocarrier densities depend on the free lifetimes τ_n and τ_p according to the relations:

$$\Delta n = f\tau_n, \quad \Delta p = f\tau_p \tag{32}$$

where f is the number of electron–hole pairs generated by light per second per unit volume. The parameter f is related to the excitation intensity $\Phi(\lambda)$ and to the absorption coefficient $\alpha(\lambda)$. The spectral response is therefore a function of λ, sensibly high approaching the band-gap light and decreasing for larger wavelengths.

By taking into account the Beer's law (Eq. (28)), f can be expressed by

$$f = \alpha\beta\Phi \tag{33}$$

with β the carrier pairs generated by each photon (typically $\beta < 1$) [45].

Photoconductivity variation $\Delta\sigma$ can be finally expressed by the relation

$$\Delta\sigma = e\beta\alpha\Phi(1-R)(1-e^{-\alpha x})(\mu_n\tau_n + \mu_p\tau_p) \tag{34}$$

where R is the reflection coefficient and x the penetration depth. A strict correlation thus exists between $\alpha(h\nu)$ and $\Delta\sigma(h\nu)$.

For photon energy $h\nu > E_G$, a high absorption region exists, where light is principally absorbed on the surface; hence, the photoconductivity $\Delta\sigma$ is controlled by the surface carrier lifetime. The intermediate region $h\nu \approx E_G$ is characterized by a still strong absorption, although photoconductivity is now controlled by the bulk lifetime, with a maximum occurring when $\alpha \approx 1/d$, where d is the thickness of the sample. For $h\nu < E_G$, light penetrates deeply into the material and bulk lifetime still controls the photoconductivity, which decreases by orders of magnitude, as absorption coefficient does [46].

Below band gap light can excite extrinsic transitions, where the extrinsic absorption coefficient is proportional to the density of absorbing centers N_t by the relation

$$\alpha = \sigma^{opt} N_t \tag{35}$$

where σ^{opt} is the optical capture cross section of the center [43].

Corresponding to the photon energy exciting extrinsic and intrinsic transitions, photoconductivity spectra show peaks whose height can be correlated with the density of defect-related deep levels.

Light penetration depth is proportional to α^{-1} [47,48]; therefore, it can be inferred that different spatial regions are investigated during the wavelength variation: below-band-gap light crosses the whole sample and the spectra refer to bulk; approaching the near-band-gap value, characterized by a high absorption, the exploration refers to the near-surface region. This, which would be a drawback in bulk investigation, can be easily overcome with nanostructures by orienting opportunely the light normally with respect to column axis to cross the nanometric diameter.

Figure 15 shows a typical spectrum obtained in thin films of GaN where the band-to-band transition and the defect-related bands are visible [49,50].

In spectral photoconductivity (SPC) a light source is used, which covers a wide photon energy range so as to probe the whole forbidden gap. Light, pulsed by a chopper, enters a monochromator and is focused onto the sample. Current measurements are carried out by a lock-in amplifier (Fig. 16) and the variations of current signal due to the release of photogenerated carriers from deep traps evidences the deep center-to-band transitions.

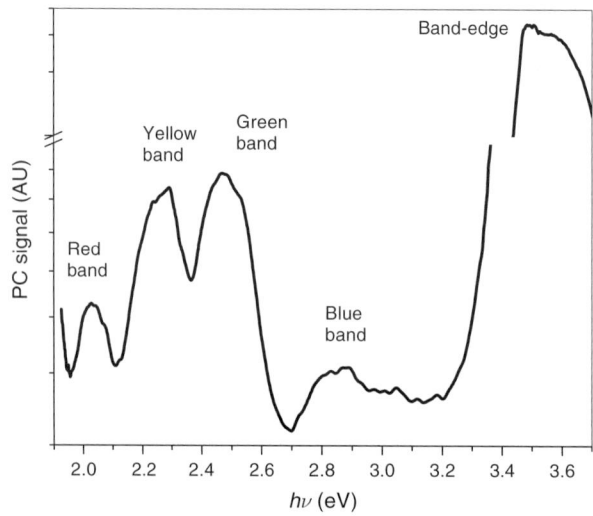

Fig. 15. Typical SPC spectrum referring to thin layers of GaN HVPE grown on sapphire.

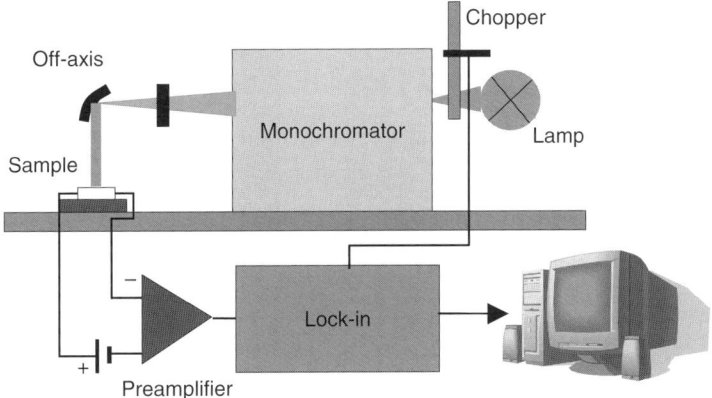

Fig. 16. Experimental setup for SPC measurements.

Appropriate bias conditions and chopper frequency dramatically increase the signal-to-noise ratio, taking into account the speed of response, strictly related to the rise time (the steady-state value of photoconductivity after turning on the light) and the decay time (the decrease to dark values after turning off the light) [51]. Speed of response is usually decreased in presence of traps, which introduce additional delay due to trap filling and emptying.

The choice of the light source is also essential to measure significant signals, as generally $\Delta\sigma \ll \sigma_0$. The lamp spectral response must therefore be high and uniform enough in the selected range of wavelengths, to excite as much transitions as possible, avoiding at the same time high injection conditions.

Because excitation intensity $\Phi(\lambda)$ strongly depends on wavelength, due to lamp and monochromator spectral response, photoconductivity signal $\Delta\sigma(\lambda)$ is usually normalized to the photon flux $\Phi(\lambda)$, hence giving the value of responsivity (photoconductivity per unit of excitation intensity):

$$I_{PC} = \frac{\Delta\sigma}{\Phi_{ph}} \tag{36}$$

4. Case study: electrical characterization of GaN nanowires

To examine the efficiency of some of the experimental techniques described in this chapter, we finally analyze the case of GaN nanowires, a challenging benchmark, taking into account the large forbidden gap.

GaN-based nanostructures and quantum devices recently widened the range of applications toward nanolasing, nanotransistors, and nanofluidic biochemical sensors [52]. The ever-increasing interest for gallium nitride and its wide applications has promoted intensive investigations and strong efforts devoted to the study of the material and the modeling of its based devices [53,54].

Details on the physical mechanisms acting in the bulk material are still under discussion, as well as the role played by the well-known defect-related yellow band [51]. The knowledge concerning GaN-based nanostructures is still at its infancy and even more confusing.

In this section, electrical investigation of GaN nanowires by means of dark conductivity and UV photoconductivity [22], DLTS [27], spectral photoconductivity [55], and surface photovoltage [56] will be presented. The application studies reported here are the first, to our knowledge, efficiently applied to GaN nanostructures.

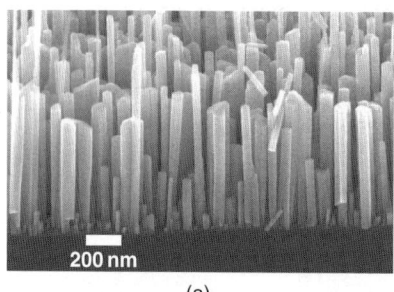 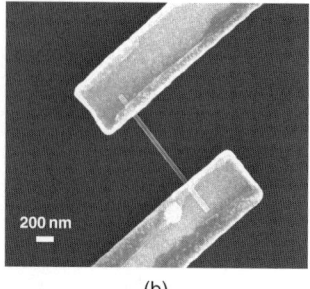

(a) (b)

Fig. 17. (a) SEM micrograph of PAMBE-grown GaN nanowires, (b) singly contacted nanowires (reprinted with permission from Ref. [22]. Copyright 2005 American Chemical Society).

The present study refers to the electrical characterization of GaN nanowires. Nanowires have been grown by radio frequency plasma-assisted molecular beam epitaxy (MBE) on Si (111) substrates in N-rich stoichiometry. The as-grown sample shows a wide distribution of hexagonal wires with diameters ranging between 20 and 500 nm and lengths from 0.3 up to 2 µm [22,57].

Nanowires were subsequently released from the native silicon by exposure to an ultrasonic bath and deposited on a SiO_2/Si substrate. Ti/Au ohmic contacts were patterned by electron beam lithography for electrical characterization [22] (Fig. 17).

4.1. Dark and photoconductivity studies

Calarco et al. [22] carried out I–V (in dark and under illumination) and photocurrent time decay measurements, interpreting the size dependence by a surface pinning barrier model.

Three main results are experimentally obtained in this work:

1. Dark current is negligible for thin wires (up to 80 nm) while showing a SCL behavior at biases larger than 0.1 V for thicker wires.
2. Photoconductivity is size dependent: the photocurrent remains at a relatively high level as long as the column diameters exceed the critical value around 80–100 nm, and decreases dramatically for wires of diameter below 80 nm.
3. The time decay measurements of photocurrent after UV illumination is also size dependent: longer for thick wires and negligible for wires thinner than 100 nm.

Current–voltage measurements in dark or under UV illumination were carried out showing regimes of ohmic and SCLC as explained in Section 2.1.1 (Fig. 18).

Surface states have been shown to induce a Fermi level pinning especially in III–V compounds, dramatically affecting the conductivity of nanostructures. In n-type GaN, the Fermi level at surface is pinned at 0.5–0.7 eV below the conduction band at low and moderate Ga/N ratios and has been assigned to Ga dangling bonds, while it is pinned at 1.8 eV above the valence band for high Ga/N ratios, due to Ga–Ga bonds [58].

The Fermi level pinning causes a depletion charge layer corresponding to the whole wire thickness or to a part of it, depending on the wire diameter. Depending on the extent of surface space charge layer, with respect to the wire diameter, photogenerated opposite carriers can be more or less spatially separated.

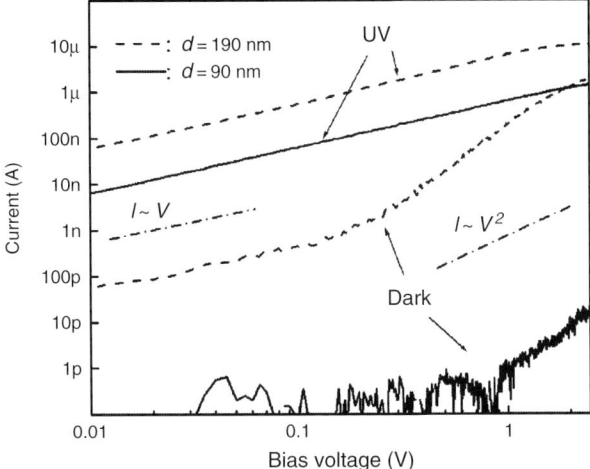

Fig. 18. Dark and UV conductivity behavior in GaN nanowires (reprinted with permission from Ref. [22]. Copyright 2005 American Chemical Society).

The measurements showed that photocurrent depends on wire diameter, demonstrating that the main mechanism controlling photoconductivity is the surface recombination of non-equilibrium carriers.

The photocurrent shows an ohmic behavior for wires larger than 100 nm due to the thin conductive channel present in these samples.

The model proposed by Calarco et al. [22] is reproduced in Fig. 19: the Fermi level pinning at the surface bends upward the semiconductor bands. It is based on the calculus of the depletion space charge layers, evaluated on the order of 50–100 nm.

From the I–V curves it is shown that, for diameters up to 200 nm, the space charge-limited dark conductivity is dominating, while the increase in conductivity due to the photogenerated carriers allows for an ohmic behavior in nanowires larger than 100 nm.

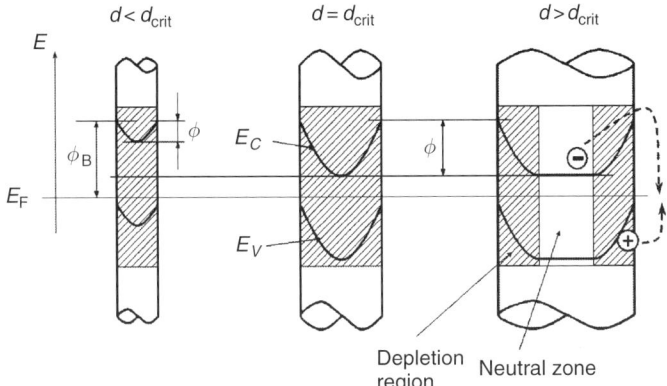

Fig. 19. Model of carrier recombination in differently sized nanowires (reprinted with permission from Ref. [22]. Copyright 2005 American Chemical Society).

The model shows the recombination mechanism occurring in differently sized wires:

(a) thin wires (diameter <80 nm) are characterized by full depletion, hence explaining their insulating character; moreover, they show a reduced band curvature due to the small dimensions and therefore a reduction of the surface electron–hole recombination. The photocurrent decays strongly with decreasing barrier height, i.e., with decreasing column thickness.
(b) For wires of diameter around the critical value of 80 nm, the sample is still fully depleted but the barrier height is on the order of the surface pinned Fermi level as calculated by experimental [59] and theoretical [14] studies.
(c) The persistent photocurrent for columns with diameters above 100 nm remains at a relatively high level. The recombination mechanism is mainly due to surface traps as electrons prefer the inner part of the column, whereas holes tend to move to the surface, hence inhibiting the recombination of non-equilibrium carriers. Electrons would have to surpass the barrier at the surface for surface recombination.

The time decay of the persistent photocurrent is essentially due to surface recombination, where holes are pushed to the surface due to the surface band bending and electrons have to overcome the corresponding surface barrier to recombine.

The stationary photocurrent, calculated from the generation and the surface recombination rate balance gives an exponential trend of the recombination rate r in the form

$$r \approx \exp-\left(\frac{\Phi}{kT}\right) \qquad (37)$$

where Φ corresponds to the Fermi level pinning value (0.55 eV) for columns larger or equal to the critical diameter d_{crit} of 80 nm, while it decreases with diameter for thinner wires. Hence, for $d > d_{crit}$ (with d_{crit} minimum diameter for a fully depleted column), a constant barrier height with N_D as donor concentration causes a photocurrent proportional to column diameter, as is observed in Fig. 20.

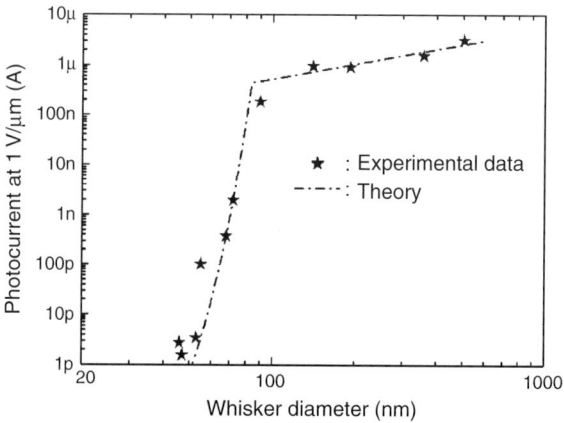

Fig. 20. Photocurrent trend as a function of nanowire diameter (reprinted with permission from Ref. [22]. Copyright 2005 American Chemical Society).

For thinner columns ($d < d_{\text{crit}}$), the recombination barrier Φ depends on the square of the column diameter d following the equation

$$\Phi = \frac{eN_D d^2}{16\varepsilon\varepsilon_0} \qquad (38)$$

The authors fitted the PC values vs diameter reported in Fig. 20, and obtained a good matching with the estimated values of donor concentration N_D and of surface barrier Φ. In conclusion, this model of surface electron–hole pair recombination excellently describes the unusual behavior of size-dependent persistent photocurrents in GaN nanocolumns. The observed interplay between column diameter and space charge extension can be of extreme importance with respect to optoelectronics and sensor technology applications.

4.2. DLTS study on GaN nanorod p–n junctions

Very recently, the real possibility of carrying our DLTS on single MBE-grown GaN nanowire p–n junction was demonstrated [27]. GaN nanorods were grown on Si (111) substrates with different dopings: Si-doped GaN showed a free electron concentration of 3×10^{18} cm^{-3} while Mg doping showed a free hole concentration of 1×10^{17} cm^{-3}. E-beam lithography was used for Ti/Au electrodes patterning, which confirms the ohmic behavior as already found in Ref. [22]. Before carrying out DLTS, the authors spatially localized the p–n junction by cathodoluminescence experiments (see Chapter 7).

Previous current–voltage characteristics have shown a strong temperature dependence of current, suggesting that thermoionic emission over the barrier is the dominant mechanism of charge transport. The large reverse bias suggested the presence of deep levels.

DLTS was carried out by amplifying the signal opportunely, and the temperature was raised from 77 to 300 K. Figure 21 shows the corresponding DLTS peak at about 240 K at emission rate equal to 11.71 s^{-1}.

From Arrhenius analysis, they found an energy level of E_C–0.40 eV and capture cross section σ on the order of 10^{-16} cm^2. This kind of research also demonstrates the big potentiality of DLTS as investigation tool for nanometric devices.

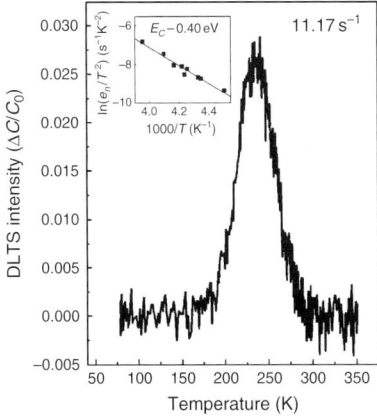

Fig. 21. DLTS spectrum of the p-n GaN nanorod and Arrhenius plot of the corresponding energy level in the inset (reused with permission from Ref. [27]. Copyright 2006, American Institute of Physics).

4.3. SPS and SPC measurements and Franz–Keldysh effect evidence

The work by Cavallini et al. [56] is complementary to two earlier publications [22,55] by the same groups and widens the understanding of the influence of size and geometry on fundamental optoelectronic properties of GaN nanowires.

The observation of sub-band-edge "tails" in the spectral response of the surface photovoltage signal (for large ensembles of vertically oriented wires) and of the photoconductivity (for single wires cast flat on a substrate) has been attributed to a Franz–Keldysh effect, arising from a built-in field due to surface Fermi level pinning, as confirmed by theoretical simulations. Although the Franz–Keldysh effect is well known in the literature of thin-film semiconductors, one might expect it to be very common in nanowires [60]. Franz–Keldysh effect is a physical phenomenon associated with the influence of a high electric field on the near-band-edge absorption [46]. Franz [61] and Keldysh [62] predicted that the electric field may cause a red shift of the absorption edge, giving rise to the presence of an absorption tail for band-to-band transitions.

SPS measurements were performed on the large ensemble of nanowires (an SEM picture of this type of sample is shown in Fig. 17(a)) under the SPS electrode area (about 3–4 mm^2). Figure 22(a) and (b) shows the typical SPS spectrum obtained in the band-edge region [56].

The SPS signal V_S significantly increases when the optical excitation energy varies from 3.3 to 3.5 eV in the band-edge absorption region, characteristic of a band-to-band transition of an n-type semiconductor [30]. The increase in V_S corresponds to the minority carrier (holes) accumulation caused by the optical generation of electron–hole pairs near to the surface: the electrons move toward the bulk, due to the electric field in the surface SCR, while holes are collected at the surface barrier.

For direct band-gap materials like GaN, the following relationship holds [63–65]:

$$h\nu \times V_S \propto \left(h - E_{\text{gap}}\right)^{1/2} \tag{39}$$

where E_{gap} is the energy gap and $h\nu$ the photon energy of the optical excitation.

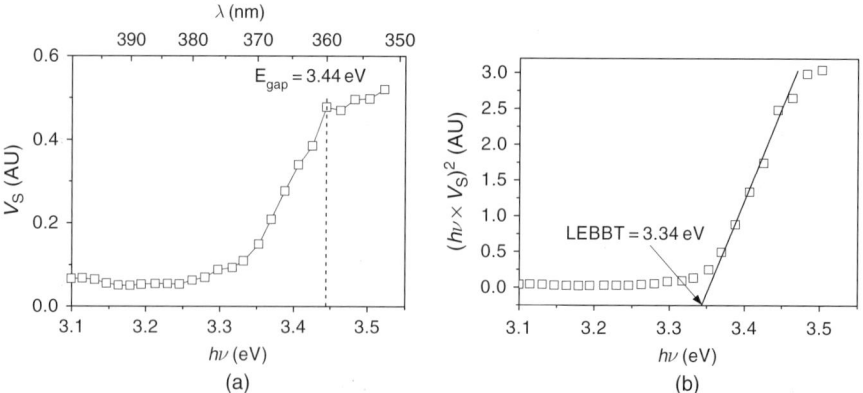

Fig. 22. (a) SPS of the as-grown nanowire ensamble where E_{gap} is estimated by the knee, (b) the fitting curve of the linear region $(h\nu \times V_S)^2$ vs $h\nu$ is shown (reprinted with permission from Ref. [56]. Copyright 2007 American Chemical Society).

The energy gap E_{gap}, evaluated from the intercept of the linear fitting of $(V_S \times h\nu)^2$ as a function of $h\nu$ (Fig. 22(b)) [66] is equal to 3.34 ± 0.03 eV, lower than that usually reported in the literature. The band-gap transition is affected by a large band tail, reasonably due to the significant contribution of sub-band-gap absorption, dramatically affecting the slope of and, in turn, the intercept of the linear fit [67].

The energy gap was hence estimated by the energy corresponding to the knee in the SPS spectrum of Fig. 22(a), i.e., $E_{gap} = 3.44 \pm 0.05$ eV, and the value 3.34 ± 0.03 eV is therefore determined as the lowest energy band-to-band transition (LEBBT) due to the band tails.

In light of the measured size-dependent properties of nanowires [22] already described in Section 4.1, the authors investigated a possible dependence of ΔE_{SPS} on the thickness of nanowires.

SPS cannot be carried out on singly contacted nanowires of different thickness, as the macroscopic SPS electrode cannot be capacitively coupled to a single nanowire.

To this aim, SPC measurements were carried out on singly contacted nanowires whose results are shown in Fig. 23 for four samples of different diameter.

Spectral PC measurements were carried out at bias maintaining similar conditions of operation for all measurements in differently sized nanowires.

As observed in Fig. 23, the sub-band-edge contribution clearly increases with nanowire diameter. By recalling the equation of photoconductivity (Eq. (34) in Section 3.3) it can be observed that, for nanowires whose diameter ranges on the order of hundreds of nanometers, the term $(1-e^{-\alpha x})$ related to the penetration depth can be reasonably considered constant in the band-edge region; hence, photoconductivity becomes proportional to the absorption coefficient α. The SPC absorption tails (Fig. 23(b)) have been therefore well fitted by the exponential law

$$\text{SPC} \sim \alpha \sim \exp\left(\frac{h\nu}{\Delta E}\right) \quad (40)$$

with ΔE tail width value deduced by fitting. As already explained in the model [22] in Section 4.1, in GaN nanowires a strong electric field exists in a depletion region due to the Fermi level pinning. This surface depletion region of high electric field can have a significant contribution in sub-band photoeffects due to Franz–Keldysh effect, especially in the case of NWs with a high surface/volume ratio than for compact layers.

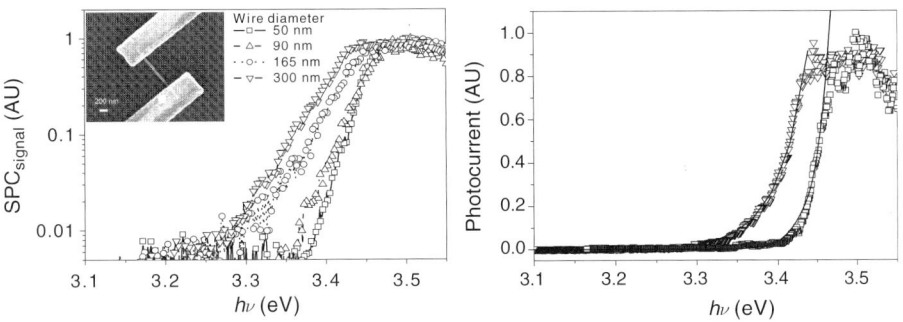

Fig. 23. (a) SPC spectra of four different nanowires (diameters are shown in the legend). (b) SPC spectra in linear scale of the thinner (50 nm) and thicker (300 nm) nanowires and exponential fitting (reprinted with permission from Ref. [56]. Copyright 2007 American Chemical Society).

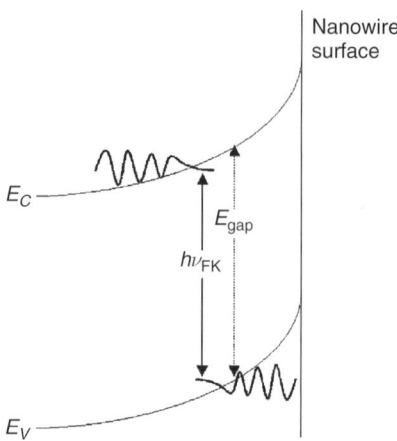

Fig. 24. Sketch of the band structure of a n-type semiconductor surface: the wave function of an electron in the conduction band and a hole in the valence band are characterized by tails in the forbidden band gap induced by the internal electric field (Franz–Keldysh effect): the band-to-band transition occurs at an energy $h\nu$ lower than the energy gap (reprinted with permission from Ref. [56]. Copyright 2007 American Chemical Society).

Under a constant electric field, the free electron wave function modifies into an Airy function that exponentially decreases in the band-gap region. As a result, a sub-band-gap optical absorption occurs, which can be interpreted as a photon-assisted tunneling through the forbidden band region, with photon energy smaller than E_{gap} (Fig. 24).

In the sub-band region, the Franz–Keldysh absorption coefficient vs $h\nu$ can be approximated by an exponential function [68, 69]

$$\alpha(h\nu) \sim \exp\left(-\left|\frac{h\nu - E_{\text{gap}}}{\Delta E}\right|^{3/2}\right) \tag{41}$$

By the fitting of curves in Fig. 23, the single exponential tail was fitted following Eq. (40), by inferring the values of ΔE for different wires. ΔE is theoretically described by the equation [43]

$$\Delta E = \frac{2}{3}\frac{(e\hbar\xi)^{2/3}}{(m^*)^{1/3}} \tag{42}$$

where ξ, m^*, e, and $\hbar$ are the electric field, effective mass, electron charge and Planck's constant ($h/2\pi$), respectively.

The cylindrical model [22], shown in Fig. 19, referring to the four thicknesses considered was hence taken into account to simulate the relative values of ξ and ΔE.

The experimental values of the exponential tail energy ΔE show a very good agreement with the calculated trend, confirming the presence of the Franz–Keldysh effect.

While SPS measurements refer to the whole distribution, hence giving an average value dominated by the contribution of thicker and shorter wires, which is the majority of the wires present in the sample, SPC determines the strong dependence of the band absorption tail as a function of wire diameter, explainable by the Franz–Keldysh effect induced by the electric field in the depletion region towards the wire surface. The experimental absorption tails fit well to a single exponential function $\exp(h\nu/\Delta E)$, suggesting a strong influence of the Coulomb (exciton) interaction on the Franz–Keldysh effect.

4.4. Spectral photoconductivity study of singly contacted GaN nanowires

In this section a very recent application of photoconductivity to nanostructures is presented, where this kind of investigation is applied for the first time to GaN nanowhiskers [55]. Due to their high surface-to-volume ratio, a detailed study of nanowire defects is very fruitful to get a deeper insight into the character of the PC bands usually detected in microscopic samples.

The photoconductivity spectra show, besides the band-gap-related transition, deep levels corresponding to the yellow, green, and blue bands. A strong spatial localization of specific photocurrent peaks has been observed, indicating that the defects responsible for such transitions are distributed inhomogeneously along the column growth direction.

Several nanowhiskers with diameters on the order of 100 nm, hence just above the critical diameter (see Section 4.1), have been analyzed with the three-finger configuration as reported in Fig. 25(a). Two electrodes have been prepared at the ends and one approximately in the center, and this three-contact configuration allows for distinguishing the signal contributions from the bottom part, containing interface-related defects, from those of the top part of the nanostructure.

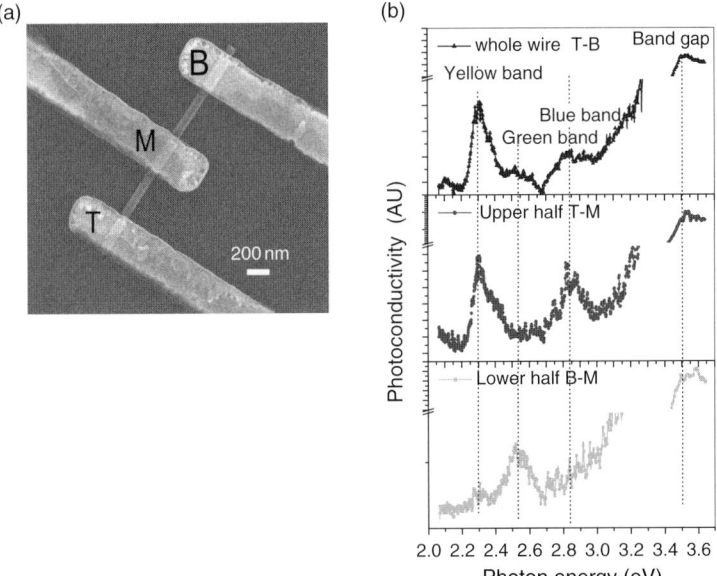

Fig. 25. (a) SEM image of a typical GaN nanowisker, the electrode T is located at the top, electrode B is located at base, electrode M is approximately in the middle. (b) Photocurrent spectra referring to: whole nanowire (electrodes T-B); upper half (electrodes T-M) and lower half (electrodes B-M) with respect to the growth direction. The y-axis is interrupted to account for the different orders of magnitude between near-band edge and defect-related band signals (reprinted with permission from Ref. [55]. Copyright 2006 American Chemical Society).

Besides the band-gap-related transition, other bands in the visible range are clearly distinguished at around 2.30 eV (yellow band), 2.52 eV (green band), and 2.85 eV (blue band).

Spectral photoconductivity does select responses coming from differently located volumes of the nanowires, just by conveniently choosing the electrodes: in this way the active volume contributing to the signal can be straightforwardly identified, differently from other techniques, such as micro-PL, whose spatial resolution is on the order of the sample size. The local investigation on the two sections of the nanowires might thus give a deeper insight into the origin of the photocurrent bands and their link with defects referring to interface, bulk, or surface.

From the two spectra relevant to the top part (T-M) and to the bottom part (B-M) of the nanowire (Fig. 25(b)), it can be observed that the amplitude of the green and yellow bands differ sensibly. Green band dominated the lower half while yellow band dominates upper half and whole wire-related spectra. In addition, we systematically observe a degradation of the signal-to-noise ratio in spectra carried out in the lower parts of the nanowires, explained by taking into account the influence of interface-related defects on the current transport characteristics.

Although the density of defects is not straightforwardly calculated from these measurements, owing to the many unknown parameters, we can however estimate their "impact" by analyzing the defect-to-band-gap ratio as reported in the histogram shown in Fig. 26.

Yellow band amplitude is slightly affected by the position along the column, while the green band signal increases by approximately one order of magnitude. By a similar approach we can quantitatively measure the relative weight between the yellow and the green bands in all the nanowires analyzed, observing that the yellow/green signals ratio ranges between 2 and 5 in the upper half, and between 0.1 and 0.4 in the lower half. We can definitely assess that the green band transition is strongly localized in the lower part of the nanowire, and is thus mainly related to defects originating at the interface with the silicon substrate. On the contrary, the yellow band, whose contribution is homogenously distributed along the nanocolumn, is the dominant feature in the upper part, in agreement to its presumed surface-related character. The spectrum referring to the whole wire reflects an average behavior, which takes into account both interface- and surface-related contributions.

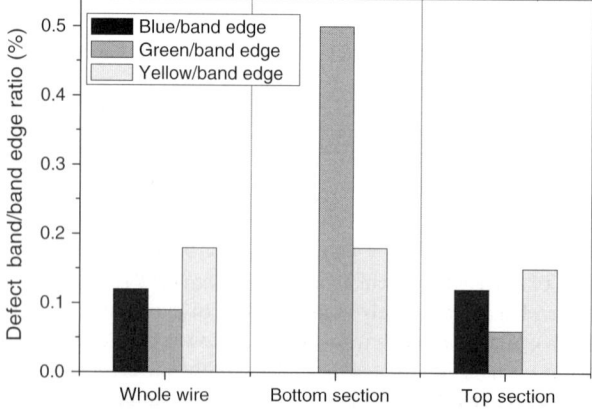

Fig. 26. Defect-to-band-edge peaks ratio percentage referring to the spectra shown in Fig. 26.

5. Summary and future development

In recent years, the synthesis and study of nanostructures has become a major interdisciplinary research area, owing to their numerous applications in different fields. One-dimensional nanostructures, in particular, represent a forefront for nanoelectronics as well as for optoelectronic and biological applications [70].

The aim of this chapter is to give a brief survey on the widely used characterization methods for investigation of electrical properties of semiconductors, in view of their application in the study of hetero- and nanostructures.

Current–voltage, capacitance–voltage, and DLTS studies are considered necessary to obtain an detailed insight of the electrical behavior of materials and devices. Moreover, light-induced spectroscopy such as photoconductivity and surface photovoltage techniques can be very powerful tools to explore the energy gap by encompassing some of the limits of pure electrical characterization.

These methods are widely used in the study of bulk and of heterostructures, while their application to nanostructures is still at its infancy and further improvements to the sensitivity of techniques are under development. Indeed, collecting electrical signal from nanostructures has been considered as a challenge for current experimental technology and for metrology in particular [71–73]. Excellent ideas were developed in the past decade to improve the capability of techniques, and further developments are expected to efficiently extend the range of investigation to the nanoscale [74] level.

Moreover, as shown in the case studies presented here concerning GaN nanowires, the use of standard techniques can be extended to nanostructures by conveniently implementing the experimental setup.

Acknowledgments

The authors thank Dr Raffaella Calarco (IBN1-Forschungszentrum Juelich, Germany) and Dr Daniela Cavalcoli (Physics Department of the University of Bologna) for the useful discussions.

References

[1] P. Blood and J.W. Orton, The Electrical Characterization of Semiconductors: Majority Carriers and Electron States, Academic Press (London, UK) 1992.
[2] H.J. Queisser and E. Haller, Science, 281 (1998) 945.
[3] E.H. Rhoderick and R.H. Williams, Metal-Semiconductor Contacts, Oxford Science Publications, Clarendon Press (Oxford, UK) 1988.
[4] M.A. Lampert and P. Mark, Current Injection in Solids, Academic Press (New York, USA) 1970.
[5] M.A. Lampert, Phys. Rev., 103 (1956) 1648.
[6] S.M. Sze, Physics of Semiconductor Devices, John Wiley and Sons (New York, USA) 1981.
[7] C.R. Crowell and S.M. Sze, Solid-State Electron., 9 (1966) 1035.
[8] D.W. Palmer, Growth and Characterization of Semiconductors, edited by R.A. Stradling and P.C. Klipstein, Adam Hilger (Bristol, UK) 1990, IOP Publishing.
[9] W. Shockley and W.T. Read, Jr, Phys. Rev., 87 (1952) 835.
[10] R.N. Hall, Phys. Rev., 87 (1952) 387.
[11] F. Dubecký and B. Olejníovà, Defect Control in Semiconductors, edited by K. Sumino, Elsevier Science Publishers B.V (North Holland) p. 1599, 1990.

[12] L. Polenta, PhD Thesis, University of Bologna, 1998.
[13] D.V. Lang, J. Appl. Phys., 45 (1974) 3023.
[14] D.V. Lang, J. Appl. Phys., 45 (1974) 3014.
[15] G.L. Miller, D.V. Lang, and L.C. Kimerling, Annu. Rev. Mater. Res., 7 (1977) 377.
[16] D.C. Look, Electrical Characterization of GaAs Materials and Devices, Wiley (New York) 1989.
[17] G.L. Miller, J.V. Ramirez, and D.A.H. Robinson, J. Appl. Phys., 46 (1975) 2638.
[18] L.C. Kimerling, IEEE Trans. Nucl. Sci., NS-23 (1976) 1497.
[19] Z.-Q. Fang, D.C. Look, and L. Polenta, J. Phys.: Condens. Matter, 14 (2002) 13061.
[20] L. Polenta, Z-Q. Fang, and D.C. Look, Appl. Phys. Lett., 76 (2000) 2086.
[21] D. Cavalcoli and A. Cavallini, Phys. Rev. B, 56 (1997) 10208.
[22] R. Calarco, M. Marso, T. Richter, et al., Nano Lett., 5 (2005) 981.
[23] R. Tu, L. Zhang, Y. Nishi, and H. Dai, Nano Lett., 7 (2007) 1561.
[24] K. Schmalz, I.N. Yassevich, H. Rücker, et al., Phys. Rev. B, 50 (1994) 14287.
[25] J.W. Kim, G.H. Song, and J.W. Lee, Appl. Phys. Lett., 88 (2006) 182103.
[26] F. Rossi, M. Pavesi, M. Meneghini, et al., J. Appl. Phys., 99 (2006) 053104.
[27] Y.S. Park, C.M. Park, C.J. Park, et al., Appl. Phys. Lett., 88 (2006) 192104.
[28] A. Einstein, Ann. Physik, 17 (1905) 132.
[29] F. Berz, Surface Physics of Phosphors and Semiconductors, edited by C.G. Scott and C.E. Reed, Chapter 3, Academic Press (London, UK) 1978.
[30] H. Lüth, Solid Surfaces, Interfaces and Thin Films, 4th ed., Springer (Berlin, Heidelberg) 2001.
[31] L. Kronik and Y. Shapira, Surf. Sci. Rep., 37 (1999) 1.
[32] L. Kronik and Y. Shapira, Surf. Interface Anal., 31 (2001) 954.
[33] A. Rohatgi, R. Sudharsanan, S.A. Ringel, et al., Proc. 20th IEEE Photovoltaic Specialists Conf., IEEE, New York, p. 1477 (1988).
[34] D. Cavalcoli, A. Cavallini, M. Rossi, and S. Pizzini, Semiconductors, 41 (2007) 421.
[35] W.H. Brattain, Phys. Rev., 72 (1947) 345.
[36] W.H. Brattain and J. Bardeen, Bell System Tech. J., 32 (1953) 1.
[37] C.G.B. Garrett and W.H. Brattain, Phys. Rev., 99 (1955) 376.
[38] A.M. Goodman, J. Appl. Phys., 32 (1961) 2550.
[39] J.F. Muth, J.H. Lee, I.K. Shmagin, et al., Appl. Phys. Lett., 71 (1997) 2572.
[40] P. Blood and J.W. Orton, The Electrical Characterization of Semiconductors: Measurements of Minority Carrier Properties, Academic Press (London, UK) 1992.
[41] R.A. Smith, Semiconductors, 2nd ed., Cambridge University Press (London) 1978.
[42] W. Smith, Nature, 7 (1873) 303.
[43] B. Gudden and R. Pohl, Z. Physik., 23 (1922) 417.
[44] R.H. Bube, Photoconductivity of Solids, John Wiley and Sons (New York) 1960.
[45] S.M. Ryvkin, Photoelectric Effects in Semiconductors, Consultant Bureau (New York) 1964.
[46] R.H. Bube, Photoelectronic Properties of Semiconductors, Cambridge University Press (New York, USA) 1992.
[47] J.J. Pankove, Optical Processes in Semiconductors, Dover Publications Inc. (New York, USA) 1971.
[48] S.H. Wemple and J.A. Seman, Appl. Opt., 12 (1973) 2947.
[49] A. Castaldini, A. Cavallini, L. Polenta, et al., J. Phys.: Condens. Matter, 14 (2002) 13095.
[50] A. Castaldini, A. Cavallini, and L. Polenta, Appl. Phys. Lett., 87 (2005) 122105.
[51] N.V. Joshi, Photoconductivity: Art, Science and Technology, Marcel Dekker Inc. (New York, USA) 1990.
[52] T. Kuykendall, P.J. Pauzuaskie, Y. Zhang, et al., Nature Mater., 3 (2004) 524.
[53] Properties, Processing and Applications of Gallium Nitride and Related Semiconductors EMIS edited by J.H. Edgar, S. Strite, I. Akasaki, H. Amano, and C. Wetzel, Published by Institution of Engineering and Technology 1999.
[54] H. Morkoç, Nitride Semiconductors and Devices, Springer-Verlag (Berlin, Heidelberg, Germany) 1999.

[55] A. Cavallini, L. Polenta, M. Rossi, et al., Nano Lett., 6 (2006) 1548.
[56] A. Cavallini, L. Polenta, M. Rossi, et al., Nano Lett., 7 (2007) 2166.
[57] R. Calarco, R.J. Meijers, R.K. Debnath, et al., Nano Lett., 7 (2007) 2248.
[58] D. Segev and C.G. Van de Walle, Europhys. Lett., 76 (2006) 305.
[59] M. Kočan, A. Rizzi, H. Lüth, et al., Phys. Status Solidi B, 234 (2002) 773.
[60] V. Perebeinos and P. Avouris, Nano Lett., 7 (2007) 609.
[61] W. Franz, Z. Naturforsch., 13 (1958) 484.
[62] L.V. Keldysh, Sov. Phys.-JETP, 7 (1958) 788.
[63] B. Adamowicz and J. Szuber, Surf. Sci., 247 (1991) 94.
[64] V.G. Litovchenko and V.G. Popov, Sov. Phys. Semicond., 16 (1982) 472.
[65] A. Kalnitsky, S. Zukotynski, and S. Sumski, J. Appl. Phys., 52 (1981) 4744.
[66] L. Kronik and Y. Shapira, Surf. Sci. Rep., 31 (2001) 954.
[67] O. Ambacher, W. Rieger, P. Ansmann, et al., Solid State Commun., 97 (1996) 365.
[68] H. Haug and S.W. Koch, Quantum Theory of the Optical and Electronic Properties of Semiconductors, 2nd ed., World Scientific Publishing Co. Pte. Ltd. (1994).
[69] H.Y. Peng, M.D. McCluskey, Y.M. Gupta, et al., Appl. Phys. Lett., 82 (2003) 2085.
[70] Y. Xia, P. Yang, Y. Sun, et al., Adv. Mater., 15 (2003) 353.
[71] E.M. Vogel, Nature Nanotechnol., 2 (2007) 25.
[72] M. Van Rossum, Mater. Sci. Eng. B, 20 (1993) 128.
[73] R. Agarwal and C.M. Lieber, Appl. Phys. A, 85 (2006) 209.
[74] P. Kim and C.M. Lieber, Science, 286 (1999) 2148.

4

Strain and composition determination in semiconducting heterostructures by high-resolution X-ray diffraction

Claudio Ferrari and Claudio Bocchi

IMEM-CNR, Parco Area delle Scienze 37/A, 43010 Fontanini-Parma, Italy

Abstract In this chapter, the basics of the methods for measuring the strain and the composition in two- and zero-dimensional structures by means of high-resolution X-ray diffraction techniques and laboratory X-ray sources are presented, with the aim of introducing these techniques to the reader. The main physical properties for defining the strain state of a heterostructure are given and the basics of the elasticity theory for cubic and hexagonal crystals are also introduced. The X-ray diffraction method for determining the composition in semiconducting alloys is explained, allowing to conclude that the lattice parameter measurement method is one of the most accurate way to determine the composition, provided that the composition versus the lattice parameter dependence is known. The comparison between composition values obtained from X-ray diffraction method and that determined by other analytical techniques has allowed to measure a deviation from the linear Vegard law in several semiconducting alloys. The experimental determination of the deviation of the Vegard law in the InGaAs alloy is reported.

The methods of asymmetric diffraction geometries and reciprocal lattice maps to characterize lattice-matched and lattice-mismatched heterostructures are briefly introduced.

Some of the most cited theories describing the strain release in semiconductor heterostructures are introduced, even if a semi-empirical approach has to be used to fit the experimental data. By using a method similar to that proposed by Tersoff, the theory is extended to composition graded heterostructures, which are of great interest for obtaining virtual substrates or strain engineered heterostructures.

The theoretical models are then compared with experimental results for GaAlSb/GaSb single heterostructures and InGaAs/GaAs composition graded heterostructures.

Finally, the kinematical theory of X-ray scattering from quantum dot (QD)-based heterostructures is briefly introduced and discussed. The independent determination of strain and composition profiles in QD heterostructures is a very complex task as laboratory X-ray sources do not allow to change the X-ray wavelength. The characterization of heterostructures containing QDs is based on the comparison between simulations of X-ray scattering based on finite element calculations of the heterostructure containing the dots and reciprocal lattice maps obtained in different symmetrical and asymmetrical geometries. A satisfactory agreement between experimental and simulated reciprocal lattice maps of samples containing several stacks of InAs/GaAs QD is reported.

Keywords X-ray characterization of heterostructures, strain measurement, composition, measurement, strain release models, composition graded heterostructures, X-ray characterization of quantum dot heterostructures

1. Introduction

The X-ray diffraction profiles measurements for determining the composition and strain in semiconducting heterostructures have the advantage of being not completely destructive and also reliable in comparison with other characterization techniques.

A preliminary requirement of X-ray-based techniques for accurate measurements in semiconducting heterostructures, which often have the highest crystalline quality among all crystals, is the use of well-collimated and monochromatic X-ray sources with divergence of the order of a few arc seconds and wavelength dispersion $\Delta\lambda/\lambda$ better than a few 10^{-4}. In perfect crystals intense diffracted beams are formed only if the incident X-rays fulfil or are within a few arc seconds from the Bragg condition with respect to a given set of lattice planes.

In high-resolution diffractometers, X-ray beams are conditioned by X-ray optics based on single or multiple crystals, and goniometers with accuracies of 10^{-4} degrees both on the sample and the detector rotation axes are used. By this technique, it is by far more convenient to measure relative differences in the peak position in relatively narrow angular ranges with respect to a substrate diffraction peak used as a reference, rather than absolute measurements of lattice parameter and lattice deformation, which require more sophisticated procedures.

In the present chapter, the application of high-resolution X-ray diffraction (HRXRD) techniques for the basic characterization of semiconductor heterostructures will be introduced and presented, starting from the simple two-dimensional heterostructures up to some zero-dimensional or QD structures. Some basic considerations about the elastic theory applied to cubic as well as hexagonal crystals will also be introduced.

2. Lattice-mismatched and pseudomorphic heterostructures

2.1. Lattice misfit, lattice mismatch, strain and stress

In the following, the basic definitions of the physical quantities used to describe strained heterostructures are introduced. The lattice *misfit* m is the relative difference of the lattice parameter of a material with respect to the one chosen as a reference, often the substrate material in a heterostructure:

$$m = \frac{a_f - a_s}{a_s} \tag{1}$$

where a_f is the free lattice parameter of the given material for the film with respect to the lattice parameter a_s of the reference material. This definition is valid for materials having a cubic lattice and defined by a single parameter a. For instance, the lattice misfit between InAs ($a = 6.0585$ Å) and GaAs ($a = 5.6536$ Å) used as a reference is $m = 7.16 \times 10^{-2}$.

The definition of lattice *mismatch* $= \Delta a/a_s$ is similar to definition (1) but is essentially related to the actual lattice parameter difference in the layers forming a heterostructure. In the general case, the strain induced in the film changes the lattice from cubic to tetragonal or to a lower symmetry lattice. In the simplest case, we can distinguish between parallel $\Delta a^{\parallel}/a_s$ and perpendicular $\Delta a^{\perp}/a_s$ lattice mismatches, with reference to the interface between film and substrate:

$$\frac{\Delta a^{\parallel}}{a_s} = \frac{a_f^{\parallel} - a_s}{a_s}, \quad \frac{\Delta a^{\perp}}{a_s} = \frac{a_f^{\perp} - a_s}{a_s} \quad (2)$$

The *strain* ε is the relative deformation of the lattice due to the presence of an applied force, i.e., a stress. In the case of a cubic lattice, we can define a perpendicular and a parallel strain:

$$\varepsilon^{\parallel} = \frac{a_f^{\parallel} - a_f}{a_f}, \quad \varepsilon^{\perp} = \frac{a_f^{\perp} - a_f}{a_f} \quad (3)$$

It is easy to demonstrate that, in the limit $m \ll 1$,

$$\frac{\Delta a^{\parallel}}{a_s} = m + \varepsilon^{\parallel} \quad (4)$$

Finally, the *stress* σ is the applied force to the lattice, which induces the lattice strain or deformation. In the general case, the stress and the strain in a solid are expressed by tensors. Considering a small cubic volume of the material, σ_{ij} is the force per unit area along the direction i applied on a surface whose normal is oriented along the j direction and ε_{ij} is the relative shift along the i direction of that volume surface.

2.2. Basic elastic theory for cubic lattice heterostructures

In the framework of the linear elasticity theory (Hooke's law), there is a linear relationship between the components of the strain ε_{ij} and of the stress σ_{kl} given by the fourth-order rank tensor c_{ijkl}:

$$\sigma_{ij} = c_{ijkl} \times \varepsilon_{kl} \quad (5)$$

In the case of cubic lattices, the symmetry allows to reduce the 81 constants c_{ijkl} to only three independent *elastic* constants c_{11}, c_{12}, c_{44}, and Eq. (5) is reduced to (see Ref. [1]):

$$\begin{cases} \sigma_{xx} = c_{11}\varepsilon_{xx} + c_{12}\varepsilon_{yy} + c_{12}\varepsilon_{zz} \\ \sigma_{yy} = c_{12}\varepsilon_{xx} + c_{11}\varepsilon_{yy} + c_{12}\varepsilon_{zz} \\ \sigma_{zz} = c_{12}\varepsilon_{xx} + c_{12}\varepsilon_{yy} + c_{11}\varepsilon_{zz} \\ \sigma_{ij} = c_{44}\varepsilon_{ij}, \quad i \neq j \end{cases} \quad (6)$$

We now consider the case of cubic lattices and the growth of epitaxial layers along a <001> direction, coincident with our z direction. In such case, the lattice of the epitaxial layer is deformed to adapt, also partially, its lattice parameter along the interface (x and y directions). The situation is depicted in Fig. 1. For such *biaxial strain* $\sigma_{zz} = 0$, $\sigma_{xx} = \sigma_{yy}$, $\sigma_{ij} = 0$ for $i \neq j$, and we obtain from Eq. (6),

$$\varepsilon_{zz} = -2\frac{c_{12}}{c_{11}}\varepsilon_{yy} \quad \text{or} \quad \varepsilon^{\perp} = -2\frac{c_{12}}{c_{11}}\varepsilon^{\parallel} \quad (7)$$

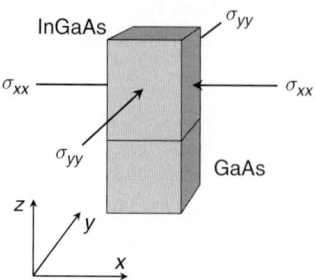

Fig. 1. Scheme of the applied stress σ and of the lattice deformation in a strained InGaAs/GaAs heterostructure.

In this simplified case, the two quantities $\varepsilon^\perp$ and $\varepsilon^\parallel$ allow the complete strain characterization of the film and are easily determined by a couple of suitable X-ray measurements (see Section 3). In the most general case, as for instance for the growth on off-cut substrates, x, y and z do not coincide with crystallographic directions and in general $\sigma_{ij} \neq 0$. In this case, to determine all the strain components X-ray measurements performed with different geometries are needed (see, for instance, Ref. [2]).

X-ray measurements do not directly provide the strain value of the crystal lattice, the measurable quantities being the lattice mismatches $\Delta a^\perp/a_s$ and $\Delta a^\parallel/a_s$. From Eq. (7) we can now obtain a relationship between lattice mismatch components and misfit m with respect to the substrate:

$$\varepsilon^\perp = -2\frac{c_{12}}{c_{11}}\varepsilon^\parallel$$

$$a^\perp - a_f = -2\frac{c_{12}}{c_{11}}\left(a^\parallel - a_f\right) \tag{8}$$

$$a^\perp - a_s = -2\frac{c_{12}}{c_{11}}\left(a^\parallel - a_s\right) + (a_f - a_s)\left(1 + 2\frac{c_{12}}{c_{11}}\right)$$

Dividing all members by a_s, we obtain

$$\frac{a_f - a_s}{a_s} = m = \frac{a^\perp - a_s}{a_s}\frac{c_{11}}{c_{11} + 2c_{12}} + 2\frac{a^\parallel - a_s}{a_s}\frac{c_{12}}{c_{11} + 2c_{12}} \tag{9}$$

which can be expressed in terms of the Poisson ratio $\nu = c_{12}/(c_{11} + c_{12})$ for a cubic crystal and z coincident with a $<001>$ direction:

$$m = \frac{\Delta a^\perp}{a_s}\frac{1-\nu}{1+\nu} + \frac{\Delta a^\parallel}{a_s}\frac{2\nu}{1+\nu} \tag{10}$$

Relation (10) is the basic equation for the strain and composition characterization of heterostructures for cubic lattice materials. From the misfit value m and from Eq. (4)

Table 1
List of the C_{11} and C_{12} elastic constants for some common semiconducting materials and the calculated Poisson ratios ν for the (100) direction

Compound	$C_{11}/10^{11}$ dyne/cm^2	$C_{12}/10^{11}$ dyne/cm^2	ν
AlAs	12.5	5.34	0.299
AlSb	8.765	4.341	0.331
GaN	29.6	13.0	0.305
GaP	14.12	6.523	0.306
GaAs	11.81	5.32	0.310
GaSb	8.834	4.026	0.318
InP	10.22	5.76	0.360
InAs	8.329	4.526	0.352
InSb	6.472	3.645	0.360
AlP	13.2	6.30	0.323
Si	16.57	6.393	0.278
Ge	12.853	4.826	0.272

the parallel $\varepsilon^{\parallel}$ and perpendicular $\varepsilon^{\perp}$ strain values are easily determined. Finally, in the case of a semiconducting alloy A_xB_{1-x} the composition x can be obtained if the relationship between composition and lattice parameter is known. This will be discussed in detail in Section 4, but it is worth noting here that in Eq. (10) the Poisson ratio ν, which has values close to 1/3 for most semiconducting materials, is also composition dependent [3,4], and that a more exact solution of Eq. (10) is obtained by a recursive method. We also note that the use of the Poisson ratio is only valid for isotropic material. For a cubic lattice, it can only be applied for high symmetric directions as <001>, <011> and <111>, but the Poisson ratio may be different along different directions. Table 1 [5] reports the elastic constants and the derived Poisson ratios for some semiconducting materials along the <001> direction.

2.3. *Basic elastic theory for hexagonal lattice heterostructures*

In the case of a hexagonal lattice the independent elastic constants are ε_{11}, ε_{12}, ε_{13}, ε_{44} and ε_{66}. We assume that the z direction is aligned with the c-axis of the hexagonal cell. The relationships between stress σ and strain ε is then expressed by [6]

$$\begin{cases} \sigma_{xx} = c_{11}\varepsilon_{xx} + c_{12}\varepsilon_{yy} + c_{13}\varepsilon_{zz} \\ \sigma_{yy} = c_{12}\varepsilon_{xx} + c_{11}\varepsilon_{yy} + c_{13}\varepsilon_{zz} \\ \sigma_{zz} = c_{13}\varepsilon_{xx} + c_{13}\varepsilon_{yy} + c_{33}\varepsilon_{zz} \\ \sigma_{xy} = c_{44}\varepsilon_{xy} \\ \sigma_{xz} = c_{66}\varepsilon_{xz} \end{cases} \quad (11)$$

In the case of an epitaxial growth, we have a *biaxial strain* $\sigma_{zz} = 0$, $\sigma_{xx} = \sigma_{yy}$, $\sigma_{ij} = 0$ for $i \neq j$, so from Eq. (11) we obtain

$$\varepsilon_{zz} = -2\frac{c_{13}}{c_{33}}\varepsilon_{yy} \quad \text{or} \quad \varepsilon^{\perp} = -2\frac{c_{13}}{c_{33}}\varepsilon^{\parallel} \quad (12)$$

From Eq. (12) assuming that $c_f/a_f = c_s/a_s$, which is the ratio between c and a parameters, is the same for the epitaxial layer and for the substrate irrespective of the composition, we can obtain a relationship between misfits m_c and m_a of the epitaxial layer along the c- and a-axis of the hexagonal cell, respectively, and the lattice mismatches $\Delta c/c_s$ and $\Delta a/a_s$ measurable by X-ray diffraction:

$$m_c + 2\frac{c_{13}}{c_{33}} m_a = \frac{\Delta c}{c_s} + 2\frac{c_{13}}{c_{33}}\frac{\Delta a}{a_s} \tag{13}$$

Here again assuming c/a constant for the substrate and epilayer, $m_c = m_a$ and Eq. (13) becomes

$$m_c\left(1 + 2\frac{c_{13}}{c_{33}}\right) = \frac{\Delta c}{c_s} + 2\frac{c_{13}}{c_{33}}\frac{\Delta a}{a_s} \tag{14}$$

Analogous to Eq. (10) for cubic lattice materials, Eq. (14) can be used to calculate the misfit m_c of the epitaxial layer from the measurable quantities $\Delta c/c_s$ and $\Delta a/a_s$ and by using Eq. (4), the strain values ε_c and ε_a, can be derived. The assumption of c/a constant is quite good in the case of nitride alloys, with variations of the order of 1% [7]. In any case, for accurate measurements, the dependence of the elastic constants c_{13} and c_{33}, as well as the ratio c/a on the composition of the alloy must be taken into account. Numerical methods must then be employed for more precise calculations (see Ref. [8] and references therein).

3. X-ray diffraction profiles of semiconducting heterostructures

High-resolution X-ray diffraction techniques are well-established methods for the structural characterization of high-quality crystalline materials [9–13]. Double- and triple-axis diffractometers with multiple-reflection beam conditioners are widely used for the measurements of the lattice deformation, composition, thickness and interface roughness of epilayers, and are particularly appropriate for studying compound semiconductor heterostructures, known to made up of high-quality crystal layers. The analysis of X-ray diffraction profiles makes possible the accurate determination of the strain tensor components in real heteroepitaxial systems, where the following strain status of the layers can be distinguished: pseudomorphic (strained), metamorphic (partially relaxed) and unstrained (fully relaxed), the last condition being only the theoretical end limit of the strain relaxation [13]. The mechanisms underlying the strain relaxation will be analysed in Section 4.5; the application of X-ray diffraction methods for the characterization of pseudomorphic and metamorphic heterostructures is the aim of the following sections.

3.1. Summary of basic scattering geometries

In an X-ray diffraction experiment a set of crystal lattice planes (hkl) is selected by the incident conditions and the lattice spacing d_{hkl} is determined through the well-known Bragg's law. The Bragg diffraction from planes parallel or inclined by an angle φ with respect to the

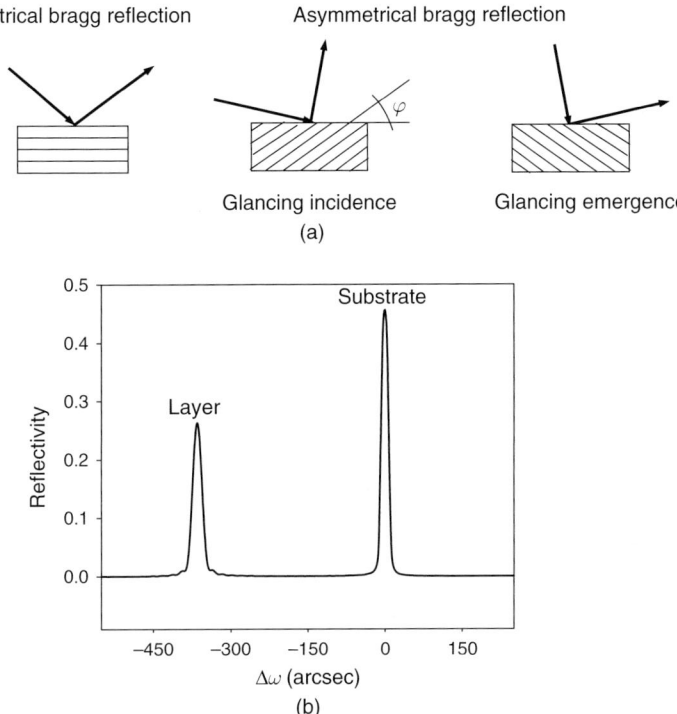

Fig. 2. (a) Scheme of symmetric and asymmetric Bragg diffraction geometries with the two possible incident conditions for an asymmetric reflection. (b) Example of an X-ray diffraction profile from a single layer heterostructure as achievable by either ω scan or ω–2θ scan mode. The angular separation $\Delta\omega$ between substrate and layer peaks is proportional to the mismatch components.

crystal surface is defined symmetrical ($\varphi = 0$) or asymmetrical ($\varphi \neq 0$), respectively (Fig. 2(a)). The crystal surface is the entrance and exit reference plane for the X-ray beams in Bragg scattering geometry. For this reason and because incident and diffracted beams make the same angle with the lattice planes, the Bragg diffraction from a crystal is commonly called *Bragg reflection*.

Let ω be the incidence angle with respect to the sample surface of a parallel and monochromatic X-ray beam; by rocking a crystal through a selected angular range centred on the Bragg angle of a given set of lattice planes a diffraction intensity profile $I(\omega)$ is collected. Dealing with a single layer heterostructure, the intensity profile will show two main peaks (Fig. 2(b)), corresponding to the diffraction from the same lattice planes (*hkl*) in the layer and the substrate, respectively. The angular separation ($\Delta\omega$) of the peaks accounts for the difference Δd_{hkl} between the layer and substrate lattice spacing.

The Bragg diffraction is a coherent and elastic scattering phenomenon with a momentum transfer between incident and scattered radiation, and the intensity distribution of an X-ray scattering experiment is plotted in reciprocal space (RS) (the space of wave vectors). From the momentum conservation principle the Bragg law in the RS becomes $\mathbf{Q} = \mathbf{k}_s - \mathbf{k}_i = \mathbf{h}_{hkl}$, where $\mathbf{h}_{hkl}$ is the reciprocal lattice vector with $|\mathbf{h}_{hkl}| = 2\pi/d_{hkl}$; $\mathbf{Q} = \mathbf{k}_s - \mathbf{k}_i$ is the scattering vector (momentum transfer) and $\mathbf{k}_{s,i}$ with $|\mathbf{k}_{s,i}| = 2\pi/\lambda$ are the scattered and incident wave vectors, respectively;

λ is the X-ray wavelength. A well known and useful way for representing the Bragg law in RS is given by the sphere of reflection (Ewald sphere). For a set of lattice planes hkl the Bragg condition is satisfied when the reciprocal lattice point hkl falls on the surface of the sphere.

Two possible scan modes (Fig. 3(b)) can be performed to measure the intensity profile $I(\omega)$: (i) the ω scan when the detector is fixed in $2\theta_B$ position (θ_B being the Bragg angle) and ω is changed by rotating the sample on the diffractometer axis (rocking curve); and (ii) ω–2θ scan if also the detector is rotated but twice as fast as the sample (for each increment $\Delta\omega$, $\Delta 2\theta = 2\Delta\omega$). The paths in the RS described by the tip of the scattered wavevector $\mathbf{k}_s$ for the two scan modes are shown in the mixed representation of Fig. 3(b).

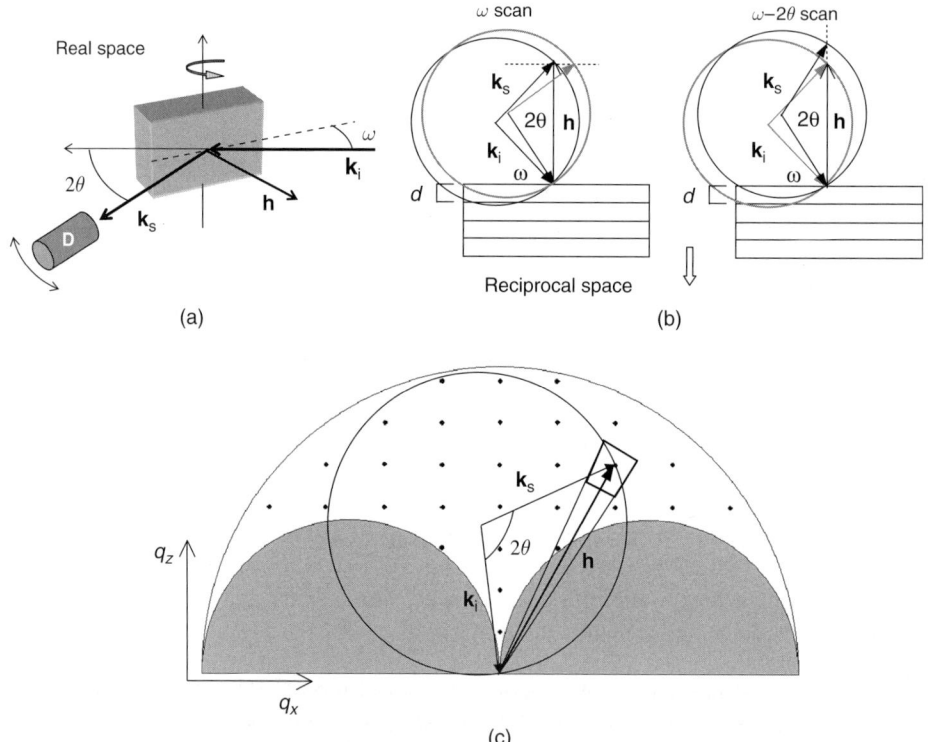

Fig. 3. (a) Scheme of an experimental diffraction measurement in real space; (b) mixed representation of the sample movements in reciprocal space by Ewald sphere construction with the real incident condition on the sample lattice planes. $\mathbf{h}$ is a reciprocal lattice vector (normal to the lattice planes); $\mathbf{k}_{i,s}$ are the incident and scattered wave vectors, respectively. The section of Ewald sphere is shown for both ω and ω–2θ scan. The dashed lines represents the movement in the reciprocal space (points from which the scattered radiation is collected) associated to the different scan modes. (c) Example of reciprocal space section showing accessible nodes for Bragg reflection measurements. $q_{x,z}$ are the reciprocal space co-ordinates. The outer semicircle is defined by the maximum 2θ angle of the diffractometer. The two inner regions defined by semicircles mark off lattice nodes which are accessible only in transmission (Laue) geometry. The area around a node showed in the figure is covered by combining ω and ω–2θ scans (reciprocal space map).

3.2. Strain determination in pseudomorphic heterostructures

In heteroepitaxial systems, the lattice parameter of each layer is modified compared to its ideal free-standing value due to the strain induced by the misfit. In the case of a *pseudomorphic* (coherent) growth, the lattice of a layer is matched in the growth plane to that of the substrate under the effect of a biaxial stress and free to get deformed along the growth direction. The strain will result compressive or tensile depending on the sign of the misfit (see Eq. (1)). Only the component of the mismatch perpendicular to the growth plane $(\Delta a/a_s)^\perp$ differs from zero $((\Delta a/a_s)^\parallel = 0)$, while both the in-plane and perpendicular strain components assume their highest values (maximum deformation). For cubic systems with the substrate [001] oriented, a tetragonal distortion of the layer lattice cells occurs. The unstrained lattice parameter a_f of the layer, which is important for determining e.g. the composition of alloys, is then related to $(\Delta a/a_s)^\perp$ by the simple expression (from Eq. (10))

$$m = \frac{a_f - a_s}{a_s} = \frac{1-\nu}{1+\nu}\frac{a_f^\perp - a_s}{a_s} = \frac{1-\nu}{1+\nu}\left(\frac{\Delta a}{a}\right)^\perp \tag{15}$$

Because the only perpendicular lattice parameter $a_f^\perp$ is needed, the mere symmetrical Bragg reflections are sufficient for pseudomorphic structures characterization.

In epitaxial systems, the mismatch components are determined from the angular separation $(\Delta\omega)$ between the Bragg peaks of layer and substrate obtained by the rocking curve. Assuming the sample surface oriented (nearly parallel to a crystallographic plane), let $\omega = \theta \pm \varphi$ be the incident angle on the surface, θ the incident angle on the lattice planes and φ the asymmetry angle. The exact general expression for the angular separation $\Delta\omega$ is given by [14]

$$\Delta\omega = \Delta\varphi + \Delta\theta = \mp\tan^{-1}\left[\tan\varphi\frac{1+(\Delta d/d)^\perp}{1+(\Delta d/d)^\parallel}\right] \pm \varphi$$

$$+ \sin^{-1}\left\{\sin\theta_B\left[\frac{\sin^2\varphi}{(1+(\Delta d/d)^\parallel)^2} + \frac{\cos^2\varphi}{(1+(\Delta d/d)^\perp)^2}\right]^{1/2}\right\} - \theta_B \tag{16}$$

where the upper and lower signs are relative to the glancing incidence and glancing emergence geometries, respectively, when asymmetric reflections are selected. $\Delta\omega$ is the sum of two terms: $\Delta\theta$ depends on the difference in the lattice spacing d_{hkl}, and $\Delta\varphi$ is the difference in [hkl] orientation due to the layer cell deformation (Fig. 4). Furthermore, the surface of the sample is frequently misoriented (miscut angle β) with respect to the lattice planes and even for small angles β the effect must be considered. To overcome this problem, it is sufficient to repeat each measurement after a rotation by 180° around the normal to the sample surface (azimuthal rotation Φ). By averaging the peak separation $<\Delta\omega>$ of each pair of scans with the azimuthal angle Φ differing by π, the effect of the miscut angle is eliminated.

For symmetric reflections, the asymmetry angle is $\varphi = 0$, and Eq. (16) reduces to

$$\Delta\omega = \Delta\theta = \sin^{-1}\left(\frac{\sin\theta_B}{1+(\Delta d/d)^\perp}\right) - \theta_B \tag{17}$$

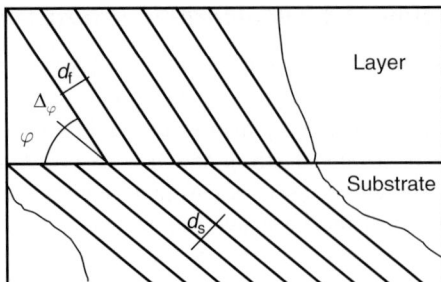

Fig. 4. Scheme of a strained single layer heterostructure. Due to the layer cells deformation, the same asymmetric lattice planes in the substrate and in the layer are no longer parallel. $\Delta\varphi$ is the difference in the [*hkl*] orientation.

where θ_B is the Bragg angle of the substrate. $\Delta\theta$ depends at the same time on the composition and strain of the epilayer $d = d(x)$, as well as the Poisson ratio, being $v = v(x)$. Moreover, if the layer is very thin, $\Delta\theta$ is also influenced by thickness effects [15]. Finally, taking into account that the layer composition is frequently unknown with the required accuracy, the unstrained perpendicular lattice parameter of the layer a_f is not directly computable from $\Delta\theta$. The analysis of the diffraction curves must then be performed by a best-fit simulation procedure accounting for the apparent value of the peaks separation [16].

The calculations are based on the theoretical approach developed by Takagi [17] and Taupin [18] for elastically deformed crystals, which constitutes a generalization of the dynamical X-ray diffraction theory [19]. It allows to calculate the ratio R (reflectivity) of diffracted to incident X-ray beam amplitudes for the general case of a structure with an elastic strain gradient perpendicular to the surface (multilayers stacking or single layer with composition gradient) and it can be also applied when asymmetric reflections are selected.

This theoretical approach provides two main advantages: (i) it is quite general; (ii) it is possible to make use of an analytical solution of the basic Takagi–Taupin differential equation for the reflectivity R when $\Delta d/d$ is a constant. Also in the case of any strain gradient perpendicular to the surface, it is possible to slice the part of the crystal affected by this gradient in several slabs, each slab having a constant lattice parameter but differing with the value of the others. In this way, for each slab it is possible to apply the analytical solution to calculate the reflectivity that, at the exit surface, can be determined following iterative methods.

This procedure is commonly utilized for pseudomorphic structures characterization, and innumerable works were published during the last 30 years dealing with diffraction profile simulations. Commercial programmes based on the Takagi–Taupin equation are also available today.

An example of a quantitative analysis of a pseudomorphic single $GaAs_{1-x}P_x$ layer grown by molecular beam epitaxy (MBE) on a GaAs 001 oriented substrate is given in Fig. 5.

In pseudomorphic multilayer systems the waves scattered from different layers interfere with each other producing very complicated diffraction profiles. In spite of this, if all the stacked layers are coherently grown, a complete characterization of the structure is still possible by using the simulation procedure summarized above.

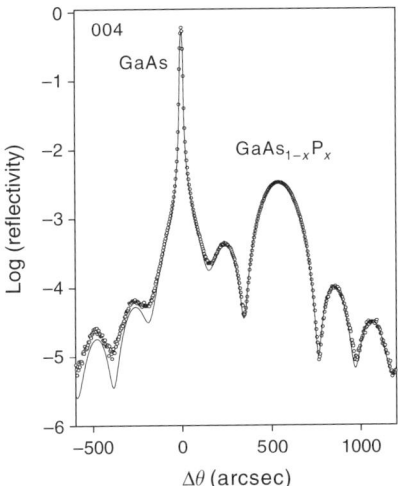

Fig. 5. 004 CuKα_1 experimental (dots) and calculated (solid line) X-ray diffraction profiles from a GaAs$_{1-x}$P$_x$/GaAs pseudomorphic heteroepitaxy. A good simulation of the experimental profile was obtained by using a calculation programme developed on the basis of the theoretical approach described in the text. The data of the fit were the following: $(\Delta d/d)^{\perp} = -4.1 \times 10^{-3}$; the layer thickness: $t = 91$ nm. From the perpendicular mismatch component, it was possible to determine the unstrained lattice parameter of the layer and its composition ($x = 0.06$) through the Vegard's law (see Section 4).

3.3. Investigation of extremely thin layer or interfaces in semiconductor heterostructures

The extinction length of X-rays in semiconductor materials is of the order of several microns and therefore due to the weak diffracted intensity the detailed X-ray characterization of very thin layer in the nanometre range is difficult. There are nevertheless remarkable exceptions when (a) the thin layer is embedded in thick layers and (b) when the thin layer is replicated several times as in a superlattice.

In the case of a single thin embedded layer, as for a δ-doped layer or a single quantum well, the diffraction profile does not normally show a direct diffracted peak from the layer, but the presence of a thin mismatched layer introduces a lattice shift between the thick cladding layers and this produces relevant effects on the diffraction profiles. In this way, the interface abruptness has been investigated by Ferrari et al. [20] in GaAlAs/GaAs and InGaAs/InP heterostructures. Tapfer et al. [21] have shown that the effect on the rocking curve is proportional to the product $(\Delta d/d)t$, where $\Delta d/d$ is the lattice mismatch of the thin layer in the scattering direction and t is the thickness of the layer. A sensitivity in the submonolayer range can be achieved in the case of InAs layers embedded in GaAs [22]. The sensitivity to the product $(\Delta d/d)t$ rather than on the single parameters $\Delta d/d$ and t also means that thin layers having the same product $(\Delta d/d)t$ have similar rocking curves and are in principle not distinguishable. Ferrari et al. [22] and Bocchi et al. [23] have shown that in the case of positively mismatched layers a sensitivity to the single parameters t and $\Delta d/d$ can be achieved by comparing diffraction profiles measured in different geometries. In this way, the In segregation length in MBE-grown InAs single monolayers was obtained with remarkable accuracy [23].

In the case of periodical heterostructures, the X-ray diffraction profiles exhibit superlattice peaks in the vicinity of the substrate diffraction peak. These peaks correspond to the Fourier component of the lattice parameter profile in the superstructure, so that a sensitivity at the interface shape in the angstrom range in superstructures can be achieved.

Different InGaAs/InP and InP/InGaAs interface abruptness in InGaAs/InP superlattice was studied by Lamberti et al. [24]. Using a similar method, the unexpected InAsP interface formation during the growth of InGaAs/InP superlattices could be measured and correlated with growth parameters, such as growth interruption time and growth temperature [25]. In the case of multiple similar interfaces, the chemical sensitivity of the rocking curve analysis on composition profiles is by far better of those obtainable by simple transmission electron microscopy (TEM) analysis, unless mathematical analysis of TEM images at atomic resolution is performed (Chapter 5 of this book).

3.4. Metamorphic structures: the method of asymmetric reflections

Beyond a given critical thickness, the growth of a heterostructure changes from pseudomorphic to metamorphic regime and partial strain relaxation occurs. Therefore, both the mismatch components $(\Delta a/a)^{\perp,\|}$ differ from zero (now $a_f^{\|} \neq a_s$) and their accurate determination is required for calculating the strain status and the free-standing lattice parameter (a_f) of the layer (Eq. (10)). A well-assessed method based on asymmetric Bragg reflections [26] gives the possibility to measure the lattice parameters both perpendicular and parallel to the interface between the layer and the substrate.

By selecting an asymmetric reflection, the lattice plane spacing d_{hkl} can be separated into the parallel ($d^{\|}$) and perpendicular ($d^{\perp}$) components to the sample surface (Fig. 6(a)), which is required for the determination of mismatch components. As seen previously for pseudomorphic heterostructures, the separation $\Delta\omega$ between the layer and the substrate peaks has to be measured from the diffraction profiles, but dealing with asymmetric reflections also the incidence conditions strongly influence $\Delta\omega$. By an appropriate experimental procedure, all contributions to the separation of the peaks can be taken into account.

In metamorphic structures, the partial or total strain relaxation is accompanied by the onset of misfit dislocations (MDs). The presence of these extended defects at the layer–substrate interface produces a certain misorientation of the lattice cells (tilt, mosaic disorder) with a consequent broadening of both the substrate and the layer peaks in the diffraction profiles. Nevertheless, the peak separation is still detectable with some accuracy, even if the fitting of the diffraction profiles based on the dynamical diffraction theory for perfect crystals is no longer possible.

An example of the method is given in Fig. 6(b). Let $\Delta\omega_1$ be the angular separation between layer and substrate peaks in a rocking curve of an asymmetric reflection measured in glancing incidence (ω^-) condition, and $\Delta\omega_2$ the same quantity obtained in glancing emergence (ω^+) geometry; then the difference $\Delta\theta$ in the lattice spacing d_{hkl} and the difference $\Delta\varphi$ in $[hkl]$ orientation are given by [26]

$$\Delta\theta = \frac{\Delta\omega_1 + \Delta\omega_2}{2}, \quad \Delta\varphi = \frac{\Delta\omega_1 - \Delta\omega_2}{2} \tag{18}$$

Without loss of generality, an example of a complete *measurement procedure* in the case of a cubic [001] oriented heteroepitaxial system, is given in the following. Eight asymmetric

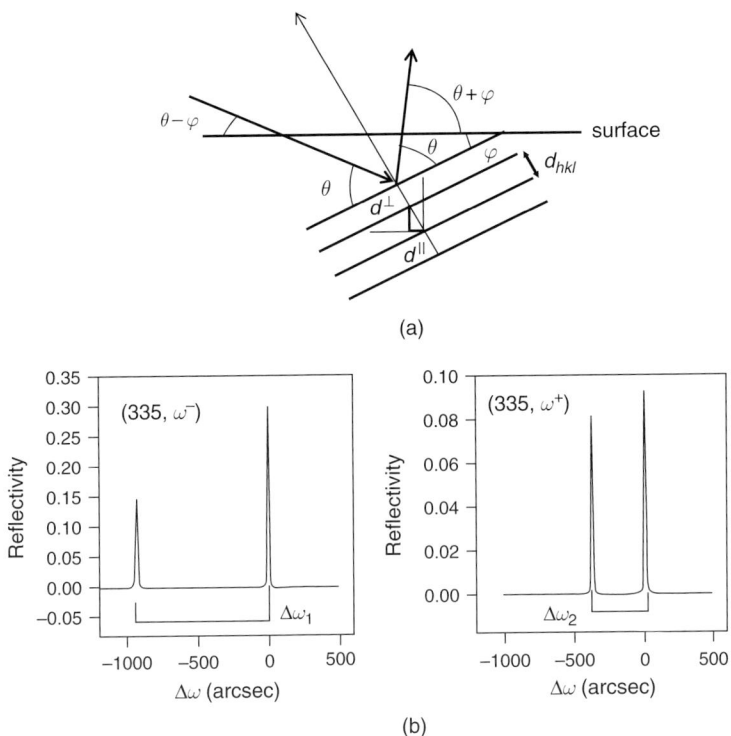

Fig. 6. (a) Scheme of an asymmetric reflection geometry. θ is the incident angle on the *hkl* lattice planes; φ is the angle between the lattice planes and the surface; d_{hkl} is the plane spacing; $d^{\parallel}$ and $d^{\perp}$ are the parallel and perpendicular spacing components to the surface, respectively. (b) Example of the asymmetric reflections method showing two diffraction profiles from 335 asymmetrical lattice planes of a zinc blende [001] oriented single layer heterostructure. By using the grazing incidence (ω^-) and grazing emergence geometries (ω^+), the two different peaks separation $\Delta\omega_1$ and $\Delta\omega_2$ are obtained.

hhl reflections with the same Miller indices, in both glancing incidence (ω^-) and glancing emergence (ω^+) geometries, have to be measured. The choice of the asymmetric reflection is limited by the sample orientation, and by the need for a high asymmetry angle to enhance the parallel component of the spacing d_{hkl}, but not too weak diffracted intensity. A good option for 001 oriented cubic structures could be the use of 335 ($\varphi = 40.32°$) or 224 ($\varphi = 35.26°$) reflections. An unconventional measurement scheme is given in Ref. [27] with the choice of the extremely asymmetric 551 ($\varphi = 81.5°$) reflection.

Four of the *hhl* reflections must be recorded on the same scattering plane (ω^-, ω^+ with $\Phi = 0, \pi$) and the other four reflections after a 90° azimuthal rotation (ω^-, ω^+ with $\Phi = \pi/2$, $3\pi/2$). In this way, it is possible to detect the diffraction profiles with the two in-plane [110] directions lying in the scattering plane. The perpendicular $(\Delta d/d)^{\perp}$ and the two parallel $(\Delta d/d)^{\parallel}$ components of the mismatch along the [110] directions have therefore been calculated from Eq. (16) using the average peak separation $\langle\Delta\omega\rangle$ of each pair of reflections with the azimuthal angle Φ differing by π (to avoid the miscut angle influence).

Different values of the parallel $(\Delta d/d)^{\parallel}$ components of the mismatch along the two [110] directions in the plane of the interface are often obtained in metamorphic samples.

This asymmetry in the strain release is associated with a different MD density along the two in-plane directions and the assumption of a pure tetragonal deformation of the epilayer cell can no be longer satisfied.

The presence of MDs also causes lattice plane tilts and deviations from the orthogonality of the principal axes of the epilayer lattice cell. The strain tensor components can be obtained by this method also for the general case of an arbitrary deformation (tetragonal → triclinic) of the epilayer lattice cell. In these cases, it is necessary to consider all the strain tensor components (ε_{ij}) and the rotation tensor components (r_{ij}) of the cell of the layer with respect to the substrate unit cell. [28–30].

3.5. The method of reciprocal lattice maps

In a conventional single- or double-crystal diffraction experiment, all the radiation scattered by the sample is integrated along the acceptance angle of the receiving slit of the detector. It follows an overlap of the different contributions to the intensity diffraction profile. Mosaic disorder, bending, composition and strain gradients broaden the intensity distribution in the RS along different directions. A reciprocal space mapping (RSM) of the scattered intensity, obtained by combining ω and ω–2θ scan modes (see Fig. 3(c)), enables to separate these contributions [31].

A high-resolution triple-axes diffractometer with an analyser crystal is required for RSM measurements. The analyser is placed on the third axis and probes the intensity scattered from the sample with a resolution in the RS comparable to its reflection domain (a few seconds of arc). This makes it possible to separate the scattered radiation directions, and increases the accuracy in the mismatch components determination of heteroepitaxial systems [32].

The reciprocal lattice is the Fourier transform of a real lattice. Each node (hkl) in the reciprocal lattice condenses the properties of a set of lattice planes (hkl) of the real space lattice. The direction of the reciprocal lattice vector corresponds to the normal to the real space set of planes, and its magnitude is equal to the reciprocal of the interplanar spacing of the real space planes.

The reciprocal lattice of a heteroepitaxial system is a superimposition of the layer and substrate lattice. When the layer is elastically deformed at the first growth stage, its reciprocal lattice lines up to the substrate nodes along the surface normal direction. As the relaxation takes place, the layer nodes move to the positions corresponding to the unstrained lattice. At the end of the relaxation process, the crystallographic directions of the two lattices must be aligned ($\Delta\varphi = 0$). Under the general assumption that Hook's law is obeyed during the relaxation, the path covered by the lattice nodes between the initial and final point of the process is a straight line. A schematic of this behaviour is shown in Fig. 7. The strain status of the layer is then "simply" determined by the position of its reciprocal lattice nodes. Like in a single scan mode where the substrate peak position is taken as the reference, also for RSMs the distances are measured from the substrate nodes.

Collecting a two-dimensional intensity map around a lattice node corresponding to an asymmetric reflection, $\Delta\omega$ (peak separation) can be obtained from the respective positions of the layer and substrate. The conversion of a peak intensity position (ω, θ) in RS co-ordinates (q_z, q_x) is given by

$$q_x = R[\cos(\omega) - \cos(2\omega' - \omega)], \quad q_z = R[\sin(\omega) - \sin(2\omega' - \omega)] \tag{19}$$

where R is the Ewald sphere radius ($R = |\mathbf{k}_i| = 2\pi/\lambda$); $2\omega'$ corresponds to any arbitrary position of the detector and $2\omega' = 2\theta_B$ when the Bragg condition is satisfied.

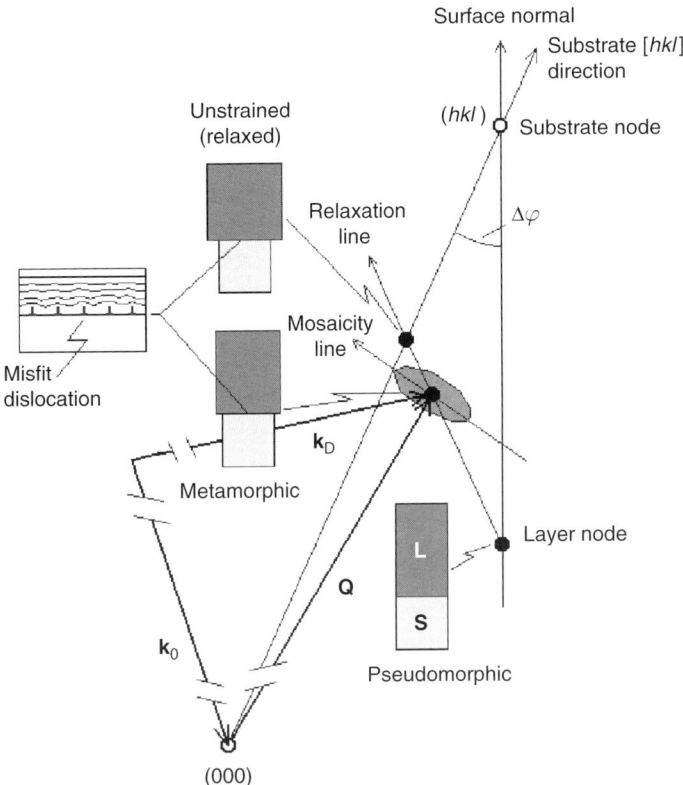

Fig. 7. Scheme of the representative layer lattice node (*hkl*) evolution in reciprocal space depending on the strain regime. When the structure is pseudomorphic, the strain is maximum and the layer lattice is aligned with that of the substrate along the surface normal. Starting the relaxation (metamorphic), the *hkl* node moves along a straight line connecting the pseudomorphic position with that one associated to the free reciprocal lattice of the layer (full relaxation) with the crystallographic directions parallel to those of the substrate ($\Delta\varphi = 0$). The broadening due to the mosaic disorder is also shown. The mosaic spread is observed along the direction perpendicular to the scattering vector Q (mosaicity line). k_0 and k_D are the incident and diffracted wave vectors, respectively.

Figure 8 deals with a study of InGaAs/GaAs epitaxial layers grown with different composition and thickness to follow the strain relaxation process in this system. RSMs around the 224 substrate and layer reciprocal lattice nodes of two different samples are shown in Fig. 8(a) and (b). The iso-intensity contours are plotted in RS co-ordinates (q_z, q_x). The map in Fig. 8(a) points out that the growth of the layer in the first sample was pseudomorphic with the nodes aligned parallel to the surface normal, whereas in Fig. 8(b), the layer is in a metamorphic state with its representative lattice point shifted with respect to the normal. The broadening perpendicular to the scattering vector (mosaicity) caused by the presence of MD is clearly visible in the figure. Examples of an effective use of RSM for strain characterization are reported in several papers (see, for instance, Refs [31, 2]).

According to Fewster [32], the use of RSM technique offers some advantages in the analysis of single- and multilayer systems. The most important is that the strain can be separated from tilts and mosaicity because the strain influences the intensity distribution along the layer scattering vector or surface normal, while mosaic spread or bending is observed along the

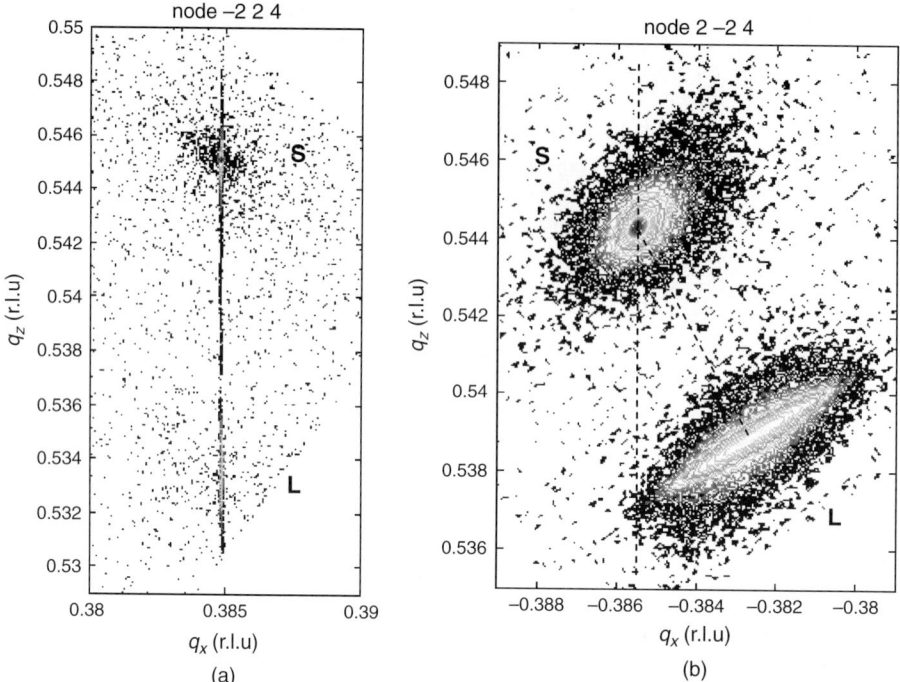

Fig. 8. Reciprocal space maps around asymmetrical nodes of the reciprocal lattice of InGaAs buffer layers grown by atomic layer MBE on [001] GaAs. (a) Pseudomorphic layer (−224 reflection). (b) Metamorphic layer (−224 reflection). Notice the alignment of substrate (S) and layer (L) nodes along the surface normal (q_z direction) in (a), and the misalignment and broadening due to the partial relaxation and mosaicity, respectively, in (b).

direction perpendicular to the same vector. Time consumed for collecting RSMs with an optimized signal-to-noise ratio and the complexity in developing a fitting calculation for a RSM are the main disadvantages of the method.

4. Determination of the composition of a semiconducting alloy heterostructures

4.1. The Vegard law approximation

According to Section 4.2, X-ray diffraction methods can easily be employed to measure the lattice parameter or the difference of the lattice parameter with respect to a substrate used as a reference. An accuracy of 10^{-4} or better in the lattice parameter determination can be obtained without sophisticated measurements and with usual geometries. For instance, recent measurements of the Avogadro's number are based on the lattice parameter measurements of monocrystalline silicon spheres with accuracies close to 10^{-8} [33].

The strain and the composition of a semiconducting alloy can be determined accurately by X-ray diffraction methods if the dependence of the lattice parameter with the composition is known, the accuracy being mainly due to the precise knowledge of the lattice parameter–composition dependence.

In many cases a good approximation of such a dependence is given by the Vegard law [34], which assumes that in an alloy A_xB_{1-x} the lattice parameter $d(x)$ of the alloy is proportional to the stoichiometric coefficient x:

$$d(x) = xd(A) + (1-x)d(B) \tag{20}$$

From Eq. (20), the stoichiometric coefficient x is obtained:

$$x = \frac{d(x) - d(B)}{d(A) - d(B)} \tag{21}$$

If $d(B)$ is the substrate lattice parameter, the composition x can be calculated from the measured misfit $m(x)$ value:

$$x = \frac{m(x)}{m_{AB}} \tag{22}$$

where $m(x)$ is the measured misfit value of the alloy with respect to $d(B)$ according to Eqs (10) and (14) and m_{AB} is the misfit between compound A and compound B, used as a reference.

4.2. Deviation form the Vegard law in SiGe, GaAlAs, GaAlSb and InGaAs alloys

In many semiconducting binary alloys, in lack of accurate experimental data, the Vegard law is often used as a good approximation for the lattice parameter dependence on the composition.

Although the lattice parameter of an alloy can be measured and defined with the same accuracy as for a single compound crystal, extended X-ray fine absorption spectroscopy measurements in a InGaAs alloy have clearly demonstrated [35] that the bond lengths in the alloy are close to the values they assume in the compound and that on a microscopic scale the atom positions are largely modified with respect to the ideal average alloy as defined by the virtual crystal approximation (VCA) [36].

A violation of Vegard's law in covalent alloys has been theoretically predicted; simple models based on the elasticity theory predict a deviation from linear dependence, depending on the elasticity constants of the components of the alloy [37–39].

Recent theoretical works (see, for instance, Ref. [40]) have shown that the deviation from the Vegard law depends on many parameters; for instance, the difference in the atom bond length, the different atom electronegativity and the elastic constants of the compounds in the alloy. In many semiconducting alloys there is a complete miscibility in the full interval of the alloy composition and a relatively small lattice parameter difference (atom bond length) between the solute and the solvent components. Under this condition relatively small deviations from the linear dependence are expected, and the lattice parameter dependence can then be expressed by the parabolic relationship

$$d(x) = xd(A) + (1-x)d(B) + bx(1-x) \tag{23}$$

where the first two terms express the Vegard law and b is the bowing parameter accounting for the deviation from linearity. According to Eq. (23), the maximum deviation from the linear dependence occurs at $x = 0.5$ and corresponds to $\Delta a = 0.25b$.

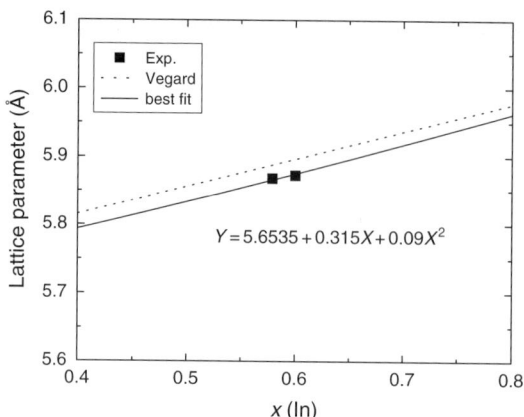

Fig. 9. Deviation from the Vegard law in InGaAs alloy as measured by X-ray absorption and electron probe microanalysis by comparison with pre-defined standards [35]. Full line: best quadratic fit with experimental data.

The determination of b requires independent techniques to measure the composition of the alloy with an accuracy comparable to that of X-rays measurements. Among the techniques used for the independent determination of the composition in the alloy, X-ray absorption, Rutherford backscattering (RBS), reflection high-energy electron diffraction (RHEED) to calibrate accurately the component fluxes during the molecular beam epitaxial growth of films, accurate measurement of density, photoluminescence, Raman spectroscopy, and inductively coupled plasma atomic emission spectroscopy have been used for of SiGe alloy [41,42], for InGaAs [43] (Fig. 9), for GaAlSb [44] and SiC at low C content [45] and for the GaAlAs alloy [46].

In the case of nitride based alloys such as, GaAsN and InGaN, the band-gap dependence with the composition is strongly nonlinear but to date there is no experimental evidence of large deviations from the Vegard law for the lattice parameter, even if they are predicted by models based on molecular dynamic model simulations (see, for instance, Ref. [47] for InGaN). A small deviation from the Vegard law, determined experimentally by X-ray diffraction and secondary ion mass spectroscopy, has been evidenced in diluted GaAsN alloys [48]. The bowing parameters from some semiconducting alloys are given in Table 2.

Table 2

List of misfit values and bowing parameters in the lattice parameter-composition dependence for some semiconducting alloys

Alloy	Compounds	Misfit	Bowing parameter (Å)	Reference
SiGe	Ge-Si	4.18×10^{-2}	-0.0021	Fabbri et al. [39]
GaAlAs	AlAs-GaAs	1.43×10^{-3}	-0.0011	Gehrsitz et al. [40]
InGaAs	InAs-GaAs	7.16×10^{-2}	-0.077	Villaggi et al. [35]
SiC	Si-C	5.23×10^{-1}	-2.4239	Berti et al. [37]
GaAlSb	AlSb-GaSb	6.46×10^{-3}	0.00105	Bocchi et al. [36]

5. Strain release in semiconducting material heterostructures

5.1. Models for strain release in semiconductor heterostructures

In semiconductor alloys the lattice parameter and the band gap can be modified by the composition in a wide range. In particular, the variation $\Delta a/a$ in the lattice parameter may range several per cents (i.e. 4.1% and 7.3% in the case of Ge_xSi_{1-x} and $In_xGa_{1-x}As$ alloys, respectively).

It is known that the lattice strain affects the electronic band structures in semiconducting materials and hence it represents a valuable tool to modify energy gaps, to shift and reverse the band edges of heavy-hole and light-hole bands, to remove band degeneracy at critical points of the Brillouin zone and to change band curvatures and hence carrier effective masses [49,50].

Examples of III–V semiconducting systems that make use of mismatched buffers heterostructures to improve device performances are (i) high electron mobility transistor [51], (ii) heterojunction bipolar transistor [52] and (iii) multi-junction solar cells that utilize a wider part of the solar spectrum, thus increasing the conversion efficiency [53]. Also SiGe-based structures have taken increasing advantages of strain engineering since the pioneering work of Abstreiter et al. [54] up to the most recent proposals [55].

At the initial growth stages all the heterostructures are characterized by the pseudomorphic condition for which the elastic energy of deformation due to the misfit is stored in the epilayer lattice. A transition from the elastically distorted configuration to the plastically relaxed one (metamorphic) occurs when a critical misfit strain is exceeded. The MDs relieve the elastic strain and the process carries on with increasing the layer thickness until the strain is reduced to its thermal residual limit value [13].

The two main points which are important for applications are

- the threshold of strain and thickness h_c at which the MDs start to form;
- the dependence of residual strain $\varepsilon(h)$ on the thickness h and initial misfit m in an heterostructure, once the threshold is overcome.

In the literature, several models were proposed to predict the critical thickness h_c and the residual strain dependence. All of them have limitations and are not able to accurately predict the residual strain dependence, as the growth and annealing conditions and, in some cases, the sample dimensions may affect the residual strain. For simplicity, we will report here the two main models that were used as basis for more sophisticated models.

The equilibrium model of Matthews and Blakeslee [56] assumes the presence of threading dislocations (TDs) from the substrate. In presence of a stress described by a tensor, a force per unit line $\vec{F}$ is exerted on existing dislocations. The dislocations can easily move if dislocation lines and the Burgers vectors belong to the easy glide planes as {111} planes in fcc crystals. As shown in Fig. 10 the threading segments of the dislocations belonging to a (111) plane can glide along a [110] direction.

The force per unit length F_x parallel to a (111) glide plane and along a [110] direction exerted to a TD line with Burgers vector $\vec{b}$ in a strain field expressed by the stress tensor $\vec{\tau}$ is given by

$$F_\tau = \left(\vec{\tau} \cdot \vec{b} \cdot h\right)_\tau = G\frac{1+\nu}{1-\nu}bh\varepsilon \tag{24}$$

where G is the shear modulus which is connected to the elastic constants for the cubic lattice by the relationship $G = (c_{11}-c_{12})/2$, ν is the Poisson ratio, b is the modulus of the Burgers

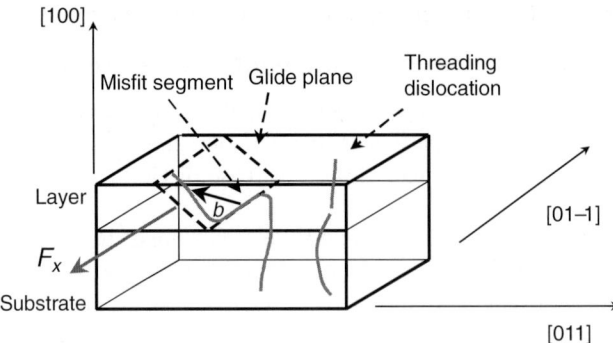

Fig. 10. Scheme of the applied shear stress acting on a threading segment of dislocation in the top layer due to the presence of a strain in the material. The force F_x acting on the threading dislocation segment in the (0–11) glide direction is indicated.

vector of the dislocation $\vec{b}$, h is the film thickness and ε is the biaxial film strain $\varepsilon = \varepsilon_{xx} = \varepsilon_{yy}$. The movement of the TD segment in the film leads to the increase of the misfit segment length and to the local reduction of the strain in the film as in Fig. 10. Opposite to that force is the tension of the misfit segment associated to the increase of the MD length:

$$F_l = \frac{Gb^2(1 - \nu/4)}{4\pi(1 - \nu)} \left(\ln\frac{h}{b} + 1 \right) \tag{25}$$

At the critical thickness h_c the two forces are equivalent:

$$h_c = \frac{b(1 - \nu/4)}{4\pi\varepsilon(1 + \nu)} \left(\ln\frac{h_c}{b} + 1 \right) \tag{26}$$

Once the critical thickness h_c is overcome, Eq. (26) gives the residual strain ε as a function of the thickness h of the layer:

$$\varepsilon = \frac{b(1 - \nu/4)}{4\pi h(1 + \nu)} \left(\ln\frac{h}{b} + 1 \right) \approx \frac{1}{h} \tag{27}$$

The main criticism to the present model is that it assumes that there are enough threading segments that are able to glide to release the strain, an assumption which is not valid of course for dislocation-free or low dislocation density materials.

The second relevant model for the calculation of the residual strain as a function of the layer thickness is the energy balance model of People and Bean [57]. The model assumes that the process governing the strain release is the kinetic barrier for the generation of new dislocations at the interface between the layer and the substrate. To generate new dislocations, the areal strain energy density E_ε (elastic strain energy per unit area of the film) should exceed a given threshold, assumed to be the self-energy density E_d associated with a dislocation. For simplicity, the dislocation has been assumed to be of screw type, even if screw dislocations cannot release the strain. The energy density per unit area E_ε and the self-energy density E_d of an isolated screw dislocation at a distance h from a free surface is given by (see Ref. [58])

$$E_\varepsilon = \frac{2G(1+\nu)}{1-\nu}\varepsilon^2 h, \quad E_d = \frac{Gb^2}{8\pi\sqrt{2}d}\ln\frac{h}{b} \tag{28}$$

where d is the film lattice parameter. From Eq. (28) the dependence of the residual strain on the film thickness is obtained:

$$h = \frac{b^2}{16\pi\sqrt{2}d}\frac{(1-\nu)}{(1+\nu)}\frac{1}{\varepsilon^2}\ln\frac{h}{b} \approx \frac{1}{\varepsilon^2} \tag{29}$$

so that a residual strain dependence on the thickness of the type $\varepsilon \approx 1/\sqrt{h}$ is obtained. It is worth noting that this model does not give any physical explanation of the mechanism of formation of new dislocations, but simply assumes that whenever a sufficient strain energy is present, new dislocations nucleate or form from multiplication processes.

Other authors tried to consider the problem of metastable states, i.e. they supposed that the strain released systems do not reach the equilibrium condition due to limited velocity of the dislocation glide even at high temperatures, as those of growth conditions [59,60]. Several observations agree with the non-equilibrium condition in partially released systems:

(a) In III–V compound layers, the threshold for the glide of dislocations is different for the [111](110) and [−1−11](011) glide systems.
(b) Much short MD segments were observed by X-ray topography in GeSi/Si-based system, because of the lower mobility of dislocations in group IV-based systems (see, for instance, Refs [61,62]).

Nevertheless, InGaAs/GaAs epilayers did not evidence significant strain release upon annealing at high temperature. The paper by Drigo et al. [63] offers a clear discussion of the proposed models and comparison with experimental data within the InGaAs/GaAs system.

5.2. Strain release in composition graded heterostructures

As pointed out by Tersoff [64], to accommodate the mismatch between the substrate and the structure grown on it, there are several advantages in growing compositionally graded buffer layers with respect to constant composition single layers or step graded structures:

(i) In composition graded structures, the misfit is maximum at the top of the structure where the MD density is minimum: thus, the generation of new MDs is favoured in regions with low MD density. On the contrary, in compositionally uniform layers, the confinement of MDs in a thin region near the interface increases the probability of dislocation repulsive interaction, which can lead to the formation of short MD segments with a high TD density. According to the model proposed by Freund [65], the repulsive interaction of MDs increases the strain necessary to permit the plastic relaxation through the dislocation glide.

(ii) The residual strain value in a graded buffer is always maximum at the top of the layer and it is much larger than that present in a uniform layer with the same average composition. This leads to an enhanced local force that drives the existing TD segments out of the crystal and reduces the TDs density.

(iii) The residual strain decreases with the depth below the surface and is minimum where a significant strain release has occurred and dislocation multiplication sources are formed [66]. This favours the strain release through the glide of the existing TDs rather than through the formation of dislocations generated by multiplication at deep sources in the structure.

5.3. The Tersoff model

The model of Tersoff [64] can be considered an extension of the equilibrium model of Matthews and Blakeslee [56] to the case of compositionally graded structures. It starts considering the energy E per unit area of a strained layer in the presence of a linear density ρ of MDs. The total energy of such a system can be approximated by

$$E = 2\lambda\rho + Kh(m - b\rho)^2 \tag{30}$$

where ρ is the linear dislocation density, h is the film thickness, $m = (a_f - a_s)/a_s$ is the film misfit (i.e. the relative difference between the lattice parameter of the unstrained film a_f and the substrate lattice parameter a_s), b the edge component in the interface of the Burgers vector of the MDs, λ is the energy per unit length of a MD and K is a constant. The first term in Eq. (30) corresponds to the elastic energy of the dislocation network and is assumed to be independent of the film thickness. In fact, the energy per unit length of a dislocation has a logarithmic dependence on the distance t from the free surface of the crystal (see, for instance, Ref. [67, p. 76]). This logarithmic dependence changes by a few per cent in a range of technological interest from a few tens of nanometres to several micrometres, so that λ can be considered as a constant with a reasonable approximation for typical thickness values. The numerical factor 2 considers a symmetrical distribution of MDs along two <110> directions. The second term gives the elastic energy of the layer due to the residual strain of the material $(m - b\rho)$, given by the misfit value reduced by the amount of relaxation $b\rho$. According to the model in Ref. [56], the equilibrium MD density ρ is that which minimizes the total energy of the system. For a single layer, Eq. (7) gives

$$\varepsilon_{\text{res}} = m - b\rho = \frac{\lambda}{bKh} \tag{31}$$

where ε_{res} is the residual strain of the layer. For comparison the equilibrium equation given by Ref. [68] for a network of 60° MDs with Burgers vectors of the type $b = a/2[110]$ is

$$\varepsilon_{\text{res}} = \frac{b(1 - \nu/4)}{4\pi(1 + \nu)h}\left(\log\frac{h}{b} + 1\right) \tag{32}$$

where ν is the Poisson ratio, a numerical factor which is very close to 1/3 for most semiconductors. Equations (31) and (32) are identical if

$$\frac{\lambda}{bK} = \frac{b(1 - \nu/4)}{4\pi(1 + \nu)}\left(\log\frac{h}{b} + 1\right) \tag{33}$$

In the following, we shall neglect the dependence on h in Eq. (33) and assume $h =$ constant ≈ 1000 nm for all the structures considered in this work. This is justified by the logarithmic dependence (the constant changes by an amount of 8% if $t = 500$ nm) and by the fact that in Eq. (33) h can be assumed as the typical distance of MDs from the free surface in a properly designed graded buffer. With this constraint the factor λ/bK becomes

$$\frac{\lambda}{bK} = 0.20 \pm 0.02 \text{ nm} \tag{34}$$

where the variation ± 0.02 nm corresponds to a thickness range of 500 to 2000 nm in Eq. (33).

Extending Eq. (30) to compositionally graded layers, the misfit $m(z)$ and the cross-sectional MD density per unit area $\rho(z)$ will depend on the distance from the interface z. In this case, Eq. (30) becomes

$$E = 2\lambda \int_0^h \rho(z)\, dz + K \int_0^h \left[m(z) - b \int_0^z \rho(\zeta)\, d\zeta \right]^2 dz \tag{35}$$

The density of dislocations per unit length and unit thickness is the function $\rho(z)$, which minimizes the energy E. Assuming a monotonic grading it is found that

$$\rho(z) = \frac{1}{b} \frac{\partial f(z)}{\partial z} \quad \text{for } z < z_c$$

and
$$\rho(z) = 0 \quad \text{for } z_c < z < h \tag{36}$$

Equation (36) states that the dislocation density profile $\rho(z)$ that minimizes the energy of the system is the one required to release completely the strain in the graded layer up to a thickness $z_c = h - h_f$ from the interface. No MDs should exist above z_c.

The results of Eq. (36) can be used to simplify Eq. (35):

$$E = \frac{2\lambda}{b} m(z_c) + K \int_{z_c}^h [m(z) - m(z_c)]^2\, dz \tag{37}$$

Finding a minimum for the energy in Eq. (37) is equivalent to finding a z_c, which obeys the following equation:

$$\int_{z_c}^h [m(z) - m(z_c)]\, dz = \frac{\lambda}{Kb} \tag{38}$$

which states that the integral of the strain $m(z) - m(z_c)$ over the thickness $h_f = h - z_c$ is a constant. Because the strain in the structure is completely released up to a thickness z_c with misfit $m(z_c)$, the growth at the top of the structure of a heterolayer with misfit $m(z_c)$ does not modify the strain profile and the layer is grown completely strain-free. In this case, the composition graded buffer layer behaves as a *virtual substrate* for the growth of a mismatched layer.

If $m(z)$ is a monotonically increasing function of z, the most strained region in the structure is always at the top and is separated by a thickness $h_f = h - z_c$ from the region containing dislocations, where sources for MD multiplication could be present.

It is also worth noting here that in graded buffers the residual strain at the surface $\varepsilon(h)$ is much higher than in the corresponding constant composition buffer layers with the same amount of average strain relaxation. Assuming an uniform layer with residual strain ε_0 and thickness $h_0 = h - z_c$, compared to a composition graded heterostructure, it follows from Eq. (38) that

$$\varepsilon_0 h_0 = \int_{z_c}^{h} [m(z) - m(z_c)] \, dz \tag{39}$$

Assuming that the grading in the buffer is chosen to obtain the same dislocation-free thickness $h_0 = h_f$, Eq. (39) states that the average of the residual strain is equivalent to ε_0. If $\varepsilon(z)$ is a monotonically increasing function of z, then $\varepsilon(h) > \varepsilon_0$. Then the formation of new MDs in the system should occur preferentially by the glide of the existing TDs in the most strained part of the structure rather than by nucleation of new dislocations from internal deep Frank–Read modified sources formed by the pinning of MDs (see, for instance, Ref. [69]), which lie in the almost completely relaxed part of the buffer.

6. Experimental results

6.1. Strain release in constant composition layers

The knowledge of the residual strain thickness dependence, as those for instance expressed by Eqs (27) and (29), is a fundamental requirement for the preparation of composition graded heterostructures with predefined profile of strain release.

Many works are devoted to the InGaAs/GaAs [63,70] and SiGe/Si materials [57]. The majority of them agree in finding a relationship between the residual strain and the thickness following approximately a $t^{-0.5}$ dependence, in better agreement with the People and Bean [57] and Mareé [71] models even if a completely satisfactory agreement with experimental data have not been obtained and some of the assumptions of the models are not completely justified.

Experimental results show that the model of Matthews and Blakeslee is able to predict correctly the critical thickness h_c at which the existing threading segments start to glide producing misfit segments. Nevertheless, this process is limited by the number of existing TDs so that a further strain release is possible only through the introduction of new dislocations or by multiplication. Once started, the residual strain is better described by the energy balance model. This is clearly reported in the paper by Bocchi et al. [72]: in Fig. 11 the measurements of residual strain versus thickness dependence in the AlGaSb/GaSb system is shown. In a range between 0.2 and 0.3 μm a very sharp change in residual strain is observed connected with a first threshold h_c for the MD generation. This is clearly due to the glide of the existing threading segments as demonstrated by the double-crystal X-ray topographs (Fig. 12) of samples corresponding to 0.2 and 0.3 μm in thickness. Once terminated, in the threading segments no further release is observed up to 0.5 μm in thickness, above which the strain decreases with a dependence $\varepsilon \sim h^{-0.5}$. Figure 11 also shows the effect of the *thermal strain*: due to a difference in the thermal expansion coefficients of the two materials a different lattice mismatch is experienced under the growth condition with respect to room temperature.

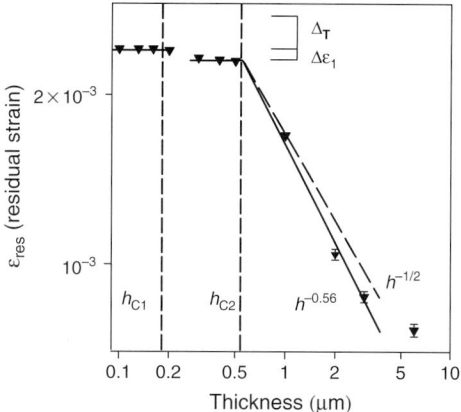

Fig. 11. Residual strain ε_{res} as a function of the layer thickness h in a series of $Ga_{0.6}Al_{0.4}Sb/GaAs$ layers. The curve (solid triangular symbols) results after the correction for thermal misfit. h_{C1} is the experimental value of a first critical thickness found to be between $0.16 < t_{C1} < 0.2$ μm.; $\Delta\varepsilon \cong 1.3 \times 10^{-4}$ is the total amount of the relaxed elastic strain for thicknesses increasing from t_{C1} to t_{C2} and $\Delta_T = -3.28 \times 10^{-4}$ is the thermal misfit correction. A second critical thickness h_{C2}, for a new nucleation mechanism of misfit dislocations, has been found to be slightly larger than 0.5 μm. The theoretical $t^{-0.5}$ (dashed line) and the experimental $t^{-0.56}$ (solid line) dependences of the strain release are drawn [44].

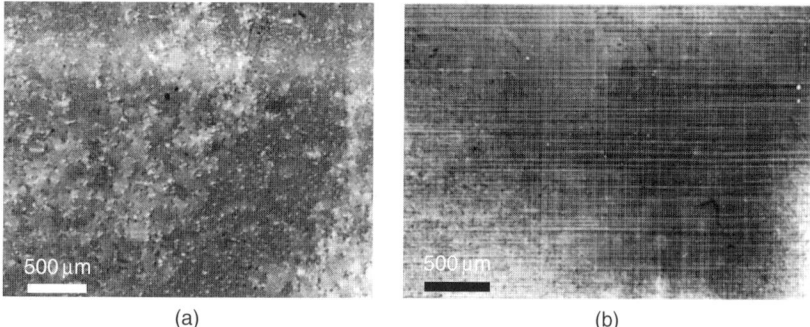

Fig. 12. Double-crystal 622 X-ray topographs of GaAlSb/GaSb samples 0.16 and 0.2 μm in thickness, respectively. In (a) only threading dislocations are visible, whereas in (b) the majority of dislocations have formed a MD line, demonstrating that the Matthews et al. mechanism [50] is applicable but only a small part of the strain is released in this way [72].

6.2. Strain release in composition graded heterostructures

The residual strain $\varepsilon(h)$ at the surface, the dislocation-free thickness h_f and the distribution of MDs in different composition graded profiles of InGaAs/GaAs samples, grown by MBE, have been studied by Lazzarini et al. [66] using TEM, RBS and high-resolution diffraction techniques. The layers were grown on (001) semi-insulating GaAs substrates by MBE by varying the growth temperature in the range 400–500°C. The composition profiles were designed according to Bosacchi et al. [73] and the predictions of the semi-empirical model described by Salviati et al. [74].

For a given buffer layer thickness, different composition gradings were designed to permit the growth of a strain-free $In_{0.30}Ga_{0.70}As$ layer starting from a GaAs substrate.

Cross-sectional TEM micrographs of the samples [60] have shown that the two set of MDs lying along the <110> directions parallel to the interface have a different depth distribution with the lower dislocation-free thickness t_f corresponding to dislocation lines parallel to the [1–10] direction: such an asymmetry is observed in polar materials at the beginning of the strain release process when mainly α dislocations develop along the [1–10] direction (see, for instance, Refs [75, 76]). In fact, in graded layers, the equilibrium position of new MDs is at the interface between the strained and the relaxed zone, where the dislocation density is very low, a situation similar to the early stage of MD generation in strained single layers. The different α and β dislocation density at this stage and hence the asymmetry in the strain release is attributed to the different glide velocity or nucleation rate for the two sets of dislocations [77]. The presence of an asymmetrical strain release indicates that the system is not at thermal equilibrium at the growth temperature. Nevertheless, the comparison between samples grown at 400°C and 500°C and the annealing experiments made on uniform composition InGaAs/GaAs structures by Drigo et al. [63] demonstrate that the residual strains do not change significantly with the temperature.

The main results of the paper of Lazzarini et al. [66], Bosacchi et al. [73] and Salviati et al. [74], which compare the predictions of the model of Tersoff [64] and of Drigo [63] and Dunstan [64] are as follows:

(a) the Tersoff model clearly underestimates the residual strain even if it predicts correctly the minimum dislocation-free thickness in the case of samples grown at higher temperature, i.e. closer to the thermodynamic equilibrium.
(b) the Residual strain at the surface is predicted with a good accuracy by extrapolating the $\varepsilon \sim h^{-0.5}$ dependence found in constant composition layers to graded buffer layers with the experimental determined constant k [63]:

$$\varepsilon^2 h = k = 0.0037 \pm 0.0007 \text{ nm} \tag{40}$$

The deviation in the measured residual strains from those calculated from Tersoff model can be understood on the basis of the Matthews et al. [56] model, which assumes that the strain release occurs only by the glide of existing TD segments.

In fact, even with substrates with TD densities of the order of 10^5 cm^{-2}, as in the case of poor quality GaAs substrates, and a mean MD length of 1 cm, the maximum linear MD density allowed by the substrate TDs is given by $\rho = 1/4 \times 10^5$ cm^{-1}, which corresponds to a maximum strain release of $\rho b = 5 \times 10^{-4}$. Therefore, even for accommodating low misfit values of the order of 10^{-3}, the nucleation of new dislocations is necessary and the strain release process is limited by the dislocation nucleation rate. Moreover, the Matthews curve describes correctly the threshold for the glide of existing TD segments, even if a consistent strain release occurs only at a higher threshold (see, for instance, Refs [63,70,78]).

On the basis of Eqs (15) and (19) and a simplified approach for graded layers, Salviati et al. [74] and Bosacchi et al. [73] gave an equation for the dislocation-free thickness z_c:

$$\int_{z_c}^{h} [m(z) - m(z_c)]^2 \, dz = k \tag{41}$$

where the first term is the square of the strain in the buffer layer. Equation (41) states that in a graded buffer the square of the strain integrated along the dislocation-free thickness is a constant equivalent to the case of the constant composition layers. For instance, we can calculate the dislocation-free thickness $h - z_c$ in a linearly graded composition profile with misfit profile $m(z) = m'z$:

$$\int_{z_c}^{h} m'^2 (z - z_c)^2 \, dz = \frac{m'^2}{3} (h - z_c)^3 = k \quad (42)$$

$$h - z_c = \sqrt[3]{\frac{3k}{m'^2}}$$

or to calculate the thickness h_{tot} needed to grow a strain-free layer with given mismatch m_0 with respect to the substrate. By assuming $m'z_c = m_0$,

$$h_{tot} = \frac{m_0}{m'} + \sqrt[3]{\frac{3k}{m'^2}} \quad (43)$$

The application of this formula to buffer structures gives a much better agreement with the experimental findings in InGaAs/GaAs heterostructures, except that for the asymmetry of the strain release, as already mentioned when discussing the Dunstan et al. [70] model. An example of the application of formula (43) for the growth of a fully relaxed $In_{0.2}Ga_{0.8}As$ layer is shown in Fig. 13: a linearly graded InGaAs buffer layer with composition grading of 15%(In)/μm is grown up to a In composition value of 22.5%. A fully relaxed zone is formed up to at a depth h_c at which $x = 0.2$. A 1.2-μm-thick strain-free $In_{0.2}Ga_{0.8}As$ layer is grown atop the structure, without modifying the strain profile of the underlying structure. The reciprocal lattice map of Fig. 14 shows the following:

(a) The reciprocal lattice with 335 nodes of GaAs substrate and $In_{0.2}Ga_{0.8}As$ relaxed layer appears aligned along the same 335 reciprocal lattice vector; as seen in Section 3, this indicates that the two lattices have parallel planes and thus the layer is strain-free.

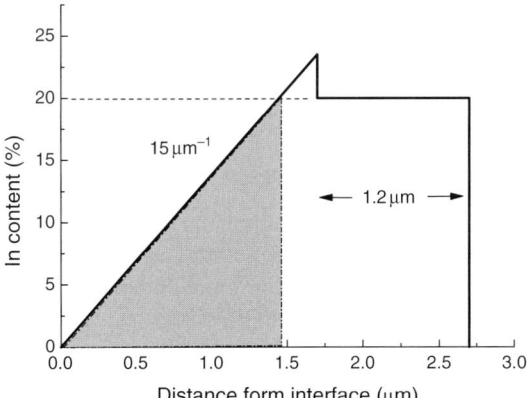

Fig. 13. Composition profile of a InGaAs/GaAs heterostructure made by a linear composition graded buffer designed in order to grow a 1.2 mm thick $In_{0.2}Ga_{0.8}As$ fully relaxed and free of misfit dislocation according to the model of Bosacchi et al. [73] and Salviati et al. [74]. The shaded area corresponds to the fully relaxed part of the composition graded buffer.

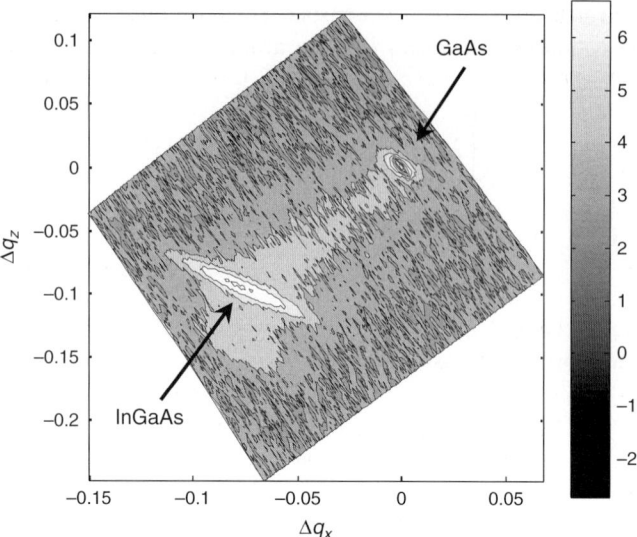

Fig. 14. X-ray reciprocal lattice map of the InGaAs structure depicted in Fig. 13 in reciprocal lattice units Δq with respect to the substrate 335 node. The GaAs substrate and the uniform composition $In_{0.2}Ga_{0.8}As$ layer have their 335 peak aligned along the same 335 direction, parallel to the map edges, demonstrating the fully relaxed state of the top InGaAs layer. The linearly graded InGaAs buffer produces a diffuse scattering intensity connecting the GaAs and the uniform composition InGaAs layer, corresponding to the fully relaxed part of the buffer. The nearly vertical diffuse scattering at low q_z values (larger lattice parameter) corresponds to the fully strained top part of the buffer.

(b) The buffer layer X-ray scattering is made by one diffuse part aligned between the two 335 nodes of substrate and layer corresponding to the full relaxed part of the buffer layer.

(c) One part of the buffer layer X-ray scattering is close to the 335 node of the $In_{0.2}Ga_{0.8}As$ layer and extends in the vertical direction. As seen in Section 3, this corresponds to a part of the buffer having a constant lattice parameter parallel to the InGaAs/GaAs interface and also corresponds to the fully strained part of the buffer layer.

7. Strain determination in zero-dimensional heterostructures

7.1. Strain as a tool and driving force

The strain theoretical studies and the experimental X-ray diffraction methods, used to determine the residual strain in heteroepitaxial structures exposed in the previous sections, may constitute a starting point for approaching more complex systems (nanostructures) for which the strain plays a leading role modifying important properties of the structures themselves.

The use of strain to tailor the electronic properties of materials requires the competence of modelling strain relaxation and the support of suitable techniques to measure it in structures matched for specific nanoelectronic devices. In this context, an example using metamorphic buffer (MB) layers is given. A basic structure configuration may consist of a substrate, a buffer layer which the lattice parameter depends on its composition and thickness through the mechanism of strain relaxation, and the active part of the structure. By controlling the lattice parameter of the MB, the mismatch between it and the upper part is changed, thus affecting the strain in the active layer with a consequent modification of its energy gap.

A significant example is represented by InAs/InGaAs QD systems in which light emission can be red-shifted to long wavelengths ($\lambda \geq 1.3\,\mu$m) [79–81] by controlling the thickness-dependent strain relaxation of InGaAs MBs [80,82].

Figure 15 shows the strain relaxation of constant-composition InGaAs MB layers as a function of their thickness in InAs/InGaAs QD structures investigated by means of X-ray diffraction (XRD) either in the reciprocal space map (RSM) or in the single ω–2θ scan modes. Raman scattering and photoreflectance were also used as a complementary investigation methods [83]. The measured values of the in-plane MB residual strain $\varepsilon(t)$ versus the thickness t of MB layers have been compared to the results of strain relaxation models that give different $\varepsilon(t)$ dependences; the experimental data confirm the validity of the energy-balance model [57,71] that foresees an approximate $t^{-1/2}$ dependence. The knowledge of the mechanism that determines the strain relaxation may allow the development of advanced epitaxial structures, where graded-composition MB layers are incorporated to take specific advantages over the constant-composition counterparts with regard to the confinement of MDs far away from the active region of the structures [73].

However, the most important example of engineering material properties by means of lattice strain is given by advanced epitaxial structures where charge carriers are zero-dimensionally confined to nanosized regions (QDs) for which the strain in a growing sample in

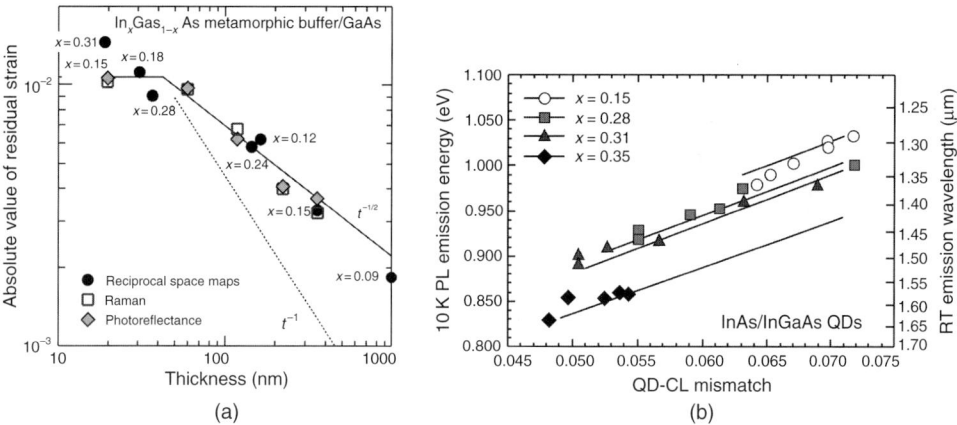

Fig. 15. (a) Absolute value of residual in-plane strain ε as a function of the thickness t of In$_x$Ga$_{1-x}$As metamorphic buffers obtained by X-ray reciprocal space map (RSM – closed circles), Raman scattering (open squares), and photoreflectance (PR – closed diamonds) measurements [83]. The continuous line shows the thickness dependence of strain both in the pseudomorphic regime (horizontal line, $x=0.15$) and in the partial relaxation one where ε can be approximated by a $t^{-1/2}$ dependence; the dotted line represents the t^{-1} behaviour foreseen by the equilibrium models. Raman and PR measurements refer to $x=0.15$ structures, while the MB compositions of structures for RSM are given in the figure. (b) PL emission energy at 10 K from InAs quantum dots embedded in In$_x$Ga$_{1-x}$As confining layers (CL) versus the mismatch between QDs and metamorphic buffer layer, for $x=0.15$ (circles), $x=0.28$ (triangles), $x=0.31$ (squares) and $x=0.35$ (diamonds). For each composition, the red shift was obtained by modifying the residual strain of the buffer layer [74].

addition to modifying the band profile of the incorporated semiconductor materials is also the main driving force leading to the formation and mutual arrangement of the nanoislands (self-organized QDs).

The fabrication of QDs takes place by depositing a layer of a material having a lattice constant very different (high misfit) with respect to that of the substrate [84,85]. At an early stage of deposition, the material grows following the layer-by-layer two-dimensional mechanism with the in-plane lattice parameter matched to that of the underlying substrate (or buffer); the resulting strained layer is termed as a wetting layer. During growth, the elastic energy stored in this layer increases. When some critical thickness is reached, this planar growth is no longer followed and the growth proceeds by the formation of three-dimensional defect-free islands due to the *elastic* relaxation mechanisms, which compete with the *plastic* release processes (accompanied by MDs) leading to *self-organized nanostructures* – Stranski–Krastanow mode [86, 87]. Plastic and elastic release mechanisms in this kind of structures have been reviewed in Ref. [88]. The islands are nanometre-sized, with heights and diameters of the order of few nanometres and few tens of nanometres, respectively. It is worth noting that by growing lattice-mismatched systems it is possible to obtain three-dimensional self-assembled coherent nanostructures without using sophisticated nanolithographic techniques. Moreover, an increased island surface density is needed for the fabrication of a number of optoelectronic devices [89] and this characteristic is achieved by growing dot multilayer (*staking*) structures. The staking is realized by repeating several times in sequence a core structure made up of a suitable spacer layer grown over a dot layer. The final structure exhibits enhanced vertical and horizontal spatial correlation of the dot positions and an improved size homogeneity [90] as required for technological application.

The knowledge of the intrinsic elastic strain field arising from the lattice mismatch between the islands and matrix material is crucial for further device modelling, as already mentioned above, the strain strongly affects the performance of optoelectronic devices. The determination of the strain in such low-dimensional systems is a difficult problem which has given a new impulse to the refinement of X-ray-based investigation techniques and to the development of calculation methods for the interpretation of the experimental results. A summary of the peculiar features of this problem will be given in the next section.

7.2. *X-ray theory for QD structures characterization: kinematical approximation*

A complete structural characterization of semiconductor QD structures requires a combination of methods to assure the clearing of the accuracy and resolution limits intrinsic to each experimental technique. Between the routinely applied methods, the X-ray-based ones are successfully suited to obtain quantitative results. In particular, X-ray diffraction is very powerful for measuring composition, strain fields and to investigate the correlations of the mutual positions of nanostructures. The exact periodic part of the electron density gives rise to sharp Bragg peaks. All non-periodic deviations from the ideal crystal structure lead to diffuse scattering in the vicinity of the Bragg peaks. Because QDs can usually exhibit different shapes or sizes, the X-ray scattering from this kind of structures is widely spread in the RS. Hence, in self-assembled QDs systems, the nanoislands are responsible for the diffuse X-ray scattering accompanying the coherent diffraction signal coming from the whole structure. The diffuse scattering depends on the difference in the scattering factors of the crystal and the dots and on the elastic deformation field in the crystal matrix. Thus, the information on the dot structure is contained only in the diffuse component of the scattered intensity. For recent reviews of X-ray scattering by self-assembled nanostructures, see, for instance, the works of Schmidbauer et al. [91], Stangl et al. [88] and Pietsch et al. [92]. The distribution of the diffusely scattered intensity can be calculated in the frame of statistical scattering theories [93,94]. To obtain information on the structure of a self-assembled QDs system, it is necessary to collect two-dimensional distribution of the diffracted X-ray intensity: RSM. Then it is possible to calculate the dependence of the diffusely scattered intensity on the reduced scattering vector $\vec{q} = \vec{K}_f - \vec{K}_i - \vec{h}$ ($\Delta\vec{K} = \vec{K}_f - \vec{K}_i$ being the scattering vector and $\vec{h}$ a vector of the reciprocal lattice, respectively) and to compare it with the experimental data. For the simulation of the diffuse scattering intensity, a kinematical approximation is usually sufficient even if extinction and refraction of the primary beam as well as any multiple scattering processes, are neglected. To improve the calculation the distorted wave Born approximation (DWBA) method has been introduced by several authors (see Ref. [92]). However, the theoretical approach must be based on a model containing the statistical description of the dot ensemble and the deformation fields due to the partial elastic relaxation of misfit strain.

Following the general treatment described by Holý et al. [94] and Darhuber et al. [95] the diffuse scattered intensity due to a large number of dots can be written as

$$I(\vec{q}) = \left\langle \sum_{\vec{R}} \sum_{\vec{R}'} E_0(\vec{q}; Z) E_0^*(\vec{q}; Z') e^{i\vec{q}\cdot(\vec{R}-\vec{R}')} \right\rangle \tag{44}$$

where the average is taken over all dot positions ($\vec{R}$). The amplitude E_0 depends also on the depth Z of the dot centre below the free surface. The positional distribution of the islands in each period of a multilayer is described by the statistical distribution function $c(\vec{R})$, unity if the point is occupied by a dot centre and zero otherwise. The average of this function is the normalized dot density $w(\vec{R}_\parallel)$ (see Refs [93,95] for details). At a given interface, the in-plane distribution of the dots can be described by the Fourier transform of the probability density:

$$w^{\text{FT}}(\vec{q}_\parallel) = \left\langle \sum_{\vec{R}_\parallel} e^{i\vec{q}_\parallel \cdot \vec{R}_\parallel} \right\rangle \tag{45}$$

where $\vec{q}_\parallel$ and $\vec{R}_\parallel$ are the in-plane components of the scattering vector and the position vector of the dots, respectively. Assuming that a relative number (m) of dots is vertically correlated (the remainder $1-m$ is not vertically correlated), Eq. (44) becomes

$$I(\vec{q}) = w^{\text{FT}}(\vec{q}_\parallel) \left[m \left| \sum_{n=1}^{N} E_0(\vec{q}; Z_n) e^{iq_z Z_n} \right|^2 + (1-m) \sum_{n=1}^{N} \left| E_0(\vec{q}; Z_n) e^{iq_z Z_n} \right|^2 \right] \tag{46}$$

This expression can be used for simulating the RSMs of the X-ray diffuse scattered intensity.

For a system of self-assembled QD embedded in a matrix and with a single QD shape function $\Omega(\vec{r})$, unity inside the dot and zero outside, the amplitude is given by [95]

$$E_0(\vec{q}) = \int d^3\vec{r} e^{i\vec{q}\cdot\vec{r}} \left[\chi_{hL} \left(e^{i\vec{h}\cdot\vec{u}(\vec{r})} - 1 \right) + (\chi_{hD} - \chi_{hL})\Omega(\vec{r}) e^{i\vec{h}\cdot\vec{u}(\vec{r})} \right] \quad (47)$$

where $\vec{u}(\vec{r})$ is the elastic displacement field due to the dot centred at point $\vec{R}$ and $\chi_{hL,hD}$ are the hth Fourier components of the polarizability at a point $\vec{r}$ of the surrounding crystal and the dot material, respectively. In turn, expanding the exponential and writing the deformation field in term of the strain tensor $\hat{\varepsilon}$ the amplitude finally becomes

$$E_0(\vec{q}; Z) \approx \chi_{hdot} \Omega^{FT}\left(\vec{q} + \vec{h}\hat{\varepsilon}\right) - \chi_{hcryst} \Omega^{FT}\left(\vec{q} + \vec{h}\hat{\varepsilon}\right) + i\chi_{hcryst} \varphi^{FT}(\vec{q}; Z) \quad (48)$$

where $\Omega^{FT}(\vec{q})$ is the Fourier transform of the shape function of a dot. The RS vector inside the dot must be corrected by $\vec{q} \to \vec{q}_i + \vec{h}_i \varepsilon_{ij}(Z)$. This correction accounts for the shift $(-\vec{h}\varepsilon)$ of the maximum of the intensity distribution due to the deformation of the dot lattice, while the shape of the distribution is influenced by the Fourier transform of the dot shape function. The function $\varphi^{FT}(\vec{q}; Z)$ depends on the analytical model chosen to describe the deformation field in and around a dot.

The calculation of the diffuse intensity distribution requires a modelling of the dot main parameters: shape, size and composition. Some of them can be obtained by others investigation techniques, e.g. dot shape and size for which cross-sectional scanning tunnelling microscopy (STM) and TEM are very powerful tools having a higher spatial resolution with respect to an X-ray probe and in-plane correlation lengths by plan-view TEM analysis. Furthermore, the elastic strain field inside and around the dots must be evaluated by using a numerical method applied to models based on either atomistic simulations or continuum elasticity theory. Finally, the results of the above two steps are used to calculate the diffuse scattering distribution by using Eq. (46) or an equivalent one. Subsequently, by the comparison with experimental X-ray intensity maps, some parameters can be refined until satisfactory agreement is achieved.

7.3. Strain calculation

However, the most important step is the calculation of the intrinsic strain field and the displacement in and around a dot. It requires to solve a three-dimensional problem for the non-trivial geometry of the QD shape. In the calculation of diffuse scattering, the elastic anisotropy and the difference in elastic constants inside and outside the dot are usually neglected. It is instead considered the relaxation of internal stresses at the free surface.

Among the commonly used methods for the solution of the strain problem, finite element analysis (FEM) [96] is one of the widely adopted method. The method has been applied to the study of different systems, e.g., Ge–GeSi/Si [97], InP/InGaP [98], InAs/GaAs [99,100], InGaAs/GaAs and InN/AlN [101,102], to mention some of the works published in recent years. The FEM is well suited for finding approximate solutions of boundary value problems for partial differential equations in finite domains. The strain field due to the constraint of epitaxy associated with the mismatched lattice parameters of the heterostructures layers is determined within the framework of continuum elasticity theory. It means that the actual sample structure is replaced by an elastic continuum and the strain distribution is obtained by solving the elasticity equilibrium equation or by minimizing the elastic energy stored in the sample. The structure is spatially described with a mesh (Fig. 16), which could be more

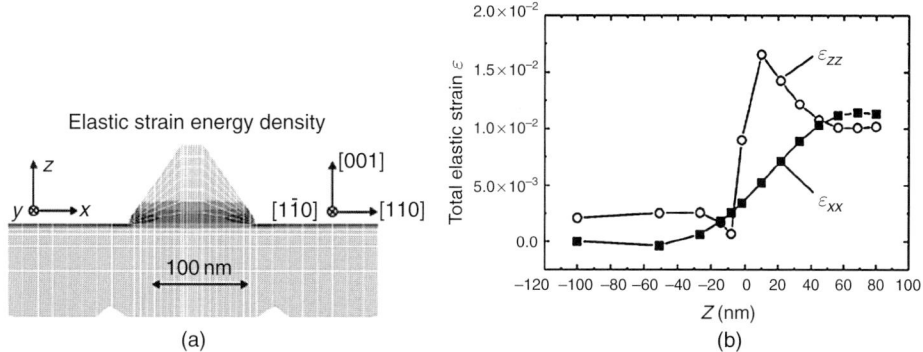

Fig. 16. Finite element calculation for a SiGe pyramid grown on Si (001). (a) Elastic strain energy density in the 110 plane through the middle axis of the dot. The white lines represent the used FEM mesh. (b) Strain tensor components $\varepsilon_{xx,zz}$ along z-axis (growth axis) [103].

refined near free surfaces and in regions where the mismatch between adjacent layers is larger [103]. It has been demonstrated that the comparison with calculation based on atomistic simulations (the other used methodological approach) shows discrepancies between the two techniques only in regions close to the interfaces [100]. For a review of different calculation approaches, see Ref. [89, pp. 763–769].

7.4. Strain effect on reciprocal space maps

In Eq. (47), the strain field in a layer is evidenced by the product $\vec{h} \cdot \vec{u}(\vec{r})$. In principle, this makes it possible to distinguish the influence of the strain on the distribution of diffracted intensity in RSMs [104]. The distance between the peaks of a layer and the substrate is determined by

$$\Delta q_{x,z} = -h_{x,z}\Delta_{xx,zz} \tag{49}$$

where

$$\Delta_{xx,zz} = \frac{a_{x,z} - a_s}{a_s}$$

are the mismatch components with respect to the substrate; $a_{x,z}$ are the average lattice parameters of the epilayer structure parallel and perpendicular to the sample surface, respectively, and a_s is the lattice parameter of the substrate. If a superlattice with period D instead of an epilayer is considered, a sequence of peaks, labelled by $SL_n, \ldots, SL_1, SL_0, SL_{-1} \ldots$, appears along q_z; SL_0 is the zeroth-order satellite corresponding to the contribution of a layer with thickness and composition equal to the total thickness of the superlattice and the weighted mean composition, respectively. The distance between the peaks along q_z is inversely proportional to the superlattice period $\Delta q_z = 2\pi/D$, and the width of the peaks is inversely proportional to the total thickness of the superlattice ($t = nD$, where n is the number of periods). The distance between the substrate peak and the SL_0 is determined by the average strain in the superlattice:

$$\Delta q_z = -h_z \langle \Delta_{zz} \rangle \tag{50}$$

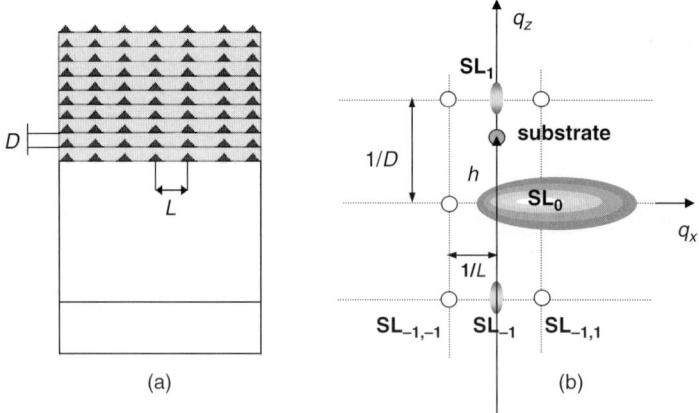

Fig. 17. (a) Scheme of self-assembled quantum dots superlattice of period D and lateral array of the dots with period L. (b) Schematic of intensity distribution around a reciprocal lattice point for coplanar HRXRD experiment. Only the SL_1, SL_0 and SL_{-1} superlattice satellites are shown. The double labelled satellites ($SL_{m,p}$ – open circles) are relative to the lateral dots array pattern.

The translation symmetry group of the resulting structure in the RS contains the translational symmetry of the crystal lattice, the vertical periodicity of the superlattice and the lateral periodicity of the structure.

The crystal lattice periodicity is represented by the three-dimensional reciprocal lattice, with the diffraction vector $\vec{h}$ as one of its translational vectors. An RSM is then collected in the vicinity of $\vec{h}$. The vertical superlattice periodicity causes superlattice satellite nodes SL_m (m as an integer) along the direction (q_z) of the surface normal unit vector with period $1/D$. In the case of a two-dimensional dots array, the lateral satellites are arranged in a two-dimensional reciprocal array and the satellites can be thus labelled by the indices (m, p) (Fig. 17). If the lateral pattern is not perfectly periodic, the width of the lateral maxima increases with p and is proportional to the root mean square deviation of the dot positions. Information on the lateral order of the dots is contained in the Fourier transform of the in-plane correlation function of the dot positions. This function gives rise to maxima (lateral satellites) in the diffuse intensity arranged periodically in the q_x direction. The in-plane strain affects the lateral position of the diffuse maxima close to non-zero order coherent satellites. The maximum of the diffuse intensity around SL_0 is shifted in the direction of positive q_x. This is due to the strain field always present in such structures and in the case of a lateral structure perfectly periodic, the strain field is periodic with the same period. As a result, the lateral positions and the widths of the lateral intensity maxima are not affected by the strains. The strains modify only the shape of the envelope functions of these maxima. In the lateral direction, the maximum of the envelope function is shifted from the substrate maximum by $-h_x \varepsilon_{xx}$ and if the strain is homogeneous, the intensity distribution along q_z is shifted by $-h_z \varepsilon_{zz}$.

7.5. Experimental methods and results

Structural investigation of QD-embedded systems can be carried out by using different X-ray scattering and diffraction methods, which yield complementary and/or additional information to that obtained with transmission and cross-sectional electron microscopy

(TEM, STM) and atomic force microscopy (AFM); the last one applies only in the case of uncapped QD single layer. Several scattering geometries can be applied:

(i) The conventional coplanar diffraction in high-resolution (HRXRD) set-up with analyser crystal (triple axes) for collecting RSMs around suitable reciprocal lattice nodes (symmetric and asymmetric Bragg reflections). This geometry is particularly sensitive to the composition and strain.
(ii) X-ray reflectivity (XRR) [105] and grazing-incidence small-angle X-ray scattering (GISAXS) [106,107] for probing the intensity distribution (maps) close to the (000) reciprocal lattice node. Due to the very small incidence and scattering angles, these methods are sensitive to shape, size and correlation properties of QDs and are insensitive to the crystal lattice and strain fields. For these reasons, both techniques are widely applied for the morphology of surfaces and interfaces investigation. Furthermore, they represent the ideal complementary methods to the HRXRD because their results are not affected by the strain. Consequently, it is possible to obtain the main dot parameters from XRR or GISAXS experiments and to use them to separate the strain effect on the diffracted intensity distribution in the RS. The strong disadvantage is because of the need of a very high-power primary X-ray beam, which limits the employment of the grazing incidence techniques with the use of a synchrotron radiation source.
(iii) The grazing incidence diffraction (GID) [108] that takes place on lattice plane perpendicular to the surface of the structure. Therefore, the GID technique is particularly suitable for the in-plane strain components determination. As in the previous case, it is a very useful diffraction geometry, complementary to coplanar HRXRD measurements; however, this technique also needs synchrotron radiation sources when applied to the study of small volume samples. Then, a strong sensitivity limitation of XRD and XRR methods occurs when the diffracting volume is very small and it makes the characterization of structures with an uncapped single dots layer on the surface by using conventional X-ray sources impossible. Reviews of nanostructure characterization by grazing incidence methods are provided in in Chapter 10 by Metzger et al. and Chapter 11 by Proietti et al.

During the last decade, a large number of QD systems have been investigated by different authors and by using all the techniques listed above (see Ref. [82] and references therein). Most of the studies have dealt with InAs/GaAs- and GeSi/Si-based structures, the former being one of the first QD semiconductor system obtained via the Stranski–Krastanow growth mode [84].

The 7% misfit between InAs and GaAs is larger than for most other semiconductor systems and the critical thickness for islands formation is below 2 MLs. Due to the strong elastic relaxation, this system promoted a remarkable number of strain studies. Furthermore, the difference between the atomic scattering factors of the constituent atoms, weak in GaAs and larger in InAs, allowed to distinguish In-rich from Ga-rich regions by suitable X-ray diffraction geometries, making possible the investigation of segregation and/or diffusion phenomena inside the islands and their correlation with the deposition rate and temperature. The effects on islands size and density of spacer thickness and composition have been studied in structures capped with InGaAs instead of GaAs. As reported at the beginning of this paragraph, the strain of InAs QD embedded in InGaAs layers is also reduced resulting in a decrease of the energy gap. The latter effect allows to achieve a significant red-shift of the QD emission. Figure 18 shows an example of characterization by RSM simulation of

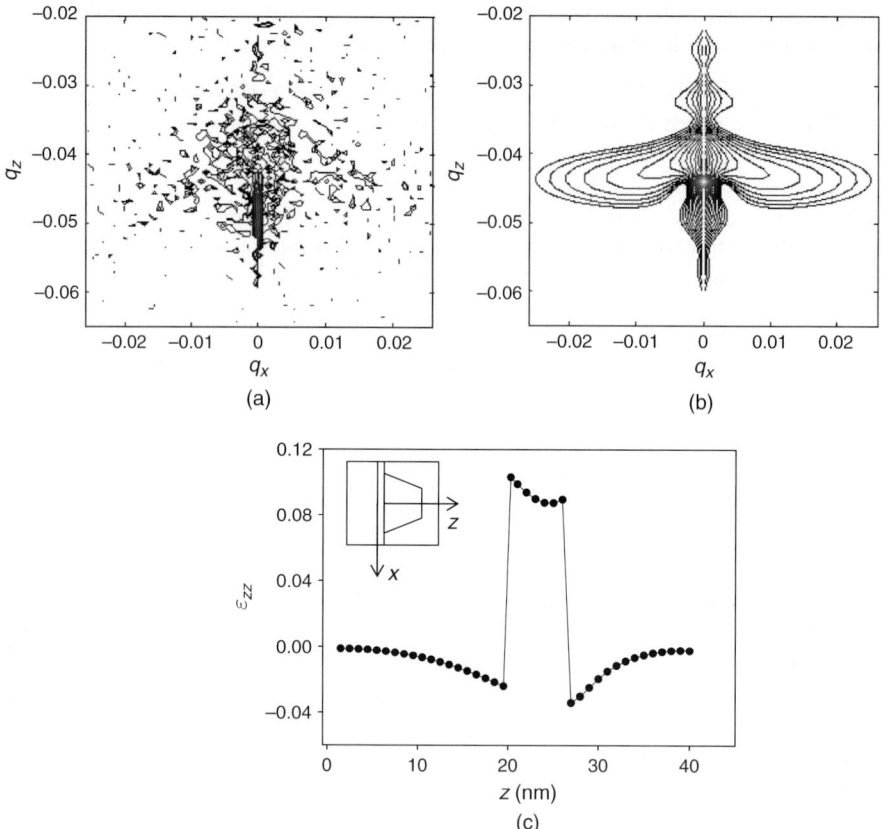

Fig. 18. Experimental (a) and simulated (b) 004 symmetrical reciprocal space maps from InAs/GaAs QD staking with InAs/GaAs period repeated eight times. The calculation were focused on the diffuse maximum around the SL_0 satellite peak. The substrate iso-intensity contours are out of the figure [109]. (c) Axial strain ε_{zz} versus the z co-ordinate along a line passing through the centre of the dot as obtained from FEM analysis. The inset shows a schematic of an embedded QD. This strain profile was used for the map simulation (b).

a 001-oriented InAs/GaAs QD superlattice sample prepared by MBE growth technique with eight layers of InAs QD separated by GaAs spacer layers 100 Å thick. RSMs were collected around reciprocal 004, 224 and 404 lattice nodes by using a high-resolution diffractometer with a conventional X-ray source. The experimental intensity distribution around the SL_0 peak relative to the 004 GaAs substrate node is reported in Fig. 18(a). The simulated RSM (Fig. 18(b)) was calculated by using the FEMAP program by assuming the ε_{zz} strain component behaviour inside the dot shown in Fig. 18(c). The same analysis was carried out on all the RSMs to determine the in-plane component of the strain and to obtain the main parameters of the structure. The dot shape considered was a truncated cone with average ratio between the base radius (r) and the height (h), $r/h = 2$. No evidence of in-plane correlation between the dots was found.

Many effects have been observed also in GeSi on Si system. Some of them like Ge diffusion and the effect of the Si cap on dot size and shape are similar to those exposed above for InAs/InGaAs or /GaAs QD systems. A good summary of the properties and main applications of various materials systems is reported in Ref. [88].

8. Summary and future perspectives

In this chapter the basics of the methods of measuring the strain and the composition in two- and zero-dimensional structures by means of HRXRD techniques are presented. A rapid overview of the basic models employed to predict the strain release occurring during the epitaxial growth of lattice-mismatched materials is also given in the chapter.

The use of compositionally graded buffer layers for the growth of mismatched heterostructures appears as a promising strategy for the integration of mismatched semiconducting materials on the same substrate. The most important factors, which make convenient the use of graded buffers, are:

(i) the decrease of TD density on the surface with respect to constant composition layers,
(ii) the possibility of accommodating high mismatch values still avoiding three-dimensional growth modes, which introduce a high density of TDs.

Although important results in terms of structure quality have been obtained, a physical model able to predict accurately the strain and the MD distribution is not available. To date, the degree of strain release is somehow material-dependent, and in many cases recipes for obtaining a given strain value are used.

It is shown that the X-ray method is accurate, non-destructive and reliable for determining the composition through the measurement of the lattice parameter, even if there is an increasing evidence of relevant deviations from the Vegard law as far as new accurate measurements of the lattice parameter dependence are performed. This field is open for studies of many new semiconducting alloys.

Although a significant progress has been made both in experimental and calculation approaches, the strain and composition determination in QD structures remains a very complex problem if conventional X-ray sources are used, mainly due to the limitation in brilliance of the X-ray beam and the impossibility of selecting the wavelength. New opportunities for studying nanostructured materials in the laboratory may arise in the future by the use of new focussing X-ray optics, as those based on capillary optics or on two-dimensional focussing.

The availability of using synchrotron sources gives rise to other opportunities; in Chapter 10 several examples are given in which strain and composition profiles of QDs and other periodic structures are obtained using anomalous scattering and well-collimated and intense beams, even if measurements at synchrotron must be considered more for fundamental studies rather than to be a standard tool for semiconductor characterization.

References

[1] C. Kittel, Introduction to Solid State Physics, John Wiley & Sons, New York (2005), Chapter 3.
[2] S. Milita and C. Ferrari, Vuoto: Scienza Tecnologia 24 (1995) 38.
[3] C. Bocchi, C. Ferrari, P. Franzosi, et al., J. Cryst. Growth 132 (1993) 427.
[4] B.K. Tanner, A.G. Turnbull, A. Stanley, et al., Appl. Phys. Lett. 59 (1991) 2272.
[5] Landolt-Bornstein, Semiconductors, Vol. 17, subvol. A Springer, Berlin (1987).
[6] O. Ambacher, J. Majewski, C. Miskys, et al., J. Phys.: Condens. Matter 14 (2002) 3399.

[7] L. Gorgens, O. Ambacher, M. Stutzmann, et al., J. Appl. Phys. Lett. 76 (2000) 577.
[8] A. Krost, G. Bauer, and J. Woitok, High Resolution X-ray diffraction, in: Optical Characterization of Epitaxial Semiconductor Layers, Springer-Verlag, Berlin, Heidelberg (1996) p. 287.
[9] C.R. Wie, Mater. Sci. Eng. 13 (1994) 1.
[10] P.F. Fewster, Semicond. Sci. Technol. 8 (1993) 1915.
[11] T. Picraux, B.L. Doyle, and J.Y. Tsao, in: Semiconductors and Semimetals, T.P. Pearsall (ed.), Academic Press, New York, (1991) pp. 139–220.
[12] B.K. Tanner, J. Cryst. Growth 99 (1990) 1315.
[13] C. Bocchi, S. Franchi, A. Bosacchi, et al., Appl. Phys. Lett. 71 (1997) 1549.
[14] M. Servidori, F. Cembali, R. Fabbri, and A. Zani, J. Appl. Phys. 25 (1992) 46.
[15] P.F. Fewster and C.J. Curling, J. Appl. Phys. 62 (1987) 4154.
[16] M.A.G. Halliwell, M.H. Lyons, and M.J. Hill, J. Cryst. Growth 68 (1984) 523.
[17] S. Takagi, Acta Crystallogr. 15 (1962) 311.
[18] D. Taupin, Bull. Soc. Fr. Miner. Crist. 87 (1964) 69.
[19] A. Authier, Dynamical Theory of X-ray Diffraction, in: ICUR Monographs on Crystallography, Vol. 11, Oxford University Press Inc., New York (2001).
[20] C. Ferrari and P. Franzosi, J. Appl. Phys. 65 (1989) 1544.
[21] L. Tapfer and K. Ploog, Phys. Rev. B 40 (1989) 9802.
[22] C. Ferrari, C. Bocchi, A. Bosacchi, and S. Franchi, Mat. Sci. Eng. B 28 (1994) 183.
[23] C. Bocchi and C. Ferrari, J. Phys. D 28 (1995) A164
[24] C. Lamberti, S. Bordiga, F. Boscherini, et al., J. Appl. Phys. 76 (1994) 4581.
[25] C. Lamberti, S. Bordiga, F. Boscherini, et al., J. Appl. Phys. 83 (1998) 1058.
[26] K. Ishida, J. Matsui, T. Kamejima, and I. Sakuma, Phys. Status Solidi A 31 (1975) 255.
[27] C. Bocchi, A. Bosacchi, C. Ferrari, et al., J. Cryst. Growth 165 (1996) 8.
[28] Yu. P. Khapachev and F.N. Chukhovskii, Sov. Phys. Crystallogr. 34 (1989) 465.
[29] L. De Caro, C. Giannini, L. Tapfer, J. Appl. Phys. 79 (1996) 4101.
[30] P.F. Fewster, J. Appl. Crystallogr. 22 (1989) 64.
[31] H. Heinke, M.O. Moller, D. Hommel, and G. Landwehr, J. Cryst. Growth 135 (1994) 41.
[32] P.F. Fewster, Appl. Surf. Sci. 50 (1991) 9.
[33] S. A. Kononogov, M. Yu. Konstantinov, and V. V. Khrushchev, Meas. Technol., 49 (2006) 311, translated from Izmeritel'naya Tekhnika 4 (2006) 3.
[34] L. Vegard, Z. Phys. 5 (1921) 17.
[35] J.C. Mikkelsen Jr and J.B. Boyce, Phys. Rev. Lett. 49 (1982) 1412; Phys. Rev. B 28 (1983) 7130.
[36] L. Nordheim, Ann. Phys. (Leipz.) 9 (1931) 606; 9 (1931) 641.
[37] B.J. Pines, J. Phys. (USSR) 3 (1940) 309.
[38] G. Fournet, Le J. Phys. Radium 14 (1953) 374.
[39] K.A. Gschneidner, G.H. Vineyard, J. Appl. Phys. 33 (1962) 3444–3450.
[40] V.A. Lubarda, Mech. Mater. 35 (2003) 53–68.
[41] P. Dismukes, L. Ekstrom, and R. J. Paff, J. Phys. Chem. 68 (1964) 3021.
[42] F.L. Fabbri, F. Cembali, M. Servidori, and A. Zani, J. Appl. Phys. 74 (1993) 2359.
[43] E. Villaggi, C. Bocchi, N. Armani, et al., Jpn. J. Appl. Phys. 41 (2002) 1000.
[44] C. Bocchi, S. Franchi, F. Germini, et al., J. Appl. Phys. 86 (1999) 1298.
[45] M. Berti, D. De Salvador, A.V. Drigo, et al., Appl. Phys. Lett. 72 (1998) 1602.
[46] S. Gehrsitz, H. Sigg, N. Herres, et al., Phys. Rev B 60 (1999) 11601.
[47] M. Ferhat and F. Bechstedt, Phys. Rev. B 65 (2002) 075213.
[48] S.Z. Wang, S.F. Yoon, W.J. Fan, et al., J. Appl. Phys. 96 (2004) 2010.
[49] P. Bhattacharya (ed.), Properties of Lattice-Matched and Strained Indium Gallium Arsenide, INSPEC, London (1993).
[50] F.H. Pollak, in: Semiconductors and Semimetals, Vol. 32, T.P. Pearsall (ed.), Academic, London (1990) p. 17.

[51] W.E. Hoke, P.J. Lemonias, J.J. Mosca, et al., J. Vac. Sci. Technol. B 17 (1999) 1131.
[52] O. Baklenov, D. Lubyshev, Y. Wu, et al., J. Vac. Sci. Technol. B 20 (2002) 1200.
[53] F. Dimorph, U. Schubert, and A.W. Bett, IEEE Electron Device Lett. 21 (2000) 209.
[54] G. Abstreiter, H. Brugger, T. Wolf, et al., Phys. Rev. Lett. 54 (1985) 2441.
[55] E.A. Fitzgerald, Mater. Sci. Eng. B 124–125 (2005) 8.
[56] J.W. Matthews and A.E. Blakeslee, J. Cryst. Growth 27 (1974) 118.
[57] R. People and J.C. Bean, Appl. Phys. Lett. 47 (1985) 322.
[58] F.R.N. Nabarro, Theory of Crystal Dislocations, Clarendon, Oxford (1967) p. 75.
[59] B.W. Dodson and J. Y. Tsao, Appl. Phys. Lett. 47 (1985) 322.
[60] B.W. Dodson and J. Y. Tsao, Appl. Phys. Lett. 49 (1986) 229.
[61] C. Ferrari, Defect and Diffusion Forum 200–202 (2002) 153.
[62] F. Romanato, E. Napolitani, A. Carnera, et al., J. Appl. Phys. 86 (1999) 4748.
[63] A.V. Drigo, A. Aydinli, A. Carnera, et al., J. Appl. Phys. 66 (1989) 3334.
[64] J. Tersoff, Appl. Phys. Lett. 62 (1992) 693.
[65] L.B. Freund, J. Appl. Phys. 68 (1990) 2073.
[66] L. Lazzarini, C. Ferrari, S. Gennari, et al., Proceedings of IX International Conference on Microscopy on Semiconducting Materials, Inst. Phys. Conf. Ser. No. 157 (1997).
[67] D. Hull and D.J. Bacon, Introduction to Dislocations, Pergamon, Oxford (1984).
[68] A.F.D. Chin and P. Battacharia, IEEE Trans. Electron. Devices 36 (1989) 2183.
[69] C.G. Tuppen, C.J. Gibbing, and M. Hickly, Appl. Phys. Lett. 56 (1990) 54.
[70] D.J. Dunstan, P. Kidd, L.K. Howard, et al., Appl. Phys. Lett. 65 (1994) 839.
[71] P.M.J. Marèe, J.C. Barbour, and J.F. van der Venn, J. Appl. Phys. 62 (1987) 4413.
[72] C. Bocchi, F. Germini, S. Franchi, et al., J. Mater. Sci. – Mater. Electron. 19 (1999) 185.
[73] A. Bosacchi, A.C. DeRiccardis, P. Frigeri, et al., F. Romanato, J. Cryst. Growth 175/176 (1997) 1009.
[74] G. Salviati, C. Ferrari, L. Lazzarini, et al., Inst. Phys. Conf. Ser. 146 (1995) 337.
[75] K.L. Kavanagh, M.A. Capano, L.W. Hobbs, et al., J. Appl. Phys. 64 (1988) 4843.
[76] M.S. Abrahams, L.R. Weisberg, C.J. Buiocchi, J. Blanc, J. Mater. Sci. 4 (1969) 223.
[77] I. Yonenaga and K. Sumino, J. Appl. Phys. 65 (1989) 85.
[78] S.J. Barnett, A.M. Keir, A.G. Cullis, et al., J. Phys. D: Appl. Phys. 28 (1995) 17.
[79] Y.C. Xin, L.G. Vaughn, L.R. Dawson, et al., J. Appl. Phys. 94 (2003) 2133.
[80] L. Seravalli, M. Minelli, P. Frigeri, et al., Appl. Phys. Lett. 82 (2003) 2341.
[81] A.E. Zhukov, A.P. Vasil'ev, A.R. Kovsh, et al., Semiconductors 37 (2003) 1411.
[82] L. Seravalli, P. Frigeri, M. Minelli, et al., Appl. Phys. Lett. 87 (2005) 063101.
[83] V. Bellani, C. Bocchi, T. Ciabattoni, et al., Eur. Phys. J. B 56 (2007) 217.
[84] L. Goldstein, F. Glas, J.Y. Marczin, et al., Appl. Phys. Lett. 47 (1985) 1099.
[85] V.A. Shchukin, N.N. Ledentsov, and D. Bimberg, Epitaxy of Nanostructures, Springer, Berlin (2003).
[86] D.J. Eaglesham and M. Cerullo, Phys. Rev. Lett. 64 (1990) 1943.
[87] D.E. Jesson, K.M. Chem, S.J. Pennycook, et al., Phys. Rev. Lett. 77 (1996) 1330.
[88] J. Stangl, V. Holý, and G. Bauer, Rev. Mod. Phys. 76 (2004) 725.
[89] S. Franchi, G. Trevisi, L. Seravalli, and P. Frigeri, Prog. Cryst. Growth Charact. Mater. 47 (2003) 166.
[90] J. Tersoff, C. Teichert, and M.G. Lagally, Phys. Rev. Lett. 76 (1996) 1675.
[91] M. Schmidbauer, M. Hanke, and R. Köhler, Cryst. Res. Technol. 36 (2002) 1.
[92] U. Pietsch, V. Holy, and T. Baumbach, High-Resolution X-Ray Scattering: from Thin Films to Lateral Nanostructures, 2nd edition, Springer-Verlag, New York (2004).
[93] M.A. Krivoglaz, X-ray and Neutron Diffraction in Non-Ideal Crystals, Springer-Verlag, Berlin (1996).
[94] V. Holý, A.A. Darhuber, J. Stangl, et al., Phys. Rev. B 58 (1998) 7934.
[95] A. Darhuber, P. Schittenhelm, V. Holý, et al., Phys. Rev. B 55 (1997) 15652.

[96] K.H. Huebner, E.A. Thornton, and T.G. Byrom, The Finite Element Method for Engineers, 3rd ed., John Wiley & Sons, New York (1994).
[97] S. Christiansen, M. Albrecht, H.P. Stunk, and H.J. Maier, Appl. Phys. Lett. 64 (1994) 3617.
[98] A. Carlsson, L.R. Wallemberg, C. Persson, and W. Seifert, Surf. Sci. 406 (1998) 48.
[99] T. Benabbas, Y. Androussi, and A. Lefebvre, J. Appl. Phys. 86 (1999) 1945.
[100] C. Pryor, J. Kim, L.W. Wang, et al., J. Appl. Phys. 83 (1998) 2548.
[101] B. Jogai, J. Appl. Phys. 88 (2000) 5050.
[102] B. Jogai, J. Appl. Phys. 90 (2001) 699.
[103] Th. Wiebach, M. Schmidbauer, M. Hanke, et al., Phys. Rev. B 61 (2000) 5571.
[104] Y. Zhuang, J. Stangl, A.A. Darhuber, et al., J. Mater. Sci.-Mater. Electron. 10 (1999) 215.
[105] M. Meduňa, V. Holý, T. Roch, et al., J. Appl. Phys. 89 (2001) 4836.
[106] M. Schmidbauer, Th. Wiebach, H. Raidt, et al., Phys. Rev. B 58 (1998) 10523.
[107] T. Metzger, V. Favre-Nicolin, G. Renaud, et al., in: Characterization of Semiconductor Heterostructures and Nanostructures, Chapter 10, p. 361, C. Lamberti (ed.), Elsevier, Amsterdam (2008).
[108] I. Kegel, T.H. Metzger, A. Lorke, et al., Phys. Rev. B 63 (2001) 035318.
[109] F. Germini, Study of InAs/GaAs stacked Quantum Dots by X-ray diffraction and reflectivity techniques, PhD thesis in Materials Science and Technology, Parma University, Italy (unpublished).

Chapter 5

Transmission electron microscopy techniques for imaging and compositional evaluation in semiconductor heterostructures

Laura Lazzarini[1], Lucia Nasi[1], and Vincenzo Grillo[2]

[1] IMEM-CNR, Parco Area delle Scienze 37/A, 43010 Fontanini-Parma, Parma, Italy
[2] Laboratorio Nazionale TASC, Area Science Park, Basovizza S.S.14 km 163, 5, 34012 Trieste, Italy

Abstract Nowadays, the advanced discovery and development of nanometric scale-structured materials with innovative properties require investigation techniques able to perform characterizations down to the atomic scale. The modern transmission electron microscopes are now able to accomplish this requirement, not only by imaging but also by obtaining chemical information with atomic resolution.

This chapter is mainly devoted to a brief introduction to the transmission electron microscopy technique: the fundamentals of the diffraction and phase contrast are explained, with regard to the defect imaging, the chemical-sensitive imaging, the high-resolution electron microscopy and the lattice fringe analysis.

At the end, some examples are discussed, to approach the reader to the problems this technique can solve in the case of quantum well structures. A typical example of the chemical-sensitive imaging and the interpretation of the strain- and defects-induced contrast is shown in InGaAs-based strain-balanced multi-quantum wells. A basic application of the lattice fringes analysis for the composition determination is described in the InGaN/GaN system, while it is shown that the In and N assessment in the InGaAsN alloy requires a combination of the diffraction contrast and the lattice fringes analysis.

Keywords transmission electron microscopy, diffraction contrast, phase contrast, high-resolution electron microscopy, chemical-sensitive imaging, composition evaluation, quantum wells

1. Introduction

1.1. The importance of the transmission electron microscopy techniques

Nano is nowadays the coolest keyword in Materials Science. The advanced discovery and development of nanometric scale-structured materials with innovative properties require investigation techniques capable of performing characterizations down to the atomic scale. The modern transmission electron microscopes (TEMs) are now able to accomplish this feat. The main reason for the use of the electron microscope resides in the superior resolution it allows due to the very small wavelength of the electrons that, depending on the accelerating voltage V (100–1500 kV), varies in the range of 4–0.3 pm, many orders of magnitude smaller than the visible light wavelength (400–700 nm). However, it is worth using electrons for many good reasons other than the extraordinary capability of imaging individual atoms in a lattice: when they interact with matter, they produce a wide range of secondary signals, giving many details on the examined samples to the materials scientists. Among these, information can be obtained on the crystal lattice parameter, the crystal structure, presence of ordering or different phases, defect nature and distribution; more importantly, chemical information from nanometric regions of the specimen with atomic resolution.

In the following, the principles of the microscope operation are described in Section 2, the fundamentals of the interpretation of the diffraction (Section 3) and phase contrast (Section 4) and a very short outline of the variety of the TEM-related techniques (Section 5). Then, in Section 6, three emblematic applications to quantum well (QW) structures of the technique are given, to approach the reader to the possibilities of the TEM and the problems it can solve.

2. Experimental: basics of the electron microscope

2.1. Basic layout and operation

A modern electron microscope usually consists of an electron source and an assembly of magnetic lenses according to the scheme in Fig. 1, where the ray path diagrams are shown in

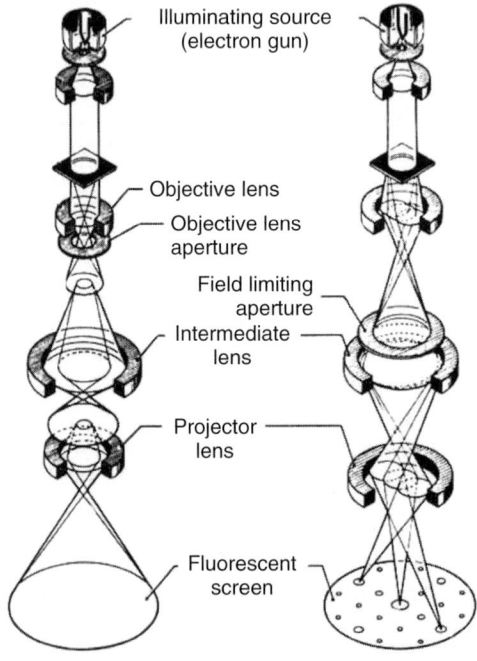

Fig. 1. Scheme of the lens assembly and the ray path diagram forming the image (left) or the diffraction pattern (right).

the cases when an image or a diffraction pattern (DP) are produced. The image or the DP is then displayed on a fluorescent screen or on a monitor.

The illuminating source produces a coherent parallel beam of electrons and must satisfy the following requirements: high brightness, small size and high stability, also of the emitted electron velocity. There are two fundamental types of sources, not interchangeable: thermionic (TI), emitting electrons when heated, or field emitters (FEs), which produce electrons when an intense electric field is applied. TI sources can be tungsten filaments or LaB_6 crystals, whereas FE sources are crystal needles with very fine tip. The sources have very different characteristics: TI sources are less monochromatic than the FE; FE produces very high current density and very small probe size, with high spatial coherence [1]. But FE TEMs are much more expensive, mainly for the UHV technology required.

The electrons are accelerated by a high potential (100–1500 kV), and the diameter and convergence angle of the beam impinging on the specimen surface can be varied by means of a condenser lens system and field-limiting apertures. The specimen is mounted on a special holder inserted within the objective lens and can be tilted, in most systems, by more than 30° around the two orthogonal axes to allow crystallographic analyses.

As the electrons are charged particles, the beam strongly interacts with matter and undergoes scattering phenomena with changes in the direction of motion and partial loss of energy. A lot of secondary signals are produced. In the crystalline specimens, the elastically diffracted electrons (Bragg diffraction) are responsible for the image contrast and also form TEM DPs that are essential in Materials Science, as they contain information on the crystal lattice parameter, the crystal structure, specimen shape, presence of ordering, identity of different phases and their orientation relationship with the matrix, spinodal decomposition, magnetic domains and similar phenomena. The analysis of the characteristic X-ray radiation, as done by X-ray energy-dispersive spectrometry, can give chemical information from a very small region of the specimen (see Section 5).

The electrons emerging from the bottom surface of the specimen recombine to form an image in the image plane of the objective lens. This is a convergent lens and all the parallel electrons (each beam having different diffraction vectors) are dispersed on the back focal plane, forming the DP. The imaging lens system must then be adjusted to have the back focal plane or the image plane of the objective lens, as the object plane for the intermediate lens in case the DP or the image, respectively, has to be seen. The DP or the image is then magnified and projected onto the viewing screen, the magnification varying between about 10^2 and 10^6 times.

The DP contains electrons from the whole illuminated specimen area. This is not very useful due to the fact that the specimen is often buckled and the transmitted beam too intense: it is a basic operation to select the small region of interest of the specimen by means of an aperture. This method is called selected area diffraction (SAD) and the aperture is inserted into the first image plane conjugate to the specimen, as it is not possible to put it at the specimen plane because of the presence of the sample. In the case of crystalline samples, any SAD pattern will contain a bright central spot and some spots arranged according to the different lattices. The electrons that will form the image are chosen by inserting an aperture into the back focal plane of the objective lens. This is the most important aperture of the TEM: if the transmitted beam only is allowed to pass through the aperture, then the image is a bright field (BF) image. If one diffracted beam is used, the image will be a dark field (DF) one.

2.2. Lens defects and resolution

The magnetic lenses are designed to produce an axially symmetric magnetic field to focus the electrons. The electron trajectories are described by the well-known paraxial ray equation. When the magnetic field axial extent is small compared to the focal length, the thin lens formula, analogue to that of geometrical optics, is obtained. The only remarkable difference is the rotation of the image, with respect to the object plane as an effect of the lens magnetic field along the optical axis.

Electrons lenses, like thin lenses do for the light, suffer from defects among which, astigmatism, chromatic and spherical aberration are the most important in relation to the objective lens, as they limit the resolution of the microscope. Astigmatism (Fig. 2(a)) occurs

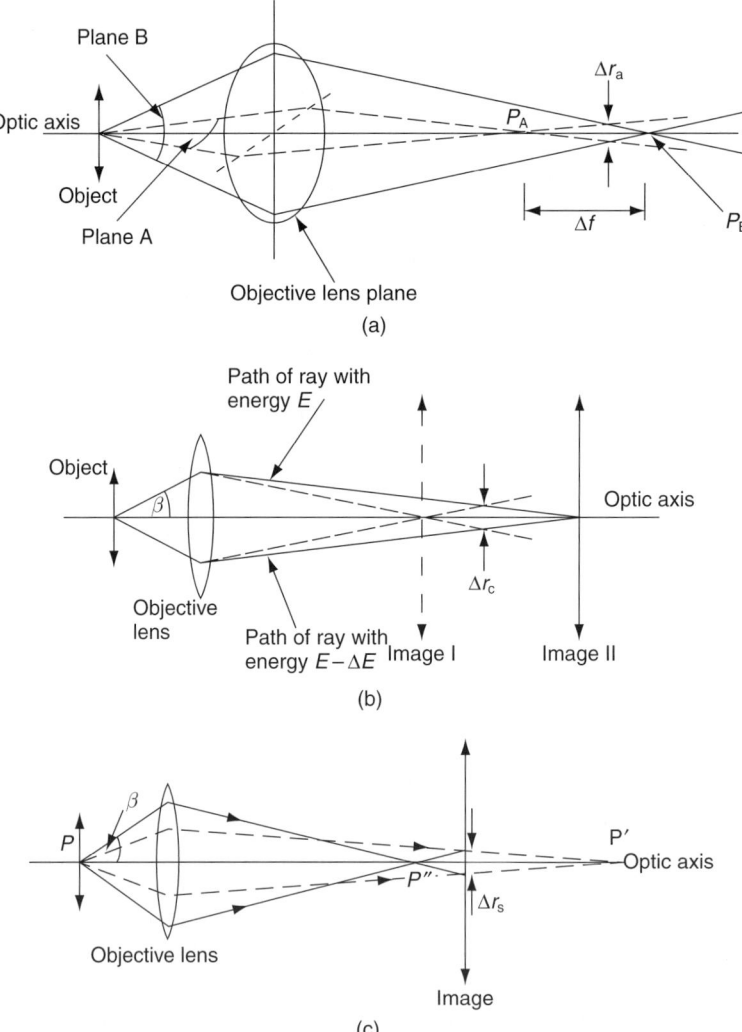

Fig. 2. Scheme of the objective lens aberrations: (a) astigmatism, (b) chromatic aberration and (c) spherical aberration.

anytime the lens presents different focal length depending on the electron path plane. Chromatic aberration (Fig. 2(b)) arises from the electron (mostly the ones emerging from the foil) energy spread, because the lens focal length varies with the electron energy. Spherical aberration (Fig. 2(c)) is due to the fact that off-axis electrons are brought to focus at different planes depending on the angle between the electron path and the optical axis. While the former two can be corrected or neglected, it is this defect that ultimately limits the resolution of modern microscopes, except one can afford to buy a very sophisticated and extremely expensive device, the C_s corrector. Normally, the spherical aberration induces in the image a confusion disc of radius

$$\Delta r_s = C_s \alpha^3 \quad (1)$$

where C_s is the spherical aberration coefficient. To maximize the instrument resolution, very small values of the lens aperture α are then needed. However, if the size of the aperture is too small, diffraction effects become a further limiting factor. According to the Rayleigh criterion, the closest distance Δr_d at which two different points can be resolved is

$$\Delta r_d = 0.61 \lambda / \alpha \quad (2)$$

where λ is the electron wavelength. This resolution increases by increasing the lens aperture that must be the smallest to minimize the spherical aberration contribution. There is an optimum aperture size that minimizes the sum of the quantities (1) and (2), which gives the maximum resolution:

$$\Delta r_{min} = \lambda^{3/4} C_s^{1/4} \quad (3)$$

The electron wavelength can be made smaller by increasing the accelerating voltage: in practice, due to voltage instability and strong damage induced in the samples, instruments with V higher than 400 kV are no more used. Typical values for C_s are 0.25–0.3 nm, but high-resolution instruments have 0.15 nm or less.

2.3. Limitations of the TEM

The above-described unique features of the TEM are unfortunately, but not surprisingly, accompanied by some drawbacks:

- Sampling – A major disadvantage of this technique is that, higher the resolution, the smaller the part of the specimen one can look at, and worse is the possibility of sampling the specimen in any one measure. The sample to be investigated must be accurately chosen to be representative of the phenomena we are interested in. It has been estimated that, as TEM is available, the total quantity of material investigated is around 0.6 mm^3! It is then evident that the TEM cannot be the only technique that should be used to characterize a material, but the complementary employment of techniques having worse resolution and better sampling is necessary.
- Interpretation of the images – The electron lenses act to increase the depth of field of the instrument so that the whole of the specimen, from the top to the bottom surface, is in focus at the same time. This is an advantage but, as the information we obtain averaged through

the whole thickness of the sample, caution must be used in interpreting the images. One single TEM image has no depth sensitivity and complementary techniques are needed for a full characterization of the specimen.
- Electron beam damage – The use of ionizing radiation can damage the samples, especially polymers and ceramics. In the case of semiconducting materials, the InP-based alloys can be affected by the electron beam with local evaporation of some species, leaving In-rich droplets. The risk of beam damage increases with the accelerating voltage and in presence of very intense electron sources that, potentially, are able to destroy any sample. So, the observation conditions must be carefully tested for each material.
- Specimen preparation – To prepare a specimen for TEM observations means that it must be thin enough to be transparent to the electrons, so that the beams of the transmitted electrons are strong enough to give on the screen or on the photographic plate an interpretable image in a reasonable time. How much thin the specimen depends on the accelerating voltage and the average atomic number of the specimen. To give an example, GaAs is transparent up to half a micron but, if you do high-resolution TEM, thicknesses <50 nm are needed. In the case of semiconducting materials, at the beginning, the sample is cut into 3-mm-diameter self-supporting disks that are thinned down to a few micrometres just in the middle by means of a dimpler instrument. Then the sample is bombarded with energetic ions and sputtered until the electron transparency in an ion milling apparatus. In general, the specimen is affected by these treatments and much care must be paid to avoid damages and to recognize the artefacts that could have been introduced. Specimen preparation is time consuming and can be considered a major drawback of the TEM technique. Again, for this reason also, the TEM cannot be used as a routine diagnostic tool.

3. Diffraction contrast in TEM images

3.1. The diffraction pattern

When the beam of electrons is accelerated through a thin crystalline specimen, several diffracted beams in addition to the transmitted beam are present at the exit side of the sample in predictable directions, which depend on the orientation and the structure of the crystal. The diffracted beams are then focused by the objective lens in its back focal plane, resulting in a spot pattern which is eventually magnified on the viewing screen.

The basic principles of the formation of SAD patterns are well described in the kinematical approximation of the electron diffraction theory, which states that the amplitudes of the scattered beams are negligible with respect to the amplitude of the incident beam that is supposed to be constant over the whole crystal. Within this assumption, no interaction between the scattered and incident beams is expected.

As the distances electron source–specimen and specimen–objective lens are very large as compared to specimen area under observation, the incident and scattered waves can be treated as plane waves with wave vector $\vec{k}$ and $\vec{k}'$, respectively.

To predict the intensity distribution in the back focal plane of the objective lens is thus equivalent to consider the phase relationship between the electron wavelets scattered from a periodic array of atoms hit by a plane wave of constant amplitude, and to calculate their amplitude and intensity at far field from the specimen (Fraunhofer wave optics model).

The solution for the diffracted wave amplitude in the direction defined by the scattering vector $\vec{h} = \vec{k}' - \vec{k}$ can be expressed as a double sum over the lattice and over the scattering centres inside the unit cell:

$$A(\vec{h}) = \sum_{\substack{j \\ \text{cell}}} f_j(\vec{h}) \exp(2\pi i \vec{h} \cdot \vec{r}) \sum_{\substack{\ell \\ \text{lattice}}} \exp(2\pi i \vec{h} \cdot \vec{r}_\ell) \quad (4)$$

where $\vec{r}_\ell$ represents the unit cell position in the lattice while the first sum represents the structure factor F_h and describes the contribution of the entire unit cell to the diffracted intensity in that direction:

$$F_{\vec{h}} = \sum_{\substack{j \\ \text{cell}}} f_j(\vec{h}) \exp(2\pi i \vec{h} \cdot \vec{r}_j) \quad (5)$$

$f(\vec{h}_j)$ being the scattering power of the individual atoms at $\vec{r}_j$.

For a perfect crystal with dimensions very large compared to cell extent, strong diffraction by crystal will occur when the waves diffused by any cell separated by a lattice vector $\vec{r}_\ell$ are in phase, i.e. when $\vec{h} \cdot \vec{r}_\ell = (\vec{k}' - \vec{k}) \cdot \vec{r}_\ell = n$, n being a relative integer. This condition is equivalent to require that $\vec{h}$ is a vector of the reciprocal lattice [2] $\vec{h} = \vec{g}$ so that Eq. (4) can be written as

$$A(\vec{h}) = \sum_{\vec{h}} F_{\vec{h}} \, \delta(\vec{g} - \vec{h}) \quad (6)$$

The DP is thus described by the reciprocal lattice nodes weighted by the corresponding structure factors. The delta functions in Eq. (6) represent the Ewald diffraction conditions. All this can be described by the Ewald sphere construction in the dual space of the direct lattice vector space, denoted in the following as reciprocal space. If a sphere with radius $|\vec{k}| = 1/\lambda$ is drawn through the origin of the reciprocal lattice, for elastic scattering ($|\vec{k}| = |\vec{k}'|$), the diffraction condition $\vec{h} = \vec{g}$ is satisfied only for the reciprocal lattice points which are cut by the Ewald sphere surface as shown in Fig. 3.

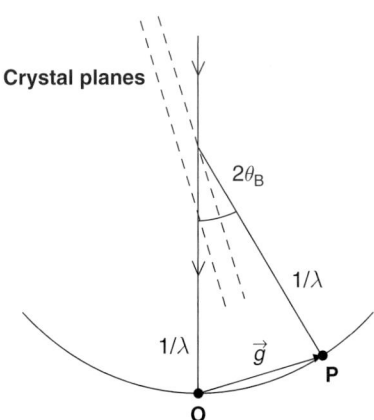

Fig. 3. Diagram showing the Ewald sphere construction. O and P are points in the reciprocal lattice.

The diffraction phenomenon can also be visualized as the reflection by the set of (hkl) crystal planes with interplanar space d_{hkl}. Setting θ_B as the semi-angle between $\vec{k}$ and $\vec{k}'$, by simple geometrical considerations, it can be written as $|\vec{h}| = 2\sin\theta_B/\lambda$. Therefore, the Ewald condition gives the well-known Bragg's law:

$$2d_{hkl}\sin\theta_B = \lambda \tag{7}$$

For the properties of the reciprocal space, the vector $\vec{g}$ is perpendicular to the diffracting planes (hkl) having magnitude $|\vec{g}| = 1/d_{(hkl)}$.

The Bragg condition is derived for an infinite crystal. The size effect of the specimen allows a tolerance around the Bragg peaks so that the reciprocal lattice points are no longer infinitely sharp, having a shape that is modulated by the Fourier transform of the finite crystal.

To interpret the SAD patterns, it is thus convenient to introduce the deviation parameter $\vec{s}$, which is a measure of how far the diffraction vector deviates from the exact Bragg condition, $\vec{h} = \vec{g} + \vec{s}$.

The specimens used in TEM observations are thin foils whose dimensions are infinite in the plane perpendicular to the beam but finite with thickness t in the direction parallel to the beam, z. The amplitude of the diffracted wave given by Eq. (4) can now be written as

$$A(\vec{h}) = \sum_{\vec{h}} F_{\vec{h}} \delta(\vec{s}_x) \delta(\vec{s}_y) \frac{\sin \pi N_z s_z}{\pi N_z s_z} \tag{8}$$

where N_z is the number of unit cells along the incident beam direction.

The modified Ewald sphere construction which takes into account for the crystal shape effect is shown in Fig. 4. The reciprocal lattice points are now elongated along the z-direction to form "relrods" perpendicular to the sample surface, whose length is about $2/t$.

Furthermore, for the accelerated electrons in a TEM, the Ewald sphere can be considered flat for the reflections of main interest, the radius being (40 Å^{-1} for 200 keV electrons) much larger than the typical reciprocal lattice spacing (0.5 Å^{-1}).

From the above considerations, one can approximate the SAD pattern from a single crystal by a magnified view of a planar section of the reciprocal lattice perpendicular to the incident beam direction, each spot corresponding to a diffracting plane in the real space. For a given

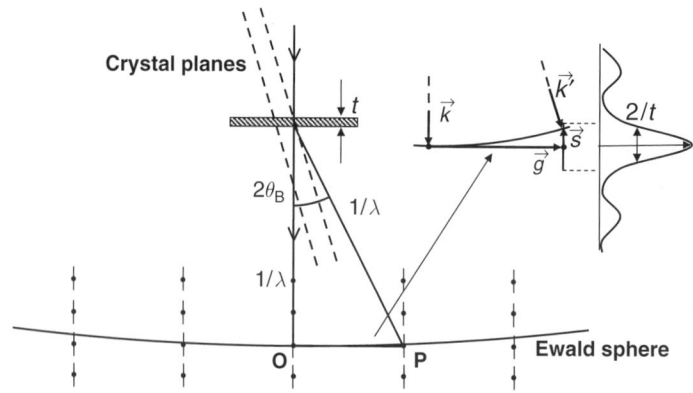

Fig. 4. Deviation from the Bragg condition due to the crystal shape along the direction parallel to the beam.

SAD pattern, the direction common to all the reflecting planes is defined zone axis (ZA), that is, the vector perpendicular to all the $\vec{g}$ vectors in the diffraction plane.

The SAD pattern analysis is an important tool in the characterization of crystalline materials. It allows identifying the crystal structure and phases, the orientation relationships between phases, the long-range ordering in crystal structures and the presence and distribution of particular defects.

3.2. TEM imaging mode: the diffraction contrast

Image contrast in the TEM arises primarily because different regions of a specimen diffract the electron beam in a different way. The scattering can change both the amplitude and the phase of the electron beam.

On the basis of the mechanisms producing contrast, there are two basic operating modes for TEM imaging, namely diffraction contrast that is related to changes in the amplitude of electron beams and phase contrast that is the result of the change in their relative phase.

To obtain images in diffraction contrast, a small objective aperture is inserted into the back focal plane of the objective lens allowing either the transmitted beam or a diffracted beam to contribute to the image formation. These imaging modes are called BF and DF, respectively, and correspond to magnified maps of the intensity distribution of the transmitted selected beam on the exit surface of the thin crystalline sample. For the main properties of the image contrast interpretation, it is assumed that only two beams contribute to the diffraction processes, that is the specimen is supposed to be tilted so that only one strong diffracted beam is excited, other than the transmitted one.

The simplest way to predict the image contrast is to consider the propagation of the incident and diffracted waves by neglecting any coupling between the two beams, in the so-called kinematical approximation [1,3]. The near-field optics approach should be applied in this case as the image is considered to form at a point P for which $\vec{r}$ is comparable with the object dimensions. In this derivation, the crystal is split into slices parallel to the surface and, for an incident plane wave, the contribution to the scattered beam with wave vector $\vec{k}'$ from a slice of thickness dz at a position $\vec{r}$ is given by

$$\frac{d\Psi_g}{dz} = \frac{i\pi}{\xi_g} \exp(2\pi i \vec{u} \cdot \vec{r}) \tag{9}$$

where ξ_g, termed the extinction distance parameter for the particular reflection $\vec{g}$ is given by

$$\xi_g = \frac{\pi \Omega \cos \theta_g}{\lambda F_g} \tag{10}$$

Ω being the volume of the crystal unit cell.

Due to the small Bragg angles in electron diffraction ($\sim 10^{-3}$ rad), the amplitude can be considered to vary only along the propagation direction z, with x and y being parameters and not integration variables [4]. The intensity at a point on the crystal bottom surface is essentially determined by diffraction events occurring in a narrow column parallel to z (column approximation), and for a thickness t it can be obtained for the scattered amplitude:

$$\Psi_g = \frac{i\pi}{\xi_g} \int_0^t \exp(2\pi i \vec{u} \cdot \vec{r}) \, dz \simeq \frac{i\pi}{\xi_g} \int_0^t \exp(2\pi i s_z z) \, dz \tag{11}$$

The scattered intensity is

$$I_g = \Psi_g \Psi_g^* = \frac{\pi^2}{\xi_g^2} \frac{\sin^2 \pi t s_g}{(\pi s_g)^2} \quad (12)$$

The above intensity is valid under kinematical approximations, when the intensity I_g of the diffracted beam is small and the decrease of the transmitted beam intensity $I_0 = 1 - I_g$ is negligible.

The agreement with experimental observations is limited to very thin samples or to reflections with large values of s_g. If the Bragg condition is exactly satisfied ($\vec{s}_g = 0$), the intensity given by Eq. (12) increases as $(\pi t/\xi_g)^2$, and the condition $I_g \ll 1$ is fulfilled only if the thickness is so small that there are not enough scattering centres to build up an important diffracted beam ($t < \xi_g/10$). If $\vec{s}_g \neq 0$, the intensity I_g oscillates with increasing t and the condition $I_g \ll 1$ can be satisfied if $s_g \gg 1/\xi_g$.

For large sample thickness in comparison with the extinction distance, or for diffracted beam with low s_g values, the intensity of the diffracted beams is not negligible in comparison to the transmitted one, kinematical theory is no longer valid and the following dynamical effects must be taken into account:

- the intensity of the diffracted beam may be comparable with the intensity of the transmitted beam
- electrons may be re-diffracted back from the diffracted into the direct beam
- the electrons are absorbed while penetrating the specimen.

Different formulations of the dynamical theory, based on wave-mechanical (Bloch wave) and wave-optical approaches, were developed [1,3] (see also Appendix).

Although the different approaches are basically equivalent for the calculation of the image contrast, each of them gives different insights into the diffraction process, allowing different convenient descriptions in particular experimental situations.

For conventional TEM imaging, in the two-beam condition, it is usually more convenient to make use of the formulation developed by Howie and Whelan [3]. Using the Fresnel optics theory, they derived a set of differential equations describing the coupling between the incident wave of amplitude Ψ_0 and a scattered wave of amplitude Ψ_g passing through a layer of thickness dz inside the crystal.

In contrast with the kinematical theory, where the amplitude of the incident beam is considered constant, it is assumed that the amplitude Ψ_0 is depleted by $d\Psi_0$ by diffraction and the amplitude Ψ_g is correspondingly increased by an amount $d\Psi_g$. These changes can be calculated using the wave Fresnel optics approach in the column approximation, resulting in a system of linear equations called the Howie–Whelan equations:

$$\begin{cases} \dfrac{d\Psi_0}{dz} = \dfrac{\pi i}{\xi_0}\Psi_0 + \dfrac{\pi i}{\xi_g}\Psi_g \exp(2\pi i s z) \\ \dfrac{d\Psi_g}{dz} = \dfrac{\pi i}{\xi_0}\Psi g + \dfrac{\pi i}{\xi_g}\Psi_0 \exp(2\pi i s z) \end{cases} \quad (13)$$

where the first term in the equations takes into account the amplitude changes due to refractive phenomena while the second term represents the Bragg scattering between the two beams.

TEM techniques for imaging and compositional evaluation

Equations (13) conserve the total intensity, i.e. $(d/dz)\left(\Psi_0\Psi_0^* + \Psi_g\Psi_g^*\right) = 0$, thus predicting that BF and DF images will be mutually complementary.

The solution for the intensity of the diffracted beam, corresponding to DF image, is given by [2,3]

$$I_g = 1 - I_0 = \left(\frac{\pi}{\xi_g}\right)^2 \frac{\sin^2(\pi t s_{\text{eff}})}{(\pi s_{\text{eff}})^2} \qquad (14)$$

where $s_{\text{eff}} = \sqrt{s^2 + 1/\xi_g^2}$ is a deviation parameter that reduces to s in the kinematical limit, that is when $s_g \gg 1/\xi_g$.

The transmitted and diffracted intensity periodically change with the two independent variables t and s_{eff}.

1. The periodic variation with the crystal thickness, which is shown in Fig. 5(a) in the case of Bragg condition $(\vec{s}_g = 0)$, gives rise to the formation of thickness fringes in wedge crystal also predicted by the kinematical theory but only for $s \neq 0$. The distance at which the diffracted beam falls to zero at the Bragg condition corresponds to ξ_g, hence, the term extinction distance already defined in Eq. (12). The period of the thickness fringes in a TEM image decreases if the specimen is tilted out of the Bragg condition for the operative reflection, with $\xi_{g\text{eff}} = \xi_g/\left(1 + s^2\xi_g^2\right)^2$ for $\vec{s}_g \neq 0$. Experimentally, the intensity of the thickness contours is observed to rapidly damp with increasing crystal thickness. The dynamical theory of image contrast accounts for this evidence if absorption effects are incorporated into the two beam equations, by replacing the parameters $1/\xi_0$ and $1/\xi_g$ by complex quantities (Fig. 5(b)).

2. The dependence of the transmitted and diffracted beams on s_{eff} is experimentally confirmed by the observation of bend contours which correspond to geometrical loci with the same s_{eff} for the operative reflection. This phenomenon occurs if the planes under diffraction are not parallel everywhere, that is when the specimen is elastically bent close to an exact Bragg position.

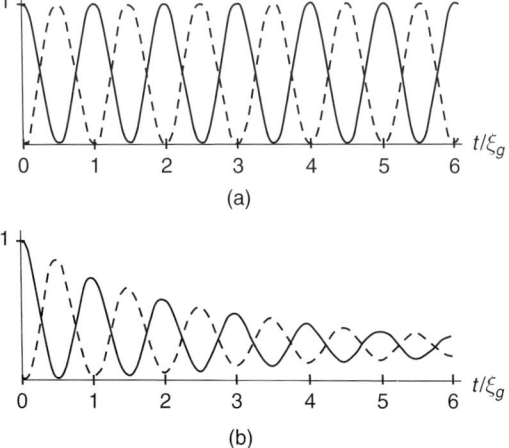

Fig. 5. Calculated transmitted (continuous line) and diffracted intensity (dashed line) along the specimen thickness in the case of (a) no absorption and (b) absorption.

3.3. Defect imaging

The theory of image contrast for a perfect crystal can be easily modified to account for the presence of defects. Deformations in crystals can be represented as the vector field $\vec{R} = \vec{R}(x,y,z)$ representing the displacement field of the unit cell from its proper position in the perfect lattice. Introducing $\vec{R}$ in the Howie–Whelan equations leads to:

$$\begin{cases} \dfrac{d\Psi_0}{dz} = \dfrac{\pi i}{\xi_0}\Psi_0 + \dfrac{\pi i}{\xi_g}\Psi_g \exp 2\pi i\left(sz + \vec{g}\cdot\vec{R}\right) \\ \dfrac{d\Psi_g}{dz} = \dfrac{\pi i}{\xi_0}\Psi g + \dfrac{\pi i}{\xi_g}\Psi_0 \exp 2\pi i\left(sz + \vec{g}\cdot\vec{R}\right) \end{cases} \quad (15)$$

The effect of a local lattice displacement on image contrast is thus equivalent to superimpose the phase factor $\exp(i\alpha)$ to the normal scattering process in a perfect crystal, with $\alpha = 2\pi\vec{g}\cdot\vec{R}$.

Different crystal defects correspond therefore to different image contrast depending on the values of the phase factor, provided $\alpha \neq 0, 2\pi n$.

If the displacement vector $\vec{R}$ is perpendicular to the operative reflection $\vec{g}$, the reflection planes are not disturbed by the defect and no contrast is produced. Such invisibility criterion is used to determine the exact nature of defects through the analysis of the phase factor $\alpha = 2\pi\vec{g}\cdot\vec{R}$ under different reflection vectors $\vec{g}$.

In the following, typical examples of the diffraction contrast expected from planar defects, dislocations and volume defects are shown and discussed.

The simplest type of planar defects is represented by the stacking faults, which are very commonly observed in close packed structures [5]. A stacking fault divides the crystal into two regions identical in spacing and orientation but translated with each other by a constant vector $\vec{R}$, as shown in Fig. 6. This planar defect can be qualitatively viewed as the boundary between two wedge-shaped crystals which are in direct contact, but with a constant displacement $\vec{R}$ along the wedge. The fringe patterns caused by the two wedges do not fit together due to the displacement, resulting in a new fringe system extending over the region of the projected stacking fault (Fig. 7).

An analytical expression for the diffracted intensity by the faulted crystal represented in Fig. 6 can be derived by substituting the constant translation vector $\vec{R}$ in Eqs (15):

$$I_g \propto \frac{1}{s^2}[A - B\cos(2\pi st')] \quad (16)$$

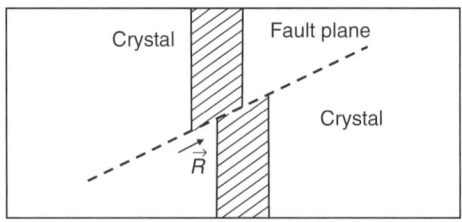

Fig. 6. Column displacement along a fault plane.

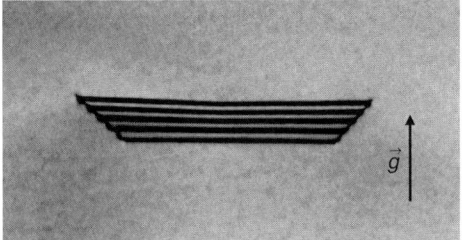

Fig. 7. Contrast from a stacking fault in GaAs. The fringes extend over the projection of the stacking fault on the plane of observation.

where t' is the defect depth in the thin foil, whereas A and B are constants depending on the value of the translation vector. The predicted contrast pattern from an inclined stacking fault is thus a series of cosine dark and bright fringes with period s^{-1} and intensity varying as s^{-2}, running parallel to the intersection line between the foil surface and the fault. Similar fringe contrast is obtained for other planar defects like twin boundaries, grain boundaries and antiphase boundaries. However, complex contrast analysis is required for an accurate defect identification, and the simulation of diffraction contrast image is often necessary for the characterization of real defect configurations [6,7].

Very different displacement field is associated to dislocations. In contrast to planar defects which are only characterized by a rigid shift corresponding to an abrupt change of $\vec{R}$ occurring through the defect, a dislocation is characterized by a continuous bending of lattice planes close to the defect line. Dislocations give rise to contrast because they locally distort the lattice and change the diffraction conditions, that is the effective value of $\vec{s}$. This effect can be understood by the inspection of Fig. 8 showing the bending of the reflecting planes around an edge dislocation and the qualitative effect it produces on an image. It is assumed that far from the dislocation line, the sample is slightly tilted away from the exact Bragg condition ($s > 0$), corresponding to a high-intensity background in BF image. The lattice distortion near the defect bends the reflecting planes towards perfect Bragg orientation ($s = 0$), thus increasing the intensity of the diffracted beam at the expense of the transmitted beam. BF image in the two-beam approximation will show a dark line on a bright background. Contrast analysis observations, using the invisibility criterion, enable the dislocation nature to be determined.

Dislocations are classified according to the relative orientation between the dislocation line and their Burgers vector, $\vec{b}$, which defines the lattice displacement around the defect line [5]. Special cases regard screw and edge dislocations for which $\vec{b}$ is respectively parallel and perpendicular to the dislocation direction, $\vec{u}$. The shape of the elastic field in proximity of a dislocation is quite complex but in the case of a screw dislocations the displacement field $\vec{R}$ lies along its Burgers vector. Therefore, the invisibility criterion reduces to $\vec{g} \cdot \vec{b} = 0$, thus meaning that the dislocation shows no contrast when the operation vector is normal to the dislocation line as in this case the diffracting planes are not bent by the dislocation (Fig. 9). For edge dislocations, similar geometrical considerations lead to the simultaneous realization of the conditions $\vec{g} \cdot \vec{b} = 0$ and $\vec{g} \cdot \vec{b} \wedge \vec{u} = 0$ for the dislocation to be invisible. Finally, a residual contrast is always present for mixed dislocations, as found by calculating the displacement they generate using the isotropic elasticity theory. The application of invisibility criterion to a real case is shown in Fig. 10.

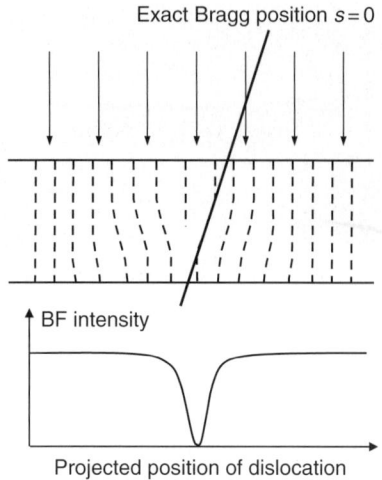

Fig. 8. Distortion of lattice planes around an edge dislocation and the schematic intensity profile of the transmitted beam along the projected position of the defect.

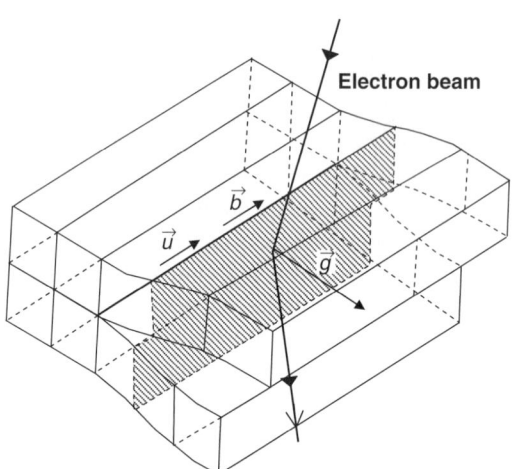

Fig. 9. Schematic diagram illustrating the distortion of lattice planes around a screw dislocation. The planes containing the dislocation direction $\vec{u}$ are not bent by the defect and diffract the electrons as if the dislocation was not present. The invisibility condition $\vec{g} \cdot \vec{b} = 0$ is satisfied.

Volume defects arising from the coherent inclusion of particles, having different structures, lattice parameters or composition, are eventually considered because the contrast features they generate are very different from the previous cases. This kind of contrast can be understood in the special case of a spherical particle, in case of continuity of lattice planes across the particle–matrix interface. Because the displacement field $\vec{R}$ can be modelled with a radial field, from the invisibility criterion it immediately follows that the corresponding contrast feature will show a line of zero contrast perpendicular to $\vec{g}$, as shown in Fig. 11.

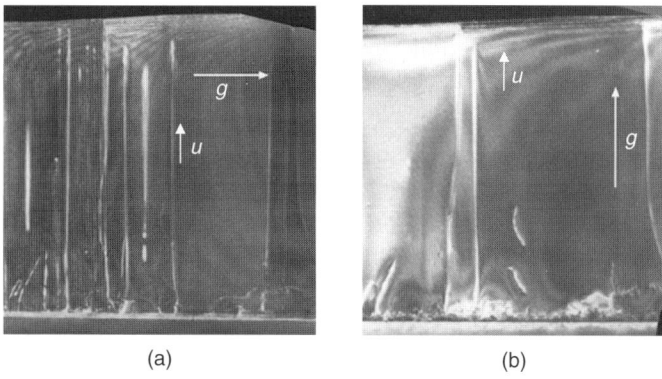

Fig. 10. The application of invisibility criterion in a GaN layer epitaxially grown on sapphire. In (a), the screw dislocations having $\vec{b}$ parallel to $\vec{u}$ are out of contrast while in (b), the edge dislocations with $\vec{b}$ perpendicular to $\vec{u}$ are invisible.

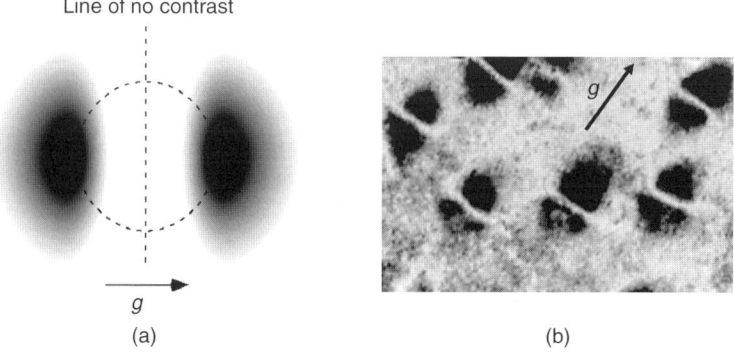

Fig. 11. (a) Calculated contrast from a coherent spherical volume particle. (b) DF image showing the observed contrast from coherent InAs quantum dots in a GaAs matrix.

3.4. The chemical-sensitive imaging

The chemical composition of the specimen can be another very important and useful contribution to the imaging contrast. The chemical dependence is essentially taken into account in the extinction distance parameter (Eq. (10)). It is intuitive that materials with different scattering factors and crystalline potentials can give rise to diffracted beams with very different intensity but many other parameters (specimen thickness, misorientation, etc.) influence this intensity. Therefore, in general, it is not easy to extract quantitative information on the composition simply by intensity measurements. It will be shown that some crystals, due to their specific lattice arrangement, select "chemical-sensitive" Bragg reflections that allow compositional investigation of the specimens.

The fundamental quantity to be considered is the structure factor F (Eq. (5)), which depends not only on the Bravais lattice, but also on the position and nature of the atoms composing the basis. The crystal structure imposes certain selection rules that determine which beams are allowed. In the following, we will treat the zinc blende lattice which is typical of many III–V semiconducting compounds (GaAs, InP and related alloys).

The zinc blende is an fcc crystal where the cation is located at (0,0,0) and the anion at (1/4,1/4,1/4). The expression for F now becomes

$$F = 4\left[f_{III} + f_V e^{i(h+k+l)(\pi/2)}\right] \quad (17)$$

This gives rise to the following selection rules:

1. $F = 0$ if h, k, l are mixed odd and even
2. $F = 4(f_{III} \pm if_V)$ if h, k, l are all odd
3. $F = 4(f_{III} - f_V)$ if h, k, l are all even and $h+k+l = 4(n+1/2)$
4. $F = 4(f_{III} + f_V)$ if h, k, l are all even and $h+k+l = 4n$

In the two-beam approximation, images are formed with only one beam (direct or diffracted) and the intensities are proportional to $|F|^2$. Case 3 is particularly interesting because the DF intensity in the kinematical approximation can be written as

$$I \propto (f_{III} - f_V)^2 \quad (18)$$

and can be very large or even very small, depending on the difference between the individual scattering factors of atoms a and b. In the case of Si or Ge diamond lattice, $f_{III} = f_V$ and these reflections are forbidden. In this sense, the reflections selected by this rule are called "chemical sensitive" and the most commonly used are of the type (002) in the DF imaging mode. Figure 12a reports the [110] projection SAD pattern of the InGaP/GaAs superlattice

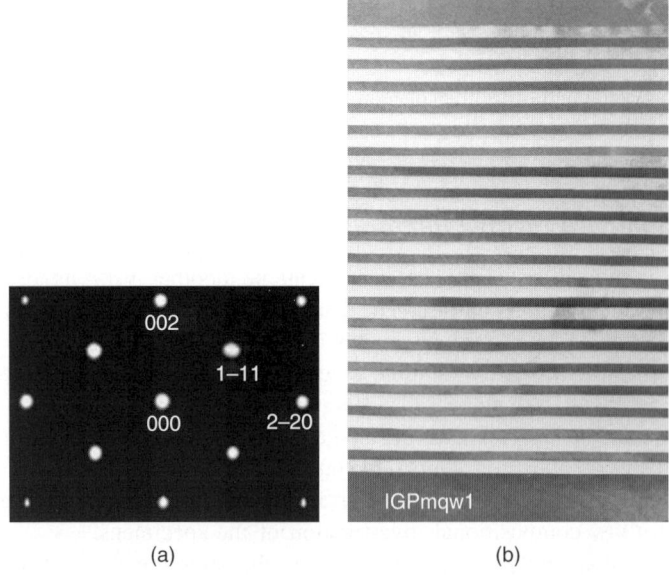

Fig. 12. (a) Diffraction pattern (011 projection) of the sample; (b) 011 cross-sectional image of a lattice-matched InGaP/GaAs MQW structure.

structure of Fig. 12(b), which shows a (002) DF image. It is obtained by appropriately tilting the sample until the direct and a (002) type reflection only are illuminated and then selecting the (002) beam: InGaP appears brighter than GaAs, as As has been replaced with the lighter P and a fraction of Ga with the heavier In, thus increasing the difference $f_{III}-f_V$ with respect to GaAs.

The quantitative evaluation of these intensities is in general made difficult by many factors like thickness, strain and dynamical effects and requires simulation procedures, the details of which are beyond the scope of this book. We mention here that, in some systems like InGaAs or GaAsN, the range of thickness and In or N content for which the kinematical approximated expression of Eq. (18) is valid is quite large. Under these conditions, the contrast between areas with different compositions is independent of thickness and can be used to evaluate the alloy concentrations when the composition of a nearby reference layer is known. Some results on specific materials systems will be illustrated in Section 6.

4. Phase contrast

4.1. Principles of the phase contrast

The above treatment of the diffraction contrast has shown that much information is contained in the single diffracted beam amplitude. A different kind of information can be obtained from a microscope using a larger objective aperture and permitting to the transmitted and the diffracted beams to interfere. For the interferometric nature of this contrast, these imaging conditions are often referred to as phase contrast as they are sensible to phase variations among the beams. Unfortunately, in general, the phase and amplitude information are entangled in a complicated mix in an interferometric image. The principle of "phase contrast" imaging is better clarified in the special but instructive case of "weak phase condition" [8] that will be illustrated hereafter. A more general discussion on many beam imaging condition, giving rise to the high-resolution transmission electron microscopy (HRTEM), will be discussed in the next section.

One refers to "weak phase conditions" when a small variation of the phase of the electron wave is the sole effect of the interaction with the sample. This condition is to a good extent realized only for very thin samples made of light elements. A simple evaluation of the phase factor introduced by the interaction with the specimen can be performed considering the difference between the wave vector in the vacuum $K_0 = \sqrt{2mE}/h$ and its value in the medium $K_m = \sqrt{2m[E - eV(r)]}/h$, where m is the electron mass, E is the beam energy, V is the potential and h is the Planck's constant. The phase difference $d\alpha$ accumulated after a thickness dz can be then written as

$$d\alpha(r) = (K_0 - K_m)\,dz \approx \frac{K_0 e}{2E} V(r)\,dz \tag{19}$$

or, for a finite thickness and considering the in plane coordinates $R = (x, y)$ and introducing the z-projection of the potential $V(R)$, the net phase accumulated by the wave is

$$\alpha(R) \approx \frac{K_0 e}{2E} V(R) z \tag{20}$$

As a consequence of Eq. (20), the beam phase variation depends on the position and is peaked at the atomic columns. For a crystal, in particular, the interaction gives rise to discrete beams as discussed above (see Section 3) whose phase is well described by the Fourier transform

$$\tilde{\alpha}(g) \approx \frac{K_0 e}{2E} \tilde{V}(g) z \qquad (21)$$

where g is the conjugated coordinate of R while $\tilde{V}(g)$ is the Fourier component of the potential $V(R)$. As the potential is periodic, the values of g are discrete and correspond to the direction of each diffracted beam. The wave function at the exit of the specimen is then described as

$$\Psi_t = \exp(i\alpha) = 1 + i\sigma V(R) + \cdots \qquad (22)$$

where $\sigma = eK_0/2E$ and the approximation to the first two terms is valid only for weak phase objects.

This sort of ideal specimen allows to study the response of optical system of the TEM in phase contrast condition. It would be desirable to be able to produce a direct image of the atomic potential with the highest resolution. It will be explained hereafter under what conditions and to what extent this is possible.

The first step necessary to produce a model describing the response of an optical system is to consider the properties of an ideal lens. Its properties, as in light optics, are to conjugate each direction of a beam in the object plane to a point in the lens focal plane (this property is at the base of the DP imaging), and to conjugate each position in the object plane to a position in the image plane. Mathematically, the first correspondence is described by the Fourier transform; the passage to the image plane is then described by an inverse Fourier transform operation. This means that the wave function Ψ_i in the image plane is, for an ideal lens,

$$\Psi_i = \mathrm{FT}^{-1}[\mathrm{FT}(\Psi_t)] \qquad (23)$$

where FT and FT^{-1} are the direct and inverse two-dimensional Fourier transform in the in-plane x, y coordinates. Because the detected intensity is just $|\Psi_i|^2$, it turns out that, for a weak phase object, the intensity is identically equal to 1. Therefore, an ideal optical system is not able to convey the phase information.

However, the presence of lens aberrations deeply modify this situation. In particular, two factors need to be considered, the defocus and the spherical aberration. The expression defocus (hereafter Δf) indicates the distance of the sample from the exact height at which the objective lens is on focus, that is, the above-described conjugation conditions are valid. This quantity can be varied either by raising or lowering the specimen or by changing the excitation of the objective lens. In an ideal electro-optical system with $\Delta f = 0$, the different trajectories departing from the object plane converge to a single point in the image plane. Under these conditions, the phase that accumulates in each path is the same but, as one moves above or below the image plane, the trajectories traversing the lens at different angles θ (Fig. 13(a)) are characterized by different phase. This angle-dependent phase shift is [9].

$$\chi_f(\theta) = -\frac{2\pi}{\lambda}\left(\frac{1}{2}\Delta f \theta^2\right) \qquad (24)$$

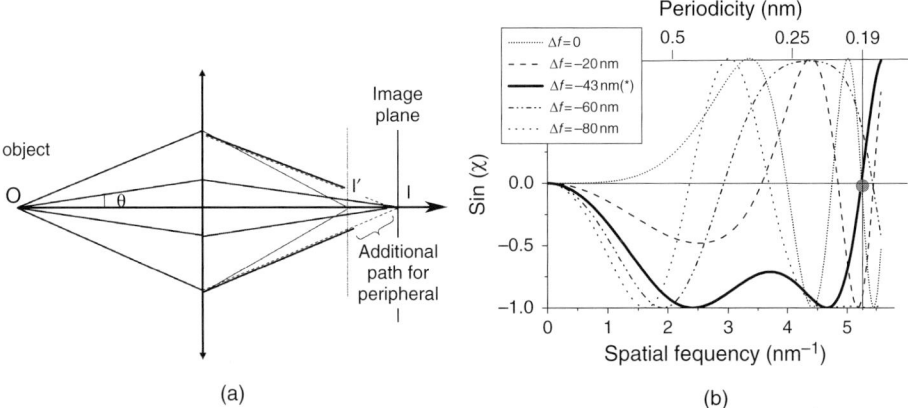

Fig. 13. (a) Schematic representation of an aberration affected optical system. For the rays close to the optical axis, the conjugation occurs in I while at I' trajectories at higher angle. The additional path of the external rays in I are due to spherical aberration. (b) Contrast transfer function as a function of the spatial function for different values of the defocus Δf ($Cs = 0.5$ mm). The Scherzer condition is indicated by an (*). The first zero indicating the resolution of the instrument is evidenced.

where λ is the electron wavelength. However, even at exact focus condition, the non-ideality of the lens causes an additional phase shift for the trajectories forming larger angles with the optical axis. This phase factor is equal to

$$\chi_s(\theta) = \frac{2\pi}{\lambda}\left(\frac{1}{4}C_s\theta^4\right) \qquad (25)$$

If k is an in-plane spatial frequency, i.e. a conjugated variable of the in-plane position vector R in the object plane, $\theta = \lambda k$ for the trajectories close to the optical axis. The overall phase shift can then be written as

$$\chi(k) = \pi\lambda k^2\left(0.5 C_s \lambda^2 k^2 - \Delta f\right) \qquad (26)$$

Considering such a phase factor, each Fourier coefficient of the wave function is modified by a factor

$$T(k) = A(k)\exp i\chi(k) \qquad (27)$$

where the top-hat function $A(k)$ describes the objective aperture limiting the maximum spatial frequency k_{max} ($A(k) = 1$ for $k < k_{max}$ and 0 otherwise).

Therefore it is found that

$$\Psi_i = FT^{-1}[T(k)\cdot FT(\Psi_t)] = h(r)\otimes\Psi_t(r) \qquad (28)$$

with $h = FT^{-1}(T)$. This leads to the following general expression for the final intensity:

$$I = |h(r)\otimes\Psi_t(r)|^2 \qquad (29)$$

At this point, the recourse to the weak phase approximation allows to simplify this expression. Using Eqs (22) and (29) yields

$$\begin{aligned} I = |1 \otimes h(r) + i\sigma V \otimes h(r)|^2 + \cdots &\approx 1 + 2\sigma V \\ \otimes [ih(r) - ih^*(r)] + \cdots &= 1 + 2\sigma V \otimes [h_{\text{WP}}(r)] \end{aligned} \qquad (30)$$

where the * denotes the complex conjugation and $h_{\text{WP}}(r) = ih(r) - ih^*(r)$ is the point spread function for the weak phase object. If this expression assumes real values different from 0, it is possible to have an image of the potential. More explicitly, the convolution can be written in the Fourier space where the Fourier transform H_{WP} is simply $H_{\text{WP}} = A(k)\sin(\chi(k))$. The image Fourier transform is then written as

$$\tilde{I}(k) = \delta(k) + 2\sigma\tilde{V}(k)H_{\text{WP}}(k) \qquad (31)$$

This means that, depending of the values of Δf and C_s, different spatial frequencies of the atomic potential are actually transferred in the final image.

The optimal condition has been established by Scherzer [10] by setting $\Delta f = \sqrt{1.5 C_s \lambda}$. Figure 13(b) shows, for example, the transfer function for the Scherzer conditions, $C_s = 0.5$ mm and an accelerating voltage of 200 kV. The H_{WP} for different defoci are also shown for comparison. It is apparent that some spatial frequencies are better imaged for defoci different from optimal. The resulting images, however, may be difficult to interpret if the low-frequency components, containing most of the more intuitive information on the structure, are not well transferred. In this sense, the Scherzer condition represents a good compromise between resolution and interpretability. The first zero of the transferred function at the Scherzer condition defines the microscope interpretable resolution. This corresponds to a real space distance of $d_s = 0.64(C_s \lambda^3)^{1/4}$: in the present case, this corresponds to $d_s = 0.19$ nm.

4.2. High-resolution transmission electron microscopy

The previous discussion also justifies why this technique is called HRTEM, as this working mode allows the highest resolution in a TEM.

Unfortunately, almost all the objects we are interested in cannot be described as "weak phase objects". Real object imaging violates the above treatment for two reasons. One is inherent to the imaging system below the sample: for a strong beam–specimen interaction the expression (29) cannot be simplified by an approximate linearization [8]. Due to the non-linearity of this expression, the "exit wave function" below the sample has a complicated relation with the final intensity.

However, even if the exit wave function were precisely known, a second difficulty arises as the exit wave function cannot be directly interpreted in terms of crystal potential. This is due to the character of the electron–specimen interaction: the electron beam undergoes a continuous series of scattering events that completely modify its shape [11]. In other words, the exit wave formation cannot be traced back to a single scattering event and the shape of the potential cannot be easily extracted.

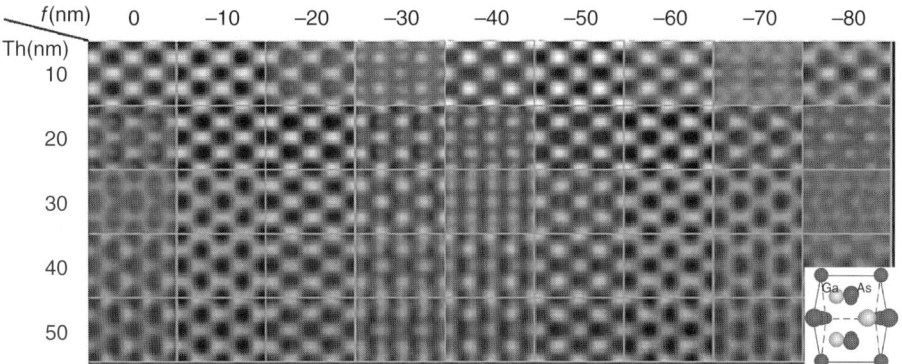

Fig. 14. Simulation of HRTEM images for different values of defocus and thickness for GaAs in the [110] zone axis ($C_s = 0.5$ mm).

As a consequence, HRTEM images show a complex dependence on both the specimen thickness and defocus condition. Figure 14 is a tableau of simulated HRTEM images of GaAs for a beam at the [110] direction for various sample thickness and defocus conditions [12,13].

In particular, Fig. 14 shows that the maxima of intensity do not always coincide with atomic column positions: conversely, in these positions, minima and maxima alternate by varying thickness and defocus. All the images in the pattern have however in common a fact of non-negligible importance. The basic spatial distance between two repeated features is the same as in the projected crystal: somehow this information is maintained throughout the image formation process.

In particular, in HRTEM imaging mode the crystal is typically oriented in a low index orientation ZA [7]. This is in a highly symmetric condition: a low-index ZA is characterized by the alignment of the atomic columns along the beam direction. Such alignment favours the concentration of a consistent fraction of the wave function along the atomic columns. This phenomenon is sometimes referred to as "channelling". As an effect of the channelling, the exit wave function will be characterized by a strong atomic contrast with the same periodicity of the crystal.

To understand the channelling, a useful approximation has been introduced by Van Dyck [14] in the case of the exact ZA orientation. He assumed that the fast electron of the beam, in the direction of propagation, behaves as a classical electron with velocity

$$v = \frac{z}{t} = \frac{\hbar K_0}{m} \qquad (32)$$

The time-dependent Schrödinger in the x, y plane simplifies as

$$-\frac{\hbar}{i} \frac{\partial}{\partial t} \Psi(R, t) = H_R \Psi(R, t) \qquad (33)$$

with

$$H_R = -\frac{\hbar^2}{2m} \Delta_R - eV(R, t) \qquad (34)$$

where Δ_R is the Laplacian operator acting in the plane coordinate $R=(x,y)$ perpendicular to z. From expressions (32)–(34) one obtains

$$\frac{\partial}{\partial z}\Psi(R,z) = \frac{i}{4\pi k}\left(\Delta_R + \frac{2me}{\hbar^2}V(R,z)\right)\Psi(R,z) \qquad (35)$$

The in-plane propagation is completely separable from the z propagation and the projected potential $V(R)$ can be used instead of $V(R,z)$. The in-plane propagation component is therefore ruled by the special Hamiltonian with eigenvalues E_n (also called transverse energy) and eigenstates ϕ_n. Interestingly, this Hamiltonian allows for a few two-dimensional bound states. The physical meaning of a bound state is a lateral confinement of the beam propagation on atomic column, which is just the channelling.

It can be demonstrated that these states are the dominating electron scattering at moderate thickness. To understand this, the overall solution $\Psi(R,z)$ in terms of all ϕ_n eigenstates can be written as

$$\Psi(R,z) = \sum_n C_n \phi_n \exp\left(-i\pi\frac{E_n}{E_0}kz\right) \qquad (36)$$

where E_0 is the electron total energy. The coefficients C_n can be determined using the boundary condition of continuity on the upper surface. After a bit of algebra, the same expression can be rewritten as

$$\Psi(R,z) = 1 + i\pi\frac{eV(R)}{E_0}k_0 z + \sum_n C_n \phi_n \left[\exp\left(-i\pi\frac{E_n}{E_0}k_0 z\right) - 1 + i\pi\frac{E_n}{E_0}\right] \qquad (37)$$

In practice, the first two terms are the weak phase approximation while the third term accounts for the rest of the beam propagation. This term becomes important for thicker samples. Indeed this factor is 0 if the exponential is developed at the first order and only those states for which $|E_n| \geq E_0/K_0 z$ will significantly contribute. Moreover, this condition favours the bound states as these are usually characterized by a large negative transverse energy. This is at the origin of the spatial localization of the information in HRTEM and of the possibility to transfer the crystal periodicity to the image.

The information on the lattice periodicity is also conserved in the imaging process performed by the lens system below the sample: the non-linear imaging expressed by Eq. (29) maintains the periodicity of the exit wave function but adds some spurious ones. The HRTEM image can indeed be considered as the sum of the linear contribution as discussed in the previous section plus a non-linear part [8,15].

The former derives from the interference of the diffracted beams with the transmitted beam, and the latter is the effect of the interference of the diffracted beams with each other. When varying defocus or thickness, the two contributions vary their sign and their relative importance giving rise to the periodical image inversions and to different HRTEM patterns [16] as seen in Fig. 14.

For this reason, to interpret an HRTEM image in terms of the crystal structure, it is necessary to make large use of image simulations. In many cases a trial and error or a fitting procedure is used. Imaging parameters like defocus, sample thickness as well as structural parameters like the local composition are systematically varied to obtain the best match to experimental images.

In some favourable cases, from some features of an image, it is possible to directly deduce important parameters and, especially, the chemical composition. This is the case of the spherelite structure for which the chemical sensibility of the {200} reflections can be exploited to derive the chemical information. Obtaining an interference image of the sole {200} and transmitted beam and with the aid of computer analysis, Rosenauer et al. [17] were able to determine directly both the sample thickness and local composition in an alloy.

Finally, a recently developed technique has made it possible to retrieve directly the exit wave function [18]. This is usually achieved by systematically varying defocus and consequently the transfer function. From each acquired image, it is possible to get the piece of information better transferred for the given defocus. All the pieces are then recombined together to obtain the exit wave function. Some extra calculations are, however, necessary to remove the non-linearity effects [19]. The exit wave function can be directly interpreted in terms of the investigated structure, getting rid of the lens system effects. This is also the starting point of new research efforts to retrieve the full crystal structure directly by the exit wave function. However, this subject is beyond the aim of this book.

4.3. *The lattice fringes analysis for strain determination*

If the aim of the experiment is to extract the chemical information, it is sometimes not necessary to calculate the wave function at the exit of the sample. The fact that the periodicity of the phase contrast image is the same as of the crystal is often a piece of information sufficient to infer about the chemical composition in a material. This is especially true in semiconductor heterostructures made by materials with different lattice parameters. The necessity to match the lattice parameters in the plane parallel to the interface directly leads to the formation of strain fields. If the layer with lattice parameter different from the substrate is thick enough, such strain is relieved by the formation of dislocations.

In this context, it is however more interesting to analyse the case where a thin layer of few nanometres of lattice-mismatched material is inserted, along the growth direction z, in a matrix of otherwise uniform material. This is the typical case of the QW structures. From the point of view of the elastic strain field, this situation is highly symmetric as the system is completely space invariant in the x, y direction of the interface plane. In this case, it is easy to demonstrate that the strain is concentrated in the QW region: the QW material unit cell gets deformed in the z direction to compensate the unit cell volume variation it had undergone to adapt to the host material. For example, in the case of a cubic unit cell, the distortion of the unit cell transforms the symmetry from cubic to tetragonal; therefore, this phenomenon is referred to as "tetragonal distortion".

An HRTEM image will show a periodicity variation of the atomic fringes in the z direction when passing from the substrate to the QW material. Applying the classical elasticity theory, the variation in lattice parameters amounts to (see Chapter 4)

$$\frac{a_\perp - a_s}{a_s} = -\frac{1-\nu}{1+\nu}\frac{a_0 - a_s}{a_s} \tag{38}$$

where a_s is the substrate lattice parameter, a_0 and ν are the bulk lattice parameter and the Poisson ratio of the QW material while $a_\perp$ is the new distorted lattice parameter. If the QW material is an alloy of two compounds A and B, the Vegard [20] empirical law states that a_0 can be written as the combination of the two bulk parameters a_A and a_B:

$$a_0 = a_A + x(a_B - a_A) \tag{39}$$

In this case, Eq. (38) states a direct proportionality between the x molar fraction of the alloy and the lattice parameter variation. This is generally used in experiments to extract the chemical composition. In addition, if not specified otherwise, it will be assumed that the alloy extreme compound A is just the same material of the substrate: this results in a further simplification of Eq. (38) to

$$\frac{a_\perp - a_A}{a_A} = x \left[\left(-\frac{1+\nu}{1-\nu} \right) \frac{a_B - a_A}{a_A} \right] \tag{40}$$

In spite of the relative simplicity of the principle on which this method is based, some complication may arise in experiments. The first problem to deal with when using this technique is that the lattice parameters must be determined with an accuracy better than 1%. It may appear as a very challenging problem as long as the ultimate resolution is that of the single image pixel (typically 5–10% of the lattice parameters). Fortunately, a noticeable mathematical property can be used in this context: by means of the centroid technique (averaging of the coordinates of neighbouring positions with weight given by the corresponding intensities) or parabolic fit on intensity profiles around a maximum, the position of a maximum can be determined with a far better accuracy. The additional information from the intensity in the pixels around the maximum is the reason of the improved accuracy: the better the intensities known (i.e., the better is the signal-to-noise ratio), the better is the maximum position known [21].

As a direct application of this principle, it is possible to determine the position of the maxima of the atomic fringes and to extract their distance. This has been done in the literature by different authors [21,22]. In particular, Rosenauer has successfully used this technique to characterize the composition variation in quantum dots [23] and QW [24] systems.

Another interesting approach, we just mention here, is to consider the periodicities in the Fourier space. The method consists in filtering out all the periodicities except one and performing an inverse Fourier transform. The phase component of the resulting complex image gives information about local displacements of atomic planes. This kind of analysis, which has been used with a spectacular success in the analysis of the strain field around a dislocation [25], is slightly less accurate for the measurement of strain fields with abrupt discontinuity as in an ideal QW.

Both analysis techniques have to afford a problem of physical origin that can affect lattice fringes analysis, related to the extreme small thickness in one transverse direction the specimen must have (few tens of nanometres) to be suitable for an HRTEM experiment. The possibility to relax strain in different directions alters the distribution of the strain with increasing deviation from the simple relation of Eq. (40).

Theoretical investigations using classical elasticity theory have revealed that the key parameters is the ratio R between the sample thickness and the QW width [26]. It is easy to see that if $R \gg 1$, more precisely $R \geq 5$, the strain in the QW approaches the perfect one-dimensional case for which Eq. (40) is valid. When R is of the order of 1, the strain profile starts to show an unphysical dip at the centre of the QW that becomes increasingly important for decreasing values of R.

It can be shown that the lattice distortion in a totally relaxed layer can be evaluated as

$$\frac{a_\perp - a_A}{a_A} = x \left[(-(1+\nu)) \frac{a_B - a_A}{a_A} \right] \qquad (41)$$

For typical values of $\nu \sim 0.3$, the reduction of the projected lattice parameter is of 60%. For this reason, it is important to acquire the HRTEM images for lattice fringes analysis in a section of the specimen where it is sufficiently thick. For example, for a 8-nm-wide QW this means a foil thickness of 40 nm that, in a common 200 kV microscope, is not quite the best condition for HRTEM imaging. It is therefore a challenging problem to choose the optimal condition for the fringe spacing measurement.

Finally, an additional problem may arise due to the imperfections of the optical system. Hytch [27] has found that, for example, it is unwise to use tilted condition for lattice fringes analysis due to the objective lens imperfection effects. Moreover, he suggested to consider the distortion effect on the final image due to imperfections of the projector lenses [28]. Projector lens distortions have a minor but not totally negligible effect on distortion maps. Fortunately, by acquiring an HRTEM image in identical lens conditions also in a undistorted area (e.g. the substrate far away from the QW) it is possible to subtract the latter effect. This method has permitted an accuracy in the strain field determination of 0.03 Å [25].

5. Other techniques

Diffraction contrast and HRTEM techniques are only two of the very many possibilities offered by modern microscopes. In particular, the improvements in the electron optics have made it possible to create focused beams with lateral size of the order of magnitude of 1 Å [29]. Scanning transmission electron microscope (STEM) exploits this capability giving chemical information with the same and often better lateral resolution as the HRTEM.

An STEM machine is equipped with a scanning coil that moves the beam on the sample and a series of detector at different positions. For each beam position the signal of each detector is collected, and the final image is obtained assigning to each position the corresponding intensity.

The standard detectors attached to an STEM are localized in the diffraction plane. They usually integrate the intensities arising from angles ranging from a few to hundreds of milliradians. They are mainly distinguished in the BF and DF detectors with a terminology that reminds single beam diffraction contrast. Indeed a BF detector includes the transmitted beam while DF does not [1]. DF detectors are usually annulus shaped with the centre on the optical axis to integrate isotropically in all directions. The best and unrivalled resolution in STEM is obtained using an annular DF detector with a large inner aperture angle [29] being the state-of-the-art resolution 0.7 Å. This technique can be referred to as HAADF (high-angle annular dark field) or simply "Z contrast". This latter definition points out that this technique shows a high sensitivity to the atomic number Z of the elements in the sample [30]. Indeed at high angle the screening effects of the atomic electrons are small: the scattering factor for each atom is close to that of the pure Rutherford (or Mott) scattering. Moreover, it can be demonstrated that, under these conditions, each atom scatters almost incoherently from the other [31]. As a result, the overall intensity on a HAADF detector, at least for thin samples, is simply proportional to Z^α with $\alpha \sim 2$. The incoherent nature of Z contrast has also an additional effect: in contrast to diffraction contrast intensity, the Z-contrast intensity is less affected by small distortion of the crystal lattice [32].

These characteristics make this technique very close to an ideal mass thickness mapping of specimens. It is therefore a very useful technique to characterize interfaces and compositional variation in nanostructures.

In conclusion, HAADF allows atomic resolution images of crystalline samples as HRTEM. The reason for this high spatial resolution are, again, to be found in the nature of the beam–sample interaction. As in the case of HRTEM, the beam can couple with in-plane bound states of the two-dimensional crystal potential and undergo channelling [33,34]. Moreover, the large integration over different directions, typical of HAADF, introduces a large indetermination in the diffraction direction. In quantum mechanics, an indetermination in momentum is repaid by a better spatial localization and, therefore, an enhancement of channelling effects [35]. As a result, images show the characteristic fringes typical of the projection of the crystal structure.

Figure 15 shows a simulation for a through defocus, thickness series analogous to that shown in Fig. 14 for HRTEM. The main difference with HRTEM is that, due to the incoherent image formation, the atomic fringes do not undergo contrast reversal: the intensity maxima always correspond to atomic column position. For this reason, HAADF provides unambiguous information on the atomic positions in extended defects and interfaces [30].

HAADF is often combined with other techniques among which it is worth citing EELS (electron energy loss spectroscopy) and EDX (energy-dispersive X-ray spectroscopy). A typical HAADF + EELS set-up is shown schematically in Fig. 16(a).

The EELS deploys the transmitted part of the beam while the electron scattered at high angle are collected by the HAADF detector, allowing to simultaneously acquire the two signals or to select with atomic resolution the position where EELS spectra are acquired.

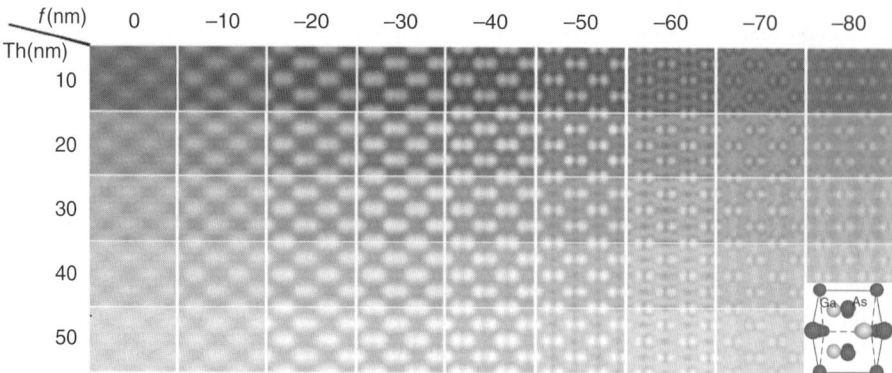

Fig. 15. Simulation of HAADF images for different values defocus and thickness for GaAs in the [110] zone axis (for simulation details, see Ref. [12]).

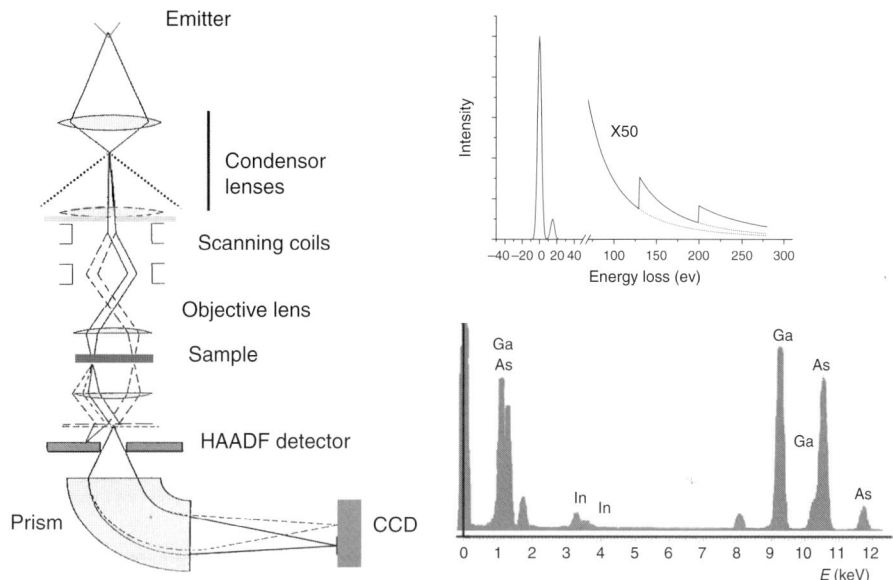

Fig. 16. (a) STEM HAADF + EELS schematic system (b) Schematic representation of an EELS spectrum: each edge is characteristic of an atomic shell of a species in the sample. (c) Example of EDX spectrum if an InGaAs alloy: each peak is characteristic for a transition in a given material.

Both EELS and EDX are based on the inelastic beam–sample interaction and, one of the most important effects of the interaction with atoms is the energy transfer to electrons in the atomic inner shells. These are then promoted to higher states or to the continuum. These processes are characterized by distinctive activation energies, well recognizable in the energy spectrum of the transmitted electrons. This is at the origin of the ability of EELS to distinguish different materials and compositions. The EDX, on the contrary, is based on the detection of the X fluorescence. After the promotion of an inner shell electron to higher states, a deep level of the atom is left free and a rearrangement of electrons is necessary. Such rearrangement implies the emission of X-ray photons with energy peculiar to the different chemical species.

Schematic examples of EDX and EELS spectra are compared in Fig. 16(b) and (c). The EDX spectra are formed by several peaks, often more than one corresponding to each element. Conversely, the EELS spectrum can be divided into different regions as a function of the lost energy ΔE. The most important for chemical characterization is the so-called high-loss spectrum ($\Delta E > 50\,eV$). In this region, the spectrum comprises several contribution characterized by a sharp rise and a low decrease towards higher energy losses. Each of these triangular shapes is called "edge" and is characteristic of the energy absorption by electrons at a precise atomic level. This idealized triangular shape is only found in the energy absorption from isolated hydrogen atoms while real edges usually have more complex shapes.

EDX spectra are typically detected by means of a liquid nitrogen-cooled semiconductor detector positioned above the sample at an angle of about 20° off the vertical. EELS spectra are often collected by conveying the transmitted beam on a special magnetic lens that bends the electron trajectory by about 90°. This kind of lens acts as an electron prism as the actual bending depends on the electron energy. It transfers the electrons from its object plane to its image plane where the electrons are dispersed according to their energies. A charge-coupled device (CCD) detector is then used to collect the signal for each energy independently.

By scanning the beam position on the specimen, it is possible to get an EELS or EDX two-dimensional map for each energy or equivalently a spectrum for each beam position. Both these techniques can give quantitative information if the appropriate cross-sections and experimental details like the sample thickness are known in detail. EELS is better suited for the detection of low-Z elements, while EDX is often limited to materials with high Z.

While EDX is mainly meant to give chemical information, EELS is a more versatile tool. It can be demonstrated that it is sensible to both initial and final electronic states of the electron transition and that the details of the spectrum for each shell absorption are determined by the local electronic environment. Moreover, many spectroscopic studies based on EELS can give similar information comparable with X-ray absorption spectroscopy [1].

However, compared to X-ray techniques, the EELS in a (S)TEM is limited by a poorer resolution of the order of 0.5 eV, for a FE source TEM. On the contrary, the energy resolution of EDX is of about 120 eV [3], which is sufficient to separate most of the common characteristic peaks.

As far as the spatial resolution is concerned, EELS cannot usually benefit from the large angles integration typical of HAADF since a restricted angular range in the forward direction is conventionally used. This means that the channelling and the related signal localization is comparatively lower. Nevertheless, a resolution of 2–3 Å [35] has been demonstrated by means of state-of-the-art STEM machines. In contrast, EDX resolution is comparatively much worse as it is strongly affected by the beam spreading depending on the sample thickness [3].

Finally, it is worth mentioning that energy filters can be used in combination with TEM in non-scanning mode. In this case, the images are acquired for different positions, but a fixed energy, in one shot. By comparison of images at different energies, it is possible to extract the chemical information and evaluate the spectra for each point [36].

6. Application of the TEM technique to semiconductor heterostructures

6.1. Defects and compositional contrast in strain-balanced InGaAs multi-quantum wells

In the following, a typical application of the chemical-sensitive imaging and the interpretation of the strain- and defects-induced contrast will be given in the case of a novel system for photovoltaic applications, which combines InGaAs-based strain-balanced multi-quantum wells (MQWs) with compositionally graded InGaAs/GaAs virtual substrate. The cells were intended to absorb at around 1 eV. Virtual substrates with predetermined lattice parameter at the top of the buffer layer and MD profiles were designed according to the semi-empirical model on strain relaxation for the InGaAs/GaAs system proposed by Romanato et al. [37]. InGaAs p–i–n junctions were deposited on these virtual substrates, containing strain-balanced MQWs in the intrinsic region. Compositions and thickness of the InGaAs/InGaAs MQW layers were chosen to meet strain balance condition with respect to the top of buffer layers according to the design proposed by Ekins-Daukes et al. [38].

The results obtained by XRD studies allow to conclude that the top of the structure is almost completely relaxed so that MDs were found to be strictly confined at the buffer/GaAs substrate

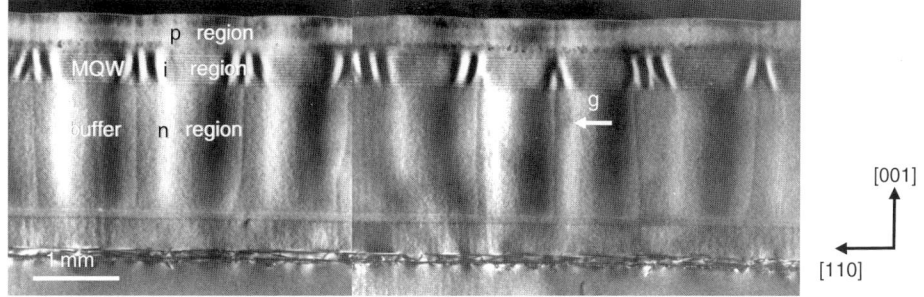

Fig. 17. 220 Dark field cross-sectional image of the complete structure.

interface in a few nanometres thick layer as shown in the cross-sectional DF TEM ($g = 220$) micrograph of the whole structure (Fig. 17). The stress fields associated with the underlying misfit dislocation network cause a roughening of the epilayer surface, resulting in a cross-hatched (CH) surface with a periodicity of about 1 µm. Two features are worth noting: a regular coarse contrast modulation all along the structure with the same wavelength of the CH pattern and a wavy fine contrast modulation in the MQW at the CH valleys. All the bright and dark regions were found to invert contrast when the sign of the operating $g = 220$ reflection was reversed, indicating the presence of a typical strain-induced contrast pattern. The coarse contrast modulation extending over the whole structure consists of regular bright and dark columnar regions, which invert contrast across the middle of each ridge of the CH surface morphology. This evidences that the CH pattern is accompanied with stress concentration regions that elastically relax when thin foils for TEM observations are made. In contrast, fine strain modulation extends only in the MQW region in correspondence to the CH valleys and develops in closed circular loops due to the intersection of the two orthogonal [1–10] and [110] CH striations (Fig. 18). It must be ascribed to lateral thickness modulation of the MQW layers as

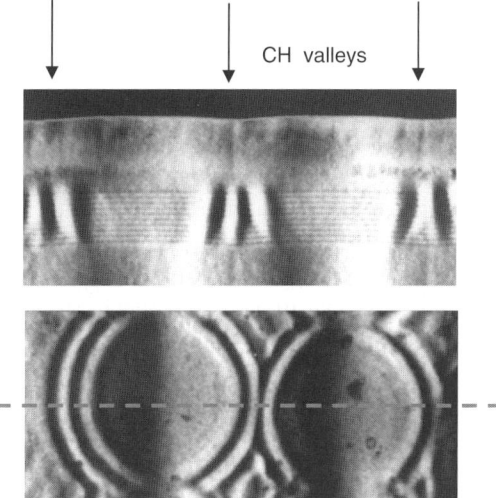

Fig. 18. The upper picture shows a cross-sectional image taken along the line in the plane view of the bottom picture.

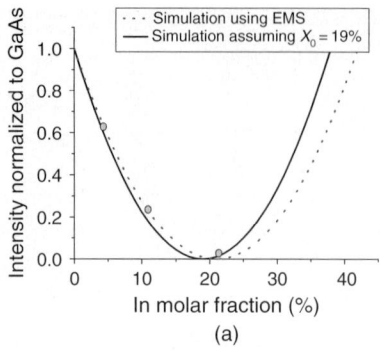

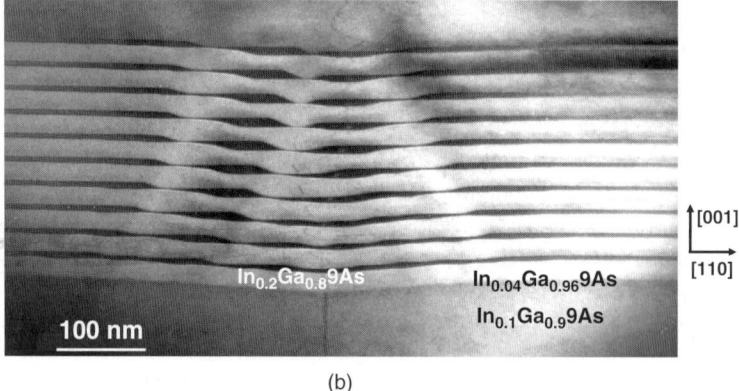

Fig. 19. (a) EMS simulation and fit of the InGaAs 200 type reflection intensity as a function of the In content, grey circles corresponding to the experimental In concentration of the layer, well and barrier. (b) 200 DF image of the corrugated QW region showing the "vertical well" formed in the CH valley.

clearly illustrated in the $g = 200$ dark-field cross-sectional TEM picture of Fig. 19(b). The ribbon-like shape of the layers can be considered as the "finger print" of the step-bunching phenomena, which typically occur during the growth of crystals on a vicinal surface [39]. In the present case, the MQW interface morphology develops by bunching of steps originally present at the CH valleys, where the growing surface presents the maximum slope. The 200 reflection dramatically enhances the contrast difference due to different In concentration in the layers of the structure (see Section 3.4).

A closer inspection of the image in Fig. 19(b) reveals the presence of a so-called vertical QW, which is usually ascribed to a preferential In-rich growth in the bottom of the valley [40]. The graph in Fig. 19(a) reports the EMS [41] simulation and a recent fit of experimental data [42] of the InGaAs 200 type reflection intensity as a function of the In content: if we plot on this graph the points corresponding to the values of In composition of the buffer layer, the well and the barrier as they have been measured by HRXRD, we obtain a perfect agreement with the contrast they exhibit in the picture. The vertical well "colour" being brighter than the well and darker than the barrier, we can deduce that the vertical well is In-rich, but the In content should not be higher than 40%.

6.2. An example of lattice fringes analysis for the InGaN/GaN system

A basic application of the lattice fringes analysis for determining the composition in the InGaN/GaN system is described in the following. InGaN QWs are the basis for light-emitting diodes (LEDs) in the blue and green spectral region [43]. One of the interesting features in InGaN alloys with In concentration $x < 0.1$ is the high exciton localization [44]. One possible explanation of this phenomenon is the In concentration fluctuations. For this reason, it is extremely important to be able to test the compositional homogeneity on the atomic scale. In one of the first articles of this kind, this was performed by means of lattice fringes analysis and compared with EELS maps [45]. We report here on the details of the lattice fringes analysis.

The two analysed samples comprised one single and one multiple quantum well of InGaN grown by metallorganic chemical vapor phase epitaxy onto dislocation-free GaN single crystals. In the case of the single QW the upper barrier was composed by $Al_yGa_{1-y}N$ with nominal composition $y = 15\%$. The single QW and the MQWs have nominal In concentration of 20% (QW) and 10% (MQW). The nominal QW thickness was 3.5 nm.

The HRTEM experiment was performed on a Philips CM300 UT with LaB_6 filament and a point resolution 0.17 nm operated at 300 keV. HRTEM images have been acquired on a negative and digitized by means of an off-line camera. Digital analysis has been performed by means of the software DALI [21].

The lattice fringes analysis explained in Chapter 4 can be suitably applied to hexagonal structures. Eq. (41) remains valid now with c being the lattice parameter in the growth direction. In the specific case of InGaN,

$$\frac{c - c_{GaN}}{c_{GaN}} = x\left[\left(-\frac{1+\nu}{1-\nu}\right)\frac{a_{InN} - a_{GaN}}{a_{GaN}}\right] \quad (42)$$

The actual values of the Poisson ratio ν and of the lattice parameters for the InN are not precisely known. This error can actually strongly affect the final estimation [46]. The used values are summarized in Table 1.

An example of the HRTEM images used for the analysis of the MQW is shown in Fig. 20 together with an enlargement of a detail. The image is obtained in the [11–20] ZA[1]. A scheme of the atomic structure as viewed in this projection is also shown. The image pattern resembles the atomic structure when the Ga positions only are considered. Indeed the instrument resolution does not permit to observe the N and Ga as separated maxima.

Table 1

Lattice parameters and the Poisson ratios for binary compounds alloying in InGaN and AlGaN compounds

	GaN	InN	AlN
a (nm)	0.3189	0.354	0.3112
c (nm)	0.5185	0.5705	0.4982
ν	0.15	0.39	0.23

---

[1] Details on the four indexes notation can be found in Ref. 7, p. 292 (Appendix 2).

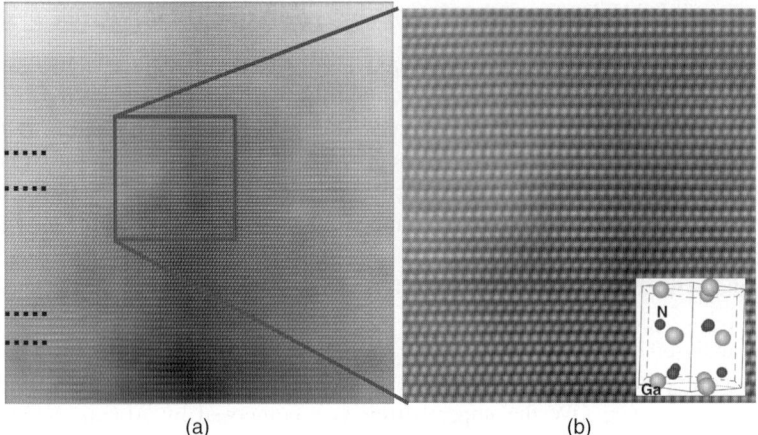

Fig. 20. Example of HRTEM image used for the lattice fringes analysis of InGaN alloys. The lines are guide for the eyes to indicate the QWs position. The enlargement in (b) shows a detail of (a) compared with the unit cell structure.

The application of the DALI to this image yields a point-to-point map of the lattice distortion that can be directly interpreted, according to Eq. (42), as an In composition map (Fig. 21(a)). In this figure, the composition values have been represented with different grey levels as indicated by the legend. The grey level indicated by the label 0% has been taken as a reference and corresponds to the GaN average lattice parameter in the barrier regions.

The darker zones correspond effectively to the MQWs as indicated in Fig. 20. The compositional contrast is hardly visible in Fig. 20 while the strain mapping is more reliable in evidencing the effective QW position. Residual strain appears in the upper left part of the image but it is more probably an artefact due to a contrast change in the HRTEM image. Indeed HRTEM images of III–nitride wurtzite structures can be strongly affected by small

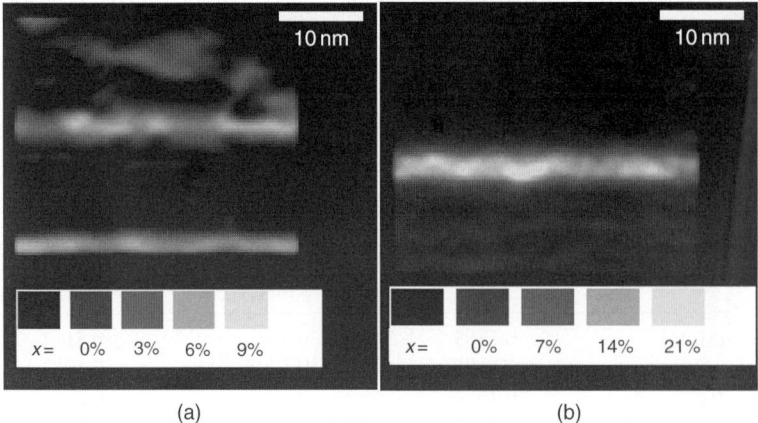

Fig. 21. In composition maps in $In_xGa_{1-x}N/GaN$ structure. Different grey levels correspond to different In compositions as described by the legends. Figure (a) refers to the MQW sample while (b) refers to the single QW structure.

variation of imaging condition that can happen even in the same image (small local tilt from exact ZA condition or small sample thickness variations). Artefacts due to digitization and noise appear even in the nominally pure GaN layers and have been reduced by a lowpass filter. The final image is affected by a 1% uncertainty in composition. From the image inspection, we can therefore state that the average composition is of about $x = 10\%$ with fluctuation down to 5% and up to 13%.

The same procedure has been applied to the case of the single QW sample to obtain the map in Fig. 21(b). Evidently, the sample presents a much higher uniformity and the average composition amounts to $x = 20\%$ with fluctuation between 18% and 23%.

The AlN barrier on the top of the QW is also visible as a darker area. The average Al composition in this layer was here evaluated as $y = 14\%$.

6.3. Combining diffraction contrast and lattice fringes analysis: In and N assessment in the InGaAsN system

6.3.1. Method

In comparison to the case of ternary systems studied in the previous section, the application of DF and lattice fringes analysis to the quaternary $In_xGa_{1-x}As_{1-y}N_y$ represents a problematic case. Both the methods when used alone fail in giving reliable chemical information on this system as two variables, the In and N composition, need to be evaluated from a single measurement. It is therefore natural to try to combine the two techniques to obtain a simultaneous characterization for both In and N composition [47]. However, due to the complications of this kind of analysis, it is necessary to clarify some details of the evaluation procedure and some InGaAsN system peculiarities.

First, it must be noticed that, in the case of a quaternary alloy, Eq. (41) is slightly more complicated and the tetragonal distortion ε can be written as

$$\varepsilon = \frac{a_\perp - a_{GaAs}}{a_{GaAs}} = -\left(\frac{1-\nu}{1+\nu}\right)\frac{a_{InGaAsN} - a_{GaAs}}{a_{GaAs}}$$
$$= -\left(\frac{1-\nu}{1+\nu}\right)\frac{(a_{InAs} - a_{GaAs})x + (a_{GaN} - a_{GaAs})y + ((a_{InN} - a_{GaN})(a_{InAs} - a_{GaAs}))xy}{a_{GaAs}}$$
(43)

In the same formula, a_{GaAs}, a_{InAs}, a_{InN} and a_{GaN} refer to the bulk lattice parameters. Their values and the elastic constants are shown in Table 2.

Substituting the values in Eq. (43), it is easily found that the tetragonal distortion increases with the In concentration [x] but decreases with N concentration [y]. Therefore, for each value of In composition, there exists a value of y yielding the same tetragonal distortion. This is given by

$$y = \frac{1}{\gamma + \delta x}\left[\left(\frac{1-\nu}{1+\nu}\right)\alpha\varepsilon_\perp - \beta x\right]$$
(44)

where $\alpha = a_{GaAs}$, $\beta = a_{InAs} - a_{GaAs}$, $\gamma = a_{GaN} - a_{GaAs}$ and $\delta = ((a_{InN} - a_{GaN})/(a_{InAs} - a_{GaAs}))$. For this material system, as δ is a relatively small number, the relation between x and y in Eq. (44) is almost linear (Fig. 22(a)) and this results in a certain simplification of the analysis.

Table 2

Lattice parameters and the Poisson ratios for binary compounds alloying in InGaAsN compounds

	GaAs	InAs	GaN (cubic)	InN (cubic)
a	0.56535	0.60585	0.4503	0.498
ν	0.311	0.3521	0.387	0.342

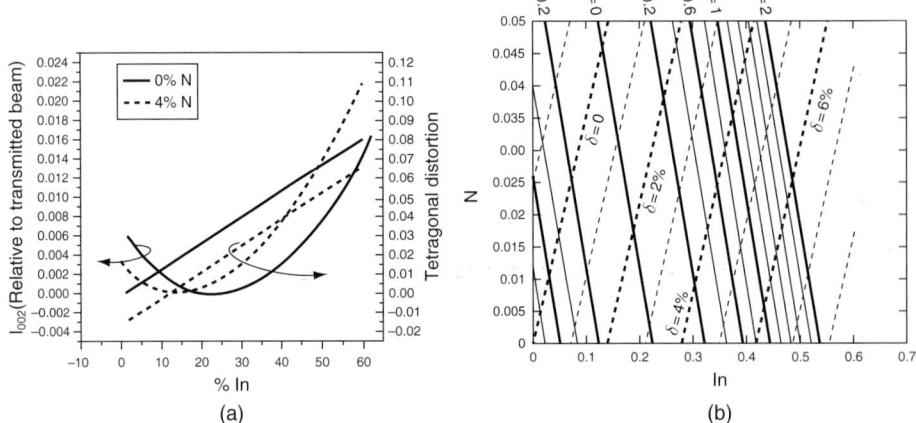

Fig. 22. (a) Tetragonal distortion and dark field 200 intensity as a function of the In composition for N compostion equal to 0% (straight lines) and 4% (dashed line). (b) Locus of the points with the same values of tetragonal distortion or same dark field (200) intensity as a function In and N compositions.

The other piece of information arises directly from the measurement of the intensity in dark-field conditions using the chemical-sensitive (200) reflection. Unfortunately, as already stated in Sections 4 and 6.1, such quantity shows a non-linear dependence on the composition (see, e.g. Fig. 22(a)). In detail, the image intensity at a point can be calculated as a function of the local In and N composition by considering the scattering factors for cation and anion:

$$\begin{cases} f_{\text{III}} = x(f_{\text{In}} - f_{\text{Ga}}) + f_{\text{Ga}} \\ f_{\text{V}} = y(f_{\text{N}} - f_{\text{As}}) + f_{\text{As}} \end{cases} \quad (45)$$

The intensity for a {200} reflection is, as mentioned, to a large extent proportional to the square of their difference (see Eq. (18)).

The dependence on compositions can be written as

$$I_{200} \propto (ax + by + c)^2 \quad (46)$$

where $b = f_{\text{As}} - f_{\text{N}}$, $a = f_{\text{In}} - f_{\text{Ga}}$ and $c = f_{\text{Ga}} - f_{\text{As}}$. Then, for a given N concentration, the intensity I_{200} depends on the square of the In composition (Fig. 22).

Using the scattering factor values as calculated by the simulation program EMS [41], the resulting curve shows an intensity minimum ($I = 0$) for an In composition $x_0 = 21\%$ if $y = 0$, but such minimum shifts to smaller values of x when y increases. Due to the parabolic profile in Fig. 22(a), two sets of concentrations give rise to the same value of I_{200}. In particular, for a given value of x, the same intensity I_{200} is obtained if y is one of the two values of y^+ and y^-:

$$y^+ = \frac{\sqrt{I-c}}{b} - \frac{a}{b}x \tag{47a}$$

$$y^- = \frac{-\sqrt{I-c}}{b} - \frac{a}{b}x \tag{47b}$$

Even if the intensity dependence of the composition is not linear, the locus of the compositions with the same intensity is a linear function of y vs x.

Some recent articles have risen some doubts on the scattering factor values, as values of $x_0 = 19\%$ have been found [42] (see also Section 6.1 for comparison). In practice, independent from the exact value of the numeric constants, the two expressions (Eqs (47a) and (47b)) can be coupled together with Eq. (44) to form two sets of linear equations with x and y as unknowns.

The solution can also be found graphically by inspection of Fig. 22(b) showing the loci of points in a x, y graph having the same value of tetragonal distortion or intensity I_{200}. For a given set of experimental values of ε and I_{200}, the composition can be obtained as the crossing point of the corresponding lines. The only ambiguity remains between the two solutions y^+ and y^- that determine the two values of x^+ and x^-. As these values are usually very different, this ambiguity can be removed provided that the composition range is even roughly known.

6.3.2. Experimental results

Experiments have been performed testing the validity of this method. Here a brief account is given mainly as an example of analysis.

The experiments have been performed on samples prepared in cross section in the projections [011] and [001] with low-temperature ion milling. TEM analysis has been performed using a Philips CM300UT operated at 300 kV. DF (200) micrographs were taken close to the [001] pole while HRTEM analysis was done in both [001] and [011] ZA.

Particular care was taken to avoid strain relaxation caused by thin foil effects. The DF images were acquired directly with a CCD mounted on the TEM. The intensity was normalized to have the level 0 corresponding to the background intensity and the level 1 corresponding to the GaAs intensity. HREM micrographs acquired on photographic negatives have been scanned off-line prior to be analysed by means of the DALI program.

All the analysed samples were grown by MBE at a temperature of 450°C on a GaAs substrate and comprised five QWs of InGaAsN with different In and N composition. The nominal QW width was of 7 nm.

In particular, in Fig. 23 an example of analysis is shown on a QW with nominal In and N content of 35% and 1.7%, respectively. Figure 23(a) and 23(b) shows an HREM image of the sample along with the relevant tetragonal distortion profile obtained by DALI lattice fringes analysis. Figure 23(c) is a DF image of the same sample, the profile below (Fig. 23(d)) being the normalized intensity in the QW region. In Fig. 23(c), the dark lines on the QW sides correspond

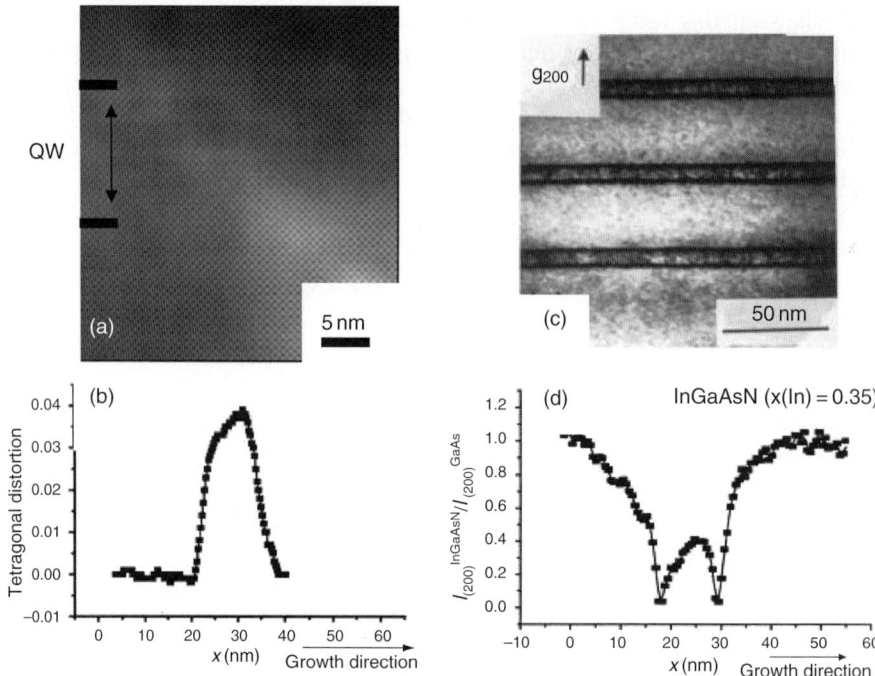

Fig. 23. Example of analysis of the sample with 35% In and 1.7% N. HRTEM image (a) and corresponding, laterally averaged, tetragonal distortion profile evaluation (b). Dark field (200) image of the same sample (c) and relevant normalized intensity profile in the zone of interest (d).

to the condition of no intensity or, at least, to the minimum intensity condition. Indeed, the QW interfaces are graded so that the composition assumes that all the intermediate values included the ones for which $f_{III} = f_V$. Such minimum intensity condition is fulfilled for compositions of the order of $x = 10$–15% and $y = 2\%$, which correspond to about half of the nominal composition value in the middle of the QW.

By evaluating the DF intensity and the tetragonal distortion at the centre of the QW, it is possible to apply the above-described methodology, so as to obtain the compositions of In, $x = 34 \pm 2$, and N, $y = 2.5 \pm 0.5$. These values are reasonably similar to the nominal ones, thus confirming the validity of the methodology. It is also possible to perform the same analysis point by point, paying attention to perform the correct matching of the two profiles. The result is shown in Fig. 24(a): the In and N composition profiles are completely disentangled and directly comparable. These curves indicate that the In concentration profile is quite symmetric (as effectively observed in InGaAs/GaAs QWs grown at low temperatures) while N concentration is highly asymmetric.

This result can be explained in terms of the growth modality: radio frequency used to produce atomic N is turned on prior to the actual QW growth and opening of the N shutter; however this favors the incorporation of the residual N is present in the growth chamber. On the contrary, the shutdown of the radio frequency determines an abrupt stop in the N incorporation visible at the upper interface.

Figure 24(b) and (c) is a further example of disentangled composition profiles for QWs with different In (respectively 20% and 10%) and N (1.7%) composition. It has to be stressed again

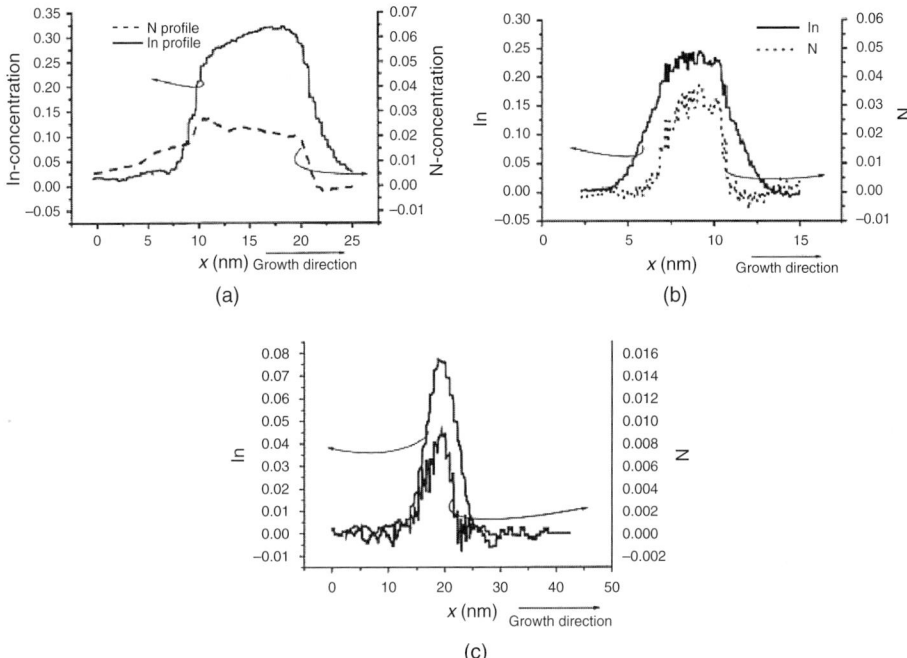

Fig. 24. Composition profiles for In and N obtained by the combined analysis of tetragonal distortion and dark field for the QW with a nominal composition of 1.7% N and 35% (a), 20% (b) and 10% (c) In.

that the results are in good agreement with the nominal values and that the main features of the chemical profiles are common to all the analysed samples. In particular, the In profile symmetry and the abrupt N composition drop at the upper interface. This is a further confirmation of the reliability of this analysis method.

Appendix: Simulation of TEM images

The weak phase approach or even the simplified approach to ZA channelling theory described by van Dyck (see Section 4.2) is only qualitative approximations. The necessity to obtain quantitative information by different TEM techniques (HRTEM, STEM, etc., but also by other techniques like convergent beam diffraction) has been the driving force for developing simulation programs able to predict the experimental results.

Depending on the technique under study and on the nature of the structure to be simulated, different approaches have been used. The most popular are "Bloch waves" [1,3,48] and "Multislice" [9,49,50] simulations. Both have been employed with small modification in the simulation of HRTEM images [8,16], STEM images [51,52] and diffraction [53,54].

The Bloch wave theory is based on the assumption of a periodical structure and is therefore especially suited for homogeneous crystal structures. One of the advantages of the "Bloch wave" theory is that it permits also analytical solutions and a simplified treatment of special cases. The Howie and Wehlan equations (see Section 3.2) for a perfect crystal can be deduced, for example, as a special case of the Bloch wave theory. However, the results are quite more general as this theory can be easily used to predict channelling or to quantitatively evaluate the pattern of HREM images.

In this Appendix, some of the basic idea of Bloch wave methods and Multislice method are briefly explained, with reference to the plane wave illumination case. The Bloch wave theory produces a solution of the stationary Schroedinger equation of the electron in the crystal:

$$\nabla^2 \psi(r) + \frac{8\pi^2 me}{\hbar^2}\left(E + \frac{\hbar^2}{2me}\sum U_g e^{i2\pi g \cdot r}\right)\psi(r) = 0 \qquad (A1)$$

where the periodic crystal potential has been decomposed into Fourier components U_g. The solution is searched by decomposing the wave function in eigenstates.

As a consequence of the Bloch theorem, such states have the same periodicity of the crystal and are called Bloch states. These Bloch states are practically the same used in the solid-state theory of electrons in a crystal [2]. The spectrum of eigenvalues is therefore characterized by a continuum of states with a defined pseudo-momentum k grouped in discrete number of bands indicated here by j.

It is therefore natural to write each state as

$$b^{(j)}(r) = \sum_g C_g^j\left(k^{(j)}\right) e^{i2\pi\left(k^{(j)}+g\right)r} \qquad (A2)$$

as done in solid state.

The C factors are the coefficients that describe the Bloch eigenstates in terms of reciprocal lattice vectors: they are the Fourier transform of the shape of the Bloch states in each single unit cell. The only difference with solid-state electrons is that these electrons are characterized by a large velocity in the beam direction, which can simplify the mathematical expressions.

The Schrödinger equation, rewritten on this basis, becomes an algebraic system of equations with unknowns given by the factors C_g^j:

$$\left\{-\left|k^{(j)}+g\right|^2 + K^2\right\}C_g^{(j)} + \sum_{h\neq 0} U_h C_{g-h}^{(j)} = 0 \qquad (A3)$$

This homogeneous equation can be solved to determine the $C_g^j(k)$ factors.

If the incident wave is a single plane wave as in TEM experiments, a single state with a definite k is excited for each Bloch band j. Therefore, the total wave function inside the crystal is simply described as

$$\psi(r) = \sum_j \varepsilon^j b^j\left(r, k^j\right) \qquad (A4)$$

where the coefficients ε for each Bloch wave are determined by the boundary condition of phase matching between the impinging plane wave and the waves inside the crystal.

Figure A1 (a) shows an example of the most important Bloch states excited in a GaAs unit cell by a plane wave along the [110] direction. The first states are localized on the atomic columns and are responsible for channelling, other are more localized between columns. These first states alone account for 99% of the wave function. At the entrance surface of the crystal, these Bloch waves sum up according Eq. (A4) to match the impinging plane wave. Inside the crystal, each of them evolves with z in a different way, their interference producing the final wave function at the exit surface of the specimen. Outside the crystal, the states are re-projected to eigenstates of the vacuum Hamiltonian, namely plane waves propagating in different

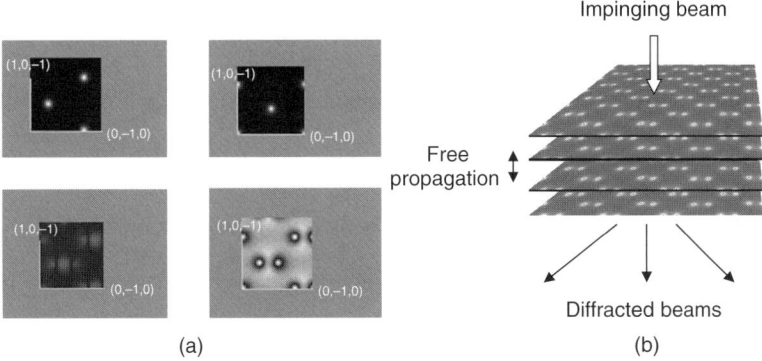

Fig. A1. (a) Spatial representation of the wave amplitude of the Bloch states in GaAs that are most excited by a plane wave in the [110] direction. (b) Scheme of the Multislice simulation: the potential is reduced to two-dimensional slices and free propagation occurs between slices.

directions given by the reciprocal lattice vectors of the crystal. These diffracted and transmitted beams produce the final intensity on the screen.

The core of the Bloch waves problem is to solve the system of Eq. (A3). If only two beams and two Bloch waves are present, the solution is the Howie and Wehlan equation system. Approximate solutions are also possible in exact ZA conditions if the effect of only a few of the most excited states is considered: this is the channeling simplified approach case introduced in Section 4.2.

For all the other cases simulations are used, Eq. (A3) must be solved numerically: this is usually performed by classic matrix-based algorithms (see, e.g., Ref. [55]). The boundary conditions are then applied at the top surface extracting the value of the excitation of each Bloch wave.

At this point, the wave function is completely determined as well as the amplitude and phase of the beams at the exit of the sample. Finally, the effect of the optical system is accounted for. The procedure is based on Fourier transforming the wave function at the exit of the sample, applying the multiplicative effect of the transfer function of the objective lens on each component (see Section 4.1) and then back Fourier transforming the final wave function. The intensity will be just its square modulus.

The Multislice approach is more general as it can be also used to describe the TEM images for defected crystals. Indeed it does not require the discreteness of the Fourier spectrum of the wave function implicit in the Bloch wave formulation. This has as a drawback a usually larger computing time.

The idea of the simulation is that the weak phase approximation, which is generally a bad approximation for the overall sample beam interaction, can be still used if the specimen is divided in consecutive potential slices (see Fig. A1 (b)). Each slice contains the potential projected for a thickness Δz that produces a change of the phase of the sample as described in Section 4.1. Therefore, after the interaction at the ith slice, the wave function results in

$$\psi_i(\bar{r}, z) = \psi_{i-1}(\bar{r}, z) e^{i\sigma V(\bar{r})_i} \qquad (A5)$$

where $\psi_i(\bar{r}, z)$ is the wave function after the ith slice, $\bar{r}$ is the in-plane x, y coordinate and V is the slice projected potential.

After the interaction with each slice, the beam is considered to propagate as in free space for the distance Δz to the next slice. While the interaction with the potential is a multiplicative operation, the propagation into free space is accounted for by convolution with spherical waves or more often with their parabolic approximation. After this propagation, the wave function results in

$$\psi_i(\bar{r}, z + \Delta z) = e^{2\pi i \Delta z/\lambda}[\psi_i(\bar{r}, z) \otimes p(r, \Delta z)] \tag{A6}$$

where the propagator p can be rewritten in its parabolic approximation as [9]

$$p(r, \Delta z) = \frac{1}{i\lambda\Delta z}\exp\left[\frac{i\pi}{\lambda\Delta z}r^2\right] \tag{A7}$$

Due to these assumptions, the Multislice presents itself as an alteration of multiplication in real space and convolution that are better performed in Fourier space. As a result of the high-speed implementations of the fast Fourier transform, this algorithm produces very accurate results in a reasonable time if the sample is not too thick.

References

[1] D.B. Williams, C.B. Carter, Transmission Electron Microscopy, Plenum Press, NewYork, 1996.
[2] N.W. Ashcroft, N.D. Mermin, Solid State Physics, Holt-Saunders International Editions, Philadelphia, 1976.
[3] P. Hirsh, A. Howie, R.B. Nicholson et al., Electron Microscopy of Thin Crystals, Robert E. Krieger, Malabar, FL, 1977.
[4] L. Reimer Transmission Electron Microscopy. Physics of Image Formation and Microanalysis, Springer-Verlag, Berlin, 1989.
[5] D. Hull, D.J. Bacon, Introduction to Dislocations, 3rd Edition, International Series on Materials Science and Technology, Pergamon Press, Oxford, 1984, Vol. 37.
[6] M.H. Loretto, Electron Beam Analysis of Materials, Chapman & Hall, London, 1984.
[7] J.W. Edington, Practical Electron Microscopy in Material Science, Van Nostrand Reinhold Company, New York, 1976.
[8] P. Buseck, J. Cowley, L. Eyring, High-Resolution Transmission Electron Microscopy and Associated Techniques, Oxford University Press, Oxford, 1988.
[9] E.J. Kirkland, Advanced Computing in Electron Microscopy, Plenum, New York, 1998.
[10] O. Scherzer, J. Appl. Phys. 20 (1949) 20.
[11] P. G. Merli and M. V. Antisari (Eds.), Proc. Int. School on Electron Microscopy, in Materials Science, (World Scientific, Singapore, 1992) p. 193.
[12] E. Carlino, V. Grillo, Arch. Met. Mat. 51 (2006) 23.
[13] R.W. Glaiser, A.E.C. Spargo, D.J. Spargo, Ultramicroscopy 27(1989) 35.
[14] D. VanDyck, Ultramicroscopy 64 (1996) 99.
[15] K. Ishizuka, Ultramicroscopy 5 (1980) 55.
[16] D. Stenkamp, W. Jäger, Ultramicroscopy 50 (1993) 321.
[17] A. Rosenauer, U. Fisher, D. Gerthsen, A. Förster, Ultramicroscopy 72 (1998) 121.
[18] W. Coene, M.O. Debeeck, D. Van Dyck, Phys. Rev. Lett. 69 (1992) 26.
[19] M.O. DeBeeck, D. VanDyck, W. Coene, Ultramicroscopy 64 (1996) 167.
[20] P.Yu, M. Cardona, Fundamentals of Semiconductors, Springer-Verlag, Berlin, 1996, p. 182.
[21] A. Rosenauer, T. Remmele, D. Gerthsen et al., Optik 105 (1997) 99.
[22] R. Bierwolf, M. Hohenstein, F. Philipp et al., Ultramicroscopy 49 (1993) 273.

[23] A. Rosenauer, U. Fisher, D. Gerthsen, A. Förster, Appl. Phys. Lett. 71 (1997) 3868.
[24] D. Gerthsen, E. Hahn, B. Neubauer et al., Phys. Stat. Sol A 177 (2000) 145.
[25] M.J. Hytch, J.L. Putaux, J.M. Penisson, Nature 423 (2003) 270.
[26] L. De Caro, A. Giuffrida, E. Carlino, L.Tapfer, Acta Crystallogr. A 53 (1997) 168.
[27] M.J. Hytch, T. Plamann, Ultramicroscopy 87 (2001) 199.
[28] F. Hue, C.L. Johnson, S. Lartigue-Korinek et al., J. Electron Microsc. 54 (2005) 181.
[29] P.E. Baston, N. Dellby, O. Krivanek, Nature 418 (2002) 61.
[30] S.J. Pennycook, Advances in Imaging and Electron Physics, Vol. 123, Academic, New York, 2002.
[31] P.D. Nellist, S.J. Pennycook, Ultramicroscopy 78 (1999) 111.
[32] S. HillYard, J. Silcox, Ultramicroscopy 56 (1995) 6.
[33] G.R. Anstis, D.Q. Cai, D.J.H. Cockayne, Ultramicroscopy 94 (2003) 309.
[34] J. Broeckx, M. Op de Beeck, D. Van Dyck, Ultramicroscopy 60 (1995) 71.
[35] A.R. Lupini, S.J. Pennycook, Ultramicroscopy 96 (2003) 313.
[36] T. Walther, A.G. Cullis, D.J. Norris, M. Hopkinson, Phys. Rev. Lett. 86 (2001) 2381.
[37] F. Romanato, E. Napolitani, A. Carnera et al., J. Appl. Phys. 86 (1999) 4748.
[38] N.J. Ekins-Daukes, J. Zhang, D.B. Bushnell et al., Proc. 28th IEEE Photovoltaic Specialists Conf., Anchorage, Alaska, USA, 2000.
[39] H.M. Cox, D.E. Aspnes, S.J. Allen et al., Appl. Phys. Lett. 57 (1990) 611.
[40] K. Leifer, S. Mautino, H. Weman et al., IoP Conf. Ser. 180 (2003) 131.
[41] P. Stadelmann, Ultramicroscopy 51 (1985) 389.
[42] K. Leifer, P.A. Buffat, J. Cagnon et al., J. Cryst. Growth 237–239 (2002) 1471.
[43] S. Nakamura, T. Mukai, M. Senoh, Appl. Phys. Lett. 64 (1994) 1687.
[44] S. Chichibu, T. Azuhata, T. Sota, S. Nakamura, Appl. Phys. Lett. 69 (1996) 4188.
[45] M. Albrecht, V. Grillo, J. Borysiuk et al., Inst. Phys. Conf. Ser. 169 (2001) 267.
[46] S. Stepanov, W.N. Wang, B.S. Yavich et al., MRS Int. J. Nitride Semicond. Res. 6 (2001) 6.
[47] V. Grillo, M. Albrecht, T. Remmele et al., J. Appl. Phys. 90 (2001) 3792.
[48] H. Bethe, Ann Phys 87 (1928) 55.
[49] J.M. Cowley, A.F. Moodie, Acta Crystallogr. 10 (1957) 609.
[50] D.F. Lynch, M.A. O'Keefe, Acta Crystallogr. A. 28 (1972) 536.
[51] S.J. Pennycook, D.E. Jesson, Phys. Rev. Lett. 64 (1990) 938.
[52] E.J. Kirkland, R.F. Loane, J. Silcox, Ultramicroscopy 23 (1987) 77.
[53] J.M. Cowley, J.C.H. Spence, Ultramicroscopy 3 (1979) 433.
[54] J.C.H. Spence, J.M. Zuo, Electron Microdiffraction, Plenum, New York, 1992.
[55] W.H. Press, S.A. Teukolsky, W.T. Vetterling, B.P. Flattery, Numerical Recipies in C, Cambridge University Press, Cambridge, 1992.

6

Accessing structural and electronic properties of semiconductor nanostructures via photoluminescence

Stefano Sanguinetti[1], Mario Guzzi[1], and Massimo Gurioli[2]

[1] L–NESS and CNISM, Dipartimento di Scienza dei Materiali, Universitá degli Studi di Milano Bicocca, Via Cozzi 53, I-20125, Milano, Italy
[2] L.E.N.S. and CNISM, Dipartimento di Fisica, Universitá degli Studi di Firenze, Via Sansone 1, I–50019, Sesto Fiorentino, Italy

Abstract Photoluminescence (PL) is one of the most widely diffused experimental techniques for the characterization of semiconductor nanostructures (in particular quantum wells (QWs)) and for the study of their electronic properties. PL allows to study a number of interesting intrinsic effects in QWs, like the exciton binding energy increase due to carrier localization, the homogeneous broadening of the exciton recombination line, the carrier–carrier interaction, and so on. In addition, the analysis of the radiative recombination spectra of nanostructures can help in the characterization of the structure, providing information on the interface morphology and on the quality of the materials. In this chapter the information that can be gained from PL spectra is critically summarized with the aim of providing a reference scheme for the characterization of nanostructures through PL measurements. The discussion in the chapter is supported by a wide and detailed bibliography, in which basic textbooks, review articles, and research papers are included in order to provide the reader with up-to-date information on the application of PL to the study of optical and structural properties of semiconductor QWs.

Keywords photoluminescence, semiconductor nanostructures, quantum wells, excitons, interfaces, carrier dynamics

1. Introduction

The photoluminescence (PL) is a nondestructive spectroscopic technique commonly used for the study of intrinsic and extrinsic properties of both bulk semiconductors and nanostructures.

A general discussion of the principles of this technique and of its applications to bulk semiconductors can be found in the still very actual textbooks by H. B. Bebb and E. W. Williams [1] and by J. I. Pankove [2] and in the review papers by P. J. Dean [3] and by L. Pavesi and M. Guzzi [4]. For a general introduction of PL applications to the study of quantum confined structures, and in particular of quantum wells (QWs), the reader is referred to the recent book by M. Grundmann [5] and to the review articles by M. A. Herman and coworkers [6] and K. K. Bajaj [7], to whom we will frequently refer in the following. Finally, in a recent work by the authors of the present chapter [8] the applications of PL to the quantum dot (QD) study are discussed.

PL consists in the radiation emitted by a crystalline or amorphous solid or by a nanostructure as a consequence of optical excitation; in particular, it derives from the radiative recombination processes of photoexcited electron–hole pairs (e–h pairs). The wide diffusion of this technique in the field of semiconductor nanostructures is motivated by the fact that it allows obtaining general information on the electronic properties and, which is of particular interest, on the quality of the nanostructures. The study of the PL spectra dependence on external parameters, such as sample temperature, energy and intensity of the exciting radiation, and applied fields (electric field, magnetic field, pressure), helps obtaining these information.

As shown in Ref. [9], complementary data on the sample quality can be gained using a similar technique, the cathodoluminescence (CL), in which the generation of e–h pairs in excess respect to the equilibrium density is obtained exciting the material or the nanostructure by an accelerated electron beam. The description of CL and of its applications to the study and to the characterization of nanostructures is the subject of Chapter 7 of this volume [10]. Finally, for sake of completeness, we cite the electroluminescence (EL), a process in which light is emitted as a consequence of electrical carrier injection in a semiconductor or in a structure.

PL spectroscopy is an experimental technique with some well-known advantages. It is a spectroscopic characterization technique, in that it provides energy resolved information; it is non-destructive, it provides information mainly on the minority carrier properties, and of consequence it is complementary to the electrical characterization techniques. In principle, it is easy to use, because it does not require any particular sample preparation or treatment. Finally, simple and relatively cheap equipments are needed for a rapid sample characterization. In the following section, we will present the basic structure of the experimental apparatuses for routine PL measurements.

However, some PL disadvantages should be mentioned. PL provides information mainly on the radiative recombination processes; information on non-radiative recombination processes can be obtained only in an indirect way through the analysis of the radiative recombination efficiency and of its dependence on external parameters, such as sample temperature and excitation intensity. Furthermore, it could be very difficult to get quantitative information on some material properties (such as the defect or impurity content) through PL measurements.

In the case of nanostructures, and in particular of QWs, PL spectroscopy is widely used for the analysis of the relations between the structure quality and the epitaxial growth conditions, for both molecular beam epitaxy (MBE) and metal–organic chemical vapor deposition (MOCVD) grown samples [6]. In particular, as it will be shown in detail in the following, the features of the PL spectra can provide information on the structural properties of the interfaces and on the quality of the well and barrier materials, including the effects of the strain induced by the lattice mismatch [11,12]. Important information can be obtained also on the intrinsic properties, such as valence and conduction band discontinuities, exciton level energy and exciton binding energy, and recombination kinetics.

The target of this chapter is to critically summarize the information that can be gained from PL spectra of QWs to provide a reference scheme for the characterization of nanostructures and, in particular, of QWs through PL measurements. A general discussion of properties and applications of QWs can be found in Refs [13–15]. This chapter is structured as follows. In Section 2 the experimental systems and techniques used for PL measurements, both in continuous wave (CW) and in time-resolved (TR) configurations, are described. In Section 3 the properties of the electronic states and the physics of the recombination process in QWs are summarized. Then, Section 4 deals with a detailed illustration of the PL applications in the characterization of the QW structural properties, in particular in the evaluation of the interface quality. In this section, a synthetic discussion of the carrier dynamics, thermalization and lifetime, is also presented. Finally, Section 5 contains a summary of the presented topics.

Due to space limitations, only the properties of QWs are discussed. The discussion of the properties of another important class of quantum-confined structures, the QDs has been omitted; the interested reader is referred to Ref. [8] and to the references quoted therein. The discussion of some interesting aspects of the radiative recombination in QW structures, which are less directly related to the applications of PL for the nanostructure characterization, are not included. In particular, we decided not to discuss the PL measurements in external fields and the high excitation effects, and thus the stimulated emission in QWs.

As for what concerns the materials systems, we will refer mainly to GaAs/AlGaAs and to InGaAs/GaAs QWs, because they have been widely studied due to their diffuse applications. When needed, other materials systems will be referred to.

2. Experimental systems and techniques

PL measurements are usually performed using two different excitation conditions. When an excitation constant in time is used, the material can reach a stationary nonequilibrium state. The PL measured in these conditions, the so-called continuous wave PL (CW-PL), provides information mainly on the structure of the electronic levels participating in the radiative recombination process. However, when the PL is excited using a pulsed excitation with light pulses of suitable duration, the so-called time-resolved PL (TR-PL) is obtained and the excited state lifetime, which provides information on carrier dynamics and recombination kinetics, can be measured.

This section is divided into two parts: in the first subsection the main features of the experimental set-up for CW-PL are illustrated and the second subsection is devoted to a synthetic presentation of the experimental techniques used in TR-PL measurements.

2.1. Continuous wave photoluminescence

A synthetic description of techniques and instrumentation used for CW-PL measurements and a detailed bibliography on the subject can be found in Ref. [6].

The scheme of a typical experimental set-up for routine CW-PL measurements is reported in Fig. 1(a). The sample is excited by absorption of photons with energy higher than the energy of the optical gap of the barrier material (nonresonant excitation) or below it, therefore injecting the carriers directly in the quantum structure (resonant excitation). These two excitation conditions allow to get different kind of information on the optical and physical properties of the nano-structure [16] and are obtained using different excitation sources with different wavelengths. In general, gas discharge lasers, such as Ar^+, Kr^+, or He–Ne lasers, solid-state lasers, such as the Nd:YAG laser, semiconductor lasers, and tunable lasers are used. In the past years, organic dye lasers were widely diffused as tunable light sources; recently they tend to be substituted by tunable Ti:Sapphire lasers.

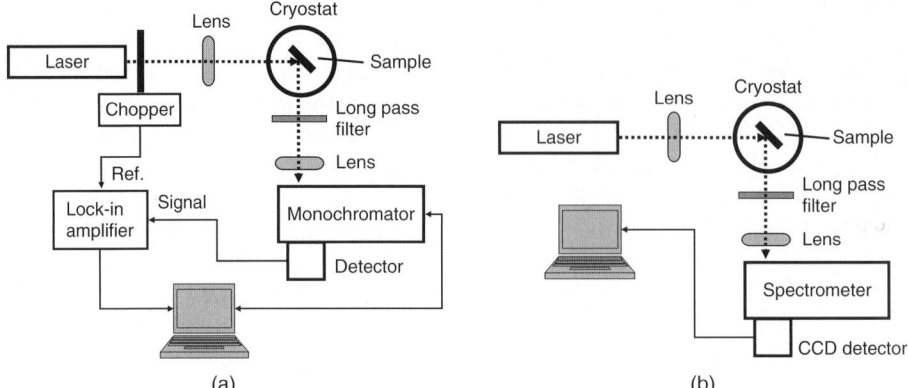

Fig. 1. Schemes of typical experimental set-ups for routine CW-PL measurements making use, for the PL detection, of photomultiplier tubes or semiconductor detectors (a) or of a CCD detector (b).

Two different schemes can be used for the spectral analysis and detection of the radiation emitted by the excited sample.

The luminescence can be spectrally analyzed using a grating monochromator and detected using photomultiplier tubes or semiconductor detectors, depending on the emitted photon energy. The monochromator wavelength tuning is usually performed with stepper motors driven by PCs through suitable programs. The laser radiation may be chopped by a mechanical chopper, as shown in Fig. 1(a), to allow the signal amplification by a lock-in amplifier [17]; Alternatively, photon counting techniques are used, as well. For the signal recording PCs are used, which make use of commercial or home-made programs.

In recent years, the increasing diffusion of image detectors, such as the silicon charge-coupled devices (CCD) or InGaAs diode arrays, favored a wide application of multichannel detection techniques in the PL spectroscopy (Fig. 1(b)). This detection scheme is characterized by a remarkable reduction in measuring time with respect to the previous one, sometimes accompanied by a limited loss of spectral resolution. In multichannel spectroscopy, the emitted light is spectrally analyzed with grating spectrometers and all the emitted spectrum is measured at the same time by the image detector, CCD or diode array. The sensitivity of cooled silicon CCD detectors is frequently as high as to allow few photon detection, and thus the use of silicon CCD coupled to grating spectrometers is, for many routine applications, advantageous with respect to the use of photomultiplier tubes coupled to monochromators. The main limitation of CCD detectors is their spectral response. For the detection of photons with energy lower than the silicon energy, gap-cooled InGaAs diode arrays can be used. In the case of InGaAs, the signal-to-noise ratio is markedly lower than for silicon CCD detectors, due to the lower energy gap of the ternary semiconductor. However, InGaAs diode arrays coupled to grating spectrometers are, for many routine applications, advantageous with respect to the use of single semiconductor detectors coupled to monochromators. In the case of multichannel detection, the detector signal is directly recorded, without any amplification stage; usually, the recording occurs on a PC, via suitable acquisition programs.

It has to be remarked that the spectral shape of the measured PL usually differs from the shape of the spectrum emitted by the sample, due to the energy dependence of the spectral response of the optical system, including the monochromator or spectrometer, the single or multichannel detector, and the optical components. Therefore, the optical system response should be measured, to get spectrally corrected PL spectra.

The elevated sensitivity of CCD detectors has favored in recent years the diffusion of the so-called micro-PL (μPL) measurements. In the μPL measurements, usually performed with confocal microscopy techniques [18], the PL is measured with micrometric, or even sub-micrometric, spatial resolution. Recently, near-field spectroscopies have been developed and used to probe the PL from QW with lateral resolution below 100 nm. The principle consists in collecting the emitted fields at very short distance from the emitter thus overcoming the diffraction limit [19]. The success of the scanning near-field optical microscopy (SNOM) was given by the development of methods to control the probe-sample distance, which permitted the simultaneous recover of the surface morphology in addition to the optical properties [20]. Different approaches were used for the technology, such as tapered optical fiber, metalized optical fiber [19], and apertureless SNOM [21].

The routine PL measurements on semiconductors and semiconductor quantum structures are commonly performed at cryogenic temperatures, typically few degrees Kelvin, or as a function of the temperature. Samples are then mounted with a strain-free mount in a cryostat. Two types of cryostats are widely diffused: the immersion cryostats, in which samples are directly immersed in the cryogenic liquid, usually liquid He, or in its vapor, and the cold finger cryostats, usually a closed-cycle refrigerator, in which samples are cooled by conduction.

Complementary information can be provided by the so-called photoluminescence excitation (PLE) spectroscopy. In a PLE measurement the PL intensity is measured at a given wavelength, while scanning the wavelength of the exciting light, and thus the PLE spectrum gives the CW-PL intensity at that wavelength as a function of the exciting photon energy. A tunable exciting source, such as a tunable laser or a lamp coupled with a monochromator, is used for these measurements. The PLE spectroscopy may provide a convenient and easy way to obtain the absorption spectrum without any sample preparation, such as sample thinning or selective etching. In particular, PLE reflects the absorption in nanostructures where the absorption is quite small; in bulk the incoming light beam is fully absorbed and PLE spectra may also reflect modulations in the carrier relaxation efficiency [22].

Fourier transform (FT) spectroscopy, which is usualy applied for transmission and reflection measurements in the infrared, in recent years has also been proposed for PL and PLE measurements (see Ref. [23] and references quoted therein). This is due to the well-known advantages of the FT spectroscopy with respect to the dispersive spectroscopy. A critical discussion of the FT approach to PL measurements as well as of its limitations is presented in Ref. [23].

2.2. Time-resolved photoluminescence

The availability of laser sources with a large range of wavelength tunability and pulse duration and energy has allowed in the last decades a strong development of TR spectroscopies. After the excitation by an ultrashort pulse of a semiconductor structure, the photoexcited carriers undergo several stages of relaxation before they return to the initial condition. The investigation of these relaxation processes allows to get information on the intrinsic time constant ruling the carrier dynamics. TR spectroscopic techniques with the appropriate time resolution are then largely exploited to follow, in real time, the relaxation paths of the photoexcited carriers. For a general review, we refer to Ref. [24].

Ultrashort laser technology is mainly based on the mode-locking method, in which a large number of longitudinal modes of the laser cavity, in active systems with large gain bandwidth, are locked in phase. Synchronously pumped organic dye lasers, invented at the end of the 1970s, were the first generation of ultrashort pulse lasers. Enormous progress in tunability, stability, and quality of the laser pulses has been achieved with the development of Ti:Sapphire lasers. Light pulses shorter than 10 fs have been realized directly from the laser oscillator without the need of laser compression, and the development of optical parametric oscillators and amplifiers has recently increased the tunability range of ultrafast lasers.

Different detection techniques can be used for TR PL spectroscopies. Among them, the most widely used is based on streak camera, where the PL photons, focused on a photocathode, extract photoelectrons that are deflected by time-modulated electrostatic fields to be streaked, after multichannel plate amplification, across a phosphorescent screen. The pulsed electron beam impinging onto the screen at different position transfers into a spatial distance the information on the emission time and picosecond resolution can be achieved. Higher time resolution, only limited by the pulse duration, can be achieved by non-linear techniques like the parametric up-conversion. In this case the PL signal is combined in a nonlinear crystal, such as KTP or $LiIO_3$, with part of the laser pulse suitably delayed. Frequency sum generation can occur only during the laser pulse gate, and its intensity as a function of the delay of the laser pulse directly gives the PL time evolution.

3. Electronic and optical properties of quantum structures

3.1. Quantum-confined structures

Advanced epitaxial growth techniques, in particular the MOCVD and the MBE [25], characterized by very rapid technological improvement, opened the way to the growth of innovative semiconductor heterostructures. Thanks to these growth techniques, a capability of interface control and of composition changes with the precision of one atomic layer, thus improving dramatically the qualities of the heterostructures.

Important parameters for the heterostructure design are the difference in the energy gaps of the semiconductors constituting the heterostructure (Fig. 2) and the band alignment, which give rise to the valence and conduction band discontinuity at the heterointerface. The two main different types of band alignment are reported in Fig. 3. In the case of type I band alignment (Fig. 3(a)), the conduction band edge and the valence band edge of the smaller gap material are, respectively, lower and higher than the corresponding edges of the wider gap material. On the contrary, for type II interfaces (Fig. 3(b)) a staggered band alignment occurs for both the valence and the conduction band edge.

The measure and the modeling of the band alignment is a crucial task for the understanding and the design of semiconductor heterostructures. More details on this important topic can be found in Chapters 2 and 12 of this book [26,27] and in Refs [28–30].

Of particular interest is the possibility to spatially confine the carriers, thanks to the spatial variation of the composition in the heterostructure, because it adds further degrees of freedom to the band gap engineering, that is to the possibility to tailor the conduction and valence band profiles to have a structure with given physical properties or to get a device with certain functionality [31]. In fact, the spatial variation of composition gives rise to a spatial variation of the electron energy and thus to potential wells that may be used to control the carrier

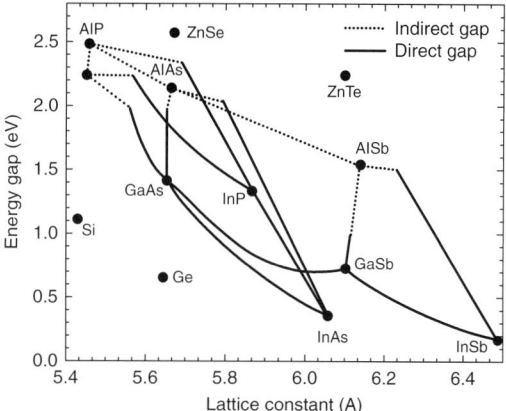

Fig. 2. Energy gap at 300 K as a function of the lattice constant. Full lines and dotted lines indicate direct gap and indirect gap ternary alloys, respectively.

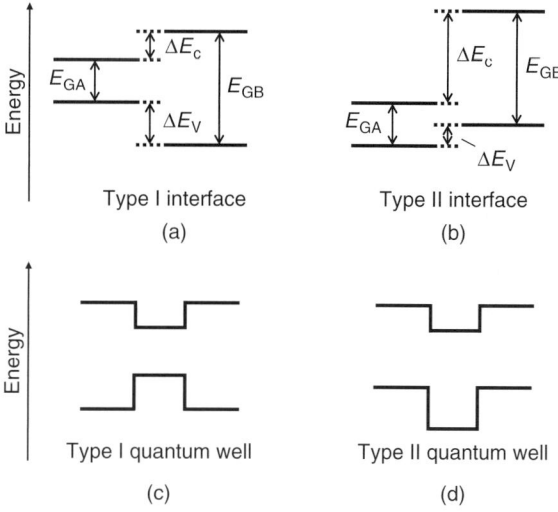

Fig. 3. Band alignment in type I (a) and type II (b) heterointerfaces and band structure of type I (c) and type II (d) quantum wells.

motion and therefore to modify the electronic and optical properties of the system. This spatial confinement can be obtained in one, two, or three directions, depending on the design of the heterostructure.

When the extent of the spatial confinement, at least in one direction, becomes comparable to the De Broglie carrier wavelength (usually of the order of few tens of nm), quantum mechanical effects occur. The particle confinement energy in a potential well depends on the spatial dimension of the well and on the height of the confining potential barriers. Thus, thanks to a suitable tuning of the structure dimensions and the barrier height, it is possible to

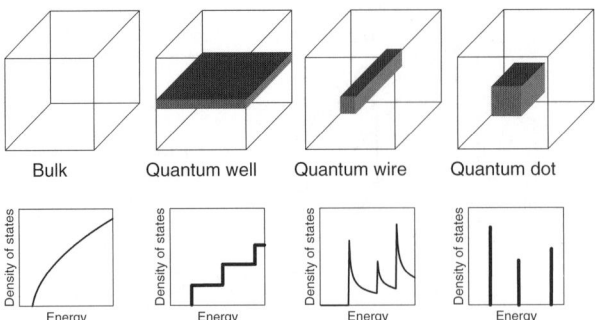

Fig. 4. Sketch of the geometry and of the density of states of a bulk material and of nanostructures in which the carrier confinement is in one (quantum well), two (quantum wire) or three dimensions (quantum dot).

modulate the electron and hole confinement energy, and thus to modulate the so-called optical gap of the nanostructure.

Nanostructures are classified on the basis of the number of dimensions in which the carriers are confined or, alternatively, free to move. Nanostructures in which the carrier confinement is in one, two, or three dimensions are sketched in Fig. 4. The structure in which carriers are confined in one dimension is a QW: it consists of a thin layer of a semiconductor with low energy gap embedded in a material with higher energy gap, and the confinement occurs in the direction perpendicular to the interfaces. In a quantum wire (QWr) the carrier motion is confined in two dimensions: a QWr consists in a long and thin rod of a semiconductor with low energy gap embedded in a material with higher energy gap. In this structure carriers are free to move in the direction of the rod only. Finally, in a QD the carrier confinement is in the three spatial directions: a QD consists of a low energy gap semiconductor nanocrystal embedded in a material with higher energy gap.

Of relevant interest for their potential technological optoelectronic applications are nanostructures made with semiconductor pairs which present a type I band alignment. In fact, as shown in Fig. 3, in type I QWs electrons and holes, localized in the lower energy gap material, are confined in the same region of the structure. This increases the wave function overlap and, of consequence, the absorption coefficient and the recombination efficiency of the active structure. On the contrary, in type II QWs electrons and holes are localized in different materials, and thus the superposition of the wave function is reduced. In the following discussion we will refer to type I nanostructures only.

The physical properties of semiconductor nanostructures and their potential applications in the field of electro-optics are determined by their electron and hole density of states (DOS), which depend on the dimensionality of the nanostructure. A sketch of the DOS for QWs, QWrs, and QDs is reported in Fig. 4. One of the main reasons of interest for optoelectronic applications of nanostructures stems from the possibility to concentrate the electron and hole DOS in a restricted energy interval nearby the ground state. Of consequence, the use of quantum confined structures as active layer in semiconductor lasers results in devices with better performance with respect to the conventional heterostructure lasers. For instance, it can be shown that the optical gain of a QW structure is about a factor 2 larger than that of a bulk semiconductor; theoretically, a further factor 10 is gained passing from a QW structure to a QD structure (Fig. 5).

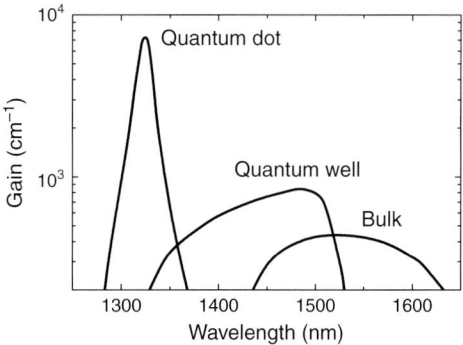

Fig. 5. Calculated gain spectra at 300 K for QD (cube of 10 nm side), QW (width of 10 nm) and bulk $Ga_{0.47}In_{0.53}As/InP$ (data from Ref. [32]).

3.2. Electronic states in quantum-confined structures

3.2.1. Single-electron states

As reported in every quantum mechanics textbook [33], the confinement of the electrons (or of the holes) in a small space region, whose dimensions are comparable to the De Broglie wavelength of the particle, induces the quantization of the electronic states from the continuum to a ladder of localized states whose energy spacing depends on the spatial extension of the confinement box.

The more widespread method for calculating the electronic energy level scheme in QWs is the effective mass approximation (EMA) [34]. In its simplest form, EMA calculates the conduction and valence band eigenvalues by solving the standard time-independent Schrödinger equation for electrons and holes with the effective masses of the relevant particle and the confining potential given by the band edge energy profile of the heterostructures [35]. A more general form of EMA requires the expansion of the electronic wave function on a set of bulk conduction and valence eigenstates [22]. The main advantages in the EMA approach are due to its simple formulation, which in some cases allows for an analytic solution and, as a consequence, for the possibility of a simple physical picture of the underlying phenomena connected with the quantum confinement. In the literature more complex and precise methods for the calculation of the QW quantum-confined states are available, namely tight-binding [36,37], pseudopotential [38], and density functional (see Chapter 2 of this book [26]) approaches.

The validity of EMA theory for the analysis of semiconductor nanostructures is based on two considerations. The first is that most of the materials which constitute the QW heterostructure display similar band structures. In fact, the more widely used semiconductor heterostructures for optoelectronic applications are made by III–V compounds with direct band gap, both in the barriers and in the well. In addition, the periodic part of the Bloch functions [39], at the relevant band edges, does not differ very much from one material to the other. The second point is connected to the fact that the relevant energy states are close to the band edges.

The relevant band edges of the III–V semiconductor QWs belong to the Γ ($k=0$) point of the Brillouin zone and, more specifically, have the Γ_6 (conduction band), Γ_8 (light and heavy hole valence bands), and Γ_7 (split–off valence band) symmetries [40]. Then, the QW eigenstates can be expanded as

$$\Psi_i^{EMA}(\mathbf{r}) = \sum_{n=1}^{N_B} \left\{ \int_{BZ} d\mathbf{k}_\| dk_z C_n^{EMA}(\mathbf{k}_\|, k_z) \exp\left[i(\mathbf{k}_\|\mathbf{r}_\| + k_z z)\right] \right\} u_{n,\Gamma}(\mathbf{r})$$

$$= \sum_{n=1}^{N_B} f_n(\mathbf{r}) u_{n,\Gamma}(\mathbf{r}) \quad (1)$$

where $u_{n,\Gamma}(\mathbf{r})$ are the bulk Bloch functions of the nth band at Γ point, $\mathbf{k}_\| \equiv (k_x, k_y)$ and $\mathbf{r}_\| \equiv (r_x, r_y)$ are the two-dimensional wave and position vectors, respectively. The integral is, in principle, extended to the whole Brillouin zone, although, as already mentioned, the $C_n^{EMA}(\mathbf{k}_\|, k_z)$ are different from zero only in a small region around the Γ point.

Taking into account the small nonparabolicity of the conduction and valence bands in III–V semiconductors, we can then decouple the Hamiltonian of the system in a sum of Schrödinger-like equations for each relevant band separately. The electron and hole states in the QW can be then found by solving the Schrödinger-like equations for the envelope functions of the respective bands [22]. Within these approximations, the nth band envelope eigenfunction

$$f_n(\mathbf{r}) = \int_{BZ} d\mathbf{k}_\| dk_z C_n^{EMA}(\mathbf{k}_\|, k_z) \exp\left[i(\mathbf{k}_\|\mathbf{r}_\| + k_z z)\right] \propto f_n(z) \exp(i\mathbf{k}_\|\mathbf{r}_\|) \quad (2)$$

will be the product of a plane wave in the direction parallel to the interface, due to the fact that the particles are free to move in the well plane with momentum $\hbar \mathbf{k}_\|$, and of a function in the z direction, solution of the equation

$$\left\{ -\frac{\hbar^2}{2} \frac{\partial}{\partial z} \left[\frac{1}{m^*(z)} \frac{\partial}{\partial z} \right] + V_n(z) \right\} f_n(z) = E f_n(z) \quad (3)$$

where $V_n(z)$ is the z profile of the Γ_6 conduction band in the case of electrons and of the Γ_8 valence band for heavy and light holes (see Fig. 6). The Ben Daniel–Duke boundary conditions [41] impose the continuity of the wave function ($f(z)$) and of the probability flux ($[1/m^*(z)\partial f(z)/\partial z]$) at the heterostructure interface.

In the following, we will treat only the case of square, type I, QWs (see Fig. 6). Usually V_n is set to zero in the well region, so that the eigenvalues of Eq. (3) give the increase (decrease) in energy, with respect to the band edge, of the electron (hole) states. Different heterostructure geometries can be found in Refs [42,43].

The solution of Eq. (3) can be easily calculated in the case of the *infinitely deep well approximation* (IDWA), as the wave function must be zero in the confining layers and at the interfaces. This gives, for the nth band, a ladder of confined eigenstates, with energies

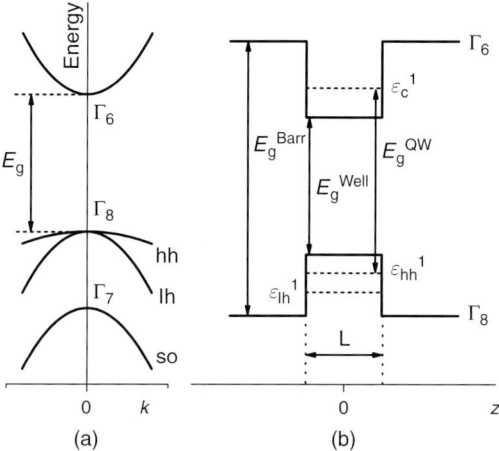

Fig. 6. (a) Typical band structure of a direct gap III–V compound semiconductor (e.g., GaAs) in the vicinity of the center of the Brillouin zone. (b) Band edge profiles of the conduction and valence (heavy and light hole) in a type I undoped QW.

$$\epsilon_n^i = i^2 \frac{\pi^2 \hbar^2}{2m^* L^2} \qquad (4)$$

where L the well width and i is an integer that identifies the quantized levels. IDWA is a very crude approximation that has problems when dealing when the band alignment leads to poorly confined carriers.

When the finite height of the confining potential barrier (ΔV_n), due to the band alignment of the constituent semiconductors, is taken into account, the IDWA results are modified along the following lines:

1. The eigenenergies are reduced with respect to the values ϵ_n^i in IDWA.
2. The localized state wave functions penetrate into the barrier region.
3. The number of confined states is finite.

A comparison between IDWA and realistic EMA predictions using Eq. (3) is reported in Fig. 7. The dependence of the lowest electronic state energy on the QW width L is reported in Fig. 7(a) for $\Delta V_n = 250$ meV and in the IDWA. For large well widths ($L > 30$ nm), the two curves almost coincide. In the case of extremely small well widths, $\epsilon_n^1 \cong \Delta V_n$, thus making the particle just slightly confined in the QW (see Fig. 7(a)), and the wave function strongly penetrates in the barriers. This is shown in Fig. 7(b), where the comparison of the full width at half maximum (FWHM, Ξ) of the square of the envelope function $|f_n^1(z)|^2$ as obtained from EMA and IDWA is reported. A marked dependence of the FWHM on L is observed, which clearly shows the localization of the electronic wave function in the z direction and its quasi-two-dimensional character. However, for well widths below 2.5 nm, the EMA wave function shows a dramatic increase of the FWHM, due to the large spread into the barriers, and for $L \approx 0$, the wave function is no more localized in the QW recovering its three-dimensional behavior.

It is worth noticing that attached to each quantized state in the z direction there is a two-dimensional momentum space (see Eq. (2) and Fig. 8). In other words, the introduction

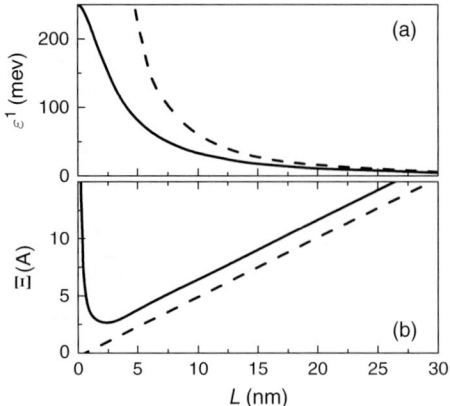

Fig. 7. (a) Lowest QW electronic state energy ϵ^1 as a function of the QW width L for $\Delta V_n = 250$ meV (continous line) and IDWA (dashed line). (b) Lowest QW electronic state wave function FWHM (Ξ) as a function of the QW width L for $\Delta V_n = 250$ meV (continous line) and IDWA (dashed line). The used effective mass is that of the GaAs conduction band ($m_e^* = 0.067\, m_0$).

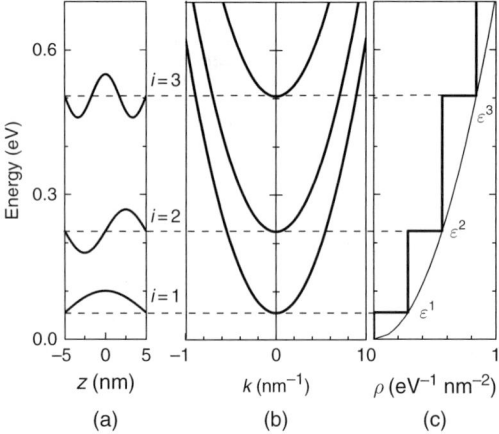

Fig. 8. (a) IDWA potential well with energy levels and wave functions. (b) Total energy including the in-plane kinetic energy of each subband. (c) Steplike density of states of an IDWA QW. The QW thickness is 10 nm. The used effective mass is that of the GaAs conduction band. The thin curve in panel (c) is the density of states of unconfined bulk electrons.

of the QW, by breaking the symmetry along the z direction, splits the n band into a set of sub-bands belonging to different quantized state along the z direction. Each state in the sub-band is identified by four quantum numbers, namely the n band index, the i sub-band index (corresponding to the number of the quantized state along the z direction), and the two momentum values parallel to the interface (k_x and k_y). The DOS $\rho_n^i(E)$ of each sub-band is a step function of the energy:

$$\rho_n^i(E) = m/(\pi\hbar^2)\theta(E - \epsilon_n^i) \qquad (5)$$

3.2.2. Optical transitions in a quantum well

Within the EMA framework, it is possible to determine the selection rules that give the allowed transitions between the QW states. The transition probability between two QW states can be derived, using the Fermi golden rule:

$$W_{j,i} = \frac{2\pi}{\hbar}\left(\frac{eE_0}{m_0\omega}\right)^2 |\langle j|\mathbf{e}\cdot\hat{\mathbf{p}}|i\rangle|^2 \delta(E_i - E_j - \hbar\omega) \qquad (6)$$

where $\mathbf{e}$ and E_0 are the electric field polarization and intensity, $\hat{\mathbf{p}}$ the momentum operator, and $\hbar\omega$ is the photon energy. When the transition takes place between an electron and a hole state confined in a QW, the initial (electron) and the final (hole) state can be written, in the EMA framework, as

$$\psi_i \equiv A^{-1/2} f_{c,i}(z) \exp(i\mathbf{k}_\|^{c,i}\mathbf{r}_\|) u_{c,\Gamma}(\mathbf{r}) \qquad (7)$$

$$\psi_j \equiv A^{-1/2} f_{v,j}(z) \exp(i\mathbf{k}_\|^{v,j}\mathbf{r}_\|) u_{v,\Gamma}(\mathbf{r}) \qquad (8)$$

so that the matrix element in Eq. (6) takes the form:

$$\langle j|\mathbf{e}\cdot\hat{\mathbf{p}}|i\rangle = p_{cv} g \delta(\mathbf{k}_\|^{c,i} - \mathbf{k}_\|^{v,j}) \int dz f_{v,j}^*(z) f_{v,i}(z) \qquad (9)$$

Here $p_{cv}g = \langle u_{v,\Gamma}|\mathbf{e}\cdot\hat{\mathbf{p}}|u_{c,\Gamma}\rangle$ is the optical oscillator strength of conduction to valence electron transitions and g is the spin–orbit factor, which takes into account the selection rules for optical transitions between states with different angular momentum (e.g., no transition is possible between conduction and heavy hole states with the emission of a photon polarized in the z direction) [22]. The possible recombination transitions, which involve conduction and valence band derived sub-bands, are thus vertical in the $\mathbf{k}_\|$ space. The oscillator strength of such transitions is controlled by the overlap $\int dz f_{v,j}^*(z) f_{v,i}(z)$ between the envelope functions of the initial and final states. The overlap integral assumes a simple form in the case of a type I IDWA-QW:

$$\int dz f_{v,j}^*(z) f_{v,i}(z) = \delta_{i,j} \qquad (10)$$

In other words, the interband transitions in IDWA-QW are permitted between electron and hole states with the same principal quantum number: $\Delta n \equiv (i-j) = 0$. In the case of QWs with finite confinement potentials, this rule is no more strictly followed, but, anyway, transitions with $\Delta n \neq 0$ are very weak. For symmetric QWs, transitions with $\Delta n = \pm 2m+1$, with $m = 0, 1,\ldots$ are strictly forbidden. A schematic of the permitted transitions in a type I symmetric QW is reported in Fig. 9.

3.2.3. Excitons

Excitons are fundamental electronic excitations in semiconductor crystals associated to bound states of e–h pairs [39]. As in the case of the hydrogen atoms, the Coulomb attraction leads to

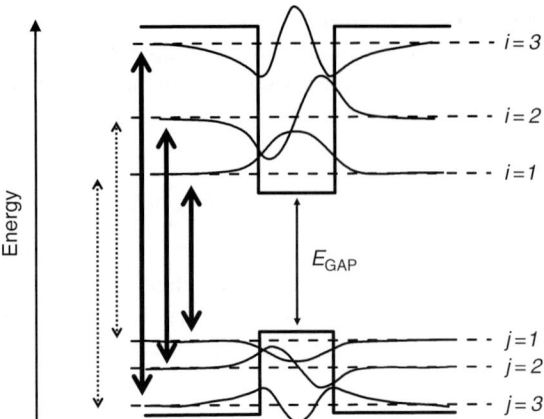

Fig. 9. Transitions between bound states in the valence and conduction bands of a QW. Only the envelope functions are shown. Countinous lines indicate strong transitions ($\Delta n = 0$), and dashed lines mark weaker, but allowed, transitions.

bound levels, characterized by discrete quantum numbers with energy $-R^*/n^2$ below the continuum level. The exciton binding energy and the exciton Bohr radius (a_B^*) in bulk semiconductors are:

$$R^* = \frac{2m_{re}^* e^4}{\hbar^2 (8\pi\epsilon)^2}$$
$$a_B^* = \frac{4\pi\epsilon\hbar^2}{m_{re}^* e^2} \qquad (11)$$

where m_{re}^* is the reduced mass of the e–h pair ($1/m_{re}^* = 1/m_e + 1/m_h$) and ϵ the permittivity of the semiconductor. Therefore, the fundamental optical transition energy occurs at a photon of energy $h\nu = E_g - R^*$.

Because in usual semiconductors $a_B^* \approx 10$ nm, wave functions and energies of excitons are quite modified in a QW whose thickness is usually of the order or smaller of the Bohr radius a_B^*. In the limiting case, that is when the exciton can be considered exactly two-dimensional ($L \ll a_B^*$), the exciton binding energy is four times larger than in the bulk ($R_{2D}^* = 4R^*$), while the effective exciton radius is reduced by a factor two ($a_{2D}^* = a^*/2$).

An exciton in a QW of finite thickness is described, within the EMA approach, by the two particle Schrödinger equation:

$$\left[-\frac{\hbar^2}{2m_e^*}\nabla_{r_e}^2 - \frac{\hbar^2}{2m_h^*}\nabla_{r_h}^2 - \frac{e^2}{\epsilon_0 |r_e - r_h|} + V_e(r_e) + V_h(r_h) \right] \Psi(r_e, r_h) = E\Psi(r_e, r_h) \qquad (12)$$

where $V_a(r_a)$ ($a = e, h$) are the spatially varying confining potential for electron and holes, respectively. The fundamental excitonic transition energy is then given by

$$E_{QW}^{exc} = E_g + E_{c,1} + E_{hh,1} - R_{QW}^* \qquad (13)$$

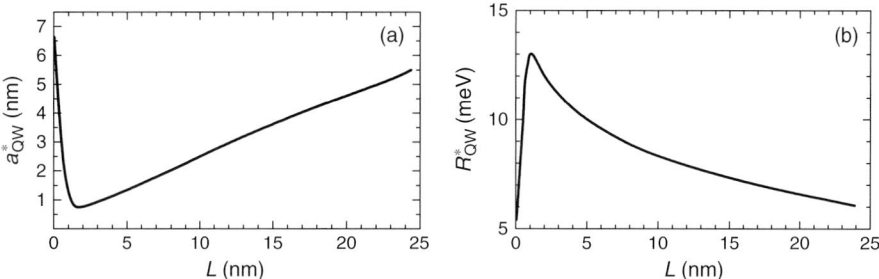

Fig. 10. Heavy hole exciton extension in the z direction (a) and heavy hole exciton binding energy (b) for GaAs/Al$_{0.4}$Ga$_{0.6}$As QWs (data from Ref. [46]).

where E_g is the energy gap of the well material, $E_{c,1}$ and $E_{hh,1}$ are confinement energies of electrons and heavy holes, and R^*_{QW} is the exciton binding energy. The exciton binding energy R^*_{QW}, oscillator strength, and Bohr radius depend on the well width [44]. Binding energies and oscillator strengths are first increased as the well width is reduced, due to the smaller e–h separation. Values even larger than the two-dimensional limit can be observed in GaAs/AlAs QW due to the concomitant effect of the dielectric mismatch [47]. The large binding energy makes the excitons in QW very stable and they dominate the PL spectra even at room temperature. For narrow wells the wave function penetrates into the barriers (see Section 3.2.1), and the exciton binding energy decreases toward the bulk value [48]. The calculated dependencies of the ground state R^*_{QW} and a^*_{QW} on the well width are reported in Fig. 10 [46].

As far as the radiative decay of excitons is concerned, it is worth noting that, in the bulk, the exciton is a coherent elementary excitation over the whole crystal, with a macroscopic transition dipole moment. Then, due to the translational symmetry of the system, the exciton can only interact with a photon that has the same wave vector: as a result, the polariton, i.e., a hybridized mode of exciton and photon, is formed [49]. The decay of the polariton in the bulk crystal is possible only by a leak through the surface of the crystal or by radiative and nonradiative recombination at crystal imperfections. Experimentally, long PL decay times of the order of few nanoseconds have been measured in bulk GaAs [50]. In a two-dimensional QW, the breaking of the translational invariance dramatically modifies this result. The QW exciton can decay super-radiantly in a monodimensional continuum of states, through its macroscopic transition dipole moment. Predicted decay times are of the order of few tens of picosecond, depending on the well width [51,52].

3.2.4. Excitons in disordered media

Even planar semiconductor QWs are unavoidably characterized by structural defects arising in the fabrication process. This disorder affects the optical response of the lowest-lying excited states, particularly the exciton. Although disorder is usually considered an unwanted feature [53], disorder breaks the in-plane translational symmetry and can lead to a dramatic change in the physical behaviour of excitons, producing multiple scattering and localization [54]. This behaviour has several implications in the optical response of the system. In particular, it produces an inhomogeneous spectral broadening that is often used as a figure of merit of the interface quality.

Now some definitions are introduced for supporting the following discussion; to this aim we follow mainly Refs [7] and [55]. The epitaxial growth process takes place in the so-called layer-by-layer mode, but usually a new layer starts growing before the surface of the previous layer is completed [56]. For this reason, the interface between two materials A and B will be not well defined on an atomic scale, but will be characterized by a structural, or morphological, disorder on this scale. The term *interface roughness* refers to this kind of disorder. Following Ref. [55], the interface between a material A and a material B is defined as *abrupt* if the transition between A and B occurs in one atomic layer on all the interface. This means that the deposition of the material B after termination of the epitaxial growth of the material A does not perturb the atomic configuration of material A. An abrupt interface is defined *atomically smooth* on a length scale d if the steps on the surface of the material A do not have lateral extent larger than d. It is also useful to classify the lateral steps on the basis of their lateral dimensions: a step with lateral dimension larger than the exciton radius a^*_{QW} is usually called *island*, whereas lateral step dimensions much lower than a^*_{QW} is called *microroughness*.

An interface between two materials A and B can also be classified as *direct* or *inverted* [57]. In fact, the interface roughness of QWs grown by epitaxy depends also on the diffusivity of the atoms of the two materials which form the interface. Just to fix the ideas, let us consider the case where a cation of one of the two materials constituting the interface has a large diffusivity; this is the case of Ga at the GaAs/AlAs interface. Taking the growth direction as a reference, the two interfaces of a QW behave in a different way and have different denominations. In a GaAs/AlGaAs QW, the normal or direct interface is that in which the alloy AlGaAs is grown on GaAs and the inverted one is that in which GaAs is grown on AlGaAs. A schematic of normal and inverted interfaces is reported in Fig. 11. In general, the interface in which the segregating element is in the upper layer is called *inverted*, whereas the one in which the segregating element is in the lower layer is called *normal* or *direct* [58]. It has been shown that, due to the diffusion process, the inverted interface is rougher than the normal or direct one (see Ref. [59] and references quoted therein). This asymmetry can be evidenced with transmission electron microscopy (TEM) [59] and with spectroscopic techniques, in particular PL.

In fact, interface disorder results in an in-plane fluctuation of the electron and hole confinement energy in the QW which induces an inhomogeneous spectral broadening of the emission [53]. However, disorder can affect not only the potential but also the kinetic

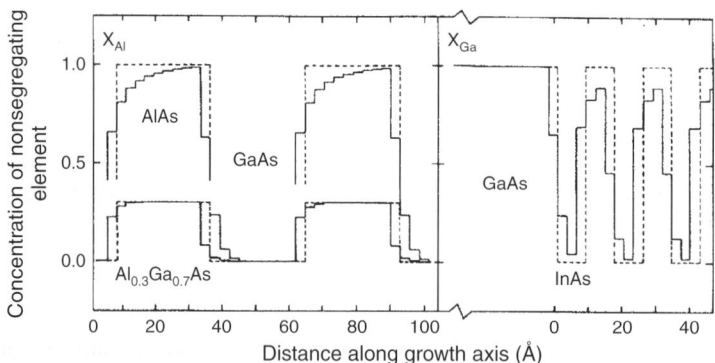

Fig. 11. Sketch of the profile of the composition for AlGaAs/GaAs QWs. The difference between *direct* and *inverted* interfaces is evidenced (reprinted figure with permission from Ref. [58]. Copyright (1989) by the American Physical Society).

energy of the exciton moving along the QW plane. The exciton then behaves as a massive particle subject to a disordered potential, giving rise to spatially localized eigenstates of the center of mass (COM). A complete review of the theory of exciton states and their optical properties in disordered QWs can be found in Ref. [54].

In most QWs of average quality, the exciton binding energy is significantly larger than the confinement energy fluctuations due to interface disorder. This suggests that disorder has only a weak influence on the e–h relative motion in the exciton state. The solution $\Psi(\mathbf{r}_e, \mathbf{r}_h)$ of the exciton Schrödinger Eq. (12) can be therefore factorized into the confinement electron and hole wave functions (along the growth direction) and the in-plane wave functions for the relative and COM exciton motion [60]. The presence of disorder principally affects the exciton COM part of the wave function, giving rise to exciton states with a spatially localized COM motion. The presence of exciton COM localized states has strong effects on the optical DOS. The first effect is, naturally, to induce an inhomogeneous bradening of the emission lines. However, the emission spectrum, even in the presence of a Gauss-distributed random potential, shows an asymmetric lineshape, with a tail on the high energy side [61,54]. The exciton COM localization lengths are distributed over a broad range and show a clear energy dependence with a dramatic rise as a function of energy across the inhomogeneously broadened exciton line [62].

An interesting effect induced by the wave function localization in disordered media is the spatial and energetic correlations of the wave functions as well as the corresponding correlations of energy levels [63]. In simple words, two localized exciton states, due to quantum correlation effects, cannot be simultaneously close in energy and space. Therefore, two localized wave functions having almost identical energies will, in general, be localized in different spatial areas. Such quantum correlation effects, called "level repulsion", can be detected in the autocorrelation of near-field exciton emission spectra. Evidences of level repulsion effects are reported in the literature [63–65] showing the localized nature of QW excitons in the presence of a disordered interface.

3.3. Physics of the recombination process

In a PL experiment, e–h pairs are photoexcited by photon absorption. In case of nonresonant excitation, when the photon energy is larger than the energy gap of the barrier material, the direct absorption in the nanostructure can be neglected, due to its very low thickness. The e–h pairs in the barrier material are excited in nonequilibrium states; then they diffuse and relax into a quasi-equilibrium distribution and at the same time recombine radiatively and non-radiatively. In QW systems, the confined states in the well are the lowest states and therefore carriers eventually relax on them; this process is sometimes also quoted as carrier capture.

Then, different processes can be distinguished in the PL process: e–h pair excitation by photon absorption, e–h thermalization and diffusion, and e–h pair recombination. These are now shortly discussed. After the absorption, the spatial dependence of the photoexcited carrier density is given by the well-known Beer's law:

$$G(x) = [1 - R]\alpha I(0) \exp(-\alpha x) \qquad (14)$$

where x is the coordinate normal to the sample surface, R is the surface reflectivity of the sample, $I(0)$ is the incident light intensity, and α is the absorption coefficient. The α and R values depend on the excitation energy and for the most common semiconductor materials are tabulated in handbooks [66].

The penetration depth of the exciting radiation, i.e., $1/\alpha$, depends on the nature, direct or indirect, of the energy gap E_G of the barrier and on the difference between the energy of the absorbed photon $\hbar\omega_{exc}$ and E_G. For indirect gap semiconductors such as Si, Ge, or AlAs, the penetration depth is of the order of few micrometers; on the contrary, in direct gap semiconductors the penetration depth varies from a fraction of micrometer for near-band-gap excitation to few tens of nanometers for excitation energies much larger than E_G. In Eq. (14) it has been assumed that the diffusion is negligible, which is very often a wrong assumption. Diffusion tends to spread the carriers over a volume of the order of the diffusion length (usually several micrometers).

In nonresonant excitation conditions, $\hbar\omega_{exc}$ is larger than the barrier energy gap and usually the carriers have a kinetic energy, larger than the lattice thermal energy, which is mostly given to the electrons, due to their smaller effective mass with respect to holes. Carriers tend to relax their excess kinetic energy by reaching a thermal equilibrium with the lattice. Different steps can be distinguished during the thermalization process. In the first step, the carriers thermalize among themselves reaching a common temperature that is higher than the lattice temperature. In the second step these hot carriers lose their excess energy by emitting optical phonons, thus cooling down toward the lattice temperature. These first two steps take a time of the order of some tens of picoseconds after the excitation. The third step is the final carrier cooling and consists in the emission of acoustic phonons. During this last step, which is in the hundreds of picoseconds range, the carriers start to recombine, both radiatively and nonradiatively. Obviously, in QWs the main contribution to the radiative recombination is given by the electronic states confined in the nanostructures. The thermalization steps are discussed in more detail in Ref. [4] and in the references quoted therein. Further details on the thermalization process in nanostructures will be discussed in the following sub-section 4.2.

During the thermalization process, the carriers diffuse away from the surface inside the sample; the spatial profile of the photoexcited carriers is modified, with an increase of the width of the region in which excited carriers are present. This also helps the carrier capture into the QW, where they recombine. However, due to the presence of the thermalization and diffusion processes in the barrier, the PL efficiency in semiconductor nanostructures depends also on the properties and on the quality of the barrier material [67].

On the other hand, in resonantly excited nanostructures the exciting photon energy is lower than the energy gap of the barrier material and thus photon absorption generates e–h pairs directly into the nanostructure. The total absorption rate is then reduced, due to the nanometric thickness of the absorbing material, the diffusion process is limited to the nonconfined spatial directions, and the thermalization proceeds only through the states confined in the nanostructure. One of the main advantages of the resonant excitation is therefore to bypass the carrier capture and the possible nonradiative recombination in the barrier layer [67].

Once the photoexcited carriers have reached the ground state, the PL lifetime τ_{tot} is given by the contributions of the radiative and nonradiative lifetime (τ_r and τ_{nr} respectively), by the relation [2]:

$$\frac{1}{\tau_{tot}} = \frac{1}{\tau_r} + \frac{1}{\tau_{nr}} \tag{15}$$

As a consequence, the radiative efficiency is given by τ_{tot}/τ_r and usually strongly depends on temperature and excitation power density. The carrier lifetime in QW structures will be discussed in detail in the following sub–sections 4.2 and 4.3.

3.4. Quantum well emission

The luminescence of undoped QWs at low temperatures consists of a quite narrow line (see Fig. 12 where only a limited energy range is reported; please note also the logarithmic scale of the intensities) with different spectral characteristics and brightness which make it different from the impurity-related emissions observed in the constituent bulk materials [1,53,68]. In fact, the emission in high-quality undoped QW is mainly due to intrinsic exciton recombination [53] and several reasons concur to this:

1. The carrier capture in QW at low temperature is extremely efficient due to the efficient thermalization mechanisms (Section 3.3) and to the localized nature of the QW states (Section 3.2.1).
2. The exciton is thermally stable (Section 3.2.3).
3. The free exciton radiative lifetime increases due to quantum confinement in two dimensions [44–46] (Section 3.2.3).
4. Impurity gettering may occur during the multilayer growth, which reduces the nonradiative recombination center density [69].

Evidence of quantum size effect is clearly shown in Fig. 12. As the QW width decreases, the emission undergoes to a clear blue-shift. This is mainly because the electron and hole confinement energy in QWs is strongly determined by the width of the QW. The doublet clearly visible in the emission of the $L = 22.2$ nm QW is attributed to single-exciton (high-energy structure) and bi-exciton (low-energy structure) recombinations [70–72].

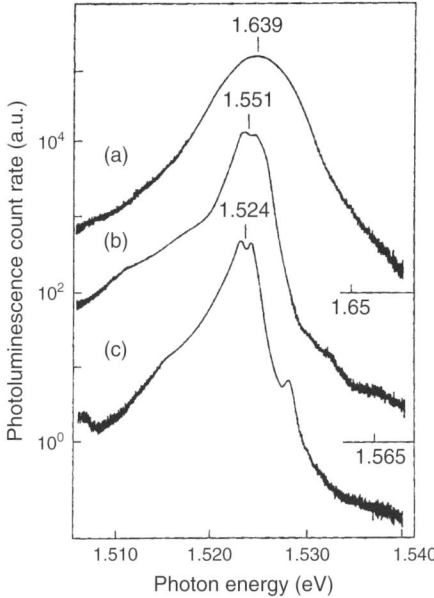

Fig. 12. PL from QW with varying thickness: (a) 5.1 nm, (b) 10.2 nm, and (c) 22.2 nm (reprinted figure with permission from Ref. [14]. Copyright (1991) by Elsevier).

At low temperature the main source of the PL line broadening is the inhomogeneous broadening introduced by the interface disorder (see Section 4.1). As the temperature increases, exciton–phonon scattering events introduce another source of PL line broadening [73]. The phonon-driven homogeneous increase of the emission linewidth with the temperature follows the law [74]:

$$\Gamma_{\text{hom}}(T) = \Gamma_{\text{hom}}(0) + \gamma_{\text{AC}} T + \gamma_{\text{LO}} n_{\text{LO}}(T) \tag{16}$$

where the term linear in the temperature is due to exciton scattering from acoustic phonons, and the term involving the Bose function of the LO-phonons $n_{\text{LO}}(T)$ is due to interactions with LO-phonons. The homogeneous linewidth al zero temperature (which may be due to impurity scattering) is usually quite small [74].

Linewidth-broadening coefficients of ≈ 4 meV/K for acoustic phonons and ≈ 17 meV for optical phonons have been reported in GaAs/Al$_{0.4}$Ga$_{0.6}$As QWs [74]. Contradictory results are reported in the literature on the well-width dependence of these coefficients. A nearly constant value was found by Gammon et al. [74] and even a decreasing value when decreasing the well width is reported by Schultheis et al. [75], while Hillmer et al. [76] measure exciton mobilities indicating an acoustic-phonon scattering that is decreasing with increasing the well width.

4. Photoluminescence characterization of quantum wells

In fact, PL is an extremely powerfull tool for characterising QW properties. Many intrinsic effects in QWs can be accessed via PL, like band offsets [77], exciton binding energies [47], trions [78], biexcitons [71], quantum-confined Stark effect [79,80], homogeneous broadening [73,74], etc. In addition, through PL measurements it is possible to get detailed informations on the structural disorder, defect content, and nonradiative recombination channels of a QW system. In other words, via PL measurements it is possible to characterize the overall QW quality, from both electronic and structural side. In the following we will focus our discussion on the QW quality caracterization by PL, in agreement with the book plan.

4.1. Interface evaluation

In the previous section, it has been shown that the width of the structure has an important role in determining the exciton recombination energy in semiconductor QWs. This is because the electron and hole confinement energy in QWs is determined by the height and the width of the potential well. Therefore, as it will be discussed in detail in the following, the quality, and in particular the flatness, of the interfaces has important effects on PL spectra of QWs and thus on the performances of devices. PL and PLE are frequently used in place of other more complex and expensive techniques such as the TEM and X-ray diffraction. An overview of the applications of PL in the characterization of QW heterointerfaces is reported in the review papers by Bajaj [7] and Herman and coworkers [6].

In high-quality QW structures, the low temperature recombination is dominated by exciton transitions, and thus PL and PLE spectra provide information on a sample volume comparable with the exciton volume, which is usually on the nanometer scale. Therefore, PL and PLE measurements provide a simple way for getting useful information on the QW quality, and in particular on the disorder present at the QW heterointerfaces, as evidenced for the first time in Ref. [53].

4.1.1. Linewidth and shape

As shown by Eq. (13), the $n=1$ exciton recombination energy in a QW is directly related to the properties of the materials constituting the structure and to the morphological characteristics of the confining structure. In fact, the energy gap of the barrier and QW materials determine the amplitude of the confining potential, and thus the confinement energies, and the energy gap of the QW material directly enters Eq. (13). Furthermore, the electron and heavy hole confinement energies are also determined by the QW width (see Section 4.1). In the characterization of the QW morphology, it is commonly assumed that the exciton binding energy E_{exc}(QW) does not depend on the QW width fluctuations; in fact, the E_{exc}(QW) value is a small fraction, of the order of 1%, of the QW energy gap E_G, and thus its dependence on the QW width fluctuations can be neglected [6].

The modeling of the effects of the interface roughness on the shape of the exciton PL spectrum in QWs has been discussed in a number of papers. In the following, the conclusions of these works are summarized; for a detailed discussion of the experimental data and of the theoretical models, which in some cases allow also quantitative conclusions, the reader is referred to the review papers by Bajaj [7], Herman and coworkers [6], and Zimmermann and coworkers [54].

In the analysis of the information that can be deduced on the interface properties from the linewidth and shape of the PL spectra, it is useful to take into account two different situations:

1. When the barrier and/or the QW materials are semiconductor alloys, the alloy disorder may give an important contribution to the FWHM of the exciton recombination, as widely discussed in the case of bulk semiconductors [7,81];
2. When the growth of the structure is interrupted at the interfaces, the interface quality is deeply modified [9]. In fact, the QW heterointerfaces are not precisely defined, having, in general, asperities of the order of one monolayer. It has been proved that growth interruption up to several tens of seconds before the interface formation produces an effective smoothing of the interface roughness [82,83].

We introduce the following discussion summarizing the results of a simple bimodal model for the interface roughness [84], which considers the presence of both nanoroughness and islands; the model can account for the main experimental features of the PL spectra. The nanoroughness with a length scale smaller than the exciton diameter does not influence the exciton transition energy, the QW width being defined by the average, or effective, position of the interface, but it is predicted to play a fundamental role in determining the exciton COM localization [54]. On the contrary, if in the QW extended regions are present which differ in effective thickness by one or more monolayers, the PL spectrum appears structured, being composed of different lines separated in energy by a quantity corresponding to a monolayer thickness variation.

A schematic of the interface disorder is reported in Fig. 13. We start discussing the situation in which alloy disorder effects are not present in the barrier and in the QW material.

Without growth interruptions, interfaces with a short range disorder on a scale much lower than the exciton diameter are expected. In this case a inhomogeneously broadened lineshape (see Section 3.2.4) for the exciton recombination is found, with a FWHM related to the interface disorder [6,54,85]. Experimental data on QWs and superlattices support this conclusion [86–89].

As a consequence of growth interruptions, the interface microscopic structure completely changes. It has been reported in different papers (see Ref. [7] and references quoted therein) that

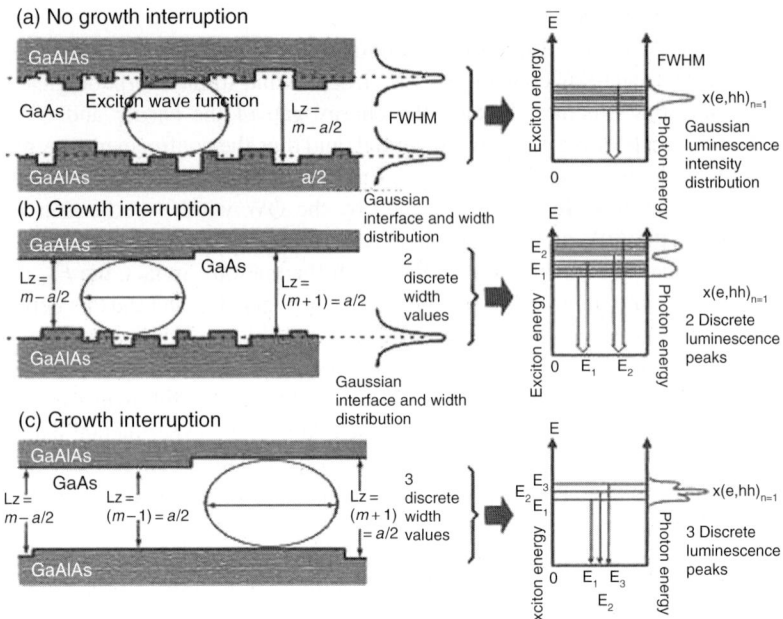

Fig. 13. Sketch of the interface disorder in a AlGaAs/GaAs QW; (a) QW grown without growth interruption; (b) QW in which, as a consequence of growth interruption, the GaAs surface (normal interface) is smoother than the AlGaAs surface (inverted interface); (c) QW grown with a long growth interruption at the interfaces: smoothing of both interfaces results. The energy states and the corresponding PL spectra are also sketched for each case (reprinted figure with permission from Ref. [6]. Copyright (1991) by the American Institute of Physics).

growth interruptions play an important role in the improvement of the interface quality, due to the formation of large dimension islands [87]. This is usually attributed to the cation diffusion on the interface during the growth interruption.

For the analysis of the effects of the growth interruption, it should be taken into account that, as discussed before, usually the two QW interfaces are not equivalent. The effects of growth interruption at the interfaces of GaAs/AlAs QWs is discussed in detail in Ref. [87]; these results can be summarized as follows (Fig. 14). In samples grown without interruptions, where the interface roughness is fully averaged over the exciton area, the PL spectrum is composed of a single peak with Gaussian shape. In the case of interruption at the lower interface similar results are found; in general, for the inverted interface the growth interruption is not expected to give rise to a significant reduction of the interface roughness. On the contrary, if the growth interruption is performed at the upper interface, the PL spectrum changes and becomes composed of two components of Gaussian shape in agreement with the fact that growth interruption at the normal interface allows the formation of islands of sizable dimensions. Finally, in samples where the growth has been interrupted at both interfaces for long times, two or three well-resolved components with Lorentian shape are found: in this case the islands become sufficiently large to give rise to discrete well-resolved levels. This is due to the fact that the smoothing of the inverted interface is slower than the one of the direct interface [6].

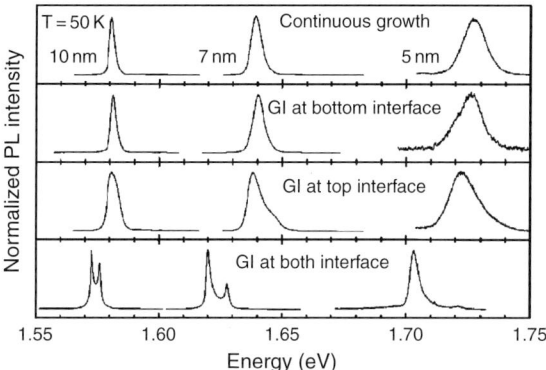

Fig. 14. PL spectra at T = 50 K of three GaAs/AlAs QWs 5 nm, 7 nm, and 10 nm thick. The samples were grown with 120 s growth interruption at the different interfaces as indicated (reprinted figure with permission from Ref. [87]. Copyright (2000) by the American Physical Society).

When well-resolved components are observed in the PL spectra of QWs, the thermalization of excitons in regions characterized by lower energy can be evidenced (Fig. 15); the relative intensity of the high energy component, corresponding to the narrower region of the QW, increases for increasing temperature [6].

We move now to the analysis of the effects of the alloy disorder on the shape, and in particular on the FWHM, of the QW-PL spectrum. Detailed models have been proposed to take into account the combined effects of the alloy and interface disorder [7,90]. We summarize in the following the discussion proposed in Ref. [90] for the analysis of linewidth of PL spectra in InGaAs/GaAs QWs; in this case the alloy disorder characterizes the material constituting the QW. In Ref. [90] it is shown that the dependence of the exciton FWHM on the QW thickness is well described only when interface roughness and alloy disorder are taken into account. In wide QWs, where the exciton wave function is fully localized in the well, the experimental FWHM is fully described by the disorder contribution alone. On the contrary, for narrow QWs the experimental data can be accounted for by considering also the effect of the interface roughness.

In the case of GaAs/AlGaAs QWs, the alloy constitutes the barrier material; in this case, the alloy disorder is expected to play a relevant role in narrow QWs, because in this case the exciton wave function extends in the barrier material [91].

4.1.2. Stokes shift

In the previous section, it has been shown that the FWHM of the exciton recombination in QWs provides a qualitative but easy indicator of the effects of the interface roughness and of the alloy disorder in the well and/or in the barrier material on the structure quality.

A further relevant feature of the optical spectra of QWs is the difference between the energy of the maximum of the absorption spectrum, or more commonly of the PLE spectrum, and that of the emission spectrum, the so-called Stokes shift (SS) (Fig. 16). The SS is commonly considered, together with the FWHM of the PL spectrum, an indicator of the QW quality; in particular, the presence of a large SS is a negative indicator of the QW quality.

It has been shown that the values of FWHM and SS are related to each other, in that they are determined by the same factors characterizing the QW morphology. Very interesting is

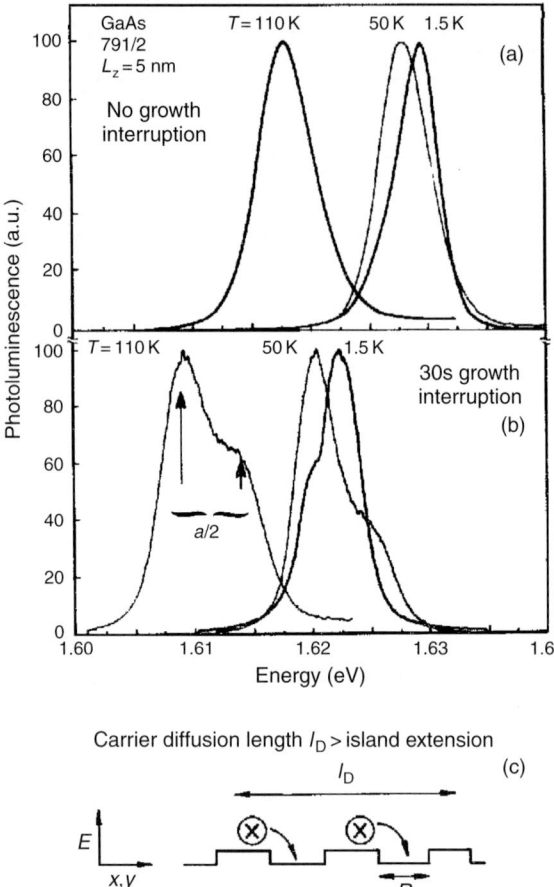

Fig. 15. PL spectra of two similar QWs grown with (b) and without (a) interruption at the interfaces measured at different temperatures. The carrier thermalization process, schematically shown in (c), explains the doublet structure present in the high-temperature spectra in (b) (reprinted figure with permission from Ref. [6]. Copyright (1991) by the American Institute of Physics).

the first paper on this subject [93], in which FWHM and SS data from III–V and II–VI QWs are collected and compared. It is shown that a general trend exists, independent of the material system: the SS is proportional to the FWHM over at least two orders of magnitude, as shown in Fig. 17. The origin of the SS is explained as due to the presence of QW width fluctuations, which extend over a length scale larger compared with the exciton diameter. The absorption spectrum reflects the probability distribution of the QW width, and the shift of the PL peak is due to the exciton relaxation by phonon emission in the largest regions of the QW before recombination. In the same paper [93], the experimental data (see Fig. 17) are compared with a theoretical calculation based on a simple model that describes the fluctuations of the effective well width with a Gaussian random function and neglects the alloying disorder effects. The model gives a constant ratio of the SS to the FWHM, whose value (0.553) is in fairly good agreement with the experimental results (0.60).

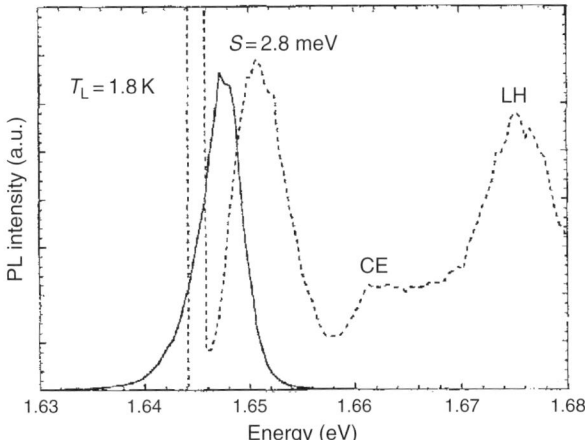

Fig. 16. Comparison between the PL (full line) and the PLE (dashed line) spectra of a 4 nm GaAs/AlGaAs QW at $T = 1.8$ K. A SS of 2.8 meV is observed. Note that the continuum edge and the light hole exciton recombination are resolved in the PLE spectrum (reprinted figure with permission from Ref. [92]. Copyright (1994) by the American Physical Society).

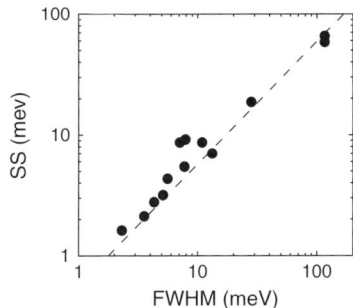

Fig. 17. Literature data of SS and FWHM in different QW systems. The line shows the theoretical prediction with a slope $\gamma = 0.553$ [93] (the experimental data come from Refs [94–101]).

Later on, this model has been criticized by Gurioli and coworkers [102], who have shown that in good quality QWs the SS can be simply explained as a consequence of the thermal population of the exciton band inhomogeneously broadened by interface roughness and alloy disorder. From this hypothesis, the authors deduce a quadratic dependence of the SS on the FWHM of the exciton PLE spectrum with a coefficient depending on the effective carrier temperature T_c: these conclusions are in very good agreement with experimental data on GaAs/AlGaAs single QWs of different widths.

An exhaustive study on the SS has been later reported by Polimeni and coworkers [103], who measured and discussed the relationship between the SS and the FWHM for the heavy hole free exciton in a large set of InGaAs/GaAs QWs with different In molar fraction and therefore with different optical quality. The results of this work show that experimental data follow the previsions of one model or the other depending on the values of SS and FWHM, i.e., on the sample quality. In samples characterized by FWHM values lower than about 6 meV, i.e., in samples with low In molar fraction, the exciton thermalization process in the broadened exciton band dominates and the dependence of the SS on the FWHM is quadratic, in agreement with the model proposed by Gurioli and coworkers [102]. On the contrary, in samples characterized by values of FWHM larger than about 6 meV experimental data show a linear dependence of SS on the FWHM, thus following the model by Yang and coworkers [93]. A slightly different conclusion has been reported by Aït-Ouali and coworkers [104]. These authors, analyzing the SS of InGaAs/InP and InGaP/InP multiple QWs, concluded that the two contributions of thermalization and localization are always contained in the SS, independently of its value.

4.1.3. Localization

Spatial exciton localization is one of the main consequences of the heterointerface roughness; evidence of exciton localization, as discussed in the preceding sections, can be found in the FWHM of the PL and PLE spectra of QWs and in the SS between absorption (or excitation) and recombination spectra.

Strong spatial exciton localization manifests itself also in the shape of the PL spectra, through the presence of multiplets, in general doublets, with an energy separation consistent with difference in the QW thickness of one monolayer, the so-called monolayer splitting (see Sections 4.1.1 and 4.1.2 and Refs [6,55]).

Spectroscopic measurements with high spatial and spectral resolution directly evidence the exciton localization. These measurements show that the PL band of a QW, with FWHM of the order of tens of meV due to the inhomogeneous broadening, is due to the superposition of a number of very narrow lines, with FWHM of the order of few μeV due to the recombination of excitons localized at the interface fluctuations [105]. These effects are more evident in narrow QWs. The regions at which excitons localize have been also called naturally formed QDs [105]. The exciton localization is due to the fact that electrons and holes can be localized into energetically favorable QW regions that are a monolayer wider than the surrounding regions. If these regions have typical dimensions of the order of 10^2 nm, they can laterally confine the carriers, modifying their confinement energy. The recombination of excitons localized in these regions results in narrow lines with FWHM of the order of few μeV. The different lateral dimensions of these regions give rise to recombination lines of different energy which contribute in forming the inhomogeneously broadened PL band, typical of the recombination in QWs.

4.2. Carrier dynamics and recombination kinetics

Most of the PL experiments are performed using nonresonant excitation (in the barrier) in which the continuum states are initially excited. The photoexcited e–h pairs thermalize (see Section 3.3), and the pairs with low kinetic energies emit acoustic phonons to form excitons at large in-plane wave vectors $\mathbf{K}_\parallel$ in less than 20 ps. These excitons interact among themselves and with phonons relaxing to $\mathbf{K}_\parallel = 0$ excitons which can radiate. The initial dynamics of nonresonantly excited luminescence in intrinsic samples is dominated by this relaxation process and shows a slow rise of several hundreds of picoseconds [106–108]. In samples exhibiting localization effects, relaxation through localized states plays a major role and also modifies the radiative recombination kinetics [109]. During the process of formation of excitons and relaxation to $\mathbf{K}_\parallel = 0$, excitons also undergo spin relaxation process [110].

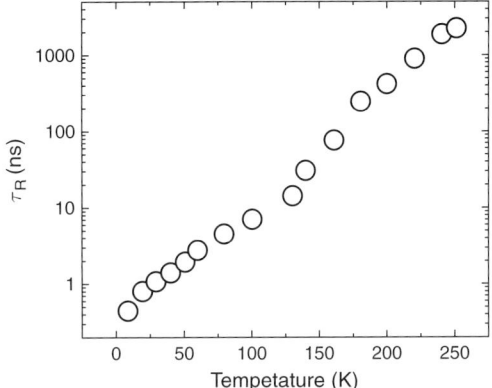

Fig. 18. Temperature dependence of the radiative recombination time (Eq. (17)) extracted from the combined measurements of $\tau_L(T)$ and $I_L(T)$ for a 7 nm GaAs/AlGaAs QW (data from Ref. [112]).

As already reported in Section 3.2.3, in QWs the breaking of the translational invariance along the z (growth) direction strongly modifies the nature of the polariton [51,52] giving rise to a very short lifetime; predicted decay times are of the order of few tens of picoseconds, depending on the well width [111] for excitons having $\mathbf{K}_\parallel \leq \mathbf{K}_{ph}$, where $\mathbf{K}_{ph}$ is the photon wavevector. At the same time excitons with $\mathbf{K}_\parallel > \mathbf{K}_{ph}$ are optically inactive, because they cannot annihilate in a photon due to the $\mathbf{K}_\parallel$ conservation law. Despite the theoretical predictions, all the nonresonant experiments have reported long decay times for the exciton PL [16,112]. In addition, a strong temperature dependence of the radiative decay time τ_R, which exceeds the µs already at 200K, has been observed in GaAs/AlGaAs QWs [112] against a theoretical prediction of a linear dependence of τ_R with a slope of $\approx$35 ps/K [52] (see Fig. 18).

The understanding of the QW PL lifetime is strictly related to the thermalization between the different states involved in the recombination kinetics, which are:

1. Free excitons at $\mathbf{K}_\parallel \leq \mathbf{K}_{ph}$, which have very short decay time ($\approx$25 ps) [51,52].
2. Free excitons at $\mathbf{K}_\parallel > \mathbf{K}_{ph}$, which are dark due to the conservation of the $\mathbf{K}_\parallel$ in the emission of a photon.
3. Exciton trapped at interface defects, thus implying localization of the exciton COM wave function. The radiative lifetime is predicted to increase by nearly one order of magnitude as a consequence of the reduction of the exciton coherence length [113]. Localized excitons emit at energies lower than free excitons.
4. Free carrier recombination, following the exciton thermal ionization [114,115]. The spectral position of this channel is on the high-energy side of the exciton PL band.

The interplay between all these possible recombination channels is quite complicated and strongly depends on the temperature. At low temperature the disorder tends to localize the excitons. This is usually demonstrated by the presence of a sizable SS between the absorption and the emission peaks. The absence of a SS does not necessarily imply the free exciton nature of the emission band [92]. The effect of localization of excitons at the interface defects is the increase of the exciton radiative lifetime due to the difference in the coherence length imposed by the interface defects [113]. This essentially explains the discrepancy between experiment and theoretical predictions, and, in general, the overall behaviour of the exciton recombination at low temperature depends on the relative thermal population of the localized/free exciton states [92]. Nevertheless in very high quality samples it is expected that the free exciton nature of the recombination will play a relevant role. At the same time, the intrinsic radiative lifetime of the excitons with $\mathbf{K}_\| \leq \mathbf{K}_{ph}$ is quite difficult to be detected due to the thermalization with dark excitons states at $\mathbf{K}_\| > \mathbf{K}_{ph}$, which determines a linear increase of the radiative lifetime with a slope of the order of 35 ps/K. Experimentally, this linear increase of the PL lifetime has been observed by careful selection of the excitation wavelength to inject cold excitons in the QW system and therefore selecting free or localized states [16]. Strictly resonant excitation has been also used to observe the super-radiant emission from free excitons at $\mathbf{K}_\| \leq \mathbf{K}_{ph}$ [116]. This experiment is also strictly related to the resonant Rayleigh scattering topics, which have been longly debated [117] in the literature and whose discussion is well outside the scope of this contribution [118].

At high temperature, the recombination kinetics has to take into account the exciton ionization. Although the binding energy of excitons, R^*_{QW}, is enhanced in QWs (in GaAs/AlGaAs QWs $R^*_{QW} = 10$ meV [115]), as shown in Section 3.2.3, the thermal ionization of excitons and the converse exciton formation are processes always active. At each temperature, the excitons will establish a dynamic equilibrium with free electrons and holes [114]. It is also worth noting that, at low temperature, the carrier temperature is always larger than the lattice one, thus modifying the expected exciton band dependence on the temperature [92]. In general, there are two paths along which radiative recombination of free excitons takes place. The first is the direct exciton decay through a polaronic state, and the second is the exciton dissociation in free carriers followed by e–h radiative recombination. A plot on the logarithmic scale of the PL spectrum evidences the presence of free carrier recombination in the QW emission (Fig. 19) at room temperature. The ratio of exciton to free carrier recombination increases as the temperature is lowered. A two-dimensional law of mass action describes the coexistence of excitons and free carriers [115]. The increase, with increasing temperature, of the weight of the free carrier recombination fraction induces a temperature dependence of the radiative lifetime. In fact, the exciton radiative decay is

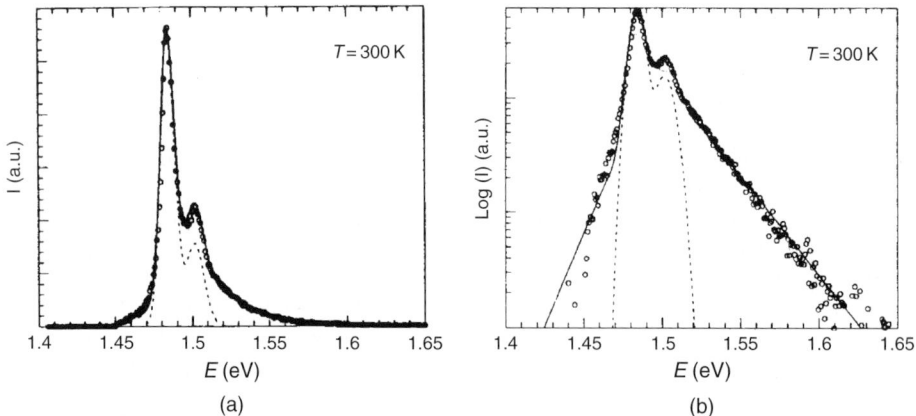

Fig. 19. PL emission of a 7 nm QW of GaAs/$A_{l0.3}Ga_{0.7}$As at $T = 300$ K. (a) Plot in linear scale; (b) semilogarithmic scale. The contribution arising from exciton recombination is shown as a dashed line (reprinted figure with permission from Ref. [115]. Copyright (1990) by the American Institute of Physics).

much faster than the free carrier recombination time. A model taking into account the exciton-free carrier equilibrium distribution gives a rather accurate description of the τ_R values at high temperatures [112,114].

Finally at very high temperature, the thermally activated nonradiative channels start to play a role. This topic will be discussed in the next section.

Many other informations can be obtained by careful analyses of the recombination kinetics, such as the distinction between impurity incorporation and interface roughness [112], the exciton localization on QW grown in misoriented substrates [115], the intervalley scattering of carriers [119], and the Γ–X state mixing [120]. Recently, great interest has been addressed to spatially indirect excitons with electrons and holes confined in different layers (type II QWs, see Fig. 3(d)), which show a recombination lifetime in the hundreds nanosecond scale due to the small e–h overlap [121]. The long recombination lifetime of indirect excitons promotes accumulation of these Bose particles in the lowest energy states and allows the photoexcited excitons to cool down to very low temperatures. This has led to the report of the achievement of dilute two-dimensional gas of indirect excitons which are statistically degenerate [122]. Even if the claim of exciton condensation has been debated in the literature [123], many experimental data show the creation of excitons with very large coherence length corresponding to a very narrow spread of the exciton momentum distribution, much smaller than that for a classical exciton gas [124].

4.3. Nonradiative decay channels

TR spectroscopy, together with the temperature dependence of the integrated radiative intensity can be used as sensitive probe of defect incorporation at the interfaces. In fact, from the PL decay time $\tau_L(T)$ and integrated radiative intensity $I_L(T)$, we can extract the radiative and nonradiative recombination times $\tau_R(T)$ and $\tau_{NR}(T)$ according to

$$\tau_R(T) = I_0 \frac{\tau_L(T)}{I_L(T)} \tag{17}$$

$$\tau_{NR}(T) = I_0 \frac{\tau_L(T)}{I_0 - I_L(T)} \approx \tau_L \tag{18}$$

where I_0 is a normalization factor depending on the absolute radiative efficiency at low temperatures. Usually the radiative efficiency of the QW systems is rather low, thus justifying the assumption $I_0 - I_L(T) \approx I_L$. The absolute value and the temperature dependence of τ_{NR} gives a direct measure of the strength of the nonradiative decay channels active in the QW system, thus allowing a direct measure of the effect of the change of the growth parameters on the defect incorporation (see Fig. 20) [125].

At high temperatures, $I_L(T)$ undergoes a strong quenching, which is attributed to the thermal emission of carriers from the QW into the barrier. Usually, the slope of the Arrhenius plot of $I_L(T)$ at high temperatures gives access to the activation energy of the process [127,128]. However, the strong temperature dependence of the radiative time τ_R in QWs [126] (see Section 4.2) nullifies this approach. Indeed it is the temperature dependence of τ_L that contains the information about the nonradiative recombination time τ_{NR}. From the Arrhenius plot of τ_L, it is possible to derive the activation energy of the thermal emission process. Activation energies much smaller than the e–h-pair confinement energies but in good

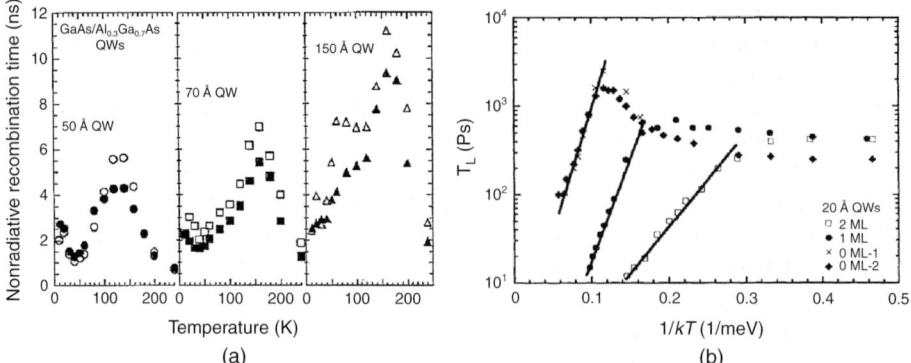

Fig. 20. (a) Temperature dependence of the nonradiative recombination time as calculated from Eq. (18), choosing an absolute radiative efficiency of 0.9 at $T = 4$ K. Open and closed circles refer to two different samples (reprinted figure with permission from Ref. [125]. Copyright (1991) by the American Institute of Physics). (b) Arrhenius plot of the PL decay time τ_L versus $1/KT$ of 2 nm GaAs/AlAs/AlGaAs double barrier quantum wells. The straight lines are the best fit to the high temperature slopes from which it is possible to extract the activation energies of the nonradiative process (reprinted figure with permission from Ref. [126]. Copyright (1992) by the American Physical Society).

agreement with the unipolar escape out of the wells of the less confined species of carriers have been found [126]. It is worth noticing that, in non–steady state conditions, the free carrier escape is more efficient than that of excitons. The e–h correlation, however, plays a major role in CW conditions, thus modifying the activation energy of the escape process which thus becomes close to the e–h-pair confinement energies [129].

5. Summary

PL is one of the most widely diffused experimental techniques for the characterization of semiconductor nanostructure and for the study of their optical and electronic properties. In particular, PL allows to study a number of interesting intrinsic effects in QWs, such as the band offsets in the heterostructure, the splitting between the heavy hole and the light hole valence band induced by the dimensionality reduction, the strain in pseudomorphic QWs, the exciton binding energy increase due to carrier localization, the homogeneous broadening of the exciton recombination line, the carrier–carrier interaction, and so on.

Furthermore, the analysis of the radiative recombination spectra of QWs can help in the characterization of the structure, providing information on the interface morphology and the quality of the materials of the QW and the barriers. In this chapter, the information that can be gained from PL spectra is critically summarized with the aim of providing a reference scheme for the characterization of nanostructures and, in particular, QWs through PL measurements.

After a presentation of the basic structure of the experimental systems and techniques used for PL measurements, both in CW and in TR configurations, the first part of the chapter is devoted to explain the basic concepts of the PL process in semiconductor quantum structures. The characteristics of quantum-confined structure are shortly presented and, then, the properties of electron states in quantum-confined structures are introduced, starting from the discussion of the single-electron states and the optical transitions in QWs. Then, the properties of excitons are discussed, because the PL in QWs is dominated by the exciton recombination. Finally, to support the understanding of the information that can be gained from PL spectra on the quality of quantum-confined structures, the physics of the recombination process in QWs and the effects of the disorder on the electron states and on the optical transition are summarized.

The second part of the chapter deals with a detailed illustration of the PL applications for characterizing the structural properties and for evaluating the optical properties of QWs. The possibility of extracting from the features of PL spectra information useful for interface evaluation is then discussed in detail. In this last part of the chapter, the carrier dynamics and the recombination kinetics are synthetically presented.

The discussion in the chapter is supported by a wide and detailed bibliography, in which basic textbooks, review articles, and research papers are included to provide the reader with up-to-date information on the application of PL to the study of optical and structural properties of QWs.

References

[1] H. Bebb, E. Williams, Photoluminescence i: Theory, in: R. Willardson, A. Beer (Eds), Transport and Optical Phenomena, Vol. 8 of Semiconductors and Semimentals, Academic Press, New York, 1972, p. 181.
[2] J. I. Pankove, Optical Processes in Semiconductors, Dover, New York, 1975.
[3] P. Dean, Progr. Cryst. Growth Charact. 5 (1982) 89.
[4] L. Pavesi, M. Guzzi, J. Appl. Phys. 75 (1994) 4779.
[5] M. Grundmann, The Physics of Semiconductors, Springer, Berlin, 2006.
[6] M. Herman, D. Bimberg, J. Christen, J. Appl. Phys. 70 (1991) R1.
[7] K. K. Bajaj, Mater. Sci. Eng. R 34 (2001) 59.
[8] M. Guzzi, S. Sanguinetti, M. Gurioli, in: H. S. Nalwa (Ed.), Encyclopedia of Nanoscience and Nanotecnology, American Scientific Publishers, 2004, p. 735.
[9] D. Bimberg, J. Christen, T. Fukunaga et al., J. Vac. Sci. Technol. B 5 (1987) 1191.
[10] G. Salviati, F. Rossi, N. Armani et al., in: C. Lamberti (Ed.), Characterization of Semiconductor Heterostructures and Nanostructures, Chapter 7, Elsevier, Amsterdam, 2008, p. 227.
[11] E. P. O'Reilly, Semicond. Sci. Technol. 4 (1989) 121.
[12] D. Colombo, E. Grilli, M. Guzzi et al., J. Lumin. 121 (2006) 375–378.
[13] R. Dingle (Ed.), Semiconductor and Semimetals, Springer, Berlin, 1987.
[14] C. Weisbuch, B. Vinter, Quantum Semiconductor Structures, Academic Press, Boston, 1991.
[15] P. Bhattacharya (Ed.), Properties of III-V Quantum Wells and Superlattices, Vol. 15 of EMIS Datareviews, INSPEC, London, 1996.
[16] J. Martinez-Pastor, A. Vinattieri, L. Carraresi et al., Phys. Rev. B 47 (1993) 10456.
[17] J. H. Scofield, Am. J. Phys. 62 (1994) 129.
[18] K. Okamoto, A. Kaneka, Y. Kawakami et al., J. Appl. Phys. 98 (2005) 064503.
[19] E. Betzig, J. K. Trautman, T. D. Harris et al., Science 251 (1991) 1468.
[20] E. Betzig, P. L. Finn, J. S. Weiner, Appl. Phys. Lett. 60 (1992) 2484.
[21] F. Zenhausern, Y. Martin, H. K. Wickramasinghe, Science 269 (1995) 1924.
[22] G. Bastard, Wave Mechanics Applied to Semiconductor Heterostructures, Les Editions de Physique, Les Ulis, 1988.
[23] A. Bignazzi, E. Grilli, M. Radice et al., Rev. Sci. Instrum. 67 (1996) 666.
[24] J. Shah, Ultrafast Spectroscopy of Semiconductors and Semiconductor Nanostructures, Springer, Berlin, 1996.
[25] M. A. Herman, H. Sitter, Molecular Beam Epitaxy, Springer, Berlin, 1996.

[26] M. Peressi, A. Baldereschi, S. Baroni, in: C. Lamberti (Ed.), Characterization of Semiconductor Heterostructures and Nanostructures, Chapter 2, Elsevier, Amsterdam, 2008, p. 19.
[27] G. Margaritondo, in: C. Lamberti (Ed.), Characterization of Semiconductor Heterostructures and Nanostructures, Chapter 12, Elsevier, Amsterdam, 2008, p. 441.
[28] G. Margaritondo, Rep. Prog. Phys. 62 (1999) 765.
[29] A. Franciosi, C. G. Van de Walle, Surf. Sci. Rep. 25 (1996) 1.
[30] C. Lamberti, Surf. Sci. Rep. 53 (2004) 1.
[31] L. Esaki, IEEE J. Quantum Electron. QE 22 (1986) 1611.
[32] M. Asada, Y. Miyamoto, Y. Suematsu, IEEE J. Quantum Elelctron. QE22 (1986) 1915.
[33] R. Eisberg, R. Resnick, Quantum Physics of Atoms, Molecules, Solids, Nuclei and Particles, J. Wiley and Sons, New York, 1974.
[34] L. J. Sham, Superlattices Microstruct. 5 (1989) 335.
[35] J. H. Davies, The Physics of Low-Dimensional Semiconductors, Cambridge University Press, Cambridge, 1998.
[36] Y. Fu, K. A. Chao, Phys. Rev. B 43 (1991) 12626.
[37] D. A. Contreras-Solorio, V. R. Velasco, F. García-Moliner, Phys. Rev. B 48 (1993) 12319.
[38] A. Franceschetti, A. Zunger, Appl. Phys. Lett. 68 (1996) 3455.
[39] K. Seeger, Semiconductor Physics – An Introduction, Vol. 40 of Solid–State Sciences, Springer, Berlin, 1985.
[40] P. Y. Yu, M. Cardona, Fundamentals of Semiconductors, Springer, Berlin, 1996.
[41] D. J. Ben Daniel, C. B. Duke, Phys. Rev. 152 (1966) 683.
[42] C. Lamberti, Comput. Phys. Commun. 93 (1996) 53.
[43] C. Lamberti, Comput. Phys. Commun. 93 (1996) 82.
[44] R. C. Miller, D. A. Kleinman, W. T. Tsang, A. C. Gossard, Phys. Rev. B 24 (1981) 1134.
[45] G. Bastard, E. E. Mendez, L. L. Chang, L. Esaki, Phys. Rev. B 26 (1982) 1974.
[46] M. Grundmann, D. Bimberg, Phys. Rev. B 38 (1988) 13486.
[47] M. Gurioli, J. Martinez-Pastor, M. Colocci et al., Phys. Rev. B 47 (1993) 15755.
[48] R. L. Greene, K. K. Bajaj, D. E. Phelps, Phys. Rev. B 29 (1984) 1807.
[49] J. J. Hopfield, Phys. Rev. 112 (1958) 1555.
[50] G. W. p't Hooft, W. A. J. A. van der Poel, L. W. Molenkamp, C. T. Foxon, Phys. Rev. B 35 (1987) 8281.
[51] E. Hanamura, Phys. Rev. B 38 (1988) 1228.
[52] L. C. Andreani, F. Tassone, F. Bassani, Solid State Commun. 77 (1991) 641.
[53] C. Weisbuch, R. Dingle, A. Gossard, W. Wiegmann, Solid State Commun. 38 (1981) 1709.
[54] R. Zimmermann, E. Runge, V. Savona, in: T. Takagahara (Ed.), Quantum Coherence, Correlation, and Decoherence in Semiconductor Nanostructures, Academic Press, London, 2003, p. 89.
[55] R. F. Kopf, E. F. Schubert, T. D. Harris, R. S. Becker, Appl. Phys. Lett. 58 (1991) 631.
[56] P. Petroff, A. Gossard, W. Wiegmann, A. Savane, J. Cryst. Growth 44 (1978) 5.
[57] C. Gourdon, I. V. Mashkov, P. Lavallard, R. Planel, Phys. Rev. B 57 (1998) 3955.
[58] J. M. Moisson, C. Guille et al., Phys. Rev. B 40 (1989) 6149.
[59] A. R. Smith, K. Chao, C. K. Shih et al., Appl. Phys. Lett. 66 (1995) 478.
[60] R. Zimmermann, Jpn. J. Appl. Phys. 34 (1995) 228.
[61] R. Zimmermann, E. Runge, Phys. Status Solidi (a) 164 (1997) 511.
[62] V. Savona, Unpublished.
[63] F. Intonti, V. Emiliani, C. Lienau et al., Phys. Rev. Lett. 87 (2001) 076801.
[64] G. von Freymann, U. Neuberth, M. Deubel et al., Phys. Rev. B 65 (2002) 205327.
[65] C. Lienau, F. Intonti, T. Guenther et al., Phys. Rev. B 69 (2004) 085302.
[66] E. D. Palik (Ed.), Handbook of Optical Constants of Solids, Academic Press, London, 1985.
[67] S. Sanguinetti, D. Colombo, M. Guzzi et al., Phys. Rev. B 74 (20) (2006) 205302.
[68] D. D. Sell, S. E. Stokowski, R. Dingle, J. V. DiLorenzo, Phys. Rev. B 7 (1973) 4568.
[69] P. M. Petroff, C. Weisbuch, R. Dingle et al., Appl. Phys. Lett. 38 (1981) 965.

[70] R. T. Phillips, D. J. Lovering, G. J. Denton, G. W. Smith, Phys. Rev. B 45 (1992) 4308.
[71] J. C. Kim, D. R. Wake, J. P. Wolfe, Phys. Rev. B 50 (1994) 15099.
[72] T. W. Steiner, A. G. Steele, S. Charbonneau et al., Solid State Commun. 69 (1989) 1139.
[73] J. Lee, E. S. Koteles, M. O. Vassell, Phys. Rev. B 33 (1986) 5512.
[74] D. Gammon, S. Rudin, T. L. Reinecke et al., Phys. Rev. B 51 (1995) 16785.
[75] L. Schultheis, A. Honold, J. Kuhl et al., Phys. Rev. B 34 (1986) 9027.
[76] H. Hillmer, A. Forchel, S. Hansmann et al., Phys. Rev. B 39 (1989) 10901.
[77] R. C. Miller, A. C. Gossard, D. A. Kleinman, Phys. Rev. B 32 (1985) 5443.
[78] A. Esser, E. Runge, R. Zimmermann, W. Langbein, Phys. Rev. B 62 (2000) 8232.
[79] D. A. B. Miller, D. S. Chemla, T. C. Damen, Phys. Rev. B 32 (1985) 1043.
[80] E. E. Mendez, F. Agulló-Rueda, J. M. Hong, Phys. Rev. Lett. 60 (1988) 2426.
[81] E. F. Schubert, E. O. Gobel, Y. Horikoshi et al., Phys. Rev. B 30 (1984) 813.
[82] H. Sakaki, M. Tanaka, J. Yoshino, Jpn. J. Appl. Phys. 24 (1985) L417.
[83] M. Tanaka, H. Sakaki, J. Yoshino, Jpn. J. Appl. Phys. 25 (1986) L155.
[84] C. A. Warwick, W. Y. Jan, A. Ourmazd, T. D. Harris, Appl. Phys. Lett. 56 (1990) 2666.
[85] J. Singh, K. K. Bajaj, S. Chaudhuri, Appl. Phys. Lett. 44 (1984) 805.
[86] D. Bimberg, D. Mars, J. N. Miller et al., J. Vac. Sci. Technol. B 4 (1986) 1014.
[87] K. Leosson, J. R. Jensen, W. Langbein, J. M. Hwam, Phys. Rev. B 61 (2000) 10322.
[88] F. Genova, A. Antolini, L. Francesio et al., J. Cryst. Growth 120 (1992) 333.
[89] C. Lamberti, S. Bordiga, F. Boscherini et al., J. Appl. Phys. 83 (1998) 1058.
[90] A. Patané, A. Polimeni, M. Capizzi, F. Martelli, Phys. Rev. B 52 (1995) 2784.
[91] D. C. Bertolet, J. Hsu, K. M. Lau et al., J. Appl. Phys. 64 (1988) 6562.
[92] M. Gurioli, A. Vinattieri, J. Martinez-Pastor, M. Colocci, Phys. Rev. B 50 (1994) 11817.
[93] F. Yang, M. Wilkinson, E. J. Austin, K. P. O'Donnel, Phys. Rev. Lett. 70 (1993) 323.
[94] F. Yang, B. Henderson, K. P. O' Donnel et al., Appl. Phys. Lett. 59 (1991) 2142.
[95] G. Bastard, C. Delalande, M. H. Maynadier et al., Phys. Rev. B 29 (1984) 7042.
[96] O. Brandt, H. Lage, K. Ploog, Phys. Rev. B 45 (1992) 4217.
[97] E. Göbel, Excitons in Confined Systems, Springer, Heidelberg, 1988.
[98] J. Hegarty, L. Goldner, M. D. Sturge, Phys. Rev. B 30 (1984) 7346.
[99] M. Pistol, X. Liu, Phys. Rev. B 45 (1992) 4217.
[100] K. Shahzad, D. J. Olego, C. G. van de Walle, D. A. Cammack, J. Lumin. 46 (1990) 109.
[101] T. Taguchi, Y. Yamada, Mater. Res. Soc. Symp. Proc. 161 (1990) 199.
[102] M. Gurioli, J. Martinez-Pastor, A. Vinattieri, M. Colocci, Solid State Commun. 91 (1994) 931.
[103] A. Polimeni, A. Patané, M. Grassi Alessi et al., Phys. Rev. B 54 (1996) 16389.
[104] A. Aït-Ouali, J. L. Brebner, R. Y. Yip, R. A. Masut, J. Appl. Phys. 86 (1999) 6803.
[105] D. Gammon, E. S. Snow, B. V. Shanabrook et al., Phys. Rev. Lett. 76 (1996) 3005.
[106] T. C. Damen, J. Shah, D. Y. Oberli et al., Phys. Rev. B 42 (1990) 7434.
[107] R. Eccleston, R. Strobel, W. W. Rühle et al., Phys. Rev. B 44 (1991) 1395.
[108] P. Roussignol, C. Delalande, A. Vinattieri et al., Phys. Rev. B 45 (1992) 6965.
[109] J. P. Bergman, P. O. Holtz, B. Monemar et al., Phys. Rev. B 43 (1991) 4765.
[110] T. C. Damen, K. Leo, J. Shah, J. E. Cunningham, Appl. Phys. Lett. 58 (1991) 1902.
[111] L. C. Andreani, A. Pasquarello, Phys. Rev. B 42 (1990) 8928.
[112] M. Gurioli, A. Vinattieri, M. Colocci et al., Phys. Rev. B 44 (1991) 3115.
[113] D. S. Citrin, Phys. Rev. B 47 (1993) 3832.
[114] B. K. Ridley, Phys. Rev. B 41 (1990) 12190.
[115] M. Colocci, M. Gurioli, A. Vinattieri, J. Appl. Phys. 68 (1990) 2809.
[116] B. Deveaud, F. Clérot, N. Roy et al., Phys. Rev. Lett. 67 (1991) 2355.
[117] M. Gurioli, F. Bogani, S. Ceccherini, M. Colocci, Phys. Rev. Lett. 78 (1997) 3205.
[118] V. Savona, E. Runge, Phys. Status Solidi (b) 234 (2002) 96.
[119] B. Deveaud, F. Clérot, A. Regreny et al., Phys. Rev. B 49 (1994) 13560.
[120] B. Chastaingt, M. Gurioli, P. Borri et al., Phys. Rev. B 55 (1997) 2393.

[121] L. V. Butov, A. L. Ivanov, A. Imamoglu et al., Phys. Rev. Lett. 86 (2001) 5608.
[122] L. V. Butov, A. C. Gossard, D. S. Chemla, Nature 418 (2002) 751.
[123] R. Rapaport, G. Chen, D. Snoke et al., Phys. Rev. Lett. 92 (2004) 117405.
[124] S. Yang, A. T. Hammack, M. M. Fogler et al., Phys. Rev. Lett. 97 (2006) 187402.
[125] M. Gurioli, A. Vinattieri, M. Colocci et al., Appl. Phys. Lett. 59 (1991) 2150.
[126] M. Gurioli, J. Martinez-Pastor, M. Colocci et al., Phys. Rev. B 46 (1992) 6922.
[127] J. D. Lambkin, D. J. Dunstan, K. P. Homewood et al., Appl. Phys. Lett. 57 (1990) 1986.
[128] G. Bacher, H. Schweizer, J. Lovac et al., Phys. Rev. B 43 (1991) 9312.
[129] G. Bacher, C. Hartmann, H. Schweizer et al., Phys. Rev. B 47 (1993) 9545.

7

Power-dependent cathodoluminescence in III–nitrides heterostructures: from internal field screening to controlled band-gap modulation

Giancarlo Salviati[1], Francesca Rossi[1], Nicola Armani[1], Vincenzo Grillo[2], and Laura Lazzarini[1]

[1]*IMEM-CNR, Parco Area delle Scienze 37/A, 43100 Parma, Italy*
[2]*Laboratorio Nazionale TASC-INFM-CNR, Area Science Park, Basovizza S.S. 14 km 143, 5, 34012 Trieste, Italy*

Abstract The aim of this chapter is to show mainly the potentiality of the cathodoluminescence (CL) technique and to what extent it can be applied in the study of III–nitrides heterostructures. A quite recent and unusual application of the CL technique, namely power-dependent CL, will be presented, exploiting and discussing the influence of different injection power conditions, from low to high injection regimes (the latter occurring for $n \gg n_0$ free carrier concentration, with typical values ranging from about 10^{16}–10^{20} injected e–h pairs cm^{-3}).

It has therefore been decided to abbreviate the basics of the technique, which can be easily found in dedicated textbooks, in favor of a deeper discussion of the experimental examples and of some peculiarities of the technique.

Keywords cathodoluminescence, nitrides, internal fields screening, electron beam writing, band-gap modulation, injection power, point defects, random-walk modelling, in-depth resolution

1. Short introduction to cathodoluminescence spectroscopy

Cathodoluminescence (CL) is the emission of photons as the result of the interaction between energetic electrons and matter. In semiconductors, the excitation by a highly energetic electron beam generates electron–hole (e–h) pairs within a specimen volume. The excess carriers then thermalize and diffuse inside the material and finally recombine, either by nonradiative processes or by photon emission.

CL spectroscopy is a powerful method, which allows one to study local variations in composition, impurity concentration and doping, the influence of different phases and crystal defects on the optical emissions, etc. A spectral acquisition of the intensity as a function of wavelength and a mapping of the intensity and wavelength distributions can be performed. Time-resolved CL spectroscopy with a time resolution on the picosecond scale can also be carried out. All the aforementioned modes are conducted by combining optical spectroscopy and electron microscopy with submicrometric lateral (in-plane) and in-depth resolutions. A wide range of applications is possible for different classes of materials, from semiconducting bulk, hetero- and nanostructures [1] to rare-earth compounds [2], and from diamond to geological specimens [3,4].

CL investigations can be performed by means of different types of electron beam sources, like high-energy electrons (100–400 keV) inside a transmission electron microscope (TEM) (see, for instance, Refs [5,6]), low-energy electrons (1–40 keV) inside a scanning electron microscope (SEM), or very low-energy electrons (tenths of volts) in the case of a scanning tunnel microscope equipment (see, for instance, Refs [7,8]). The lateral resolution of the technique is strongly influenced by the beam/sample interaction volume, which is mainly determined by the electron energy. For example, in the case of TEM–CL the reduced thickness of the typical TEM foil combined to the high energy of the electrons and the very small diameter of the beam allow to reduce the lateral resolution of about one order of magnitude with respect to the SEM mode.

The intrinsic limitation in CL lateral resolution due to the generation/recombination volume is overcome by combining the peculiarity of atomic force microscopy and SEM techniques. The experimental setup developed for this new technique, scanning near-field cathodoluminescence (SNCLM), is described in Refs [9,10]. In brief, as for CL in near-field conditions, the vicinity of the tip–sample "contact" area is irradiated homogeneously with primary electrons and the resulting luminescence is picked up directly above the recombination centers by the SNOM probe with a probe–sample distance of a few nanometers. The AFM tip is used to diffract the evanescent waves at the surface, while the light collection is carried out by means of an optical fiber. The CL excitation is performed just below the tip by a conventional SEM focused electron beam. The authors demonstrated an improvement in spatial resolution of at least one order of magnitude with respect to the classical "far-field" CL. This is possible because the SNCLM lateral resolution is not energy-transfer dependent, but it is mainly determined by the tip dimensions.

Compared to other luminescence techniques like photoluminescence (PL) (see, e.g., Chapter 6, this book [11]) and electroluminescence (EL), there are some peculiar aspects associated to the nature of the excitation processes that mark the CL: e.g., the generation and recombination volumes and rates (see Sections 1.2 and 1.4). However, the basic luminescence mechanisms responsible, both in bulk materials and quantum confined structures, for the radiative transitions are the same for the different forms of excitation (Fig. 1). We can distinguish between the two types of transitions:

(a) *Intrinsic emissions* that are due to recombination of electrons and holes across the fundamental energy gap, by interband transitions from the bottom of the conduction band to the top of the valence band. In the spectral region close to the energy gap, we can detect *free excitons or bound excitons*, with one of the carriers localized at an impurity center.

(b) *Extrinsic luminescence* that is due to radiative transitions involving states in the band gap, shallow or deep, mainly due to native defects and impurities complexes acting as donor or acceptor centers. Different processes of recombination between free carriers and trapped carriers can take place, basically indicated as free-to-bound (e.g., donor-to-free-hole, $D°h$; free-electron-to-acceptor, $eA°$) and donor-to-acceptor pair (DAP) transitions.

A detailed analysis of the nature, transition probability, and behavior under varying experimental conditions (temperature, excitation power, etc.) of the aforementioned optical transitions is a wide and more general subject, which is beyond the purpose of this

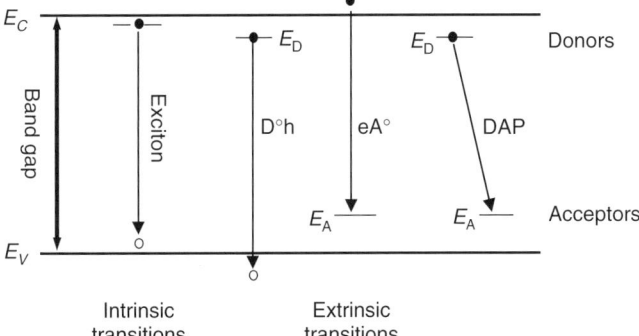

Fig. 1. Schematic diagram of radiative transitions in a direct band-gap semiconductor.

short treatment. For a complete and in-depth discussion, see, for instance, Pankove [12, Chapter 6], Pavesi-Guzzi [13] and Yacobi-Holt [14, Chapter 3].

In CL analysis performed under steady-state conditions (see Section 1.1), after radiative carrier recombination occurs, the emitted photons propagate inside the material and the fraction coming out from the sample is finally collected and detected. The intensity of the recorded CL signal (L_{CL}) can be expressed [14,15] as

$$L_{CL} = \int_V F(z) \frac{\Delta n(r)}{\tau_{rr}} dV \quad (1)$$

where $\Delta n(r)$ is the stationary excess carrier density at position r, τ_{rr} (see Section 1.2.2 and Appendix) is the radiative recombination lifetime, and F is a function defining the amount of photons actually leaving the surface as a function of the generation coordinate. (In this expression a translation invariance of the optical response in the x–y direction has been assumed, therefore F is just a function of the z coordinate.)

Equation (1) states that, given the number of photons emitted by radiative carrier recombination inside the material, the intensity of the recorded CL signal depends (through F) [16] on factors accounting for:

(i) processes of self-absorption of photons by the material along their path;
(ii) internal losses due to scattering/reflection of the light at interfaces, and finally at the surface;
(iii) CL collection/detection system response.

Factor (iii) must consider experimental setup parameters like the overall collection efficiency of the light acquisition system, including also the fraction of solid angle covered above the specimen, the losses due to all the optical elements, the transmissive efficiency of the monochromator, the signal amplification factor and the detector quantum efficiency and yield varying with frequency.

1.1. The CL experimental setup in the scanning electron microscope

The more diffuse CL setup uses an SEM beam as a probe. The principles of the SEM are very well known and will not be summarized here (for a deep treatment of the SEM principles, see, e.g., Ref. [17]). In a typical CL experimental setup, the only peculiarity to be taken into account, as it influences to a different extent the integrated intensities of deep and band-to-band-type CL emissions, is that the electron beam dwells on each single point of the sample for a time τ, which depends on the scan rate (in our case $\tau = 67 \times j$ ns at a TV/j scan rate, $j = 1,2,4,8,16,\ldots$). Comparing the electron beam dwell time with average carrier lifetimes of near band-edge (NBE) transitions in III–V(N), we can assume to be in stationary conditions. A steady-state emission spectroscopy is therefore performed.

The CL signal coming out from the sample can be collected in two main ways: placing an optical fiber close to the analyzed sample or using a mirror, of ellipsoidal or parabolic shape, positioned exactly above the sample surface. A light guide is used for signal transport to the detection system, outside the microscope: a monochromator for light dispersion (Fig. 2) and a set of detectors with related signal processing systems. The intensity of the cathode ray tube (CRT) is modulated by one of the signals recorded, coming from the detector of secondary and backscattered electrons or from the CL mirror (i.e., visible or infrared detectors), to form an image. The specimen temperature can be controlled by a cryostat between 6 and 300 K (flux of liquid helium or liquid nitrogen are used to reach 6 or 77 K). The entire SEM column is kept in vacuum at 10^{-6}–10^{-7} torr by rotary and turbomolecular pumps.

A recent improvement of the dispersion/detection system is the parallel detection mode that uses a photodiode [18] or array of diodes in spite of the standard photomultiplier as detector (Fig. 3). This configuration has given a strong impulse to the use of the CL

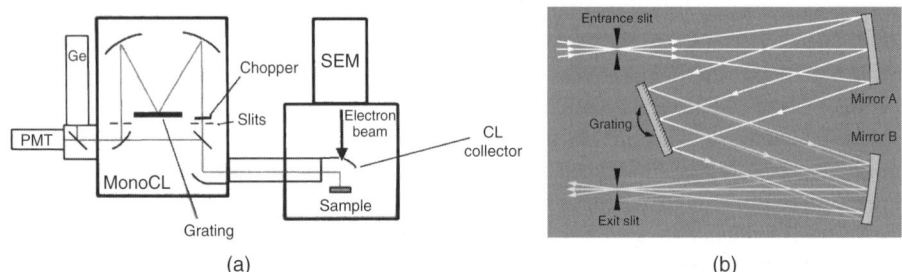

Fig. 2. (a) Schematic path of the light from sample to the detectors, (b) diagram of the light paths inside a typical Czerny–Turner monochromator using a dispersion grating.

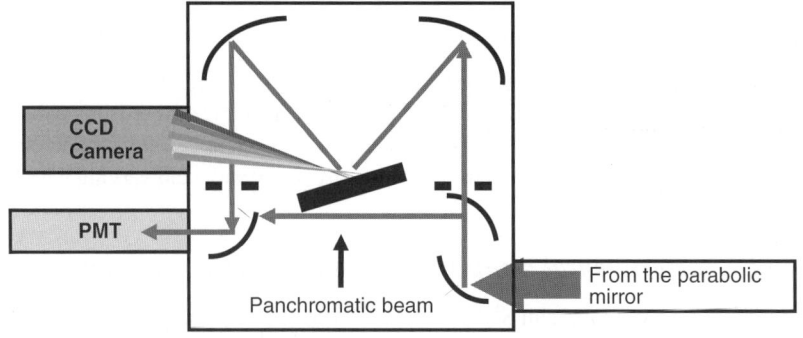

Fig. 3. Scheme of the CL setup for parallel detection system using a modified Czerny–Turner monochromator and a CCD camera as visible light detector.

technique as a suitable characterization technique, as it allows very fast acquisition times (typically from 0.5 to 40 s) thereby minimizing the damage effects induced by electron beam irradiation. This is particularly valuable for the study of emission properties and emission decay rates in beam sensitive materials as organic crystals [19]. In addition, an ad hoc software able to control the electron beam movements and the CL signal acquisition permits to work in the so-called "spectral imaging" mode. This CL mode enables to acquire up to four analogues and four pulse inputs simultaneously, to choose pixel image size in X and Y dimensions with a pixel dwell time from 400 ns to 400 ms per pixel (http://www.gatan.com/sem/digiscan2.php). Alternatively n-spectra ($1 \leq n \leq 256$) may be acquired by dividing the array via software into vertical tracks [20].

The spectral resolution of a CL system is determined by the luminescence efficiency of the material and by the monochromator parameters. In particular, its focal length, the density of the dispersion grating and the width of the entry and exit slits, improving for longer focal length, higher density of the grating and narrower slits.

As a general trend, accurate CL studies require a low specimen temperature to separate contiguous emission peaks, due to the linewidth narrowing and the increase in intensity usually obtained by lowering the temperature.

In our work, we used a Cambridge 360 Stereoscan SEM with a tungsten filament (resulting beam size on the sample surface typically ranging between a few microns and a few tens of nanometers), equipped with a Gatan MonoCL2 system. The spectra as well the panchromatic and monochromatic images have been acquired using a dispersion system, equipped with three diffraction gratings and a system of a Hamamatsu multialkali photomultiplier and a couple of liquid-nitrogen-cooled (Ge and InGaAs) detectors. This experimental setup provides a spectral resolution of 2 Å and a detectable 200–2200 nm (0.6–6.2 eV) wavelength range. Three kinds of CL analyses can be performed with this standard setup:

(i) Spectral analysis, with a spectral resolution dependent on the working parameters but typically less than 1 nm (min. 0.1 nm) for the studies reported here.
(ii) Analysis of the spatial distribution of the luminescence by panchromatic and monochromatic CL maps, allowing to carry out correlated spectral and spatial studies. Complementary morphologic SEM information from submicroscopic regions allows, for example, to spatially localize nonradiative centers.
(iii) A time-dependent analysis on a timescale of tens of seconds. The time evolution of a given transition intensity and the dynamic change of the relative integrated intensities of different emissions give indication of point defects migration, breaking of complexes and/or other damage effects, intentionally induced in a controlled way by both by ex situ thermal treatments or by the in situ electron beam irradiation.

It has to be noted that, in dedicated CL systems, time-resolved analyses are possible. Differently to the more standard time-resolved PL studies employing a pulsed laser, the major difficulty in performing time-resolved CL analyses is having a short pulsed electron beam, with a subnanosecond period. This is achieved in two different ways:

(a) By blanking the electron beam of the SEM by means of a short pulsed electric field (up to 1–2 GHz); the time resolution of this kind of analysis is around tenths of nanoseconds [21].
(b) An alternative intriguing method to achieve a time resolution in the picosecond range is obtained by re-designing the electron beam source and by generating the primary electrons from a gold target hit by an "excimer" picosecond pulsed laser [22].

In the first case, a conventional detecting system synchronized to the electron beam is enough to have good results, while in the case of the picoseconds system, the use of a high sensitive streak camera, typical for time-resolved PL analyses, is necessary. For picosecond time-resolved analysis a modified system is required, while for studies on the nanosecond scale a commercial beam blanking system is usually suitable.

1.2. Carrier generation, diffusion, and modeling

1.2.1. Generation

Multiple excitation processes are produced by interaction of the electron beam (primary electrons) with a semiconducting sample [14]. In the framework of the CL analysis, it is relevant to consider, as strictly related to the signal formation and origin, the process of e–h pair generation, the lateral and in-depth size of the generation volume and the spatial distribution of the generated carriers.

Carrier generation was the first topic to be extensively studied in the literature, owing to the contemporary development of other SEM-related techniques such as electron beam-induced current (EBIC), backscattering electron analyses, and energy-dispersive X-ray microanalysis (EDS), which required an accurate modeling of the probe-sample interaction [23,24]. Because a CL system can be attached to both an SEM and a TEM (or even scanning transmission electron microscope, STEM) [25–27], the accurate description of the electron beam characteristics for both types of microscopes is the fundamental "first step" to build a simulation of general validity. In addition, the peculiarities of the samples studied in these instruments (thin foils for TEM and thick materials for SEM) and the different trajectories of the primary electrons through these specimens must be taken into account. Additionally, the simulation should also consider the possibility of a largely defocused or of a narrow beam. However, the details of the electro-optical system determining the probe are, in most of the cases, not necessary when a moderate convergence angle is used. Indeed, in many cases, a Gaussian intensity distribution can be assumed [23] and the only parameter that matters in simulations is the width, σ, of such Gaussian profile.

Most of the work is to calculate the electron trajectories in the material and to evaluate the energy dose released to the crystal through inelastic scattering. In the case of the SEM, a large number of phenomenological descriptions of the beam–sample interactions have been given. The mean electron path length (Bethe range [17]) in the material depends on energy-releasing processes, which sets the kinetic energy of the electrons to zero, initially accelerated at a voltage of the order of some keV. In particular, there is a maximum effective penetration depth (Grün range, R), function of the beam energy (E_b), which the primary electrons reach, at which the e–h pairs can still be generated. This is in general parameterized as $R \propto (E_b)^\alpha$. In particular, Kanaya and Okayama [28] derived the following phenomenological expression well in agreement with experimental results for a wide range of materials:

$$R = \left(\frac{0.0276 A}{\rho Z^{0.889}}\right) E_b^{1.67} \quad (2)$$

where A is the atomic weight, Z the atomic number, and ρ (in g/cm^3) the density of the material.

Because these relations work only in the case of uniform samples made of a single material, an approach successfully describing the generation volume in presence of compound heterostructures would be necessary. This is usually accomplished by the numerical

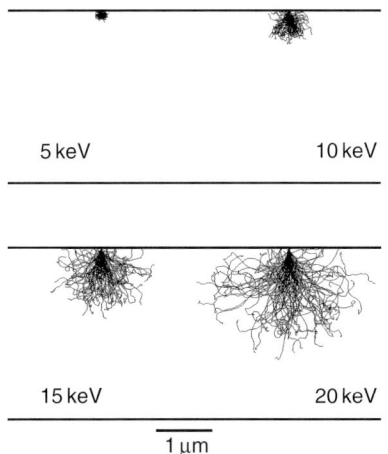

Fig. 4. Graphical 2D representation of electron trajectories in a GaN layer (ALLES program – see Appendix).

solution of the generation problem through a classical Monte Carlo simulation (MCS) [23,29]. Inside the material, primary electrons undergo a series of stochastic scattering events, both elastic and anelastic. Rather than the exact single electron path, the distribution of electron trajectories, significant for CL and SEM analyses, can be estimated (Fig. 4). The structure of MCS, in spite of a large number of improvements, is still unchanged: an electron is considered to move unperturbed along its trajectory between the two elastic scattering events and to undergo a continuous slowdown due to inelastic scattering. This approach is justified because the inelastic scattering, in most cases, produces little deflections in the electron direction, while elastic scattering (including the quasi-elastic thermal diffuse scattering) produces high-angle deflections with negligible energy losses. Additionally, the medium is supposed to be homogeneous and the number of inelastic events is supposed to be sufficiently high, at a typical length scale of the mean free path between two elastic collisions, so that the average over the energy loss can be considered and the fluctuations ignored.

A good "recipe" for a MCS must contain as "ingredients" a description of

1. the total elastic cross section, $\sigma_{elastic}$;
2. the differential elastic cross section, $d\sigma/d\Omega$ (Ω being the solid angle);
3. the average electron energy loss, dE, after a distance, ds; the quantity, dE/ds, is often referred to as stopping power.

The difficulties of this formulation are hidden in the correct description of the factors $\sigma_{elastic}$, $d\sigma/d\Omega$, and dE/ds as a function of energy.

The first description of the elastic cross section is based on the Rutherford cross section (both integrated and differential) with a screening factor [23]. However, it is known that the Rutherford cross section can give inaccurate results as it does not account for relativistic effects and, for example, the Mott cross section should be alternatively used [30,31]. Nevertheless, the Rutherford cross section with some "ad hoc" corrections continues to be used, as it speeds up the calculation procedures and simplifies the algorithm implementation [32]. One point that may surprise in the use of Rutherford or Mott cross sections is that the resulting description of the electron–material interaction is completely incoherent; usually no effects like diffraction or channeling need to be considered. Actually, so far no channeling-related effect has ever been invoked in CL: in practice, the diffusion of the carriers often masks any channeling effect on the beam broadening. Moreover, CL measurements often require a sample thickness of at least a few hundreds of nanometers for which channeling effects start weakening.

The description of the stopping power is somehow even more crucial as it is the main factor determining the penetration range of the electron in a material. It has to account for plasmon generation, inner shell ionization and conduction electron excitation. In particular, the inner shell ionization is difficult to be described, at least when considering that the opening of each single inelastic channel is strongly energy dependent. Nevertheless, it is customary to use the approximate Bethe formula [33]

$$\frac{dE}{ds} = -785 \frac{\rho Z}{AE} \ln\left[\frac{1.166E}{J}\right] \text{ eV Å}^{-1} \qquad (3)$$

where E is the instantaneous energy of the incident electron in eV, s is the path length in Å, ρ is the material density in g/cm^3, Z the atomic number, A is the atomic weight, and J the mean ionization potential. Many efforts have also been made to correct this formula, for example, by adapting the value of the mean ionization potential or even of the effective atomic number Z, mainly in the low-energy regime [34]. The detailed description of this parameterization is however beyond the aims of this chapter.

The real interest in MCS for CL analyses is how to simulate the distribution of the e–h pairs in the sample. Unfortunately, this is affected by an indetermination due to the application of the continuous slow down. In MCS, an electron releases a certain amount of energy between two elastic scatterings, but it is unclear exactly at which point it does it. The usually adopted approximated solution is to consider the middle point between the two consecutive elastic scattering events. By performing the average over different simulated electron trajectories, a good description can usually be obtained for the generation volume and the dose distribution. However, this method shows its inaccuracy near the surface at length scales shorter than the elastic mean free path. Indeed, in the region close to the injection point at the surface, the primary electron covers a large distance before the first elastic scattering occurs as it is still highly energetic. In this case, the assumption of a large energy release and carrier generation localized at a single point half way to the first scattering is unrealistic: it creates a too large inhomogeneity in the generation with no physical meaning. For this reason, it is sometimes necessary to correct the basic mechanisms if a more accurate generation function needs to be calculated [35].

Once the dose is known, the generation volume and the injected carrier distribution can be determined. The number of carriers is largely determined as the number of e–h pairs generated by the interaction of the primary beam with the material. The number of electrons in the primary electron beam is usually negligible, compared to the generated e–h pairs, because the former is at least two orders of magnitude lower for beam energies typically larger than 1 keV. To determine the connection between the energy loss of the beam and the e–h pair generation is a priori a difficult task. It is known that, for example, a plasmon should decay into two excitons plus an amount of energy lost in thermal agitation [36], but for other processes, the fraction of e–h pairs is more difficult to be estimated. An empirical evaluation of the number of generated e–h pairs per incident electron, G, is usually performed by using

$$G = \frac{E_b}{E_I}(1-\gamma) \tag{4}$$

where γ is the backscattered rate and E_I is the ionization energy [37]. A rule of thumb estimation assigns E_I as

$$E_I = 2.67 E_{gap}\,(\mathrm{eV}) + 0.87\,\mathrm{eV} \tag{5}$$

where E_{gap} is the energy gap of the material.

Like the energy deposition, the spatial distribution of the e–h pairs is strongly nonuniform (that plays a fundamental role in heterostructures where each layer is differently injected) and is determined by a local generation rate. Monte Carlo methods provide both the lateral and in-depth doses. In the case of uniform materials, a universal depth–dose function ($g(z)$, see Fig. 5), representing the number of pairs generated per electron of energy E per unit depth and per unit time, has been also proposed by Everhart and Hoff [38] as a polynomial in terms of a normalized depth z/R:

$$g(z) = \frac{1}{R}\left(0.6 + 6.21\frac{z}{R} - 12.4\left(\frac{z}{R}\right)^2 + 5.69\left(\frac{z}{R}\right)^3\right) \tag{6}$$

It can be seen that as the dissipated energy density in the generation volume is concentrated, the lower is the voltage, up toward the beam impact point.

The effects of beam energy, sample thickness and material density on the size and shape of the carrier generation distribution can be studied directly [17]. Roughly, for low electron energy values ($E_b < 5$ keV), high Z materials ($Z > 20$) and thick specimens ($t > 1$ μm) the distribution is found to be a small hemisphere centered at the surface. As the energy/density ratio increases, the generation volume shape becomes spherical and larger and is centered below the surface. For highest energy/density ratios the distribution is pear-shaped and is much larger; for thin

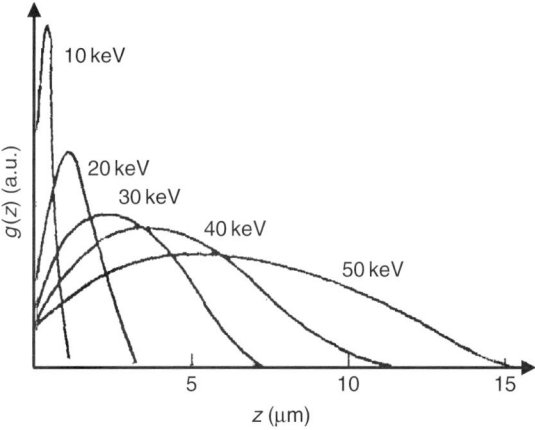

Fig. 5. Depth-dose function (from Ref. [14]).

samples the shape is conical, because it corresponds to the initial pear neck, and much smaller. Various contributions such as electron–electron scattering, hole–hole scattering, electron–hole scattering, electron–phonon interactions and hole–phonon interactions, mix together and should be taken into account in simulation programs for a quantitative CL analysis.

Actually, as the CL signal is determined by the position and concentration of the excess carriers at the recombination time, the details of this distribution related to the generation by primary electrons need to be corrected taking into account first the thermalization and later the diffusion of carriers. Thermalization processes [39,40] are strongly active, all the more compared to PL experiments [13], because the excitation source is the energetic (keV) electron beam, resulting in excess carriers of energy far above the conduction band edge in the case of electrons and in the bottom of the valence band for holes. Briefly, the thermalization process can be divided into three steps: (i) mutual carrier thermalization among the carriers themselves at a temperature higher than that of the host lattice (hot carrier distribution); (ii) further cooling down to the lattice temperature by emission of optical phonons (after the excitation, the two steps require some picoseconds depending on the specific material due to the reciprocal scattering); and (iii) final cooling by emitting acoustic phonons (0.1–1 ns time) with radiative and nonradiative recombination of the carriers. Carriers diffuse until they recombine in a different region of the sample, therefore modifying the initial excitation distribution and extending the volume from which the CL emission originates. The detailed analysis of diffusion-related effects and modeling approaches is reported in the following.

1.2.2. Diffusion

In the low-injection regime, it has been demonstrated [41] that it is sufficient to describe the diffusion of the minority carriers to understand the distribution of the CL recombination. The master equation for steady-state diffusion is the differential equation [42]

$$D\nabla^2 n - \frac{n}{\tau} = -g \qquad (7)$$

where D is the diffusivity of the material (eventually accounting for ambipolar diffusion [43]), n the excess minority carrier density, τ the lifetime of the carriers, and g the generation function.

Moreover, the diffusion length (L) can be defined as

$$L = \sqrt{D\tau} \qquad (8)$$

where D can be linked to the mobility as

$$D = \frac{\mu k T}{e} \qquad (9)$$

A quantity called diffusion velocity (v_d) can be also introduced as

$$v_d = \frac{L}{\tau} \qquad (10)$$

The specific solution depends on the boundary conditions, and on the shape of the generation volume.

In two dimensions, the solution is a zeroth-order modified Bessel function, but can be approximated [44] for large r as

$$n \propto \frac{e^{-(r/L)}}{\sqrt{r}} \qquad (11)$$

Expression (11), the solution to Eq. (7), can be used, for example, to describe the diffusion inside a quantum well (QW); in a general case, however, it is difficult to apply this kind of "unlimited volume" boundary conditions and the relative solutions. The main reason is that the geometry of CL injection necessarily imposes a different kind of boundary conditions to the diffusion equation (see Eq. (7)). In particular, the effect of at least the surface on which the electron beam impinges should be considered. The boundary condition at surfaces is usually written as

$$D\frac{\partial n}{\partial z}\bigg|_{z=z_s} = sn \qquad (12)$$

where s is the surface recombination velocity. For practical reasons, a dimensionless recombination velocity is also introduced as $S = s\tau/L$.

The physical origin of this boundary condition is the strong perturbation that a surface induces in a crystal: even under ideal conditions, the fact that part of the atoms at the surface do not have all the electronic orbital saturated (dangling bonds), introduces electronic states different from those of bulk. Part of these bonds can be saturated by rearranging the bonds and atoms in the first 1–2 monolayers of the surface (surface reconstruction), while the exposure to the environment results in a contamination of the surface favored by the presence of the dangling bonds [45]. All these factors determine the presence of point defects in proximity of the surface that creates deep electronic levels acting as traps and nonradiative recombination centers. For this reason, a significant fraction of the carriers diffusing toward the surface recombines while the remaining part diffuses back into the material.

Unfortunately, the effective value of s cannot be determined for a specific material, as it depends on the sample surface properties, like oxidation and surface reconstruction, which can vary with time and with treatment type (i.e., passivation by means of the deposition of a thin layer material saturating the surface dangling bonds). Moreover, the value of s can be expected to vary as a function of the carrier density at the surface. First of all, at large carrier density, the deep centers can be partly saturated if those states are metastable (with lifetimes of the order of μs). Further, as an indirect consequence of the surface-localized states, a distortion of the band profile is often observed in proximity of the surface causing electrical charges trapping and a local modification of the density of states. The net effect is a local change in the energy position of the Fermi level to reach the thermodynamic equilibrium (this is sometimes referred to as Fermi level pinning).

In this case, the band alignment is modified and the carriers close to the surface experience a curved band profile equivalent to the alteration produced by an electrostatic field. The effect of such a field is to modify Eq. (7) in a region of a depth z_d from the surface (the depletion zone, some nm wide) obtaining the following equation:

$$D\nabla^2 n + C\frac{dn}{dz} - \frac{n}{\tau} = -g \qquad (13)$$

where $C = -\mu(\partial U/\partial z)$, U being the effective potential accounting for the surface band bending. It has been shown [46,47] that, in practice, the surface recombination can be obtained as a function of the surface charge, and that in regions deeper than z_d, the presence of the upper depletion zone can be accounted for by a modification of the actual surface recombination velocity. Moreover, it can be demonstrated [48] that, as a consequence of the partial screening of the surface fields, the surface recombination should actually depend on the carrier density. Unfortunately, a value of s depending on the carrier density is a mathematical complication, as it destroys the linearity of Eq. (8), making it very difficult to find a rigorous general solution. For this reason, in most cases, a constant recombination velocity is considered at least in a limited range of injection densities. In this case, an analytical solution can still be given by a method analogous to the image charge method in electrostatics.

If the surface recombination velocity is sufficiently high ($S \gg 1$), the lateral spread of the distribution of the diffusing carriers, and therefore the total carrier density, can be influenced significantly mostly for carriers close to the surface. Because the lower part of the sample in a classical SEM–CL experiment is practically unlimited, diffusion to distances of the order of the diffusion length is still possible in all the remaining directions away from the surface.

While the problem of surfaces has been extensively studied, only a few papers take into account the influence of the interface between the two materials. This kind of problem occurs quite often when semiconductor heterostructures have to be analyzed. In this case, the different interfaces are often aligned parallel to the surface. From an analytical point of view, the problem of the interfaces can be considered similar to the surface problem. The main difference is that, except for the losses due to the interfacial recombination (e.g., tunneling to deep traps at the interface), the carriers crossing the interfaces are not lost but continue diffusing through the adjacent material and can also diffuse back.

The rigorous description in terms of boundary conditions is not straightforward: a solution was given in a general case of interfaces by Chen [49], imposing a discontinuity in the derivative of the carrier density at the interface.

A simple rule of thumb prediction is that the probability to overcome the energy barrier ΔE of the band discontinuity is given by

$$P = e^{-(\Delta E/kT)} \tag{14}$$

A complete calculation, using a 1D geometry for a source at only one side (B side in Eq. (15)) of the interface and accounting for the density of states, predicts the current at the interface to be [50]

$$j = env \frac{N_A}{N_B} e^{-(\Delta E/kT)} \tag{15}$$

where v is the velocity of the carriers near the interface on A side, usually considered as the diffusion velocity v_D, and the constants N_A and N_B are the density of states in the conduction (or valence) band at the two sides of the interface. It is easy to see that without barrier, the current becomes simply $j = env_D$ and the boundary condition just imposes the continuity of the derivative of n at a selected plane.

So far, a few CL simulations have been attempting to model the effect of QWs, quantum wires (QWire) or quantum dots (QDs) on CL spectra. In these three cases of low-dimensional structures (LDSs), there are some similarities, in particular because the carrier diffusion inside those confining structures is limited, with respect to the bulk material, to one (QWire) or two (QW) directions. This effectively changes the solution of the diffusion equation inside the structure. As, however, the confining structures are part of a three-dimensional matrix, the interaction between these two systems often needs to be accounted for. In particular, it is necessary to consider that carriers are captured inside an LDS and then possibly re-emitted through the surrounding material by thermal activation. These two processes are ruled by the relevant characteristic capture (τ_c) and re-emission (τ_r) times. The difficulty in modeling these processes explains why little has been done so far in this direction. Considering, for instance, the case of a QD or a QWire characterized by a very long re-emission time, they can be described like a point defect or a dislocation. The main noticeable difference will be that the relevant CL signal is now coming from LDS rather than from the bulk. The case of a QW can be treated in the hypothesis that no carriers can cross the QW area without being captured and that again no thermal re-emission is possible. This case can be treated as a sort of surface with a large recombination velocity. When both the capture and re-emission processes have to be considered, a significant quantity is the relative probability of the two processes.

As for the re-emission probability, Schneider and Von Klitzing [51] used for a QW the formula

$$\frac{1}{\tau} = \frac{1}{w}\left[\frac{kT}{2\pi m^*}\right]^{1/2} \exp\left[-\frac{\Delta E}{kT}\right] \quad (16)$$

where m^* is the effective mass, k the Boltzmann constant, w the QW width, and ΔE the energy difference between the levels in the QW and in the barrier.

As for the carrier capture process, it has a weaker dependence on temperature and depends on the relaxation dynamics through emission of optical phonons and Auger electrons. Brum et al. [52] show, for example, an oscillating dependence of such capture time on the QW width. In any case, the average value is of the order of a picosecond or less [53].

The case of a QD is, in principle, simpler as no diffusion needs to be considered inside the structure. This suggests that the defect-like approach can be used considering an effective capture time

$$\tau_{\text{eff}} = \tau_c \left(1 + \frac{\tau_{\text{LDS}}}{\tau_r}\right) \quad (17)$$

where τ_{LDS} is the lifetime inside the low-dimensional system; in this case the QD.

Actually, the realistic case of Stranski–Krastanov QDs is more difficult to be simulated because a QW-like layer (wetting layer) surrounds the QDs, acting as carrier reservoir. In this case, the use of analytical methods is probably completely unfeasible.

1.2.3. Numerical methods

The results reported in the quoted references so far are mainly deduced by analytical calculations; however it is clear that each approach shows some limitations, especially when the generation volume should be accurately described or a perturbation of the crystal symmetry should be inserted into the calculation. The modeling of the generation volume as a sphere is well suited for an analytical solution of the combined injection and diffusion processes, but such an approximation yields too crude results. On the contrary, the analytic expression of the depth–dose function $g(z)$ in Eq. (6) only considers effects along the z direction. The only realistic solution for CL simulations is therefore the numerical approach. The most efficient and direct solution was introduced by Pey et al. [54], who modeled the diffusion by means of finite differences. This kind of program is probably the fastest and the most precise one available in the CL literature, with one important drawback: it becomes increasingly difficult to model interfaces, confining structures and all cases, where one cannot simply apply the master diffusion equation (Eq. (7)). Therefore, we developed a multipurpose software able to account for all different problems at the same time. This simulation program ("ALLES"), which has been used for the experiments reported in this work, is briefly described in the Appendix.

1.3. Lateral, in-depth, and analytical resolution

The *lateral* resolution in CL imaging can be roughly defined as the minimum detectable distance between the two regions presenting different CL intensity. In the SEM–CL the spatial imaging resolution depends mainly on the size of the recombination volume (generation volume broadened for the diffusion length) of e–h pairs inside the material, entailing also a dependence on the diffusion length of minority carriers. A second factor affecting the resolution is the electron probe diameter on the specimen surface, which depends on the beam current following a power law [55]. A typical value of the lateral resolution of about 100 nm in far-field conditions can be reached as a lower limit in suitable working conditions, for instance, on III–V semiconducting quantum confined heterostructures [22,56]. A different topic is represented by free-standing nanoribbons, wires or belts, etc., as single nanowire spectroscopy can be achieved by CL. In this case, the lateral resolution is only determined by the width and thickness of the nanoribbon. CL spectroscopy and imaging of nanobelts of 50 nm in diameter can also be easily performed [57,58].

The *in-depth* analysis is a CL peculiarity which permits to vary, in a single experiment, the electron penetration range through a controlled beam energy variation. A proper value of the beam current at each voltage must be calculated to balance the variation of the explored volume by an adequate dose change [59]. This procedure allows to avoid artifacts in detecting in-depth inhomogeneities, which could affect the shape, the energy position and the intensity of the emission bands (in summary, the analytical resolution: see below). The achievable resolution, Δt, related to the depth–dose function, is of the order of 100 nm [21] or lower in controlled conditions, in particular for LDSs, where a quantum confinement of the carriers is achieved.

As for the CL *analytical* resolution, let us introduce it as the local sensitivity as a function of lateral or in-depth resolutions, the sensitivity being the minimum detectable variation of chemical and physical properties, like dopants, impurities, point and extended defects, compositional variations and internal fields etc. As an example of the outer sensitivity limit in our experimental setup, we have reported how adding 1 monolayer of GaN in GaN/AlN stacked self-assembled QDs resulted in an average energy shift of the CL emission energy of about 150 meV [60].

1.4. Role of excitation conditions

The dependence of CL on excitation conditions and materials properties is a wide field whose detailed discussion is beyond the scope of this chapter. Here only two major points affecting the CL signals will be shortly discussed. A wider description of that dependence can be found in Chapter 4.4 of Ref. [14].

1.4.1. CL emission efficiency, energy peak position and lineshape broadening

The possibility of varying in a single experiment and in a controlled way, the injection power for any given generation–recombination volume is the main peculiarity of the SEM–CL technique. Beam current, beam voltage and focusing conditions, during prolonged electron beam irradiations at high beam currents, can determine significant changes in peak intensity, energy values and lineshape of the CL spectra, with respect to low-injection conditions.

The CL emission efficiency, L_{CL}, generally depends on both the beam current and the voltage and, according to Ref. [14], and references therein, can be written as

$$L_{CL} = f(I_b)(V - V_0)^n \qquad (18)$$

where V_0 is a "dead voltage," accounting for surface nonradiative recombinations (Section 1.2.2), and $1 \leq n \leq 2$. The value of V_0 is affected mainly by the properties of materials and primary beam brilliance; in the specific case of very efficient GaN layers, with our experimental setup, V_0 was found to be around 0.25 keV [61]. $f(I_b)$ is a power function of the beam current ($f(I_b) \propto I_b^m$), which assumes values partly dependent on the type of transition but strongly influenced by a large number of variable parameters. In general [62], a sublinear ($m < 1$) dependence of $f(I_b)$ has been reported for extrinsic DAP and deep level-related emissions, whereas a linear ($m = 1$) or superlinear ($m > 1$ and in particular $m = 2$ in GaN layers), dependence has been observed for the near-band-edge emission intensity [63,64]. In addition to defect concentration, e–h pair density and injection rates, there is evidence of beam size and focusing-related effects on absolute and relative intensities and energy position of the CL bands [65,66]. Roughly, defocusing the electron beam on the sample surface without changing the e-beam energy, results in a broadening of the beam size and in a modified carrier distribution inside the generation volume which can be quantitatively determined only using complicated experimental procedures [66]. However, as the influence of materials on defocusing effects has also to be considered, results of general validity are not yet provided.

To get reliable CL analyses, self-absorption effects should also be taken into consideration as they cause noticeable modifications of the relative intensity and of the energy peak baricenter in the outcoming CL spectra [16,67].

As for the band emission energy values, for specific transitions, they depend on the carrier density, hence on the excitation conditions. A typical example is the DAP recombination that shifts to higher energies as the excitation intensity increases independent of the excitation source. This follows from the reciprocal dependence of the peak energy on the pair separation, r, and on the reduction in the transition probability with increasing r [68]. Concerning CL, the aforementioned effect is much stronger due to the wider excitation range achievable. An outstanding case of the advantages offered by CL spectroscopy in nitrides is the carrier recombination in nanostructures, where an inverse quantum confined Stark effect (QCSE) is obtained by e–h pair generation. This subject will be extensively discussed and experimentally demonstrated in Sections 3.1 and 3.3.

As already mentioned, CL is particularly useful for in-depth studies, provided the electron generation and recombination volumes can be properly simulated [16]. However, as primary electrons accelerated at keV energies in general activate all the luminescence mechanisms of the material, whereas PL emission is limited by the laser frequency, the main point is the different role of thermalization processes in PL and CL experiments and the different energy level population. In fact, the electron beam excitation entails stronger energy transfer to the lattice and phonon emission, and the temperature of the excess carriers remains usually higher. All these factors produce a linewidth broadening of the CL bands compared to PL transitions.

Furthermore, for most of the generation conditions, the e–h pair density is higher in CL than in PL and carriers populate more energetic levels of the semiconductor. This also causes a strong dependence on the scan rate of the transitions involving levels with longer carrier lifetime, because the kinetics of the filling or emptying of traps can be commensurate with the raster of the electron beam. If the scan rate is reduced or even the spot mode is used (in this configuration the e-beam is spatially steady in a defined surface specimen position), the electron beam dwells on the same area and these levels are saturated earlier. For example, this is observed for the yellow band in GaN, whose integrated intensity relative to the near-band-edge peak is minimum in the spot mode and maximum in standard *TV* scan, and strongly exceeds that detected in PL experiments [69].

Finally, the influence of electron irradiation on the specimen temperature has to be considered. The electron beam-induced sample heating has been estimated by Myhajlenko et al. [70] on the basis of the Vine–Einstein model [71], predicting a maximum temperature rise under steady-state conditions from

$$\left(\frac{P}{a}\right)_{\text{eff}} = 4.27 \int_{T_{sub}}^{T} k(T)\, dT \tag{19}$$

where P is the beam power (in W), a is the beam radius (in cm), $k(T)$ is the thermal conductivity (in W/cm/K), and $(P/a)_{\text{eff}}$ is the effective beam power corrected for the penetration. Typical temperature rises of about 150 K were achieved for ZnSe with substrate temperature of 300 K, $E_b = 30$ keV, $I_b = 10\,\mu\text{A}$ and beam diameter of $3\,\mu\text{A}$. Reducing the substrate temperature to 100 K, a temperature rise of 30 K was estimated. It has to be noted that by increasing the temperature, the shape of both intrinsic (edge emission) and extrinsic emission bands cannot be entirely described by a Gaussian function of symmetrical shape. Actually, the shape around the peak is asymmetric with high- and low-energy exponential tails [72].

Furthermore, artifacts on extended defects CL contrast can also be induced by local specimen heating. For instance, contrast inversion at a dislocation in GaAs:Se substrate has been evidenced in Ref. [73] by increasing the electron beam current from 10^{-6} to 10^{-5} A. The effect has been explained in terms of localized heating leading to enhanced nonradiative recombination at the dislocation line and to a decrease of the CL signal.

2. Cathodoluminescence on III–nitrides heterostructures

2.1. Properties of III–N and dilute nitrides

The group of compound semiconductors containing nitrogen (III–N and III–V–N dilute nitrides) GaN, AlN, InN, and their ternary (e.g., InGaN, AlGaN, GaAsN) and quaternary

(InAlGaN, InGaAsN) alloys have attracted an increasing interest as strategic materials for a wide range of industrial applications and device production [74,75]. They are suitable systems for applications in optoelectronics (light-emitting diodes (LEDs) and lasers) and high-power electronics (e.g., high-electron-mobility transistors), due to the potential given by: the large tunability of direct band gaps with alloy composition, from about 0.8 to 6.2 eV, the high peak and saturation drift velocity, the existence of polarization-induced effects [76], such as the formation of high-density low-dimensional charge gases.

While dilute nitrides present the zinc blende structure like the majority of III–V compounds, because of the low N concentration in the alloys, the other III–N semiconductors are characterized by the wurtzite crystal structure. In luminescence studies of hetero- and nanostructures based on hexagonal nitrides, internal polarization fields play the leading role. The wurtzite crystal structure in these polar semiconductors determines a nonvanishing spontaneous polarization (SP) and the onset of a piezoelectric polarization in presence of strain [77]. In particular, for biaxial strain the nonzero component of the piezoelectric polarization is along the growth axis

$$P_3^{PZ} = 2\varepsilon_1(e_{31} - e_{33}C_{13}/C_{33}) \quad (20)$$

where C_{13}, C_{33} are the elastic constants, e_{31}, e_{33} are the piezoelectric constants and ε_1 is the strain ($\varepsilon_1 = (a-a_0)/a_0$, relative change of the basal plane lattice parameter with respect to the relaxed crystal). The values of the piezoconstants (Table 1) of the III–N binaries are up to 10 times larger than for conventional III–V and II–VI compounds, in particular those of AlN which are the largest known so far among the tetrahedrally bonded semiconductors. In ternary alloys, a nonlinear dependence on the alloy composition is observed for both the spontaneous and the piezoelectric polarizations [78].

For basal-plane strains typical of epitaxial wurtzite nitride multilayers, the SP is comparable to the piezoelectric polarization and the two components can sum up or subtract each other depending on strain and polarity. In AlGaN/GaN systems SP effects are dominant, because AlN has an SP about three times larger than GaN and a relatively small (~2.5%) lattice mismatch to it. In contrast, InGaN/GaN structures are mostly influenced by piezoelectricity, because InN has nearly the same SP as GaN but a large lattice mismatch (~11%)

Table 1

Lattice parameters, energy gap, spontaneous polarization, elastic constants, piezoelectric constants, and dielectric constant of binary hexagonal nitrides (values are taken from Refs [80,81])

Parameters	GaN	AlN	InN
a (Å, RT)	3.189	3.112	3.545
c (Å, RT)	5.185	4.982	5.703
$E_G(0)$ (eV)	3.510	6.23	~0.8
SP (C/m^2)	−0.034	−0.090	−0.042
C_{13} (GPa)	106	108	92
C_{33} (GPa)	398	373	224
e_{31} (C/m^2)	−0.34	−0.53	−0.41
e_{33} (C/m^2)	0.67	1.50	0.81
ε (ε_0)	10.28	10.31	14.61

to it. It is worth noting that for layers under compressive strain, a compensation is achievable in alloys with appropriate composition [79].

In heterostructures as multiple quantum wells (MQW) or superlattices (SL), polarization differences at interfaces (or at the surface) cause interface (surface) charge densities with typical values of a few $10^{13}\,\mathrm{cm}^{-2}$ [82]. Electrostatic fields up to a few MV/cm are produced and rule the carrier dynamics. To estimate the value of the field in the jth layer of a MQW or SL system under assumptions of periodic boundary conditions and no free carrier screening, the formula

$$E_j = \frac{\sum_k (P_k - P_j) l_k / \varepsilon_k}{\varepsilon_j \sum_k l_k / \varepsilon_k} \qquad (21)$$

has been proposed [79], where the sum runs over all the structure and P_k, l_k, and ε_k are the total transverse polarization, the thickness, and the dielectric constant of the k layer. A field exists in both QW and confining barrier, typically of opposite sign.

The band profile is strongly modified by the built-in fields [83]. In confined structures, an effectively triangular confinement potential is obtained and the carriers are localized at the interfaces, opposite for e^- and h^+, where the bound polarization-induced charge is located (Fig. 6). Strong changes are observed concerning the optical properties, mainly the quantum confined Stark effect (QCSE) and consequent anomalies in recombination dynamics. A Stark-like red shift in recombination and absorption energies and a suppression of the oscillator strength, due to the spatial separation of electron and hole wave functions are produced. Therefore, optical transitions in LDSs, for example, between the electron and hole ground states in a QW, usually occur well below the band-gap energy, are slowed down and made less intense. The transport properties are also strongly modified. In particular, the formation of a bidimensional electron gas in n-type heterostructures, where free electrons accumulate at interfaces having positive polarization-induced charges, has been of special interest for the fabrication of devices as high electron mobility transistors (HEMTs).

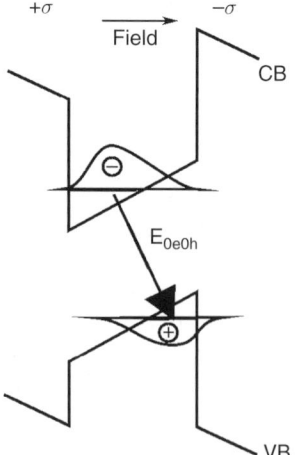

Fig. 6. Schematic band-edge profiles and fundamental electron and hole wave functions in case of QCSE in a QW system.

These processes are very sensitive to the carrier distribution. The value of the built-in fields can be reduced by different screening mechanisms, which allow a recovery of oscillator strength and a blue back-shift of the transition energies in a permanent or transient mode. A partial permanent screening of the bound polarization charge is achieved mainly through extrinsic carriers provided by doping, oppositely charged interface defects, or in the case of surfaces charges adsorbed from the environment. The transient screening can be obtained through free carrier generation by electrical or optical (e.g., PL, CL) injection. We present this process in the following section as for CL excitation.

Dilute nitrides are a new class of semiconductor materials, which attracted the interest of the scientific community over the past 5 years because of their puzzling physical properties and their great potential for applications. Indeed, in $GaAs_{1-x}N_x$, the substitution of As with N atoms leads to dramatic changes in the materials electronic, optical and structural properties [75]. For x as small as 0.01, the energy gap decreases by ~ 200 meV, the electron effective mass doubles, a tensile strain develops when $GaAs_{1-x}N_x$ is grown on GaAs, and new modes appear in the infrared absorption spectra. All these N-induced effects stem from the large difference in size and electronegativity between the N and the As atoms, which breaks severely the lattice translational symmetry, leads to a mixing between different conduction bands of the crystal, and causes sizeable local lattice distortions. Some of these features render this material suitable for a variety of applications, ranging from telecommunications through fiber-optic cables, to multijunction solar cells, heterojunction bipolar transistors, and terahertz applications.

2.2. Why doing cathodoluminescence in semiconductors containing nitrogen?

Multilayers and nanostructures based on nitride alloys exhibit optical emissions spanning a wide spectral range and having an inhomogeneous spatial distribution. In this context, the CL is more flexible than conventional PL, as it allows a simultaneous excitation of all the recombination channels (e.g., yellow luminescence in GaN – Section 1.4), a selected penetration depth and the controlled generation on a given 3D volume in a single experiment. Due to the very high luminescence efficiency of III–N, the CL spectra can be obtained with a spectral resolution comparable to PL (0.2 nm). In these conditions, the peculiarities of CL can be applied without loss of spectral details.

The typical feature of CL spectra is to show the NBE emissions as well as broad bands due to shallow and deep levels related to dopants (e.g., Mg, Si), intrinsic point defects (e.g., V_{Ga}, V_N), and complexes (e.g., involving H – see Section 3.3). In particular, in LDSs such as QWs and QDs [60], transitions between electron and hole quantized levels are observed, where the ground states and one or more excited states are involved depending on the generated carrier density and the injection conditions.

Finally, CL takes a leading role in presence of polarization fields, as it provides an effective controlled field screening, up to almost a complete recovery of the flat-band conditions. This screening process is a function of the excitation power and can be extensively studied in power-dependent experiments, performed by varying the injection current at fixed voltage (see Section 3.1).

Further, in the high-injection conditions, it is possible to modify in a controlled way the crystal band gap of hydrogenated dilute nitrides in the spatial region where the electron beam is steered (see Section 3.2).

3. Power-dependent cathodoluminescence: three examples

3.1. Internal field screening in InGaN QWs to restore flat band conditions

In presence of internal fields, the generation of free carriers results in a screening of the bound polarization charge and effectively reduces the field strength [84]. The efficiency of the process depends on the sheet carrier density and distribution. Considering, for example, a QW, at low excitation power the sheet density has a marginal effect and the electrostatic field remains uniform over the whole QW. As the generated carrier density increases, the field gets progressively screened, in a way depending on the well width. An inverse QCSE is produced, driving a recovery of oscillator strength (recombination rate increased by several orders of magnitude) and a blue back-shift of transition energies. However, the field is not screened abruptly, but dies off gradually with an effective screening length [85], due to the larger spatial extension of the screening charge as compared to the polarization charge. This entails that for QWs much wider than the screening length a complete recovery of the zero-field conditions cannot be reached even in the high-power limit.

The above-mentioned mechanisms concerning free carrier screening have a general validity, independent of the particular nature of excitation and generation processes. Power-dependent CL spectroscopy has specific strengths, as it allows to follow, mainly by monitoring the intensity and the energy position of the emission peaks, the field evolution for close variations of the carrier density. This is obtained by varying in a controlled way the beam current, and a wide range can be explored. Compared to time-integrated (TI) PL experiments, conventionally used for these studies, the main advantage of the CL methodology is indeed to cover an excitation range extending to about four orders of magnitude, up to carrier densities comparable to polarization charges. This highlights that the efficient field compensation is achievable only by CL in high-injection conditions.

We applied the power-dependent CL methodology to the analysis of InGaN/GaN MQWs. In this system, the engineering of the emission wavelength by simply selecting the In composition and the well width is known to fail [86]. The quantum energy levels and the carrier dynamics are indeed mainly controlled by the built-in electrostatic field, dependent on many structural parameters such as growth design, doping, surface/interface states, and alloy composition [81]. In addition to the QCSE, In clustering, notably relevant at $x > 0.2$, can occur and induce strong carrier localization in QD-like regions [87,88]. Moreover, remarkable changes in the optical properties can result from rather small compositional inhomogeneities, as the energy gap difference between GaN and InN (its narrow band gap of about 0.8 eV has been recently proved [89]) is very large.

To reduce the effects related to inhomogeneities in In distribution, we analyzed structures based on five period $In_xGa_{1-x}N$/GaN MQWs with low In content ($0.025 < x < 0.065$), grown on a 1 μm-thick GaN template by MOCVD using (0001) sapphire substrates. The structural parameters, as assessed by high-resolution X-ray diffraction (HRXRD) and TEM analyses [90], are reported in Table 2. The existence of major compositional fluctuations affecting these alloys can be ruled out on the basis of the structural characterizations.

A research focused on the role of internal fields and their interplay with carrier generation was carried out [91]. The QW-related optical transition was excited at increasing power

Table 2

Summary of the structural parameters of three different five period $In_xGa_{1-x}N$/GaN MQW samples as determined by HRXRD and TEM measurements. A final GaN cap layer, 33 nm thick in #MQW02 and 20 nm thick in #MQW03 and #MQW04, is omitted

Sample	Period (nm)	QW thickness (nm)	Composition $In_xGa_{1-x}N$
MQW02	44.0	3.5	$x = 0.060 \pm 0.004$
MQW03	28.3	2	$x = 0.060 \pm 0.005$
MQW04	27.5	1÷1.5	$x = 0.0375 \pm 0.025$

densities, in PL[1] and CL studies. From spectral analysis, the peak evolution was assessed, allowing to determine the dependence of the transition energy on the excitation conditions. An almost rigid blue shift of the QW emission was revealed (see the representative spectra in Fig. 7), mainly as a result of the internal field screening by the generated carriers on the wide investigated range (10^8–10^{12} carriers cm^{-2} in a QW, estimated as a function of the CL excitation conditions by using our software ALLES). The blue-shift is accompanied by a continuous increase of the QW emission intensity due to the gradual recovering of the oscillator strength. It must be stressed that the CL emission at the lowest injection (centered at 3.07 eV) takes place between the PL peaks for the limit conditions of low (centered at 3.064 eV) and high (3.08 eV) excitation. This underlines that the two techniques identify two complementary power regimes and can be joined in the intermediate excitation region. We point out that, due to the effect of the screening, huge variations of the optical matrix element are found. Consequently, for a fair comparison of the energy shifts, the spectra have been normalized.

A systematic power-dependent analysis, which provided the complete QW emission energy evolution, was performed. The case of the 3.5 nm QW sample is considered in Fig. 8, where the values obtained from PL studies (filled symbols) have been combined with the CL ones (open symbols). An overall blue-shift of about 66 meV (16 meV across the PL excitation range and 58 meV in the CL range) is obtained at $T = 6$ K in this structure, up to a saturation tendency. As for the methodology, a significant result was achieved, that is CL demonstrates to be effective in reaching the most efficient field screening.

The dependence of the blue-shift amount on the QW width at fixed composition was also evidenced. This is clear by comparing the results for samples 02 and 03, as reported in Fig. 9. In CL the shift decreases, as expected, from about 60 to 13 meV by decreasing the well thickness from 3.5 to 2 nm. As shown in the figure, the experimental results are in very good agreement with simulations based on a theoretical model where a self-consistent solution of the Schrödinger–Poisson equations is coupled to a rate equation model [91] to calculate the electronic states and the optical transition rates. In detail, the agreement was obtained considering a nonradiative time constant equal to 3 ns, confirmed by time-resolved PL data, and a value of the internal field of about 0.6 MV/cm.

---

[1] For PL experiments (courtesy of A. Vinattieri), the second harmonic of a picosecond self-mode locked Ti:sapphire laser pumped by a CW Ar$^+$ laser was used, varying the excitation power in the range 0.05–30 mW and tuning the energy of the exciting photons between 3.3 and 3.5 eV.

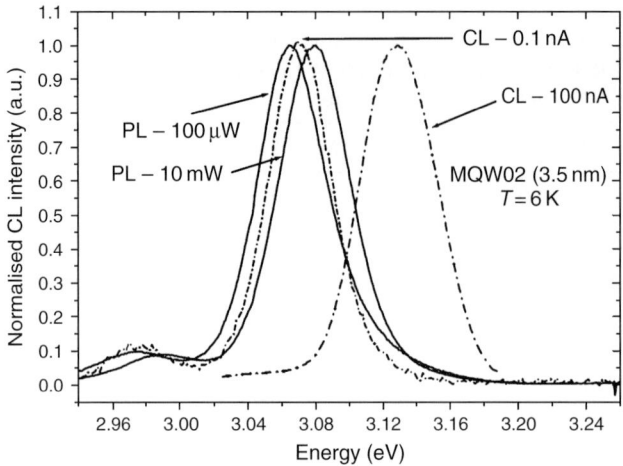

Fig. 7. PL (solid lines) and CL (dotted lines) spectra obtained in the limit of minimum and maximum injection power (PL: 0.1 and 10 mW, respectively, CL: $I_B = 0.1$ and 100 nA, respectively) on sample MQW 02 at $T = 6$ K. The peak intensity has been normalized to unity.

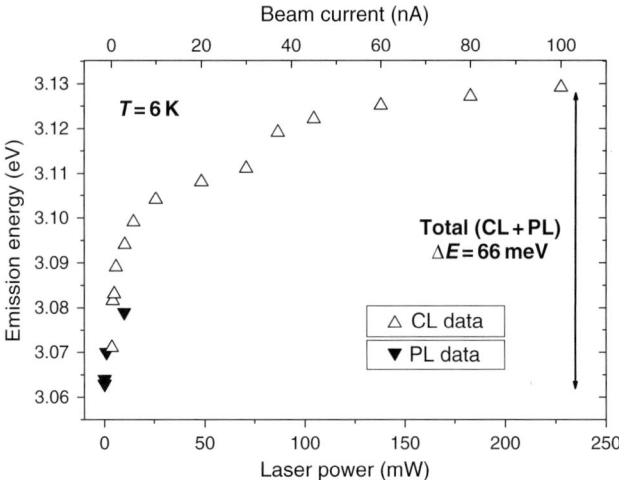

Fig. 8. Excitation dependence of the QW peak position for sample MQW 02 at $T = 6$ K. Filled triangles: PL data at 0.050, 0.100, 1, and 10 mW, respectively. Open symbols: CL peak position in a current range 0.1–100 nA at $E_B = 10$ keV. The measured total blue shift is about 58 meV in CL and 16 meV in PL, respectively.

The increase in the QW emission energy for decreasing well width, as obtained from the CL results in the high-injection limit, is noteworthy. In such conditions, any effect arising from possibly different internal fields is minimized. This confirms the existence of a two-dimensional quantum carrier confinement in these InGaN nanostructures. Finally, no shift was found for sample 04 (not shown here) in the overall range investigated in TI PL and up to 2 nA in the CL analysis. Later, an 11 meV shift was obtained in CL at higher beam currents. This suggests that an almost complete free carrier screening of the built-in field is already achieved at

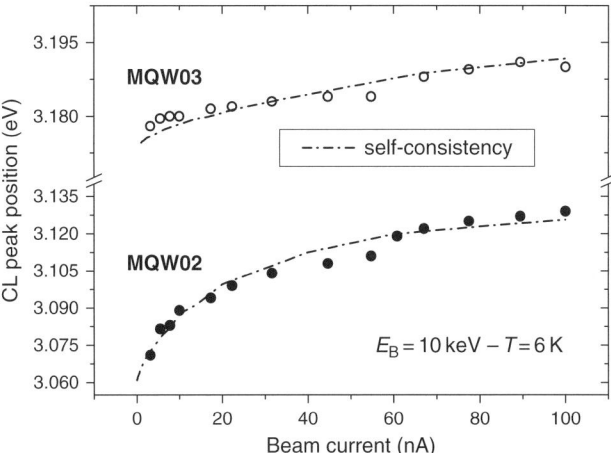

Fig. 9. Blue shift of the QW-related CL peak for samples with QW width of 3.5 (MQW02) and 2 nm (MQW03) (symbols: experimental data, dotted-dashed lines: theoretical model).

low-injection conditions and the observed shift at the highest densities is more likely due to band-filling effects. A small field value in this sample, and the consequent observed initial absence of peak shift, is consistent with its very low QW thickness and In composition, which can also account for the QW emission energy well above those of the other specimens.

3.2. Controlled in-plane band-gap modulation in hydrogenated dilute nitrides

In modern epitaxial growth techniques, the control of the electronic and optical properties of a semiconductor heterostructure *along the growth direction* is easily achieved via layer-by-layer deposition of materials with different chemical composition and thickness [92,93]. On the contrary, the control of those material properties *in the growth plane*, which is required to fabricate one-dimensional or zero-dimensional nanostructures, is not easy to attain. That limits the freedom of modulating the in-plane optical properties of a heterostructure, where materials with different band gaps and refractive indexes are comprised on a same chip. Two main methods are used for the in-plane control of the physical properties:

(i) "Top-down" methods based on lithographic processes, which give a lateral modulation of the material band gap by chemical removal of parts of the specimen, allow to grow highly uniform QDs, wires, and rings, but result in a poor optical quality [94],
(ii) "Bottom-up" methods based on nanometer-sized aggregates forming spontaneously by self-assembly in highly strained heterostructures [95,96], but that are in turn affected by a lack of control of the spatial arrangement.

A third route to the engineering of the electronic properties in the growth plane of a semiconductor can be achieved by exploiting the effect hydrogen has on dilute nitrides, such as $GaAs_{1-x}N_x/GaAs$ (and $GaP_{1-x}N_x/GaP$) [97]. In these material systems, the replacement of a tiny fraction ($x \sim 1\%$) of arsenic (phosphorus) atoms by nitrogen atoms leads to highly nonlinear effects in the electronic properties of the host lattice [75,98]. These include a giant reduction in the band-gap energy and a deformation of the conduction band structure, which render this material of high potential for telecommunications through fiber-optic cables [99], multijunction solar cells [100], heterojunction bipolar transistors [101] and terahertz applications [102]. Previous experiments have shown that postgrowth irradiation of $GaAs_{1-x}N_x$ with atomic hydrogen leads to a complete reversal of the drastic band-gap reduction, as well as of other material parameters, caused by nitrogen incorporation [103–107]. Here, by deposition of metallic masks on and subsequent hydrogen irradiation of $GaAs_{1-x}N_x$, we create a planar heterostructure with zones having the band gap of a $GaAs_{1-x}N_x$ well surrounded by GaAs-like barriers (metallic wire experiment). In this chapter, an alternative way based on the controlled electron beam irradiation of hydrogenated samples is presented. By focusing an energetic electron beam on the surface of hydrogenated $GaAs_{1-x}N_x$, we remove hydrogen atoms off their passivation sites, thus leading to a controlled decrease of the crystal band gap in the spatial region where the electron beam is steered.

The possibility of focusing electrons on a small area (beam diameter ~ 10–100 nm) gives an insightful tool for investigating the interaction of hydrogen locally with point defects [108,109]. In particular, it was shown that an energetic (10–50 keV) beam of electrons can break the bonds that hydrogen forms with silicon impurities in GaAs through an electronic excitation of the Si–H complexes [109]. Here, an unpatterned $30 \times 40\,\mu m^2$ rectangular area of a $GaAs_{1-x}N_x$ sample treated previously with hydrogen was scanned by an intense electron beam having current $I_b = 400$ nA and energy $E_b = 5$ keV. Figure 10(a) shows two monochromatic CL images obtained at $T = 5$ K around the sample region where the rectangle was irradiated (*writing process*) by the intense electron beam. To collect the CL signal (*reading process*) a reduced beam current ($I_b = 10$ nA or lower) was used to avoid any further modification of the carrier potential. A typical irradiation time of 40 s is enough for obtaining a complete recovering of the $GaAs_{1-x}N_x$ band-gap value before H irradiation. The images in Fig. 10(a) were collected with the monochromator energy set at $E_{det} = 1.415$ eV (left image) and 1.465 eV (right image). Therefore, the two bright areas (each complementary to the other) correspond to photons having energy equal to that of the band gap of the as-grown (left) or hydrogenated (right) sample at $T = 5$ K. A spectral analysis of the light emitted from the electron-irradiated rectangle and from the area outside is shown in Fig. 10(b). The CL spectrum of hydrogenated plus electron-irradiated $GaAs_{1-x}N_x$ (middle line, energy gap equal to 1.415 eV) coincides with that of the as-grown lattice (topmost line) and differs markedly from that of the hydrogenated sample (bottommost line, energy gap equal to 1.465 eV). In other words, the rectangular area designed by the intense (writing) electron beam acts in all respects as a potential well for carriers.

The temporal evolution of the CL emission intensity under electron beam irradiation is shown in Fig. 11. The NBE GaAs characteristic emission is not affected by the "writing" process and is not reported. The curves reported in Fig. 11 have been obtained recording the CL intensity with an energy window $\Delta E \ll$ FWHM of the monitored CL band.

These findings demonstrate that the electron beam has broken the bonds of the nitrogen–dihydrogen complexes responsible for the dramatic changes in the electronic properties of the crystal [97]. Such breaking most likely results in a displacement of H atoms from the nitrogen–dihydrogen complex but not in hydrogen outdiffusion from the lattice. We point out that after irradiation with the electron beam the sample was cycled between $T = 5$ K and room temperature several times. Nevertheless, the same CL images and spectra were recorded after each cycle, thus demonstrating the thermal stability of the H displacement process.

The data shown so far indicate that spatially selective implantation/displacement of hydrogen defines a rather sharp *vertical interface* between planar zones having different

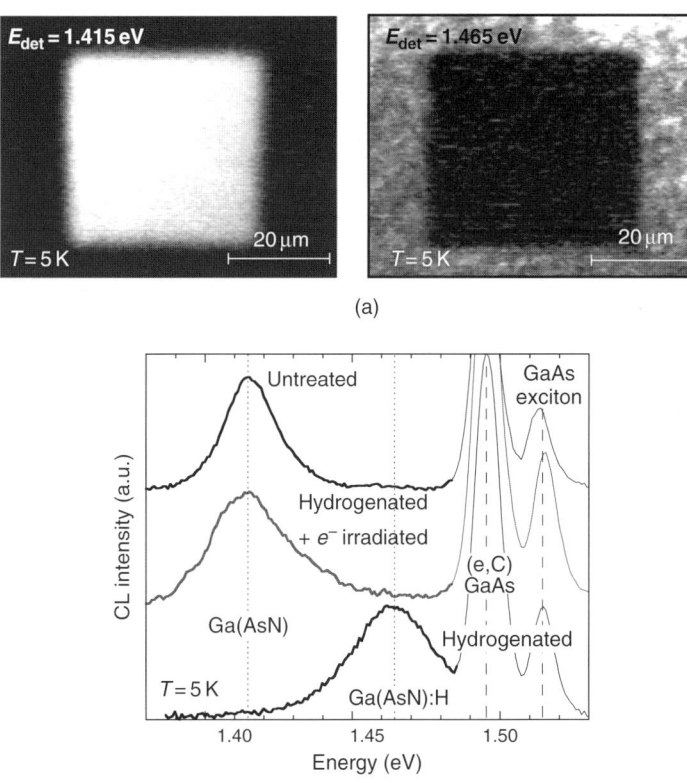

Fig. 10. (a) Low-temperature (T = 5 K) CL images acquired after sweeping a hydrogenated GaAs$_{1-x}$N$_x$ sample with an intense electron beam having current $I_b = 400$ nA and energy $E_b = 5$ keV (writing operation). The CL images were recorded with a less intense electron beam having $I_b = 10$ nA and $E_b = 20$ keV (reading operation). Left image: The monochromator energy E is set equal to that of the photons emitted from the GaAs$_{1-x}$N$_x$ well region. Right image: Same as left but $E = 1.465$ eV, that is the band-gap energy of the hydrogenated GaAs$_{1-x}$N$_x$ barrier region. (b) CL spectra recorded at $T = 5$ K in the region outside (bottommost line) and inside (middle line) the rectangular area swept by the intense electron beam. The CL spectrum of the GaAs$_{1-x}$N$_x$ sample before H irradiation is also shown (topmost line) for comparison purposes. Dotted vertical lines indicate carrier recombination from the GaAs$_{1-x}$N$_x$ region (either hydrogenated or not). Dashed vertical lines indicate carrier recombination from the GaAs band-gap exciton and free electron to carbon impurity levels ((e,C)).

band-gap energies. The abruptness of the crossover region between H-free and hydrogenated areas can be assessed by CL [110,111,14]. We model the impinging electron beam as having a Gaussian section with standard deviation $\sigma = 100$ nm, which was estimated by Monte Carlo simulations performed in GaAs under the given experimental conditions. By setting the origin in the impinging point of the electron beam, the number of carriers diffused in a point (x, y) of the plane perpendicular to the beam is given by

$$n(x,y) = C \int_{-\infty}^{\infty} dx_1 \int_{-\infty}^{\infty} dy_1 \, e^{-\left(\sqrt{(x-x_1)^2+(y-y_1)^2}/L_D\right)} e^{-\left((x_1^2+y_1^2)/2\sigma^2\right)} \tag{22}$$

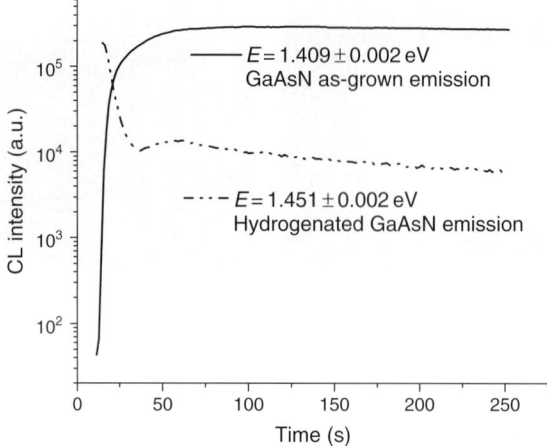

Fig. 11. Dynamical behavior of the "writing" process. The N–H dissociation and consequently the restoring of the original band gap of GaAsN is complete in about 20–30 s in these "writing" conditions. In particular, an I_b of about hundreds of nanometers is necessary for all the structures studied (threshold effect).

where C is a constant, the first exponential function represents the typical diffusive profile of the excited carriers with diffusion length L_D and the Gaussian function represents the electron beam section. A one-dimensional representation of $n(x,y)$ is shown in Fig. 12(a). The CL intensity corresponding to photons emitted from the whole well region after impingement of the electron beam in x_0 is then given by

$$I_{CL}(x_0) = C \int_{-\infty}^{\infty} dy_2 \int_{x_b}^{x_b+w} dx_2 \, n(x_0 - x_2, y_2) \qquad (23)$$

where x_b indicates the position of the border of the wire having thickness w. A fitting of $I_{CL}(x_0)$ to the CL intensity profile measured in the metallic wire experiment is shown in Fig. 12(b).

We found $L_D = 0.66\,\mu\text{m}$, which compares well with values of the carrier diffusion length reported in III–V semiconductors [112]. A similar analysis performed at room temperature gives $L_D = 1.8\,\mu\text{m}$, consistently with the temperature increase of the sample. This also implies that the spatial sharpness of the edge between the H-diffused and the H-free part of the sample ought to be better than $0.6\,\mu\text{m}$, namely, the carrier diffusion length at low temperature. The outcome of an analogous study of the vertical interface quality for the electron-irradiated rectangle shown in Fig. 10 is displayed in Fig. 12(c) (in this case $w = \infty$). A carrier diffusion length equal to $0.62\,\mu\text{m}$ is found. This value nearly coincides with that derived in the case of the metallic wire and confirms that our present capability of determining the sharpness of the in-plane band-gap profile is limited by carrier diffusion, only.

The method presented here for in-plane engineering the band gap of $GaAs_{1-x}N_x$ can be readily extended from the near-infrared to the visible spectral range by using $GaP_{1-x}N_x/GaP$

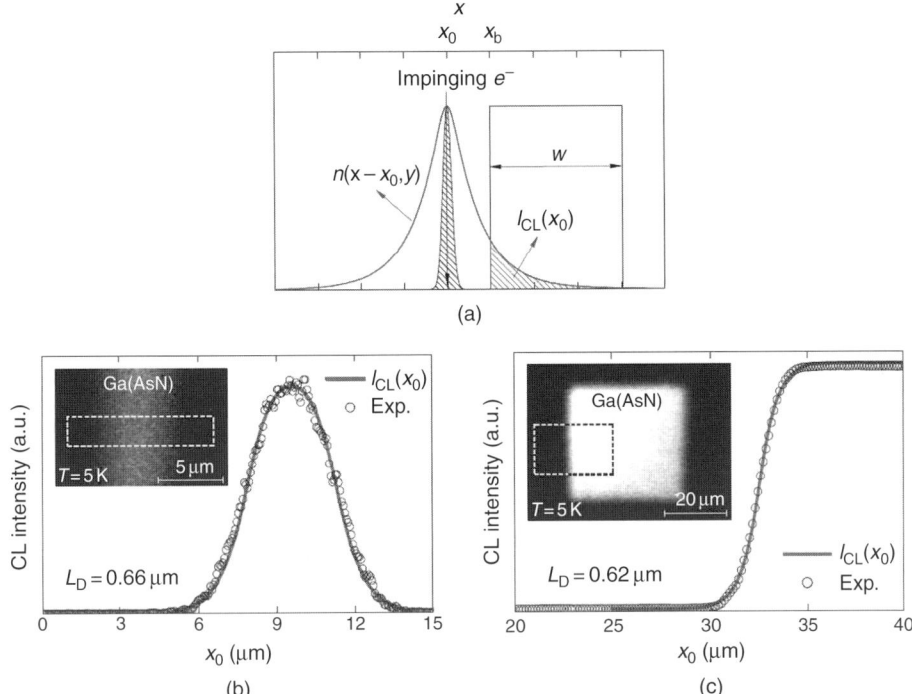

Fig. 12. (a) The black hatched area in the center indicates the impinging electron beam in x_0, the solid line is the carrier diffusion profile $n(x-x_0,y)$ in steady-state conditions. The rectangular box indicates the electron-irradiated region. The black hatched area on the right defined by the overlap between the electron-irradiated region and the carrier diffusion profile is proportional to the CL intensity due to the photons emitted from the GaAs$_{1-x}$N$_x$ region, $I_{CL}(x_0)$. (b) Fitting of the CL intensity profile across the GaAs$_{1-x}$N$_x$ wire (open circles) by the function $I_{CL}(x_0)$ defined in the text (solid line). The estimated carrier diffusion length, L_D, is equal to 0.66 µm. The inset reproduces the monochromatic CL image (E = 1.4 eV) of the region beneath the wire. The CL intensity data were obtained after a 2 µm vertical binning applied to the inset region defined by the dashed line rectangle. c) Fitting of the CL intensity profile across the rectangular region obtained by electron irradiation of hydrogenated GaAs$_{1-x}$N$_x$ (open circles) by the function $I_{CL}(x_0)$ defined in the text (solid line); $L_D = 0.62$ µm. The inset reproduces the CL image shown in Fig. 10(a). A 15 µm-vertical binning has been applied to the inset region defined by the dashed line rectangle.

heterostructures, whose emission wavelength can be varied from red to green by hydrogen irradiation [99,113]. In Fig. 13 the effects of electron beam irradiation on a GaPN:H sample is reported. Also in this case a complete recovering of the prehydrogenation conditions is achieved.

The modulation of the material band gap described here should affect the refractive index, n, which increases as the N concentration increases (and the band gap decreases) in GaAs$_{1-x}$N$_x$ [114]. Because n should be larger in as-grown GaAs$_{1-x}$N$_x$ than in the hydrogenated material, selective incorporation/displacement of hydrogen would allow to design integrated optical circuits. In particular, optical elements like planar waveguides and optical couplers working in the telecommunication wavelength range could be implemented using the methods presented in this chapter.

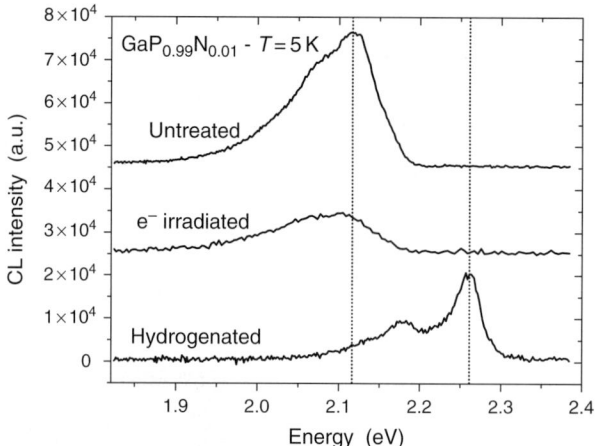

Fig. 13. CL spectra obtained at $T = 5\,K$, $E_b = 5\,keV$, $I_b = 400\,nA$ at 1000× after 30 s of irradiation on a GaPN:H specimen. The spectra have been shifted for clarity.

In conclusion, by either spatially resolved displacement of H from hydrogenated GaAs$_{1-x}$N$_x$ by direct electron beam writing, we are able to modulate on a small scale the band gap of a semiconductor heterostructure in its growth plane. In perspective, we should be able to attain such a modulation on a nanometer scale as to lead to quantum confinement effects on carriers.

3.3. Electron beam-induced aging of InGaN-based LEDs

A procedure of high-energy electron beam irradiation (HEEBI) can be effectively exploited in the analysis of LEDs based on III–N heterostructures. We performed in the SEM irradiation experiments correlated to a comprehensive device failure analysis [115,116], aimed at identifying the role of extended and point defects interactions [117,118], contact degradation [119] and pure annealing effects [120] in the worsening of device performance during electrical aging treatments. We will summarize here the main results of our research on MOCVD-grown blue LEDs with single InGaN QW-active region (see the schematic structure in Fig. 14), showing how changes in the concentration and distribution of point defect complexes involving the p-dopant and H impurities are responsible for the device degradation, under stress-induced joined electrical and thermal effects.

A set of samples was submitted to accelerated life tests at DC forward currents ranging from 20 to 100 mA [121]; some devices were intentionally left without heat sink, so that elevated junction temperatures were achieved during tests. The electrical (current–voltage (I–V) and capacitance–voltage (C–V) characteristics) and optical (EL and CL spectra) characterizations performed at different aging stages suggested a main role of self-heating in the LED performances and showed an electrothermal threshold effect at a current of 100 mA without a heat sink (theoretical current density of about 110 A/cm^2; junction temperature of about 380°C). For this reason, we concentrated on the electrical aging at 100 mA, which induces dramatic changes in the conduction mechanisms (steep voltage increase in the I–V curves as a "break" process [115]) and in the shape and intensity of the luminescence spectral response.

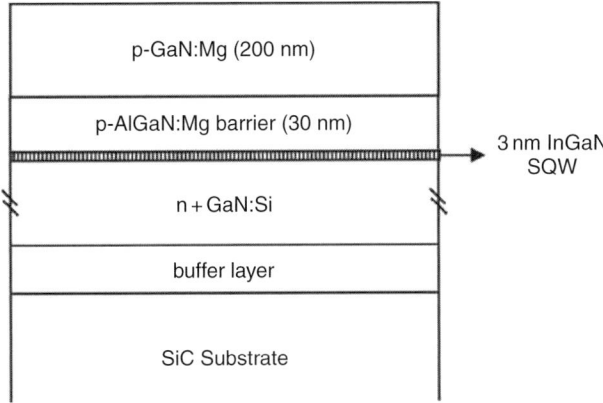

Fig. 14. Schematic cross section of a device hetero structure. Square chips with 250 μm side, mounted on TO46 packages were analyzed. The ohmic contacts are made by a thin Pt semitransparent layer with a golden pad for bonding wire (top), and a bi-component conductive resin ensuring both good electrical and thermal conductivity (bottom). A Si_3N_4 passivating layer is settled all over the chip.

In stressed devices, the QW-related emission peak undergoes a strong efficiency loss, due to an increase of nonradiative centers concentration during the aging process, but its shape and energy position remain unchanged. The most relevant variation is the onset of a broad band at about 3.08 eV (D band, see Fig. 15), detected in the whole device area; a complete quenching of this band is observed above 200 K, indicating its donor–acceptor pair nature. This band was ascribed to shallow acceptor levels resulting from the formation of Mg-related metastable complexes (possibly Mg–H_2), as a consequence of a combined electrical and thermal effect [116].

The evidence of a permanent quenching of the D band under exposure to highly energetic electron beam in the SEM, demonstrating the metastable nature of the relevant Mg complexes, has been observed. HEEBI was performed by scanning the electron beam for 30 min–1 h at a pixel clock of 15 MHz on a 25×25 or 40×40 μm² wide area, at accelerating voltage 12–16 keV and currents ranging from 50 to 120 nA. Here we report on the time decay of the CL intensity during this treatment and its dependence on the beam current and on the size of the irradiated region (Fig. 16).

A linear decrease in the optical emission was found on the largest area (open triangles in Fig. 16), whereas the decay became exponential with two time constants ($250\,s < t_1 < 750\,s$ and $t_2 = 1500\,s$) reducing the size of the irradiated regions. The slope of the linear curve is clearly lower compared to the exponential decays, so evidencing the electron beam irradiation is more efficient in the second mode, due to the higher achieved local injected power density. As for the existence of two time constants, the value t_2 is nearly independent of the beam current at fixed beam voltage and was supposed to be related to a matting of the passivating layer of the LED. On the contrary, t_1 decreases with the beam current, suggesting a correlation with the dissociation of point defect complexes. Our findings can be compared to literature results, such as the CL decays presented by Gelhausen et al. [122], who reported a similar decay on massive p-type GaN layers but with smaller time constant values. The efficiency of the InGaN QW in draining the generated carriers and the presence of interlayers interfaces in our devices could account for the discrepancy in the temporal quenching of the Mg-related band.

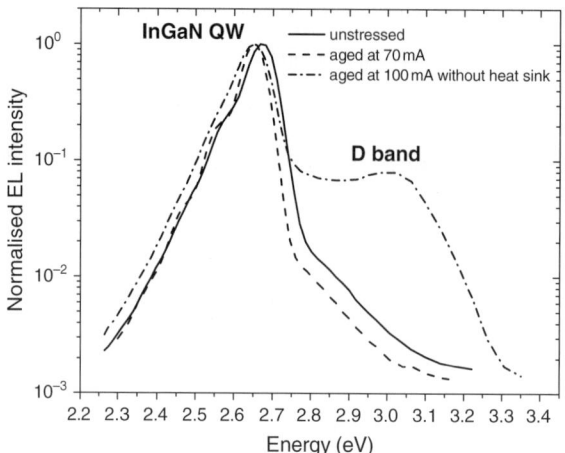

Fig. 15. Comparison between low-temperature ($T = 100$ K) EL spectra at 10 mA, before (solid line) and after (dashed-dotted curve) stress at 100 mA for 3 h without heat sink. The spectrum after stress at 70 mA (dashed line) is also shown for comparison. The intensity has been normalized to the QW emission peak.

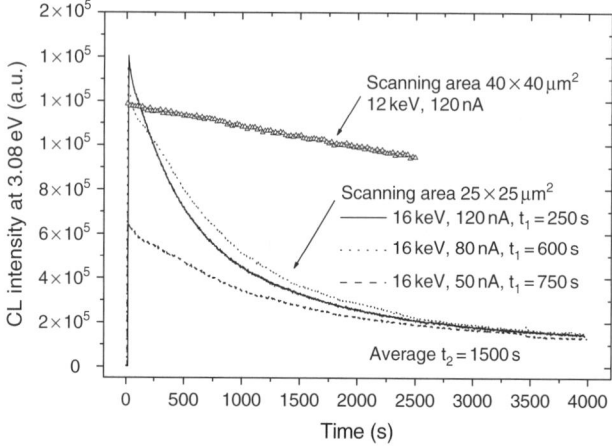

Fig. 16. Evolution in time of the D band during the HEEBI treatment at $T = 77$ K. Decays of the CL intensity at $E_B = 12$ keV $- I_B = 120$ nA on a 40×40 μm^2 wide area (straight solid line) and at $E_B = 16$ keV $- I_B = 50$ (dashed line), 80 (dotted line), 120 nA (solid) on a 25×25 μm^2 area are reported.

The effect of the HEEBI treatment and its interaction with D band was further studied by repeating the irradiation experiments on an unstressed sample. The exposure to the electron beam up to currents of 2 μA ($E_B = 16$ keV) did not induce the onset of the D band. This result confirms that the physical mechanism leading to the creation of the new Mg–H$_2$ complexes is activated by the coexistence of both adequately high junction temperature and density of low-energy carriers, typical of the electrical aging.

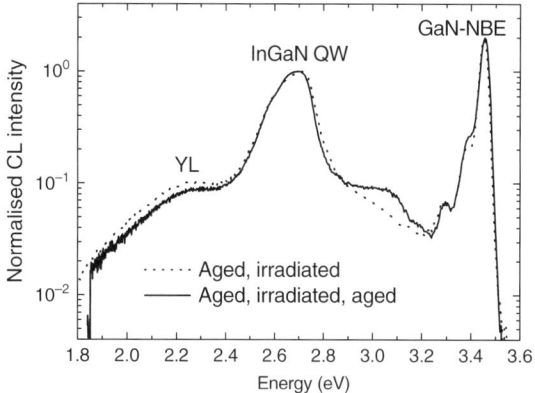

Fig. 17. Comparison between the CL spectra at $T = 77$ K before (dotted curve) and after (solid line) the repeated electrical aging. The spectra are collected on the same $30 \times 40\,\mu m^2$ wide area at $E_B = 16$ keV, $I_B = 11$ nA; the curves are normalized to the QW intensity.

The HEEBI quenching of the D band was achieved on the whole device area, by scanning adjacent regions of about $25 \times 25\,\mu m^2$. The reversibility of the process was then studied by carrying out a further electrical aging (at 100 mA for 30 min), to check if the removed D band could be formed again by the DC forward stress. The results are reported in Fig. 17.

CL spectra acquired before and after the further electrical aging on the same local area (far from the metallic pad) differ only in the energy range between 2.9 and 3.2 eV. After the first aging and irradiation, the main contribution comes from a band centered at about 2.93 eV, possibly ascribed to Mg–N–H complexes acting as acceptors, detected in all the CL spectra of the LEDs after "break" and not quenched by HEEBI. Instead, it is plain from the plot the D band was almost completely removed. On the contrary, in the CL curve after the second aging the D band reappears with a barycenter shifted to about 3.03 eV.

This behavior confirms that the irradiation reduces the D band optical intensity by breaking the Mg-related complexes, whereas the electrical stress restores the D band creating them again. It is worth noting that, while the D band was completely removed in the CL spectra after irradiation, in the EL spectra it reduced its intensity without disappearing, because the irradiation did not "break" the complexes in the device volume below the metallic pad.

4. Conclusions

This chapter is mainly devoted to discuss the advantages of using the SEM–CL technique under high-injection power regimes in nitrides compounds.

A short summary of the CL basics and peculiarities has been first discussed.

Carrier generation diffusion and recombination processes have been reviewed and modeled. The difference between lateral and in-depth resolution as well as the analytical sensitivity have been discussed.

The role of excitation conditions and their effect on the main advantages and drawbacks with respect to PL has also been described.

Finally, three emblematic examples on the advantages of employing power-dependent CL in counterbalancing internal fields in InGaN QWs, modulating the band-gap values in hydrogenated dilute nitrides and assessing metastable point defect complexes in nitrides-based LEDs have been presented and discussed.

In particular, the free carrier screening of the polarization fields in low In content ($x < 0.07$) In$_x$Ga$_{1-x}$N/GaN multiple QWs has been studied by combining CL and TI PL on a large excitation density range. The blue shift of the QW emission in samples with almost the same In fraction decreases from about 50 to 17 meV by decreasing the well width from 3.5 to 2 nm. The peak energy position is well reproduced as a function of the carrier density in the framework of a theoretical model, where a self-consistent solution of Schrödinger and Poisson's equation is coupled to a time-dependent simulation.

Furthermore, CL induced spatially resolved displacement of hydrogen from nitrogen passivation sites in hydrogenated GaAs$_{1-x}$N$_x$ by direct electron beam writing, allowed to modulate on a small scale the band gap of a semiconductor heterostructure in its growth plane. In perspective, we should be able to attain such a modulation on a nanometer scale, which would lead to quantum-confining effects on carriers.

Finally, a permanent quenching of an emission band at 3.08 eV (D band) due to Mg-related metastable complexes (possibly Mg–H$_2$) under exposure to highly energetic electron beam in the SEM has been observed on-line by CL studies on InGaN QW-based LEDs submitted to electrical aging. The observation demonstrated the metastable nature of the Mg complexes. The same experiments performed on unstressed samples revealed the D band had an irreversible nature.

Acknowledgments

Thanks are due to the Italian Ministry of Foreign Affairs (MAE) for partially supporting the research activity on LEDs through the Italian–Japanese Project of Great Relevance "High electron mobility HEMTs and high brightness LEDs."

Appendix: "ALLES" software for the simulation of CL processes in semiconductors

A full CL simulation usually has a number of problems and different situations to deal with. Even if it was possible in principle, so far no multipurpose software able to account for all different problems at the same time has been developed. However, this has been attempted with the software *ALLES* [16,123]. *ALLES* is able to describe the injection by a built-in Monte Carlo simulation, as described in Section 1.2.1, or as an alternative it is possible to import the injection matrix provided by other programs. The main feature of the program is the description of carrier diffusion. Indeed, diffusion is numerically simulated by means of the "random walk" (RW) approach. Finally, the optical absorption can also be considered, with some limitations, to be solved in future implementations.

In comparison to the previous simulation schemes like the one proposed by Pey et al. [54], the use of the RW simulation presents advantages and disadvantages. RW is a much slower process than a direct numerical approach to the differential equations, but it permits a more intuitive implementation of many processes. Recently, the use of a RW approach has also been proposed independently by other authors [124] to solve the analogous diffusion problem in SEM charge collection techniques. However, in their approach they limit themselves to the study of dislocations, while the generality of the present approach is better appreciated in CL simulations. For example, the introduction of different processes for the carrier dynamics is possible and can be directly interfaced with rate equations. Moreover, the diffusion in structures of different dimensionality and with complex geometry can be described more naturally.

The program has been written in the C language and a graphical interface *gALLES* has been added using Borland C++ [125]. This allows for easy changes to relevant parameters and to build different structures. In addition, up to three parameters can be scanned automatically to explore a complete set of parametrical dependencies of CL variables.

A.1. The random walk approach

The RW approach solves the diffusion equation (Eq. (7)) treating the diffusion and the recombination of carriers as separate steps. In particular, ALLES is based on a continuum space approach as a time discretization with a fixed time step dt rather than a space discretization is applied.

The first step consists of placing the excess carriers in the initial position prior to diffusion, according to the generation function, which is calculated as an injection matrix from the injection program. Numerically, the random positioning of each individual particle according to the generation function distribution is obtained by means of the so-called double extraction method [126], using the injection matrix as a probability function. After this, the following steps are performed:

1. Displace the particle from the previous position by an amount $\vec{\delta d}$ where $\vec{\delta d}$ is a random number normally distributed in three dimensions. The distribution has the width $2\sqrt{Ddt}$ and is centered at $\vec{\delta d} = 0$.
2. Calculate the recombination time τ and the recombination probability p for each step as a function of the calculated position as

$$p(r) = 1 - e^{-(dt/\tau(r))} \approx \frac{dt}{\tau(r)} \quad (A1)$$

3. Extract a random number R between 0 and 1. If $R > p(r)$, the particle recombines and the simulation starts for a new particle positioned according to the injection matrix. If $R < p(r)$, the particle does not recombine and the loop is restarted from point (1).

If, in the same region, more than one recombination channel is to be considered, the effective lifetime τ_{tot} is calculated as

$$\frac{1}{\tau_{tot}} = \sum_{channels} \frac{1}{\tau_i} \quad (A2)$$

For each recombination event, a second random extraction [16] can be used to decide which channel was responsible for recombination.

This simple scheme allows, in principle, to account for diffusion and recombination of carriers with a temporal accuracy dt and a spatial resolution of the order of $dx = \sqrt{Ddt}$. This means that the parameter dt should be accurately chosen to be smaller than the characteristic time of any possible process in the sample. Moreover, the corresponding dx should be smaller than the typical lateral dimension of the simulated structures. This condition is most important as this determines the overall simulation time.

The other determining factor for evaluating the total computation time is the number of particles used for the simulation (N_S); this depends on the quantity to be measured and on the accuracy with which it has to be known. Some thousands of particles are typically sufficient in most cases.

The RW approach is suitable for both transient and steady-state CL simulation. To obtain a steady-state density of carriers, the position of the particle during its diffusion is sampled at constant time intervals and stored in a carrier density matrix. The numerical value of the matrix entry is increased by 1 for each particle that is seen in the corresponding position. The final number of counts will be proportional to the carrier density at that point. The time intervals do not coincide just with the single step dt, but the sampling for the density matrix is performed on longer time intervals. Each electron path is different and a single representative position for each trajectory should be used, in principle, in the computation of the steady-state density. Of course, as the number of considered trajectories is limited, the resulting distribution is affected by statistical fluctuation around the exact distribution. However, if more than one point of the same trajectory is used for the density matrix calculation, the fluctuations turn out to be artificially correlated, with the result that the convergence to the exact distribution is slower. The best compromise must be considered for each diffusion condition. Fig. A1 shows an example of diffusion trajectory and sampling for the carrier density matrix. If the sampling time between two adjacent events is too short (top left dots in Fig. A1(a)), the samplings can be considered correlated (i.e., if both the events are taken into account, it is like the same point is counted twice). Considering only the correct sampling (filled dots), it is possible to properly fill up the density matrix: the contribution of the selected trajectory is evidenced in Fig. A1(b).

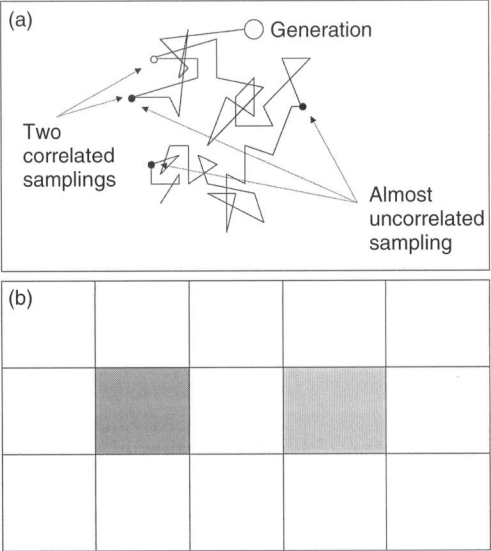

Fig. A1. (a) Example of diffusion trajectory and sampling for the carrier density matrix. If the time between two samplings is too short (top left dots), they can be considered correlated. (b) Example of a piece of carrier density matrix spatially corresponding to the above trajectory (only full dots have been considered). Gray levels from white to black correspond to an increasing number of samplings in the given pixel. In this case, two in the dark gray pixel and one in the light gray pixel.

This position of the sampling is important as the density matrix obtained by summing the sampling for different particles is a realistic estimation of the density inside the material. Indeed at steady state, the density is given by

$$n(r) = g_0 \int_{T=-\infty}^{T_0} \tilde{n}(r, T - T_0) \, dT = g_0 \int_{t=0}^{\infty} \tilde{n}(r, t) \, dt \quad \text{(A3)}$$

which is the integral of the density of diffused carriers (here $\tilde{n}(r,t)$ indicates the density normalized to 1, reached after diffusion for a time t) that had been continuously generated (with a total rate g_0) from the time $-\infty$ till the time T_0 of observation. A simple mathematical substitution demonstrates that this density is equivalent to the carrier density produced by particles generated at a time $t=0$ and whose diffusion density is sampled at times between 0 and ∞, that is just what the simulation calculates.

The program also provides two methods to evaluate the number of recombined electrons in a given channel. One is more rigorous and obvious and consists of simply recording the type of recombination at the time the particle is recombined. The second is based on the fact that the number of particles recombining in a given region (n_{rec}) where a recombination channel with a recombination time τ is active, is supposed to be proportional to the steady-state density of carriers in that region:

$$n_{rec} \propto \frac{n}{\tau} \quad \text{(A4)}$$

This expression could be easily modified to describe a nonlinear dependence of the number of generated phonons on the carrier density. This is the case, for example, of excitonic or localized states recombination. However, it should be noted that such non-linearity implies a certain degree of inconsistency with Eq. (7), as it involves a dependence of the recombination lifetime, in the given channel, on the local carrier density. If this were the dominating recombination channel, this would imply that Eq. (13) becomes a nonlinear diffusion equation, whose solution would be quite complicated. However, as in many cases the nonlinearity is not dramatic, or competing recombination channels, with constant recombination time, dominate the recombination, these effects can often be neglected. It is often possible to use for diffusion a constant recombination time selected on the basis of the average injection regime.

A.2. Random walk with surfaces and interfaces

The main difficulty in a RW approach is how to include the boundary condition on the first derivative, as typically required for surfaces and interfaces (Eq. (12)).

The boundary condition, with s a finite number, imposes that a fraction p of the particles impinging on a surface during a discrete time step dt recombines. Considered in a time interval Δt, the number of recombined particles would roughly be $n_{surf} = (\Delta t/dt) \times p$. However, the time dt is arbitrary so that p must necessarily contain a compensation term canceling this spurious dependence of n_{surf} on dt. The dependence on position complicates the discussion slightly but does not alter the statement that p is necessarily dependent on dt.

The solution for this complex problem was given by Holloway [127], who produced the formula for the recombination probability as

$$p = \left[1 - \exp\left(-2S^2 \frac{Ddt}{L^2}\right)\right]^{1/2} \tag{A5}$$

Thanks to this formula, the effect of a flat surface can be easily incorporated.

In the case of interfaces, the situation may appear similar, but the fact that carriers can exist and even be generated on both sides of the interface makes the situation quite different. Carriers impinging on an interface have a probability

$$p = ce^{-(\Delta E/KT)} \tag{A6}$$

of overcoming it, where ΔE is the energy barrier at the interface and c is a constant which can be evaluated as N_a/N_b [50]; in many cases, this ratio is assumed to be of the order of 1.

An energy barrier $\Delta E > 0$ means, for example for electrons, that the conduction band of the material beyond the interface is located at a higher energy than the starting material. Conversely, one defines $\Delta E < 0$ as the energy difference for the same interface approached from the opposite direction. The probability of overcoming a barrier with $\Delta E < 0$ is considered to be 1.

In the RW simulation, whenever a particle impinges on an interface (or a surface) a random number R ($0 < R < 1$) is extracted: if $R > p$ the particle is reflected (diffused back for a surface), otherwise it is transmitted (or recombined for a surface) and diffuses in the new material according to its diffusivity.

As it will be seen in the following sections, this approach has been tested to be independent of the discretization time dt.

A.3. Random walk and carrier dynamics in confining structures

The advantage of the RW approach is more evident in the case of confining structures. In these cases, the possibility to account for the coupled diffusion between a "bulk" material and a nearby LDS is hardly achievable by common analytical or numerical approaches.

The inclusion of such effects is made considerably easier by the fact that these passages are ruled by transition rate as described in Section 3.4; the probability of each process can be written simply as

$$p = \frac{dt}{\tau} \tag{A7}$$

where $1/\tau$ is the rate of the specific process.

Once a particle is trapped in, for example a QW, a check is performed by extraction of random numbers R to be checked against the probability that particles are either recombined or emitted to the barrier, for each stage of the RW loop. In the case of capture, it is often lengthy, even if still possible, to adopt this kind of scheme. The reason is that the necessarily small QW thickness (w, some nanometers) requires a small sampling of dx and therefore dt. The alternatives we propose can be of two kinds:

1. to artificially increase the value of w to some tens of nonometers and to increase the capture time accordingly to have $w/\tau_c = $ const.;
2. to consider an effective capture time $\tau_{eff} = \tau_c(w/dx)$ accounting for the fact that, in a jump with a too coarse average step dx, the QW could be overlooked. Because this may happen with a probability of roughly w/dx, this correction factor should be applied to the capture probability.

Both these possibilities have been tested in simulations, giving a reasonable approximation of the exact approach. The second correction has been applied mostly to QWs, while a correction similar to correction 1 has sometimes been applied to the case of QDs.

Finally, it is useful to underlie that in the case of QDs, the procedure is even further simplified: as no lateral diffusion is possible inside the QD, the sole rate equation [128] is sufficient to describe the QD behavior. Once a carrier is captured inside a QD, it is considered to remain there for an average time

$$\tau_{TOT} = \left(\frac{1}{\tau_{LDS}} + \frac{1}{\tau_r}\right)^{-1} \tag{A8}$$

where τ_{LDS} is the lifetime inside the QD and τ_r is the re-emission time. A random number extraction is then used to evaluate whether the particle is recombined or thermally reemitted to the barrier.

The simulation also permits, in the case of QDs, the definition of an array of QDs and eventually to consider the CL signal arising from a subset of the array separately. This simulates the typical situation of the monochromatic CL mode, for which the accurate selection of the emission wavelength could, in principle, permit to isolate single QDs.

References

[1] A. Gustafsson, J. Microsc. 224 (2006) 72.
[2] M. Gasgnier, Phys. Stat Sol A 114 (1989) 11.
[3] R. Heiderhoff, R. M. Cramer, O. V. Sergeev, and L. J. Balk, Diamond Relat. Mater. 10 (2001) 1647.
[4] S. Boggs and D. H. Krinsley, Application of Cathodoluminescence Imaging to the Study of Sedimentary Rocks, Cambridge University Press, Cambridge, 2006.
[5] D. Nakaji, V. Grillo, N. Yamamoto, and T. Mukai, J. Electron Microsc. 54 (2005) 223.
[6] M. Albrecht, H. P. Strunk, I. Grzegory, et al., J. Appl. Phys. 92 (2002) 2000.
[7] M. Markmann, A. Zrenner, G. Bohm, et al., Phys. Stat Sol A – Appl. Res. 164 (1997) 301.
[8] A. Olkhovets, S. Evoy, and H. G. Craighead, Surf. Sci. (including Surf. Sci. Lett.) 453(1–3) (2000) L299.
[9] M. Troyon, D. Pastré, J. P. Jouart, and J. L. Beaudo, Ultramicroscopy 75 (1998) 15.
[10] R. M. Cramer, V. Ebinghaus, R. Heiderhoff, and L. J. Balk, J. Phys. D: Appl. Phys. 31 (1998) 1918.
[11] S. Sanguinetti, M. Guzzi, and M. Gurioli, in Characterization of Semiconductor Heterostructures and Nanostructures, Chapter 6, C. Lamberti, Ed., Elsevier, Amsterdam, 2008, p. 191.
[12] J. I. Pankove, Optical Processes in Semiconductors, Prentice-Hall, Englewood Cliffs, NJ, 1971.

[13] L. Pavesi and M. Guzzi, J. Appl. Phys. 75 (1994) 4779.
[14] B. G. Yacobi and D. B. Holt, Cathodoluminescence Microscopy of Inorganic Solids, Plenum Press, New York, 1990.
[15] W. Hergert and L. Pasemann, Phys. Stat Sol A 85 (1984) 641.
[16] V. Grillo, F. Rossi, N. Armani, et al., in Beam Injection Based Nanocharacterization of Advanced Materials, G. Salviati, T. Sekiguchi, S. Heun, and A. Gustafsson, Eds., Research Signpost, www.transworldresearch.com, 2007.
[17] L. Reimer, in Scanning Electron Microscopy: Physics of Image Formation and Microanalysis, Springer Series in Optical Sciences, Vol. 45, Springer-Verlag, Berlin, 1985.
[18] J. Christen, T. Riemann, F. Bertram, et al., Phys. Stat Sol C 0 (2003) 1795.
[19] Day, John C. C., Ph.D. Thesis, University of Bristol, UK, 1988. Source: Dissertation Abstracts International, Volume: 50-05, Section: B, p. 1992.
[20] C. Trager-Cowan, P. G. Middleton, and K. P. O'Donnell, MIJ-NSR Volume 1, Article 6 (1996).
[21] A. Gustafsson, M. E. Pistol, L. Montelius, and L. Samuelson, J. Appl. Phys. 84 (1998) 1715.
[22] M. Merano, S. Sonderegger, A. Crottini, et al., Nature 438 (2005) 479.
[23] R. J. Hawryluk, A. M. Hawryluk, and H. I. Smith, J. Appl. Phys. 45 (1974) 2551.
[24] P. Hovington, D. Drouin, and R. Gauvin, Scanning 19 (1997) 1.
[25] R. J. Graham, T. D. Moustakas, and M. M. Disko, J. Appl. Phys. 69 (1991) 3212.
[26] T. Mitsui, N. Yamamoto, T. Tadokoro, and S. J. Ohta, J. Appl. Phys. 80 (1996) 6972.
[27] M. Albrecht, V. Grillo, J. Borysiuk, et al., Inst. Phys. Conf. Series 169 (2001) 267.
[28] K. Kanaya and S. Okayama, J. Phys. D: Appl. Phys. 5 (1972) 43.
[29] D. C. Joy, Monte Carlo Modeling for Electron Microscopy and Microanalysis, Oxford University Press, New York, London, 1995.
[30] Z. Czyżewski, D. O'Neill Mac Collumn, and A. Romig, J. Appl. Phys. 68 (1990) 3066.
[31] D. Drouin, P. Hovington, and R. Gauvin, Scanning 19 (1997) 20.
[32] D. Drouin, R. Gauvin, and D. C. Joy, Scanning 16 (1994) 67.
[33] H. Bethe, Ann. Phys. 5 (1930) 325.
[34] P. Hovington, D. Drouin, and R. Gauvin, Scanning 19 (1997) 29.
[35] G. Han, M. Khan, Y. Fang, and F. Cerrina, J. Vac. Sci. Technol. B 20 (2002) 2666.
[36] M. E. Bachlecher, W. Macke, H. M. Miesemböck, and A. Schinner, Physica B 168 (1991) 104.
[37] C. A. Klein, J. Appl. Phys. 39 (1968) 2029.
[38] T. E. Everhart and P. H. Hoff, J. Appl. Phys. 42 (1971) 5837.
[39] Hong Ye, G. W. Wicks, and P. M. Fauchet, Appl. Phys. Lett. 75 (2000) 1185.
[40] L. Rota and P. Lugli, Semicond. Sci. Technol. 7 (1992) B180.
[41] M. Boulou and D. Bois, J. Appl. Phys. 48 (1977) 4713.
[42] K. Seeger, Semiconductor Physics, Springer, New York – Wien, Chap. 5, 1973.
[43] M. L. Cone and R. L. Hengehold, J. Appl. Phys. 54 (1983) 6346.
[44] V. Grillo, A. Genseki, N. Yamamoto, and Y. Watanabe, Surf. Interface Anal. 35 (2003) 40.
[45] P. Y. Yu and M. Cardona, Fundamentals of Semiconductors, Springer, Berlin, 1996, p. 451.
[46] A. Djemel, A. Nouiri, and R. J. Tarento, J. Phys.: Condens. Matter 12 (2000) 10343.
[47] A. Djemel, R. J. Tarento, J. Castaing, et al., Phys. Stat Sol A 101 (1987) 425.
[48] A. Nouiri, A. Djemel, and R. J. Tarento, Microelectron. Eng. 51 (2000) 151.
[49] G. Chen, J. Appl. Phys. 97 (2005) 83707.
[50] C. Van Opdorp and G. W. 't Hoof, J. Appl. Phys. 52 (1981) 3827.
[51] H. Schneider and K. V. Klitzing, Phys. Rev. B 38 (1988) 6160.
[52] J. A. Brum and G. Bastard, Phys. Rev. B 33 (1986) 1420.
[53] R. Kersting, R. Scwedler, K. Wolter, et al., Phys. Rev. B 46 (1992) 1639.
[54] K. L. Pey, D. S. H. Chan, and J. C. H. Phang, Scanning Microsc. 9 (1995) 355; K. L. Pey, J. C. H. Phang, and D. S. H. Chan, Scanning Microsc. 9 (1995) 367.
[55] R. F. W. Pease and W. C. Nixon, Adv. Imag. Elect. Phys. 133 (2004) 195.
[56] M. Merano, S. Sonderegger, A. Crottini, et al., Appl. Phys. B 84 (2006) 343.

[57] G. Salviati, L. Lazzarini, M. Z. Zha, et al., Phys. Stat Sol A 202 (2005) 2963.
[58] T. Sekiguchi, J. Hu, and Y. Bando, J. Electron Microsc. 53 (2004) 203.
[59] V. Grillo, N. Armani, F. Rossi, et al., Inst. Phys. Conf. Ser. 180 (2003) 565.
[60] G. Salviati, F. Rossi, N. Armani, et al., J. Phys.: Condens. Matter 16 (2004) S115.
[61] G. Salviati, F. Rossi, N. Armani, et al., Inst. Phys. Conf. Ser. 169 (2001) 251.
[62] P. J. Dean, Prog. Cryst. Growth Charact. Mater. 5 (1982) 89.
[63] M. R. Phillips, H. Telg, S. O. Kucheyev, et al., Microsc. Microanal. 9 (2003) 144.
[64] S. O. Kucheyev, M. Toth, M. R. Phillips, et al., Appl. Phys. Lett. 79 (2001) 2154.
[65] A. Castaldini, A. Cavallini, L. Polenta, et al., J. Phys.: Condens. Matter 14 (2002) 13095.
[66] U. Jahn, S. Dhar, O. Brandt, et al., J. Appl. Phys. 93 (2003) 1048.
[67] K. Knobloch, P. Perlin, J. Krueger, et al., MRS Internet J. Nitride Semicond. Res. 3 (1998) 4.
[68] L. Pavesi and M. Guzzi, J. Appl. Phys. 75 (1994) 4779.
[69] F. Rossi, Interplay between structural and optical properties in semiconducting N-based heterostructures for optoelectronic applications, Ph.D. Thesis in Materials Science, University of Parma, 2004.
[70] S. Myhajlenko, W. K. Ke, and B. Hamilton, J. Appl. Phys. 54 (1983) 862.
[71] J. Vine and P. A. Einstein, Proc. IEEE 111(1964) 921.
[72] B. G. Yacobi, S. Datta, and D. B. Holt, Philos. Mag. 35 (1977) 145.
[73] L. J. Balk, E. Kubalek, and E. Menzel, in Scanning Electron Microscopy, O. Johari and R. P. Becker, Eds, IIT Research Inst., Chicago, 1976, p. 257.
[74] H. Morkoç, Handbook of Nitride Semiconductors and Devices, Physics and Technology of GaN-Based Optical and Electronic Devices, ISBN 978-3-540-46930-8, Vol. 3, Springer Verlag, Berlin, 2007.
[75] M. Henini, Ed., Dilute Nitride Semiconductors, Elsevier, Amsterdam, The Netherlands, 2005.
[76] O. Ambacher, J. Majewski, C. Miskys, et al., J. Phys.: Condens. Matter 14 (2002) 3399.
[77] F. Bernardini and V. Fiorentini, Appl. Surf. Sci. 166 (2000) 23.
[78] F. Bernardini and V. Fiorentini, Phys. Stat. Sol. A 190 (2002) 65.
[79] F. Bernardini and V. Fiorentini, Phys. Stat. Sol. B 216 (1999) 391.
[80] I. Vurgaftman and J. R. Meyer, J. Appl. Phys. 94 (2003) 3675.
[81] A. Zoroddu, F. Bernardini, P. Ruggerone, and V. Fiorentini, Phys. Rev. B 64 (2001) 045208.
[82] V. Fiorentini, F. Bernardini, and O. Ambacher, Appl. Phys. Lett. 80 (2002) 1204.
[83] V. Fiorentini, F. Bernardini, F. Della Sala, et al., Phys. Rev. B 60 (1999) 8849.
[84] F. Della Sala, A. Di Carlo, P. Lugli, et al., Appl. Phys. Lett. 74 (1999) 2002.
[85] V. Fiorentini, F. Bernardini, F. Della Sala, et al., Phys. Rev. B 60 (1999) 8849.
[86] A. Hangleiter, Phys. Stat. Sol. C 0 (2003) 1816.
[87] S. Chichibu, T. Sota, K. Wada, and S. Nakamura, J. Vac. Sci. Technol. B 16 (1998) 2204.
[88] K. P. O'Donnell, R. W. Martin, and P. G. Middleton, Phys. Rev. Lett. 82 (1999) 237.
[89] J. Wu, W. Walukiewicz, W. Shan, et al., J. Appl. Phys. 94 (2003) 4457.
[90] F. Rossi, N. Armani, C. Ferrari, et al., Inst. Phys. Conf. Ser. 180 (2003) 277.
[91] N. Armani, F. Rossi, C. Ferrari, et al., Superlattices Microstruct. 36 (2004) 615.
[92] F. Capasso, Science 235 (1987) 172.
[93] C. Weisbuch and B. Vinter, Quantum Semiconductor Structures, Academic Press, San Diego, USA, 1991.
[94] Ch. Gréus, L. Butov, F. Daiminger, et al., Phys. Rev. B 47 (1993) 7626.
[95] D. Bimberg, M. Grundmann, and N. N. Ledentsov, Quantum Dot Heterostructures, Wiley, Chichester, UK, 1998.
[96] A. Polimeni, M. Henini, A. Patanè, et al., Appl. Phys. Lett. 73 (1998) 1415.
[97] G. Ciatto, F. Boscherini, A. Amore Bonapasta, et al., Phys. Rev. B 71 (2005) 201301.
[98] I. A. Buyanova and W. M. Chen, Eds, Physics and Applications of Dilute Nitrides, Taylor & Francis, New York, USA, 2004.
[99] G. Steinle, F. Mederer, M. Kicherer, et al., Electron. Lett. 37 (2001) 632.

[100] S. R. Kurtz, A. A. Allerman, E. D. Jones, et al., Appl. Phys. Lett. 74 (1999) 729.
[101] R. J. Welty, H. Xin, C. W. Tu, and P. M. Asbeck, J. Appl. Phys. 95 (2004) 327.
[102] A. Ignatov, A. Patane', O. Makarovsky, and L. Eaves, Appl. Phys. Lett. 88 (2006) 032107.
[103] A. Polimeni, G. Baldassarri Höger von Högersthal, M. Bissiri, et al., Phys. Rev. B 63 (2001) 201304.
[104] G. Baldassarri Höger von Högersthal, M. Bissiri, A. Polimeni, et al., Appl. Phys. Lett. 78 (2001) 3472.
[105] A. Amore Bonapasta, F. Filippone, P. Giannozzi, et al., Phys. Rev. Lett. 89 (2002) 216401.
[106] A. Polimeni, G. Baldassarri Höger von Högersthal, F. Masia, et al., Phys. Rev. B 69 (2004) 041201.
[107] A. Polimeni, G. Ciatto, L. Ortega, et al., Phys. Rev. B 68 (2003) 085204.
[108] M. Pavesi, M. Manfredi, G. Salviati, et al., Appl. Phys. Lett. 84 (2004) 3403.
[109] S. Silvestre, D. Bernard-Loridant, E. Constant, et al., Appl. Phys. Lett. 77 (2000) 3206.
[110] J. Christen, M. Grundmann, and D. Bimberg, J. Vac. Sci. Technol. B 9 (1991) 2358.
[111] A. Petersson, A. Gustafsson, L. Samuelson, et al., Appl. Phys. Lett. 74 (1999) 3513.
[112] L.-L. Chao, G. S. Cargill III, M. Levy, et al., Appl. Phys. Lett. 70 (1997) 408.
[113] A. Polimeni, M. Bissiri, M. Felici, et al., Phys. Rev. B 67 (2003) 201303.
[114] G. Leibiger, V. Gottschalch, B. Rheinländer, et al., J. Appl. Phys. 89 (2001) 4927.
[115] M. Pavesi, F. Rossi, and E. Zanoni, Semicond. Sci. Technol. 21 (2006) 138.
[116] M. Pavesi, G. Salviati, N. Armani, et al., Appl. Phys. Lett. 84 (2004) 3403.
[117] T. Egawa, H. Ishikawa, T. Jimbo, and M. Umeno, Appl. Phys. Lett. 69 (1996) 830.
[118] F. Manyakhin, A. Kovalev, and A. E. Yunovich, MRS Internet J. Nitride Semicond. Res. 3 (1998) 53.
[119] D. L. Barton, M. Osinski, P. Perlin, et al., Microelectron. Reliab. 39 (1999) 1219.
[120] S. Nakamura, N. Iwasa, M. Senoh, and T. Mukai, Jpn. J. Appl. Phys., Part 1, 31 (1992) 1258.
[121] F. Rossi, N. Armani, G. Salviati, et al., Superlattices Microstruct. 36 (2004) 859.
[122] O. Gelhausen, H. N. Klein, M. R. Phillips, and E. M. Goldys, Phys. Stat. Sol. B 239 (2003) 310.
[123] V. Grillo, N. Armani, F. Rossi, et al., Inst. Phys. Conf. Ser. 180 (2003) 565.
[124] M. Ledra and N. Tabet, J. Phys. D. 38 (2005) 3845.
[125] www.borland.com.
[126] W. H. Press, S. A. Teukolsky, W. T. Vetterling, and B. P. Flannery, "Numerical Recipes" in C++, 2nd edition, Cambridge University Press, 2002, p. 290 (http://www.nr.com).
[127] H. Holloway, J. Appl. Phys. 62 (1997) 3241.
[128] K. Akiba, N. Yamamoto, A. Genseki, and Y. Watanabe, Phys. Rev. B 70 (2004) 1653222.

8

Raman spectroscopy

Daniel Wolverson

Department of Physics, University of Bath, Bath BA2 7AY, UK

Abstract An introduction to the theory of Raman scattering and the effects of symmetry on the selection rules governing the scattering process is followed by a brief review of developments in experimental technique. An overview of some recent results in the application of Raman spectroscopy to the study of low-dimensional semiconductors is then given, with some comments on the likely development of the field.

Keywords Raman, vibrational spectroscopy, microscopy, quantum theory of Raman scattering, magneto-Raman scattering

1. Introduction

Raman scattering is the inelastic scattering of light quanta (photons) by some excitation of a material. Most often, the excitation implied by the term "Raman scattering" is a vibrational mode of the atoms of the material and, as is appropriate for the semiconductor context of this book, this chapter will concentrate mainly on the vibrational excitations (specifically, the optical phonons) of crystalline materials and their heterostructures. However, many other types of excitation can interact with photons in a material and thus may also lead to the inelastic scattering of light; examples that will be mentioned include plasmons and localised and collective electronic modes (scattering by charge density fluctuations), and spin flips (scattering by spin density fluctuations), whilst examples that will be omitted include magnons, spin waves and the vibrational excitations of amorphous (non-crystalline) semiconductors. After a short discussion of the theoretical and experimental fundamentals of inelastic light scattering, we shall follow the application of Raman spectroscopy to epitaxial layers of semiconductor alloys, doped semiconductors, heterointerfaces and then to quantum wells, wires and dots, with references in most cases to more specialist reviews.

The scattering of light by sound waves in solids was predicted by Brillouin [1] in 1922 and the quantum-mechanical theory was developed by Smekal [2] in 1923; its observation was reported in 1928 by Sir C.V. Raman [3, 4] and also, independently, by Landsberg and Mandelstam [5]. Initially, the importance of Raman's observation was as a success of the emerging theory of quantum mechanics but, with the invention of lasers, Raman spectroscopy rapidly became a standard tool for solid-state physics, for chemistry and for materials science whilst still providing renewed challenges for quantum theorists at every stage of its development. The terms Brillouin and Raman scattering are now widely accepted to denote light scattering from acoustic and optical phonons respectively; we shall not discuss Brillouin spectroscopy here (a review has been given by Pine [6]).

1.1. The semi-classical theory of Raman scattering

It is assumed that the reader is familiar with the concepts of phonons in crystalline materials at at least the level of the diatomic linear chain model of undergraduate physics textbooks, e.g. Ref. [7], including the concept of the Brillouin zone, longitudinal and transverse acoustic and optical phonons (LA, TA, LO, TO), their dispersion curves, and the qualitative patterns of their atomic displacements in space. Here, the origins of Raman scattering are first treated within a classical model that will give insight into the polarisation selection rules (and, thus, the experimental geometries that may be used to identify what excitation gives rise to a particular spectroscopic signal). A reader who only wishes to find out quickly "what Raman can do" may jump to Section 3, treating LO, TO, etc. simply as labels for different vibrational modes with, in general, different frequencies and dispersion curves. A reader who requires an advanced text should consult the *Light Scattering in Solids* volumes that are part of the series *Topics in Applied Physics* [8].

All tutorial introductions to Raman scattering adopt an approach similar to the present one, which follows that of Ref. [9]. We represent the interaction between light and matter by Eq. (1), which gives the polarisation **P** (a vector) which is induced by light (of electric field vector **E**) incident on the material:

$$\mathbf{P} = \varepsilon_0 \chi \mathbf{E} \qquad (1)$$

Here, χ is the dielectric susceptibility of the material (χ may be a function of the frequency of the incident light and is a second-rank tensor, χ_{jk}). The electric field of the incident light is oscillatory (angular frequency ω_I, wave vector $\mathbf{k}_\mathrm{I}$) and, if χ_{jk} also contains an oscillatory component due to a modulation of the material properties by an excitation such as a phonon (angular frequency ω_ph, wave vector $\mathbf{q}$), then it is clear already from the product on the right-hand side of Eq. (1) that the emitted light (whose amplitude is proportional to **P**) will contain sidebands at sum and difference frequencies $\omega_\mathrm{I} \pm \omega_\mathrm{ph}$. These sidebands are the Raman-scattered light waves; conventionally, the terms with $+$ and $-$ signs are referred to as "anti-Stokes" (blue-shifted) and "Stokes" (red-shifted) scattered light, respectively.

In order to gain a more detailed understanding, we can write

$$\mathbf{E}(\mathbf{r},t) = \mathbf{E}(\mathbf{k}_\mathrm{I},\omega_\mathrm{I}) \cos(\mathbf{k}_\mathrm{I}\cdot\mathbf{r} - \omega_\mathrm{I} t) \qquad (2)$$

for the incident light and

$$\mathbf{u}(\mathbf{r},t) = \mathbf{u}(\mathbf{q},\omega_\mathrm{ph}) \cos(\mathbf{q}\cdot\mathbf{r} - \omega_\mathrm{ph} t) \qquad (3)$$

for the displacement due to the phonon in question. On the assumption that the modulation of the susceptibility by the displacement **u** is small, χ_{jk} may be expanded in a Taylor series in **u** (where j, k, l, m run over the spatial co-ordinates x, y, z, and where a summation over repeated indices is implied):

$$\chi_{jk}(\mathbf{k}_\mathrm{I},\omega_\mathrm{I}) \approx \chi_{jk} \left(\frac{\partial \chi_{jk}(\mathbf{k}_\mathrm{I},\omega_\mathrm{I})}{\partial u_l}\right)_{\mathbf{u}=0} u_l + \left(\frac{\partial^2 \chi_{jk}(\mathbf{k}_\mathrm{I},\omega_\mathrm{I})}{\partial u_l \partial u_m}\right)_{\mathbf{u}=0} u_l u_m + \cdots \qquad (4)$$

We may now substitute Eqs (2) to (4) into Eq. (1) to obtain an explicit result for the time dependence of **P**. All terms in the resulting expression are of potential interest, representing other optical processes such as Rayleigh (elastic) scattering and optical absorption, but here we extract only the term representing first-order Raman scattering (which is the term proportional to the second term on the right-hand side of Eq. (4)):

$$p_j(\mathbf{r},t,\mathbf{u}) = \frac{1}{2}\varepsilon_0 \left(\frac{\partial \chi_{jk}(\mathbf{k}_\mathrm{I}\omega_\mathrm{I})}{\partial u_l}\right)_{u=0} u_l(\mathbf{q},\omega_\mathrm{ph}) E_k(\mathbf{k}_\mathrm{I},\omega_\mathrm{I})$$
$$\times \left\{\cos\left[(\mathbf{k}_\mathrm{I}+\mathbf{q})\cdot\mathbf{r} - (\omega_\mathrm{I}+\omega_\mathrm{ph})t\right] + \cos\left[(\mathbf{k}_\mathrm{I}-\mathbf{q})\cdot\mathbf{r} - (\omega_\mathrm{I}-\omega_\mathrm{ph})t\right]\right\} \quad (5)$$

This part of the induced polarisation indeed contains cosine terms oscillating at sum (anti-Stokes) and difference (Stokes) frequencies. Furthermore, Eq. (5) shows that there is a transfer of momentum between the incident photons (momentum $\hbar\mathbf{k}_\mathrm{I}$) and the phonon mode (crystal momentum $\hbar\mathbf{q}$) so that the scattered photons have energy and momentum given (after multiplication by $\hbar$) by the following

$$\omega_\mathrm{S} = \omega_\mathrm{I} \pm \omega_\mathrm{ph}$$
$$\mathbf{k}_\mathrm{S} = \mathbf{k}_\mathrm{I} \pm \mathbf{q} \quad (6)$$

as is required by the conservation of energy and momentum. The phonon dispersion curve (ω_ph versus $\mathbf{q}$) of the particular material under study and the experimental conditions together determine the values of ω_ph and $\mathbf{q}$ that can be observed in a Raman spectroscopy experiment (we shall see several examples of this below). Typically, the photon momenta are small on the scale of the Brillouin zone and so $\mathbf{q}$ is small; thus, at least in a pure bulk semiconductor, only phonons near the zone centre are probed (but see Section 3.1).

The prefactor appearing on the right-hand side of Eq. (5) is normally referred to as the Raman tensor, R, and expresses a further property of the material; it determines the amplitude of the scattered waves for a given vibrational mode l:

$$R_{jk,l} = \left(\frac{\partial \chi_{jk}(\mathbf{k}_\mathrm{I},\omega_\mathrm{I})}{\partial u_l}\right)_{u=0} u_l(\mathbf{q},\omega_\mathrm{ph}) \quad (7)$$

With this definition, we can write the intensity I of the scattered beam (polarised in the direction of unit vector $\mathbf{e}_j$) that arises from incident light (polarised in direction $\mathbf{e}_k$) when scattered by the lth mode in the following, more compact form:

$$I \propto \left|\mathbf{e}_j^\mathrm{S} \cdot R_{jk,l} \cdot \mathbf{e}_k^\mathrm{I}\right|^2 \quad (8)$$

If the lth mode is degenerate, a summation over all $R_{jk,l}$ corresponding to the set of degenerate modes is required [9]. It is beyond the scope of this review to discuss how the components of R are determined; it is enough to say that group theory may be used to establish, for a given crystal symmetry, which components of R are non-zero, and which of the non-zero components take the same values [10,11]. Thus, Raman spectroscopy using polarised excitation and detection yields information on crystal symmetry and on any deviations from ideality.

1.2. The quantum theory of Raman scattering

Before providing examples of the applications that arise from the above considerations, we shall look briefly at two aspects of the quantum mechanical nature of the Raman phenomenon that are not apparent from the classical treatment. These are effects due to (i) resonance and (ii) phonon statistics.

Resonant effects lead to a dependence of the intensity and form of the Raman spectra on the incident photon energy. Such phenomena were analysed by Loudon [12,13] and are discussed extensively elsewhere (for example, Refs [9,14,15]) and are only summarised here, though the use of resonant excitation is important in many of the practical examples that will follow.

In a quantum picture, the Raman process involves the excitation of the electronic system by the photon to a higher-energy state (which may be either *real*, that is, an eigenstate of the electronic system, or *virtual*). The interaction with the phonon arises due to any form of coupling between the electronic system and the lattice vibrations.

Following Ref. [13], these processes may be represented either using Feynman diagrams (Fig. 1) or a band structure diagram (Fig. 2). Both depict the following three events, or "vertices": (1) the generation of an electron–hole pair (Fig. 1(a)) or exciton (Fig. 1(b)) in state $|n\rangle$ in the crystal via the absorption of a photon; (2) the creation or annihilation of a phonon, leaving the electronic system in state $|m\rangle$ and (3) the recombination of the electron–hole pair, returning the electronic system to its original ground state $|i\rangle$ (taken as the zero of energy) and generating the scattered photon. The requirements imposed by the conservation of energy and momentum are evident from Fig. 2, whilst the Feynman diagram of Fig. 1 gives a systematic means of calculating via perturbation theory the scattering intensity in terms of the electron–radiation and electron–lattice coupling strengths H_{ER} and H_{EL}. The resulting expression (for example, Ref. [14]) contains six terms, corresponding to the six permutations of the vertices of Fig. 1(a) and (b); of these, the following term will be a dominant one as the incident or the scattered photon energies approach those of the real electronic states $|n\rangle$ or $|m\rangle$, respectively:

$$I(\omega_S, \omega_I) \propto \left| \sum_{n,m} \frac{\langle i|H_{ER}(\omega_S)|m\rangle \langle m|H_{EL}|n\rangle \langle n|H_{ER}(\omega_I)|i\rangle}{[\hbar\omega_I - E_n + i\Gamma_n][\hbar\omega_S - E_m + i\Gamma_m]} \right|^2 \delta(\hbar\omega_I - \hbar\omega_S - \hbar\omega_{ph}) \qquad (9)$$

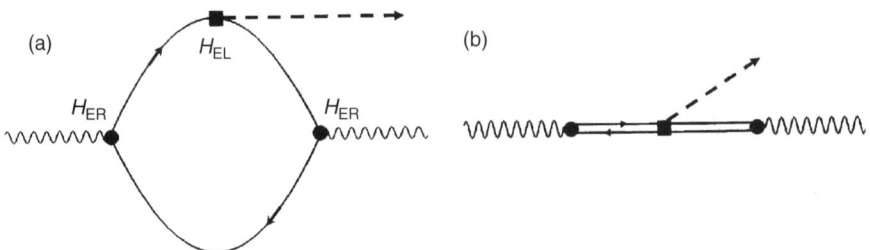

Fig. 1. Feynman diagrams representing phonon Raman scattering, following Loudon [13]. Time flows from left to right. Wavy lines: photons; solid lines: electrons and holes; dashed lines: phonons; solid circles: electron–radiation interaction vertex; solid squares: electron–lattice interaction vertex. Intermediate states: (a) electron–hole pair; (b) exciton.

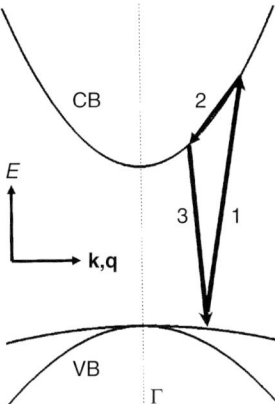

Fig. 2. The scattering process shown in Fig. 1(a), now represented as transitions on the diagram showing the band structure near the point (**k** = 0) of a zinc blende semiconductor. The transitions are numbered 1–3 in the sequence in which they appear (from left to right) in Fig. 1(a) and in the text.

because, under these conditions, one of the two terms in the denominator becomes small (the finite linewidths $\Gamma_{n,m}$ of the transitions however ensure that the scattering intensity remains finite). For a discussion of the derivation of this expression, see Ref. [15]. It is also possible for *both* terms in the denominator of Eq. (9) simultaneously to become small when the splitting in energy of the electronic intermediate states equals the energy of the phonon or other excitation; this is termed "double resonance".

The significance of resonant Raman scattering (RRS) is firstly, from Eq. (9), that an appropriate choice of excitation energy will enhance the signal strength. This is vital in many applications of Raman spectroscopy. Secondly, the use of different excitation energies may allow one to obtain Raman spectra selectively from different parts of an inhomogeneous sample (for instance, different single quantum wells in a multiple quantum well structure). In fact, the photon energy at which resonance occurs may itself be the information of interest as a probe of the electronic density of states of the sample. In this case, a plot of Raman scattering intensity versus incident photon energy (that is, of $I(\omega_l)$, called a "resonance profile") is often presented. Finally, it should be recalled that we have so far restricted our discussion to the first-order Raman scattering process, that is, the process arising from the term $\partial\chi/\partial u$ in Eq. (4); however, the quantum-mechanical description of second- and higher-order processes is analogous to Eq. (9) and shows that they too are enhanced by resonant excitation.

The second aspect of the quantum mechanical nature of the Raman process that should be mentioned is that the probability of annihilating a phonon in an anti-Stokes process depends on the temperature of the sample, as the probability of occupation of a given phonon state (energy ω_{ph}) obeys Bose–Einstein statistics. Therefore, Stokes and anti-Stokes scattering intensities I_S and I_{AS} are not equal (in contradiction to what Eq. (5) suggests) and, in fact, their ratio under non-resonant conditions is

$$\frac{I_{AS}}{I_s} = \left(\frac{\omega_l + \omega_{\text{ph}}}{\omega_l - \omega_{\text{ph}}}\right)^4 \exp\left(-\frac{\hbar\omega_{\text{ph}}}{kT}\right) \qquad (10)$$

an expression which is sometimes employed in determining temperature via Raman spectroscopy. Note that, if photon counting techniques are employed (see Section 2), the measured signal is a photon flux, rather than an intensity, and the fourth power in the prefactor of Eq. (10) should be replaced by a third power. This relationship does not generally hold close to resonance as, as is implied by Eq. (9), the resonances of Stokes and anti-Stokes processes occur at different incident photon energies. Equation (10) may also be used when considering spin or charge density excitations, rather than phonons, though the temperature involved is then that of the electronic system and is not necessarily equal to that of the lattice under typical experimental conditions.

1.3. Examples of selection rules

Before discussing selected "state-of-the-art" applications of Raman spectroscopy, we shall give some relatively simple examples to illustrate the above points. First, we look at Raman spectra of two semiconductor materials and at how one may identify the phonon modes giving rise to the lines in their Raman spectra. We choose GaAs and GaN as important representatives of semiconductors having zinc blende and wurtzite crystal structures and thus exemplifying cubic and hexagonal symmetry, respectively. Being polar materials, these show a splitting in energy of the LO and TO phonons, [7], and therefore show relatively rich Raman spectra.

The Raman tensor was introduced in Eqs (7) and (8) and we shall now demonstrate its use. We stated in Section 1.1 that group theory allows one to predict which vibrational modes of a given crystal structure have the correct symmetry properties to participate in Raman scattering. In the technical language of group theory and for the first-order Raman scattering discussed above, the "Raman-active" modes are all those which correspond to the irreducible representations of the crystal point group that transform in the same way under the point group symmetry operations as the products of the Cartesian co-ordinates $xx \ldots zz$ and $xy \ldots zx$ [16]. The important consequence of this for our present purposes is that, conventionally, the vibrational modes are given the same labels as the irreducible representations to which they correspond. These labels depend only on the symmetry of the scattering material (though alternative labelling conventions exist, as shown in Table 1).

In the case of zinc blende crystals (e.g. bulk GaAs, point group T_d), the zone-centre ($\mathbf{q} \sim 0$) optical phonons active in Raman scattering have the symmetry type labelled Γ_4 (or, equivalently, Γ_{15} or T_2). These modes are three-fold degenerate (as x, y and z are equivalent in the cubic zinc blende lattice) and the Raman tensors corresponding to the x, y, and z-polarised modes of the degenerate set of Γ_4 modes are as follows:

$$R^{\Gamma_4}_{jk,x} = \begin{pmatrix} 0 & 0 & 0 \\ 0 & 0 & d \\ 0 & d & 0 \end{pmatrix}, \quad R^{\Gamma_4}_{jk,y} = \begin{pmatrix} 0 & 0 & d \\ 0 & 0 & 0 \\ d & 0 & 0 \end{pmatrix}, \quad R^{\Gamma_4}_{jk,z} = \begin{pmatrix} 0 & d & 0 \\ d & 0 & 0 \\ 0 & 0 & 0 \end{pmatrix} \quad (11)$$

Here, the suffices $x \ldots z$ indicate the direction of the polarisation $\mathbf{p}$ of the mode. As $\mathbf{q}$ is not strictly zero and because there is a splitting in energy between modes depending on whether $\mathbf{p}$ and the propagation direction $\mathbf{q}$ are parallel (LO) or orthogonal (TO), the LO

Table 1
Alternative conventions for the labelling of irreducible representations and their associated vibrational modes (after Ref. [15])

Crystal point group	Koster [17] notation	Bouckaert, Smoluchowski and Wigner [18] notation	Molecular notation
T_d (zinc blende)	Γ_1	Γ_1	A_1
	Γ_2	Γ_2	A_2
	Γ_3	Γ_{12}	E
	Γ_4	Γ_{15}	T_2 (or F_2)
	Γ_5	Γ_{25}	T_1 (or F_1)
C_{6v} (wurtzite)	Γ_1	–	A_1
	Γ_2	–	A_2
	Γ_3	–	B_1
	Γ_4	–	B_2
	Γ_5	–	E_1
	Γ_6	–	E_2

and TO modes form two separate threefold degenerate Γ_4 sets, each described by the Raman tensors given in Eq. (11) though with different constants d_{LO} and d_{TO}.

In the case of wurtzite crystals (e.g., GaN, point group C_{6v}), x and y are equivalent to each other but are normal to the non-equivalent direction of the sixfold rotation axis z. The three Raman-active symmetry types are usually labelled A_1 (which is non-degenerate and corresponds to displacement in the z direction, defined along the high-symmetry sixfold rotation axis) and E_1 and E_2 (each twofold degenerate). The corresponding Raman tensors are

$$R^{A_1}_{jk,z} = \begin{pmatrix} a & 0 & 0 \\ 0 & a & 0 \\ 0 & 0 & b \end{pmatrix}, \quad R^{E_1}_{jk,x} = \begin{pmatrix} 0 & 0 & c \\ 0 & 0 & 0 \\ c & 0 & 0 \end{pmatrix}, \quad R^{E_1}_{jk,y} = \begin{pmatrix} 0 & 0 & 0 \\ 0 & 0 & c \\ 0 & c & 0 \end{pmatrix}$$

$$R^{E_2}_{jk,1} = \begin{pmatrix} d & 0 & 0 \\ 0 & -d & 0 \\ 0 & 0 & 0 \end{pmatrix}, \quad R^{E_2}_{jk,2} = \begin{pmatrix} 0 & -d & 0 \\ -d & 0 & 0 \\ 0 & 0 & 0 \end{pmatrix} \quad (12)$$

Note that the two E_2 modes cannot be labelled in simple terms by x, y and z; see Ref. [19] for a description of the E_2 polarisation properties.

As a first example, we consider a backscattering Raman experiment from the (001) surface of a GaAs crystal. The incident light travels along z and the scattered light along $-z$ (often written $\bar{z}$); the incident and detected polarisation states are, as in Eq. (8), $\mathbf{e}_I$ and $\mathbf{e}_S$, respectively. Given the transverse nature of light in free space, we can assert that $e_z = 0$ in both cases. From Eq. (6), momentum conservation implies that the phonon involved must also propagate parallel to z; if it is transversely polarised (TO), its polarisation direction must be x or y, and if it is longitudinal (LO), the polarisation direction must be z. We can now deduce which phonon modes may be observed; evaluating Eq. (8) for two of the zinc blende Raman tensors from Eq. (11) gives:

$$R^{\Gamma_4}_{jk,x}: \quad \begin{pmatrix} e^S_x & e^S_y & 0 \end{pmatrix} . \begin{pmatrix} 0 & 0 & 0 \\ 0 & 0 & d \\ 0 & d & 0 \end{pmatrix} . \begin{pmatrix} e^I_x \\ e^I_y \\ 0 \end{pmatrix} = \begin{pmatrix} e^S_x & e^S_y & 0 \end{pmatrix} . \begin{pmatrix} 0 \\ 0 \\ de^I_y \end{pmatrix} = 0$$

$$R^{\Gamma_4}_{jk,z}: \quad \begin{pmatrix} e^S_x & e^S_y & 0 \end{pmatrix} . \begin{pmatrix} 0 & d & 0 \\ d & 0 & 0 \\ 0 & 0 & 0 \end{pmatrix} . \begin{pmatrix} e^I_x \\ e^I_y \\ 0 \end{pmatrix} = \begin{pmatrix} e^S_x & e^S_y & 0 \end{pmatrix} . \begin{pmatrix} de^I_y \\ de^I_x \\ 0 \end{pmatrix} \neq 0$$

(13)

The first expression shows that there will be no signal detected from x-polarised (TO) phonons. Inspecting the second expression of Eq. (13) shows that if the incident light is purely x-polarised $\left(e^I_y = 0 \right)$, then a signal from z-polarised (LO) phonons may be observed using y-polarised detection. The experimental geometry is conveniently represented in the Porto notation $k_I(e_Ie_S)k_S$; for example, we can summarise the results of Eq. (13) by saying that TO phonons cannot be detected in experiments of types $z(xy)\bar{z}$ or $z(xx)\bar{z}$ and that LO phonons may be detected in $z(xy)\bar{z}$ but not $z(xx)\bar{z}$. For practice in such calculations, and a more thorough introduction to the use of group theory, we recommend the problems at the end of Chapter 7 of Ref. [15].

For our second example, we consider wurtzite crystals (such as GaN and ZnO, where the c-axis is now defined as z). In a similar backscattering experiment from the (0001) face, one can use Eq. (12) to make similar predictions:

$$R^{A_1}_{jk,z}: \quad \begin{pmatrix} e^S_x & e^S_y & 0 \end{pmatrix} . \begin{pmatrix} a & 0 & 0 \\ 0 & a & 0 \\ 0 & 0 & b \end{pmatrix} . \begin{pmatrix} e^I_x \\ e^I_y \\ 0 \end{pmatrix} = \begin{pmatrix} e^S_x & e^S_y & 0 \end{pmatrix} . \begin{pmatrix} ae^I_x \\ ae^I_y \\ 0 \end{pmatrix} \neq 0$$

$$R^{E_1}_{jk,x}: \quad \begin{pmatrix} e^S_x & e^S_y & 0 \end{pmatrix} . \begin{pmatrix} 0 & 0 & c \\ 0 & 0 & 0 \\ c & 0 & 0 \end{pmatrix} . \begin{pmatrix} e^I_x \\ e^I_y \\ 0 \end{pmatrix} = \begin{pmatrix} e^S_x & e^S_y & 0 \end{pmatrix} . \begin{pmatrix} 0 \\ 0 \\ ce^I_x \end{pmatrix} = 0$$

(14)

$$R^{E_2}_{jk,1}: \quad \begin{pmatrix} e^S_x & e^S_y & 0 \end{pmatrix} . \begin{pmatrix} d & 0 & 0 \\ 0 & -d & 0 \\ 0 & 0 & 0 \end{pmatrix} . \begin{pmatrix} e^I_x \\ e^I_y \\ 0 \end{pmatrix} = \begin{pmatrix} e^S_x & e^S_y & 0 \end{pmatrix} . \begin{pmatrix} de^I_x \\ -de^I_y \\ 0 \end{pmatrix} \neq 0$$

$$R^{E_2}_{jk,2}: \quad \begin{pmatrix} e^S_x & e^S_y & 0 \end{pmatrix} . \begin{pmatrix} 0 & -d & 0 \\ -d & 0 & 0 \\ 0 & 0 & 0 \end{pmatrix} . \begin{pmatrix} e^I_x \\ e^I_y \\ 0 \end{pmatrix} = \begin{pmatrix} e^S_x & e^S_y & 0 \end{pmatrix} . \begin{pmatrix} -de^I_y \\ -de^I_x \\ 0 \end{pmatrix} \neq 0$$

Here, we see that A_1(LO) and E_2 modes will be detected in an experiment of type $z(xx)\bar{z}$ but that only the E_2 modes will appear for crossed optical polarisations, $z(xy)\bar{z}$. Figure 3 demonstrates this; in the top panel, we see that the A_1(LO) and E_2 modes are observed for unpolarised detection (which one can write as $z(xx)\bar{z} + z(xy)\bar{z}$; sometimes $z(uu)\bar{z}$ is used). We leave it as an exercise to the reader to verify that the results of the bottom panel, for experiments

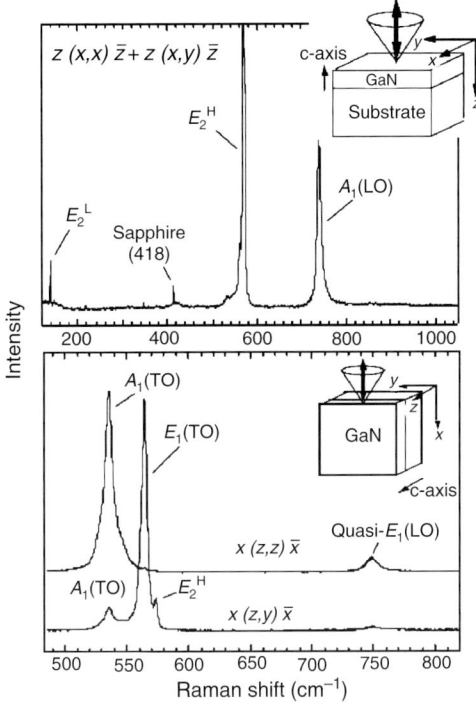

Fig. 3. Typical Raman spectra of hexagonal GaN observed at different scattering geometries. The inset shows schematically the directions of the incident and scattered light (reproduced with permission of the Institute of Physics (UK) from Ref. [20]).

$x(zz)\bar{x}$ and $x(zy)\bar{x}$, are also as expected. The figure nicely demonstrates the motivation for the rather awkward experiment $x(zy)\bar{x}$ in which the epitaxial layer (whose normal is the crystalline c-axis and defines the z direction) is viewed edge-on; it provides a means of obtaining a strong Raman signal from the E_1(TO) mode.

Table 2 summarises for both crystal systems a few of the experimental geometries and the types of phonon, which may be observed; in this table, the cubic crystal directions [011] and

Table 2
Examples of experimental geometries (expressed in the notation described in the text) and the phonons which may be observed in those geometries for crystals of zinc blende and wurtzite symmetries

Geometry	Zinc blende	Wurtzite
$x(yy)\bar{x}$	None	A_1(TO), E_2
$x(zz)\bar{x}$	None	A_1(TO)
$x(zy)\bar{x}$	LO	E_1(TO)
$z(yy)\bar{z}$	None	A_1(LO), E_2
$z(xy)\bar{z}$	LO	E_2
$y'(z'z')\bar{y}'$	TO	

[0$\bar{1}$1] are represented by y' and z' respectively (these are relevant to backscattering from zinc blende (011)-type surfaces, for example). Extensive tables are given elsewhere (e.g., Refs [15,20–22] and tables of Raman tensors for all point groups can be found in Refs [13,23].

2. Experimental methods

Definitions of many technical terms and descriptions of different illumination geometries are usefully provided in a recent IUPAC report [24]. Here, we focus on recent developments with specific reference to studies of semiconductors. For example, we shall not discuss Fourier transform Raman spectroscopy despite its importance in other areas of Raman spectroscopy [21] because of the limited resolution and restricted choice of excitation wavelengths that it offers.

The major components of the "classic" experimental system for Raman spectroscopy are an excitation laser and a high-resolution spectrometer. The spectrometer is a two-stage instrument with the two diffraction gratings arranged so that the effects of their dispersions are additive; typical double grating instruments have a focal length of 75 cm, an entrance f-number of around eight, a resolution of around 0.03 nm and a high stray light rejection, of around 10^{-12}; such a good rejection of stray light is essential because the Raman scattering process is generally weak and as the sample is illuminated by intense laser light at a very close-lying photon energy. Such a system is usually equipped with a GaAs-based photon-counting photomultiplier with background level of a few counts per second in the dark. Depending on the strength of the signals, a scan over a range of $\sim 100\,\mathrm{cm}^{-1}$ (the spectroscopist's traditional energy unit is $1\,\mathrm{cm}^{-1}$ to $\sim 0.124\,\mathrm{meV}$) will require minutes to hours for a good signal-to-noise ratio (SNR). A review of various sample types, the optical arrangements appropriate for them and the resulting light collection efficiencies is given in Ref. [21].

2.1. Multichannel detection

The use of array detectors brings a massive advantage in terms of the speed of data acquisition and charge-coupled devices (CCDs) are now used as detectors in most commercial Raman spectrometers; for details of how CCDs work and of their use in spectroscopy, see Ref. [25]. An array detector having ~ 1000 simultaneously illuminated elements will be able to acquire a spectrum with a given SNR in a tiny fraction of the time that the scanning system above will require, despite the minor disadvantages that (i) the quantum yield of a CCD pixel is sometimes poorer than that of a photomultiplier and (ii) there is a new noise source associated with the read-out process of the CCD. However, to illuminate an array detector, the exit slit of the conventional spectrometer described above must be removed (and the central slit widened) and this seriously degrades its stray light rejection. For this reason, holographic notch filters (specific to a single laser wavelength) are often used to provide rejection of the stray excitation light, followed by a single-grating stage and CCD. Such a spectrometer generally has fewer optical components and therefore a lower loss that a double grating system, so that the improvement in SNR may be even more than the multichannel advantage alone suggests. The nearest possible approach to the laser line with a holographic filter is typically ~ 50–$100\,\mathrm{cm}^{-1}$.

Where tunable excitation is necessary, a triple-grating spectrometer is often used in which the first two gratings are arranged so that their dispersions cancel (termed subtractive dispersion); the final dispersing stage is as before but the subtractive double-grating stage plays the role of a *tunable* notch filter, giving much increased loss but greater versatility; in addition, the cut-off of such a "filter" can be sharp enough that signals within a few cm^{-1} of the laser line can be recorded. Other more efficient variants of tunable filters are also available (e.g., the Renishaw NExT filter, which gives a cut-off at about 5 cm^{-1} from the laser in the visible spectrum).

Recently, CCDs have been developed in which impact ionisation is used to provide amplification of the charge accumulated in each pixel, in a manner similar to the process exploited in avalanche photodiodes. In such "electron-multiplying" CCDs (EMCCDs), the read-out noise that is intrinsic to a CCD becomes insignificant in comparison to the charge associated with a single photon event, so that true photon counting becomes possible. At the lowest light levels, EMCCDs are therefore very promising for spectroscopic applications. However, the gain of the amplification process fluctuates randomly with time; this is unimportant at low light levels as photon counting only requires that each signal pulse passes a set threshold but, at high light levels, this can represent both a new noise source and a limitation on linearity.

2.2. Microscopy

The use of array detectors has facilitated the combination of Raman spectroscopy with conventional optical microscopy, directing the laser beam via a beam splitter into the microscope objective, and then collecting the scattered light via the same objective and delivering it to a spectrometer (for a general review, see Ref. [21]). The scattered light is collected over a large solid angle so the collection efficiency of such a system is good, and the technical challenge of matching spectrometer and microscope optics has been met by several manufacturers. One benefit is the ability to correlate Raman spectra with position on the optical image of the sample surface ("Raman mapping"), and Raman microscopes (with long working length objective lenses) are useful in work with, e.g., diamond anvil pressure cells or helium flow cryostats. Several of the examples of Raman spectroscopy below employed microscopy simply for convenience because of its reproducible alignment and efficient light collection.

A typical lateral spatial resolution is around 10^{-6} m (1 µm) and the use of a confocal microscope design (reference), in which a spatial filter confocal with the sample restricts the collection region to about half a micron above and below the focal plane of the objective, defines a sample volume of, very roughly, 10^{-18} m^3. The lateral resolution may be improved by use of a solid immersion lens [26].

2.3. Near-field techniques

The diffraction-limited resolution in the far field, of around one micron, is still not sufficiently small for studies of truly nanoscale systems. To obtain better spatial resolution, near-field techniques must be used and the usual starting point for the near is via modification of the operation of a far-field Raman microscope. Three approaches will be mentioned here; these involve the use of metal apertures, scanning near-field optical microscopy (SNOM or NSOM), and metal tip-enhanced Raman scattering (TERS). The "best-yet" resolution in Raman spectroscopy already reached the single molecule level in 1997 [27] using surface-enhanced Raman scattering, SERS, to which TERS is related, but typical resolutions of near-field systems are 100–200 nm. The SERS and TERS depend, at least partially, on the local enhancement of the electric field of the incident light that can result in the region of a strongly curved metal surface (provided, often, by a metal nanoparticle or nanoscale tip).

One approach is to deposit an opaque metal layer (typically ~100 nm of Al) in which electron beam lithography is used to make apertures (ranging from tens down to tenths of a micron in diameter [28]). By use of several apertures separated by greater than the far-field resolution, the microscope can be used to find an aperture above a suitable part of the lateral distribution of quantum dots (QDs). This technique has been successful in studies of QDs, allowing a variety of single QD spectra to be obtained.

In SNOM, no sample processing of this type is required. Instead, the tip of a metal-coated and tapered optical fibre provides a sub-wavelength sized aperture; the fibre tip is scanned over the sample surface at about 10 nm above it, this distance being held constant using techniques similar to those of scanning tunnelling microscopy (STM) [29–31]. The excitation laser light may be delivered via the fibre and the scattered light collected via a far-field Raman microscope. Alternatively, the laser light may be focused conventionally on the surface or via attenuated total reflection for back illumination of the sample [32] and the fibre-optic probe may be used for the light collection. In both cases, the tip diameter (of about 20–500 nm) determines the lateral resolution.

2.4. Coherent and time-domain techniques

With the advent of lasers able to generate very short light pulses (with durations of a few picoseconds or less), it became possible in the 1970s to use Raman scattering processes to investigate the dynamics of phonon processes and to determine phonon lifetimes. Typically, optic phonons can decay into combinations of acoustic phonons with a lifetime of a few picoseconds [33,34]. An example of such an optic phonon lifetime measurement in diamond is provided in Fig. 4.

The complexity and cost of ultrafast time-resolved techniques and the very fundamental nature of the information they provide mean that they can hardly be regarded as "characterisation" techniques and so a lengthy description is not appropriate here. A good review has been provided by Laubereau in Ref. [36].

Briefly, two light pulses are used, with a time delay between them that may be varied (a typical measurement shows intensity as a function of this time delay, Fig. 4). The first ("pump") pulse (momentum $\hbar\mathbf{k}_1$) excites an optic phonon mode ($\hbar\mathbf{q}$) and the delayed second ("probe") pulse ($\hbar\mathbf{k}_2$) then interacts with what remains of that excited phonon population to produce coherent Raman scattered light. The pump and probe beams and light collection directions are determined by the condition that the same phonon mode (with the same frequency *and* propagation direction) is addressed by the pump and the probe beams [36]. If the material under study shows birefringence, this can be exploited to achieve this "matching" condition. Collection of the scattered light is very efficient because the scattered light is coherent and forms a beam with a well-defined direction in space; we return to this concept in Section 3.4.

2.5. Extreme conditions: high pressure

Because Raman spectroscopy only requires optical access to a sample and this may be through robust, common window materials, it is a useful probe of structure and composition for samples in extreme environments. For a homogeneous bulk sample, the obvious externally controllable parameters are temperature (T) and hydrostatic pressure (p). We shall only discuss Raman spectroscopy under pressure but, at a fundamental level, variation of T or

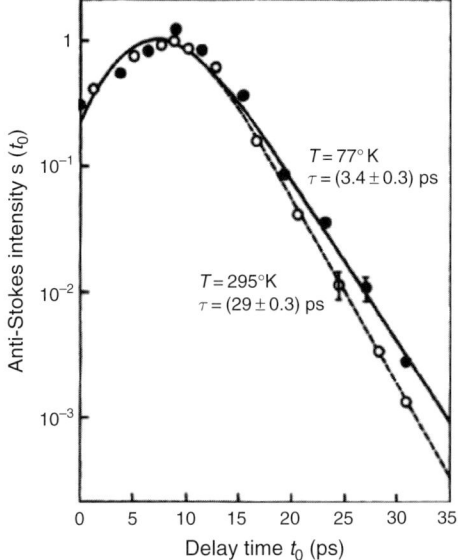

Fig. 4. TO phonon lifetime in diamond at room temperature and at 77 K. Data at 77 K reveal the charateristic relaxation time of a TO phonon into TA and LA phonons, where occupation numbers of the final states are small. The data at room temperature show a TO phonon lifetime consistent with the linewidths observed in conventional (continuous wave) Raman scattering experiments (reproduced with permission from Ref. [35]. Copyright (1971) by the American Physical Society).

p allows one to probe details of the interatomic potential and its anharmonicity; phase changes at high pressure may also be revealed.

Our first example is that of GaN under pressure (Fig. 5). The sample is a triangular flake of GaN, held in a pressure-transmitting liquid in a hole (diameter $\sim 200\,\mu$m) drilled through a thin metal plate ("gasket") that is compressed between the small flat facets at the tips of two approximately conical diamonds. A view of the sample from above is shown in Fig. 5. The diamonds both create the high pressure (which is greatly magnified compared to the pressure

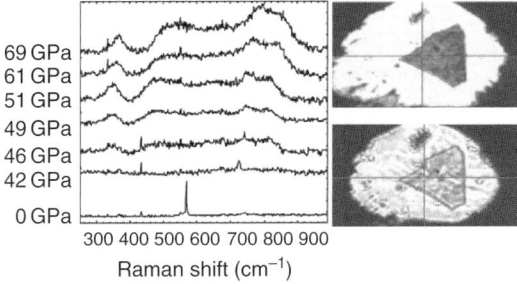

Fig. 5. A series of Raman spectra of GaN recorded at pressures up to 69 GPa, showing the change of phase from wurtzite to rocksalt. The images show the sample under back illumination before the phase transition (at 3 GPa) and after it is completed (at 62 GPa) (reproduced with permission from Ref. [39]. Copyright (2004) by the American Physical Society).

applied to their back surfaces because of the taper of the diamonds) and they serve as the windows for the experiment. The pressure is determined by measuring the pressure-induced shift in the wavelength of the photoluminescence (PL) of a small ruby crystal placed next to the sample; p ranges up to some tens of gigapascals. This is a standard arrangement for Raman and other optical spectroscopies under high pressure; for recent reviews, see Ref. [37,38].

A phase change in the sample is easily recognised in Fig. 5 as the sample becomes much less transparent to visible light above pressures of $p \sim 42$ GPa. Above this pressure, the broad bands seen in the Raman spectrum (Fig. 5) allow one to identify the new crystalline structure as rocksalt (though first-order Raman scattering is forbidden for the rocksalt structure, disorder relaxes the selection rules and, as will be seen later, the Raman spectrum then reflects the shape of the phonon density of states) [39].

It can be seen in Fig. 5 that the phonon frequencies shift continuously with pressure. As a second example, the sign and rate of change of frequency with pressure for several phonon modes of wurtzite ZnO is demonstrated in Fig. 6, which shows clearly that these shifts depend

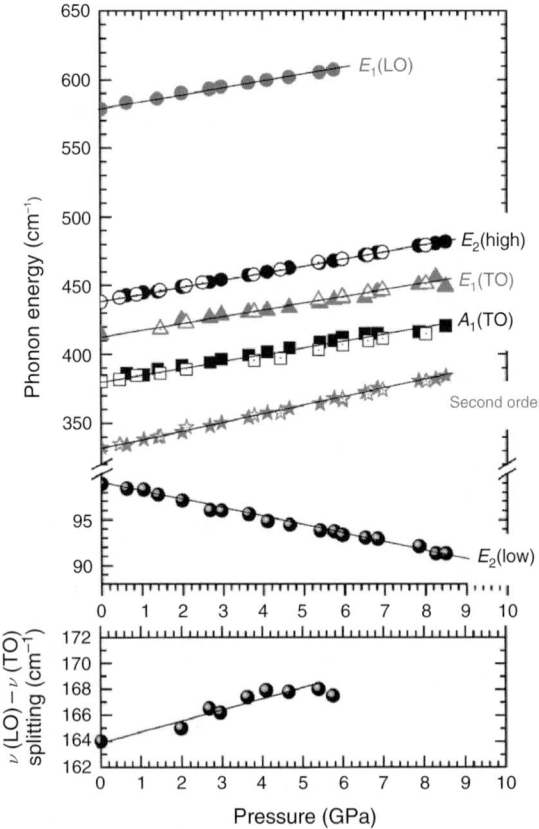

Fig. 6. Pressure dependence of the optic phonon energies in wurtzite ZnO. Bottom: the LO–TO splitting as a function of pressure (reproduced with permission from Ref. [42]. Copyright (2002) by the American Physical Society).

on the particular phonon mode. Thus, pressure-dependent measurements provide a more stringent test of theoretical calculations of phonon modes than a comparison with the predicted phonon frequencies only at ambient pressures. In both the examples of this section, the experimental data were compared to calculations using ab initio methods that are rapidly becoming standard tools for the experimentalist; see also Chapter 2 booking Ref. [40] and, for a review, see Ref. [41].

It is clear that fundamental information about the phonons of bulk semiconductors can be obtained via pressure-dependent studies. This is true also of semiconductor heterostructures and quantum systems [43]. For example, a semiconductor epilayer still attached to some substrate material will experience a biaxial stress even in a notionally isotropic hydrostatic pressure experiment if the compressibility of the epilayer differs from that of the substrate, and this effect has occasionally been exploited [44].

3. Applications

For the remainder of this chapter, we give examples of recent results in the characterisation of semiconductor heterostructures by Raman spectroscopy, moving from issues relating to sample "quality" through to probes of new effects in quantum structures. We conclude with some remarks about Raman spectroscopy applied to the investigation of semiconductor devices under operational conditions.

3.1. Epilayers: composition, doping and strain

As all the preceding chapters make clear, alloy semiconductors form the basis of many semiconductor heterostructures, the classic example being $Al_xGa_{1-x}As$ as a barrier material for GaAs quantum wells. A significant challenge for characterisation techniques is therefore to investigate the state of the alloy material. Even in an alloy where the cations are randomly (and therefore isotropically) mixed, Raman scattering is sensitive to the presence of more than one cation type because of the breakdown of long-range order and, therefore, the relaxation of the momentum selection rule of Eq. (6). In alloy semiconductors, therefore, Raman scattering often reveals signals arising from phonons at finite q; this can be recognised by comparison of the spectrum with the expected phonon density of states, now taking into account also the acoustic phonons. Disorder due to other types of disruption of the crystal symmetry will also result in phonons far from the Brillouin zone centre being detected. This provides a means of assessing damage in, for instance, ion implantation [45]. As an example of "disorder-activated Raman scattering", we consider microcrystalline $Ga_{1-x}Mn_xN$, which shows such structure emerging even at quite low Mn concentrations, $x \sim 0.005$ [46] (bands b–d of Fig. 7; in the spectral region where they appear, the spectrum is completely flat in the absence of significant disorder). The degree to which Mn substitutes randomly for Ga in this compound is a key question for its possible application as a ferromagnetic semiconductor in spintronics (see also Chapter 2 of this book [40]).

The zone-centre modes themselves are also affected by alloying, as one would expect. Typically, for a material with two cation types $A_{1-x}B_xC$ one observes either (i) an interpolation of the LO and TO frequencies between their corresponding values for the end members of the composition range, AC and BC, (termed "one-mode" behaviour) or (ii) the co-existence of independent phonon modes for the AC and BC sub-lattices within the alloy ("two-mode" behaviour). In the latter case, exemplified by $Al_xGa_{1-x}As$, the frequencies of the phonon

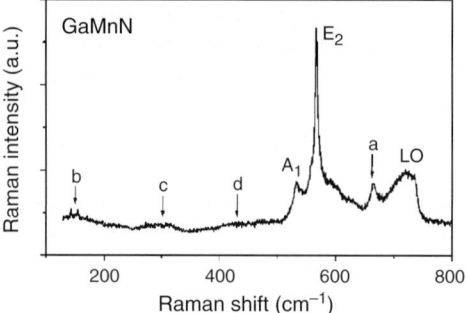

Fig. 7. Raman scattering from zone centre phonons (A_1, E_2) together with peaks arising from the phonon density of states, obtained for microcrystalline $Ga_{1-x}Mn_xN$ with Mn concentration ~ 0.005 (reproduced with permission from Ref. [46]. Copyright (2001) by the American Institute of Physics).

modes of AC (the "AC-like" modes) tend to the vibrational frequency of an isolated impurity of species A in BC as the concentration of A becomes very small. The issues affecting whether a system displays one- or two-mode behaviour have been discussed in detail recently with special reference to the case of $Ga_xIn_{1-x}P$ [47]. In either situation, a measurement of the phonon frequencies can yield an estimate of the composition of the alloy though, it should be noted, in general the uncertainty in this estimate is usually high and other techniques (e.g., X-ray diffraction, XRD, to determine the lattice parameter [48]) are generally superior. A model based on the one-dimensional linear diatomic chain applied to a material with "average" properties [49] has been widely used, and very successful, in describing the variation of phonon frequency with composition in the two-mode case. This model has been adapted in many ways: to deal with quaternary alloys, wurtzite materials, and the full three-dimensional phonon dispersion [50]. In Fig. 8, we show recent results for wurtzite $Al_xGa_{1-x}N$, which represents a

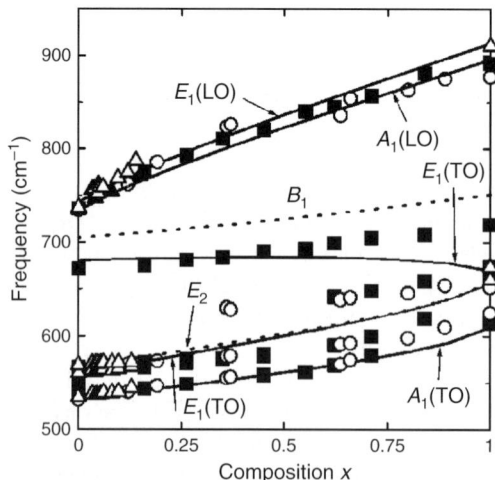

Fig. 8. Calculated (lines) and measured (symbols) phonon frequencies in 2H $Al_xGa_{1-x}N$ (reproduced with permission from Ref. [50]. Copyright (2000) by the American Physical Society).

technologically important but rather complicated case; however, one can see that the LO modes in this case clearly show one-mode behaviour [50]. The scatter of the experimental data gives an impression of the usefulness of phonon data for determination of composition; an accuracy of no better than a few per cent is realistic, though this may still be valuable where other techniques may fail (for instance, in probing small volumes of material such as are present at partially inter-diffused heterointerfaces, Section 3.2).

Another issue affecting the "quality" of alloy semiconductors is segregation of the components into separate phases. If this occurs over micron or larger length scales, Raman microscopy can be valuable to identify precipitates; if it occurs on a smaller scale, Raman spectra may show a superposition of the spectra of the two materials. For example, the possible existence of In-rich precipitates in the alloy $In_xGa_{1-x}N$ is of importance as the existence of In-rich QDs offers one possible (and rather controversial) explanation for the high radiative efficiency of this alloy (which forms the active layer of most blue light-emitting diodes, LEDs). In Fig. 9, a peak (labelled S) is seen between the TO and LO modes of cubic $In_xGa_{1-x}N$ layers (with $x = 0.19, 0.33$) and this is attributed to the presence of In-rich regions with $x \sim 0.8$ [51]. In this work, RRS (Section 1.2), PL and XRD data were combined to reach these conclusions.

Another example from the field of nitride semiconductors is given by the ability of Raman scattering to detect hexagonal (wurtzite) inclusions in nominally cubic material, for example in cubic GaN down to the level of about 1% [52]. One earlier example of the success of Raman spectroscopy in detecting the formation of new phases is the case of spontaneous ordering in $Al_xIn_yGa_{1-x-y}P$ (a semiconductor of importance in red LEDs); this system has a well-known tendency to form a new crystalline phase, effectively a short-period superlattice, in which the cations are not randomly mixed. Raman spectroscopy enables one to quantify the degree of ordering, for instance, in $Al_{0.5}In_{0.5}P$ [53]. Such effects may prove also to be an issue in the nitride semiconductors [54].

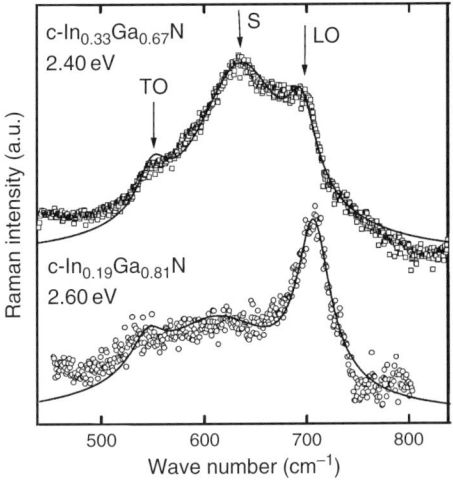

Fig. 9. Raman spectra of cubic $In_xGa_{1-x}N$ for two values of x; S indicates a mode assigned to an In-rich phase with $x \sim 0.8$ (reproduced with permission from Ref. [51]. Copyright (2000) by the American Physical Society).

Our final example of compositional inhomogeneity comes from recent work on quaternary alloys where the anions are also partially substituted, in this case, $Al_xGa_{1-x}As_yN_{1-y}$. The interest in the mixed (As, N) semiconductors arises from the decrease of the band gap on the introduction of a small amount of nitrogen, which has opened up the possibility of new optoelectronic devices in the infrared region (see also Chapters 5 and 9). Here, Raman spectroscopy has revealed a strong tendency of the Al to bond preferentially to N atoms, leading to an alloy which is locally highly non-random (for only 5% Al content, it is found that almost no N remains in the expected configuration, $-Ga_4N$). This may be seen in Fig. 10 by the rapid extinction of the GaN-like LO mode and the growth of the AlN-like LO mode as Al is introduced. This observation has important implications for attempts to gain a fundamental understanding of the mechanism by which the band gap is modified [55].

In summary, we see that Raman spectroscopy has a useful role to play in detecting microstructures that may *spontaneously* form in materials that are expected, or hoped, to behave as isotropic components of a larger-scale semiconductor heterostructure.

In addition to intrinsic compositional variations of alloy semiconductors, dopant and other impurities break the crystal symmetry and thus lead to disorder-induced Raman scattering. More importantly, however, isolated impurities show Raman scattering from localised vibrational modes which may provide a useful diagnostic tool. Figure 11 shows the case of the nitrogen local mode in $GaAs_{1-x}N_x$ for low N content; this spectrum [56] incidentally provides another beautiful example of the disorder-activated transverse acoustic (DATA) and longitudinal acoustic (DALA) phonon scattering mentioned earlier (Fig. 7).

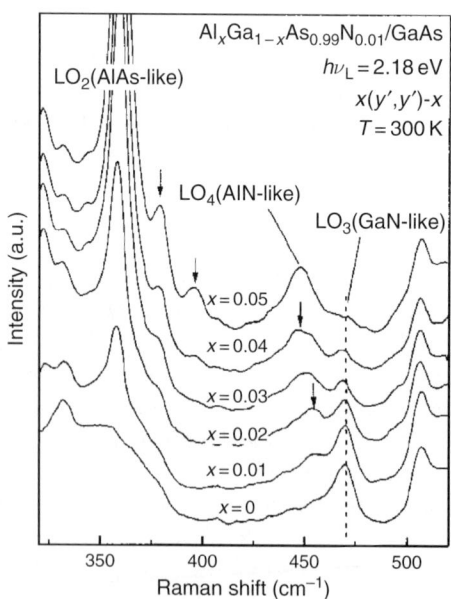

Fig. 10. Raman spectra of $Al_xGa_{1-x}As_{0.99}N_{0.01}$ with Al concentrations as indicated. Modes arising from the simultaneous presence of Al and N are marked by vertical arrows (reproduced with permission from Ref. [55]. Copyright (2002) by the American Institute of Physics).

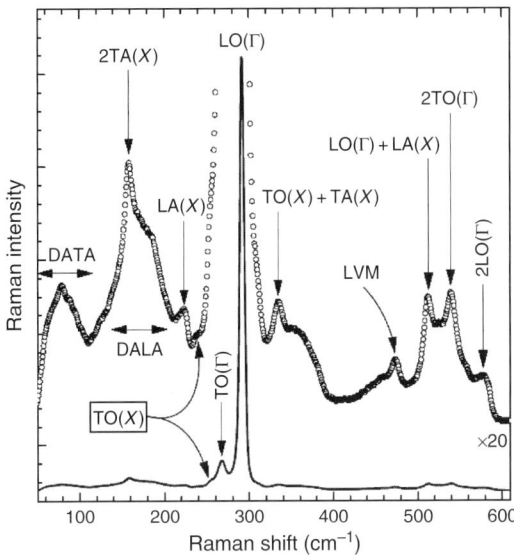

Fig. 11. Raman spectrum of $GaAs_{1-x}N_x$ ($x = 0.5\%$) showing the local vibrational mode (LVM) of nitrogen. The upper spectrum shows the lower spectrum magnified vertically by a factor of 20 (reproduced with permission from Ref. [56]. Copyright (2002) by the Institute of Physics (UK)).

Another particularly important application is that of Raman scattering from hydrogen in semiconductors which has been widely applied, as has infrared absorption, to monitor the passivation of dopants (for instance, Mg acceptors in GaN) via the distinctive, relatively high frequency vibrational mode of a single hydrogen atom bonded to a point defect at about 2000–3000 cm^{-1}. In GaN, it has proved possible to observe the removal of hydrogen upon annealing and its subsequent re-absorption, and to correlate this with the degree of activation of the Mg acceptors; Raman spectroscopy provides a good tool to monitor the degree of hydrogen passivation [57]. A very rich Raman spectrum was observed for hydrogen-implanted silicon in the frequency range of the bond-stretching modes of hydrogen, shown in Fig. 12, and lead to the identification of a range of vacancy-hydrogen clusters [58]. The changes of these modes on annealing were compared to electron spin resonance data (which helps identify defects through their charge state and symmetry) and this comparison was key to the interpretation of the spectra. Another example of the success of Raman spectroscopy in identifying point defects is provided by the observation of carbon dimers in GaAs [59] and AlAs [60], where first principles calculations were employed to explore a range of possible substitutional and interstitial sites for the complex, leading to the conclusion that it is located on an As site.

The carriers introduced by dopants also lead to two new types of Raman signal; firstly, it is possible to observe Raman scattering between the electronic levels of a neutral dopant impurity and thus to determine its energy level structure in some detail. In terms of semiconductor characterisation, this may be useful for identifying residual impurities by reference to known impurity depths as was done, for example, for the case of residual donors such as In, Ga and Cl in epitaxial ZnSe [61].

Secondly, it is found that high electron or hole concentrations interact with the long-range electric fields of optic phonons to produce coupled plasmon–phonon modes, giving a shift and

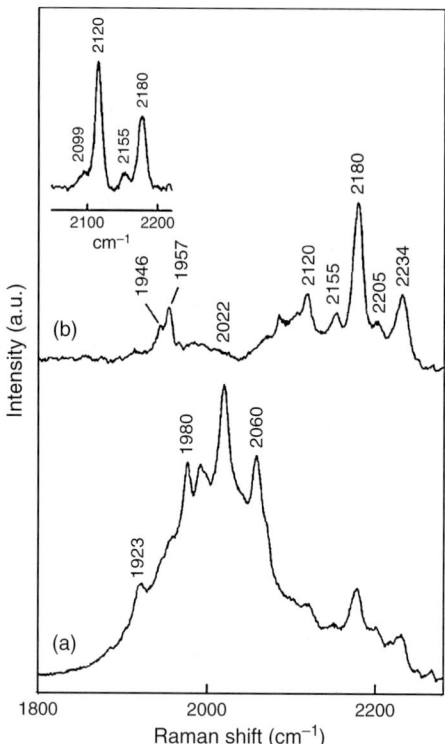

Fig. 12. Raman spectra measured at room temperature on a H_2-implanted Si sample: (a) as-implanted sample; (b) after annealing at 400°C for 2 min. Spectra are offset vertically for clarity (reproduced with permission from Ref. [58]. Copyright (2001) by the American Physical Society).

damping of the phonon mode, which is dependent on the carrier concentration in a simple way, so that Raman spectroscopy may be used to determine carrier concentrations where direct electrical measurements are not possible. An illustration of Raman spectra in the coupled plasmon–phonon regime is given in Fig. 13, for GaAs:Mn [62] and other such results have been presented [63]; the aim of introducing Mn is to introduce magnetic effects but Mn^{2+} also forms an acceptor in GaAs and the resulting hole concentration may play a key role in the origin of its ferromagnetism [64,65].

The final perturbation to be considered here is the strain induced by the mismatch of lattice parameters in many heterostructures (see also Chapter 4). Raman spectroscopy has been very widely used to monitor strain, as the shifts of phonon frequencies with strain are well-established for many materials. Usually, the frequency shift is proportional to the biaxial strain; the constant of proportionality contains the elastic constants and deformation potential constants of the material. In ternary alloys, it may be difficult to distinguish between strain and composition effects (a particularly difficult case is $In_xGa_{1-x}As$, where the two effects can compensate). A recent example is provided by Fig. 14, from Ref. [66], for the case of GaN on various buffer layers on SiC; the Raman shift of the GaN E_2 mode shows a clear sensitivity to the type of buffer layer. Here, a careful comparison of X-ray diffraction, Raman and PL data yielded values for the deformation potential constants and a simple correlation with the bound exciton PL line. We shall return to the question of strain in device structures in Section 3.7.

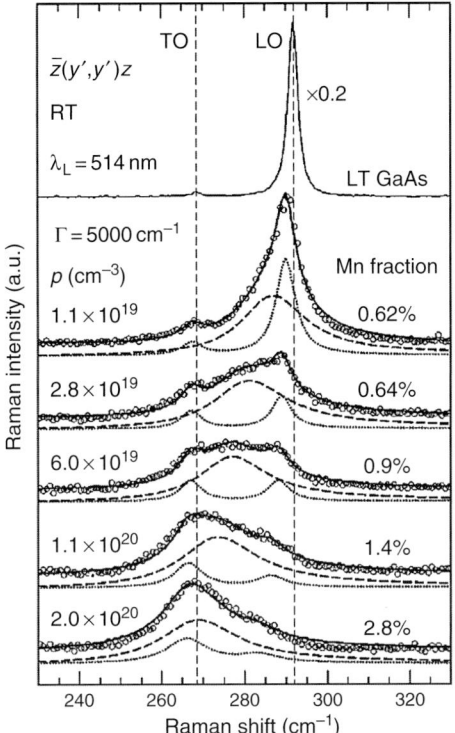

Fig. 13. Room temperature Raman spectra (open circles) recorded in a backscattering configuration at the (001) surfaces of $Ga_{1-x}Mn_xAs$ layers with different Mn fractions x. The hole densities p were estimated from a full line-shape analysis (solid lines) including scattering by coupled plasmon–LO phonon modes (dashed lines). Additional contributions (dotted lines) arise from the symmetry forbidden TO phonon and from the LO phonon in the depletion layer (reproduced with permission from Ref. [62]. Copyright (2002) by the American Physical Society).

3.2. Surfaces and interfaces

Raman spectroscopy has also yielded useful results in the study of surfaces and of heterointerfaces between the constituent layers of low-dimensional structures. We shall look first at the study of free surfaces and of epitaxial growth on them where, for instance, surface reconstructions can lead to surface-specific vibrational states which Raman spectroscopy can detect, as shown in Fig. 15 (Raman spectroscopy is performed in situ in the UHV growth chamber via suitable windows).

In situ Raman studies of the epitaxial growth process can be carried out in the same type of experimental system. In this case, the deposition of Sb on the InP surface leads to the appearance of new modes as the first Sb monolayer (ML) forms [68]. These results have lead to the rich field of the use of Raman spectroscopy for the monitoring of semiconductor growth [69]; recently, results have also been reported for the II–VI semiconductor system BeTe [70].

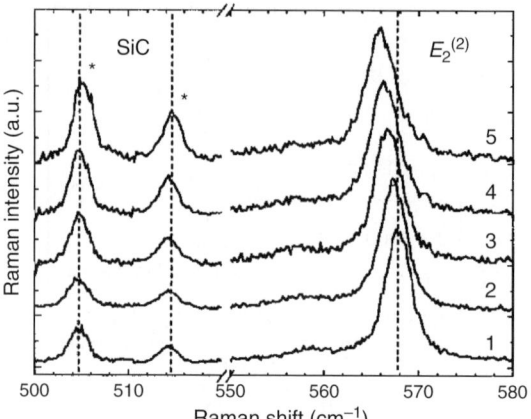

Fig. 14. Raman spectra near the E_2 mode for GaN layers grown on SiC with a range of $Ga_xAl_{1-x}N$ and AlN buffer layers producing a variety of strain states which result in the different positions of the E_2 peak (samples 1 and 2 had $Ga_xAl_{1-x}N$ buffer layers, samples 3 and 4 had AlN buffers, and sample 5 had no buffer layer). The Raman modes of the SiC substrate are also indicated (*) and provide an unshifted reference point (reproduced with permission from Ref. [66]. Copyright (1997) by the American Institute of Physics).

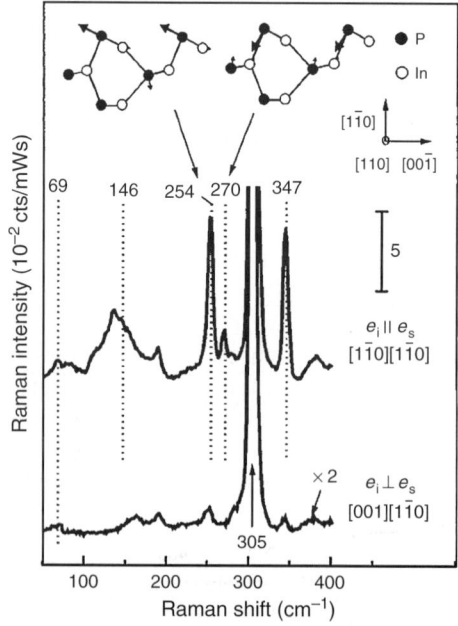

Fig. 15. Raman spectra of the clean InP (110) surface. The calculated atomic displacements for two surface modes are shown at the top of the figure (reproduced with permission from Ref. [67]. Copyright (1997) by the American Physical Society).

Turning now to buried interfaces between the heterostructure layers, recent work has shown that ab initio calculations of the vibrational states give a good insight into the displacements at internal hetero-interfaces [71]; in a heterostructure of type AB/CD (that is, with no common elements in alternate layers), this is particularly important as the interfaces, which can determine the symmetry reduction of the electronic states, offer the possibility for several different types of interface compound (e.g., AD, BC). There are many examples in the literature of the detection of relatively large-scale quantities of unwanted interface compounds during epitaxial growth, e.g., Ga_2Te_3 in the growth of ZnTe on GaSb [72], but it is particularly interesting that Raman spectroscopy under favourable conditions can be sensitive enough to monitor the interface state even in "high-quality" structures and to reveal a wealth of detail, showing, for example, clear differences between CdTe-type and BeSe-type interfaces for CdSe layers or dots embedded in BeTe [71,73] (Fig. 16).

Resonant Raman spectroscopy has likewise revealed significant inter-mixing at the interfaces of a single $GaN/Ga_xAl_{1-x}N$ quantum well, as shown in Fig. 17, where a shifting and broadening of the $A_1(LO)$ GaN phonon near $732\,cm^{-1}$ is observed for the quantum wells of narrowest nominal thickness; to avoid the Raman spectra of the barrier materials dominating the spectra, the use of excitation (3.54 eV) near resonance with the quantum well transition was essential [74]. As the quantum wells become thicker, one sees the emergence of the phonon mode characteristic of GaN (also shown in Fig. 3) and its counterpart for the $Ga_xAl_{1-x}N$ barrier, at higher energy (as expected from Fig. 8).

The resonant nature of this scattering process underlines the important point of Section 1.2; in an inhomogeneous system, even a small change of excitation photon energy is likely to select slightly different regions of the sample under study. Thus, other forms of optical spectroscopy (absorption, reflection, PL) must often be coupled with Raman spectroscopy to design a Raman experiment with the maximum sensitivity to the required part of a heterostructure.

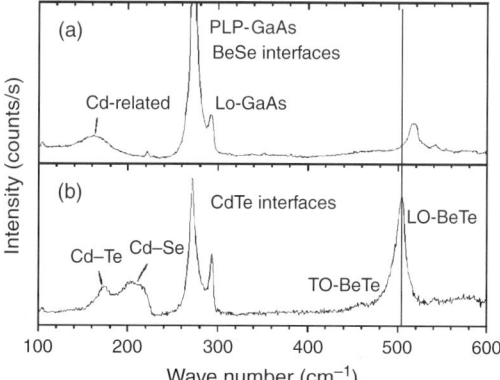

Fig. 16. Raman spectra of BeSe- and CdTe-type interfaces for CdSe layers embedded in BeTe barriers, providing evidence for interdiffusion in the BeSe case and relatively sharp interfaces in the second case (reproduced with permission from Ref. [71]. Copyright (2004) by the American Physical Society).

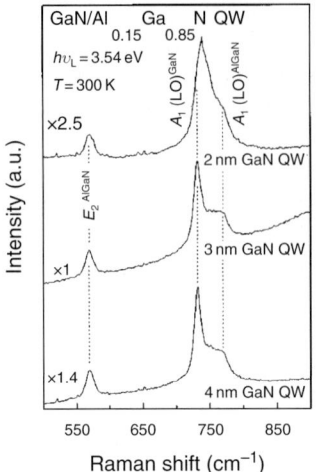

Fig. 17. Resonant Raman spectra of single GaN quantum wells at room temperature, showing a shift and broadening of the GaN A_1(LO) mode for the narrowest quantum well (reproduced with permission from Ref. [74]. Copyright (1997) by the American Institute of Physics).

3.3. Quantum wells and superlattices

In the previous section, we saw that the ability of Raman spectroscopy to investigate the composition and homogeneity of compound semiconductors can be useful as a diagnostic tool for semiconductor heterostructures and, particularly, the imperfections of their interfaces. However, even in "ideal" semiconductor heterostructures, new vibrational modes are introduced which are either intrinsic to superlattices or quantum wells or are made detectable by the symmetry-breaking they induce. These include "zone-folded" acoustic phonons [75], confined optic phonon modes [76] and interface modes [77].

We consider first the case of the acoustic phonons. In a superlattice, the longer-range periodicity d_{SL} along the superlattice normal gives rise to a reduced Brillouin zone width (π/d_{SL}) in that direction, and both acoustic and optic phonons far from $\mathbf{q}=0$ are mapped into this smaller Brillouin zone and become accessible via Raman scattering. This has been observed in many materials and a striking recent example is provided by the acoustic phonons of a superlattice of Ge dots in Si as displayed in Fig. 18; the inset shows the reduced Brillouin mini-zone (left) and the "unfolded" bulk-like dispersion (right). The momentum transfer in the scattering process, Eq. (6), is no longer small compared to the Brillouin zone width and in fact lies quite close to the new zone edge (π/d_{SL}), at the position indicated by the vertical dashed line in the inset of Fig. 18. The points on the unfolded dispersion curve are derived from the experimental data [78].

The dispersion of the superlattice acoustic modes along the superlattice axis may be calculated from the classical theory of elastic media [77,79]. Their dispersion typically shows only small gaps between successive branches at the new Brillouin zone boundary and centre (not visible in Fig. 18) and the size of these gaps depends on the mismatch in acoustic impedance of the two layers. Recently, time-resolved coherent Raman scattering has been used to probe these modes in the time domain, where beating effects between modes of the type shown in Fig. 18 can be seen (see Ref. [80] and references therein).

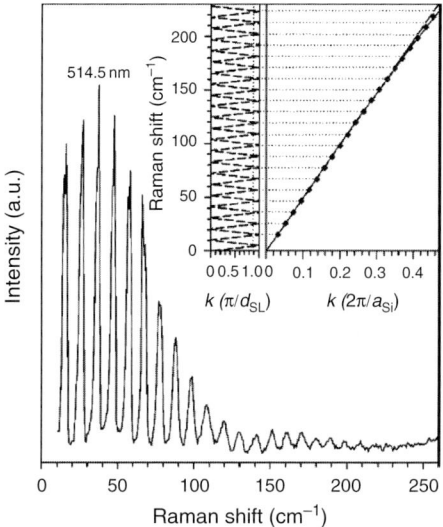

Fig. 18. Raman spectrum of zone-folded acoustic phonons in a modulation-doped Si/Ge quantum dot superlattice structure. The inset shows the calculated dispersion of LA phonons in the superlattice mini-zone and extended Brillouin zone schemes (reproduced with permission from Ref. [78]. Copyright (2004) by the American Institute of Physics).

In Section 3.4, we shall consider some examples of Raman scattering in which the sample is placed in a magnetic field. However, it is appropriate to mention here that resonant magneto-Raman scattering has been applied to study in detail the acoustic phonon Raman spectra and dispersion, particularly in GaAs/AlAs superlattices and single quantum wells [79]. The role of the magnetic field in this system is to quantise the electron and holes in Landau orbits, giving rise to strong variations in the Raman scattering intensity (for both optic and acoustic phonons). The interpretation of the resonance behaviour of the Raman spectra then gives detailed information on the electronic states, the electron–phonon interactions and, in Fig. 19, the acoustic phonon dispersion. This figure shows the broad "wing" of the laser line commonly observed in Raman spectra and sometimes dismissed as stray light leakage; in fact, under resonant conditions, this wing is strongly enhanced, pointing to its origin from disorder-induced RRS. Furthermore, as Fig. 19 shows, a complicated structure is superimposed on the smooth wing and this structure can be interpreted as arising from acoustic phonon scattering (of the type discussed above) but also more complicated features ("intensity anomalies") due to the band gaps in the acoustic phonon dispersion and also to crossing points of the LA and TA dispersions [79,81].

The behaviour of the optic phonons is often quite different from that of the acoustic phonons. As the optic phonon dispersion is relatively weak, there may be very little or no overlap in energy of the dispersion curves of the two layer materials. Under these circumstances, the optic phonons of each layer cannot propagate into the other layer; they thus form standing waves confined to a substantial extent in their respective layers (somewhat analogous to electrons confined in a semiconductor quantum well). This is well-established in the case of GaAs/AlAs superlattices, for example, where calculations of the atomic displacements confirm almost complete phonon confinement [82]. Because the thickness of the confining

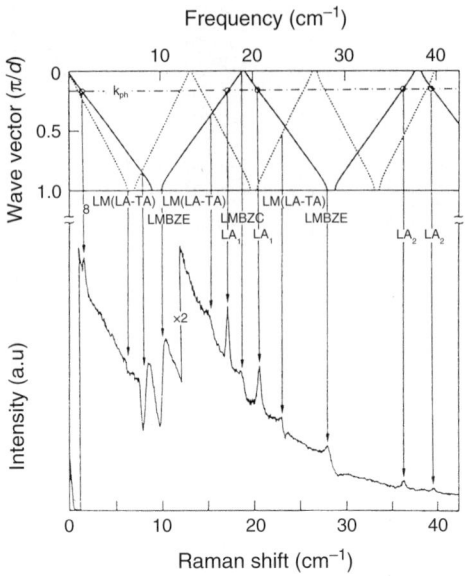

Fig. 19. Intensity anomalies and folded phonon doublets in the Raman spectrum of a [001]-oriented GaAs/AlAs multiple quantum well compared to the acoustic phonon dispersion (reproduced with permission from Ref. [81]. Copyright (1992) by the American Physical Society).

layer is generally known, one can then relate the energy and wavelength of the confined phonon standing wave to those of the unconfined phonon of the same wavelength, and thus one can make deductions about the form of the bulk dispersion curve [77]. Conversely, one can assume this knowledge and use the Raman data to infer the layer thicknesses. Even in cases where the optic phonon dispersions of the two layers overlap, confinement effects can occur, for example, in GaAs/Ga$_x$Al$_{1-x}$As [83], and this has been observed even in the unusual case of superlattices consisting of alternate Ge isotopes [84]. It has become conventional to index the confined LO phonons as LOn where the confined mode has $n-1$ nodes; an example is given in Section 3.5.

The above discussion assumes a backscattering experiment, as this is the most convenient experimental situation, with the superlattice normal aligned along the light propagation direction, (Fig. 3, top). However, there are various means to probe the dispersion of the optic modes of superlattices; for example, polishing a series of bevelled edges allows backscattering experiments with different amounts of momentum transfer in the superlattice plane [85]. The results also give insight into the three-dimensional dispersion of the superlattice optic phonons, which evolve from the LO and TO phonons of the constituent layers into the so-called interface modes with a characteristic dispersion which depends on the layer thicknesses and bulk phonon modes, and which gives four phonon energies between the two pairs of bulk LO and TO values, Fig. 20. Thus, angle-dependent studies may be necessary to identify, for example, whether a new mode in this spectral region arises from an unexpected composition or compound, or is an intrinsic interface mode of the intended heterostructure.

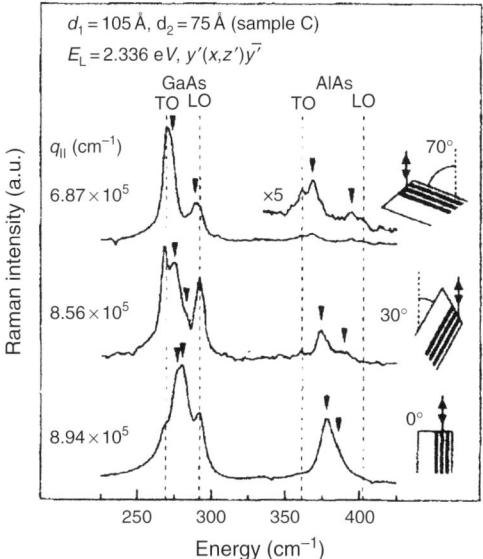

Fig. 20. Raman spectra of a GaAs/AlAs superlattice for different values of the wave vector component $q_\parallel$ in the plane of the superlattice. The arrows mark the interface modes and their shift with momentum transfer indicates their dispersive nature (reproduced with permission from Ref. [85]. Copyright (1992) by the American Physical Society).

We have already seen the importance of resonant excitation above (Figs 17 and 19). For quantum wells or superlattices, the excitation may be resonant with well or barrier states (discussed in Chapters 2 and 3) and this enables one to probe selectively the excitations of either. An early example of a resonance profile of the LO phonon Raman signal from a GaAs/Ga_xAl_{1-x}As heterostructure is shown in Fig. 21, where several peaks in the resonance profile can

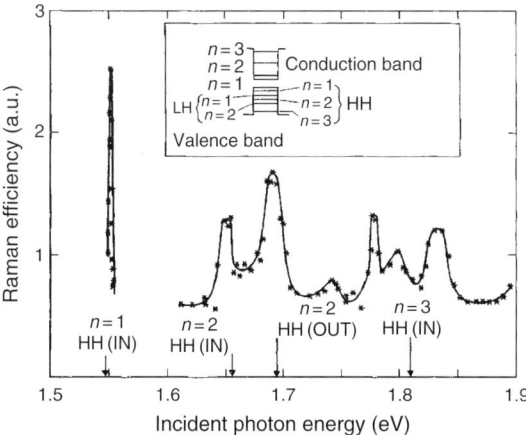

Fig. 21. Resonant Raman profile for a GaAs/Al_xGa_{1-x}As multiple quantum well structure; the arrows indicate exciton energies calculated using a single quantum well model (reproduced with permission from Ref. [86]. Copyright (1983) by the American Physical Society).

be identified with quantum-confined electronic transitions of the quantum well [86] or even the electronic states of the superlattice that lie above the confining barrier potential in energy and are therefore delocalised [87]. In some circumstances, the resonance profile of the Raman signal can be used to confirm the association between a particular layer and particular features in the electronic density of states (the latter being detected via PL or reflectivity measurements).

3.4. Spin effects

In view of the growing interest in the manipulation of spin in semiconductors, and therefore in techniques that can probe, for instance, the state of magnetic impurities such as Mn in semiconductor hosts, it is worth giving a brief view of some recent results in magneto-Raman scattering.

The simplest case is that of a conduction band electron (spin $S = 1/2$); under the action of a magnetic field B, the degeneracy between its two spin states is lifted and their separation in energy becomes $\Delta E = g_e \mu_B B$ (a Zeeman splitting). Here, the gyromagnetic ratio ("g-factor") g_e is modified from the free electron value by spin-orbit effects via the semiconductor band structure and is therefore an "effective" g-factor, in the same way that the effective mass is modified from the free electron mass. Raman scattering involving a transition of the electron between the two spin states (a "spin-flip") is possible and the Raman shift, typically now dependent on magnetic field, allows one to measure g_e. If the electron is confined in a quantum well, its g-factor is changed from the bulk semiconductor value, and becomes anisotropic. If the electron is bound to an impurity (for instance, a neutral donor) then its g-factor is sometimes observed to be modified by an amount which depends on the donor species and energy separation from the conduction band edge. If the electronic state in question is localised on a *magnetic* impurity, then the g-factor may be determined mainly by the nature of the impurity (the commonest example being the $3d^5$ state of Mn^{2+}, for which $g_e = 2.00$ independent of the host semiconductor). Finally, if the semiconductor shows strong bulk magnetic effects (for instance, due to the paramagnetism of the Mn ions, or carrier-induced ferromagnetism), then spin-flip Raman scattering by the charge carriers gives a measure of those effects. In all these cases, the resonant intermediate state in the Raman scattering process is found still to be an excitonic state of the host semiconductor. This implies that Raman scattering can probe selectively the spin parameters of a centre in a given layer of a quantum heterostructure.

Similar remarks apply to holes though the $J = 3/2$ valence band of zinc blende semiconductors, which is fourfold degenerate at $\mathbf{k} = 0$, may show a lifting of its degeneracy even in zero magnetic field if there is any lowering of symmetry (for instance, the splitting of light and heavy holes due to strain). In general, one cannot therefore speak of a simple g-factor, but must solve a spin Hamiltonian H, often of the following form:

$$H = 2\kappa\mu_B \mathbf{J} \cdot \mathbf{B} + \Delta_c \left(J^2 - \frac{1}{3} J(J+1) \right) \quad (15)$$

in which Δ_c is a zero-field splitting, κ is a parameter analogous to g_e and $\mathbf{J}$ is the spinor representing the valence band spin states; the energies are obtained as the eigenvalues of, here, a 4×4 matrix for the basis $m_J = (3/2, 1/2, -1/2, -3/2)$. This procedure is analogous to that used to model electron spin resonance spectra [88] (see also Chapter 13).

As an example of impurity-bound carriers (in a non-magnetic semiconductor), we consider acceptor-bound holes in multiple quantum well structures, for which the energies of the lines in the Raman spectrum are plotted as a function of applied magnetic field in Fig. 22 [89]. The acceptor fine structure arising from the magnetic field and quantum confinement effects (which appear here in a way that is formally similar to a strain splitting) is indicated in the inset and is described by Eq. (15). The Raman splittings correspond to the energy separations between pairs of the levels shown in the inset, thus giving a measurement of both the zero-field splitting Δ_c and the constant κ. Figure 22 furthermore shows that the magnitude of Δ_c differs for acceptors associated with different quantum wells.

The use of a microwave field in addition to the laser field allows one to induce coherent transitions between the carrier spin states [90,91]. The magnetic field is scanned and, when the microwave quantum energy matches the Zeeman splitting, a degree of coherence develops between all three levels involved in the scattering process, Fig. 23(b). The Raman-scattered light then forms a beam (similar to the situation discussed in Section 2.4) co-propagating with the reflected laser light. Given a fast enough photodiode, heterodyne detection may be used to detect the presence of the scattered beam as its beating with the reflected beam regenerates a microwave signal which may itself be detected by mixing with a signal from the original microwave source.

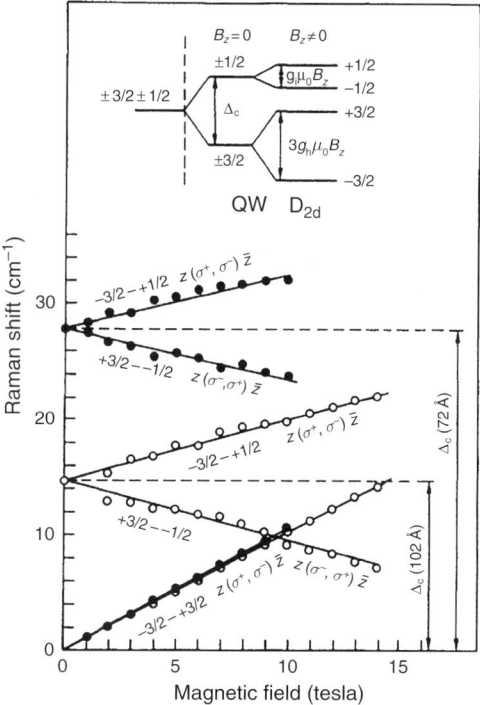

Fig. 22. Spin-flip Raman scattering of acceptor-bound holes in two GaAs/Al$_x$Ga$_{1-x}$As multiple quantum wells for light propagation and a magnetic field B_z normal to the layer plane, showing the combined effects of magnetic field and "crystal field" on the acceptor level structure (which is displayed in the inset for the cases of zero and finite magnetic field) (reproduced with permission from Ref. [89]. Copyright (1994) by the American Physical Society).

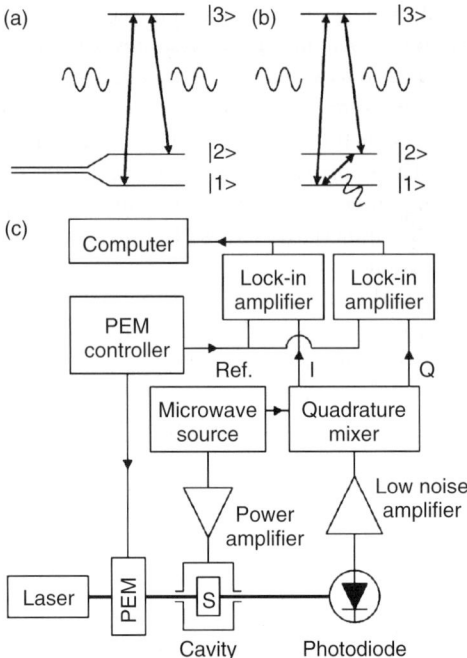

Fig. 23. Experimental set-up for coherent Raman-detected electron spin resonance. S represents the sample at 1.5 K in a microwave cavity within a superconducting magnet (reproduced with permission from Ref. [91]. Copyright (2001) by the American Physical Society).

The final output exists only when the spin resonance condition is satisfied and gives information analogous to electron spin resonance (with absorption- and dispersion-like components of the spin resonance signal), but with the added feature that the Raman process is, as before, resonant with an intermediate excitonic state, giving selectivity to specific layers and types of centre as well as high sensitivity. The results of Fig. 24 show the a signal at a field

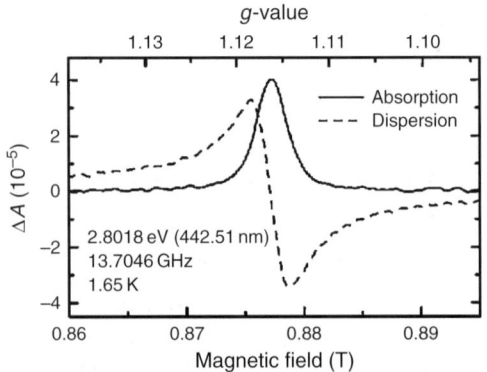

Fig. 24. Coherent Raman-detected electron spin resonance of donor-bound electrons in nominally undoped epitaxial ZnSe, excited in resonance with the donor-bound exciton (reproduced with permission from Ref. [91]. Copyright (2001) by the American Physical Society).

corresponding to the g-factor of conduction band electrons in ZnSe; the high resolution compared to that of spectrometer-based Raman scattering led to the demonstration of a monotonic dependence of the g-factor on excitation energy which no other technique could reveal [91].

This technique has recently also been applied to the cases of donor-bound electrons in single ZnSe quantum wells and Mn^{2+} ions in CdTe single quantum wells; in each case, it was confirmed that selective excitation of Raman-detected spin resonance in an individual quantum well was possible.

3.5. Quantum dots and wires

The sensitivity of Raman scattering can be enough to detect phonon Raman scattering from only a few MLs of material, so that the structural properties of QDs can be probed. As an example, we show results for the growth of GaSb on GaAs [92]. For deposition of an average coverage of 1.8 MLs of material, only the Raman signals of the GaAs substrate were observed, Fig. 25. Atomic force microscopy (AFM) also shows no dot-like features on the surface. However, when 3 ML of GaSb were deposited, AFM reveals structures with a surface density of $\sim 10^{10}\,cm^{-2}$ and dimensions $\sim 30 \times 280$ Å. The Raman spectrum (lower trace in Fig. 25) shows a signal at $234\,cm^{-1}$, between the LO and TO phonon frequencies of GaSb (238 and $228\,cm^{-1}$, respectively). Thus, it is inferred that this signal arises from dots of binary GaSb composition. By contrast, the deposition of AlSb gave rise to dots whose composition was shown to be an ternary alloy $Al_xGa_{1-x}Sb$. Thus Raman spectroscopy can give insight into quantum confinement effects in dots, as it can be difficult otherwise to distinguish quantum confinement shifts of electronic transition energies from shifts that are merely due to alloying.

The effects of localisation on the phonon modes themselves is also potentially interesting; we have seen already that optic phonon confinement in a superlattice can give rise to the appearance in the Raman spectra of modes with $\mathbf{q} \neq 0$ that are not normally detected, and this is the case also for dots, with the difference that this effect is now dependent on the dot size and shape rather than a single simple parameter. A distribution of dot sizes generally gives rise to a Raman spectrum containing a distribution of phonon energies around the optic phonon energy, which may be probed selectively via resonant Raman.

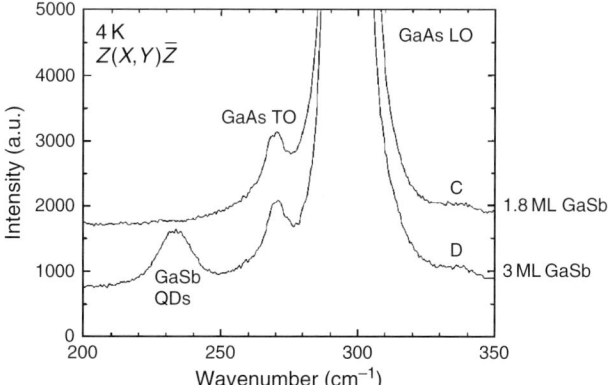

Fig. 25. Raman spectra (at 4 K) of (C) 1.8 ML and (D) 3 ML of GaSb deposited on GaAs (reproduced with permission from Ref. [92]. Copyright (1996) by the American Institute of Physics).

For instance, we take as a very clear example the case of CdSe dots embedded in a ZnSe matrix; this is a case where high-resolution transmission electron microscopy (TEM) has shown that a substantial degree of intermixing can occur [93], and the results of Raman spectroscopy (by contrast, a simpler and non-destructive technique) are in agreement; Fig. 26 shows that, as the excitation photon energy is changed to excite resonantly dot regions with different electronic transition energies, the phonon Raman peak evolves smoothly from that of the ZnSe barrier to that of a $Cd_xZn_{1-x}Se$ alloy [73].

For quantum wires, a principal question is the anisotropy of their properties in the radial and axial directions; polarised micro-Raman spectroscopy has been used to investigate this in several cases; clearly, it is necessary to be able to detect signals from a single wire for such a study. Figure 27 shows the Raman spectra with polarisations parallel and perpendicular to a single GaN quantum wire. Comparison with TEM studies showed that Raman spectra could distinguish between wires with principal axes corresponding to either the crystallographic a- or c-axis.

The picture is thus emerging that Raman spectroscopy in the case of nanostructured samples is certainly a powerful technique, but is most useful when combined with other forms of optical spectroscopy to understand the electronic transitions that determine the resonant effects and combined with direct structural probes to determine size, shape and even composition independently. With sufficient experience of a given material system, Raman spectroscopy can then provide a reliable diagnostic for, e.g., QD growth.

The ultimate detection limit in the semiconductor context is a single dot or scattering centre. Several single dot measurements have been achieved and we show in Fig. 28(E) an example of the resonant Raman spectra of a GaAs-based QD obtained using an aperture-based near-field technique as discussed in Section 2.3 and compared to the ensemble spectrum obtained in a conventional backscattering experiment, Gammon. For a single dot, one can see confined optic phonon modes indexed LO_n as discussed earlier (Section 3.3).

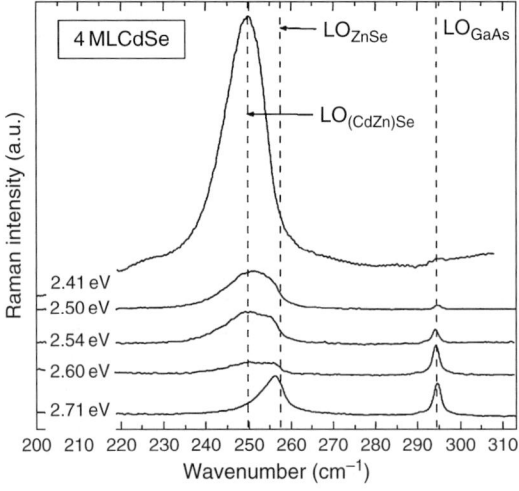

Fig. 26. Raman spectra of $Cd_xZn_{1-x}Se$ quantum dots, embedded in ZnSe for a series of laser lines with photon energies between 2.41 eV and 2.71 eV. Nominal CdSe thickness: 4 ML. Sample temperature = 77 K (reproduced with permission from Ref. [73]. Copyright (2005) by Wiley).

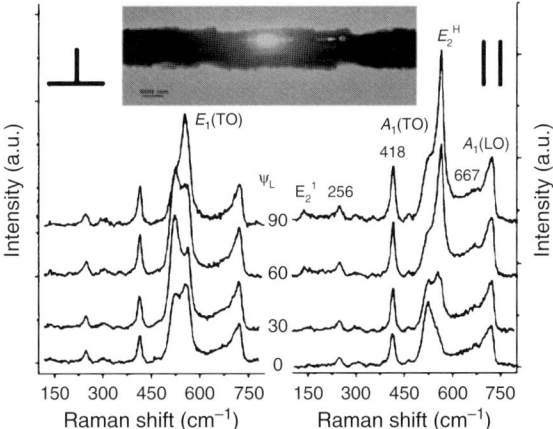

Fig. 27. Raman spectra of a single GaN nanowire of diameter 500 nm (see inset) whose axis was shown by transmission electron microscopy to be oriented along the wurtzite c-axis. The detection polarisation was either parallel ($\parallel$) or perpendicular ($\perp$) to the excitation polarisation and the wires were rotated with respect to the excitation polarisation by angle ψ_L (reproduced with permission from Ref. [94]. Copyright (2006) by the American Physical Society).

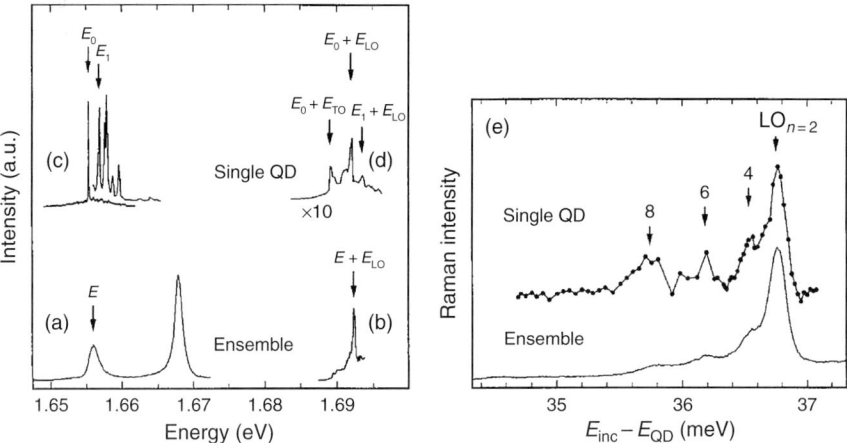

Fig. 28. GaAs/AlAs emission spectra demonstrating single quantum dot Raman spectroscopy. (a) An ensemble PL measurement through a 25-μm aperture. (b) Ensemble PL excitation spectrum (25 μm). (c) and (d) PL and PL-excitation spectra, respectively, obtained through a 0.8-μm aperture. The PL spectrum shows a single-QD PL line at E_0. The excitation spectrum shows sharp electronic resonances within the first 10 meV and Raman resonances starting about 34 meV above the luminescing state. (e) The top spectrum is a high-resolution LO–phonon spectrum from a QD, obtained from the intensity of the QD emission as a function of the energy difference between the QD (E_{QD}) and the laser (E_{exc}) as the laser is scanned. The bottom (ensemble) spectrum is the resonant Raman spectrum obtained in the conventional way from the 25-μm aperture (reproduced with permission from Ref. [95]. Copyright (1997) by the American Association for the Advancement of Science).

The correspondence of the LO_n phonon Raman spectra between single dot and ensemble spectrum points to the fact that the dominant phonon confinement effect arises from confinement in the growth direction only, the dots being more extended laterally than vertically.

3.6. Nanostructures: nanotubes and non-epitaxial nanocrystals

Although this review is focussed on conventional semiconductor heterostructures, Raman spectroscopy is widely applied in the characterisation of many other types of nanostructured materials. For instance, a great deal of work has been carried out on free-standing nanoparticles based on, for example, CdSe [96], PbS [97], ZnO [98] or Si [99], and semiconductor QDs embedded in glassy matrices [100]. The strong phonon confinement effects and new surface or interface geometries present challenges in terms of the theoretical modelling of the vibrational modes of small clusters or nanoparticles [101]. Sometimes relatively simple interpretations of the Raman spectra are possible, as in the case of mesoporous silica structures with pores partially filled with binary semiconductors, where Raman spectra indicated clearly the LO phonon modes of the semiconductor [102]. Elsewhere, the interpretation of Raman spectra has needed detailed first-principles analysis, as in the case of one-dimensional quantum-confined wires of –Ti–O–Ti– formed within a microporous silicate matrix (the material termed "ETS-10"). Here, the signature of the –Ti–O–Ti– chains is a mode (at $725\,cm^{-1}$) which arises from –Ti–O–Ti– bond stretching [103].

As well as semiconductor nanostructures based on inorganic matrices, the nanoscale combination of semiconductors with organic templates or matrices has also yielded promising results. One example is provided by organic–inorganic superlattices, where extremely high barriers for electron confinement lead to very large electron–hole exchange contributions to the exciton fine structure, probed by magneto-Raman spectroscopy as in Section 3.4 [104]. Another is at the interesting limit of very small semiconductor-like clusters (e.g., Zn_4O_{13}) coordinated by organic groups. In that case, optical spectroscopy suggested the transfer of optical excitations between the organic and inorganic parts of the clusters, whilst Raman spectroscopy was able to identify at least one vibrational mode characteristic of the ZnO cluster [105].

One particularly well-defined nanosystem for which Raman spectroscopy has rapidly become an essential characterisation tool is that of carbon nanotubes [106,107]. The vibrational modes of nanotubes are amenable to calculation [108] and Raman spectra may be obtained from single nanotubes with good SNRs. One of the most important vibrational modes involves the symmetrical radial expansion and contraction of the nanotube, the radial breathing mode (RBM). The frequency of this mode is indicative of the tube structure, and can be used to determine the integer values n, m that specify the diameter and chirality of the nanotube [106]. Figure 29, from Hartschuh [109], shows the RBM Raman signals of single nanotubes with single walls and, furthermore, shows how the frequency of the vibrational mode correlates with the energy of the fluorescence emitted from the same region.

3.7. Devices and thermal properties

Because phonon frequencies are functions of temperature (due to anharmonicity), mapping of a given phonon frequency over the surface of an operating semiconductor device may give an indication of the temperature distribution over the device [110]. This is of crucial importance in modelling device performance and lifetime. Likewise, the strain (discussed in Section 3.1)

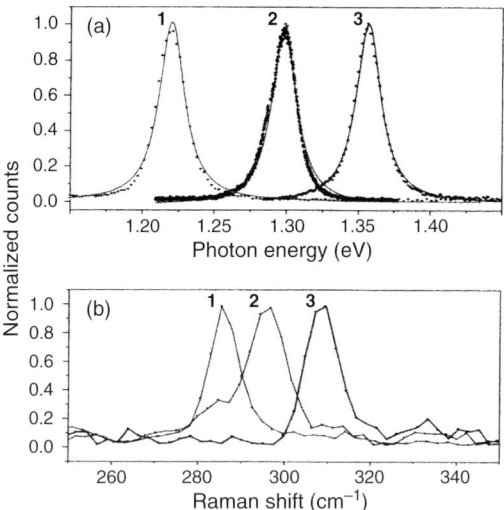

Fig. 29. Fluorescence and Raman spectra of single single-walled carbon nanotubes. (a) Fluorescence spectra were detected for three different sample positions, labelled 1–3. Solid lines are fits with single Lorentzian line-shape functions. (b) Raman spectra obtained from the positions described in (a). The Raman shift corresponds to scattering from the nanotube RBM (reproduced with permission from Ref. [109]. Copyright (2003) by the American Association for the Advancement of Science).

may be mapped to estimate the deformations generated in a layer by its neighbours [111]. The advantages of Raman microscopy in these contexts are that the method is non-destructive, it can be performed on a device under a range of operating conditions and, for temperature mapping, it has a superior diffraction-limited spatial resolution (typically around 1 μm) to infrared thermal imaging. Here, we shall give just one example of each application of Raman microscopy.

One of the most famous applications of Raman microscopy for strain mapping is in the area of silicon devices; for example, the very successful one-dimensional strain mapping of Si underneath a Si–N stripe has attracted a great deal of attention, with manufacturers of commercial Raman spectroscopy systems using this as a benchmark demonstrator for the capabilities of their systems. The experimental results need careful interpretation and must be simulated via finite element methods for a proper understanding [112,113] (Fig. 30).

Temperature mapping has been applied to many types of device, including III–nitride-based heterostructure field effect transistors (HFET) [114–116] and $In_xGa_{1-x}As/Al_xGa_{1-x}As$ semiconductor lasers (giving the ability to determine the facet temperatures, which must be minimised to maximise device lifetimes). In several of these applications, an attractive feature of Raman microscopy is that one can choose the excitation wavelength to control the penetration of the excitation into the sample, thus obtaining a high-resolution surface-specific temperature probe. For instance, UV excitation was used for the nitride and green excitation for the arsenide semiconductors to obtain data over a depth of ∼100 nm from the device surface. In Fig. 31, we show an example of thermal mapping along a line crossing the gate structure of a multi-finger $Al_xGa_{1-x}N/GaN$ HFET. As with the example of strain mapping, a detailed numerical simulation of the data is useful to extract the maximum information.

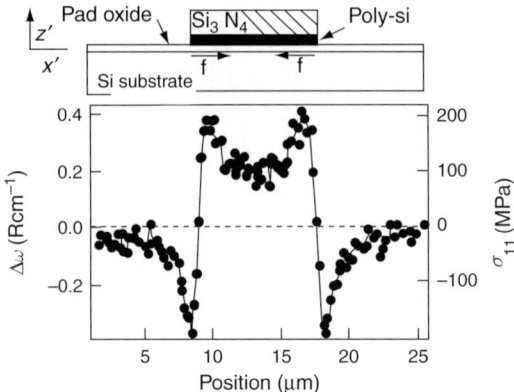

Fig. 30. Raman shift $\Delta\omega$ and corresponding stress σ in silicon near and beneath a 9.4-mm-wide Si_3N_4/polycrystalline Si line. A schematic drawing of the sample is shown at the top of the figure (reproduced with permission from Ref. [112]. Copyright (1996) by the American Institute of Physics).

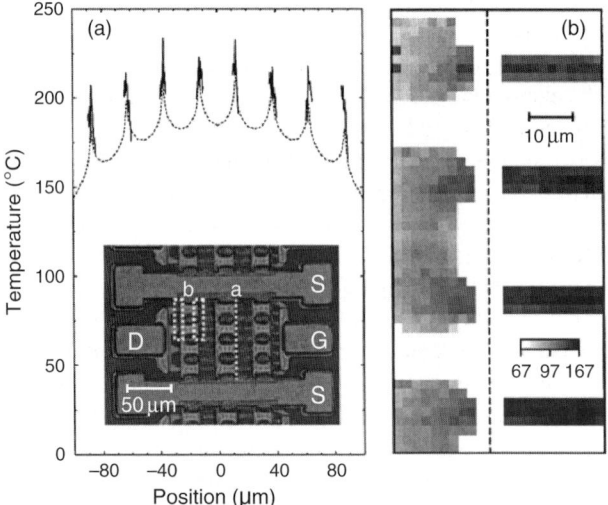

Fig. 31. (a) Temperature line scan recorded along a line perpendicular to the gate fingers near the central airbridge, (b) temperature map in the vicinity of the drain feed structure of a multifinger $Al_xGa_{1-x}N$/GaN heterostructure field effect transistor (HFET). The dotted line in (a) shows simulation results. The inset in (a) shows a photograph of the multifinger $Al_xGa_{1-x}N$/GaN HFET with the area of recording of (a) and (b) indicated by dotted lines (reproduced with permission from Ref. [115]. Copyright (2002) by the American Institute of Physics).

4. Summary and future perspectives

This chapter has aimed to give an overview of the types of scientific problems in semiconductor characterisation to which Raman spectroscopy can make a useful contribution. The examples provided represent only a tiny fraction of the "case studies" that one could find and this is to be expected of such a versatile and well-established technique. This review has drawn on the recent literature, so some famous examples are omitted which are easily found in earlier reviews.

Firstly, as CCD detectors become cheaper, more efficient and less noisy, and solid-state lasers steadily replace gas lasers, it seems likely that Raman spectroscopy of the conventional type (backscattering with fixed excitation energy, far-field sampling), which is perfectly adequate for much materials characterisation, will become progressively cheaper and more widely used as a process monitoring tool.

Secondly, it is clear that a great deal of attention recently has focussed on nanostructures, and the "single quantum dot" and "single nanotube" limits have now been reached for some materials systems. This activity is likely to continue to be important whilst the development and optimisation of preparation techniques for new types of nanoparticle remains an active area. At the same time, one can expect Raman spectroscopy to be applied more widely to semiconductor heterostructures and semiconductor nanoparticles in their "working environments", which span the vast range from traditional planar device structures to the free-standing CdSe nanoparticles used, for example, as in vivo fluorophores.

It also appears that much progress still remains to be made in the combination of plasmonics (that is, the control of light fields via their interaction with the plasmon modes of *metal* nanostructures) with Raman spectroscopy for the near-field sampling required to study of *semiconductor* nanostructures. This has been hinted at above with the use of metal apertures for near-field Raman microscopy, though there it is assumed that the metal mask plays a passive role in defining the illuminated area. There has been promising work done on the local enhancement of the excitation electric field via the use of a sharp metal tip (leading to TERS [117]) provoked by the widespread application of the SERS that typically uses colloidal gold or silver particles to increase Raman signals for materials in aqueous solution. The manipulation of the light field via the techniques of plasmonics is probably the key to the maximisation of the usefulness of Raman spectroscopy in the growing area of nanotechnology.

Finally, we note that several open-source packages for performing ab initio local density functional calculations are available (see also Chapter 2 in Ref. [40]) and, as we have seen above, many research groups working in Raman spectroscopy now regard the detailed modelling of vibrational states at the atomic level as a routine procedure. Packages such as PWSCF [118] allow one to calculate vibrational densities of states and non-RRS intensities [41] and it is likely that the demand from "experimentalists" for such tools will continue to grow.

Acknowledgements

Support from the EPSRC, BBSRC, EU-INTAS 03-51-5266 and the Royal Society is gratefully acknowledged. I am very grateful to Professor J. John Davies for his encouragement and for his helpful comments on this review.

References

[1] L. Brillouin, Annales des Physique, 17 (1922) 88.
[2] A. Smekal, Naturwissenschaften, 11 (1923) 873.
[3] C. Raman, Indian J. Phys., 2 (1928) 37.
[4] C. Raman and K. Krishnan, Nature, 121 (1928) 501.
[5] G. Landsberg and L. Mandelstam, Naturwissenschaften, 16 (1928) 557.

[6] A.S. Pine, *Light Scattering in Solids I*, M. Cardona, Ed., Springer-Verlag, Berlin, 1983, Vol. 8.
[7] C. Kittel, *Introduction to Solid State Physics*, John Wiley and Sons, Inc., New York, 1996.
[8] M. Cardona and G. Guentherodt, *Topics in Applied Physics*, Springer-Verlag, Berlin, 1983–2007, Vols 8, 50, 51, 54, 66, 68, 75, 76, 108.
[9] C. Hamaguchi, *Basic Semiconductor Physics*, Springer-Verlag, Berlin, 2001.
[10] J.L. Birman, *Theory of Crystal Space Groups and Lattice Dynamics*, Springer-Verlag, Berlin, 1984.
[11] G. Burns and A.M. Glazer, *Space Groups for Solid State Scientists*, Academic Press, New York, 1978.
[12] R. Loudon, Proc. R. Soc. A, 275 (1963) 218.
[13] R. Loudon, Adv. Phys., 50 (2001) 813.
[14] R.M. Martin and L.M. Falicov *Light Scattering in Solids 1*, M. Cardona, Ed., Springer-Verlag, Berlin, 1983, Vol. 8.
[15] P.Y. Yu and M. Cardona, *Fundamentals of Semiconductors*, Springer-Verlag, Berlin, 1996.
[16] J.P. Elliott and P.G. Dawber, *Symmetry in Physics*, Macmilllan, London, 1979.
[17] G.F. Koster, *Solid State Physics*, Academic Press Inc., New York, 1957.
[18] L.P. Bouckaert, R. Smoluchowski and E. Wigner, Phys. Rev., 50 (1936) 58.
[19] K. Yee, K. Lee, E. Oh et al., Phys. Rev. Lett., 88 (2002) 105501.
[20] H. Harima, J. Phys. Condens. Matter, 14 (2002) R967.
[21] D.J. Gardiner and P.R. Graves, *Practical Raman Spectroscopy*, Springer-Verlag, Berlin, 1989.
[22] I. De Wolf, J. Jimenez, J. Landesman et al., *Catalogue of optical and physical parameters*, 'Nostradamus' project SMT4-CT-95–2024, EUR, 18595 (1998).
[23] H. Kuzmany, *Solid-State Spectroscopy*, Springer-Verlag, Berlin, 1998.
[24] B. Schrader and D. Moore, Pure Appl. Chem., 69 (1997) 1451.
[25] J. Sweedler, K. Ratzlaff and M. Denton, *Charge-transfer Devices in Spectroscopy*, VCH Publishers, Inc., New York, 1994.
[26] C. Poweleit, A. Gunther, S. Goodnick and J. Menendez, Appl. Phys. Lett., 73 (2006) 2275.
[27] K. Kneipp, Y. Wang, H. Kneipp et al., Phys. Rev. Lett., 78 (1997) 1667.
[28] D. Gammon, E.S. Snow, B.V. Shanabrook et al., Phys. Rev. Lett., 76 (1996) 3005.
[29] E. Betzig, P. Finn and J. Weiner, Appl. Phys. Lett., 60 (1992) 2484.
[30] J.K. Trautman, J.J. Macklin, L.E. Brus and E. Betzig, Nature, 369 (1994) 40.
[31] C.L. Jahncke, Appl. Phys. Lett., 67 (1995) 2483.
[32] M. Futamata and A. Bruckbauer, Chem. Phys. Lett., 341 (2001) 425.
[33] J. Menendez and M. Cardona, Phys. Rev. B, 29 (1984) 2051.
[34] A. Debernardi, S. Baroni and E. Molinari, Phys. Rev. Lett., 75 (1995) 1819.
[35] A. Laubereau, D. von der Linde and W. Kaiser, Phys. Rev. Lett., 27 (1971) 802.
[36] R.R. Alfano, *Semiconductors Probed by Ultrafast Laser Spectroscopy*, Academic Press, Orlando, 1984.
[37] M. Cardona, Phys. Status Solidi B, 241 (2004) 3128.
[38] M. Cardona, High Pressure Res., 24 (2004) 17.
[39] M.P. Halsall, P. Harmer, P.J. Parbrook and S.J. Henley, Phys. Rev. B, 69 (2004) 235207.
[40] M. Peressi, A. Baldereschi and S. Baroni, *Characterization of Semiconductor Heterostructures and Nanostructures*, Chapter 2, C. Lamberti, Ed., Elsevier, Amsterdam, 2008. p. 19.
[41] S. Baroni, S. de Gironcoli, A. Dal Corso and P. Giannozzi, Rev. Mod. Phys., 73 (2001) 515.
[42] F. Decremps, J. Pellicer-Porres, A.M. Saitta et al., Phys. Rev. B, 65 (2002) 092101.
[43] S.H. Tolbert and A.P. Alivisatos, Ann. Rev. Phys. Chem., 46 (1995) 595.
[44] B. Rockwell, H.R. Chandrasekhar, M. Chandrasekhar et al., Phys. Rev. B, 44 (1991) 11307.
[45] W. Limmer, W. Ritter, R. Sauer et al., Appl. Phys. Lett., 72 (2006) 2589.
[46] M. Zajac, R. Doradzinski, J. Gosk et al., Appl. Phys. Lett., 78 (2001) 1276.
[47] O. Pages, A. Chafi, D. Fristot and A.V. Postnikov, Phys. Rev. B, 73 (2006) 165206.

[48] C. Ferrari and C. Bocchi, *Characterization of Semiconductor Heterostructures and Nanostructures*, Chapter 4, C. Lamberti, Ed., Elsevier, Amsterdam, 2008, p. 101.
[49] I.F. Chang and S.S. Mitra, Adv. Phys., 20 (1971) 359.
[50] H. Grille, C. Schnittler and F. Bechstedt, Phys. Rev. B, 61 (2000) 6091.
[51] V. Lemos, E. Silveira, J.R. Leite et al., Phys. Rev. Lett., 84 (2000) 3666.
[52] H. Siegle, L. Eckey, A. Hoffmann et al., Solid State Commun., 96 (1995) 943.
[53] L.Y. Huang, C.H. Chen, Y.F. Chen et al., Phys. Rev. B, 66 (2002).
[54] L.M.R. Scolfaro, Phys. Status Solidi A, 190 (2002) 15.
[55] T. Geppert, J. Wagner, K. Kohler et al., Appl. Phys. Lett., 80 (2002) 2081.
[56] A.A. Mascarenhas and M.J. Seong, Semicond. Sci. Technol., 17 (2002) 823.
[57] H. Harima, T. Inoue, S. Nakashima et al., Appl. Phys. Lett., 75 (1999) 1383.
[58] E.V. Lavrov, J. Weber, L. Huang and B.B. Nielsen, Phys. Rev. B, 64 (2001) 035204.
[59] J. Wagner, R.C. Newman, B.R. Davidson et al., Phys. Rev. Lett., 78 (1997) 74.
[60] B.R. Davidson, R.C. Newman, C.D. Latham et al., Phys. Rev. B, 60 (1999) 5447.
[61] P. Blanconnier, J.F. Hogrel, A.M. Jean-Louis and B. Sermage, J. Appl. Phys., 52 (1981) 6895.
[62] W. Limmer, M. Glunk, S. Mascheck et al., Phys. Rev. B, 66 (2002).
[63] M.J. Seong, S.H. Chun, H.M. Cheong et al., Phys. Rev. B, 66 (2002) 033202.
[64] T. Dietl, H. Ohno, F. Matsukura et al., Science, 287 (2000) 1019.
[65] K.C. Ku, S.J. Potashnik, R.F. Wang et al., Appl. Phys. Lett., 82 (2003) 2302.
[66] V. Davydov, N. Averkiev, I. Goncharuk et al., J. Appl. Phys., 82 (1997) 5097.
[67] K. Hinrichs, A. Schierhorn, P. Haier et al., Phys. Rev. Lett., 79 (1997) 1094.
[68] M. Hunermann, J. Geurts and W. Richter, Phys. Rev. Lett., 66 (1991) 640.
[69] V. Wagner, D. Drews, N. Esser et al., J. Appl. Phys., 75 (2006) 7330.
[70] V. Wagner, J. Wagner, S. Gundel et al., Phys. Rev. Lett., 89 (2002).
[71] T. Muck, J.W. Wagner, L. Hansen et al., Phys. Rev. B, 69 (2004) 245314.
[72] M.P. Halsall, D. Wolverson, J.J. Davies et al., Appl. Phys. Lett., 60 (1992) 2129.
[73] J. Geurts and V. Wagner, Phys. Status Solidi B, 242 (2005) 2644.
[74] D. Behr, R. Niebuhr, J. Wagner et al., Appl. Phys. Lett., 70 (1997) 363.
[75] C. Colvard, R. Merlin, M. Klein and A. Gossard, Phys. Rev. Lett., 45 (1980) 298.
[76] R. Enderlein, D. Suisky and J. Roseler, Phys. Status Solidi B, 165 (1991) 9.
[77] B. Jusserand and M. Cardona, *Light Scattering in Solids V*, M. Cardona and G. Guentherodt, Eds, Springer-Verlag, Berlin, 1989, Vol. 32.
[78] P. Tan, D. Bougeard, G. Abstreiter and K. Brunner, Appl. Phys. Lett., 84 (2004) 2632.
[79] T. Ruf, *Phonon Raman scattering in Semiconductors, Quantum Wells and Superlattices*, Springer-Verlag, Berlin, 1998.
[80] A. Bartels, T. Dekorsy, H. Kurz and K. Köhler, Phys. Rev. Lett., 82 (1999) 1044.
[81] V.F. Sapega, V.I. Belitsky, T. Ruf et al., Phys. Rev. B, 46 (1992) 16005.
[82] H. Ruecker, E. Molinari and P. Lugli, Phys. Rev. B, 45 (1992) 6747.
[83] A. Fasolino, E. Molinari and J. Maan, Phys. Rev. B, 39 (1989) 3923.
[84] J. Spitzer, T. Ruf, M. Cardona et al., Phys. Rev. Lett., 72 (1994) 1565.
[85] R. Hessmer, A. Huber, T. Egeler et al., Phys. Rev. B, 46 (1992) 4071.
[86] J. Zucker, A. Pinczuk, D. Chemla et al., Phys. Rev. Lett., 51 (1983) 1293.
[87] J. Zucker, A. Pinczuk, D. Chemla et al., Phys. Rev. B, 29 (1984) 7065.
[88] A. Abragam and B. Bleaney, *Electron Paramagnetic Resonance of Transition Metal Ions*, Dover, New York, 1970.
[89] V.F. Sapega, T. Ruf, M. Cardona et al., Phys. Rev. B, 50 (1994) 2510.
[90] R. Romestain, S. Geschwind, G.E. Devlin and P.A. Wolff, Phys. Rev. Lett., 33 (1974) 10.
[91] S.J. Bingham, J.J. Davies and D. Wolverson, Phys. Rev. B, 65 (2002) 155301.
[92] B. Bennett, B.V. Shanabrook and R. Magno, Appl. Phys. Lett., 68 (1996) 958.
[93] M. Strassburg, V. Kutzer, U.W. Pohl et al., Appl. Phys. Lett., 72 (1998) 942.
[94] T. Livneh, J. Zhang, G. Cheng and M. Moskovits, Phys. Rev. B, 74 (2006) 035320.

[95] D. Gammon, S.W. Brown, E.S. Snow et al., Science, 277 (1997) 85.
[96] L. Brus, Appl. Phys. A, 53 (1991) 465.
[97] K.K. Nanda, S.N. Sahu, R.K. Soni and S. Tripathy, Phys. Rev. B, 58 (1998) 15405.
[98] M. Rajalakshmi, A.K. Arora, B.S. Bendre and S. Mahamuni, J. Appl. Phys., 87 (2000) 2445.
[99] J. Zi, H. Buscher, C. Falter et al., Appl. Phys. Lett., 69 (1996) 200.
[100] L. Saviot, B. Champagnon, E. Duval and A.I. Ekimov, Phys. Rev. B, 57 (1998) 341.
[101] G. Faraci, S. Gibilisco, P. Russo et al., Phys. Rev. B, 73 (2006) 033307.
[102] F.J. Brieler, M. Fröba, L. Chen et al., Chem. Eur. J., 8 (2002) 185.
[103] A. Damin, F.X. Llabres i Xamena, C. Lamberti et al., J. Phys. Chem. B, 108 (2004) 1328.
[104] K. Ema, K. Umeda, M. Toda et al., Phys. Rev. B, 73 (2006).
[105] S. Bordiga, C. Lamberti, G. Ricchiardi et al., Chem. Commun. (2004) 2300.
[106] A. Jorio, R. Saito, J.H. Hafner et al., Phys. Rev. Lett., 86 (2001) 1118.
[107] A.M. Rao, A. Jorio, M.A. Pimenta et al., Phys. Rev. Lett., 84 (2000) 1820.
[108] E. Richter and K.R. Subbaswamy, Phys. Rev. Lett., 79 (1997) 2738.
[109] A. Hartschuh, H.N. Pedrosa, L. Novotny and T.D. Krauss, Science, 301 (2003) 1354.
[110] S. Jain, Semicond. Sci. Technol., 11 (1996) 641.
[111] E. Bonera, M. Fanciulli and G. Carnevale, J. Appl. Phys., 100 (2006) 033516.
[112] I. De Wolf, H. Maes and S. Jones, J. Appl. Phys., 79 (1996) 7148.
[113] I. De Wolf, J. Raman Spectrosc., 30 (1999) 877.
[114] I. Ahmad, V. Kasisomayajula, M. Holtz et al., Appl. Phys. Lett., 86 (2005) 173503.
[115] M. Kuball, S. Rajasingam, A. Sarua et al., Appl. Phys. Lett., 82 (2002) 124.
[116] M. Kuball, Surf. Interface Anal., 31 (2001) 987.
[117] W. Sun and Z. Shen, Ultramicroscopy, 94 (2003) 237.
[118] S. Baroni, A. Dal Corso, S. de Gironcoli et al., http://www.pwscf.org/.

Characterization of Semiconductor Heterostructures and Nanostructures
Edited by C. Lamberti
© 2008 Elsevier B.V. All rights reserved
Doi: 10.1016/B978-0-444-53099-8.00009-9

9

X-ray absorption fine structure in the study of semiconductor heterostructures and nanostructures

Federico Boscherini

Department of Physics and Consorzio Interuniversitario per la Fisica della Materia, University of Bologna, Bologna, Italy

Abstract X-ray absorption fine structure (XAFS) is a powerful tool in the study of the local atomic environment in condensed matter. It has been often applied to the study of semiconductor heterostructures and nanostructures, significantly contributing to their characterization at the local level and to the understanding of the relation between atomic structure and physical properties.

This chapter begins with an introduction to XAFS as a tool for the determination of local structure; the physical origin of the fine structure is illustrated and the present understanding of X-ray absorption spectra in the framework of multiple scattering theory is outlined. The second section contains a description of the diverse experimental setups and detection schemes, which can be used in the field of semiconductor science; an effort has been made to make this section both as complete and up-to-date as possible, so that it can serve as a useful reference, also outside the field of semiconductor physics. In the third section, a review of the use of XAFS to study semiconductor heterostructures and nanostructures is presented, covering bond length variations in strained heterostructures, local atomic environments in nanostructures (Ge islands, embedded nanoparticles, porous Si, and related systems), nitride heterostructures, and thin films, and finally dilute alloys heterostructures, that is dilute nitrides and dilute magnetic semiconductors.

Keywords X-ray absorption fine structure (XAFS), synchrotron radiation, nanostructures, heterostructures, local atomic structure

1. Introduction to X-ray absorption spectroscopy for local structural studies

In this section an introduction to X-ray absorption fine structure is reported. After a review of some basic phenomenology, the origin of the fine structure in X-ray absorption spectra is described. An outline of multiple scattering (MS) theory is presented and, in this framework, the extended and near-edge regions of the X-ray absorption spectrum are discussed.

1.1. Outline and origin of the fine structure in X-ray absorption spectra

In the X-ray and vacuum ultraviolet (VUV) spectral regions, photoelectric absorption is the dominant interaction mechanism between electromagnetic radiation and matter. An *X-ray absorption spectrum* is a measurement of the absorption coefficient of a sample, $\mu(\hbar\omega)$, or of a quantity to which it is directly proportional, as a function of the energy of the impinging photons $\hbar\omega$. One can determine (conceptually and experimentally) the absorption coefficient by measuring the flux of monochromatic photons incident on and transmitted by a sample of thickness x, I_0, and I_T, respectively; these quantities are related by

$$I_T(\hbar\omega) = I_0(\hbar\omega) \, e^{-\mu(\hbar\omega)x} \tag{1}$$

The product $\mu(\hbar\omega)x$ (a quantity proportional to the *absorbance* commonly used in VUV and visible spectroscopy) can be obtained by inverting Eq. (1). The overall absorption coefficient of a sample is related to the atomic concentrations and absorption cross sections of the constituent atoms, ρ_i and $\sigma_i(\hbar\omega)$, respectively, by

$$\mu(\hbar\omega) = \sum_i \rho_i \, \sigma_i(\hbar\omega) \tag{2}$$

X-ray absorption spectra exhibit sharp *absorption edges* which occur at the energies at which excitation of particular core levels becomes energetically allowed. The energies of absorption edges are characteristic of each element; for example, excitation of electrons from 1s, 2s, $2p_{1/2}$, and $2p_{3/2}$ states correspond to the K, L_I, L_{II}, and L_{III} absorption edges, and so on, according to the established X-ray nomenclature.

The presence of *fine structure* in X-ray absorption spectra has been known since 1920; this fine structure shows up as a modulation of $\mu(\hbar\omega)$ at energies greater than the absorption edge, the relative magnitude of which can be as high as ~10% and decays with increasing energy from the edge itself. Until the 1970s the field was plagued by contradictory experimental data and uncertain physical interpretation. Two decisive factors subsequently contributed to asserting XAFS as a valuable *structural tool*: the availability of brilliant and broadband synchrotron radiation sources and the development of an approximate but simple and useful point scattering theory. The "modern" era of XAFS is usually traced to the paper by Sayers et al. [1]. Some historical reviews have been published recently [2,3] and the interested reader is referred to them for details on the evolution of the technique; the introductory paragraph of the paper by Filipponi [4] contains a compact historical summary.

The physical origin of XAFS in the framework of scattering theory can be understood quite easily. Consider an atom bound in a molecule or solid. In the one-electron approximation, the core level photoelectric absorption process is described in terms of the initial state consisting of the impinging photon of energy $\hbar\omega$ plus the electron in the core atomic orbital characterized by a wave function ψ_i with a binding energy E_B and a final state consisting of a core hole plus an excited electron characterized by a wave function ψ_f with an energy E_f. Depending on the photon energy, the electron can be excited to a discrete and bound final state or to a continuum and unbound final state. The final energy of the photoelectron is dictated by energy conservation,

$$E_f = \hbar\omega - E_B \tag{3}$$

In the dipole approximation, the atomic absorption cross section (just as for electronic transitions in the optical and UV region) is proportional to the square modulus of the matrix element of the position operator between the initial and final states:

$$\sigma(\hbar\omega) \propto \left|\langle\psi_i|\hat{\varepsilon}\cdot\vec{r}|\psi_f\rangle\right|^2 \tag{4}$$

where $\hat{\varepsilon}$ is the unit vector defining the direction of the electric field of the exciting X-ray beam and $\vec{r}$ is the position operator. Let us consider only continuum final states. If the atom were not embedded in a matrix (e.g., a monoatomic gas) the final state wave function would be, at sufficiently high energies, that of an outgoing spherical wave and no fine structure would result. The relation between wave number ($k = 2\pi/\lambda$, with λ the wavelength) and kinetic energy for such a free electron is

$$k = \frac{\sqrt{2mE_f}}{\hbar} = \frac{\sqrt{2m(\hbar\omega - E_B)}}{\hbar} \tag{5}$$

where the second equality uses Eq. (3). If the atom is bound in a molecule or solid the photoelectron wave function will be modified by the presence of the surrounding atoms' potential; one describes this modification as due to the *scattering* (one or more times) of the photoelectron by the surrounding atoms. The final state wave function is now the outgoing spherical wave plus the scattered waves; the scattering process due to the presence of two neighbors to the excited atom is pictorially represented in Fig. 1. Depending on the

Fig. 1. Pictorial representation of the photoelectron scattering process. The solid lines represent the crests of the spherical wave fronts of the photoelectron wave function and their thickness is roughly proportional to the amplitude. The atom at the top of the figure is the photoexcited one (the central atom) while the two atoms at the bottom scatter the photoelectron. In the represented case, scattering gives rise to an enhancement of the wave function at the site of the central atom and therefore to an increase of the cross section with respect to the isolated atom case.

photoelectron wave number and on the relative atomic arrangement, the scattering process may lead to an enhancement or a decrease of the amplitude of the final state wave function close to the core level orbital and thus to an increase or decrease of the cross section described by Eq. (4). This is an *interference* effect between the outgoing and scattered wave functions; the scattered field is the electron wave function while the scattering "objects" are the neighboring atoms. As for any interference effect (e.g., in optics) the position of maxima and minima in reciprocal space depends on the wavelength of the scattered field and on the geometry of the scattering object (e.g., slits, grating, or atomic lattice). In the present case, for a given atomic structure, as the wave number depends on the photon energy via Eq. (5), the cross section varies with photon energy, exhibiting characteristic oscillations which are quasi-periodic as a function of the wave number. The structural sensitivity stems from the dependence on the relative atomic positions between absorbing ("central") atom and scattering (neighboring) atoms. In Fig. 2 we report a typical XAFS spectrum, in this case of crystalline GaAs at the Ga-edge: the presence of fine structure modulations is clearly visible.

The preceding discussion easily allows us to illustrate the common separation of an XAFS spectrum into extended and a near-edge region. If the energy of the photoelectron is high it will be only weakly affected by the neighboring atoms' potential; usually, it will be scattered a very limited number of times by the neighboring atoms and single scattering (SS) processes will be dominant. This is the so-called extended X-ray absorption fine structure or EXAFS region. In this region one defines the EXAFS spectrum $\chi(k)$ as

$$\chi(k) = \frac{\mu(k) - \mu_0(k)}{\mu_0(k)} \tag{6}$$

where $\mu_0(k)$ is the absorption coefficient in the absence of neighboring atoms. The EXAFS spectrum for GaAs at the Ga-edge is shown in the top inset of Fig. 2.

As the photoelectron energy decreases, closer to the absorption edge, photoelectron scattering by the neighboring atoms' potential will become increasingly important and MS processes will give a greater contribution to the fine structure. This spectral region, which extends from the edge to roughly 50 eV above it, is commonly referred to as X-ray absorption

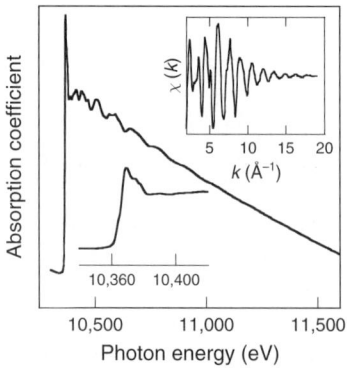

Fig. 2. Ga K-edge X-ray absorption spectrum of GaAs. The top inset reports the EXAFS function, $\chi(k)$, while the bottom inset highlights the XANES spectral region.

near-edge structure (XANES) or near-edge X-ray absorption fine structure (NEXAFS). It must be stressed that the distinction between the two regions is largely arbitrary and that there is no difference in the basic physical mechanism which gives rise to the fine structure; however, the analysis methods for EXAFS and XANES are different and thus it is reasonable to maintain the nomenclature. In the next section the separation between the two spectral regions will be discussed in more quantitative terms in the framework of MS theory.

In the past 30 years, XAFS has evolved into a reliable *tool* which is able to provide a quantitative measurement of the local structure in condensed matter and has been applied to fields ranging from solid-state physics to cultural heritage materials. A number of review papers [5–10] and books [11,12] on XAFS have been published and the reader is referred to them for a thorough description of experimental, theoretical, and data analysis aspects.

The purpose of this chapter is to report on the application of XAFS to the subject matter of the present book, semiconductor heterostructures and nanostructures. After description of theoretical and experimental aspects of XAFS, a review of applications and results is reported. The use of synchrotron radiation techniques for the characterization of semiconductor heterostructures has been recently reviewed by Lamberti [13]. Fine structure is also observed in electron energy loss spectra (e.g., in the transmission electron microscope) and this is also a powerful tool in semiconductor research offering spatial resolution of the order of 1 nm (but generally a lower signal-to-noise ratio which does not allow recording extended spectra, especially for absorption edges of high Z atoms); interested readers are referred to the review by Spence [14].

1.2. Multiple scattering theory framework and key approximations

In this section the basis of XAFS in the framework of MS theory [15–21] is briefly outlined, with the objective of providing a guide for the reader [4].

Photoelectric absorption is treated [22] in time-dependent perturbation theory to first order in the perturbation $\vec{A} \cdot \vec{p}$ (the scalar product of the vector potential of the radiation field and the electron momentum operator). Application of Fermi's golden rule gives the transition rate between discrete initial and final many electron states, Ψ_i and Ψ_f, respectively, as

$$W_{i \to f} = \frac{2\pi}{\hbar} \sum_f \left| \left\langle \Psi_i \left| \sum_j \frac{e}{m} \vec{A}(\vec{r}_j) \cdot \vec{p}_j \right| \Psi_f \right\rangle \right|^2 \delta(E_f - E_i - \hbar\omega) \quad (7)$$

where the index j covers all the electrons in the system and the label f identifies all final states compatible with the energy conservation dictated by the Dirac delta function. This intractable problem is greatly simplified in the *one-electron approximation*, in which it is assumed that only one electron is involved in the process, the others being unaffected; many-body corrections are added a posteriori (see below). In this approximation, it is assumed that one electron makes a transition between a localized core atomic orbital (e.g., the 1s orbital), described by a wave function ψ_c, and a final state described by a wave function ψ_f; in ab initio calculations, the presence of the core hole can be taken into account in calculation of ψ_f. A further common simplification is the *dipole approximation*, which neglects the spatial variation of the vector potential across the extent of the core level orbital. The photoelectric absorption cross section for transitions to continuum final states can be written [23] in these approximations as

$$\sigma(\hbar\omega) = 4\pi^2 \alpha \hbar\omega |\langle \psi_c | \hat{\varepsilon} \cdot \vec{r} | \psi_f \rangle|^2 \rho(E_f) \tag{8}$$

where $\rho(E_f)$ is the density of the final states, the energy of which is given by energy conservation as $E_f = E_c + \hbar\omega$, and (in SI units) $\alpha = e^2/4\pi\varepsilon_0\hbar c \cong 1/137$ is the fine structure constant, which is thus seen to determine the order of magnitude of the cross section. Important consequences of the dipole approximation are the *selection rules* which determine, on the basis of symmetry, which final states are allowed for a given initial state; for a free atom they are

$$\Delta\ell = \pm 1$$
$$\Delta m_\ell = 0, \text{ linearly polarized radiation} \tag{9}$$
$$= \pm 1, \text{ circularly polarized radiation}$$

where ℓ is the orbital angular momentum quantum number and m_ℓ determines its projection on the quantization axis. Thus, for a K absorption edge, the transition will be to p states.

One must now treat the modification of the final state due to the presence of neighboring atoms, that is, to the scattering of the photoelectron by their potential. Because the principal interaction of the photoelectron is with the core level electrons of surrounding atoms, the *muffin tin approximation* for the scattering potential is commonly adopted. This consists in spherically averaging the potential around each atom and adopting a constant interstitial potential in between. In this approximation, the calculation of the absorption cross section is simplified into an MS problem of the final state wave function by a collection of spherically symmetric scattering centers.

Under the previously stated assumptions, it has been demonstrated [16–18] that for a randomly oriented polycrystalline sample the polarization-averaged cross section for a transition to a final state of angular momentum ℓ is

$$\sigma(\hbar\omega) = \frac{\sigma_0(\hbar\omega)}{(2\ell+1)\sin^2(\delta_\ell^0)} \text{Im}\left\{\sum_{m_\ell} \left[(\mathbf{I} - \mathbf{TG})^{-1}\mathbf{T}\right]_{L,L}^{0,0}\right\} \tag{10}$$

where $\sigma_0(\hbar\omega)$ is the atomic cross section. Each element of the matrices which appear in Eq. (10) is identified by four indices: i, j running over the sites of the atoms surrounding the central one and L, L' (where $L = \{\ell, m_\ell\}$) being angular momentum indexes. $\mathbf{I}$ is the unit matrix while $\mathbf{T}$ and $\mathbf{G}$ are the atomic scattering and propagator matrices in a local basis, respectively. The $\mathbf{T}$ matrix is diagonal in the atomic site and angular momentum indexes, $T_{L,L'}^{i,j} = \delta_{ij}\delta_{LL'} e^{i\delta_\ell^i} \sin\delta_\ell^i$ where δ_ℓ^i is the phase shift for the partial wave of angular momentum ℓ centered on the ith atom. The $\mathbf{G}$ matrix is composed of null diagonal blocks in the site indexes i, j $\left(G_{L,L'}^{i,i} = 0\right)$ and describes the free propagation of the electron from site i to site j. Thus, the $\mathbf{T}$ matrix depends on the atomic composition of the sample while the $\mathbf{G}$ matrix depends on the geometrical arrangement of the atoms.

At sufficiently high energies above the edge (say 50 eV) the formal matrix expansion

$$\mathbf{T}(\mathbf{I} - \mathbf{GT})^{-1} = \mathbf{T}(\mathbf{I} + \mathbf{GT} + \mathbf{GTGT} + \mathbf{GTGTGT} + \cdots) \qquad (11)$$

in Eq. (10) is convergent; this leads to the *multiple scattering (MS) series* whereby the cross section can be written as the sum of a limited number of terms, each relative to a particular scattering path. This is the EXAFS region in which the cross section can be written as

$$\sigma(\hbar\omega) = \sigma_0(\hbar\omega)\left\{1 + \sum_{i\neq 0}\chi_2^{0i0} + \sum_{(i\neq j)\neq 0}\chi_3^{0ij0} + \cdots\right\} \qquad (12)$$

where 0 identifies the central atom, the subscripts indicate the number of scattering "legs" and the superscripts indicate the scattering path, starting and ending on the central atom via atoms i and j. Scattering paths with two and three legs are pictorially represented in Fig. 3; scattering paths with two legs are usually termed SS paths while those with three or more legs are termed MS paths. An alternative decomposition of the cross section in the EXAFS region into irreducible n-body signals has been used by Filipponi et al. [21].

The matrix expansion Eq. (11) is not convergent in the near edge, or XANES, region. Here a calculation of the spectrum implies the inversion of the matrix (**I-TG**); this is the full multiple scattering (FMS) regime because an infinite number of scattering paths contribute to the XAFS cross section.

1.3. EXAFS or extended X-ray absorption fine structure

Sufficiently far from the edge, in the EXAFS region, the MS series is convergent and this leads to a compact expression for the EXAFS function (Eq. (6)) which is used for quantitative structural refinement. Before proceeding, we must take into account some effects which have been neglected up to now. Despite the success of a one-electron picture, many body effects do have an appreciable influence on the fine structure and it is necessary to introduce some corrections to improve agreement between theory and experiment, the most important of which we now briefly discuss. The first correction is due to the relaxation of the wave function of electrons not directly involved in the transition and leads to a reduction of the amplitude of $\chi(k)$ by a factor

$$S_0^2 = \left|\left\langle \Psi_{N-1}^i \middle| \Psi_{N-1}^f \right\rangle\right|^2 \qquad (13)$$

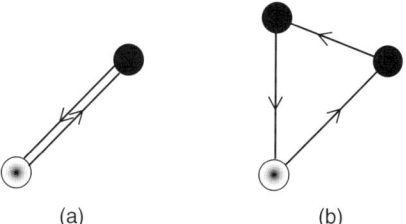

(a)   (b)

Fig. 3. Pictorial representation of paths with (a) two and (b) three scattering legs. The excited atom is highlighted at the bottom left.

where Ψ^i_{N-1} and Ψ^f_{N-1} are the many-body wave functions of the "passive" electrons before and after excitation of the "active" electron; the value of this amplitude reduction factor is usually between 0.7 and 0.9. The second correction is due to the limited range of the excited electron with the matrix (at most ~ 10 Å) and to the limited lifetime of the core hole (~ 1 fs); both have the effect of damping the contribution to $\chi(k)$ from distant atoms and this is, in fact, the origin of the local sensitivity of XAFS. One may model these two effects with an effective electron mean free path $\lambda(k)$, the value of which is energy dependent and ranges from a few angstroms to ~ 10 Å. In up-to-date data analysis schemes these effects are included in the effective scattering amplitude (see below). The third correction is due to thermal motion of the atoms. The photoabsorption process occurs on the time scale of $\sim 10^{-15}$ s, in which the atoms may be considered as fixed in their instantaneous positions; hence, a XAFS spectrum, due to a very high number of excitation processes, probes the configurational average of the atomic position. Thermal motion gives rise to the so-called Debye-Waller (DW) factor, which for a SS contribution to $\chi(k)$ involving atoms 0 and i and for motion in a harmonic potential is

$$\sigma_i^2 = \left\langle \left[\hat{r}_{0i} \cdot (\vec{u}_0 - \vec{u}_i) \right]^2 \right\rangle \qquad (14)$$

where $\langle \cdots \rangle$ denotes a configurational average, $\hat{r}_{0i}$ is the unit vector joining atoms 0 and i, and $\vec{u}_i$ is the instantaneous deviation of the position of atom i from its average position. An important point is that the XAFS DW factor depends on the degree of correlation of the vibration of the two atoms, while the DW factor which affects the intensity of XRD peaks depends only on the mean square deviation of each atom's position with respect to its equilibrium value, projected in the direction of the wave vector difference. The effect of moderate static structural disorder can also be taken into account by a supplementary DW factor.

With these approximations, the SS contribution the EXAFS function can be written in a compact form. With the central atom taken as reference, the neighboring atoms are grouped into coordination shells labeled by the index j and consisting of N_j identical atoms at the interatomic distance r_j; in the harmonic approximation for thermal motion and for a randomly oriented crystalline powder sample we have

$$\chi_2(k) = S_0^2 \sum_{j=\text{shells}} \frac{N_j f_j(r_j, k)}{k^2 R_j^2} \sin\left[2kr_j + \varphi_j + 2\delta\right] e^{-2k^2 \sigma_j^2} \qquad (15)$$

where $f_j(r_j, k)$ and φ_j are the modulus and argument of the effective scattering amplitude of atoms in shell j and δ is the phase shift due to the central atom. The scattering amplitude depends also on the interatomic distance r_j, as the curved wave nature of the photoelectron wave function is taken into account; in the *plane-wave approximation*, adopted in more elementary treatments, this dependence no longer exists. If the sample is an oriented single crystal and if, as usual, a linearly polarized photon beam is used, each atomic correlation will contribute to $\chi(k)$ with a weight w_{0i} which depends on the relative orientation of $\hat{r}_{0i}$, the unit vector joining atoms 0 and i, and the direction of polarization $\hat{\varepsilon}$; a simple expression for this weight can be obtained in the *small atom approximation* in which the spatial extension of scattering region associated to each neighboring atom is neglected with respect to photoelectron wavelength:

$$w_{0i} = 3\hat{\varepsilon} \cdot \hat{r}_{0i} \tag{16}$$

Despite the apparently crude approximation, this weighting factor is very useful to estimate linear dichroism effects.

Contributions to $\chi(k)$ due to MS paths have a form similar to Eq. (15). It can be shown that the MS contribution of order n due a set of atoms the coordinates of which are identified by a path index $\Gamma = \{\vec{r}_i\}$ and for which the total scattering path length is r_Γ are of the form

$$\chi_n^\Gamma = A_n(k, \Gamma) \sin[kr_\Gamma + \phi_n(k, \Gamma) + 2\delta] \tag{17}$$

The total EXAFS function will be the sum of all the SS and MS paths, as indicated by Eq. (12). In principle, there is an infinite number of such paths, but, in practice, their number is limited by the rapid damping of the signal for long path lengths (usually paths the length of which is greater than $\sim$10 Å have negligible amplitude) and the fact that the amplitude of MS signal quickly decreases with the order n (usually paths with $n > 4$ can be neglected). MS paths are weak in the EXAFS region but must be taken into account for an accurate structural determination; collinear paths due to three atoms in a row give rise to strong contributions to $\chi(k)$ due to scattering paths composed of three or four legs as, for example, in the case of rocksalt structures [24]. In the diamond structure, the importance of MS paths in the EXAFS region was first pointed out by Bianconi et al. [25].

Because the EXAFS function is the sum of sine functions, the argument of which is kr_Γ, a Fourier transform (FT) of $\chi(k)$ will exhibit peaks in correspondence to the path length; the FT is usually plotted as a function of the half path length which, for SS paths, is equal to the interatomic distance. In Fig. 4 we report the magnitude of the FT of the EXAFS spectrum shown in the top inset of Fig. 2, relative to GaAs at the Ga K-edge; the main contributions to the first three peaks are due to the first, second, and third coordination shells of the zinc blende structure.

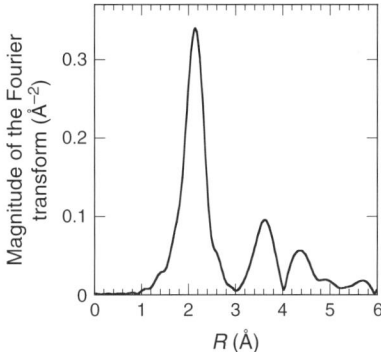

Fig. 4. Magnitude of the Fourier transform of the Ga K-edge EXAFS spectrum shown in the top inset of Fig. 2 for GaAs.

From the analysis of an EXAFS spectrum the following local structural parameters can be obtained: interatomic distances (R, typically ± 0.005 Å for the first shell), identity of neighboring atoms and their coordination number (CN, typically $\pm 10\%$) and, provided a Gaussian model for the radial distribution function is appropriate, the DW factor (σ^2, typically $\pm 5 \times 10^{-4}$ Å^2).

EXAFS analysis requires an adequate starting structural model, which can be obtained from prior knowledge of the sample. An up-to-date analysis usually is based upon a simulation of the EXAFS signal and a subsequent refinement of the local structural parameters via a least squares routine. A description of data analysis methods is beyond the scope of this paper; the interested reader is referred to existing literature and manuals. (Useful lists of XAFS software can be found in the web pages xafs.org, www.i-x-s.org, and www.esrf.fr/computing/scientific/exafs/xafs.html.)

1.4. XANES or near-edge X-ray absorption fine structure

In the near-edge spectral region, the MS series (Eq. (12)), does not converge and thus it is not possible to express the cross section as a sum of sine functions. A simulation of the XANES region requires the inversion of the matrix (**I-TG**) and it is a relatively lengthy matter in terms of computer time. The XANES lineshape is sensitive both to the details of the scattering potential (because of the low energy of the photoelectron) and to the three-dimensional position of atoms contained within a radius greater than that necessary to reproduce EXAFS. As a consequence, quantitative analysis of the XANES is not generally possible at present.

An interpretation of XANES spectra in term of *electronic structure* is also possible. In fact, referring to Eq. (8), if one neglects the energy dependence of the matrix element in a limited range of the near-edge region, one can see that the absorption cross section is proportional to the density of final states (in the presence of the core hole), selected in angular momentum by the rules given by Eq. (9) and projected on the absorbing atomic species. This kind of interpretation links XANES to electronic structure calculations [26,27] and is often used in the case of small molecules [28].

XANES finds widespread application in many different fields, in which it is often used with a "finger-printing" approach whereby the spectrum of a sample is compared to those of a number of reference compounds of known structure; for example, the XANES spectra of transition metal compounds exhibit clear and significant differences with oxidation state and coordination, the most extensive compilation of which has probably been performed in the field of mineralogy (for a specific example, see Ref. [29]). An analysis of the XANES spectrum provides information on the site symmetry of the absorbing atom (e.g., tetrahedral or octahedral) and on its oxidation state.

Because XANES is sensitive to the three-dimensional atomic geometry (while the dominant contributions to EXAFS are of a one-dimensional character) and because XANES spectra are more easily recorded than EXAFS ones, especially for dilute elements, a quantitative interpretation of XANES would be of great interest. In the field of semiconductors, we quote the recent paper by Ciatto et al. [30], in which, by making a comparison between experimental spectra and ab initio simulations, it was possible to determine the three-dimensional structure of the nitrogen–hydrogen complex responsible for the elimination of the red shift of the band gap in the dilute nitride alloy GaAsN. In an significant paper, Della Longa et al. [31] have described a structural fitting of XANES spectra for biomolecules; the extension of this method to solids would be a very important step forward in the investigation of dopants and defects in semiconductors.

Let us conclude mentioning that in the important case of the L-edges of transition metals, in which a p electron makes a transition to a narrow d band, because of the strong electron correlation effects multiplet theory is used to interpret the spectra. A useful review paper on these aspects has been written by de Groot [32]. L-edge spectra have been used to study the local structure of Mn in dilute magnetic semiconductors (DMSs), as will be described in Section 3.

2. Detection schemes and experimental setups relevant for semiconductor research

In this section, the main experimental setups that are relevant for semiconductor research are described. Of course, an essential requirement to perform an XAFS experiment is that of a brilliant synchrotron radiation source and appropriate optics, the role of which is to provide a focal spot on the sample of the required characteristics in terms of photon flux, bandpass, stability, divergence, and spot size. There has been great progress in this field in the past years and a great number of papers and books have been published, some of which are reported in Refs [33–38].

Here, we focus our attention on the experimental chamber itself and the appropriate detection schemes. The objective of any XAFS measurement is to record, as a function of photon energy, a quantity which is directly proportional to the cross section of the atom of interest in the particular environment or phase which is relevant for the experiment. A number of methods have been devised and will be described here. As for diffraction anomalous fine structure (DAFS) we refer to Chapter 11 by Proietti et al.

2.1. Transmission geometry

A transmission measurement of XAFS is the most commonly used in general. It relies on the measurement of the flux of monochromatic photons incident (I_0) and transmitted (I_T) by a sample of thickness x. Inversion of Eq. (1) gives

$$\mu(\hbar\omega)x = \ln\left[\frac{I_0(\hbar\omega)}{I_T(\hbar\omega)}\right] \tag{18}$$

where $\mu(\hbar\omega)$ is the absorption coefficient. The main requirement for this kind of measurement is that the discontinuity in μx be in the range [6] of 1–2 and that the lateral homogeneity of the sample is small enough to avoid "thickness effects" [39]. In practice, this means sample thicknesses must be in the range 1–10 μm and homogeneities must be a fraction of this; therefore, the transmission technique cannot usually be used in the soft X-ray range in which the thicknesses required would be too small to make the fabrication of a homogenous sample practical. The transmission technique can be applied when the atom of interest is not dilute or at or near the surface; its main advantage is that it generally provides data of very good signal-to-noise ratio, as it relies on the measurement of rather high photon fluxes (10^9–10^{11} photons/s). The most commonly used detector for transmission measurements are gas ionization chambers.

In the field of semiconductor heterostructures and nanostructures one is mostly interested in studying the local structure of ultrathin films, nanolayers, or low-dimensional structures on, or near, the surface. Pristine samples are thus not usually appropriate for measurements in the transmission mode, as the background due to the substrate or matrix would be too high. In a notable exception, Ridgway's group [40] has devised specific sample preparation procedures that involve the etching of the substrate and stacking of various layers of the sample of interest to obtain a final equivalent thickness appropriate for transmission measurements; accompanying this with a low measurement temperature (which reduces the DW factor, Eq. (14)) data of excellent signal-to-noise ratio were obtained by these authors for a number of cases.

2.2. Decay channels: fluorescence and electron yield

Reduction of the background absorption signal originating from the matrix can be obtained by selectively recording the intensity of an atom-specific signal. For example, one can exploit the decay of photoexcited atom, which occurs with the emission of either a fluorescence photon (this is the so-called characteristic radiation) or of an Auger electron, both of which have energies that depend on the atomic number of the photoexcited atom.

2.2.1. Fluorescence yield detection

In the fluorescence yield (FY) detection mode [41], a photon detector is used to record the fluorescence intensity as a function of energy. In the most common experimental geometry used with horizontally polarized X-rays, the detector is placed in the horizontal plane at right angles to the incident beam to minimize the intensity of elastic scattering of the impinging X-ray beam; in this geometry, the intensity of fluorescence photons, $I_f(\hbar\omega)$, of atom A embedded in a matrix M for a sample thickness d can be easily shown to be [6]

$$I_f(\hbar\omega) = I_0(\hbar\omega) \frac{\Omega}{4\pi} \varepsilon_f \frac{\mu_A(\hbar\omega)}{\mu_{tot}(\hbar\omega) + \mu_{tot}(\hbar\omega_f)\tan\theta} \left[1 - \exp\left\{-(\mu_{tot}(\hbar\omega) + \mu_{tot}(\hbar\omega_f)\tan\theta)d\right\}\right] \quad (19)$$

where ε_f is the FY, Ω is the solid angle subtended by the detector, $\hbar\omega_f$ is the energy of the fluorescence photons, θ is the angle between the impinging beam, and the sample surface and the total absorption coefficient (see Eq. (2)) is the sum of contributions due to atomic species A and that due to the matrix M: $\mu_{tot} = \mu_A + \mu_M$. Because μ_{tot} depends on μ_A, Eq. (19) is not in general proportional to the quantity one wants to measure, that is, $\mu_A(\hbar\omega)$. This proportionality is recovered in two limiting cases: thin samples (for which the series expansion of the exponential in Eq. (19) can be truncated at the term linear in d, typically below 100 nm) and for thick samples in which atom A is dilute ($\mu_{tot} \cong \mu_M$, a condition which depends on the angle θ but usually occurs below a few atomic %). If these conditions are not met, fluorescence data can be corrected for nonproportionality (also known as "self-absorption"), as has been described by various authors [42–47].

Many types of detectors for FY mode have been used, but the most common are based on solid-state Si:Li or (mostly nowadays) hyperpure Ge diodes. These detectors often are in the

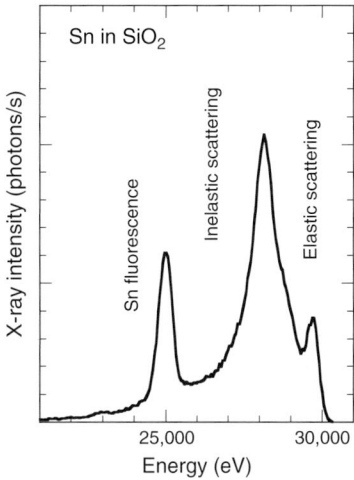

Fig. 5. Spectrum of photons scattered by a sample consisting of $\sim 10^{16}$ Sn atoms/cm^2 implanted in SiO$_2$; elastic and inelastic (Compton) scattering and Sn fluorescence are identified [50].

form of multielement arrays [48,49] and have an energy resolution in the range of 150–200 eV at 6 keV, a reasonable trade-off between counting rate and background rejection; as an example [50], the spectrum of photons emitted from a sample consisting of $\sim 10^{16}$ Sn atoms/cm^2 implanted in SiO$_2$ recorded on the GILDA beamline of ESRF [51] is reported in Fig. 5: elastic scattering, inelastic (Compton) scattering and Sn fluorescence are clearly identified. As pulse counting electronics is used, care must be taken to avoid, or correct, dead-time effects [52].

Because both the probe and the detected signals are penetrating X-rays, FY is the best choice to study dilute elements in the bulk of materials. This has led to a wide application of XAFS to the study of dopants in semiconductors. The lowest concentration which can be measured in a FY–XAFS experiment ultimately depends on the magnitude of the inelastic scattering background in the specific experimental geometry. Takeda et al. [53] have recently studied the factors which determine the lowest concentration which can be detected; they have come to the conclusion that, for the specific case of the L$_{III}$-edge of Er in GaAs or InP, the lowest dose measurable is 1–5×10^{14} atoms/cm^2. In the more general case, the lowest detectable dose or thickness depends on the absorbing atom, the matrix and the experimental geometry; XAFS spectra on thicknesses equivalent to 1–0.1 monolayers are nowadays measurable with adequate signal-to-noise ratio and even lower concentrations can be measured in specific cases [53].

2.2.2. Electron yield detection

The decay of the core hole can occur also with the emission of an Auger electron. This Auger electron creates a cascade of lower energy ("secondary") electrons due to multiple ionization and inelastic scattering processes. Various electron yield detection methods exist, depending on which electrons are detected and the detection method. The experimentally simplest method, often used in the soft X-ray region, consists of measuring the drain current using an electrometer which connects the sample to ground; this method, termed total electron yield (TEY), can be traced back to the pioneering investigation by Gudat and Kunz [54]. A variation on TEY is partial electron yield (PEY) in which only electrons of a given energy range are detected, with the use of retarding grids and electron detectors. In the hard X-ray range, cross sections are smaller than at lower energies and the most common method consists in recording the signal due to radiationless decay using a He-filled ionization chamber in which the sample acts as the anode [55–58]: He gas acts as an electron multiplier and an amplified signal is recorded from the cathode; this method is also termed conversion electron yield (CEY). Finally, the conceptually simplest method consists of directly recording of the number of Auger electrons themselves [59,60] (Auger electron yield, AEY); this is only employed in surface XAFS experiments, as it requires an electron analyzer operating in ultrahigh-vacuum conditions.

The main peculiarity of all electron detection modes is the surface sensitivity, which is guaranteed by the limited electron mean free path. Hence, these modes are particularly useful for the investigation of the near-surface region or of interfaces. Erbil et al. [61] have reported a detailed study of the depth sensitivity of TEY mode, concluding that it depends on the range of the original Auger electrons and that it generally is in the range of a few hundreds to a few thousands of angstroms. Schroeder et al. [62] have compared TEY and CEY in detail, concluding that the former is more surface-sensitive than the latter.

Spectra recorded using electron detection methods may suffer from "self-absorption effects," the origin of which is identical to that previously illustrated for FY; usually, the effect is less severe than in FY as the electron absorption coefficient is much higher than the photon absorption coefficient and thus it dominates the denominator in Eq. (19). Correction procedures have been described by Nakajima et al. [63] and Regan et al. [64].

2.3. Alternative detection methods

A number of "nonconventional" methods to detect XAFS have been proposed, each with its own advantages, limitations, and pitfalls. We briefly review the ones which have relevance to the investigation of semiconductors, leaving a discussion of results on specific systems to Section 3.

2.3.1. Optical detection

When excited by an X-ray beam, many samples emit visible or near-visible radiation. The use of this X-ray excited optical luminescence (XEOL) and the measurement of the XAFS spectrum via the resulting photoluminescence yield (PLY) have attracted considerable attention, in view of the possibility of determining the local structure of only the atoms involved in the luminescence emission, thus adding site selectivity to XAFS.

XEOL was first observed by Bianconi et al. [65] in CaF_2; their PLY-XAFS spectrum actually exhibited an inverted jump (i.e., a decrease, rather than an increase, of the PLY at the edge). The role of the attenuation lengths for the X-ray and optical photons and the sample composition in determining this behavior was subsequently understood and discussed by Goulon et al. [66] and Emura et al. [67].

A fundamental issue is the effective site selectivity of PLY-XAFS: it should be verified that the change of the PLY at the absorption edge of a particular atom is only due to absorbers in the optically active centers. In many cases, particularly in scintillators or ionic crystals, the optically active center may be different or distant from the selected absorbing atoms. In the case of a sample consisting of a mixture of ZnS–ZnSe powders, in which the two components emit at different wavelengths, site selectivity was clearly demonstrated by Pettifer and Bourdillon [68]. On the contrary, the important role that can be played by interion excitation was shown by Sonderholm et al. [69], who studied, among others, a Gd_2O_2S sample doped with 0.4% Tb; they observed a XAFS signal in the Tb luminescence yield at energies corresponding to the Gd-edge. The importance of interatomic excitations has also been stressed by Pettifer and coworkers [70]. A thorough, but quite critical, review on PLY-XAFS has been reported by Rogalev and Goulon [71]. It is clear that site selectivity in PLY-XAFS should be checked for every sample examined.

PLY-XAFS has found an important application in the study of porous Si, which emits in the visible range, and specific results will be illustrated in Section 3. Recently, PLY-XAFS has been applied in the study of ZnO nanostructured materials both at the high-energy K-edge [72] and at low energy [73].

2.3.2. Electrical detection

The electrical detection of XAFS has been investigated by some groups with the objective of determining whether the local structure of only those atoms contributing to the electrical signal due to defects could be determined, leading to another attractive site-selective version of XAFS.

Photocurrent-detected XAFS spectrum has been reported by Boehme et al. [74] and Hu et al. [75] for an Al–GaAs diode and for bulk GaAs, respectively. Their common conclusion is that these spectra contain, for these concentrated samples, the same structural information as the more traditional transmission mode ones, once corrections for sample thickness are taken into account.

Ishii et al. [76] have proposed a capacitance-detected version of XAFS, the specific system studied being a Se-doped AlGaAs–Al diode in which the semiconductor exhibits the so-called DX deep level electron trap. It is argued by these authors that, as changes in capacitance are due to electrons localized in the traps, it is possible to determine the local structure of only those atoms at a short distance from the trap itself. In fact, their spectra at the Ga-edge recorded in the fluorescence and in the capacitance mode exhibit significant differences. Ishii has also proposed variations of the capacitance method; by varying the diode bias potential a depth selectivity is proposed [77], while use of scanning capacitance probes is proposed to provide lateral resolution on the length scale of the AFM [78,79].

The site selectivity of capacitance-detected XAFS has been challenged in detail by Bollmann et al. [80]. These authors have examined samples exhibiting both the DX deep level and antisite defects; spectra measured in the fluorescence and capacitance mode *do not* show any significant differences. Also by modeling the X-ray-induced current and capacitance changes, Bollmann et al. conclude that, as the defect centers can be ionized also by electrons (directly or indirectly) generated by core ionization of atoms distant from the defect center themselves, no site selectivity results. It is clear that in order for this detection mode to be accepted by the scientific community, more work is necessary to resolve the issue of selectivity.

2.3.3. Beta environmental fine structure

We mention briefly a method which is potentially of great interest for the study of defects and dopants in semiconductors. In the context of precise measurements of the neutrino mass, it has been suggested [81] that the β decay spectrum of a nucleus belonging to an atom embedded in a crystal or molecule could be affected by a fine structure similar to that which gives rise to XAFS. In fact, the emitted β particle (an electron) will be scattered by neighboring atoms and the decay rate might be affected by the resulting interference, resulting in the so-called beta environmental fine structure (BEFS).

BEFS was detected [82] in the decay spectrum of ^{187}Re and as β-emitting nuclei include tritium, ^{14}C, and a number of metals it was realized that BEFS could potentially be of great interest in materials physics and chemistry [83]. The Re BEFS spectrum was later quantitatively analyzed [84], finding good agreement with the known crystal structure. An open issue, the angular momentum of the final state (s or p) was addressed more recently [85]. BEFS is potentially very powerful, especially for the study of the local environment of hydrogen in semiconductors (for which traditional XAFS is impossible because of the low binding energy of the K-edge) but has yet to prove to be a practical tool.

2.4. Specific experimental setups

2.4.1. Grazing incidence

The interaction between an electromagnetic wave and a sample can be described, in a macroscopic picture, by the *index of refraction*. The propagation of a wave traveling in the direction $\vec{x}$ is determined by the phase factor

$$\exp\left[i\left(\omega t - \vec{k}\cdot\vec{x}\right)\right] \tag{20}$$

where ω is the angular frequency and $\vec{k}$ is the wave vector, with the wave number being $|\vec{k}| = k = 2\pi/\lambda$ (λ is the wavelength), the velocity of propagation is $v = \omega/k$. In the X-ray range the index of refraction can be written as

$$n(\hbar\omega) = 1 - \delta(\hbar\omega) + i\beta(\hbar\omega) \tag{21}$$

where both δ and β are positive and $\ll 1$; δ and β describe dispersion and absorption processes, respectively, and, like the real and imaginary part of the permittivity, are related by the Kramers–Krönig relations. For Si at $\lambda \sim 1.2$ Å ($\hbar\omega = 10$ keV), $\delta \cong 9.8 \times 10^{-6}$ and $\beta \cong 1.5 \times 10^{-7}$. As a consequence, an X-ray beam impinging on the flat surface of a sample may undergo *total external reflection* if the angle of incidence (α, taken to be the angle between the beam's direction and the surface) is smaller than the critical angle

$$\alpha_c = \sqrt{2\delta} \tag{22}$$

For Si at $\lambda \sim 1.2$ Å, $\alpha_c \cong 4.5$ mrad. For $\alpha < \alpha_c$, the reflectivity approaches 1 and inside the sample the beam is confined to the near-surface region; compact expressions for the penetration depth Λ in the direction normal to the surface can be derived [37] for three cases (the values listed are for Si at $\lambda \sim 1.2$ Å):

$$\begin{aligned}\alpha \ll \alpha_c : \quad &\Lambda = \frac{1}{2k\sqrt{2\delta}} \approx 2.2 \text{ nm}\\ \alpha = \alpha_c : \quad &\Lambda = \frac{1}{2k\sqrt{\beta}} \approx 26 \text{ nm}\\ \alpha = 10\alpha_c : \quad &\Lambda = \frac{\alpha}{2k\beta} \approx 29 \text{ μm}\end{aligned} \tag{23}$$

The last expression in Eq. (23) is just the usual penetration length determined by the absorption coefficient ($\mu = 2k\beta$) projected on the surface normal. Moreover, it can be shown that for $\alpha = \alpha_c$ the beam intensity at the surface is four times the intensity of impinging beam. The reflectivity is a complicated function of α, δ and β, which approaches 1 for $\alpha \ll \alpha_c$ and 0 $\alpha \gg \alpha_c$. It is clear that a strong confinement of the X-ray beam in the near-surface region can be obtained by using grazing incidence.

The possibility of using grazing incidence to probe the surface region of materials was illustrated by Parratt [86]. As far as XAFS is concerned, Barchewitz et al. [87] were the first to record fine structure via the reflectivity, $R(\hbar\omega)$, and to suggest the relation with the absorption spectrum. Subsequently, Martens and Rabe [88] provided an in-depth discussion, opening the way to what has been called the "RefleXAFS" technique. The fine structure in $R(\hbar\omega)$ is related to the "usual" XAFS spectrum $\chi(k)$. However, it is not straightforward, in the general case, to obtain $\chi(k)$ from $R(\hbar\omega)$ as the latter quantity depends on both $\delta(\hbar\omega)$ and $\beta(\hbar\omega)$ in a complicated way; of course, the "usual" XAFS spectrum is contained in $\beta(\hbar\omega)$. A simplification arises sufficiently far from the edge and for $\alpha < 0.8\alpha_c$, as it can be shown [89] that

$$\mu(\hbar\omega) = \frac{1 - R(\hbar\omega)}{1 + R(\hbar\omega)} \tag{24}$$

A number of methods have been proposed to extract an XAFS spectrum from the reflectivity, each with its advantages and limitations [88–91]. Instead of measuring the reflectivity, FY or TEY may be used, of course [92]; in the case of a dilute or thin sample, a spectrum recorded in this way is directly proportional to $\mu(\hbar\omega)$ and does not require corrections [93]. Grazing incidence XAFS is a powerful tool to study surfaces and interfaces and has been used also to study semiconductor nano- and heterostructures.

2.4.2. Surface X-ray absorption fine structure

The potential of XAFS to determine the local structure of interfaces deposited in situ in an ultrahigh-vacuum environment was realized early on [59,60]. As far as use of the EXAFS spectral region is concerned, this has led to the acronym SEXAFS for "surface EXAFS." The specific advantage of surface XAFS for the study of heterostructures is the possibility of determining the atomic environment in the first few coordination shells of an excited atom in the epilayer (with an error of the order of 0.01 Å for the interatomic distances), during the formation of the interface, without exposure to the atmosphere. This information is crucial for a correct physical modeling of interface properties and now can be usefully compared to ab initio structural simulations. By exploiting the dependence of the XAFS cross section on the relative orientation of the sample and the linearly polarized synchrotron radiation beam the structure parallel and perpendicular to the growth plane can be determined [94,95]. When information on the long-range order is necessary, surface XAFS can be usefully coupled to grazing incidence X-ray diffraction (XRD). Surface XAFS has been reviewed a number of times and some of these reviews are listed in Refs [96–98].

Surface XAFS relies on detection schemes which are able to distinguish the signal originating in the interface from that originating from the bulk: TEY, PEY, or FY, often combined with grazing incidence, all of which have been described previously. Of course, the further requirement is of an ultrahigh-vacuum chamber connected to the beamline and equipped with all the instrumentation required to prepare an atomically clean and ordered surface and to deposit and characterize ultrathin epilayers. A particularly detailed description of a well-equipped surface XAFS apparatus has been reported by Oyanagi et al. [99].

2.4.3. Micro-XAFS

As brighter synchrotron sources have become available it has become possible to obtain smaller focal spot sizes and to design instruments capable of increasingly higher spatial resolution. In fact, the field of X-ray microscopy has progressed very rapidly in recent years; lateral resolutions of the order of 10 nm have been obtained with photoelectron imaging techniques, e.g., the photo emission electron microscope or PEEM (see Locatelli et al. [100] for an example of a spectromicroscopy beamline using photoelectron imaging, and Ratto et al. [100] for an example of application to Ge dots on Si).

There is a great interest in performing XAFS with high lateral resolution (micro-XAFS) as it allows a new level of description of heterogeneous samples, combining microscopy with the atomic-scale structural information obtainable from XAFS; with sufficiently small focal spots the local atomic environment of single nanostructures might be determined. The challenge is to record XAFS spectra free of systematic errors and with sufficiently good signal-to-noise ratio. At present, XAFS spectra can be recorded with spot sizes of the order of a few square micrometers, for example, on the ID22 beamline [101] of the ESRF or on the LUCIA beamline [102] of the SLS; micro-XAFS is often complemented with microfluorescence mapping. Applications of micro-XAFS to semiconductors have been relatively limited so far, but it is expected much progress will be made in this field. Micro-XAFS is particularly useful when the samples investigated exhibit lateral inhomogeneities and it is not surprising that many applications in the field of semiconductors have been on samples the growth of which has yet to be optimized, for example DMSs based on GaN, as will be discussed in Section 3. In several synchrotron radiation laboratories, efforts are on the way to combine the information which can be derived from XAFS with the nanometer-level lateral resolution which can be obtained by using atomic force microscope tips [103]. While photon-based techniques might maybe never reach the sub-angstrom spatial resolution obtainable in electron microscopy, the advances in recent years have been really dramatic as they involve many orders of magnitude improvement in spatial resolution. In the future, micro-XAFS will certainly become a powerful tool in materials science, complementing spatially averaged techniques which have been the key to the success and widespread use of X-rays for structural determination.

3. A review of the use of XAFS in the field of semiconductor heterostructures and nanostructures

In the field of materials science, new growth and deposition schemes are continuously devised, with the objective of obtaining samples and devices with novel physical properties. The role of highly sensitive and sophisticated methods such as XAFS is to provide a description of the local atomic environment which can form the structural basis for an understanding of the physical properties. In the past 20 years, XAFS has developed from an intriguing physical phenomenon to a *tool* that is used in many branches of science. In this section, an attempt is made to review the use of XAFS in the study of the local atomic environment of semiconductor heterostructures and nanostructures. This is a rather wide field and the choice of examples is partially a personal one. Not all aspects of semiconductor science are covered, for example, the important application of XAFS to the study of dopants, implanted atoms, and defects is not included.

3.1. Bond lengths in strained III–V heterostructures

The study of the effect of strain on the local atomic structure, especially on the bond lengths, is an area in which XAFS has provided key results, clarifying what initially seemed a rather confusing situation. The applicability of XAFS to study small bond length variations in (unstrained) semiconductor pseudobinary alloys was demonstrated in pioneering papers in the 1980s [104,105]. The main result of these investigations is that bond lengths exhibit a variation with relative concentration which is significantly weaker than that exhibited by the lattice parameter. In fact, the lattice parameter of a pseudobinary semiconductor alloy of composition $A_xB_{1-x}C$ *and zinc blende structure generally obeys Vegard's law,*

$$a(x) = xa_{AC} + (1-x)a_{BC} \qquad (25)$$

where a_{AC} and a_{BC} are the well-known lattice parameters of the binary compounds. It is found experimentally that the bond lengths, R_{AC} and R_{BC}, exhibit a much weaker dependence on x. This can be qualitatively understood as due to the "rigidity" of semiconductor bonds and implies the existence of local distortions, i.e., variations of the bond angles. A comprehensive theory of bond length variation in semiconductor alloys, able to reproduce existing experimental results and with predictive ability, was proposed by Cai and Thorpe [106]; a basic ingredient of their work is the assumption that there are no differences between the bond stretching and bond bending force constants in elemental tetrahedra centered on different atoms and that an average value applies in all cases. It should be noted that XAFS is the appropriate tool to study this problem because of the chemical selectivity and high resolution at the short distance scale. It is interesting to note that a pair distribution function (PDF) analysis of XRD patterns has also recently been shown to allow the determination of interatomic distances in bulk $InGa_{1-x}As_x$ samples [107]; also, it has been demonstrated [108] that ab initio structure determination of randomly oriented nanostructures is in principle feasible and that the PDF can be obtained for bulk $ZnSe_{1-x}Te_x$ from neutron diffraction data [109]. An advantage of this technique is that it combines short- and medium-scale sensitivity while a limitation is that the applicability to thin crystalline epilayers appears to be difficult.

The issue of bond lengths in thin strained semiconductor layers grew out naturally in the context of studies of bulk alloys. It is well known that, below a certain critical thickness, when a epilayer with a cubic lattice is deposited on a substrate with a different lattice parameter its unit cell will undergo a tetragonal distortion (this is termed pseudomorphic growth). The distortion is quantified by the perpendicular and parallel strains, $\varepsilon_\perp = (a_\perp - a_f)/a_f$ and $\varepsilon_\parallel = (a_\parallel - a_f)/a_f$, respectively, where a_f is the epilayer's free (unstrained) lattice parameter; for two-dimensional growth on a (001)-oriented substrate the strains are related by

$$\varepsilon_\perp = -2\frac{C_{12}}{C_{11}}\varepsilon_\parallel \qquad (26)$$

where C_{12} and C_{11} are the elastic constants. The open issue at the beginning of the 1990s was the effect of this long-range strain at the local scale, i.e., on the individual bond lengths. Is there such an effect (also in consideration of the known rigidity of bonds)? If so, in the case of an alloy, is it the same for different bonds and how does it combine with bond length variations induced by alloying itself? The problem is graphically

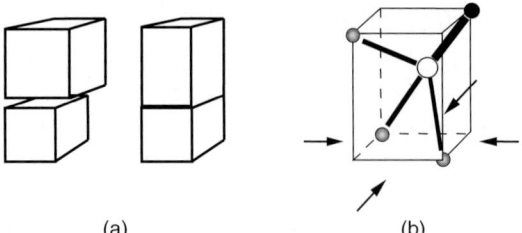

Fig. 6. (a) Tetragonal distortion of the unit cell in pseudomorphic growth, in the case of compressive strain. (b) An elementary tetrahedron for an alloy in which one of the bonds (at the top right) is shorter than the other ones, under compressive strain.

depicted in Fig. 6. XAFS is clearly the technique of choice to address the problem, also in consideration of the applicability to thin epilayers; in fact, FY was used by most groups and in some cases grazing incidence was used to enhance surface sensitivity and to reject the substrate contribution [110].

Bond length strain has been studied both for SiGe alloys deposited on Si(001) and for heterostructures based on III–V semiconductors; the clearest results have been obtained in the latter case. The issue can be tackled in two ways. For a binary epilayer or for an alloy epilayer with fixed composition, the bond lengths can be measured as a function of thickness: the effect of strain should show up as strain relaxation will occur above the critical thickness. Alternatively, for an alloy epilayer below the critical thickness, the bond lengths can be measured as a function of the composition: the strain will change as the free lattice parameter varies according to Eq. (25); in this case variations of the bond length induced by strain must be distinguished from variations due to alloying.

Initially, many authors [111–113] reported negligible variations of bond lengths in strained layers, within a quite large typical error bar of ± 0.01 or ± 0.02 Å. This confusing situation was probably due to less than optimal data acquisition and analysis and/or poor sample quality or lack of characterization. Most probably, the first clear report of a strain-induced bond length variation is that by Kuwahara et al. [114], who measured the In–As bond length in InAs and $InAs_{0.6}P_{0.4}$ layers on InP(001) (both of which are compressively strained) as a function of thickness. Their data are reported in Fig. 7: a clear compression is measured at low thicknesses while an increase is evident in a thickness range that includes the critical thickness for strain relaxation; note the reduced error bars. Later, Woicik et al. [115] obtained a similar result for InAs/GaAs(001).

A couple of years later, experiments on strained alloy layers as a function of composition allowed to gain definitive insight on strain-induced bond length variations and to propose a simple model for the effect. Romanato et al. [116] studied tensile and compressively strained $In_xGa_{1-x}As$ epilayers InP(001) in the range $0.25 < x < 0.75$ while Woicik et al. [117] studied a buried $In_{0.22}Ga_{0.78}As$ layer in GaAs, finding similar results. In Fig. 8 we report the data of Romanato et al. for $In_xGa_{1-x}As$ epilayers on InP(001), which are lattice-matched for $x = 0.53$; filled symbols are the results for the strained alloy layers while empty ones are for unstrained alloys, measured as a check. Data points on unstrained layers confirm the classical results by Mikkelsen and Boyce [104] (the dashed line is a fit to their data) while those for the strained layers clearly exhibit the effect of strain, which actually *inverts* the slope of the bond length vs concentration plots: a remarkable effect. The continuous line is the result of a model which

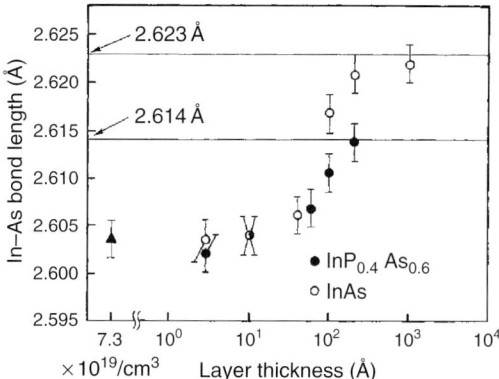

Fig. 7. In–As bond length in InAs and InAs$_{0.6}$P$_{0.4}$ layers on InP(001) (both of which are compressively strained) as a function of thickness (reprinted with permission from Ref. [114]. Copyright (1994) by the Institute of Pure and Applied Physics).

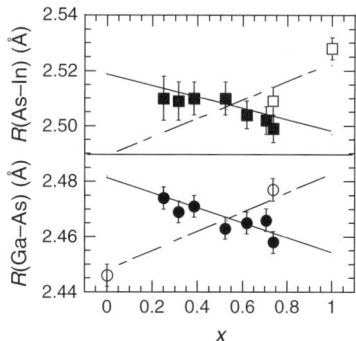

Fig. 8. In–As and Ga–As bond lengths for In$_x$Ga$_{1-x}$As epilayers InP(001) in the range $0.25 < x < 0.75$, reproducing the original results from Ref. [116]. Filled symbols are the results for the strained alloy layers while empty ones are for unstrained alloys. The dashed line is a fit to the data by Mikkelsen and Boyce while the continuous line is the result of the model as described in the text (reprinted with permission from F. Boscherini, S. Pascarelli, C. Lamberti, S. Mobilio, F. Romanato, D. DeSalvador, M. Tormen, M. Natali and A. Drigo, J. Synchrotron Rad. 6 (1999) 506. Copyright (1999) by the International Union of Crystallography).

applies to the local scale the strain tensor: to first order, the strain-induced distortion for bonds in the [111] direction is predicted to be

$$\delta r^{st} = \frac{a(x)}{2\sqrt{3}}\left(1 - \frac{C_{12}}{C_{11}}\right)\varepsilon_{\parallel} \qquad (27)$$

The effects of alloying and strain are summed linearly. As can be seen, the agreement with the data is excellent. A consequence of this model is that different bonds (e.g., In–As and Ga–As) actually respond to strain in the same way, despite the fact that bond stretching force constants are significantly different; this is not surprising as the alloying-induced variations were reproduced with average values of the force constants [106]. More refined methods to calculate bond lengths in strained layers, based on a valence force field potential, have been published by Woicik [118] and d'Acapito [119], with a similar degree of good success. The issue of the effect of strain on bond lengths can be considered as solved and measurements of variations of bond lengths in strained systems can now be used to *determine* strain, as will be described in the next section.

3.2. Atomic environment in nanostructures

XAFS has been used to determine the atomic environment in many semiconductor nanostructures. The applicability of XAFS to study the changes of the local environment in nanostructures as a function of their dimensions was recognized early in the development of the technique, as shown by the studies on metallic clusters performed in the 1980s [120–122]. These studies demonstrated that nanoparticles exhibit a contraction of bond lengths, a reduced average coordination number and an enhanced structural and vibrational disorder. XAFS is the technique of choice because its local sensitivity (which derives from the limited photoelectron mean free path and core hole lifetime) makes it extremely powerful in detecting even subtle changes in the local atomic environment; moreover, the method of analysis does not change with the degree of long-range order even in going from a dimer molecule to a perfect crystal, a situation which must be contrasted with the case of XRD, and there are no consequences of even great changes in the morphology of a nanoparticle on the analysis formalism. Here, the most important topics addressed using XAFS are reviewed.

3.2.1. Ge quantum dots and islands on Si surfaces and related systems

The possibility of altering the physical properties of semiconductors by exploiting quantum size effects has lead to the recent big interest in nanostructures. In this field, XAFS has provided key results, especially regarding the intermixing in Ge quantum dots and islands on Si. XAFS has the advantage of providing a local view of the atomic environment of the excited atom, thus providing quantitative information on intermixing and bond length strain in a direct fashion. The chemical sensitivity necessary to obtain information on intermixing can be obtained by XRD techniques only by exploiting anomalous dispersion. In this context, anomalous XRD at high momentum transfer has been successfully applied to determine composition and strain profiles of Ge islands on Si [123]; disentangling the contributions of strain, intermixing, and morphology to the XRD pattern required a significant amount of modeling. A limitation of XAFS for the ex situ study of uncapped dots and islands is that for thin samples the contribution from the surface oxide (or from other disordered phases) can be significant, which might complicate analysis; a solution is the use of a capping layer or performing an in situ experiment. In view of the advantages of XAFS in this field, a great deal of papers have been published, of which we review the most significant.

Ge islands on Si(001) are among the most studied systems. It is known that, due to the 4.2% lattice misfit, heteroepitaxial growth proceeds with the formation of a two-dimensional *wetting layer*, followed by the formation of dots which relieve the lattice strain (at the cost of an increase in surface energy); this is the so-called Stranski–Krastanov growth mode. XAFS experiments [124] on Ge dots formed on Si(001) and Si(111) were among the first to detect the presence of atomic intermixing by any technique, a phenomenon which had been neglected up to that time; other evidence for intermixing from XAFS was provided by Kolobov et al. [125] and Erenburg et al. [126] The experimental evidence is rather clear, as

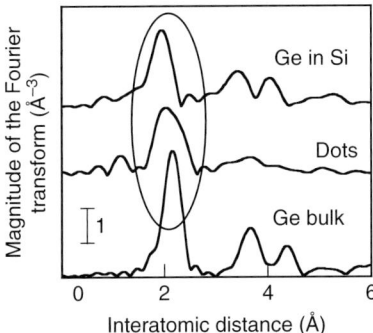

Fig. 9. Ge K-edge EXAFS spectra Ge for quantum dots on Si(001), illustrating intermixing of Si (adapted from Ref. [127]).

can be seen in Fig. 9 that compares the FT of Ge K-edge spectra for bulk Ge, for a Ge impurity in crystalline Si and for a representative sample of Ge dots on Si(001). The asymmetrical lineshape of the first peak, which is due to the first coordination shell, indicates the presence of a significant number of Si neighbors to the average Ge atom; this indication has been confirmed by a fitting of the spectra. It is now accepted that diffusion of Si atoms from the substrate reduces the strain energy associated with the heteroepitaxial growth and is a key feature of quantum dot growth. Hence, the appropriate growth mode for quantum dots is a *modified* Stranski–Krastanow one. Motta et al. [128] have also monitored the evolution of the intermixing process on the Si(111) surface, finding an enhancement of interdiffusion with increasing substrate temperature. The direct information on intermixing in quantum dots available from XAFS was also nicely illustrated by the measurements by Kovats et al. [129], who found that intermixing was inhibited by the saturation of the dangling bonds on the Si(111) surface by B atoms.

The advantages of XAFS to study the local Ge quantum dots on Si in comparison to Raman spectroscopy have been described by Kolobov et al. [130], who clearly showed that the superposition of a Raman peak of the Si substrate to that of Ge implies strong limitations of this technique in the present case. The same group also detected considerable atomic intermixing in capped Ge dots, the degree of which depends on the growth temperature of the capping layer [131] and can be eliminated by growth on SiO_2 [132]; the value of the second and third shell interatomic distances was found to indicate the strained state of the dots. The measurement of second and third shell was made possible by inclusion of the significant MS paths. The use of bond lengths to determine the degree of strain of Ge dots was also recently illustrated by d'Acapito et al. [133].

Atomic intermixing is not limited to the group IV dots and islands but was found also in those based on III–V compound semiconductors. Galluppi et al. [134] were the first to detected In/Ga intermixing in nominally pure InAs dots on GaAs by In K-edge XAS; these structural measurements were correlated to photoluminescence investigation of the optical properties. d'Acapito et al. [135] confirmed this result in a grazing incidence measurements on 1.3–3 monolayer (ML) thick samples and discussed the values of the bond lengths, finding that, also in this case, their value could provide reliable information on the state of strain of the dots. Finally, Renevier et al. [136] contrasted the case of InAs/GaAs dots to that of InAs/InP quantum wires; in the former case, the authors confirmed the previously cited results while no interdiffusion was found for the group V elements in the latter case.

3.2.2. Embedded semiconductor nanoparticles

Semiconductor nanoparticles embedded in a dielectric medium can exhibit unique optical properties and have been actively studied recently, also in view of potential applications. Variations of the physical properties derive from quantum size effects and from the increased surface-to-volume ratio in smaller particles. Among the fabrication methods proposed, ion implantation followed by thermal annealing is a very flexible means to realize nanoparticles in the near-surface region of a dielectric. We will focus our attention on studies of Ge, GaN, and II–VI nanoparticles embedded in a variety of mostly amorphous media.

The local structure of Ge nanocrystals in silica has been studied by Ridgway's group, who obtained two main results. Firstly, for mean nanocrystal sizes of 14 nm, these authors confirmed the tetrahedral coordination of Ge in the nanocrystals but showed that, compared to bulk crystalline Ge, an enhancement of both Gaussian and non-Gaussian disorder is present [137]; this study highlights the extreme sensitivity of XAFS to the details of the short range radial distribution function. Secondly, the amorphization of Ge nanocrystals in Si induced by Si ion irradiation was demonstrated to occur at a dose 100 times smaller than that required for the bulk [138]. In Fig. 10 we report Fourier transforms of Ge K-edge XAFS data for (a) polycrystalline and (b) nanocrystalline samples upon Si ion irradiation (in units of ions/cm^2); note the factor of 10 difference in the doses between the two panels. The strong damping of the second and third shell signals (peaks between 3 and 5 Å) clearly shows that preferential amorphization of the nanocrystals. The authors discussed four possible origins for this behavior: preferential amorphization at the interface, the intrinsic instability of the nanocrystals, the enhanced vacancy concentration within the nanocrystals, and ion-beam mixing.

An original route for the fabrication of GaN quantum dots in crystalline (quartz and sapphire) or amorphous (silica) dielectrics by sequential ion implantation of Ga and N followed by annealing in NH_3 or NH_3/H_2 has been reported by Borsella et al. [139] Ga and N doses of the order of 10^{17} cm^{-2} and annealing treatments at 900–1200°C for ~1 h were used. The wurtzite phase GaN dots have average dimensions of ~5 nm and their optical properties exhibit quantum confinement effects. XAFS measurements have provided important information on the local structure and its relation to the sample deposition conditions. In Fig. 11 we show the FT of Ga K-edge EXAFS spectra for bulk GaN and differently prepared samples. While an overall similarity is apparent (and confirms that the particles locally exhibit the wurtzite structure), the smaller amplitude of the second peak with respect to the first one reflects the partial incorporation of Ga atoms in the GaN clusters and indicates that a fraction of Ga atoms is also dispersed in the matrix, possibly bonded to oxygen atoms.

Studies of CdS nanoparticles embedded in glass have been reported by Hayes's group. These nanoparticles are obtained by doping a borosilicate glass in the melt with 0.1% weight CdS and subsequent quenching and annealing (at 625–900°C); with increasing annealing temperature and time, CdS particles grow in size until they finally dissolve. These particles have typical diameters ranging from 2 to 6 nm and clearly exhibit a blue shift of the optical absorption edge due to quantum confinement. XAFS at the Cd-edge showed [140] that Cd is present both in the nanoparticles and also dispersed in the glass, and that a high degree of structural disorder is present in the nanoparticles. In Fig. 12 the Fourier transforms of an as-received sample and of a sample annealed for 14 h at 700°C, compared to bulk CdS and CdO; a shift to higher distances of the first peak with annealing and the absence of structure above the first peak in the as-received sample are evident. The first and second shell interatomic distances were found [141] to be within 0.2% of the bulk value, which implies a low

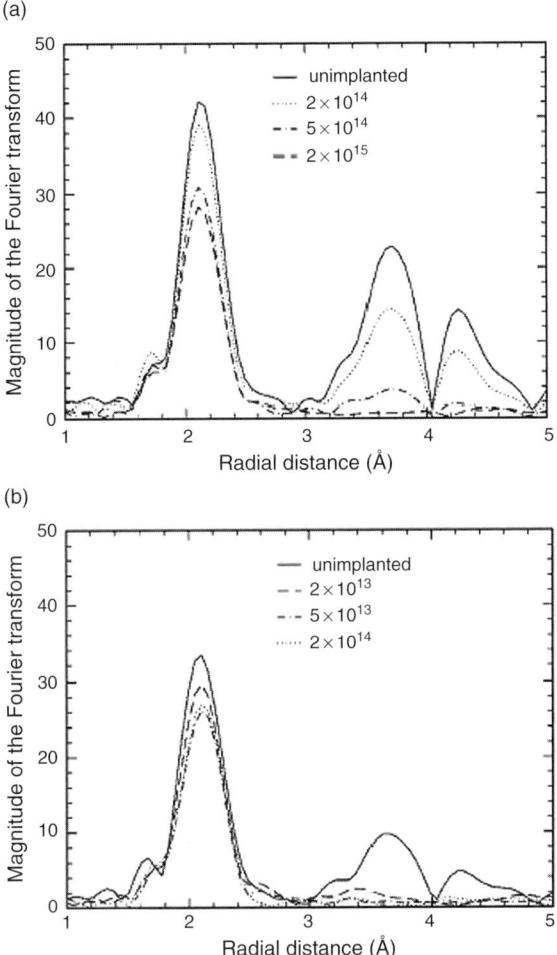

Fig. 10. Fourier transforms of Ge K-edge EXAFS data for (a) polycrystalline and (b) nano-crystalline Ge embedded in silica upon Si ion irradiation (in units of ions/cm^2); note the factor of 10 difference in the doses between the two panels (reprinted with permission from M.C. Ridgway, G. de M. Azevedo, R.G. Elliman, C.J. Glover, D.J. Llewellyn, R. Miller, W. Wesch, G.J. Foran, J. Hansen and A. Nylandsted-Larsen, Phys. Rev. B 71 (2005) 094107. Copyright (2005) by the American Physical Society).

interfacial tension, suggesting that the glass matrix is able to accommodate and relax strains, increasing the particles' stability.

3.2.3. Metallic nanostructures embedded in silica

Metallic nanocrystals embedded in amorphous SiO_2 on Si(001) have been recently investigated in view of their nonlinear optical properties, with applications ranging from optical fibers, memories, and switching devices. XAFS has been employed by several groups to study the variations of the local structure as a function of particle size and processing, and its relation to the physical properties.

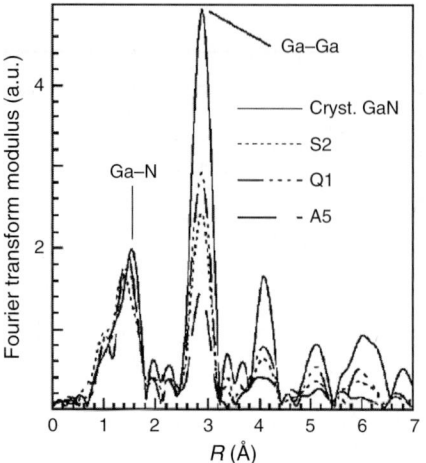

Fig. 11. Fourier transforms of Ga K-edge EXAFS spectra for differently prepared GaN quantum dots and for bulk GaN. Samples S2 and Q1 are obtained by implantation of Ga and N in amorphous silica, while for sample A5 a sapphire substrate was used (reprinted with permission from Ref. [139]. Copyright (2001) by the American Institute of Physics).

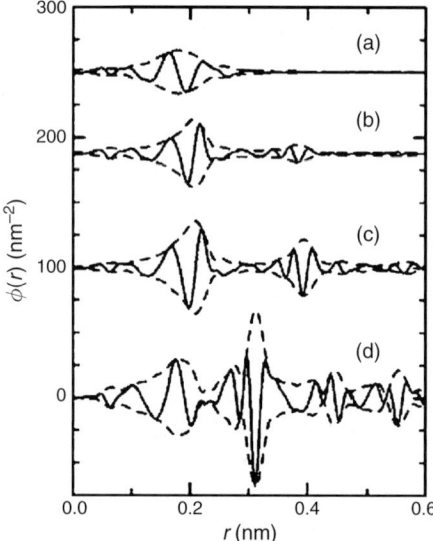

Fig. 12. Magnitude (dashed line) and real part (solid line) of the Fourier transforms of Cd K-edge EXAFS for CdS nanoparticles in glass. (a) as-received sample; (b) a sample annealed for 14 h at 700°C; (c) bulk CdS; (d) CdO (reprinted with permission from Ref. [141]. Copyright (2001) by the American Physical Society).

Spiga et al. [142] have investigated the formation and the structural properties of Sn nanocrystals produced by ion implantation in ultra thin silica films, using grazing incidence XAFS in conjunction with Mössbauer spectroscopy and transmission electron microscopy. The analysis of this system is complicated by the multiple oxidation states of Sn. Sn ion

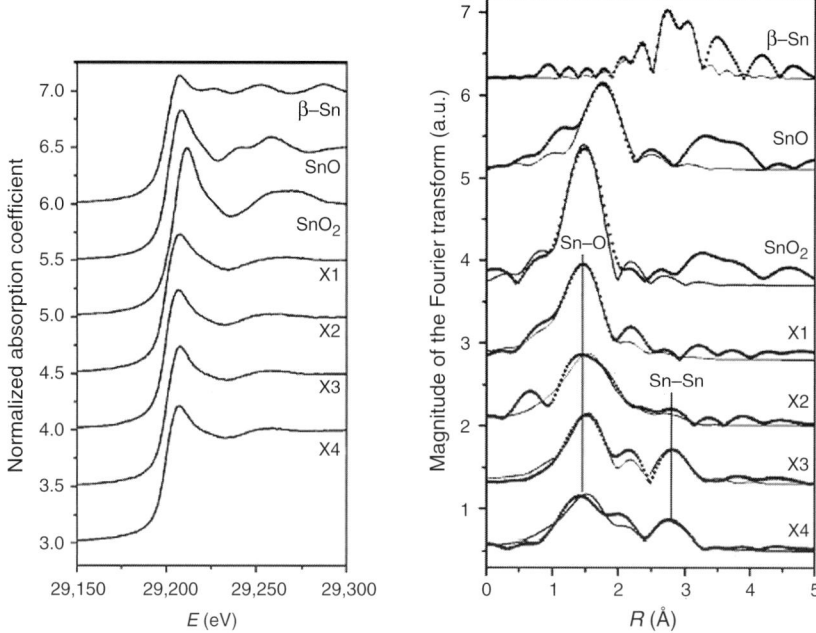

Fig. 13. XANES (left) and FT of EXAFS (right) spectra for Sn nanoparticles implanted in amorphous SiO_2 (fluence 10^{16} cm^{-2}) and reference compounds. The samples differ in the annealing conditions; X1: as-implanted; X2: 800°C, 120 s; X3: 900°C, 30 s; 900°C, 120 s (reprinted with permission from Ref. [142]. Copyright (2003) by the American Physical Society).

implantation was performed at 80 keV with a fluence of 10^{16} cm^{-2}, positioning the peak of the implantation profile in the middle of the 85-nm-thick thermally grown SiO_2 film and was followed by annealing at 800–1100°C; particle sizes were in the range 7–17 nm. It was found that in the as-implanted state all Sn ions are oxidized, with the Sn^{2+} oxidation states predominant, while annealing induces the formation of β-Sn nanoclusters. Figure 13 reports XANES and EXAFS spectra that illustrate these changes. For particle sizes less than 10 nm, a reduced coordination number and a contraction of the interatomic distances was found. The investigated annealing treatments did not lead to a complete precipitation of Sn atoms in the metallic phase, leaving a fraction of them oxidized. This study nicely illustrates the combined use of the XANES and EXAFS region of the X-ray absorption spectra; the first provides qualitative information on the oxidation state while the second measures local structural parameters. It also shows the advantage of grazing incidence to enhance the signal originating from ultrathin films.

The local structure of Au nanocrystals fabricated by ion implantation in thin amorphous silica was investigated by Kluth et al. [143] Au ions were implanted at 4.5 MeV in 2-μm-thick thermally grown SiO_2 at fluences in the range 10^{16} to 3×10^{17} cm^{-2} and were subsequently annealed in forming gas for 1 h at 1100°C. Average particle sizes ranging from 25 to 100 Å were obtained. Thanks to the high XAFS data quality, the authors clearly detected a ~0.02 Å bond length contraction in smaller particles (each bond length was measured with an error bar stated as ranging between ±0.002 and ±0.005 Å); this contraction, which is consistent with previous reports on nonembedded clusters, implies a negligible influence of the SiO_2 matrix.

The same authors also later studied the effect of irradiation with Sb ions (2.3 MeV, doses between 10^{14} and 10^{16} cm^{-2}), a method proposed to tune the properties of nanoparticles [144]. At lower irradiation doses, significant disordering is evident, with a bond length expansion; at higher doses, a significant fraction of Au atoms dissolves in the matrix and other form very small clusters (especially dimers and trimers).

3.2.4. Porous Si and related systems

Obtaining visible light emission (luminescence) from Si and fabricating a Si-based optoelectronic device has been a very active area of research in recent years. Porous Si (p-Si), typically obtained by anodic etching of a Si wafer in an electrochemical cell containing HF acid and exhibiting such luminescence, has been the object of intense research. The origin of the luminescence has been often attributed to a quantum size effect but competing interpretations attribute the luminescence to oxide-related structures on the surface of the Si. XAFS measurements, especially conducted in the PLY (or XEOL) detection mode, have played an important role in correlating the atomic structure to the physical properties and in elucidating the origin of the luminescence.

The first study by PLY–XAFS was reported by Sham's group [145]. They clearly showed that XAFS spectra in the TEY and PLY modes were different and that the former were much more sensitive to the oxide component of the sample. Their PLY-detected spectra were qualitatively similar to those of bulk Si, so it was concluded that the optically active component of p-Si locally exhibits the diamond structure. Pettifer et al. [146] studied oxidized p-Si by collecting PLY–XANES spectra using time-resolved detection: in this way they were able to distinguish the different environment of Si atoms because the fast blue component is only related to the presence of SiO_2-rich regions.

Rather extensive investigations of p-Si using XAFS have been conducted by the Trento group. In their investigations, the samples are usually freshly prepared and soaked in HF to remove the oxide component just before the measurements. A first investigation [147] was conducted in the transmission mode and allowed recording high-quality EXAFS spectra; this investigation quantitatively confirmed that the local structure of p-Si is the diamond one and detected an increase in the DW factors with porosity. Subsequently [148,149], a comparison of TEY and PLY detected spectra allowed the same group to conclude that the structures responsible for luminescence in p-Si are smaller and more disordered than the average structure of the sample. This is illustrated in Fig. 14. In the left panel, the photoluminescence spectra for two samples are reported; the higher current density sample has a higher porosity and its emission is blue-shifted. In the right panel, the corresponding FTs of the Si K-edge EXAFS spectra in the PLY mode are compared to those acquired in the TEY mode; it is clear that the TEY spectra do not change, while the PLY spectra exhibit an increasing disorder with porosity, as testified by the decreasing amplitude of the first three peaks. Finally, the same group has proposed a partial PLY detection mode [150]. According to the model in which luminescence in p-Si is due to recombination of carriers localized in quantum-confined structures emission at higher energies corresponds to smaller emitting clusters. By selecting the light emission energy, Dalba et al. were able to show that the position of the X-ray absorption edge increases linearly with the emission energy. As the X-ray absorption edge is due to a transition from a dispersion-less core state to the bottom of the conduction band, this experiment supported the stated model for the origin of luminescence in p-Si.

A field related to that of porous semiconductors is that of free-standing wires (NWs). By a variety of methods, it is possible to fabricate filamentous structures with diameters ranging

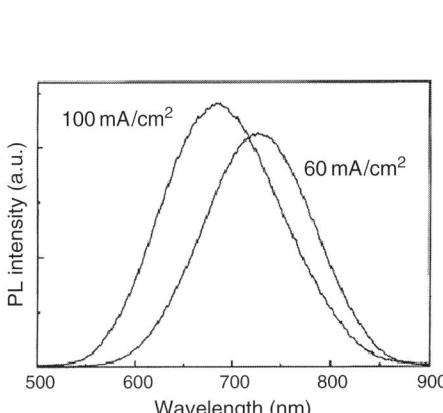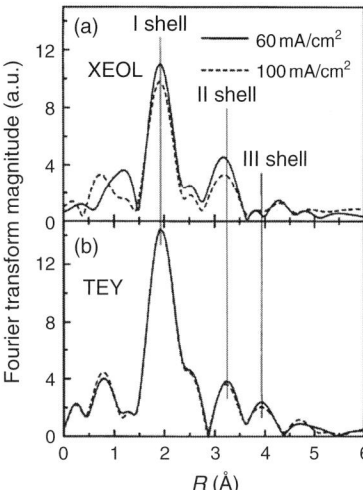

Fig. 14. In the left panel, the photoluminescence spectra for two p-Si samples are reported; the higher current density sample has a higher porosity and its emission is blue shifted. In the right panel, the corresponding FTs of the Si K-edge EXAFS spectra in the PLY mode are compared to those acquired in the TEY mode (from Ref. [149]. Copyright (1999) by Elsevier).

from a few to a few tens of nanometers and lengths up to a few hundred microns. The wires are generally covered by an oxide layer. XAFS has provided an atomistic description of the structure of these systems. In particular, Zhang et al. [151] have studied Si NWs obtained by laser ablation of a mixed Si/SiO$_2$ target. XAFS measurements at the Si K-edge of the as-deposited samples and of samples with the oxide removed showed that the core of the NW maintains the diamond local structure, with increased disorder compared to bulk Si. The same group has also more recently studied [152] Si NW using PLY detection, identifying luminescence bands originating from the quantum-confined silicon crystallites, the encapsulating silicon oxide and the silicon–oxide interface.

Finally, we mention XANES studies on silicon nanodots embedded in a SiO$_2$ film, performed by recording TEY and PLY spectra by the Trento group [153]. The comparison has allowed to propose and validate a model for the origin of the observed enhanced photoluminescence. Both experimental and theoretical results show that the interface between the silicon nanocrystals and the surrounding SiO$_2$ is not sharp: an intermediate region of amorphous nature and variable composition links the Si to amorphous stoichiometric SiO$_2$. This region plays an active role in the light-emission process.

3.3. Nitride heterostructures and thin films

GaN and related materials have been intensely studied in recent years in view of the properties which make them attractive for applications: a wide band gap (in blue–violet region), good thermal and chemical stability, and physical hardness. The most common crystal structure of nitrides is the hexagonal wurtzite one but cubic forms are also possible, albeit metastable. The wurtzite structure grows in the (0001) direction and therefore has a polar character which entails strong polarization fields (both intrinsic and piezoelectric). The cubic and hexagonal polytypes of GaN are very similar when viewed along the cubic (111) and hexagonal (0001) directions and because the formation of stacking faults is energetically favorable, stabilization of the hexagonal phase is sometimes difficult. An important problem for this class of materials is the absence of substrates with comparable lattice parameters, crystal symmetry and thermal expansion coefficients; the most commonly used substrates are Al_2O_3, SiC, and Si. XAFS investigations have contributed to clarify the local atomic structure of nitride alloys, determine the local effect of strain due to heteroepitaxial growth, investigate the relation between the relative fraction of cubic and hexagonal phases of GaN films and the choice of substrate, and finally study the fine details of the local structure of GaN.

Miyano et al. [154] studied the variation of interatomic distances in wurtzite structure $Al_xGa_{1-x}N$ alloys deposited by metal organic chemical vapor deposition (MOCVD) on sapphire, with x in the range from 0 to 0.43. The films were about 1 μm thick, so no effect of the strain due to the growth was expected; XRD showed a linear variation of the a and c lattice parameters of the wurtzite structure as a function of x. Ga K-edge EXAFS allowed the authors to measure the interatomic distances in the first and second coordination shells around Ga, finding that the first shell Ga–N bond lengths exhibit a variation with x which is much weaker than the variation of the lattice parameters. Essentially, this study extends to the wurtzite structure the classical results [104,105] on zinc blende III–V pseudobinary alloys. The effect of strain due to heteroepitaxial growth was probed in two studies of GaN ultrathin films grown by molecular beam epitaxy (MBE) on SiC substrates [155] and AlN layers [156], both of which exploited the polarization dependence of the X-ray absorption cross section (see discussion in Section 1.3) to decouple the local structure parallel and perpendicular to the growth plane. For the case of SiC substrates the growth was found to be always relaxed while the GaN on AlN films were found to be partially strained, due to the lower lattice mismatch; the degree of strain was found to depend on the substrate temperature and the ratio between the strain parallel and perpendicular to the interface was found to be in good agreement with estimates on the basis of the available elastic constants and macroscopic elasticity theory. These studies can be viewed as an extension to nitride epilayers of the previously described ones for III–V systems (see Section 3.1); it must be mentioned that nitride epilayers are intrinsically more defective than III–Vs, so an understanding of local distortions due to the heteroepitaxial growth is extremely valuable.

The effect of the substrate on the crystal structure of GaN thin films has been studied by Paloura's group [157,158]. By using the polarization dependence of N K-edge XANES spectra, these authors were able to illustrate how an $Al_2O_3(0001)$ substrate yields a wurtzite GaN film, that a Si(001) substrate yields a cubic film, and that a Si(111) substrate yields a mixed phase. The physical basis of this kind of study is that the atomic absorption cross section must exhibit (at least) the symmetry of the point group of the crystal structure in which the absorbing atom is embedded [159]. For zinc blende films the point group is T_d and the cross section is isotropic. For wurtzite films, instead, the point group is C_{6v} and the angular dependence of the cross section is

$$\sigma^{tot}(\hbar\omega, \theta) = \sigma^{isotropic}(\hbar\omega) + (3\cos^2\theta - 1)\sigma^1(\hbar\omega) \tag{28}$$

where θ is the angle between the polarization vector and the c-axis of the wurtzite structure and the two cross section terms on the right-hand side do not depend on the angle; for $\theta = 54.7°$, the so-called magic angle, the measured cross section will be equal to the isotropic one. It is therefore clear that angular-dependent measurements can distinguish between cubic

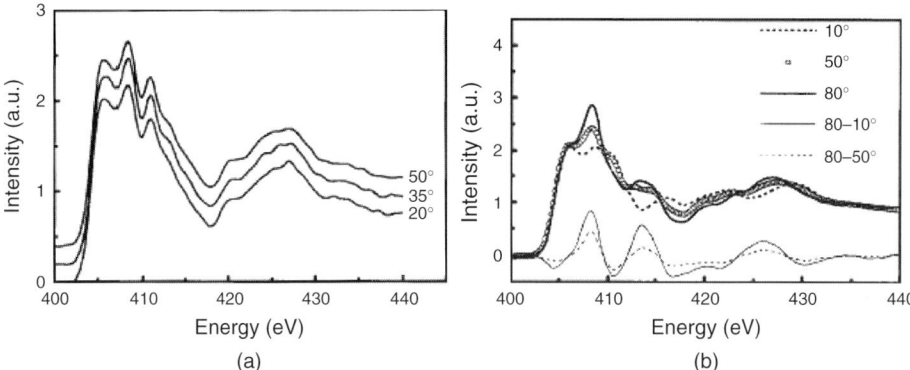

Fig. 15. Polarization-dependent N K-edge XANES spectra for GaN films deposited on (a) Si(001) and (b) Al_2O_3(0001) substrates (reprinted with permission from Ref. [158]. Copyright (1998) by the American Institute of Physics).

and hexagonal symmetry. An alternative explanation of the method, based on a molecular orbital framework, is reported in the quoted papers. In Fig 15(a) and (b) polarization-dependent XANES spectra for GaN films deposited on Si(001) and Al_2O_3(0001) substrates, respectively, are reproduced; it is quite clear that the former exhibit no polarization dependence and are thus relative to cubic films, while the latter exhibit a clear angular modulation and are thus hexagonal. These papers illustrate a general method based on XANES that can be used to determine the symmetry of the investigated crystal structure. This technique has been also extended to other nitride films [160,161].

GaN films are highly defective: along with the stacking faults that lie at the origin of the possible presence of both cubic and hexagonal phases, point defects such as N vacancies and Ga antisite defects have a relatively high concentration. This defectivity has been proposed by Katsikini et al. to cause significant structural distortions in the elementary tetrahedral building block of the material. By analysis of FY-XAFS data at both the N [162] and Ga [163] K-edges, these authors have detected a $\sim$0.25 Å elongation of a significant fraction of the four Ga neighbors to the average N atom. This distortion has subsequently been found also in ab initio structural simulations by Dimakis et al. [164] These studies once again highlight the unique sensitivity of XAFS to the fine details of the local structure.

3.4. Dilute alloy heterostructures

In recent years, there has been considerable interest in the physical properties of dilute alloy heterostructures, obtained by adding a concentration ranging from $\sim 10^{-3}$ to $\sim 10^{-2}$ of a foreign atom to the epilayer. Two materials systems will be reviewed: dilute nitrides and DMSs. In these fields, XAFS has played an important role, both as an atomic-level characterization tool to provide crucial feedback to optimize the growth procedures and also for providing an understanding of the physical properties on the basis of the local atomic structure.

3.4.1. Dilute nitrides
The addition to III–V semiconductor alloys of small quantities of an isoelectronic impurity characterized by very different electronegativity and size with respect to the constituents of the host lattice causes dramatic and unexpected changes in physical properties. For example,

the incorporation of a low concentration of nitrogen (∼0.1–1%) into GaAs, InGaAs, or GaP (the so-called dilute nitrides) leads to, among others, a giant (and counterintuitive) band-gap reduction and a reduction of the lattice parameter. These effects render dilute nitrides attractive for the fabrication of light emitters in the telecommunications wavelength range and of high-efficiency solar cells. XAFS has been used to address the relative atomic ordering in quaternary dilute nitrides and the structure of the complexes formed upon irradiation with hydrogen.

In the quaternary alloy (InGa)(AsN), the relative disposition of cations and anions on their sublattices is not uniquely determined by the atomic concentration. In fact, in the zinc blende structure each site of the anion (cation) sublattice can be occupied by In or Ga (As or N) without any limitation. Therefore, the question of the degree of atomic ordering, i.e., the relative number of each atomic bond as a function of composition, naturally arises. Specific examples of types of atomic ordering are the random case (relative number of each type of bond equal to the concentration) or short-range ordering (SRO); in the relevant case of low N concentration SRO is exhibited, in the extreme cases, by a sample having only N–In or only N–Ga bonds. For (InGa)(AsN) it has been predicted [165] that precisely this type of SRO exists, with a strong preference for In–N bonds over Ga–N ones. Moreover, SRO has been predicted to cause a significant blue shift of the band gap and hence may represent an intrinsic materials limitation, as it is the red shift that has applicative interest.

XAFS is clearly the technique of choice to address the issue of atomic ordering, due to its chemical sensistivity, and in 2003 two groups published important results in this field. Lordi et al. [166] concluded that while the as-deposited alloy has a random atomic distribution, annealing favors an increase of the relative number of In–N bonds. According to these authors, annealing brings the system toward a thermodynamically more stable state, characterized by a (not quantified) degree of SRO. Similar qualitative results were published by Uno et al. [167]. However, a quantitative measurement of the degree of SRO in as-deposited and annealed samples was not provided. A quantitative determination was instead provided by Ciatto et al. [168,169] who used combined In K-edge EXAFS and N K-edge XANES and showed that SRO does exist, but is significantly less pronounced than predicted. These authors have compared the In–N coordination number with the N concentration measured by nuclear reaction analysis. As-deposited samples exhibit a random atomic arrangement, while annealed samples show a statistically significant deviation towards SRO. The degree of SRO falls short of the maximum possible on a statistical basis and also of that predicted by Kim and Zunger [165]. This study nicely illustrates the power of XAFS to obtain quantitative information on ordering by comparing coordination numbers to average concentrations, a method well known in the study of amorphous systems [170].

Another fascinating physical properties of dilute nitrides alloys is that the dramatic changes induced by nitrogen in dilute nitrides can be reversed fully by hydrogen irradiation. In fact, XRD of hydrogenated GaAsN shows a disappearance of the diffraction peaks associated with the GaAsN epilayer and a recovery of the GaAs lattice parameter. Concomitantly, H irradiation leads to a nearly full reopening of the band gap. Such effects are technologically relevant as hydrogen is present in most growth processes and mass production steps of semiconductor devices. The nature of the hydrogen complexes responsible for these alterations of the physical properties has been addressed by a number of techniques and XAFS has played a decisive role. By combining in an original way N K-edge XANES measurements with simulations of their lineshape in the FMS framework performed on the basis of atomic coordinates obtained by ab initio simulations of the structure of the most stable N–H

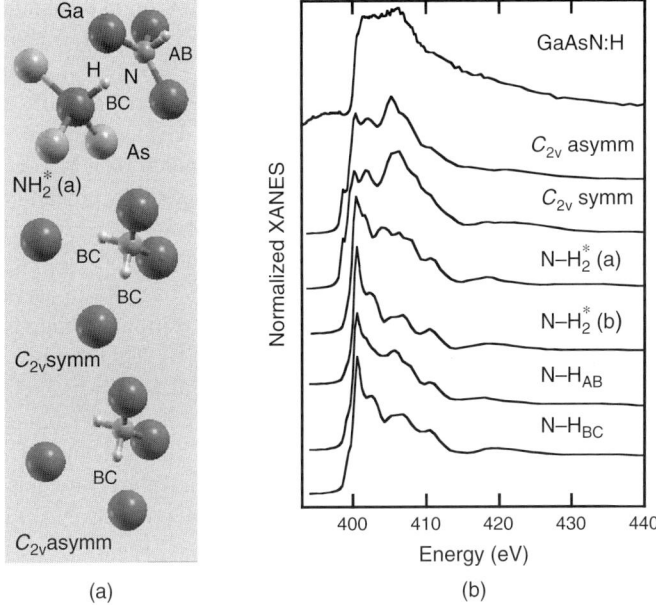

Fig. 16. (a) Sketch of some of the hydrogen complexes in GaAsN predicted to be most stable by ab initio DFT simulations: the N–H$_2^*$(a) and C_{2v} symmetric and asymmetric complexes hydrogen complexes in GaAsN. The N–H$_2^*$(b) complex has an N–H$_{BC}$–Ga–H$_{AB}$ configuration; similar "in line" configurations characterize the N–H$_{BC}$ and N–H$_{AB}$ mono–hydrogen complexes. (b) Experimental N K-edge XANES for hydrogenated GaAsN (top) and corresponding FMS simulations based on the atomic coordinates calculated via DFT in presence of the most stable monohydrogen– and dihydrogen– nitrogen complexes, some of which are sketched in Fig. 15(a) (reprinted with permission from Ref. [30]. Copyright (2005) by the American Physical Society).

complexes in hydrogenated GaAsN, Ciatto et al. [30] provided convincing evidence that dihydrogen–nitrogen complexes with C_{2v} symmetry are the most abundant. This finding contradicted all previous predictions of "in-line" N–H$_2^*$ complexes as the predominant species. In Fig. 16(a) we sketch the structures of some of the hydrogen complexes and in Fig. 16(b) the comparison between the simulations and the experiment; the best agreement is clearly found for the complexes with C_{2v} symmetry. This chapter nicely illustrates the high-energy resolution in XANES measurements which can be obtained in the soft X-ray range, the three-dimensional sensitivity of XANES, and the power of combined experimental–simulation studies (see Ref. [171] for another example).

3.4.2. Dilute magnetic semiconductors

A DMS (for a review of work up to 1988, see Ref. [172]) consists of a nonmagnetic semiconductor doped with a few percent of a magnetic ion (a transition metal, e.g., Mn or Fe). DMS thin films are typically obtained by MBE growth on an appropriate substrate. In a DMS, the ions (which ideally occupy substitutional sites in the semiconductor lattice) retain ferromagnetic alignment in the presence of spin-polarized free carriers. DMSs have been studied since the 1970s; initially much work was on II–VI-based alloys, later on III–Vs and more recently on nitrides; DMSs form the material basis of the emerging field of spintronics.

There are expectations that room temperature ferromagnetism might be obtained in DMSs. In fact, in 2000, Dietl et al. [173] predicted that room temperature ferromagnetism could be obtained by doping GaN or ZnO with a few percent of Mn. Three years later, ferromagnetism at 172 K was reported for p-type δ-doped GaAs:Mn [174] and, in 2005, the same group reported [175] a Curie temperature of 250 K in a specially designed heterostructure.

A great amount of work has been reported on DMSs, much of it is very recent and the panorama of results is continuously changing; sample deposition procedures are far from optimized and this inevitably entails that it is difficult to draw certain conclusions. For the nitride- and oxide-based systems, the reader is referred to the recent and extensive review by Chambers [176] for a materials science perspective. XAFS has played an important role in studies of DMS, both initially at the level of basic characterization and then to gain insight in the relation between local structure and physical properties. This is also due to the rich chemistry and variety of oxidation states of transition metal ions, which can be accurately characterized by the technique. Moreover, it is important to note that ferromagnetism can also be observed if ferromagnetic precipitates are present in the material,but this does not qualify as a true DMS; hence, an atomic level probe such as XAFS is of crucial value. For reasons of space, this section is limited to Mn-doped III–V semiconductors and GaN, which are probably the most intensely studied class of DMSs.

As often the case, the first studies in the field were of a preliminary character; for InAs:Mn, Krol et al. [177] reported that only a small fraction of Mn atoms was actually substitutional in the InAs lattice and most formed MnAs clusters. A few years later, using optimized sample preparation procedures and more accurate data analysis, the same group [178] clearly highlighted the role of concentration and of growth temperature: at low ($\sim$1%) concentration and growth temperature substitutional Mn was clearly detected while clusters with the local structure of MnAs precipitated at high concentrations or substrate temperatures. The origin of this behavior is the limited solubility of Mn in III–V semiconductors; to avoid the formation of MnAs inclusions, the growth temperature for Mn-doped III–Vs has to be between 200°C and 300°C, i.e., very low compared to the 500–650°C interval used for standard GaAs-based materials.

For the prototypical GaAs:Mn system more work has been reported. In this system, there is a close interplay between the local structure of Mn and the electronic properties. In fact, by using DFT simulations, Mahadevan and Zunger [179] have predicted that the formation energy of an interstitial Mn defect decreases as the Fermi level shifts toward the valence band, as can be occur as a consequence of Be codoping. Shioda et al. [180] initially reported complete Mn incorporation in substitutional sites, for concentrations $<$7% and growth temperatures quoted as 250–300°C; also, they measured a Mn–As bond length of 2.50 Å, which is 0.05 Å higher than the value for Ga–As and thus implies considerable local distortion. Later, Bacewicz et al. [181] studied samples grown between 220°C and 250°C with Mn concentrations below 4%; by a combined analysis of the XANES and EXAFS regions these authors were able to detect the presence of a significant fraction of interstitial Mn atoms; no correlation with other sample properties was reported.

More complex GaAs:Mn-based heterostructures have also been studied. Two methods have recently been proposed as alternatives to the growth of bulk GaAs:Mn alloys: the growth of digital GaAs/MnAs heterostructures, obtained by alternately depositing layers of GaAs and MnAs, and δ-doping of GaAs with Mn. The latter method consists of creating a δ-like doping profile in the growth direction by inserting in GaAs a submonolayer thick Mn layer. In both cases, the basic idea is to achieve locally higher Mn concentrations where high hole densities can be trapped. Soo et al. [182,183] have studied Mn/GaAs digital alloys. They found that in samples deposited at the low temperature of 275°C, Mn substitutes Ga and locally forms a GaMnAs alloy; upon annealing, a dramatic decrease of the first shell coordination number was found and the authors suggested that this was due to initial stages of precipitation of a MnAs phase. Later, the same group published [184] a convincing study, which combined the local structural sensitivity of Mn K-edge XAFS with an atom-specific magnetic probe such as X-ray magnetic circular dichroism (XMCD) performed at the Mn L-edge. They found a high ferromagnetic alignment of the Mn atoms and a local structure exclusively composed of substitutional Mn in the Ga sites, as in a random (GaMn)As alloy. In Fig. 17 we report Fourier transforms of Mn K-edge EXAFS data for a MnAs reference compound, a GaAs:Mn digital alloy and a (GaMn)As random alloy which clearly support this conclusion. More recently, d'Acapito et al. [185] have studied a series of Mn δ-doped GaAs samples, deposited at growth temperatures in the range 300–450°C, with and without Be codoping. In low-temperature samples, Mn was confirmed to be substitutional to Ga. Interestingly, a direct correlation between Be codoping (which lowers the Fermi level toward the valence band) and the presence of Mn interstitials was demonstrated; this study illustrates the ability of XAFS to detect relatively small fractions of defective site and to measure their local structure quantitatively.

XAFS has also made a significant contribution for the characterization of Mn-doped GaN films; again, it should be remarked that the nitrides are a more defective materials system than III–Vs and that a significant evolution in the near future is expected as the close interplay between growth and characterization will lead to improved materials. XAFS has provided important results on the site of Mn in GaN and its oxidation state.

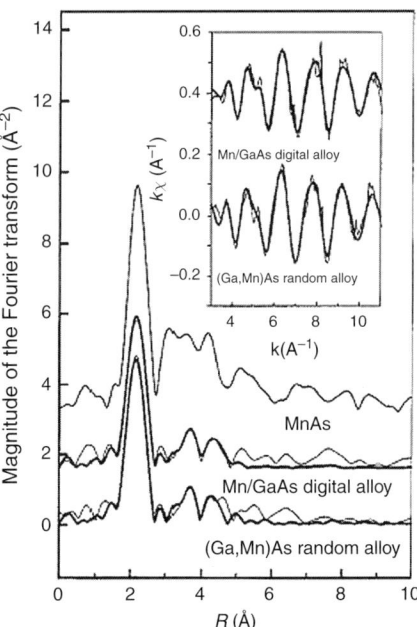

Fig. 17. Fourier transforms of Mn K-edge EXAFS data for a MnAs reference compound, a GaAs:Mn digital alloy and a (GaMn)As random alloy. The inset shows k-weighted EXAFS data (reprinted with permission from Ref. [184]. Copyright (2003) by the American Physical Society).

Soo et al. [186] studied 150-nm-thick films deposited by MBE on $Al_2O_3(0001)$ at substrate temperatures ranging between 400°C and 650°C, with a maximum Mn concentration of 2%; magnetic measurements indicated that the samples were paramagnetic. EXAFS measurements indicated that Mn occupied a substitutional Ga site with a Ga–N bond length which was found to ~ 0.07 Å greater than the Ga–N one. XANES measurements on the samples were compared to several oxide and fluoride reference compounds in which Mn has known oxidation states and it was concluded that Mn has the +2 oxidation state. While this is certainly a reasonable conclusion, it should be remarked that, ideally, such comparisons should be performed with reference compounds of known structure and composition as close as possible to the samples; it is an unfortunate situation that this is not strictly possible with Mn-doped GaN. In 2004, Thaler et al. [187] reported EXAFS data on Mn-doped GaN films grown by MBE on either $Al_2O_3(0001)$ or MOCVD GaN substrates with and without a 2-μm-thick GaN buffer layers. Samples grown at 700°C with the GaN buffer layers exhibited strong magnetization even above room temperature (but no remanence). The EXAFS measurements indicated that Mn was always substitutional, independently of the presence of the buffer layer; this result indicated the important role of the defect density of the GaN matrix in determining the magnetic behavior.

One of the concerns in the growth of Mn-doped GaN is the presence of multiple oxidation states of Mn and their spatial distribution. In this context, micro-XAFS has provided results that nicely illustrate the potential of the technique. By using an intense spot size of the order of 1 μm^2, Martínez-Criado et al. [188] performed a joint microfluorescence mapping and micro-XAFS experiment. Mapping of the Mn and Ga fluorescence intensity demonstrated their inhomogeneous distribution. Moreover, by performing maps of fluorescence intensity at specific excitation energies which correspond to XANES features of Mn in the Mn^0, Mn^{2+}, and Mn^{3+} oxidation states (chosen using as references metallic Mn, MnO, and Mn_2O_3) the authors were able to show the presence in the samples investigated of all three oxidation states, thus including metallic Mn, and their strongly inhomogeneous spatial distribution. The presence of Mn^{3+} in a substitutional site in MBE-grown samples on a GaN buffer layer up to a concentration of 5.7% was nicely demonstrated by Titov et al. [189] using high-resolution XANES measurements and their interpretation using band structure calculations. This study illustrates the usefulness of simulations in cases in which an ideal reference compound does not exist. We conclude this section by quoting a combined X-ray absorption and XMCD study at the Mn L-edge by Keavney et al. [190]; by comparison with first-principles band structure calculations, these authors propose that Mn preferentially populates Ga sites neighboring N split interstitials.

4. Summary and perspectives

XAFS has established itself as a precious tool to study the local structure in the field of materials science. The underlying theory, which has been outlined in Section 1, is well understood and a number of reliable analysis programs are now available. Owing to the many possible detection modes, most of which have described in Section 2, it can be applied to a great variety of systems and sensitivity to a particular phase, environment, or part of the sample can be enhanced on the basis of the experimental requirements. As a consequence, applications of XAFS to a wide variety of heterostructures and nanostructures, which have been reviewed in Section 3, have been widespread and have greatly contributed to the present understanding of the relation between atomic structure and physical properties.

There is no doubt that in the future XAFS will continue to be a valuable tool in this field of research. It is possible to outline specific areas in which progress is expected. The spatial resolution of synchrotron radiation techniques has greatly improved recently [100–102], and certainly further progress will occur in the near future. Micro-XAFS is expected to be very valuable in two areas: characterization of inhomogeneous samples (with possible feedback to sample growth) and study of the local atomic environment of individual nanostructures. Concerning the last point, the challenge is to record spectra of sufficient quality and energy range to allow a quantitative interpretation; a strong effort is required in this area. If this issue is resolved, it might be possible to perform "bond length maps" of nanostructures.

Time-resolved XAFS has recently emerged as a very promising field, due to the full use of the time structure of storage rings and the pump-and-probe scheme (for a review, see Ref. [191]). Differential XANES spectra of a metastable charge transfer state of an organimetallic molecule with 100 ps X-ray pulses have been reported by Saes et al. [192]. A lot of work is currently underway in synchrotron radiation laboratories and the future challenge is to use the femtosecond pulses of the LINAC-based sources. Present time resolutions range many orders of magnitude, from seconds to picoseconds [191]. In the field of semiconductor science, there have been only limited application of time-resolved XAFS, for example, studies of photo-excitation of GaAs [193] and of melting of Si [194]. More work in this area is expected, for example, to study the kinetics of annealing-induced bonding changes, crystallization, and amorphization, aggregation and growth of nanostructures in real time.

Work is in progress and interesting developments are possible also in the quantitative interpretation of the XANES region. XANES spectra contain a wealth of information, especially in view of the three-dimensional sensitivity; at present, even most of the advanced work in the solid state relies on a comparison between simulations and experiments, with semi-quantitative agreement. It would be a significant advancement to develop fitting methods, which have been devised in the field of biomolecules [31] for the quantitative determination of the three-dimensional structure of very small nanostructures, defects, and complexes in semiconductors. Finally, it must be mentioned that great progress is being made in ab initio structural simulations, with methods based on DFT theory. The accuracy of these methods and the reliability of XAFS analysis is such that joint experimental–theoretical investigations are now possible [30,171]. Simulations of the local atomic structure can be useful as a starting point for XAFS analysis, as a guide in the analysis itself and as a very useful comparison with expected results. Further progress will certainly take place in both theory and experiment and this will concern both the EXAFS and the XANES regions.

Acknowledgments

I am grateful for suggestions and critical readings of the manuscript by F. d'Acapito (Istituto Nazionale per la Fisica della Materia – Consiglio Nazionale delle Ricerche, Grenoble), P. Fons (Advanced institute for Industrial Science and Technology, Tsukuba), H.E. Mahnke (Hahn Meitner Institute, Berlin), and F. Rocca (Istituto di Fotonica e Nanotecnologie – Consiglio Nazionale delle Ricerche, Trento). My understanding of the principles, methods, and applications of X-ray absorption spectroscopy has greatly benefited from many years of close interaction with the scientific community linked to Frascati National Laboratory of Istituto Nazionale per la Fisica Nucleare.

This research has been supported in part by the University of Bologna, by Ministero dell'Università e della Ricerca (Italy) through PRIN grants, and by Istituto Nazionale per la

Fisica della Materia (INFM, Italy). Measurements at ESRF were supported in part by the ESRF public user program and by INFM; measurements at ELETTRA were supported by INFM and by the ELETTRA user program.

References

[1] D.E. Sayers, E.A. Stern, and F.W. Lytle, Phys. Rev. Lett. 27 (1971) 1204.
[2] R. Stumm von Bordwehr, Ann. Phys. (Paris) 14 (1989) 377.
[3] F.W. Lytle, J. Synchrotron Radiat. 6 (1999) 123.
[4] A. Filipponi, J. Phys.: Condens. Matter 13 (2001) R23.
[5] D. Raoux, J. Petiau, P. Bondot, et al., Rev. Phys. Appl. 15 (1980) 1079.
[6] P.A. Lee, P.H. Citrin, P. Eisenberger, and B.M. Kincaid, Rev. Mod. Phys. 53 (1981) 769.
[7] T.M. Hayes and J.B. Boyce, in H. Ehrenreich, F. Seitz, and D. Turnbull (Eds), Solid State Physics, vol 37, Academic, New York, 1982, p. 173.
[8] J.J. Rehr and R.C. Albers, Rev. Mod. Phys. 72 (2000) 621.
[9] C.R. Natoli, M. Benfatto, S. Della Longa, and K. Hatada, J. Synchrotron Radiat. 10 (2003) 26.
[10] E.D. Crozier, Nucl. Instrum. Methods Phys. Res. B 133 (1997) 134.
[11] B.K. Teo and D.C. Joy (Eds), EXAFS Spectroscopy, Plenum, New York, 1981.
[12] D.C. Koningsberger and R. Prins (Eds), Principles, Applications and Techniques of EXAFS, SEXAFS and XANES, Wiley, New York, 1988.
[13] C. Lamberti, Surf. Sci. Rep. 53 (2004) 1.
[14] J.C.H. Spence, Rep. Prog. Phys. 69 (2006) 725.
[15] W.L. Schaich, Phys. Rev. B 8 (1973) 4028.
[16] P.A. Lee and J.B. Pendry, Phys. Rev. B 11 (1975) 2795.
[17] C.R. Natoli and M. Benfatto, J. Phys. Coll. 48 (1987) C9-1077.
[18] P.J. Durham, in D.C. Koningsberger and R. Prins (Eds), Principles, Applications and Techniques of EXAFS, SEXAFS and XANES, Wiley, New York, 1988, p. 53.
[19] L. Fonda, J. Phys.: Condens. Matter 4 (1992) 8269.
[20] S.I. Zabinsky, J.J. Rehr, A. Ankudinov, et al., Phys. Rev. B 52 (1995) 2995.
[21] A. Filipponi, A. Di Cicco, and C.R. Natoli, Phys. Rev. B 52 (1995) 15122.
[22] B.H. Bransden and C.J. Joachain, Physics of Atoms and Molecules, 2nd edn, Prentice-Hall, Englewood Cliffs, 2003.
[23] E. Merzbacher, Quantum Mechanics, 3rd edn, John Wiley, New York, 1988.
[24] E. Groppo, C. Prestipino, C. Lamberti, et al., J. Phys. Chem. B 107 (2003) 4597.
[25] A. Bianconi, A. Di Cicco, N.V. Pavel, et al., Phys. Rev. B 36 (1987) 6426.
[26] E. Muller, O. Jepsen, O.K. Andersen, and J.W. Wilkins, Phys. Rev. Lett. 40 (1978) 270.
[27] P.J.W. Weijs, M.T. Czyżyk, J.F. van Acker, et al., Phys. Rev. B 41 (1990) 11899.
[28] J. Stöhr, NEXAFS Spectroscopy, Springer-Verlag, Berlin, 1992.
[29] M. Wilke, F. Farges, P.E. Petit, G.E. Brown Jr., and F. Martin, Am. Mineral. 80 (2001) 714.
[30] G. Ciatto, F. Boscherini, A. Amore Bonapasta, et al., Phys. Rev. B 71 (2005) 201301.
[31] S. Della Longa, A. Arcovito, M. Girasole, et al., Phys. Rev. Lett. 87 (2001) 155501.
[32] F. de Groot, Chem. Rev. 101 (2001) 1779.
[33] E.-E. Koch (Ed.), Handbook on Synchrotron Radiation, vol 1, North-Holland, Amsterdam, 1983.
[34] J. Baruchel, J.L. Hodeau, M.S. Lehmann, J.R. Regnard and C. Schlenker (Eds), Neutron and Sychrotron Radiation for Consensed Matter Studies, vol 1: Theory, Instruments and Methods, Springer Verlag, Heidelberg, and Les Editions de Physique, Les Ulis, 1993.
[35] D. Attwood, Soft X-rays and Extreme Ultraviolet Radiation, Cambridge University Press, Cambridge, 1999.
[36] P.J. Duke, Synchrotron Radiation, Production and Properties, Oxford University Press, Oxford, 2000.

[37] J. Als-Nielsen and D. McMorrow, Elements of Modern X-ray Physics, Wiley, New York, 2001.
[38] S. Mobilio and G. Vlaic (Eds), Synchrotron Radiation: Fundamentals, Methodologies and Applications, Conference Proceedings, vol 82, Società Italiana di Fisica, Bologna, 2003.
[39] E.A. Stern and K. Kim, Phys. Rev. B 23 (1981) 3781.
[40] M.C. Ridgway, G. de M. Azevedo, C.J. Glover, et al., Nucl. Instrum. Meth. B 218 (2004) 421.
[41] J. Jaklevic, J.A. Kirby, M.P. Klein, et al., Solid State Commun. 23 (1977) 679.
[42] J. Goulon, C. Goulon-Ginet, R. Cortes, and J.M. Dubois, J. Phys. (Paris) 43 (1982) 539.
[43] Z. Tan, J.I. Budnick, and S.M. Heald, Rev. Sci. Instrum. 60 (1989) 1021.
[44] L. Tröger, D. Arvanitis, K. Baberschke, et al., Phys. Rev. B 46 (1992) 3283.
[45] S. Eisebitt, T. Böske, J.E. Rubensson, and W. Eberhardt, Phys. Rev. B 47 (1993) 14103.
[46] P. Pfalzer, J.P. Urbach, M. Klemm, et al., Phys. Rev. B 60 (1999) 9335.
[47] R. Carboni, S. Giovannini, G. Antonioli, and F. Boscherini, Phys. Scripta T115 (2005) 986.
[48] H. Oyanagi, M. Martini, and M. Saito, Nucl. Instrum. Meth. A 403 (1998) 58.
[49] G. Derbyshire, K.C. Cheung, P. Sangsingkeow, and S.S. Hasnain, J. Synchrotron Radiat. 6 (1999) 62.
[50] S. Spiga, R. Mantovan, M. Fanciulli, et al., Phys. Rev. B 68 (2003) 205419.
[51] F. d'Acapito, S. Colonna, S. Pascarelli, et al., ESRF Newslett. 30 (1998) 42.
[52] G. Ciatto, F. d'Acapito, F. Boscherini, and S. Mobilio, J. Synchrotron Radiat. 11 (2004) 278.
[53] Y. Takeda, H. Ofuchi, H. Kyouzu, et al., J. Synchrotron Radiat. 12 (2005) 494.
[54] W. Gudat and C. Kunz, Phys. Rev. Lett. 29 (1972) 169.
[55] M.E. Kordesh and R.W. Hoffman, Phys. Rev. B 29 (1984) 491.
[56] T. Guo and M.L. den Boer, Phys. Rev. B 31 (1985) 6233.
[57] G. Tourillon, E. Dartyge, A. Fontaine, et al., Phys. Lett. A 121 (1987) 251.
[58] C.E. Bouldin, R.A. Forman, and M.I. Bell, Phys. Rev. B 35 (1987) 1429.
[59] P.A. Lee, Phys. Rev. B 13 (1976) 5261.
[60] P.H. Citrin, P. Eisenberger, and R.C. Hewitt, Phys. Rev. Lett. 41 (1978) 309.
[61] A. Erbil, G.S. Cargil III, R. Frahm, and R.F. Boehme, Phys. Rev. B 37 (1988) 2450.
[62] S.L.M. Schroeder, G.D. Moggridge, R.M. Ormerod, et al., Surf. Sci. 324 (1995) L371.
[63] R. Nakajima, J. Stöhr, and Y.U. Idzerda, Phys. Rev. B 59 (1999) 6421.
[64] T.J. Regan, H. Oldag, C. Stamm, et al., Phys. Rev. B 64 (2001) 21422.
[65] A. Bianconi, D. Jackson, and K. Monahan, Phys. Rev. B 17 (1978) 2021.
[66] J. Goulon, P. Tola, M. Lemonnier, and J. Dexper-Ghys, Chem. Phys. 78 (1983) 347.
[67] S. Emura, T. Moriga, J. Takizawa, et al., Phys. Rev. B 47 (1993) 6918.
[68] R.F. Pettifer and A.J. Bourdillon, J. Phys. C 20 (1987) 329.
[69] L. Sonderholm, G.K. Liu, M.R. Antonio, and F.W. Lytle, J. Chem. Phys. 109 (1998) 6745.
[70] R.F. Pettifer, A. Glanfield, S. Gardelis, et al., Physica B 208 & 209 (1995) 484; D.A. Hill, R.F. Pettifer, S. Gardelis, et al., J. Phys. IV (Paris) 7-C2 (1997) 553.
[71] A. Rogalev and J. Goulon, in T.K. Sham (Ed.), Chemical Applications of Synchrotron Radiation, Part II, World Scientific, Singapore, 2002, p. 707.
[72] S. Larcheri, C. Armellini, F. Rocca, et al., Superlattices Microstruct. 39 (2006) 267.
[73] R.A. Rosenberg, G.K. Shenoy, L.-C. Tien, et al., Appl. Phys. Lett. 89 (2006) 093118.
[74] R.F. Boehme, G.S. Cargill III, W. Weber, and T. Jackson, J. Appl. Phys. 58 (1985) 811.
[75] T.D. Hu, Y.N. Xie, S. Qiao, et al., Phys. Rev. B 50 (1994) 2216.
[76] M. Ishii, Y. Yoshino, K. Takarabe, and O. Shimomura, Appl. Phys. Lett. 74 (1999) 2672 and J. Appl. Phys. 88 (2000) 3962; M. Ishii, Phys. Rev. B 65 (2002) 085310.
[77] M. Ishii, Jpn. J. Appl. Phys. 40 (2001) 7129.
[78] M. Ishii, Physica B 308–310 (2001) 1153; Jpn. J. Appl. Phys. 41 (2002) 4415; Physica B 340–342 (2003) 1142.
[79] M. Ishii, N. Rigopoulos, N. Poolton, and B. Hamilton, Physica B, 376–377 (2006) 950.
[80] J. Bollmann, S. Knack, J. Weber, et al., Phys. Rev. B 68 (2003) 125206.
[81] S. Koonin, Nature 354 (1991) 468.

[82] F. Gatti, F. Fontanelli, M. Galeazzi, et al., Nature 397 (1999) 137.
[83] G. Benedek, E. Fiorini, A. Giuliani, et al., Nucl. Instrum. Methods A 426 (1999) 147.
[84] D. Pergolesi, F. Gatti, M. Raseti, et al., AIP Conf. Proc. 605 (2002) 367.
[85] C. Arnaboldi, G. Benedek, C. Brofferio, et al., Phys. Rev. Lett. 96 (2006) 042503.
[86] L.G. Parratt, Phys. Rev. 95 (1954) 359.
[87] R. Barchewitz, M. Cremonese-Visicato, and G. Onori, J. Phys. C: Solid State Phys. 11 (1978) 4439.
[88] G. Martens and P. Rabe, Phys. Status Solidi (a) 58 (1980) 415.
[89] S. Pizzini, K.J. Roberts, G.N. Greaves, et al., Rev. Sci. Instrum. 60 (1989) 2525.
[90] B. Poumellec, R. Cortes, F. Lagnel, and G. Tourillon, Physica B 158 (1989) 282.
[91] S.M. Heald, H. Chen, and J.M. Tranquada, Phys. Rev. B 38 (1988) 1016.
[92] F. d'Acapito, I. Davoli, P. Ghigna, and S. Mobilio, J. Synchrotron Radiat. 10 (2003) 260.
[93] D.T. Jiang, E.D. Crozier, and B. Heinrich, Phys. Rev. B 44 (1991) 6401.
[94] C. Lamberti, E. Groppo, C. Prestipino, et al., Phys. Rev. Lett. 91 (2003) 046101.
[95] P. Luches, V. Bellini, S. Colonna, et al., Phys. Rev. Lett. 96 (2006) 106106.
[96] P.H. Citrin, J. Phys. (Paris) C 8 (1986) 437.
[97] P.H. Citrin, Surf. Sci. 299 & 300 (1994) 199.
[98] P. Lagarde, Ultramicroscopy 86 (2001) 255.
[99] H. Oyanagi, I. Owen, M. Grimshaw, et al., Rev. Sci. Instrum. 66 (1995) 5477; H. Oyanagi, R. Shioda, Y. Kuwahara, and K. Haga, J. Synchrotron Radiat. 2 (1995) 99.
[100] A. Locatelli, A. Bianco, D. Cocco, et al., J. Phys. IV 104 (2004) 99; A.F. Ratto, A. Locatelli, S. Fontana, et al., Small 2 (2006) 401.
[101] A. Somogyi, R. Tocoulou-Tachoueres, G. Martínez-Criado, et al., J. Synchrotron Radiat. 12 (2005) 208.
[102] A.M. Flank, G. Cauchon, P. Lagarde, et al., Nucl. Instrum. Methods Phys. Res. B 246 (2006) 269.
[103] O. Dhez, M. Rodrigues, R. Felici, et al., Proc. Int. Conf. on Synchrotron Radiation Instrumentation, Deagu, Korea, 2006.
[104] J.C. Mikkelsen Jr and J.B. Boyce, Phys. Rev. Lett. 49 (1982) 1412; Phys. Rev. B 28 (1983) 7130.
[105] A. Balzarotti, A. Kisiel, N. Motta, et al., Phys. Rev. B 30 (1984) 2295.
[106] Y. Cai and M.F. Thorpe, Phys. Rev. B 46 (1992) 15872, 15879.
[107] I.-K. Jeong, F. Mohiuddin-Jacobs, V. Petkov, et al., Phys. Rev. B 63 (2001) 205202.
[108] P. Juhás, D.M. Cherba, P.M. Duxbury, et al., Nature 440 (2006) 655.
[109] P.F. Peterson, T. Proffen, I.K. Jeong, et al., Phys. Rev. B 63 (2001) 165211.
[110] M.G. Proietti, S. Turchini, J. Garcia, et al., J. Appl. Phys. 78 (1995) 6574.
[111] J.C. Woicik, C.E. Bouldin, M.I. Bell, et al., Phys. Rev. B 43 (1991) 2419.
[112] C. Lamberti, S. Bordiga, F. Boscherini, et al., Appl. Phys. Lett. 64 (1994) 1430.
[113] M.G. Proietti, S. Turchini, F. Martelli, et al., J. Appl. Phys. 77 (1995) 62.
[114] Y. Kuwahara, H. Oyanagi, R. Shioda, et al., Jpn. J. Appl. Phys. 33 (1994) 5631.
[115] J.C. Woicik, J.G. Pellegrino, S.H. Southworth, et al., Phys. Rev. B 52 (1995) R2281.
[116] F. Romanato, D. De Salvador, M. Natali, et al. in D. Cornuejols (Ed.) ESRF Highlights 1996–97, ESRF, Grenoble, 1997; F. Romanato, D. De Salvador, M. Berti, et al., Phys. Rev. B 57 (1988) 14619.
[117] J.C. Woicik, J.G. Pellegrino, B. Steiner, et al., Phys. Rev. Lett. 79 (1997) 5026.
[118] J. Woicik, Phys. Rev. B 57 (1998) 6266.
[119] F. d'Acapito, J. Appl. Phys. 96 (2004) 369.
[120] G. Apai, J.F. Hamilton, J. Stohr, and A. Thompson, Phys. Rev. Lett. 43 (1979) 165.
[121] P.A. Montano, W. Schulze, B. Tesche, et al., Phys. Rev. B 30 (1984) 672.
[122] A. Balerna and S. Mobilio, Phys. Rev. B 34 (1986) 2293.
[123] T.U. Schülli, J. Stangl, Z. Zhong, et al., Phys. Rev. Lett. 90 (2003) 066105.
[124] F. Boscherini, G. Capellini, L. Di Gaspare, et al., Appl. Phys. Lett. 76 (2000) 682.
[125] A.V. Kolobov, H. Oyanagi, K. Brunner, et al., Appl. Phys. Lett. 78 (2001) 451.

[126] S.B. Erenburg, N.V. Bausk, L.N. Mazalov, et al., Nucl. Instrum. Methods Phys. Res. A 467 (2001) 1229.
[127] F. Boscherini, G. Capellini, L. Di Gaspare, et al., Thin Solid Films 380 (2000) 173.
[128] N. Motta, F. Rosei, A. Sgarlata, et al., Mater. Sci. Eng. B 88 (2002) 264.
[129] Z. Kovats, M. Rausher, H. Metzger, et al., Phys. Rev. B 62 (2000) 8223.
[130] A.V. Kolobov, J. Appl. Phys. 87 (2000) 2926; A.V. Kolobov, H. Oyanagi, K. Brunner, et al., J. Vac. Sci. Technol. A 20 (2002) 1116; A.V. Kolobov, H. Oyanagi, A. Frenkel, et al., Nucl. Instrum. Methods Phys. Res. B 199 (2003) 174.
[131] S. Sun, S. Wei, A.V. Kolobov, et al., Phys. Rev. B 71 (2005) 245334.
[132] A.V. Kolobov, A.A. Shklyaev, H. Oyanagi, et al., Appl. Phys. Lett. 78 (2001) 2563.
[133] F. d'Acapito, M. de Seta, G. Capellini, et al., Nucl. Instrum. Methods Phys. Res. B 246 (2006) 64.
[134] M. Galluppi, M. Capizzi, F. Boscherini, et al., Appl. Phys. Lett. 17 (2001) 186.
[135] F. d'Acapito, S. Colonna, F. Arciprete, et al., Nucl. Instrum. Methods Phys. Res. B 200 (2003) 85.
[136] H. Renevier, M.G. Proietti, S. Grenier, et al., Mater. Sci. Eng. B 101 (2003) 174.
[137] A. Cheung, G. de M. Azevedo, C.J. Glover, et al., Appl. Phys. Lett. 84 (2004) 278.
[138] M.C. Ridgway, G. de M. Azevedo, R.G. Elliman, et al., Phys. Rev. B 71 (2005) 094107.
[139] E. Borsella, M.A. Garcia, G. Mattei, et al., J. Appl. Phys. 90 (2001) 4467.
[140] T.M. Hayes, L.B. Lurio, J. Pant, and P.D. Persans, Solid State Commun. 117 (2001) 627.
[141] T.M. Hayes, L.B. Lurio, J. Pant, and P.D. Persans, Phys. Rev. B 63 (2001) 155417.
[142] S. Spiga, R. Mantovan, M. Fanciulli, et al., Phys. Rev. B 68 (2003) 205419.
[143] P. Kluth, B. Johannessen, V. Giraud, et al., Appl. Phys. Lett. 85 (2004) 3561.
[144] P. Kluth, B. Johannessen, G.J. Foran, et al., Phys. Rev. B 74 (2006) 014202.
[145] T.K. Sham, D.T. Jiang, I. Coulthard, et al., Nature 363 (1993) 331.
[146] R.F. Pettifer, A. Glanfield, S. Gardelis, et al., Physica B 208, 209 (1995) 484.
[147] G. Dalba, P. Fornasini, M. Grazioli, et al., Nucl. Instrum. Methods Phys. Res. B 97 (1995) 322.
[148] G. Dalba, P. Fornasini, R. Grisenti, et al., Appl. Phys. Lett. 74 (1999) 1454.
[149] G. Dalba, N. Daldosso, D. Diop, et al., J. Lumines. 80 (1999) 103.
[150] G. Dalba, N. Daldosso, P. Fornasini, et al., Phys. Rev. B 62 (2000) 9911.
[151] Y.F. Zhang, L.S. Liao, W.H. Chan, et al., Phys. Rev. B 61 (2000) 8298.
[152] T.K. Sham, S.J. Naftel, P.-S.G. Kim, et al., Phys. Rev. B 70 (2004) 045313.
[153] N. Daldosso, M. Luppi, S. Ossicini, et al., Phys. Rev. B 68 (2003) 085327.
[154] K.E. Miyano, J.C. Woicik, L.H. Robins, et al., Appl. Phys. Lett. 70 (1997) 2108.
[155] F. Boscherini, R. Lantier, A. Rizzi, et al., Appl. Phys. Lett. 74 (1999) 3308.
[156] F. d'Acapito, F. Boscherini, S. Mobilio, et al., Phys. Rev. B 66 (2002) 205411.
[157] M. Katsikini, E.C. Paloura, and T.D. Moustakas, Appl. Phys. Lett., 69 (1996) 4206.
[158] M. Katsikini, E.C. Paloura, and T.D. Moustakas, J. Appl. Phys. 83 (1998) 1437.
[159] Ch. Brouder, J. Phys.: Condens. Matter 2 (1990) 701.
[160] K. Lawniczak-Jablonska, T. Suski, Z. Lillental-Weber, et al., Appl. Phys. Lett. 70 (1997) 2711.
[161] M. Katsikini, E.C. Paloura, M. Fieber-Erdermann, et al., J. Electron Spectrosc. Relat. Phenom. 101–103 (1999) 695.
[162] M. Katsikini, E.C. Paloura, M. Fieber-Erdermann, et al., Phys. Rev. B 56 (1997) 13380.
[163] M. Katsikini, H. Rossner, M. Fieber-Erdermann, et al., J. Synchrotron Radiat. 6 (1999) 561.
[164] N. Dimakis, G. Bunker, M. Katsikini, and E.C. Paloura, J. Synchrotron Radiat. 8 (2001) 258.
[165] K. Kim and A. Zunger, Phys. Rev. Lett. 86 (2001) 2609.
[166] V. Lordi, V. Gambin, S. Friedrich, et al., Phys. Rev. Lett. 90 (2003) 145505.
[167] K. Uno, M. Yamada, T. Takizawa, and I. Tanaka, Jpn. J. Appl. Phys. 43 (2004) 1944.
[168] G. Ciatto, F. d'Acapito, L. Grenoulliet, et al., Phys. Rev. B 68 (2003) 161201(R).
[169] G. Ciatto and F. Boscherini, J. Phys.: Condens. Matter 16 (2004) S3141.
[170] F. Boscherini, Mater. Res. Soc. Symp. Proc. 258 (1992) 217.
[171] P. Fons, H. Tampo, A.V. Kolobov, et al., Phys. Rev. Lett. 96 (2006) 045504.
[172] J.K. Furdyna, J. Appl. Phys. 64 (1988) 29.

[173] T. Dietl, H. Ohno, F. Matsukura, et al., Science 287 (2000) 1019.
[174] A.M. Nazmul, S. Sugahara, and M. Tanaka Phys. Rev. B 68 (2003) 075202.
[175] A.M. Nazmul, T. Ameniya, Y. Shuto, et al., Phys. Rev. Lett. 95 (2005) 017201 and Phys. Rev. Lett. 96 (2006) 149901.
[176] S.A. Chambers, Surf. Sci. Rep. 61 (2006) 345.
[177] A. Krol, Y.L. Soo, S. Huang, et al., Phys. Rev. B 47 (1993) 7187.
[178] Y.L. Soo, S. Huang, Z.H. Ming, et al., Phys. Rev. B 53 (1996) 4905.
[179] P. Mahadevan and A. Zunger, Phys. Rev. B 68, (2003) 75202.
[180] R. Shioda, K. Ando, T. Hayashi, and M. Tanaka, Phys. Rev. B 58 (1998) 1100.
[181] R. Bacewicz, A. Twaróg, A. Malinowska, et al., J. Phys. Chem. Solids 66 (2005) 2004.
[182] Y.L. Soo, G. Kioseoglu, X. Chen, et al., Appl. Phys. Lett. 80 (2002) 2654.
[183] Y.L. Soo, S. Wang, S. Kim, et al., Appl. Phys. Lett. 83 (2003) 2354.
[184] Y.L. Soo, G. Kioseoglou, S. Kim, et al., Phys. Rev. B 67 (2003) 214401.
[185] F. d'Acapito, G. Smolentsev, F. Boscherini, et al., Phys. Rev. B 73 (2006) 035314.
[186] Y.L. Soo, G. Kioseoglou, S. Kim, et al., Appl. Phys. Lett. 79 (2001) 3926.
[187] G. Thaler, R. Frazier, B. Gila, et al., Appl. Phys. Lett. 84 (2004) 2578.
[188] G. Martínez-Criado, A. Somogyi, A. Homs, et al., Appl. Phys. Lett. 87 (2005) 061913.
[189] A. Titov, X. Biquard, D. Halley, et al., Phys. Rev. B 72 (2005) 115209.
[190] D.J. Keavney, S.H. Cheung, S.T. King, et al., Phys. Rev. Lett. 95 (2005) 257201.
[191] C. Bressler and M. Chergui, Chem. Rev. 104 (2004) 1781.
[192] M. Saes, C. Bressler, R. Abela, et al., Phys. Rev. Lett. 90 (2003) 047403.
[193] B.W. Adams, M.F. DeCamp, E.M. Dufresne, and D.A. Reis, Rev. Sci. Instrum. 73 (2002) 4150.
[194] S.L. Johnson, P.A. Heinmann, A.M. Lindenberg, et al., Phys. Rev. Lett. 91 (2003) 157403.

Characterization of Semiconductor Heterostructures and Nanostructures
Edited by C. Lamberti
© 2008 Elsevier B.V. All rights reserved
Doi: 10.1016/B978-0-444-53099-8.00010-5

10

Nanostructures in the light of synchrotron radiation: surface-sensitive X-ray techniques and anomalous scattering

Till Metzger[1], Vincent Favre-Nicolin[2,3], Gilles Renaud[2], Hubert Renevier[2], and Tobias Schülli[2]

[1]*European Synchrotron Radiation Facility, ESRF, Grenoble, France*
[2]*CEA, Grenoble, France*
[3]*University of Grenoble, France*

Abstract The potential of X-ray scattering techniques using synchrotron radiation is described. It is demonstrated that grazing incidence (GI) conditions make X-rays near surface sensitive, which is mandatory for the characterization of semiconductor nanostructures grown in a self-organized way on or buried near the sample surface. The weak scattering from the nanostructures is thereby enhanced and can further be strengthened by using "anomalous dispersion effects", also known as resonant scattering, which makes the techniques element sensitive. When combined with in situ studies, the growth of semiconductor nanostructures can be followed in real time. The wealth of results reported in the literature is highlighted by several examples: GI small angle X-ray scattering (GISAXS) of Ge islands growth on Si(001) and the vertical stacking of GaN dots in AlN multilayers. The remaining case studies are dedicated to the combination of GI diffraction (GID) and anomalous scattering using the techniques of iso-strain scattering and multiwavelength anomalous diffraction (MAD). The concluding outlook gives a perspective on the use of the beam coherence which allows for the reconstruction of objects by phase-retrieval algorithms and thereby to accomplish a model-independent characterization of nanostructures with respect to shape, strain and composition. This technique (coherent diffraction imaging (CDI)) is currently under development at several synchrotrons around the world.

Keywords semiconductor nanostructures, synchrotron radiation, surface-sensitive techniques, grazing incidence, anomalous scattering

1. Introduction

As the characteristic dimension of semiconductor devices decreased into the regime of self-organized nanostructures with sizes of about 100 nm, advanced characterization tools are strongly needed to understand and control nanostructure properties. Nanomaterials, i.e. materials being different from its bulk phase due to their small dimension, are in the focus of intense research worldwide. The continuous shrinking of the dimension in current semiconductor devices is reaching the nanometre scale, where the performance is increasingly determined by confinement effects such as tunnelling and band structure changes induced by size, strain and composition. Recent developments in semiconductor industry are clearly targeted towards introducing nanostructures into devices.

The progress in nanoscience and nanotechnology asks for tools to characterize the structure of objects both on the mesoscopic and atomic levels. This is especially relevant in semiconductor devices based on heterostructures. Several different routes using synchrotron radiation have been pursued so far. X-ray diffraction (XRD) was very successful in determining shape, chemical composition and strain distribution within and around nanostructures, based on measurements on large nanostructure ensembles [1–5]. A comprehensive overview on the use of synchrotron radiation techniques in the characterization of strained semiconductor heterostructures and thin films is given in the review article by C. Lamberti, *Surf. Sci. Rep.*, **53** (2004) 1.

The basic problem in X-ray scattering from nanostructures is the weak scattering signal due to the small amount of material which forms the nanostructures, typically several monolayers (ML) on top of single crystalline substrates. In case of self-organized, strain-driven heterostructures, the aim is to obtain nanostructures that are dislocation free and coherently connected with the substrate, thus creating a complex mutual strain scenario, that needs to be disentangled by appropriate X-ray scattering techniques, because X-rays are highly sensitive to strain. To enhance the scattering signal from the nanostructures and suppress the strong scattering from the substrate, two techniques have been combined and will be described in this chapter: grazing incidence (GI) techniques and anomalous (or resonant) diffraction, which make X-rays surface sensitive and add chemical sensitivity, respectively.

For this chapter a group of scientists got together and report on the work done at the European Synchrotron Radiation Facility (ESRF) in Grenoble. This third generation dedicated X-ray source has the high brilliance and energy tunability indispensably needed for the investigation of nanostructures by the techniques mentioned above. The typical experimental set-ups for GI and anomalous scattering experiments can be found on the web pages of the three beamlines from which recent results will be reported here [6].

In Section 2, the scientific background for the GI small angle X-ray scattering (GISAXS) and GI diffraction (GID) will be described, in which the refraction of X-rays at an interface is of fundamental importance. The surface sensitivity at GI is based on the fact that the index of refraction for condensed matter is smaller than one and thus a critical angle for total external reflection exists. For incident angles below this critical angle, an evanescent wave is "implanted" below the surface. Its amplitude is attenuated exponentially with a typical $1/e$ penetration depth of some tens of nanometres. It is this evanescent wave which is used for all scattering experiments and which provides the surface sensitivity of GI techniques. In the second part of Section 2, the exploitation of resonant scattering techniques, such as multiwavelength anomalous diffraction (MAD) and diffraction anomalous fine-structure (DAFS) spectroscopy, is explained. Both techniques allow for chemical analysis of nanostructures.

In Section 3, examples of application of these techniques will be given based on recent publications. First it is shown that GISAXS is well suited to study shape, facets and ordering phenomena of nanostructures, where ordering can take place both laterally and in growth direction. It will be demonstrated that GISAXS is also well suited to perform in situ measurements during the growth of islands, thus revealing details of the self-organized growth process. Combined with GID measurements, which are sensitive to the atomic structure and thus the crystalline core of the islands, all important structural properties, i.e. including the strain distribution in the islands and the substrate, can be determined. The missing part, the chemical composition, is revealed by applying the simplest anomalous diffraction measurement at two X-ray energies close to the absorption edge of the island material. In most cases, the chemical composition of the islands is different from the nominally deposited material due to interdiffusion of substrate material into the islands, i.e. at high growth temperatures. It will be demonstrated that anomalous diffraction combined with "iso-strain scattering" [7] allows to determine the 3D compositional profile in Ge islands on Si. In the following subsection, the MAD technique is explained in detail and its application to the system InAs/InP and GaN/AlN multilayers is reported.

The chapter will be concluded by a summary and future perspectives for which the coherence properties and the possibility to focus the X-rays to submicron size will become important in the investigation of single nanostructured objects, which will allow for the measurement of the variation of physical parameters rather than their means and correlate their structure with the function at a particular location in a device.

2. Scientific background for grazing incidence X-ray techniques on nanostructures

2.1. General considerations

The physical properties of semiconductor heterostructures of reduced dimensions clearly depend strongly on, and are markedly affected by, their structural properties, i.e. their morphology, composition and structure at the atomic level. Their morphology includes their shape, size and organization on a substrate, and the distribution of these morphological parameters over the assembly of heterostructures, as well as the topology (steps, roughness) of the substrate surface or of the interfaces present in the sample. The atomic structure includes their strain with respect to an ideal unstrained state, their composition, possible interdiffusion between different components of the heterostructures, and very importantly, the presence of defects such as point defects, dislocations or stacking faults. It is thus essential to characterize these properties as deeply as possible, to predict and/or understand their physical properties. As discussed elsewhere in this book, these morphological and structural properties are addressed by a wealth of techniques such as laboratory XRD, transmission electron microscopy (TEM) or scanning electron microscopy (SEM), scanning tunneling microscopy (STM) or atomic force microscopy (AFM), and synchrotron radiation-based X-ray techniques. All these techniques have advantages and drawbacks, which will not be discussed in detail here, and are thus complementary to each other. The strengths of X-ray techniques are primarily the high resolution on the strain, the possibility to probe all correlation length scales, e.g. morphology and atomic structure, and the possibility to average over several square millimetres and thus basically provide statistically average information. Most of the time X-ray data can be quantitatively analysed using a simple formalism of single scattering, with no or only few assumptions. The measurements usually require no special sample environment or preparation: it is a non-destructive probe and the penetration depth can be varied between a few nanometres to micrometres, thus probing both surface or near-surface structures, as well as the bulk of the sample. The chemical contrast of a given element can be enhanced by performing anomalous scattering close to a specific absorption edge, thus yielding compositional information. Very importantly, X-ray scattering can also be applied in situ, in real environment, like in ultra-high vacuum (UHV) during the growth of nanostructures or under partial gas pressures during reactions.

However, X-ray scattering also has drawbacks. The information is obtained in reciprocal space, and thus has to be "translated" in the real world. Synchrotron radiation has to be used to characterize nanostructures because the amount of scattering material, and thus the scattered intensity, is rather small. It is extremely difficult to characterize a single nano-object because of the beam size, divergence and limitations of X-ray optics (although this field is going to develop). As only scattered intensities can be measured, the phase information is lost, which means that some model assumptions are most often mandatory. This problem can be overcome if the coherence of the synchrotron radiation is employed. The technique of coherent XRD imaging is on the rise and allows for a model-free reconstruction of small crystals and nanostructures.

2.2. Surface-sensitive X-ray scattering

To investigate heterostructures such as quantum dots (QDs) that have a small dimension perpendicular to the surface, most of the time, the incident X-ray wave vector, $\mathbf{k}_i$, is kept at a grazing angle with respect to the sample surface. This minimizes the unwanted background scattering (both elastic and inelastic) emanating from the bulk, while promoting the near-surface scattering. Figure 1 shows the schematic drawing of such measurements at grazing incident scattering geometry. It is identical to the 3D case except for the additional condition that the incident angle α_i is kept constant. The scattered beam, of wave vector $\mathbf{k}_f$, makes the scattering angle 2θ with respect to the incident wave vector. It is detected at a direction defined by slits and makes an angle α_f with respect to the sample surface and an in-plane angle δ with respect to the transmitted beam. The momentum transfer is defined as $\mathbf{Q} = \mathbf{k}_f - \mathbf{k}_i$ and is often decomposed into two components, $\mathbf{Q}_\parallel$ and $\mathbf{Q}_\perp$, respectively parallel and perpendicular to the surface. The absolute value of $\mathbf{Q}_\perp$ is a function of α_i and α_f: $Q_\perp = k(\sin \alpha_i + \sin \alpha_f)$ with $k = 2\pi/\lambda$, where λ is the wavelength.

When all angles are small, the momentum transfer is also small, typically between 0 and 1 nm^{-1}, and hence large dimensions are probed in real space. The corresponding techniques are the (i) X-ray reflectivity (XRR), specular to probe the density profile perpendicular to the surface and off-specular to probe large lateral electron density correlations (along q_x); and (ii) the GISAXS, which is used to probe the morphology parallel to the surface (along q_y) and perpendicular to it (along q_z), at intermediate length scales (typically between a few nanometres and a few hundreds of nanometres).

Large scattering angles 2θ allow probing the order at the atomic scale, with two corresponding techniques: GID, which is the typical technique to probe the crystallographic lattice of nanostructures, and surface XRD (SXRD), which is basically the same, but applied to the atomic structure determination (i.e. atomic positions) of surfaces and interfaces. When α_i and α_f are very small, $\mathbf{Q} \sim \mathbf{Q}_\parallel$, the scattering plane is nearly parallel to the surface, and diffracting lattice planes are perpendicular to it. The scattering geometry being defined by the incident beam and detector directions, one has only to rotate the sample about its surface normal to bring these lattice planes into diffraction condition, which occurs when they make an angle $(2\theta)/2$ with respect to both the incident and the scattered beam. In this way, the long-range periodicity parallel to the surface is probed. It is often useful to measure the scattered intensity as a function of $Q_\perp$, which is achieved by increasing α_f while keeping GI.

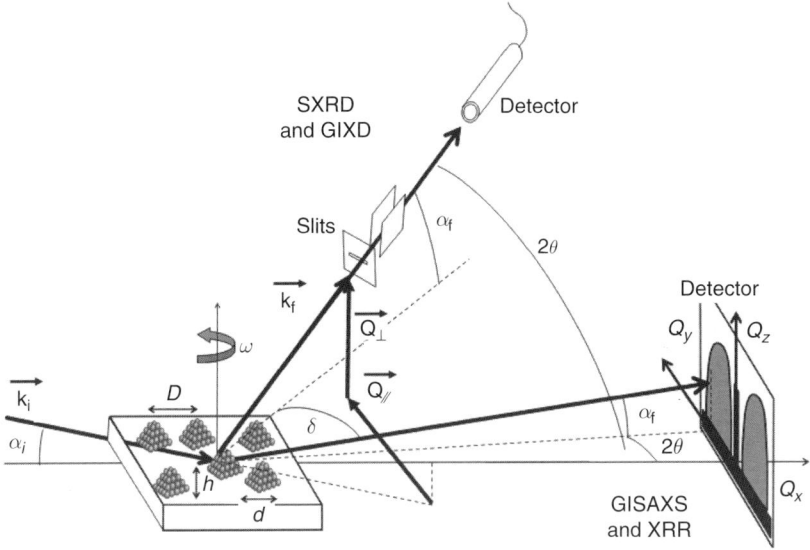

Fig. 1. Grazing incidence X-ray scattering geometry, at small angles (GISAXS and XRR), and large angles (GID and SXRD). The incident beam, of wavevector $\mathbf{k_i}$, makes a small incident angle α_i with respect to the sample surface. GISAXS and XRR probe the scattered intensity close to the specularly reflected beam. For GISAXS, a 2D detector is used, with a beam stop hiding the direct and specular beam, as well as the specular reflectivity. The intensity is recorded as a function of the two angles, 2θ and α_f, describing the in-plane and out-of-plane angles between the incident and scattered wavevectors, respectively. These angles are related to the momentum transfer coordinates, Q_x, Q_y and Q_z. The sample can be rotated by an angle ω around its surface normal. SXRD and GIXD correspond to large values of the scattering angle 2θ between $\mathbf{k_i}$ and $\mathbf{k_f}$, and hence large values of the in-plane, δ, and out-of-plane, α_f, scattering angles. The direction of the exit wavevector $\mathbf{k_f}$ is defined by slits parallel and perpendicular to the surface, behind which lies the detector, which can be punctual or linear, or even 2D. The momentum transfer coordinate parallel to the surface is directly related to the in-plane angle δ, and the out-of-plane coordinate is linked to the incident and exit out-of-plane angles α_i and α_f.

Because the incident angle is small, it is necessary to consider the effects of refraction at the surface [8]. The refractive index, n, of matter for X-rays is slightly less than unity: $n = 1 - \delta + i\beta$, with

$$\delta = \frac{1}{2\pi} \frac{e^2}{mc^2} \frac{N_a \sum_i (Z_i - f'_i)}{\sum_i A_i} \rho \lambda^2 \qquad (1)$$
$$= 2.701 \times 10^{-6} \frac{\sum_i (Z_i - f'_i)(\text{el. units})}{\sum_i A_i (\text{g/mol})} \rho(\text{g/cm}^3) \lambda^2 (\text{Å})$$

$$\beta = \frac{1}{2\pi} \frac{e^2}{mc^2} \frac{N_a \sum_i f''_i}{\sum_i A_i} \rho \lambda^2 = \mu \frac{\lambda}{4\pi} \qquad (2)$$

where N_a is Avogadro's number, and the summation is over all atomic species i. $(Z_i - f'_i)$, f''_i and A_i are respectively the scattering factor, the anomalous dispersion factor and the atomic weight of the i species, ρ is the density, λ the wavelength and μ the photoelectric absorption coefficient. Because of refraction, the transmitted beam bends towards the surface. When α_i is smaller than a critical value α_c, the beam is totally reflected, and only an evanescent wave, which decays over tens of angstroms, is present below the surface. When α_i is larger than the critical angle for total external reflection, the transmitted wave propagates into the bulk. Typical orders of magnitude are: $\delta \sim 10^{-5}$ and $\beta \sim 10^{-6}$, so that $\alpha_c \sim (2\delta)^{1/2} \sim 0.1$ to $0.5°$. Identical refractive effects occur as a function of the exit angle α_f.

The perpendicular components of the incident and emergent wave vectors are modified upon crossing the surface and become complex due to refraction and absorption:

$k'_{i,f\perp} = (2\pi/\lambda)(A_{i,f} - iB_{i,f})$, where $A_{i,f}$ and $B_{i,f}$ are given by the following expressions, valid for small incident and emergent angles:

$$A_{i,f} = \frac{1}{\sqrt{2}} \left(\sqrt{\left(\alpha_{i,f}^2 - \alpha_c^2\right)^2 + 4\beta^2} + \alpha_{i,f}^2 - \alpha_c^2 \right)^{1/2} \text{ and}$$

$$B_{i,f} = \frac{1}{\sqrt{2}} \left(\sqrt{\left(\alpha_{i,f}^2 - \alpha_c^2\right)^2 + 4\beta^2} + \alpha_c^2 - \alpha_{i,f}^2 \right)^{1/2} \tag{3}$$

and hence the perpendicular momentum transfer inside the sample $Q'_\perp = k'_{f\perp} - k'_{i\perp}$ becomes complex. The scattering depth given by

$$\Lambda = \frac{1}{\text{Im}(Q'_\perp)} = \frac{\lambda}{4\pi(B_i + B_f)} \tag{4}$$

is thus strongly affected by refraction when α_i or α_f are close to α_c. Figure 2 shows the variation of the scattering depth as a function of α_i/α_c for different values of α_f/α_c. When $\alpha_i \ll \alpha_c$ and $\alpha_f \ll \alpha_c$, the scattering depth is of the order of tens of angstroms. It rapidly increases to thousands of angstroms when α_i and α_f are larger than α_c, through a transition

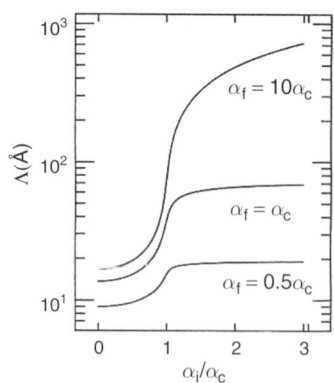

Fig. 2. Variation of the scattering depth as a function of the incident angle, for three exit angles, equal to half, 1 time and 10 times the critical angle, for a Pt surface and 1.5 Å wavelength.

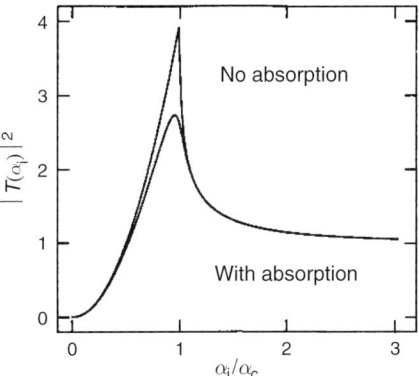

Fig. 3. Intensity transmission coefficient for a Pt surface and 1.5 Å wavelength, with and without considering absorption.

region where $\Lambda \sim 100$ Å. The incident and exit angles thus allow control of the depth contributing to a given measurement, which can be varied from about 1 to 100 nm.

The reflection and transmission coefficients for the intensity are also strongly affected by refraction, according to:

$$R_{i,f} = \frac{\alpha_{i,f}^2 - 2\alpha_{i,f}A_{i,f} + A_{i,f}^2 + B_{i,f}^2}{\alpha_{i,f}^2 + 2\alpha_{i,f}A_{i,f} + A_{i,f}^2 + B_{i,f}^2} \quad \text{and} \quad T_{i,f} = \frac{4\alpha_{i,f}^2}{\alpha_{i,f}^2 + 2\alpha_{i,f}A_{i,f} + A_{i,f}^2 + B_{i,f}^2} \quad (5)$$

The variation of the transmission coefficient as a function of $\alpha_{i,f}$ is reported in Fig. 3. Below α_c, $R_{i,f}=1$ is the regime of total external reflection. At α_c, the transmission coefficient is maximum ($T=4$). This property is sometimes used in GID by fixing α_i and/or α_f at α_c, to enhance the surface scattering.

In X-ray scattering, intensity distributions in reciprocal space, spanned by the momentum transfer (or scattering vector), are thus recorded instead of real-space images. We recall briefly the basic expression of the intensity scattered by a 3D crystal, a 2D crystal, the surface of a bulk crystal and nanostructures deposited on a substrate, with the corresponding reciprocal space schematically represented in Fig. 4. The reader is referred to standard textbooks [9,10] for a more comprehensive introduction.

Because the interaction of hard X-rays with matter is weak, the kinematical approximation of single scattering, the so-called Born approximation (BA), is valid in most cases, except for perfect crystals near Bragg scattering. The intensity scattered by a block-shaped crystal with N_1, N_2 and N_3 unit cells along the three crystal axes defined by the vectors $\mathbf{a}_1$, $\mathbf{a}_2$ and $\mathbf{a}_3$ takes the form:

$$I(\mathbf{Q}) = AF^2(\mathbf{Q}) S_{N_1}^2(\mathbf{Q} \cdot \mathbf{a}_1) S_{N_2}^2(\mathbf{Q} \cdot \mathbf{a}_2) S_{N_3}^2(\mathbf{Q} \cdot \mathbf{a}_3) \quad (6)$$

where A is a constant and

$$S_N(\mathbf{Q} \cdot \mathbf{a}_j) = \sum_{n=0}^{N-1} \exp(i\mathbf{Q} \cdot \mathbf{a}_j \cdot n), \quad j = 1, 2, 3 \quad (7)$$

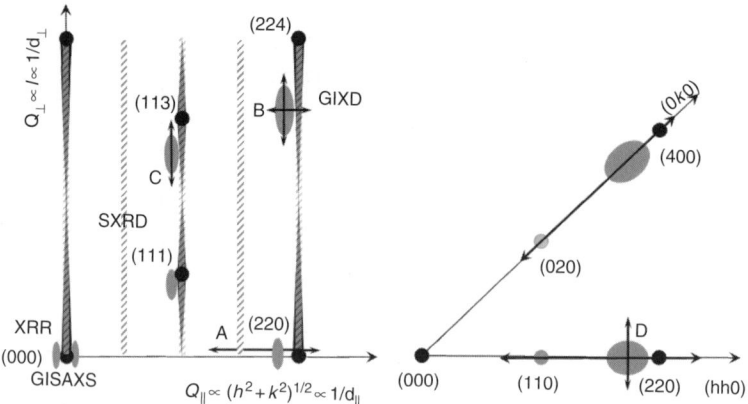

Fig. 4. Schematic representation (left side) of the reciprocal space of a 3D crystal of the diamond (Si,Ge) structure, truncated by a (001) surface, giving rise to the crystal truncation rods with maximum intensity at the bulk Bragg positions (black disks) and quickly varying intensity in between. In grey contribution from a strained 2D, two-fold reconstructed epilayers are shown, yielding flat rods of scattering (grey shaded vertical rods) at the same position as bulk CTRs because it is assumed that its lattice parameter parallel to the surface is fully strained to the one of the substrate. In case of 3D islands, diffraction peaks that are closer to the origin with respect to the substrate peaks are expected, as the overlayer is supposed to have a larger lattice parameter (note that the spacing between substrates and overlayer Bragg peaks has been enlarged for the purpose of representation). This reciprocal space pattern is typical for the Stranski–Krastanow growth of Ge islands on Si(001) for instance. Typical scans are represented by arrows. Scans A, B and C are referred to in the text. Corresponding intensity distribution in the (001) surface plane (right side).

$F(\mathbf{Q})$ is the structure factor, which is expressed as a function of atomic positions $\mathbf{r}_j$ within the unit cell as

$$F(\mathbf{Q}) = \sum_{j \text{ unit cell}} f_j \exp(i\mathbf{Q} \cdot \mathbf{r}_j) \qquad (8)$$

where f_j is the scattering factor of atom j, and

$$S_{N_j}^2(\mathbf{Q} \cdot \mathbf{a}_j) = \frac{\sin^2\left(N_j \frac{\mathbf{Q} \cdot \mathbf{a}_j}{2}\right)}{\sin^2\left(\frac{\mathbf{Q} \cdot \mathbf{a}_j}{2}\right)}, \quad j = 1, 2, 3 \qquad (9)$$

is the interference function of N_j diffracting units. The intensity is thus the product of the structure factor, which only depends on the structure within the unit cell, and the form factor, related to the shape of the crystal. In the limit of large N, the S_N function tends to a periodic array of Dirac delta functions with $\mathbf{Q}$ spacing of $2\pi/a$, i.e. the intensity is non-zero only if $\mathbf{Q} \cdot \mathbf{a}_1 = 2\pi h$, $\mathbf{Q} \cdot \mathbf{a}_2 = 2\pi k$ and $\mathbf{Q} \cdot \mathbf{a}_3 = 2\pi l$, with h, k, l integers. In other words, the intensity is non-zero only if $\mathbf{Q}$ is a vector of the reciprocal lattice of basic vectors $\mathbf{b}_1$, $\mathbf{b}_2$ and $\mathbf{b}_3$, i.e. $\mathbf{Q} = h\mathbf{b}_1 + k\mathbf{b}_2 + l\mathbf{b}_3$. When this Laue condition is fulfilled, the intensity is given by

$$I_{hkl} = AF_{hkl}^2 N_1^2 N_2^2 N_3^2 \tag{10}$$

The structure factor now takes the form

$$F_{hkl} = \sum_{j \text{ unit cell}} f_j \exp\left[2\pi i\left(hx_j + ky_j + lz_j\right)\right] \exp^{-M_j} \tag{11}$$

The summation extends over all atoms of the unit cell; f_j, (x_j, y_j, z_j) and M_j are respectively the scattering factor, fractional coordinates in the unit cell and Debye–Waller factor of atom j.

Consider now a quasi-2D crystal of finite thickness. The basic cell vector $\mathbf{a}_3$ perpendicular to the surface is chosen equal to this thickness. The diffraction is then still sharply peaked in both directions parallel to the surface, but the Laue condition on Q_3 ($=Q_\perp$) is relaxed, and the intensity is continuous in the out-of-plane direction: the reciprocal space is made of rods perpendicular to the surface plane (Fig. 4). If we still define l by $\mathbf{Q} \cdot \mathbf{a}_3 = 2\pi l$, l is taken as a continuous variable because intensity is present for non-integer values of l. The intensity is now given by

$$I_{hkl}^{2D} = AF_{hkl}^2 N_1^2 N_2^2 \tag{12}$$

The intensity variation along the rod (i.e. as a function of Q_3 or l) is solely contained in the structure factor. It is thus related to the z-coordinates of the atoms within the unit cell of this quasi-2D crystal. In general, the rod modulation period gives the thickness of the distorted layer and the modulation amplitude is related to the magnitude of the normal atomic displacements. This is the case of a reconstructed surface, for which rods are found for fractional order values of h and k, i.e. outside scattering from the bulk.

A crystal truncated by a sharp surface (or semi-infinite crystal) can be represented by the product of a step function describing the electron density variation as a function of z, the coordinate perpendicular to the surface and an infinite lattice. The diffraction pattern is then the convolution of the 3D reciprocal lattice with the Fourier transform of the step function. An infinity of Fourier components is necessary to build this latter, so that there remains non-zero intensity in between Bragg peaks as a function of l: the reciprocal space is made of rods of intensity, called crystal truncation rods (CTR), extending perpendicular to the surface and connecting bulk Bragg peaks [11,12]. They are schematically represented in Fig. 4. The intensity variation as a function of Q_3 (or $Q_\perp$ or l) can be found by replacing $S_{N_3}(Q_3 a_3)$ by $\sum_{n_3=-\infty}^{n_3=0} \exp(iQ_3 a_3 n_3)$ in Eq. (6), which gives

$$I(\mathbf{Q}) = I_{hkl}^{CTR} = \frac{\sin^2(N_1 a_1 Q_1/2)}{\sin^2(a_1 Q_1/2)} \frac{\sin^2(N_2 a_2 Q_2/2)}{\sin^2(a_2 Q_2/2)} |F_{hkl}|^2 |F_{CTR}(Q_3)|^2$$

with

$$|F_{\text{CTR}}(Q_3)| = \frac{1}{2\sin(Q_3 a_3)} \tag{13}$$

Bragg peaks are found for integer values of l, but there remains some intensity in between, even when l is a multiple of half-integer, i.e. when successive net planes scatter out of phase. At these anti-node positions, I_{hkl}^{CTR} and I_{hkl}^{2D} have comparable magnitudes, the intensity diffracted by the semi-infinite lattice is of the order of the intensity diffracted by a single ML.

If the surface is rough on an atomic scale, the step function has to be replaced by a less abrupt function, which needs fewer Fourier components to be built, so that the intensity of the CTRs between Bragg peaks is smaller than for a perfectly sharp surface. As the roughness increases, the intensity between Bragg peaks is reduced, and the result for an infinite 3D crystal is approached.

The measurement of CTRs allows the determination of the atomic structure of a surface, or of an interface between two materials. The intensity variation along CTRs between two Bragg peaks, as a function of u_l, is particularly sensitive to the difference between the bulk and surface structures.

The above expressions of the scattered intensity show that, in general, the reflections have a finite width, which is related to the finite size of the crystal. To deduce the structure factor, containing the required structural information at the atomic level, the pertinent measurement is to integrate the intensity of each reflection. This is in general done by performing a scan along one direction, while integrating along the perpendicular direction, using sufficiently opened slits, i.e. an adapted resolution function. Several corrections have next to be applied, the Lorentz correction, due to the relationship between the reciprocal space coordinates and the angular coordinates, the polarization correction, which accounts for the polarization states of the incoming and exit waves, the background correction, the normalization to the input intensity and the illuminated active area correction. The statistical error bars are also estimated during this procedure, and systematic errors are determined through measurements of several equivalent reflections, related by symmetry. This procedure can be fairly complicated, in particular for surface rods or CTRs, for which the finite width of the rods, their rapidly varying intensity or the exact surface diffractometer resolution function has to be estimated properly. The simplest features of these correction procedures may be found in many recent references [13–16].

The corrected intensities are proportional to the square of the structure factors, with an arbitrary scale factor, which will have to be determined as a free parameter (unless absolute intensities are measured). A determination of the structure on the atomic level of a reconstruction requires the quantitative measurement of as many allowed reflections as possible. Given the structure factors, standard Fourier methods of crystallography, such as Patterson function or electron-density difference function, are used.

The above expression of the scattered intensity shows that, in general, the surface diffraction peaks have laterally a finite width, which is related to the finite "domain size" parallel to the surface. In the $Q_1 (= 2\pi h/a_1)$ direction, for instance, the peaks have a finite width $\Delta h \sim 1/N_1$. The "domain size" in this in-plane direction can be approximated by $D = N_1 a_1$, and is thus simply related to the peak full-width at half-maximum (FWHM) $\Delta\omega$ by

$$D = \frac{a_1}{\Delta h} \quad \text{or} \quad \frac{2\pi}{Q_\| \Delta\omega} \tag{14}$$

where $\Delta\omega$ is the in-plane angular width of the reflection, measured by rocking the sample around its surface normal. D represents the maximum distance between two atoms that scatter coherently, i.e. the average distance between surface defects such as steps, vacancies, stacking faults, dislocations, grain boundaries, etc. that perturb the long-range atomic order, because this distance is smaller than the X-ray coherence length in this direction. This notion of "domain size" is more rigorously described by a lateral correlation function $C(r_{\parallel})$ ($C(r \to 0) = 1$, $C(r \to \infty) = 0$), defined as the probability that two atoms separated by $\mathbf{r}_{\parallel}$ scatter coherently. The lineshape of in-plane diffraction peaks corresponds to the Fourier transform of the lateral correlation function. The correlation function is in general taken as a simple analytical form: Gaussian or exponential. In the case of a Gaussian correlation function, the diffraction peak is also of Gaussian lineshape. In the case of an exponential correlation function, the lineshape is Lorentzian.

In practice, before analysing the lineshape in terms of correlation lengths, it has to be deconvoluted from the instrumental resolution function and from other contributions such as a mosaic spread or a distribution of lattice parameters parallel to the surface. This is in general done by measuring the lineshape for different orders of reflection, and in both directions, longitudinal and transversal to the momentum transfer. While a finite correlation function yields peaks of constant shape in reciprocal space, irrespective of the parallel momentum transfer value $Q_{\parallel}$, a mosaic spread or a distribution of lattice parameters yield a transverse and a longitudinal broadening proportional to $Q_{\parallel}$, respectively.

Figure 4 qualitatively shows the intensity scattered by nanostructures (or an epilayer) on top of a substrate, as a function of the in-plane and out-of-plane momentum transfers. The semi-infinite crystal yields CTRs, and the nanostructures yield wide Bragg reflections, which may be peaked at different $Q_{\perp}$ ($Q_{\parallel}$) values, if the out-of- (in-) plane unit vector differs from the bulk one. This allows independent analysis of the nanostructures structural properties, such as in-plane and out-of plane strain, composition and shape, without interference with the substrate. The strain relaxation in the nanostructures parallel and perpendicular to the surface can be analysed separately by performing respectively $Q_{\parallel}$ scans (such as scan A of Fig. 4) around in-plane Bragg peaks and $Q_{\perp}$ scans (such as scan B or C of Fig. 4) around out-of-plane Bragg peaks.

If the nanostructures, are constrained to the substrate lattice parameter parallel to the interfacial plane, the nanostructures peaks appear at the same $Q_{\parallel}$ position as the substrate CTRs.

In general, the structure and morphology of nanostructures (or an epilayer) on a substrate depend on many parameters, such as the initial structure of the substrate surface itself, the equilibrium structure of the deposited material, the balance between its surface energy, that of the substrate and the interfacial energy, the lattice parameter misfit and the growth kinetics. Because several energy terms depend on the thickness of the deposit, the structure and morphology are likely to evolve during growth. GI X-ray diffraction is one of the very rare tools that can be used in situ, for instance in UHV, to follow the evolution of the structure and morphology during growth, from sub-ML deposits to fairly thick films (up to μm in thickness). In principle, the factors influencing the overlayer growth, such as the defect density of the substrate, its temperature during growth and the incoming fluxes, can be systematically varied. The structure of the substrate, either reconstructed or relaxed, should be investigated first. Its average roughness can be deduced from analysis of the CTR-integrated intensities, and the surface domain size and average terrace width can be deduced from CTR lineshape analyses. Systematic measurements during growth may then provide the evolution of the most important parameters with the overlayer thickness. These include the interfacial distance, the average in-plane and out-of-plane lattice parameters and strain distributions, the mosaic spread and the domain size both parallel and perpendicular to the interface. In the case of a 3D growth, the domain size is directly related to the size of growing islands. These growing islands can also be studied in situ by GISAXS experiments, which can provide useful information on the growth laws. It is often energetically favourable to introduce structural defects in a growing film, in particular to accommodate the lattice parameter mismatch. GID can also be used to analyse different growth defects such as stacking faults, twins and dislocation nucleation. Indeed, stacking faults are planar defects, and thus yield rods of scattering perpendicular to the plane of the fault. Twinned crystals yield Bragg peaks that are related by mirror symmetry to those of the untwinned stacking. Finally, the nucleation of dislocations in the growing film leads to inhomogeneous deformations that may be detected through $Q_\parallel$ scans.

These kinds of studies require the sample to be in UHV, so that the necessary vacuum hardware has to be combined with an X-ray diffractometer. The main difficulty is to associate the necessary precise orientation movements of the sample with a UHV environment. Several diffractometers devoted to performing in situ GID have recently been built with these requirements.

The procedures for data collection, correction and normalization depend on the characteristics of the input beam, on the geometry of the diffractometer and on the sample characteristics (mosaic spread, domain size). While the corrections on in-plane data are simple, they may be fairly complicated for rods with large out-of-plane momentum transfer values. The reader is referred to many recent papers on the subject [13–16].

2.3. Anomalous diffraction and diffraction anomalous fine structure

We already learned in the previous section that XRD is a very powerful technique for measuring strain and correlations in hetero- and nanostructures. Nevertheless, strain is related to composition in a complicated way and the Vegard law is usually not valid for nearest neighbours interatomic distances, and it does depend also on the nanostructure shape, aspect ratio and the mutual stress which nanostructures, substrate and the matrix apply to each other. Also, one should keep in mind that only the intensity of the X-ray scattering signal is recorded, the phase is lost. The comparison with simulations, i.e. with models a priori, is needed to solve the structural problem.

The model-dependent approach can be partly overcome by exploiting the chemical sensitivity that X-ray techniques can provide. The idea is to tune the X-ray energy to the absorption edge of the atomic species that belongs to the nanostructures, i.e. to modify their atomic scattering factor by exciting the resonant scattering. This is the basic principle of anomalous (resonant) diffraction [17]. Thanks to the third-generation synchrotron sources, two complementary methods, based on X-ray resonant scattering, are nowadays available. The first one is the MAD method that allows to extract the scattering amplitude of the resonant atoms (chemical mapping in the reciprocal space). The second one is the DAFS spectroscopy that allows to recover the local environment of atoms located in an iso-strain volume selected by diffraction. As for the techniques described above, the GI geometry is used to decrease the substrate contribution to the scattering signal. Hereafter, the methods are called GIMAD (GI multiwavelength anomalous diffraction) and GIDAFS (GI diffraction anomalous fine-structure) spectroscopy. MAD consists in measuring diffraction curves (or maps in the reciprocal space) at several energies in the vicinity of the absorption edge of one element (the resonant or anomalous atom) that belongs to the nanostructures. By tuning the energy of the incoming photons to the edge energy, only the scattering factor (f_A) of that element is changed. In Section 3.4, we show how to extract the modulus of $F_A(Q) = \sum_j f^0_{Aj}(Q) c_{Aj} e^{-MQ^2_{Aj}} e^{iQ \cdot r_j}$ (resp. F_N); the partial structure factors (Thomson scattering) of all resonant (resp. non-resonant) atoms as well as the phase difference $\varphi_N - \varphi_A$; f^0_{Aj} is the Thomson scattering, $c_{Aj} = 1$ if site j is occupied by an A atom and $e^{-MQ^2_{Aj}}$ is the Debye–Waller factor. Therefore, the method provides a way to disentangle, in the reciprocal space, the resonant atoms scattering amplitude, from the substrate and/or matrix scattering (scattering amplitude deconvolution). Once the extraction is done, one can recover model-free parameters as the average strain and size of the nanostructures [18–20]. As for the composition, the knowledge of F_A, F_N and $\varphi_N - \varphi_A$ is not sufficient to conclude about the actual nanostructure chemical composition and to quantify possible atomic intermixing at the capping or substrate interface. Only in the case of an iso-strain homogeneous binary alloy, the F_A/F_N ratio gives the exact intermixing amount ($\varphi_N - \varphi_A = 0$). In such case, it is very convenient to combine MAD and "iso-strain scattering" methods. The latter was developed for as-grown QDs, basically it gives the in-plane strain of iso-strain slice above the plane of growth, providing the in-plane strain gradient as a function of height in the nanoislands [7]. The composition of a given slice at a given height above the surface can be obtained from the intensity measurements at two energies, one below the absorption edge and the other at the edge. The combination of the two methods has been successfully applied at the Ge K-edge to recover the out-of-plane Ge composition gradient in uncovered Ge/Si(001) islands [21,22] (see Section 3.2.1) and also to investigate the lateral composition of uncapped Ge domes grown on Si [23]. In general, the method is suitable for uncapped islands with a large vertical strain gradient to ensure a reasonable vertical resolution. As a matter of fact, to be suitable for devices, the nanostructures are encapsulated or embedded in a superlattice. In that case, MAD can readily give model-free information on the average size and strain [18]. However, in the most general case, the scattering signal results of the contribution of anomalous atoms and non-anomalous atoms that are strained in different way and/or located in different crystallographic sites or sample regions ($\varphi_N - \varphi_A \neq 0$) (see Section 3.4). Then, for obtaining the nanostructures composition one way is to use the finite element method, which allows to model the nanostructure shape, size, strain and composition and to simulate reciprocal space mapping as well as complementary 1D or 2D anomalous diffraction data. Nevertheless, given the huge number of atoms to be considered, modelling and calculations do not allow by now a further structural refinement. This problem can be partly overcome by the GIDAFS spectroscopy (see Chapter 10), once partial structure factors have been extracted by MAD. It consists in recording and analysing, in a quantitative way, the oscillations of the diffracted intensity vs energy, above the absorption edge, at a fixed point in the reciprocal space (fixed-Q mode). The data analysis will provide information on the composition of the nanostructures at a local atomic scale in the diffraction-selected iso-strain-volume (spatial selectivity) that contains the resonant atoms [20,24,25]. Other examples of interest of measuring DAFS oscillations in combination with MAD are among the following. Consider a layer of free standing or partially covered QDs grown on a substrate sample. Both in-plane and out-of-plane average size and strain of the nanostructures can be in principle recovered by GIMAD. However, in that case achieving absolute out-of-plane values is difficult. The main reason is that the out-of-plane diffraction data are affected by refraction, this effect is not the same for the substrate and the nanostructures, and this means that the relative position of the substrate and the nanostructure peaks (in the reciprocal space direction perpendicular to the sample surface) is not known in a direct and absolute way. Also, in GI and exit geometry, MAD cannot discriminate atomic intermixing and a mixing of in-plane iso-strained phases (the pseudomorphic materials on top and/or beneath the nanostructure). This problem can be overcome, once partial "in-plane" structure factors have been extracted by MAD, by means of GIDAFS. The incoming photon electric polarization being perpendicular to the sample surface, the GIDAFS spectroscopy mainly probes the out-of-plane local atomic environment of the resonant atom. So, the GIDAFS oscillations analysis will provide information about the out-of-plane strain accommodation of the nanostructures in the diffraction-selected iso-strain volume that contains the resonant atoms. The ultimate goal is in general to map strain and composition in three dimensions. Therefore it is of great importance to carry out anomalous diffraction 2D mappings and GIDAFS measurements together on a dedicated synchrotron beamline with a high brilliance and a high stability of beam position, diffraction geometry and energy. In this and the following chapter, we show that MAD and DAFS spectroscopy, in GI, are powerful tools to study nanostructures. DAFS can be also regarded as a meeting point technique of diffraction and absorption providing new breakthroughs for both scientific domains.

3. Application of grazing incidence techniques

3.1. Basics concepts

It is of particular interest to combine GISAXS and GID to fully characterize an assembly of nanoparticles on or close to a surface. While GISAXS yields morphological information, GID is sensitive to the structure at the atomic scale. The two techniques can also be combined in situ, during molecular beam epitaxy (MBE) growth in UHV, thus allowing to control their growth at a macroscopic scale, sometimes in real time.

Because the basics of GID have been recalled previously (see Section 2), we concentrate here on an introduction to the GISAXS technique.

The GISAXS technique has emerged in the last decade as a powerful tool to analyse the morphology and distribution of either islands on a substrate or buried particles [26–28]. The GISAXS ability to characterize granular multilayered systems containing clusters in matrices [29], implanted systems [30] as well as semiconductor QDs obtained by MBE is now well established [31,32]. Recent developments consisted of in situ GISAXS measurements performed in UHV during MBE, and with no scattering element between the synchrotron X-ray source and the detector, thus yielding a very high-intensity dynamics and background-free data [33]. This set-up provides data extended over a wide range in reciprocal space with intensity variation over several orders of magnitude, thus giving evidence to GISAXS features that could not be revealed in previous measurements. For this reason, precise quantitative analysis can be performed, as it will be shown in this chapter. In particular, the GISAXS technique has been used to characterize, in situ, the growth of metal/oxide interfaces, like Pd islands on MgO(001) [34,35], and the self-organized growth of Co clusters on the herringbone reconstruction of the Au(111) surface [36].

As shown in Fig. 1, the scattered intensity is measured by a 2D detector (most often a charge-coupled device, CCD) placed perpendicular to the incident beam. Parallel to the sample surface one measures the in-plane 2θ angle, while in perpendicular direction the scattered beam is defined by the exit angle α_f. Then, the components of the momentum transfer are given by

$$Q_x = k[\cos(\alpha_f)\cos(2\theta_f) - \cos(\alpha_i)]$$
$$Q_y = k[\cos(\alpha_f)\sin(2\theta_f)] \qquad (15)$$
$$Q_z = k[\sin(\alpha_f) + \sin(\alpha_i)]$$

Because all the angles are small, Q_x is very small and often neglected, so that, to first order, the lateral momentum transfer is proportional to the in-plane angle 2θ, and the perpendicular momentum transfer is proportional to the sum of the out-of-plane angles:

$$Q_y \sim k(2\theta_f) \text{ and } Q_z = k(\alpha_f + \alpha_i)$$

Hence, in the detector the horizontal direction probes the reciprocal space along Q_y, a coordinate parallel to the surface, and the vertical one along Q_z, the coordinate perpendicular to the surface. The probed direction along the sample surface can be chosen by rotating the sample azimuthally, corresponding to a ω rotation.

To first order [37–43], the GISAXS intensity is proportional to the form factor $F(\mathbf{Q})$ times the interference function $S(\mathbf{Q})$, where $\mathbf{Q}$ is the momentum transfer. $F(\mathbf{Q})$ and $S(\mathbf{Q})$ are the Fourier transforms of the particle shape and the island–island pair correlation function (i.e. the probability that two islands are separated by a given vector), respectively. The average particle lateral size d and height h are thus inversely proportional to the spread of the scattering parallel and perpendicular to the surface, respectively. The average island separation D is inversely proportional to the distance between the main correlation peaks along the parallel direction.

Figure 5 shows a typical GISAXS pattern obtained from Ge islands grown on an Si(001) surface. In that case, the islands have the shape of domes exposing different kinds of facets (see insets): mostly {105}, {15 3 23} and {113}. Because these facets are 2D objects in real space, their Fourier transform yields intensity rods in reciprocal space, perpendicular to the facets, the so-called facet rods. If the sample is rotated to place the incident beam parallel to the surface [1–10] direction, the detector plane becomes the (110) reciprocal plane, which contains the (113) and (−1–13) rods perpendicular to the (−1–13) and (113) facets, respectively. These rods are clearly visible on the image. Their intensity, extension and widths are characteristic of the facet sizes. GISAXS measurements for different ω angles thus allow to fully determine the shape, facets and size of the islands. However, this requires a quantitative analysis, which needs some deeper insight into the theory of small angle X-ray scattering under grazing incidence.

The intensity $I(\mathbf{Q})$ scattered by an assembly of islands, whose positions and shapes can be statistically defined, is more precisely given by the sum of a coherent and a diffuse term [38–41]:

$$I(\mathbf{Q}) = S(\mathbf{Q})\left|\overline{F(\mathbf{Q})}\right|^2 + \sum_m \Phi_m(\mathbf{Q})\, e^{-i\mathbf{Q}\cdot\mathbf{r}_m} \qquad (16)$$

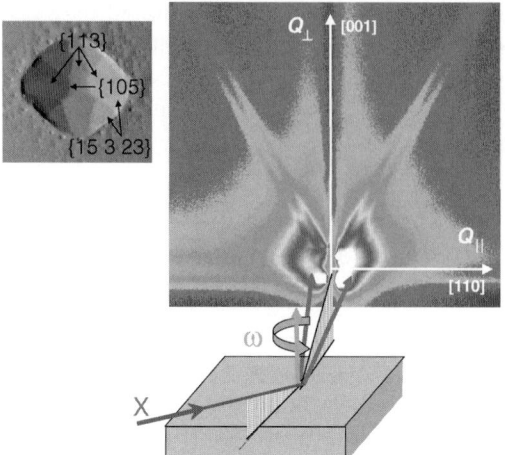

Fig. 5. Two-dimensional GISAXS pattern performed on an assembly of Ge domes (see AFM image at the upper left), with {113}, {15 3 23} and {105} facets. These facets yield intensity rods of scattering perpendicular to them. (see Colour Plate 1)

where $S(\mathbf{Q})$, the interference function [37], is the Fourier transform of the autocorrelation function with respect to the lateral positions of the islands [38]:

$$S(\mathbf{Q}_\parallel) = 1 + \rho_s \int [g(\mathbf{r}) - 1]\, e^{-i\mathbf{Q}_\parallel \cdot \mathbf{r}}\, d\mathbf{r}$$

In the BA, the island form factor $F(\mathbf{q})$ is equivalent to the amplitude scattered by an island [40] and is given by the Fourier transform of the electron density:

$$F(\mathbf{Q}) = \int_V \rho(\mathbf{r})\, e^{-i\mathbf{Q}\cdot\mathbf{r}}\, dV$$

The bar on top of $F(\mathbf{Q})$ in Eq. (16) denotes the spatial averaging. The $\Phi_m(\mathbf{Q})$ coefficient in Eq. (16), describing the correlations between island sizes, is given by [38,41]

$$\Phi_m(\mathbf{Q}) = \overline{\left[F_n(\mathbf{Q}) - \overline{F(\mathbf{Q})}\right]\left[F_{n+m}(\mathbf{Q}) - \overline{F(\mathbf{Q})}\right]^*}^n \quad (17)$$

where n denotes the average over all islands n and $F_n(\mathbf{Q})$ is the form factor of the island n. The function $\Phi_m(\mathbf{Q})$ is a measure of the correlation between the scattering amplitude of two islands separated by a vector $\mathbf{r}_m$. The summation over $\Phi_m(\mathbf{Q})$ in Eq. (16) yields the incoherent diffuse scattering arising from such correlations [41]. Two approximations are commonly used to evaluate the incoherent scattering:

1. The decoupling approximation (DA), which is appropriate for systems with small polydispersities [38,40]. It is assumed that the nature of the scatterers and their positions are completely independent, so that the partial pair correlation functions depend only on the relative positions of the scatterers, i.e. $\Phi_m(\mathbf{Q}) = 0$ for $m \neq 0$. Only the $m = 0$ term remains:

$$\Phi_0(\mathbf{Q}) = \overline{F(\mathbf{Q})^2} - \overline{F(\mathbf{Q})}^2 \quad (18)$$

2. The local monodisperse approximation (LMA), which is commonly used for polydispersed systems. Neighbouring islands are assumed to have the same shape and size, over the coherent area of the X-ray beam [42–44]. The intensity is thus the incoherent summation over monodisperse subsystems weighted by the size distribution. The intensity originating from one monodisperse domain i is

$$I_i(\mathbf{Q}) = S_i(\mathbf{Q}) \times F_i(\mathbf{Q})^2 \tag{19}$$

The measured intensity is given by the incoherent summation over the intensity I_i of all the domains and assuming that the same interference function holds for all of them:

$$I(\mathbf{Q}) = \sum_i I_i(\mathbf{Q}) \approx S(\mathbf{Q}) \times \overline{F_i(\mathbf{Q})^2} \tag{20}$$

Note that the DA and LMA are equivalent at large $\mathbf{Q}$, where the $\Phi_0(\mathbf{q})$ term decreases towards zero.

In the case of Ge island on Si(001), see Fig. 5, the assembly of islands is not particularly dense, so that the $S(\mathbf{Q})$ function is basically 1 everywhere. The GISAXS measurement is in that case only sensitive to the island shape: $I(\mathbf{Q}) \approx \overline{|F(\mathbf{Q})|^2}$. One may notice however that the facet rods of scattering are doubled, as if two images were superimposed with a vertical shift. This is because, at these low angles, the surface acts as a mirror and multiple scattering effects come into play which cannot be handled by the single scattering or BA. Instead the so-called distorted wave BA (DWBA) is applied [45,46]. Here, the form factor $F(\mathbf{Q})$ is the coherent sum of four terms [47], which represent different scattering events involving a reflection of either the incident or scattered beam on the substrate surface (Fig. 6). Each term comprises the island form factor, evaluated at different momentum transfers $\pm Q_z = \pm(k_z^f - k_z^i)$ and $\pm p_z = \pm(k_z^f + k_z^i)$, and weighted by the Fresnel reflection coefficients $R(\alpha_i)$ and $R(\alpha_f)$ of the substrate. The form factor for a supported island is expressed as

$$F(\mathbf{Q}_\|, k_i^z, k_f^z) = F(\mathbf{Q}_\|, q_z) + R(\alpha_f)F(\mathbf{Q}_\|, -p_z) + R(\alpha_i)F(\mathbf{Q}_\|, p_z) \\ + R(\alpha_i)R(\alpha_f)F(\mathbf{Q}_\|, -q_z) \tag{21}$$

These multiple scattering effects are well illustrated in Fig. 7. First, they induce a sharp intensity increase for $\alpha_f \approx \alpha_c$, called the Yoneda peak. Moreover, the "shift" between the two images clearly increases with the incident angle α_i until the critical angle α_c and then the second image gradually vanishes as the incident angle is further increased. The "two images" indeed correspond to the two first terms of Eq. (21). The rod with the smaller exit angle (α_f) comes

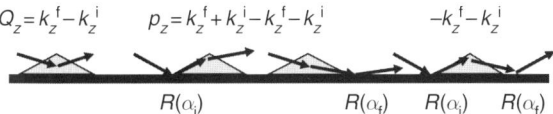

Fig. 6. Schematic representation of the four-channel scattering process. For each scattering path, the effective perpendicular wave vector transfer is given above and the corresponding weighting factors R for the amplitudes below.

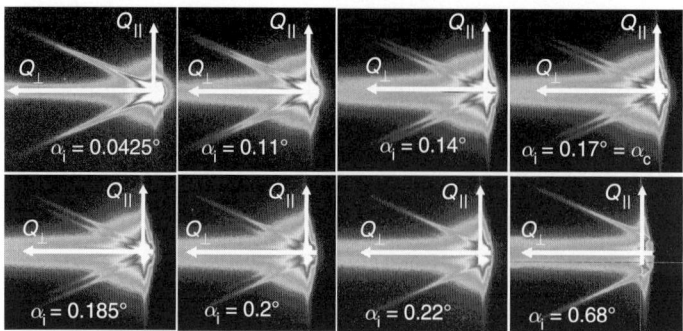

Fig. 7. Series of GISAXS pattern, recorded on an assembly of large Ge domes grown on a Si(001) surface. The azimuth setting in <110> allows for the rods from {113} facets to be recorded. Note the splitting of the rods for incident angles close to the critical angle for total external reflection, $\alpha_c = 0.17°$. (see Colour Plate 2)

from a single scattering event by the {113} facets (BA). It corresponds to the first scattering process in Eq. (21), with a Q_z component given by $Q_z = k_f^z - k_i^z = (2\pi/\lambda)[\sin(\alpha_f) + \sin(\alpha_i)]$. The other rod comes from the second scattering process in Eq. (21), which takes into account the reflection of the incident beam on the substrate surface before it is scattered by the nanoparticles. Along the Q_z direction, these two scattering rods are separated by

$$\Delta = Q_z - p_z = k_f^z - k_i^z - \left(k_f^z + k_i^z\right) = -2k_i^z = (4\pi/\lambda)\sin(\alpha_i) \quad (22)$$

Hence, this separation increases almost linearly with α_i. As long as $\alpha_i < \alpha_c$, the multiplication factor of the second term, $R(\alpha_i)$, is almost unity. When $\alpha_i > \alpha_c$, the specular reflectivity of the substrates decreases very fast as $(\alpha_i)^4$, and the second image disappears.

Finally, a quantitative analysis needs to correctly take the averaging of the form factors into account, which means that vertical and horizontal size (and possibly shape) distributions have to be introduced. We will not describe in detail here a complete quantitative analysis, for which a dedicated program is now available on the Web [37].

3.2. Examples of grazing incidence SAXS studies

3.2.1. In situ study of Ge islands grown on Si(001)

Here we will emphasize on the powerful combination of GISAXS and GID, i.e. grazing incidence X-ray scattering at small and wide angle, to fully characterize an assembly of nanoparticles. The corresponding measurements are schematically represented in Fig. 8 for the case of Ge islands on a Si(001) surface.

Figure 9 shows typical GID measurements performed (in parallel with GISAXS) during the in situ growth of Ge on Si(001) using the BM32 ESRF instrument. Some of the scans performed are radial scans as illustrated in Fig. 4 (scan labelled A), which schematically displays the corresponding reciprocal space. Figure 9(a) shows a long radial scan along the ($hh0$) direction. For the bare (2×1) reconstructed Si(001) surface, it crosses 2×1

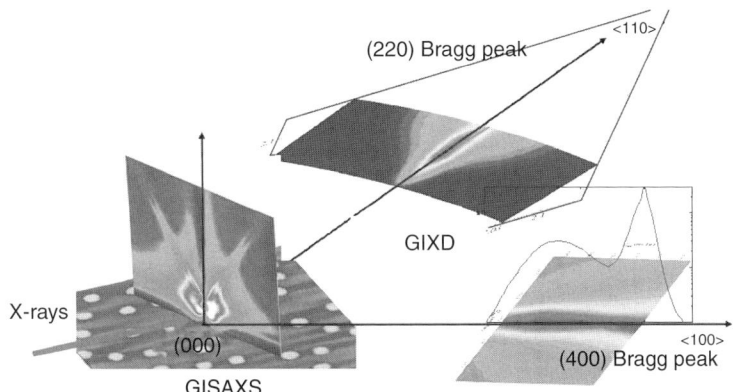

Fig. 8. Schematic representation of the reciprocal space intensity distribution of Ge islands coherently grown on Si(001), with corresponding GISAXS and GID measurements, respectively around the origin (000) and around Bragg peaks (220) and (400). (see Colour Plate 3)

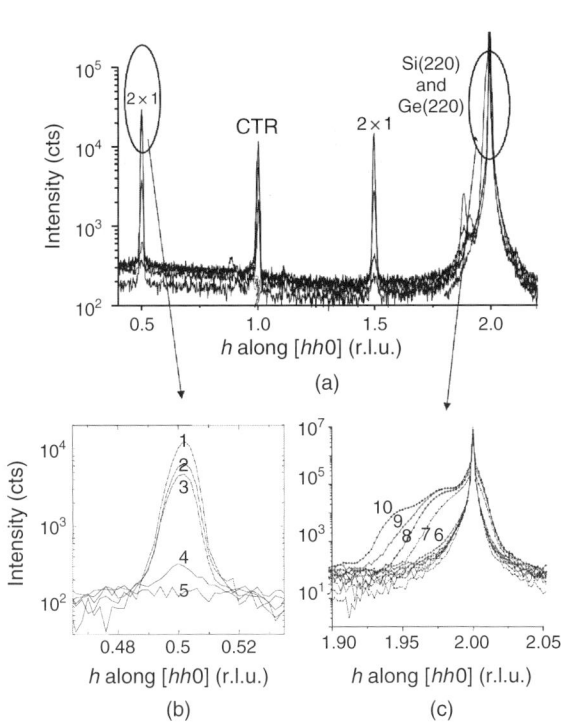

Fig. 9. GID measurements during in situ deposition: (a) In-plane radial scans along the <110> direction of Si(001), for the clean, (2 × 1) reconstructed surface, and for different amount of Ge deposited, between 1 and 5 monolayers (ML). (b) Details of the scans around the first (2 × 1) reconstruction peak at $h = 0.5$. (c) Details of the scan around $h = 2$, in which the contribution from Ge islands is visible for lower h values (fully relaxed Ge would correspond to $h = 1.92$).

reconstructions peaks for $h = 0.5$ and $h = 1.5$, the (111) CTR at $h = 1$, and the Si(220) Bragg peak at $h = 2$. These features are still present at the beginning of the Ge growth, between 0 and 5 equivalent ML. Ge is known to grow on top of Si(001) following the so-called Stranski–Krastanow (SK) process. The first 3–4 ML result in a 2D epilayer pseudomorphically strained to the substrate's lattice parameter. Beyond 4 ML, 3D growth of islands starts, with shapes evolving from "hut clusters" exposing {105} facets to pyramids with the same facets, followed by domes exposing {105}, {15 3 23} and {113} facets (see Fig. 5, inset) and finally super-domes with interfacial dislocations for thicker deposits. This island evolution can be followed simultaneously by GID and GISAXS. With GISAXS, the appearance of a small angle X-ray scattering signal is a signature of the 2D–3D transition. Then, by recording images with different azimuths, the type of islands can be identified. Simultaneously, GID also allows to follow the 2D/3D transition through the appearance of a signal of relaxed Ge on the low side of the Si(220) peak. Then, the strain state, composition and even shape of these islands can be deduced from the evolving position and shape of the Ge(220) scattering signal. These results are presented in more details in Ref. [48].

3.2.2. GISAXS analysis of vertical stacking of GaN quantum dots in AlN
The typical length scales suitable for GISAXS techniques lay on the mesoscale. The objects of interest are thus larger than atomic size but in the submicron (below the wavelength of visible light) regime. It is thus not the internal atomic structure that is addressed by this method, but rather the external shape of objects and their spatial arrangement with respect to each other. In the following example, the vertical periodicity and stacking of QDs will be considered. Ordering in such systems can be enhanced by the growth on a template or self-organized, by surface anisotropy of the elastic constants and hence a mutual influence between the QDs and their positions relative to each other [49]. In the case of QD superlattices, a vertical stacking can occur even without the presence of lateral order. This stacking is strain-mediated by the distortion field around QD in the spacer layers.

A system widely studied for its applications in the field of wide bandgap semiconductors are GaN QDs embedded in an AlN multilayer matrix. For dots that are supposed to serve as light-emitting media, a growth in multilayers means an increase in the number of dots but also can provide an improvement in size homogeneity, if self-organized ordering occurs in this process. Numerous studies have been published on the topic of the self-improvement of the order and the size homogeneity as a function of spacer layer thickness and the amount of deposited QD layers [49–51]. For such multilayered structures, the order can be studied best under GISAXS conditions. The typical distances and correlations are of the order of 100 to several 1000 Å. For a typical X-ray wavelength of $\lambda = 1$ Å, the corresponding scattering angles 2θ are to be found between 0.01 and 0.6°. In general, such experiments are carried out utilizing a 2D CCD detector that captures the whole range of interest, for the in-plane momentum transfer q_y as well as for directions perpendicular to the surface q_z, see above. In Fig. 10(a), a typical GISAXS image from a GaN/AlN multilayer is shown. In the middle of the figure, a sketch presents the structure of the multilayered sample. During the deposition of 2.5 ML of GaN, a 2D ML wetting layer grows before the QDs spontaneously form, thereby relaxing elastically the mismatch-induced elastic strain. Each of the n QD layers is covered by an AlN spacer layer. A steady state of the QD growth is reached after the deposition of 2–3 bilayers [52]. The 2D GISAXS image shows periodic intensity maxima along q_z which are substantially extended in q_y. Their distance in q_z is directly related to the spacer bilayer thickness of the multilayer. These so-called Bragg sheets, diffuse intensity of the multilayered structure, can be used to analyse the strain-driven vertical stacking order [53].

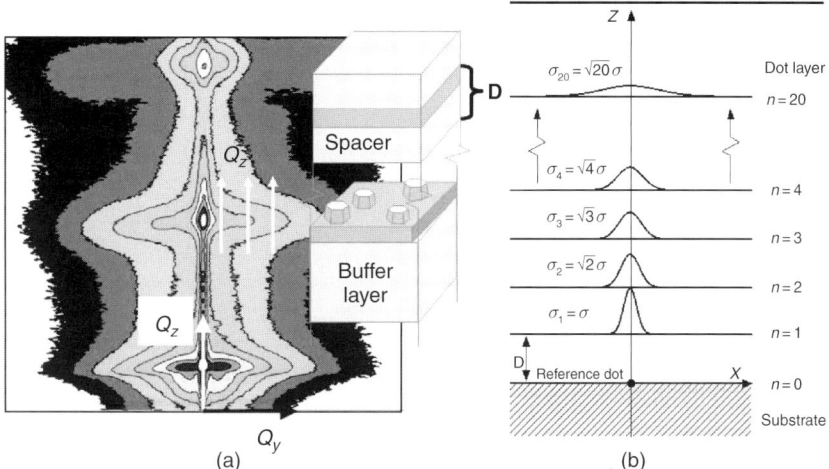

Fig. 10. (a) Typical GISAXS pattern for a multilayered sample of quantum dots separated by spacer layers. (b) Sketch of the model used to describe the random walk-like decay of the vertical order (from Ref. [53]).

The strain propagation from the GaN dots through the AlN spacer layers creates a strain-modulated surface and thus preferential nucleation sites for the successive layer of GaN dots. An analytical model of strain-driven stacking order and its effect on the scattered X-ray intensity has been developed by Kegel et al. [53]. It is based on the assumption that the strain is mediated only through one spacer layer, i.e. that every layer of QDs influences only one successive layer. Thus the probability that an island nucleates vertically aligned to the island underneath can be described in a random walk-like manner. The random displacement σ from the aligned position depends on the correlation strength and is thus certainly dependent on the strain propagation and the capping layer thickness. For the nth layer, this random walk model results in a displacement $\sigma_n = \sqrt{n}\sigma$.

As we have a vertically periodic layered structure, we expect intensity oscillations in q_z along the specular direction. Assuming a perfect vertical periodicity with a layer spacing D, it was shown in Ref. [53] that the FWHM of the intensity maxima along Q_z ($\Delta Q_z^{\mathrm{FWHM}}$) as a function of Q_x can be approximated as

$$\Delta Q_z^{\mathrm{FWHM}} = 2\frac{\sigma^2}{D}Q_y^2 + 2\frac{\mu}{D}$$

$\Delta Q_z^{\mathrm{FWHM}}$ is expected to increase as Q_y^2 and the only open parameter describing the degree of order, σ, can be quantified from the experiment. The parameter μ describes the absorption correction that has an important contribution in the grazing incidence geometry. The most interesting parameters to be varied in a growth of a sample series are the spacer layer thickness D and the number of deposited bilayers n, as in the case of strain-driven ordering, one expects an increase of order with a higher number of layers. Chamard et al. [54] published a study showing that the order increases substantially for a high number n of bilayers, even for high values of the spacer thickness D. Figure 11(a) and (b) shows the half width in Q_z as a function of Q_y along one Bragg sheet. With this parabolic dependency and the known

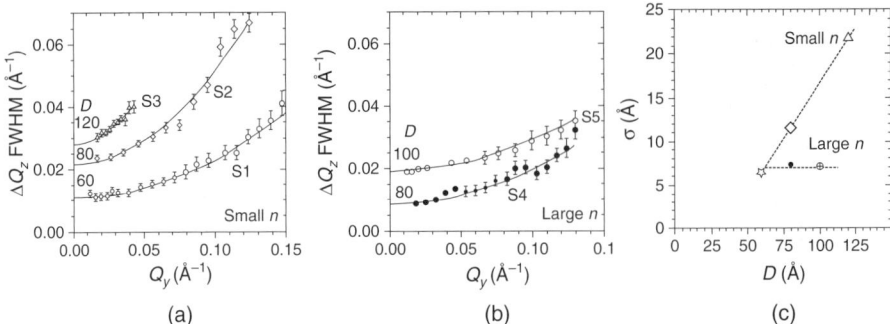

Fig. 11. (a) Full width ΔQ_z^{FWHM} of the Bragg sheets in Fig. 10(a) as a function of Q_y for three different samples with spacer thicknesses D from 60 to 120 Å and a *low* number of deposited bilayers as a function of spacer thickness D. (b) For $D = 80$ or 100 Å and a *high* number of deposited bilayers. (c) Mean displacement σ as a function of spacer thickness D.

parameters D and μ, the average lateral displacement σ can be extracted from the fits in Fig. 11(a) and (b). This average displacement between stacked QDs is plotted as a function of the spacer thickness D in Fig. 7(c). It is about 7 Å for a high number of bilayers, independent of the spacer layer thickness D. For a smaller number of periods n, this deviation still depends on the spacer layer thickness D. These results and in particular the parabolic relation between ΔQ_z^{FWHM} and Q_y served as an indirect proof of the strain-driven nature of vertical stacking in these heterostructures.

3.2.3. Concluding remarks on GISAXS and GID

The coupling of GISAXS and GID is very powerful to explore the whole reciprocal space of any kind of materials, thus fully characterizing its atomic structure as well as its morphology. In particular, the two techniques can be made sensitive to the composition by use of anomalous scattering (see, for example, Sections 3.3 and 3.4). In addition, GISAXS is very sensitive to possible ordering of nanoparticles. With increasing order narrow intensity rods located parallel to the surface plane appear, while they are rather structureless and diffuse perpendicular to it.

One of the advantages of X-ray techniques is the possibility to use them in any kind of sample environment such as at high or low temperature, in UHV as well as under gas partial pressure or during a chemical reaction. Particularly attractive are studies performed during the in situ growth of nanoparticles, which allow to characterize the very beginning of the growth, and thus accessing the kinetics as well as thermodynamics of the growth modes.

3.3. Examples of anomalous grazing incidence diffraction

3.3.1. Basic concepts

Anomalous diffraction describes the use of the energy dependence of the atomic scattering factor to separate contributions from different elements in a scattering signal. In X-ray studies of epitaxial nanostructures, one often encounters the problem of a high scattering contribution of the substrate material or the matrix, in which the structures are embedded. As the principal probe in diffraction is the lattice parameter and epitaxy necessitates similar atomic distances in different materials, this problem seems to be unavoidable. It is thus of high interest to develop an experimental method to separate both contributions. Anomalous diffraction can thus be of high interest to separate overlapping scattering signals or to analyse the concentration of an alloy. It makes use of the energy and Q dependence of the atomic scattering factor $f = f_0(Q) + f'(E) + if''(E)$.

Where $f_0(Q)$ describes the contribution of all electrons far from resonance.

As this concerns mainly outer shell electrons, f_0 depends also on the momentum tranfer $Q \cdot f'(E) + if''(E)$ are resonant corrections that become important when the X-ray energy is chosen in the vicinity of an absorption edge. As these terms originate from inner atomic shells, their momentum dependence is thus negligible for momentum transfers accessible at the corresponding absorption edge.

This energy dependence of the atomic scattering factor can, for instance, be used to determine the composition in a binary alloy. For the example of SiGe, we will demonstrate the use of applying this method at high-momentum transfers:

In the case of a (generally disordered) SiGe alloy, the scattering factor can be averaged in the form

$$f_{SiGe} = x f_{Ge} + (1-x) f_{Si} \qquad (23)$$

with x as the Ge content. As Ge with 32 electrons is a much stronger X-ray scatterer than Si with 14 electrons, the sensitivity for the Ge concentration in an anomalous diffraction experiment will be poor if we stay at low-momentum transfers. The strategy in such a measurement is to choose two energies in the vicinity of the Ge K-edge and to calculate the Ge content from the ratio of the recorded intensities [55]. As shown in Fig. 12, this ratio increases for high-momentum transfers. Thus, the anomalous scattering effects can be enhanced considerably by measuring Bragg reflections for high-momentum transfers, and hence improve the sensitivity for the composition in the SiGe system. Figure 12(a) shows the momentum dependence of the real parts of the scattering factors for Ge and Si at two energies $E_1 = 11,040$ eV and $E_2 = 11,102$ eV. The momentum-dependent behaviour of $f_0(Q)$ of Ge and Si was calculated with a code based on the parameters derived by Baró et al. [56].

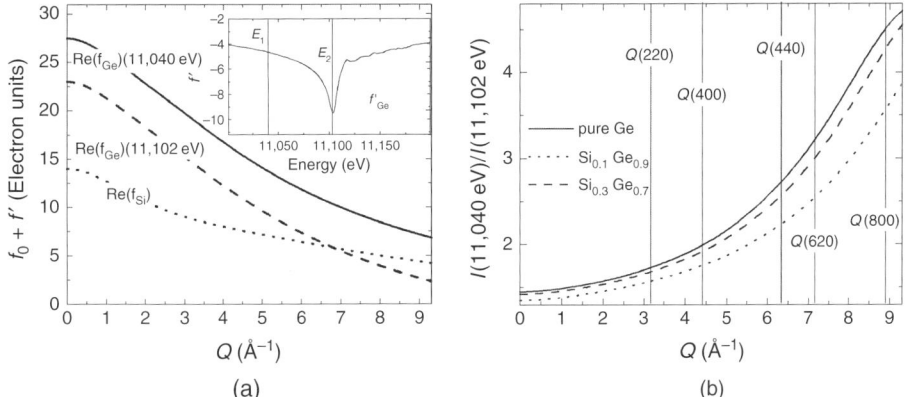

Fig 12. (a) Momentum-dependent real part of the scattering factor for Ge at $E_1 = 11,040$ eV (full line), Ge at $E_2 = 11,102$ eV (dashed) and Si (dotted). (b) Theoretical momentum dependence of the ratio of the diffracted intensities at $E_1 = 11,040$ eV and $E_2 = 11,102$ eV for pure Ge (full line), $Si_{0.1}Ge_{0.9}$ (dashed) and $Si_{0.3}Ge_{0.7}$ (dotted). Perpendicular lines mark the positions of the probed in-plane reflections.

At these X-ray energies, the corrections f' and f'' can be neglected for Si. It is clear that for high Q the Si scattering becomes more pronounced and even larger with respect to the Ge scattering. In Fig. 12(b), the intensity ratios for pure Ge, $Si_{0.1}Ge_{0.9}$ and $Si_{0.3}Ge_{0.7}$ are plotted against the momentum transfer Q. The positions of some reflections that are accessible at the Ge K-edge are indicated (as perpendicular lines). This clearly illustrates the necessity of measuring high-index reflections to improve the accuracy for the composition in these alloys.

3.3.2. Anomalous GID on Ge islands on Si(001): Iso strain scattering

In this section, we will combine anomalous diffraction and AFM to spatially resolve the composition inside SiGe islands. The sample studied consists of SiGe islands on a Si(001) surface. These were obtained via MBE deposition of 7 ML of Ge at a temperature of 600°C. An atomic force microscopy (AFM) image of the sample (Fig. 13(a)) shows a monodisperse size distribution of domes. Their base diameter is of the order of 700 Å; their height is about 130 Å. Figure 13(b) shows a line scan from the cross section indicated in (a). The scattering geometry of choice to investigate these samples via anomalous scattering is the grazing incidence diffraction technique (GID). In this geometry, only in-plane reflections can be accessed; however, it enables to suppress the scattering from the Si substrate significantly.

As the alloy composition is evaluated from a ratio between two intensities, all background contributions have to be subtracted before. It is evident that all these corrections in the data treatment introduce errors. It is therefore an advantage to use this technique to keep the background from the substrate as low as possible. The sample was illuminated at an incident angle of 0.1°, which is well below the critical angle at X-ray energies of about 11,100 eV.

To determine the Ge content in the islands by grazing incidence diffraction, two energies were chosen in the close vicinity of the Ge K-edge and 60 eV below. These two probed energies $E_1 = 11,043$ eV and $E_2 = 11,103$ eV are indicated in the inset of the graph in Fig. 12(a), where f' is presented as a function of energy. For a precise determination of the atomic scattering factors, f''_{Ge} was measured via fluorescence and f'_{Ge} was calculated using the Kramers–Kronig dispersion relation.

To demonstrate the increase of the relative anomalous effect with the momentum transfer, all accessible in-plane reflections were probed. The accessibility was limited

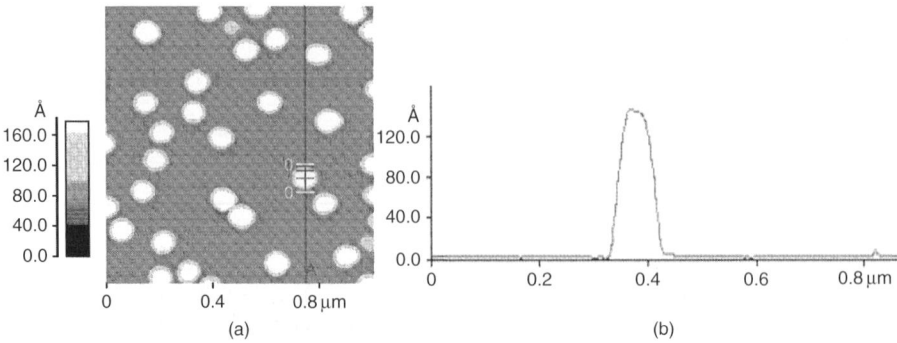

Fig. 13. (a) Atomic force microscopy of Ge islands on Si(001). (b) AFM line profile as indicated in (a).

by the energy of 11,043 eV and the maximum scattering angle of 125° of the used instrument.

All Bragg reflections as scanned in the radial direction are sketched in Fig. 14(a) together with the recorded data for each reflection. The dots correspond to the intensity at $E_1 = 11,043$ eV; the crosses represent the data at the Ge K-edge at $E_2 = 11,103$ eV. The strong Si Bragg peak on the right part of each graph is cut off, as we want to focus on a linear scale on the broad intensity distribution caused by the strained SiGe

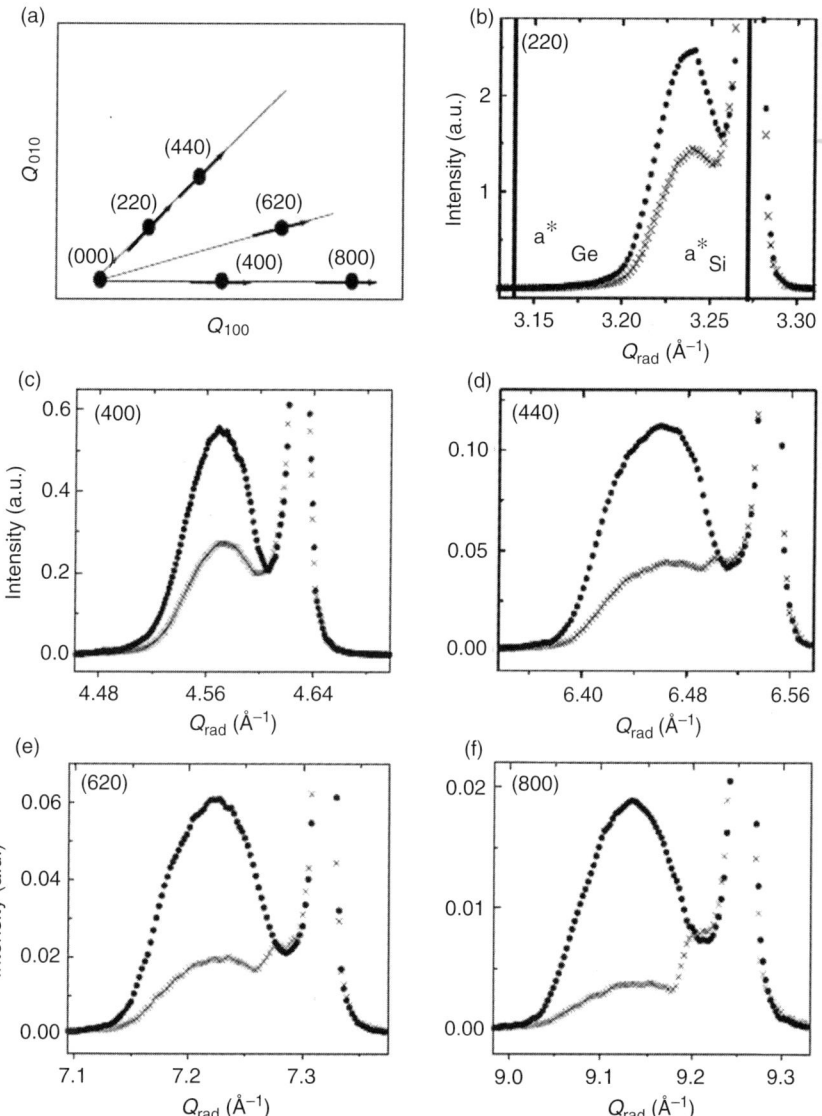

Fig. 14. All probed in-plane reflections for SiGe islands on Si(001). (a) Sketch of the radial scans across the in-plane reflections (b)–(f) Data for all radial scans at $E_1 = 11,043$ eV (dots) and $E_2 = 11,103$ eV (crosses). The vertical lines in (b) mark the Bragg points for unstrained Ge and Si.

nanocrystals. In Fig. 14(b), vertical lines mark the position corresponding to the relaxed reciprocal lattice spacings of pure Ge and pure Si. The lattice parameter distribution in the nanocrystal is much closer to Si than to the Ge position. This means that either a significant alloying has taken place, or that the islands are highly strained. As expected, the intensity ratio $I(E_1)/I(E_2)$ increases considerably for the high-index reflections.

In addition, the high-index reflections allow for a better discrimination of strain and shape. The broadening of a Bragg reflection due to the finite size R amounts to $2\pi/R$ and remains constant in reciprocal space, whereas the strain-induced broadening $\Delta a/a$ increases with Q. Therefore, if the condition

$$\frac{2\pi}{R} = \frac{\Delta a}{a} Q \qquad (24)$$

is fulfilled we can extract directly the composition for different lattice parameters with the resolution Δa. To discriminate between two regions with a difference of, for example, 0.1% in lattice parameter would require a size R of about 700 Å of the diffracting object if we probe the region of the Si(800) reflection. Evaluating the intensity ratio for the two energies E_1 and E_2 at a given momentum transfer Q_{rad}, one can now use this direct method on a high indexed reflection, to link composition and lattice parameter without any model assumptions [57–59]. As the scattering amplitude in Si is virtually the same for both energies, no elaborate normalization of the intensities is necessary.

Qualitatively, the data as presented in Fig. 14 permit already some important statements:

At the (440) reflection, a step-like feature becomes visible between the maximum of the intensity from the distorted islands and the Si Bragg position. It gets more pronounced for the reflections (620) and (800). On the right-hand side of this step, there is no observable difference in the intensities recorded at E_1 and E_2. We therefore can attribute this step as being the position of the interface between the SiGe island and the substrate of pure Si. As this location is found at a position that is 0.5% lower than the Si Bragg point, the Si is under a tensile strain of about 0.5% at the interface to the SiGe island.

Considering Eq. (16) and the base diameter of about 700 Å of our dots, we already derived our lattice parameter resolution to be 0.1% at the (800) momentum transfer. In Q-space this means that our form factor of the dot base has a width of about 0.01 Å^{-1}. This corresponds to the width of the step in Fig. 14(f). Therefore the steeping of the step for increasing momentum transfer is *purely related to the increase of the resolution*. Assuming a monotonic relaxation of the lattice parameter for increasing height above the substrate we can state that this rather abrupt change in the Ge concentration is an indication for a perfect interface between the strained Si and the island with a certain SiGe alloy composition. For our further quantitative treatment of the Ge content, we will take advantage of the attributes of the high Q-ranges and focus on the data recorded at the (620) and the (800) reflection. Considering a disordered alloy of $Si_{(1-c)}Ge_c$, the diffracted intensity at all Bragg reflections is proportional to the square of the average alloy scattering factor f. The intensity ratio for the two probed energies E_1 and E_2 then reads

$$\frac{I(E_1)}{I(E_2)} = \frac{\left| c\left(f_{0Ge}(Q) + f'_{Ge}(E_1) + if''_{Ge}(E_1)\right) + (1-c)\left(f_{0Si}(Q) + f'_{Si} + if''_{Si}\right) \right|^2}{\left| c\left(f_{0Ge}(Q) + f'_{Ge}(E_2) + if''_{Ge}(E_2)\right) + (1-c)\left(f_{0Si}(Q) + f'_{Si} + if''_{Si}\right) \right|^2} \qquad (25)$$

Except for the Ge content c, all quantities in this equation can be calculated or measured before the diffraction experiment. The measured intensity ratio is therefore the parameter that can be unambiguously linked to this Ge content c. The simplest way of the data evaluation is to interpolate a set of scans at two energies, to obtain intensity values at the same positions in reciprocal space. A numerical comparison of the intensity ratios in these points with values that were calculated then yields the composition as a function of lattice parameter.

Another point of interest is the reconstruction of the composition profile in real space. In this particular case, we can use a method introduced in Ref. [57]: the scans presented in Fig. 14 follow the radial path, and the shape and width of the signal from the nano islands is determined mainly by their lattice parameter distribution. Scans carried out in a direction perpendicular to the radial path (here, "angular scans") are therefore scans across the Fourier transform of the shape function of an iso-lattice parameter area. Note that we have already determined the composition for a fixed lattice parameter. Hence a procedure, which can link a shape function with a lattice parameter, also links shape, in this case lateral size, with the composition. On the radial scan across the (400) reflection six positions were chosen at which angular scans were recorded. These scans are presented in Fig. 15(a) and (b). The angular scans show size oscillations, proving the monodispersity of the islands. These size oscillations get narrower for higher Q positions, i.e. when we approach the reciprocal position of Si. This can be attributed to the fact that the islands have a dome-like shape. The lateral size decreases for the higher parts, which are also the most relaxed ones. It is remarkable that size oscillations are still visible at a radial point that corresponds to a position in lattice parameter space where pure Si is found. The strained parts of the substrate below the islands are also monodisperse and well defined as the islands themselves. The lateral sizes determined by the fits to the oscillations in Fig. 15(b) can be linked with certain heights inside the domes via a comparison with AFM line scans of the dome profiles. In Fig. 16 the results obtained through this procedure are presented. The AFM lineshape was corrected for the tip size broadening. The lattice parameter relaxation follows a rather continuous

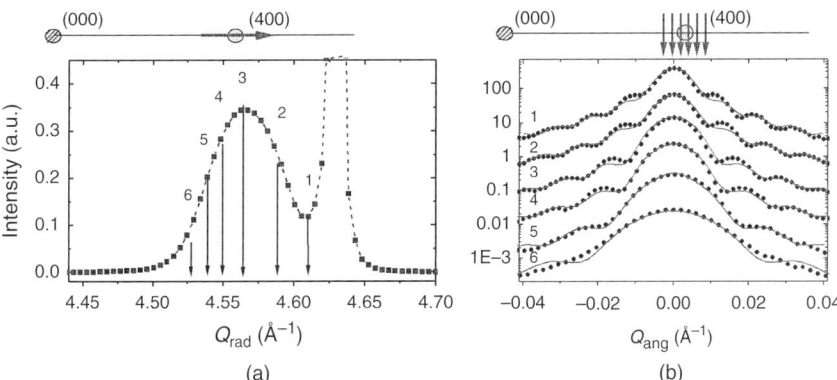

Fig. 15. In-plane map of the (400) reflection of the SiGe islands. (a) Radial scan; for the numbered positions angular scans were performed in order to determine the lateral size of a region with common lattice parameter. (b) Angular scans at the six positions marked in (a), together with fits (lines) to determine the shape function.

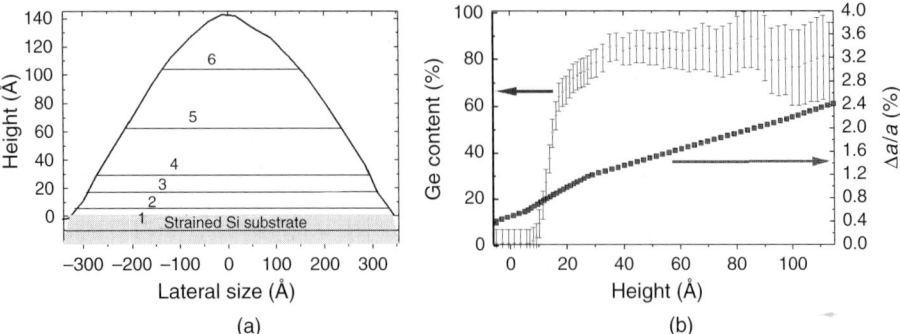

Fig. 16. Results of the real-space reconstruction. (a) The tip size corrected AFM profile is used to attribute a certain height to the lateral sizes determined from the fits in Fig. 15(b). As these lateral sizes correspond to iso-lattice parameter areas, they can be used to link directly lattice parameter and composition in (b). The squares correspond to the lattice parameter evolution with height, whereas the curve with error bars describes the composition profile as determined earlier from the (620) and the (800) reflection.

evolution whereas a rather abrupt change in the Ge concentration is observed. As still a significant alloying takes place, these results raise the question about the interdiffusion process. Interdiffusion through the substrate–island interface would blur such a sharp interface between Ge island and Si substrate. Note however that no information about lateral variations of the Ge content can be derived by this method. For a lateral resolution, a more sophisticated data treatment is required, as the relative intensities of the side maxima in the angular scans have to be fitted with an appropriate model. This was obtained by Malachias et al. [60] in the investigation of SiGe islands grown by chemical vapour deposition (CVD). Without assuming a lateral gradient and simplifying the dome base as circular in shape, the fitting of the angular scans as presented in Fig. 15(b) is performed with a Bessel function and an appropriate size distribution. In the case of the CVD-grown islands from Ref. [60], this leads to a significant underestimation of the side maxima in the angular scans. Keeping the approximation of a circular cross section but introducing a lateral concentration gradient that is a radial concentration gradient, the intensity from one iso-strain disk can be expressed as

$$I(q_a, R) = \frac{I_0}{\pi^2 R^4 |\langle f_{\mathrm{GeSi}} \rangle|^2} \left| \int_0^{2\pi} \int_0^R e^{-iq_a r \cos\theta} f_{\mathrm{GeSi}}(r) r \, dr \, d\theta \right|^2 \tag{26}$$

where f_{GeSi} stands for the concentration-averaged local atomic scattering factor of the alloy and $\langle f_{\mathrm{GeSi}} \rangle$ is the averaged atomic scattering factor for the whole iso-strain disk. Malachias et al. found good agreement with their experimental data in assuming a parabolic lateral composition profile with a Si-rich centre and higher Ge concentrations for the outermost regions. An example of radial scans with simulations for different composition profiles is shown in Fig. 17(a). The result for the complete series of angular scan is reconstructed in real space and shown in Fig. 17(b).

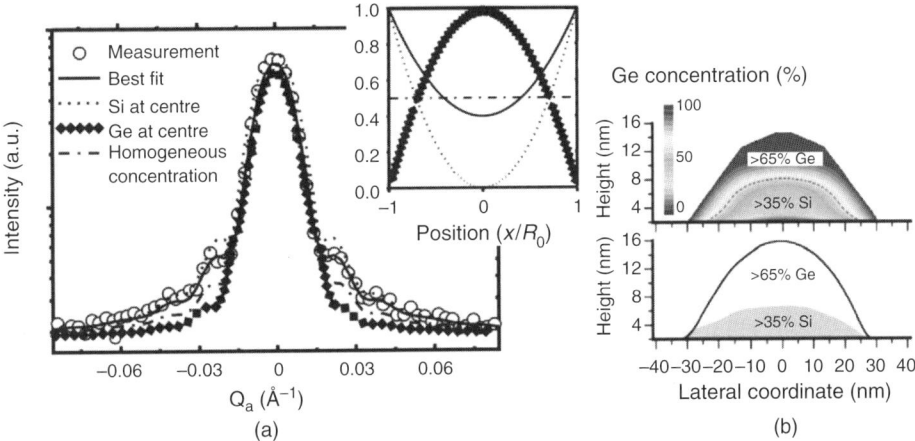

Fig. 17. (a) Angular scans from SiGe islands fitted with the assumption of different lateral concentration profiles as plotted to the right. (b) Resulting real-space chemical composition map for Ge domes (upper panel); AFM line scans taken on two Ge domes, before and after selective etch, evidencing the Si-rich core (lower panel).

These phenomena of interdiffusion differ for the different growth modes such as MBE or CVD. They are thus influenced by different kinetics rather than thermodynamics. The recent discovery of atomic ordering in SiGe islands has confirmed the importance of growth kinetics in the island formation [61]. Ordering is not observed in bulk SiGe but only in epitaxial structures grown on Si(001). It is thus a phenomenon that is purely driven by the growth kinetics.

3.4. Using multiwavelength anomalous diffraction

3.4.1. Introduction and application to buried InAs/InP and GaN/AlN

While nanostructures are generally synthesized by deposition on a surface, they are often capped for the realization of practical devices, i.e. to tailor the bandgaps. Thereby the structure is shielded from the atmosphere and oxidation is avoided. In addition, the stacking of multiple layers of dots or wires multiplies the number of "active" objects in the device and thereby their optical efficiency. This is used, for example, for GaN QDs buried in AlN (Fig. (18b)): in this case the use of a small AlN spacer thickness leads to a vertical correlation of the QDs positions and to a lateral size increase of the QD due to the strain field from the layers below.

The burying of the structures makes their study more complex as some surface techniques (AFM) are not available anymore. Moreover, during the capping process,

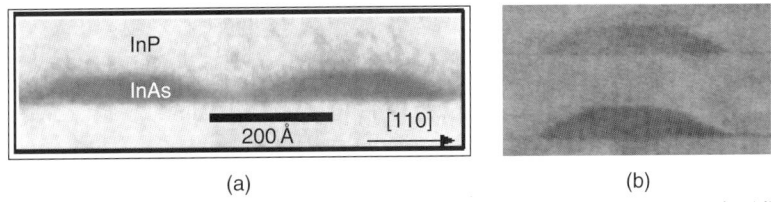

Fig. 18. (a) InAs quantum wires in InP, (b) multilayer of GaN quantum dots in AlN.

interdiffusion very often takes place between the objects and the first ML of the capping material, which can be difficult to characterize quantitatively using TEM or single-wavelength XRD. Anomalous diffraction at the absorption edge of one atom of the nano-objects (see Section 3.2), i.e. combining the spatial selectivity of diffraction with the chemical sensitivity of resonant scattering, is an ideal tool to localize selectively the atoms constituents of the nanostructures.

The most important difference with uncapped structures is that the buried objects are more strained by the capping material and therefore their scattering strongly overlap each other. In such conditions, anomalous diffraction at only one or two wavelengths is not sufficient to extract the contribution from the nanostructures. The quantitative separation of their scattering contribution can nevertheless be achieved using MAD, which we will now present.

3.4.2. Grazing incidence multiwavelength anomalous diffraction

To compute the total scattering from the material (nano and bulk), let us describe the atomic positions as $r_{i,j} = R_j^0 + r_i^0 + u_j$, where R_j^0 denotes the position of the jth unit cell in the unstrained material, r_i^0 is the position of atom i in the unit cell, and u_j is the displacement of unit cell j from this ideal position. The total scattered amplitude is the sum of the scattered amplitude from all atoms, and after factorization of the structure factor of a single unit cell, can be written as

$$A(Q) = \left[\sum_{\text{unit cell}} f_i(Q) e^{iQ \cdot r_i^0}\right] \sum_{\text{crystal}} \Omega(R_j) e^{iQ \cdot u_j} e^{iQ \cdot R_j} \qquad (27)$$

where Ω represents the shape of the object (i.e. 1 where the object is present, 0 otherwise – partial occupancy can be modelled using values in between).

From this formula one can see that the phase of the scattered amplitude depends not only on the structure factor $F(Q) = \sum f_i(Q) e^{iQ \cdot r_i^0}$, but also on the factor $\sum \Omega(R_j) e^{iQ \cdot u_j} e^{iQ \cdot R_j}$, which is the Fourier transform of the object's shape and strain field. In the case of uncapped samples, the strain relaxation is mostly a function of the height of the object and is relatively strong, resulting in an in-plane lattice parameter of the bulk, and the nanostructures sufficiently different to not overlap in reciprocal space. Therefore, the variation of the intensity with the energy close to the absorption edge is the same as for the structure factor of the bulk material, and a comparison of the intensity at two energies can be used for the determination of intermixing, as demonstrated in Section 3.2.

However, in the case of buried nano-objects, the strain relaxation is much smaller and more complicated in the nano-object, so that the scattering presents an important overlap with that of the spacer layer material (see Fig. 19(a)): the scattered amplitude is therefore the complex sum of the contribution from the different regions of the sample. Moreover, the different strain states imply different phase shifts to the scattering from the different parts of the structure, and the phase difference between the scattering from (i) the anomalous atoms and (ii) all the atoms cannot be assumed to be equal to the phase difference computed for the bulk spacer layer material. Henceforward, it is necessary to measure the diffracted intensity at more than two energies to retrieve this phase information.

This can be done with the MAD approach, which has been intensely used for more than 20 years for the structure determination of proteins and has also been used for element-selective

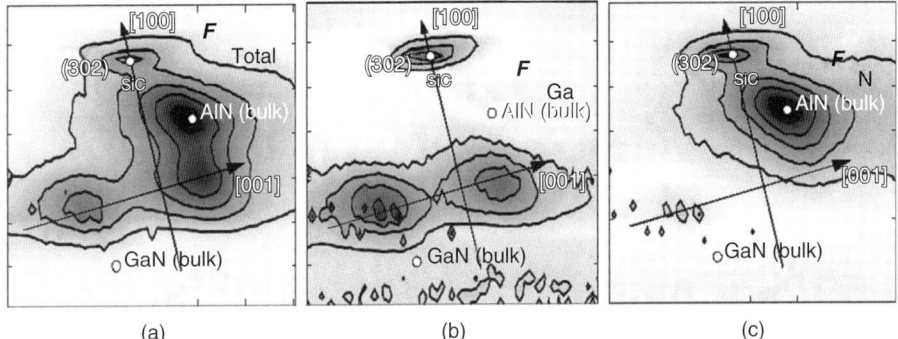

Fig. 19. (a) diffraction map close to the (302) reflection for a multilayer of GaN quantum dots embedded in AlN deposited on a SiC substrate, measured on the ID1 beamline of the ESRF. The map includes contribution from the three components of the structure: substrate (a white dot marks the (302) SiC Bragg peak), spacer layer and QDs; the position at which these diffract is related to the in-plane and out-of-plane lattice parameters and is therefore slightly different for the three parts. However, there is a very significant overlap between diffraction peaks from the QDs (GaN) and the spacer layers (AlN). The satellite peaks along the [001] direction are due to the multilayer structure. Using anomalous diffraction near the Ga K-edge, it is possible to extract the modulus of structure factor F_{Ga} from the Ga atoms only (b) and deduce (F_{Al+N}, F_N) (c), the non-anomalous atoms without any a priori knowledge on the structure. As the white dots showing the bulk positions of AlN and GaN indicate, the AlN matrix is little influenced by the QDs, and the GaN dots are compressed along [100] and stretched along [001].

diffraction (for an overview, see Ref. [17]). The scattering factor of any element can be written as (subscript A denotes the anomalous atom; in the following examples we will use A = As or A = Ga)

$$f_A(\mathbf{Q},E) = f_A^0(\mathbf{Q}) + f_A'(E) + f_A''(E) \tag{28}$$

where $\mathbf{Q}$ is the scattering vector, E is the energy of the incident photons, f^0 is the Thomson scattering factor and f' and f'' are the real and imaginary parts of the resonant scattering factor. The scattered intensity close to the absorption edge of one element A is proportional to the square modulus of the structure factor, which is written by separating the energy-dependant contributions f_A' and f_A'':

$$I = |F_T|^2 + \left|\frac{F_A}{f_A^0}\right|^2 (f_A'^2 + f_A''^2) + 2\frac{|F_A||F_T|}{f_A^0}\left(f_A'\cos(\varphi_T - \varphi_A) + f_A''\sin(\varphi_T - \varphi_A)\right) \tag{29}$$

where F_T and F_A are respectively the scattered amplitude from all atoms and from the anomalous atoms A (excluding the resonant scattering contributions), and φ_T, φ_A are the corresponding phases [17].

It should be noted that, in general, we should take into account the transmission factors T_i and T_f (see Section 2, Eq. (5)), which are also energy-dependent, but as they correspond to the air/bulk interface, the dependence around the absorption edge remains very small if the bulk does not include the resonant atom, so that we will in the following neglect these variations.

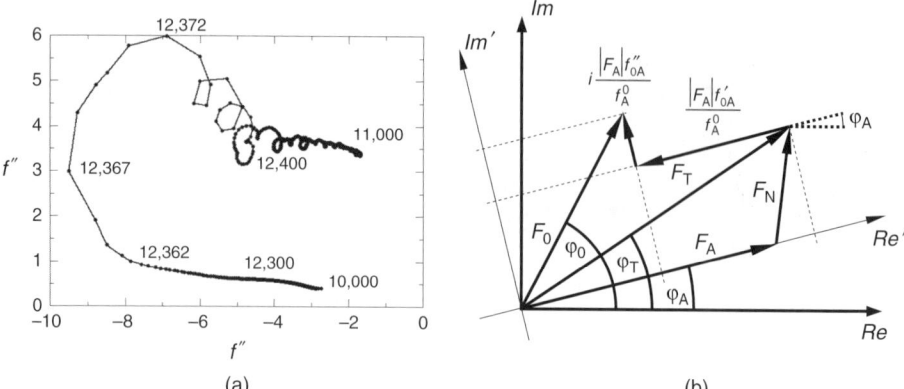

Fig. 20. (a) f′–f″ loop near the Ga K-edge. The energies used for a MAD experiment must cover the largest area possible in order to achieve the greatest phasing power and allow the most precise extraction of the anomalous scattering factor $|F_A|$ of the resonant atom. (b) Decomposition of the different contributions to the structure factor: F_T (all atoms), $F_N = F_T - F_A$ (non-anomalous atoms), F_A (anomalous atoms without their resonant contribution). The corresponding phases (φ_T, φ_N, φ_A) depend on the absolute atomic positions, i.e. both on the fractional atomic coordinates in the unit cell and on the strain field in the sample.

The three unknowns of the above equation are $|F_T|$, $|F_A|$ and $\Delta\varphi = \varphi_T - \varphi_A$. In theory, measuring the diffracted intensity at three energies is sufficient to determine them. However, to obtain a higher accuracy, it is more appropriate to use ~10 energies around the absorption edge. The optimal values for the energies can be chosen on the anomalous circle representing f'' as a function of f' in the complex plane (see Fig. 20(a)) – the larger the surface delimited by the experimental points, the higher the precision is. It is also convenient to measure a few energies far from the edge, as the variations of f' and f'' are better known in these regions, and are insensitive to small energy shifts due to the beamline optics.

$|F_T|$, $|F_A|$ and $\Delta\varphi$ are calculated by minimizing Eq. (29) using a least-squares procedure, yielding an "extracted" scattering map of the anomalous atoms only, thus allowing to study selectively the size and strain in the nano-objects without any a priori model. The scattered amplitude from the non-resonant atoms $|F_N|$ can be deduced from $|F_T|$, $|F_A|$ and $\Delta\varphi$.

As the extracted $|F_A|$ represents the contribution from the nano-objects *only*, its position, size and shape in reciprocal space can be used to determine the size, shape and average strain of the buried objects, without the need for a structural model. As shown in the following examples, a model – using an estimated shape and known elastic coefficients – can also be used to compare to the computed scattered amplitudes and refine the model.

3.4.3. First case study: InAs quantum wires in InP

With a typical photoemission wavelength around 1.55 µm, InAs quantum structures are of great interest for the possible applications for telecommunication. InAs quantum wires (QWrs) were grown by MBE using the SK growth mode, on (001) InP substrates (Fig. 18(a)). The InAs/InP lattice mismatch is around 3.2%, and leads to a strong stress during deposition, which has been found to be anisotropic and stronger along [110] than along [1–10]. The stress first relaxes along [110], which leads to the formation of QWrs. The wires are then capped by a 10–nm-thick InP layer.

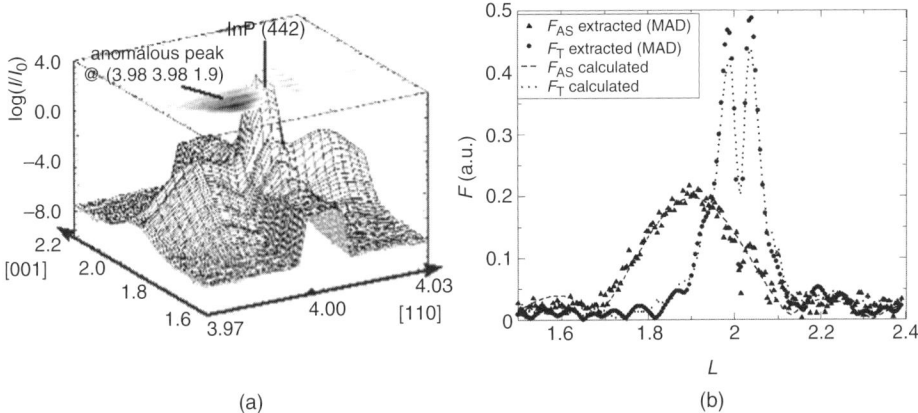

Fig. 21. (a) Experimental diffraction map around the InP (442) reflection–the projection on the upper side represents the difference between intensity recorded below (−100 eV) and at the As K-edge (b) scattered amplitude near the InAs QWrs peak along [3.98 3.98 L], extracted from the diffracted intensities measured at 10 energies near the As K-edge at 11,867 eV, along with the structure factors calculated using a FDM model of the InAS QWrs. The splitting of F_T is due to a phase shift between the scattering of InP below and above the wires.

To study the structure of the InAs wires, grazing incidence ($\alpha_i \sim 0.2°$) anomalous diffraction at the As K-edge (11,867 eV) was performed [18] at the French Collaborative Research Group beamlines BM32 and BM02 at the ESRF. Diffraction was collected close to the weak (442) reflection to enhance the anomalous signal. Figure 21 shows the diffraction map close to the reflection, as well as the extracted structure factors F_{As} and F_T.

The model-free extraction of the F_{As} peak (Fig. 21(a) and (b)) allows a direct analysis of the wires structure: its position at (3.98, 3.98, 1.90) indicates that the average out-of-plane lattice parameter is 5.3% larger than the InP lattice. The height of the wires is related to the FWHM of the F_{As} peak, and is equal to $c_{InP}/(0.9*\mathrm{FWHM}(F_{As})) = 2.54$ nm, which is close to the value found by TEM.

To further validate these values, finite difference method (FDM) simulations (Fig. 22) were performed to map the strain of the InAs wires in InP. Figure 21(b) shows the results of the best simulation, obtained by optimizing the height, width and composition of the wires. The wires were found to be 9 ML high and 22 nm wide, and the composition was pure InAs, with a weak As/P intermixing at the surface, spreading over 1 ML. To confirm the quasi-absence of interdiffusion, DAFS was also conducted and is presented in Chapter 11.

3.4.4. Second case study: GaN quantum dots in AlN, an in situ study

GaN QDs are the base for group III-nitride semiconductor devices that have been studied for their photoemission and detection properties (spanning from close-IR to UV) and used for the production of light-emitting diodes with GaN 2D quantum wells [62] and wires [63]. However, the large density of defects among group III-nitride materials leads to a reduction of emission efficiency with the temperature. To improve this, it was necessary to provide small, defect-free regions where the carriers are confined, such as QDs.

The GaN QDs presented here were grown by MBE on a 2-μm-thick AlN layer deposited on sapphire (or a 5–10 nm thick AlN layer deposited on 6H-SiC (001)). The growth follows

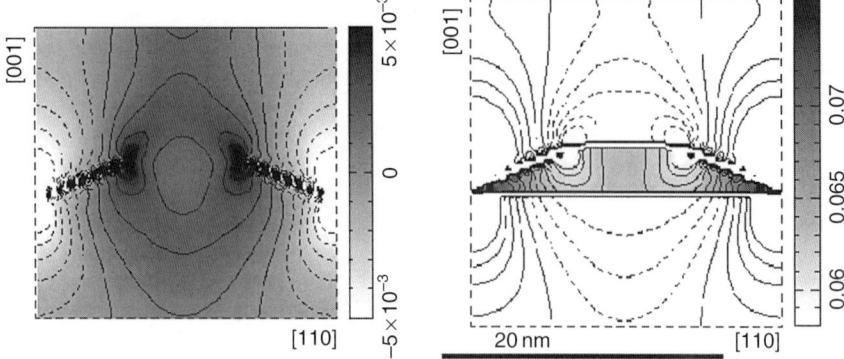

Fig. 22. (a) In-plane $\varepsilon_{xx} = (a - a_{InP})/a_{InP}$ and (b) out-of-plane $\varepsilon_{xx} = (c - c_{InP})/c_{InP}$ strains calculated using the Finite difference method around the InAs wire. Solid (resp. dashed) lines represent positive (resp. negative) contour strains. The shape of the QWr is clearly visible on the ε_{zz} map.

a modified SK mode [64]: several, 3 to 5, ML of GaN are deposited on the surface with an excess of Ga, creating a liquid layer of Ga which inhibits the formation of the QDs until the excess of Ga is evaporated. Using normal (N-rich) conditions, the QD formation would occur above the critical thickness of 2.2 ML. This growth mode allows to grow larger dots with a better control on the size distribution. Anomalous diffraction (in situ and ex situ) was used to characterize this growth mode.

The practical realization of devices implies capping of the QDs by barrier material. However, this simple operation may drastically modify the morphology and/or the optical properties of the QDs. Depending on growth conditions, capping of QDs may result in either an increase or a decrease of their size, which results from specific chemical reactions at the interface such as, for instance, dot/barrier interdiffusion in the most studied case of InAs QDs capped by GaAs or AlAs [65,66] or enhanced migration in the case of Si/Ge system [67].

As far as GaN/AlN system is concerned, which is the most recent member of the family of self-assembled QDs system [68], it has been demonstrated that capping of GaN QDs by AlN results in a size decrease due to atomic vertical exchange between Ga and Al, driven by the greater thermodynamic stability of AlN compared to GaN [69]. It should be noted that this process is limited to 2 ML and presents no interdiffusion, so that this system can be used to study directly the evolution of the capping-induced strain during the synthesis.

To investigate this dependence, we carried out an in situ study [70] of the capping of GaN QDs by AlN on the French Collaborative Research Group beamline BM32 of the ESRF, which is equipped with a MBE chamber. Samples were grown on a 6H-SiC (001) substrate at 730°C, by first depositing 8 ML of AlN, and then by growing GaN QDs using a modified SK mode, with 6 ML of GaN deposited. Then the sample was covered by n (= 0, 4, 10, 27) ML of AlN, and at each capping step the sample was studied by anomalous diffraction to extract the structure factor F_{Ga} of the Ga atoms only. To this end, we measured the diffracted intensity around the (300) reflection at 12 energies around the K-edge of Ga. To enhance the signal from the dots, a grazing incidence angle (0.15°) was used.

Figure 23 shows the extracted structure factors F_{Ga}. This shows a clear evolution for the in-plane lattice parameters of the dot, from 3.167 Å (0 ML, $\varepsilon_{xx} = -0.67\%$ relative to bulk GaN) to

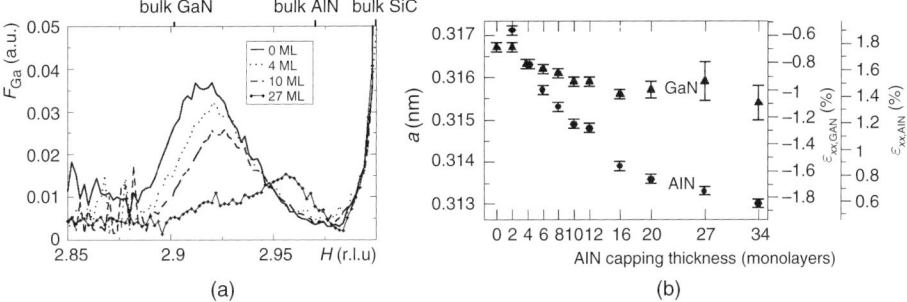

Fig. 23. (a) Extracted structure factors F_{Ga} as a function of the number of AlN monolayers deposited on the GaN QDs. The amplitude is given as a function of the in-plane reciprocal lattice unit H. $H = 3$ corresponds to the position of the SiC substrate peak, where the measured intensity (and therefore the extracted structure factor) diverges. (b) Experimental evolution of the average GaN in-plane lattice parameter, as a function of the number of capping layers deposited.

3.154 Å ($\varepsilon_{xx} = -1.1\%$). At the same time, the in-plane lattice parameter of the spacer layer material (AlN) evolves from 3.17 Å (2 ML, $\varepsilon_{xx} = 1.86\%$ relative to bulk AlN) to 3.13 Å (37 ML, $\varepsilon_{xx} = 0.58\%$). If more capping material is added (up to 100 nm), the strain remains the same in GaN (-1.1%) and AlN ($+0.6\%$) above the critical value of 30 ML (7.5 nm) deposited: this thickness value corresponds in fact to the vanishing of the strain fields (due to the buried QDs) at the surface, which is an important parameter to know to generate multiple layers of QDs with a vertical correlation of the position of the QDs between successive planes.

The production of devices with a strong photoemission efficiency requires the deposition of multiple layers of nano-object to increase the active volume in the material. Moreover, it was shown by TEM and GISAXS (see Section 3.2) that the stacking of QDs planes leads to a vertical correlation of their position, and is suspected to produce more uniform QDs assemblies [49,54,71,72].

To study this vertical correlation we followed the same GI-MAD protocol as described previously in this section to study the evolution of the QDs size and average strain, as a function of the number of QDs bilayers deposited. In this study each layer of QDs was separated from the next by 25 ML of AlN, which created a bilayer period of ~77 Å (25 ML AlN + 6 ML GaN). Again anomalous diffraction (at 12 energies around the Ga K-edge) was used to extract the structure factor of the Ga atoms only.

Figure 24 shows the results of the experiments: the most drastic changes occur when the first layer is capped, as the QDs are strained by the AlN and their size decreases. After the first layer of QDs the in-plane strain quickly stabilizes while the lateral size continues to increase until 10 layers of QDs have been deposited. This size increase is due to the correlation between successive layers: the buried QDs induce a strain field which is replicated at the AlN surface (see Fig. 3 in Ref. [54]), creating a region where it is energetically more favourable to deposit GaN (which has a lattice parameter 2.4% larger than that of AlN). Thus the creation of the QDs of the next layer preferably occurs on these strained AlN positions.

Of course, this correlation is closely related to the intensity of the induced strain field at the surface, which diminishes when the interlayer distance increases. It was found [54,72] that the critical value beyond which no vertical correlation can occur is around 30 ML of AlN (9 nm period), a value which may increase for larger dots.

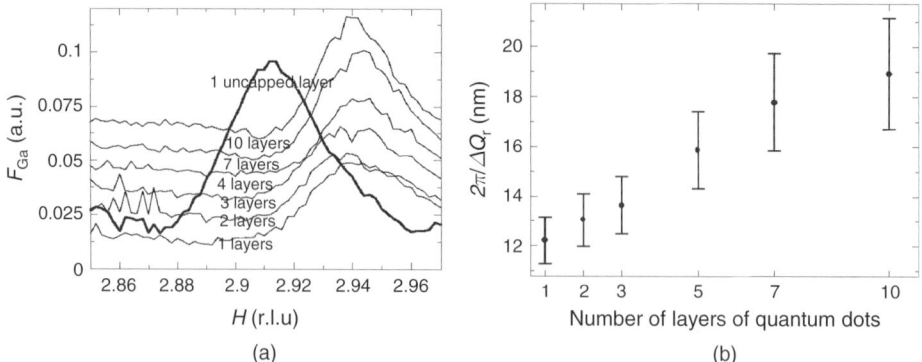

Fig. 24. (a) Extracted structure factors F_{Ga} and (b) average lateral size (computed as $D = a_{SiC}/0.9 * \text{FWHM}(F_{Ga})$) as a function of the number of QDs layers deposited in (a) the lines were vertically translated for better reading. The spacer layer thickness was $\sim 77\text{Å}$.

4. Summary and future perspective

In this chapter, we have demonstrated the application of surface-sensitive X-ray techniques at grazing incidence and the use of element-sensitive anomalous scattering to be well suited to reveal all basic structural properties of semiconductor heterostructures with nanometre dimensions. Mainly nanostructures with a 3D confinement (QDs) obtained by strain-driven, self-organized growth have been discussed. In essence, from the 3D intensity distribution in reciprocal space, we can reconstruct the 3D image in real space, concerning the shape, size, strain and composition of the islands. Inter-dot ordering, both laterally and in growth direction, can be characterized as well as atomic ordering in the part of the islands that has become an alloy, during growth, due to interdiffusion of the substrate material into the nominally pure material used to grow the nanostructures.

Recently 2D confined structures, "quantum wires", grown by self-organization in various material systems have been intensively investigated [73–76]. Structures with typical diameters of several 10–100 nm, but lengths up to several micrometres can be fabricated. Wires with built-in heterostructures in axial direction (forming QDs) have been demonstrated, as well as core-shell structures (heterostructures in radial direction). The crystal structure of long wires was shown to be different from that of bulk material. XRD experiments on large ensembles of such wires have been carried out and yielded chemical composition and strain states. Investigations on single wires would allow for clues on the growth mechanisms. X-ray scattering techniques, such as presented in this chapter, are currently being applied to reveal the complex structures of such QWrs [77,78].

Two drawbacks of the approaches described in this chapter ought to be discussed and lead to future perspectives for the application of novel X-ray techniques that can only be exploited at current third-generation synchrotrons and/or future free electron based X-ray sources, because high flux and the coherence properties of the X-ray beam will be needed. *One drawback* is that in all techniques and examples described above, real-space model assumptions for the nanostructures are needed due to the well-known "phase problem" of X-ray scattering, which does not allow for a direct reconstruction of the structures from the measured intensity distribution in reciprocal space. *The other drawback* is that the information obtained is based on measurements on large nanostructure ensembles, thus measuring their mean structure rather than the variation of individual structural properties. The knowledge on the variation of properties is especially important for structures grown by self-organization, where differences from one island to another are expected.

A possible solution for these drawbacks of conventional (incoherent) scattering experiments is offered by the use of the partial coherence of the beam and by focussing X-rays to submicron dimensions. The focussing of hard X-rays progressed to beam diameters in the 100-nm range [79] and beyond. For small beams, their partial coherence has been exploited for a model-free conversion of reciprocal space intensity distribution to the real-space sample structure; so far, this has mainly been used in coherent diffraction imaging (CDI) of almost strain-free micro- and nanocrystals [80].

In the future the combination of these different approaches, spatially resolving, submicron diffraction and CDI, offers promising tools for the characterization of *single* semiconductor nanostructures. First results on microdiffraction from rolled up nanotubes (RUNTS), which have been performed at ID01 (ESRF), have recently been published [81]. In contrast, the rather novel technique of CDI uses over-sampling for the retrieval of the phase to reconstruct the investigated objects in real space, without using an elaborate model for structure, strain and shape of a single object [82]. CDI has also been applied in the forward direction to determine the 3D morphology of objects [83–85]. In case of coherent diffraction close to Bragg reflections, additional information on strain and extended defects is obtained. It can be foreseen that CDI will be further developed to study single islands of semiconductors, coherently grown on a substrate (like SiGe dots on Si). The main issue here is to determine the model-free anisotropic strain distribution in a single island and the surrounding substrate.

Both, spatially resolved microdiffraction and CDI from low-dimensional systems will play an important role in the understanding of the structure, fabrication and functionality of many nanomaterials. The advancement of CDI will have wide-ranging applications including the investigation of self-assembled and semiconductor nanostructures, surfaces and interfaces, extended defects, granular materials and many other systems. In general, CDI can be anticipated as a versatile tool to determine the full 3D spatial arrangement of scattering objects beyond crystalline unit cells.

References

[1] M. Schmidbauer, X-Ray Diffuse Scattering from Self-Organized Mesoscopic Semiconductor Structures, Springer Tracts in Modern Physics, Vol. 199, Springer-Verlag, Heidelberg, 2003.
[2] U. Pietsch, V. Holy, T. Baumbach, High-Resolution X-Ray Scattering From Thin Films to Lateral Nanostructures, Springer Tracts in Modern Physics, 2nd edn, Advanced Texts in Physics, Springer-Berlin, Vol. 149, 2004.
[3] J. Stangl, V. Holy, and G. Bauer, Rev. Modern Phys. 76 (2004) 725–783.
[4] T.H. Metzger, T. Schülli, and M. Schmidbauer, C. R. Phys. 6 (2005) 47.
[5a] I. Kegel, T.H. Metzger, A. Lorke, et al., Phys. Rev. Lett. 85 (2000) 1694.
[5b] C. Lamberti, The Use of Synchrotron Radiation Techniques in the Characterization of Strained Semiconductor Heterostructures and Thin Films, Surf. Sci. Rep., 53 (2004) 1–97.
[6] http://www.esrf.eu/UsersAndScience/Experiments/SurfaceScience/ID01/.   http://www.esrf.eu/exp_facilities/BM2/. http://www.esrf.eu/UsersAndScience/Experiments/CRG/BM32/.
[7] I. Kegel, T.H. Metzger, A. Lorke, et al., Phys. Rev. B 63 (2001) 035318.
[8] J.D. Jackson, Classical Electrodynamics, John Wiley and Sons, New York, 1999; M. Born and E. Wolf, Principles of Optics, 5th edn, Pergamon, New York, 1975.

[9] B.E. Warren, X-Ray Diffraction, Addison-Wesley, Reading, MA, 1969; R.W. James, The Optical Principles of the Diffraction of X-Rays, Ox Bow, Connecticut, 1982.
[10] S.R. Andrews and R.A. Cowley, J. Phys. C 18 (1985) 6247.
[11] I.K. Robinson, Phys. Rev. B 33 (1986) 3830.
[12] E. Vlieg, J. Appl. Crystallogr. 30 (1997) 532.
[13] C. Schamper, H.L. Meyerheim, and W. Moritz, J. Appl. Crystallogr. 26 (1993) 687.
[14] M.F. Toney and D.G. Wiesler, Acta Crystallogr. A 49 (1993) 624.
[15] O. Robach, Y. Garreau, K. Aïd, and M.-B. Veron-Joliot, J. Appl. Crystallogr. 33 (2000) 1006.
[16] E.D. Specht and F.J. Walker, J. Appl. Crystallogr. 26 (1993) 166.
[17] J.L. Hodeau, V. Favre-Nicolin, S. Bos, et al., Chem. Rev. 101 (2001) 1843.
[18] A. Letoublon, V. Favre-Nicolin, H. Renevier, et al., Phys. Rev. Lett. 92 (2004) 186101.
[19] J. Coraux, H. Renevier, V. Favre-Nicolin, et al., Appl. Phys. Lett. 88 (2006) 153125.
[20] J. Coraux, M.G. Proietti, V. Favre-Nicolin, et al., Phys. Rev. B 73 (2006) 205343.
[21] R. Magalhaes-Panagio, G. Medeiros-Riberio, A. Malachias, et al., Phys. Rev. B 66 (2002) 245312.
[22] T.U. Schülli, J. Stangl, Z. Zhong, et al., Phys. Rev. Lett. 90 (2003) 66105.
[23] A. Malachias, S. Kycia, G. Medeiros-Riberio, et al., Phys. Rev. Lett. 91 (2003) 176101.
[24] S. Grenier, M.G. Proietti, H. Renevier, et al., Europhys. Lett. 57 (2002) 499.
[25] A. Letoublon, V. Favre-Nicolin, H. Renevier, et al., Physica B 357 (2005) 11.
[26] J.R. Levine, J.B. Cohen, Y.W. Chung, and P. Georgopoulos, J. Appl. Crystallogr. 22 (1989) 528.
[27] J.R. Levine, J.B. Cohen, and Y.W. Chung, Surf. Sci. 248 (1991) 215.
[28] A. Naudon and D. Thiaudière, J. Appl. Crystallogr. 30 (1997) 822.
[29] D. Thiaudière, O. Proux, J.S. Micha, et al., Physica B 283 (2000) 114.
[30] F. Gonella, E. Cattaruzza, G. Battaglin, et al., J. Non-Cryst. Solids 280 (2001) 241.
[31] T. Roch, V. Holy, A. Daniel, et al., J. Phys. D Appl. Phys. 34 (2001) A6.
[32] M. Schmidbauer, T. Wiebach, H. Raidt, et al., J. Phys. D Appl. Phys. 32 (1999) A230.
[33] G. Renaud, R. Lazzari, C. Revenant, et al., Science 300 (2003) 1416.
[34] C.R. Henry, Surf. Sci. Rep. 31 (1998) 235.
[35] G. Renaud, Surf. Sci. Rep. 32 (1998) 1.
[36] O. Fruchart, G. Renaud, J.P. Deville, et al., J. Cryst. Growth 237–239 (2002) 2035.
[37] R. Lazzari, J. Appl. Crystallogr. 35 (2002) 406.
[38] A. Guinier and G. Fournet, Small Angle Scattering of X-rays, John Wiley, New York, 1955.
[39] G. Porod, in: O. Glatter and O. Kratky (Eds.), Small Angle X-ray Scattering, Academic, New York, 1982, p. 37.
[40] R. Hosemann and S.N. Bagchi, Direct Analysis of Diffraction by Matter, North-Holland, Amsterdam, 1962.
[41] A. Guinier, X-Ray Diffraction in Crystals, Imperfect Crystals, and Amorphous Bodies, Dover, New York, 1963.
[42] J.S. Pedersen, J. Appl. Crystallogr. 27 (1994) 595.
[43] J.S. Pedersen, P. Vysckocil, B. Schönfeld, and G. Kostorz, J. Appl. Crystallogr. 30 (1997) 975.
[44] A. Naudon, in: H. Brumberger (Ed.), Modern Aspects of Small-Angle Scattering, Kluwer Academic, Netherlands, 1995, p. 181.
[45] G.H. Vineyard, Phys. Rev. B 26 (1982) 4146.
[46] S.K. Sinha, E.B. Sirota, S. Garoff, and H.B. Stanley, Phys. Rev. B 38 (1988) 2297.
[47] M. Rauscher, R. Paniago, et al., J. Appl. Phys. 86 (1999) 6763.
[48] T.U. Schülli, M.-I. Richard, G. Renaud, et al., Appl. Phys. Lett. 89 (2006) 143114.
[49] G. Springholz, V. Holý, M. Pinczolits, and G. Bauer, Science 282 (1998) 734.
[50] Q. Xie, A. Madhukar, P. Chen, and N. P. Kobayashi, Phys. Rev. Lett. 75 (1995) 2542.
[51] G.S. Solomon, J.A. Trezza, and J.S. Harris, Phys. Rev. Lett. 76 (1996) 952.
[52] J.-L. Rouviere, J. Simon, N. Pelekanos, et al., Appl. Phys. Lett. 75 (1999) 2632.
[53] I. Kegel, T.H. Metzger, J. Peisl, et al., Phys. Rev. B 60 (1999) 2516.
[54] V. Chamard, T.H. Metzger, M. Sztucki, et al., Europhys. Lett. 63 (2003) 268.

[55] T.U. Schülli, M. Stoffel, A. Hesse, et al., Phys. Rev. B 71 (2005) 035326.
[56] J. Baró, M. Roteta, J.M. Fernandez-Varea, and F. Salvat, Radiat. Phys. Chem. 44 (1994) 531.
[57] I. Kegel, T.H. Metzger, A. Lorke, et al., Phys. Rev. B 63 (2001) 035318.
[58] R. Magalhaes-Paniago, G. Medeiros-Ribeiro, A. Malachias, et al., Phys. Rev. B 66 (2002) 245312.
[59] T.U. Schülli, J. Stangl, Z. Zhong, et al., Phys. Rev. Lett. 90 (2003) 066105.
[60] A. Malachias, S. Kycia, G. Medeiros-Ribeiro, et al., Phys. Rev. Lett. 91 (2003) 176101.
[61] A. Malachias, T. Schülli, G. Medeiros-Ribeiro, et al., Phys. Rev. B 72 (2005) 165315.
[62] S. Nakamura, M. Senoh, S-I. Nagahama, et al., Jap. J. Appl. Phys. 35 (1996) L74.
[63] A. Kikuchi, M. Kawai, M. Tada, and K. Kishino, Jap. J. Appl. Phys. 43 (2004) L1524.
[64] N. Gogneau, D. Jalabert, E. Monroy, et al., J. Appl. Phys. 94 (2003) 2254.
[65] J.M. Garcia, G. Medeiros-Ribeiro, K. Schmidt, et al., Appl. Phys. Lett. 71 (1997) 2014.
[66] F. Ferdos, S. Wang, Y. Wie, et al., J. Cryst. Growth 251 (2003) 145.
[67] A. Hess, J. Strangl, V. Holy, et al., Phys. Rev. B 66 (2002) 085321.
[68] B. Daudin, F. Widmann, G. Feuillet, et al., Phys. Rev. B 56 (1997) 7069.
[69] N. Gogneau, D. Jalabert, E. Monroy, et al., J. Appl. Phys. 96 (2004) 1104.
[70] J. Coraux, V. Favre-Nicolin, M.G. Proietti, et al., Phys. Rev. B 73 (2006) 205343.
[71] J. Tersoff, C. Teichert, and M.G. Lagally, Phys. Rev. Lett. 76 (1996) 1675.
[72] J. Coraux, H. Renevier, V. Favre-Nicolin, et al., Appl. Phys. Lett. 88 (2006) 153125.
[73] A.I. Hochbaum, R. Fan, R. He, and P. Yang, Nano Lett. 5 (2005) 457.
[74] J. Goldberger, R. He, Y. Zhang, et al., Nature 422 (2003) 599.
[75] L. Samuelson, Materials Today 6 (2003) 22.
[76] C.P.T. Svensson, W. Seifert, M.W. Larsson, et al., Nanotechnology 16 (2005) 936.
[77] J. Eymery, F. Rieutord, V. Favre-Nicolin, O. Robach, Y.-M. Niquet, L. Frolberg, T. Martensson, and L. Samuelson, Nano Lett. 7 (2007) 2596.
[78] V. Chamard, J. Stangl, S. Labat, B. Mandl, R. Lechner and T. Metzger, J. Appl. Cryst. Accepted for publication (2008).
[79] C.G. Schroer and B. Lengeler, Phys. Rev. Lett. 94 (2005) 054802.
[80] G.J. Williams, M.A. Pfeifer, I.A. Vartanyants, and I.K. Robinson, Phys. Rev. Lett. 90 (2003) 175501.
[81] B. Krause, C. Mocuta, T.H. Metzger, et al., Phys. Rev. Lett. 96 (2006) 165502.
[82] M.A. Pfeifer, G.J. Williams, I.A. Vartanyants, et al., Nature 442 (2006) 63.
[83] J. Miao, C. Chen, C. Song, et al., Phys. Rev. Lett. 97 (2006) 215503.
[84] C. Song, R. Bergstrom, D. Ramunno-Johnson, H. Jiang, D. Paterson, M.D. de Jonge, I. McNulty, J. Lee, K.L. Wang, and J. Miao, Phys. Rev. Lett. 100 (2008) 025504.
[85] H. Jiang, D. Ramunno-Johnson, C. Song, B. Amirbekian, Y. Ohmura, Y. Nishino, Y. Takahashi, T. Ishikawa, and J. Miao, Phys. Rev. Lett. 100 (2008) 38103.

11

Grazing incidence diffraction anomalous fine structure in the study of structural properties of nanostructures

Maria Grazia Proietti,[1] Johann Coraux,[2] and Hubert Renevier[3]

[1] *Departamento de Física de la Materia Condensada, Instituto de Ciencia de Materiales de Aragón, CSIC-Universidad de Zaragoza-c. Pedro Cerbuna 12, 50009 Zaragoza, Spain*
[2] *Commissariat à l'Energie Atomique, Département de Recherche Fondamentale sur la Matière Condensée, SP2M/NRS, 17 rue des martyrs, 38054 Grenoble Cedex 9, France*
[3] *Laboratoire des Matériaux et du Génie Physique, UMR 5628, Grenoble INPMinatec, 3 parvis Louis NéelBP257, 38016 Grenoble Cedex 1, France*

Abstract In this chapter we provide a detailed view of grazing incidence diffraction anomalous fine structure spectroscopy, which has been shown to be a powerful tool in the study of the structural properties of nanostructures. We give a brief description of resonant elastic X-ray scattering and pay special attention to the major experimental and data analysis issues related to grazing incidence geometry. We show how to combine multiwavelenght anomalous diffraction and diffraction anomalous fine structure spectroscopy, in grazing incidence, to obtain strain and composition of InAs quantum wires and GaN quantum dots. This review also aims to compare extended X-ray absorption fine structure and extended DAFS results on the same samples. It gives a new insight into the complementarity of X-ray diffraction and absorption.

Keywords quantum dot, quantum wire, nanostructures, anomalous X-ray diffraction, grazing incidence DAFS, EXAFS

1. Introduction

The aim of this chapter is to illustrate the use of diffraction anomalous fine structure (DAFS) spectroscopy to study the structural properties of semiconductor nanostructures. It is closely related to Chapters 9 and 10 where extended absorption fine structure (EXAFS) spectroscopy and multiwavelength anomalous diffraction (MAD) are reported. DAFS spectroscopy consists in recording the diffraction intensity as a function of the incoming X-ray beam energy in the vicinity of an absorption edge. Like X-ray absorption fine structure (XAFS) spectroscopy, DAFS provides information about the empty electronic states [1] and the local environment of the resonant (anomalous) atom. The appearance of fine structure oscillations in the diffraction intensity was observed for the first time in the mid-1950s by Cauchois [2,3]. For years after, the applications have been sporadic [4–6], until Stragier et al. [7] gave an elegant demonstration of the utility of DAFS with a Cu single crystal. Soon after Pickering et al. [8,9] measured DAFS spectra of powdered magnetite. The reader can find extensive information about the early days of DAFS in the review articles by Sorensen et al. [10] and more recently by Hodeau et al. [11]. The very long incubation of DAFS could seem quite surprising if one thinks of the vast and massive application of extended XAFS and X-ray diffraction (XRD) to material science in the last 30 years. Indeed, DAFS provides both the advantages of X-ray diffraction and absorption; however, it is more than the addition of absorption and diffraction, it is simultaneously a *site* and *chemically* selective probe. For instance, in case of samples in which different local environments coexist, XAFS spectroscopy may fail to give pertinent information because, due to the lack of spatial selectivity, all the different environments are probed. The site/spatial selectivity of DAFS helps to solve the difficulty by selecting individual sites. Needless to say, it can be interesting both for absorption and diffraction studies by the scientific communities. One of the reasons why this technique did not take off as others, is that on a technical point of view it presents stringent experimental requirements. Essentially, one needs a very high signal-to-noise ratio, as for EXAFS, to perform a quantitative oscillations analysis on a diffraction yield that is only a very small fraction of the total one. A high brilliance beam is needed together with a high-quality diffractometer coupled to a very stable absorption-dedicated optics. Synchrotron radiation facilities, including third-generation sources, has led to the development of a number of new spectroscopic techniques, among them DAFS. Together with technical improvements, data analysis has also developed, relying on one hand on the solid support and established MAD data treatment, and, on other, on the EXAFS approach [6,7], using available and well-known codes for data simulation and analysis. Indeed, although DAFS contains contributions of both the real and imaginary parts of the complex anomalous scattering factors (XAFS is proportional to the imaginary part), they can be analyzed, in the extended region, like EXAFS [12]. An efficient program for simulating the near-edge DAFS is now also available [13]. However, one should mention an intrinsic limitation of DAFS that is the difficulty (but it is not always impossible!) to correct the data for self-absorption when measuring bulk samples (see, for instance, Refs [14,15]). Therefore, the method is definitely well suited for studying thin films or heterostructures [16–22].

In the past few years, DAFS has been applied to systems of great technological interest such as semiconductor nanostructures [23–27]. The knowledge of strain, chemical compositions, intermixing at the interfaces, i.e., structural properties on the long- and short-range scale, are of great importance to understand the growth mechanism as well as the electronic and optical properties of hetero- and nanostructures. Moreover, to be suitable for devices, the nanostructures are encapsulated or embedded in a superlattice. Capping plays a decisive role in the modification of the optical properties by modifying the strain and possibly inducing atomic diffusion. Strain is closely related to composition, shape, and aspect ratio of the nanostructures, and on the mutual stress that nanostructures, substrate, and the matrix apply to each other. X-ray diffraction is known to be a powerful tool for measuring strain fields and correlations (see Refs [28,29]). The combination of MAD (see Chapter 10 and Section 2.2) that allows to extract the scattering amplitude of the resonant and nonresonant atoms (chemical mapping in the reciprocal space), and DAFS that allows to determine the local environment of atoms located in an isostrain volume selected by diffraction is a very powerful approach to disentangle strain and compostion. In addition, X-ray diffraction is a nondestructive method that averages over many individual nanostructures and gives statistically relevant structural properties such as strain and composition. On the contrary, the structural properties of individual nanostructure could be obtained by using submicron beam size and/or coherent scattering (see, for instance, Ref. [30]).

A major technical improvement of the DAFS spectroscopy, that is extensively described in this chapter, is to perform the experiment in grazing incidence geometry (GIMAD and GIDAFS) to reduce the substrate scattering contribution, allowing to focus on the structural properties of nano-objects [31,32]. The examples that we give for GIDAFS applications indeed refer to challenging nanomaterials with strong technological impact. In particular, we show that it is possible to study strain and composition of a single nanostructure layer encapsulated by a capping layer or to follow, layer by layer the effect of capping.

The chapter is organized as follows: in Section 2 we give a brief insight on the basic theoretical principles of the technique and we describe the first-order approach to extended DAFS (EDAFS) formalism. Section 3 is devoted to the most relevant experimental aspects of the technique. In Section 4 we elucidate the role of multiple scattering (MS) and sketch the data reduction scheme. Section 5 reports examples of GIDAFS applications to GaN quantum dots (QDs) and InAs quantum wires (QWrs). We suggest to readers not familiar with quantum mechanics and diffraction formalisms to skip Sections 2 and 4 and read Sections 3 and 5 first.

2. Basic Principles
2.1. Resonant elastic X-ray scattering

2.1.1. The scattering process

In this section we underline the basic physical processes that lead to Thomson and resonant scattering [33–35]. To begin with (Fig. 1), let us consider a photon $|\mathbf{k}, \vec{\varepsilon}_\mu\rangle \equiv |\mathbf{k}, \mu\rangle$ with wave vector $\mathbf{k}$, momentum $\mathbf{p} = \hbar \mathbf{k}$, energy $E = \hbar\omega$ polarization state $\vec{\varepsilon}_\mu$ ($\mu = 1, 2$), and spin angular momentum 1, which is scattered by an atom in the fundamental state $|a\rangle$ with energy E_a. The outgoing photon is $|\mathbf{k}', \mu'\rangle$ with energy $E_0 = \hbar\omega'$ and polarization state $\vec{\varepsilon}_{\mu'}$. The direction of the transverse polarization vector $\vec{\varepsilon}_\mu$ is given by $\mu = 1,2$, which corresponds to two perpendicular directions $(\vec{\varepsilon}_\mu \perp \mathbf{k})$. In the following we will consider elastic scattering only, which means that $|a'\rangle = |a\rangle$, $E' = E$ and $k' = k$.

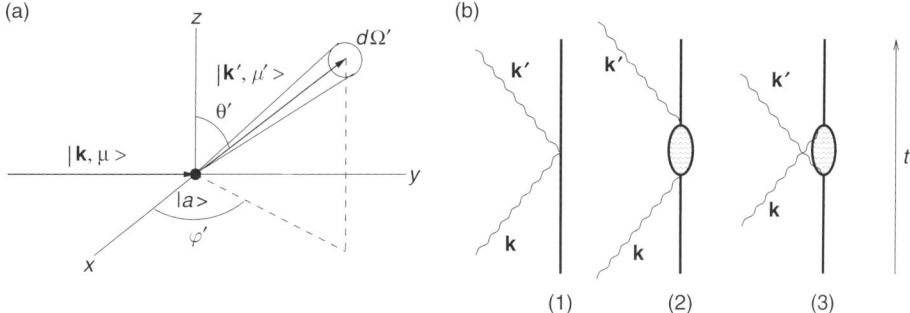

Fig. 1. (a) Atomic scattering scheme; (b) scattering process to the second-order terms of the Born series (1) simultaneous annihilation and creation of one photon (2) absorption of one photon $|\mathbf{k}, \mu\rangle$ that leads to the intermediate state $|b; 0\rangle$ followed by an emission of one photon $|\mathbf{k}', \mu'\rangle$ (resonant process) (3) emission of one photon $|\mathbf{k}', \mu'\rangle$ that leads to intermediate state $|b; \mathbf{k}, \mu, \mathbf{k}', \mu'\rangle$ with two photons followed by the absorption of one $|\mathbf{k}, \mu\rangle$ (nonresonant process).

The relevant quantity to be measured is the differential scattering cross section $d\sigma/d\Omega'$, that is, the transition probability per second, P_{fi}, from the initial state $|a; \mathbf{k}, \mu\rangle$ to the final state $|a; \mathbf{k}', \mu'\rangle$ (both states belong to a continuum), divided by the incident flux associated with the initial state.

The quantized transverse field is described by a field operator, which is given by (SI units)

$$A(r,t) = \sum_{\mathbf{k},\mu} \sqrt{\frac{\hbar}{2\varepsilon_0 \omega L^3}} \left[a_{\mathbf{k},\mu} \vec{\varepsilon}_\mu e^{i(\mathbf{k}\cdot\mathbf{r}-\omega t)} + a^\dagger_{\mathbf{k},\mu} \vec{\varepsilon}^*_\mu e^{-i(\mathbf{k}\cdot\mathbf{r}-\omega t)} \right] \quad (1)$$

where $(2\pi/L)^3$ is the elementary volume associated to vector $\mathbf{k}$, $a_{\mathbf{k},\mu}|n\mathbf{k}, \mu\rangle = \sqrt{n}|(n-1)\mathbf{k}, \mu\rangle$ and $a^\dagger_{\mathbf{k},\mu}|n\mathbf{k}, \mu\rangle = \sqrt{n+1}|(n+1)\mathbf{k}, \mu\rangle$ are the annihilation and creation operators, respectively $\left(\left[a_{\mathbf{k},\mu}, a^\dagger_{\mathbf{k}',\mu'} \right] = \delta_{\mathbf{k}\mathbf{k}'} \delta_{\mu\mu'} \right)$. In the Coulomb gauge $\Delta \cdot A(r,t) = 0$, i.e. $A = A_\perp$ is perpendicular to the wave vector $\mathbf{k}$.

If we consider only electrons (the proton mass is 1836 times greater than the electron mass) and neglect the electron spin interaction with the magnetic field, one can write the Hamiltonian $H = H_0 + V$, where H_0 and V are the unperturbed and interaction Hamiltonians, respectively:

$$H_0 = \sum_\alpha \frac{p_\alpha}{2m_e} + V_{\text{Coulomb}} + H_R \quad \text{and} \quad V = -\sum_\alpha \frac{q}{2m_e} p_\alpha \cdot A + \sum_\alpha \frac{q_\alpha^2}{2m_e}[A]^2 = H_{11} + H_{12}$$

where m_e is the electron mass and the sum is over all electrons α and $H_R = \sum_{\mathbf{k},\mu} \left(a^\dagger_{\mathbf{k},\mu} a_{\mathbf{k},\mu} + \frac{1}{2} \right) \hbar\omega$ is the radiation Hamiltonian of the quantized radiation field. The state $|a; \mathbf{k}, \mu\rangle$ is an eigenstate of H_0: $H_0|a; \mathbf{k}, \mu\rangle = (E_a + \hbar\omega)|a; \mathbf{k}, \mu\rangle$.

The initial $|a; \mathbf{k}, \mu\rangle$ and final $|a; \mathbf{k}', \mu'\rangle$ states belong to the continuum of H_0 and the transition probability per second [35] writes:

$$P_{fi} = \frac{2\pi}{\hbar} \left| \langle a; \mathbf{k}', \mu'|V|a; \mathbf{k}, \mu\rangle + \langle a; \mathbf{k}', \mu'|V \frac{1}{E_a + \hbar\omega - H + i\eta} V|a; \mathbf{k}, \mu\rangle \right|^2 \delta(E_f - E_i) \quad (2)$$

$$= \frac{2\pi}{\hbar} |\tau_{fi}|^2 \delta(E_f - E_i) \quad (3)$$

To the second-order terms of the Born series of τ_{fi}, one obtains three matrix elements:

$$\tau_{fi}^{(1)} = \langle a; \mathbf{k}', \mu'|H_{12}|a; \mathbf{k}, \mu\rangle \quad (4)$$

$$\tau_{fi}^{(2)} = \lim_{\eta \to 0^+} \sum_b \frac{\langle a; \mathbf{k}', \mu'|H_{11}|b; 0\rangle \langle b; 0|H_{11}|a; \mathbf{k}, \mu\rangle}{E_a + \hbar\omega - E_b + i\eta} \quad (5)$$

$$\tau_{fi}^{(3)} = \lim_{\eta \to 0^+} \sum_b \frac{\langle a; \mathbf{k}', \mu'|H_{11}|b; \mathbf{k}, \mu, \mathbf{k}', \mu'\rangle \langle b; \mathbf{k}, \mu, \mathbf{k}', \mu'|H_{11}|a; \mathbf{k}, \mu\rangle}{E_a + \hbar\omega - (E_b + 2\hbar\omega) + i\eta} \quad (6)$$

Where $|b; 0\rangle$ is a virtual intermediate state, $H_0|b; 0\rangle = E_b|b; 0\rangle$ and $\sum_b |b\rangle\langle b| = 1$. The three matrix elements $\tau_{fi}^{(1)}$, $\tau_{fi}^{(2)}$, $\tau_{fi}^{(3)}$ correspond to three scattering processes: (1) a simultaneous annihilation and creation of one photon ($H_{12} \propto [\mathbf{A}]^2$); (2) an absorption of one photon $|\mathbf{k}, \mu\rangle$ that leads to the intermediate state $|b; 0\rangle$ followed by the emission of one photon $|\mathbf{k}', \mu'\rangle$ (nonresonant process); and (3) the emission of one photon $|\mathbf{k}', \mu'\rangle$ that leads to intermediate state $|b; \mathbf{k}, \mu, \mathbf{k}', \mu'\rangle$ with two photons followed by the absorption of one $|\mathbf{k}, \mu\rangle$ (nonresonant process).

The total number of photon $|\mathbf{k}', \mu'\rangle$ whose energy is $E' = \hbar\omega' = \hbar\omega$ and are scattered per second in the solid angle $d\Omega'$ is

$$dn' = F_i \frac{d\sigma}{d\Omega'} d\Omega' = d\Omega' \int_0^\infty P_{fi}\rho(E')dE' = d\Omega' P_{fi}\rho(E_f) \qquad (7)$$

where $F_i = c/L^3 \, (m^{-2}s^{-1})$ is the incident photons flux associated to one incoming photon in the volume L^3; $\rho(E') = (L^3/(2\pi)^3)(\omega'^2/\hbar c^3)$ is the photon density of states; $d\sigma/d\Omega' = r_0^2 |f|^2$ is the *differential scattering cross section* (m^2) where f is the atomic scattering amplitude (scattering factor), $r_0 = q^2/4\pi\varepsilon_0 m_e c^2$ the classical electron radius ($r_0 = 2.82 \times 10^{-15}$ m, q is the electron charge). The scattering cross section can be expressed as

$$\frac{d\sigma}{d\Omega'} = r_0^2 |f^0(\mathbf{Q} = \mathbf{k}' - \mathbf{k}, \vec{\varepsilon}_\mu, \vec{\varepsilon}_{\mu'}) + \Delta f(E = \hbar\omega, \mathbf{k}, \mathbf{k}', \vec{\varepsilon}_\mu, \vec{\varepsilon}_{\mu'})|^2 \qquad (8)$$

where

$$f^0(\mathbf{Q}) = -\left(\vec{\varepsilon}_\mu \cdot \vec{\varepsilon}_{\mu'}^*\right) \sum_\alpha \int d^3r |\langle \mathbf{r}|\alpha\rangle|^2 e^{-i\mathbf{Q} \cdot \mathbf{r}} \qquad (9)$$

was obtained from the matrix element $\tau_{fi}^{(1)}$. Equation (9) shows that $f^0(\mathbf{Q})$ is the Fourier transform of the electronic density ($|\langle \mathbf{r}|\alpha\rangle|^2 = |\psi_\alpha(\mathbf{r})|^2$) in the $\mathbf{Q} = \mathbf{k}' - \mathbf{k}$ space, known as Thomson scattering ($\mathbf{Q}$ is the scattering vector). The term $\Delta f = f'(E, \mathbf{k}, \mathbf{k}', \vec{\varepsilon}_\mu, \vec{\varepsilon}_{\mu'}) + if''(E, \mathbf{k}, \mathbf{k}', \vec{\varepsilon}_\mu, \vec{\varepsilon}_{\mu'})$ is the anomalous scattering. It is the sum of a resonant Δf_{res} (complex) and a nonresonant Δf_{nres} terms obtained with $\tau_{fi}^{(2)}$ and $\tau_{fi}^{(3)}$, respectively:

$$\Delta f_{res} = \frac{1}{m_e \hbar} \sum_{\alpha,\beta} \sum_b \frac{\langle a|\mathbf{p}_\beta \cdot \vec{\varepsilon}_{\mu'}^* e^{-i\mathbf{k}' \cdot \mathbf{r}_\beta}|b\rangle \langle b|\mathbf{p}_\alpha \cdot \vec{\varepsilon}_\mu e^{i\mathbf{k} \cdot \mathbf{r}_\alpha}|a\rangle}{\omega_{ba} - \omega - i\Gamma_b/2} \qquad (10)$$

$$\Delta f_{nres} = \frac{1}{m_e \hbar} \sum_{\alpha,\beta} \sum_b \lim_{\eta \to 0} \frac{\langle a|\mathbf{p}_\beta \cdot \vec{\varepsilon}_\mu e^{i\mathbf{k} \cdot \mathbf{r}_\beta}|b\rangle \langle b|\mathbf{p}_\alpha \cdot \vec{\varepsilon}_{\mu'}^* e^{-i\mathbf{k}' \cdot \mathbf{r}_\alpha}|a\rangle}{\omega_{ba} + \omega + i\eta} \qquad (11)$$

where $\hbar\omega_{ba} = E_b - E_a$ and ω_{ba} is the Bohr frequency, $\tau_b = 4\pi/\Gamma_b$ is the lifetime of the intermediate state $|b; 0\rangle \left(e^{-iHt/\hbar}|b; 0\rangle = e^{-iE_b t/\hbar} e^{-\Gamma_b t/2}|b; 0\rangle\right)$. When $E_a + \hbar\omega \approx E_b$, $\Delta f_{res} \gg \Delta f_{nres}$. Note that a minus sign was explicitly added to the Thomson and anomalous scattering expressions because the scattered wave is π out of phase with respect to the incident wave (this ensures the attenuation of the transmitted wave in the forward direction).

Equation (2) can be used to calculate the transition probability per second from state $|a; \mathbf{k}, \mu\rangle$ to state $|b; 0\rangle$, i.e., the photoelectric absorption probability and the corresponding differential absorption cross section. To the first order of the Born series, one obtains the photoelectric transition probability per second P_{pe} and the cross section σ_{pe}:

$$P_{pe} = \frac{2\pi}{\hbar} |\langle b; 0|H_{I1}|a; \mathbf{k},\mu\rangle|^2 \delta(E_b - E_a - \hbar\omega) \tag{12}$$

$$\sigma_{pe} = \frac{4\pi c}{\omega} r_0 \sum_{\alpha,\beta} \sum_b \langle a|\mathbf{p}_\beta \cdot \vec{\varepsilon}_\mu^* e^{-i\mathbf{k}\cdot\mathbf{r}_\beta}|b\rangle \langle b|\mathbf{p}_\alpha \cdot \vec{\varepsilon}_\mu e^{i\mathbf{k}\cdot\mathbf{r}_\alpha}|a\rangle \delta(E_b - E_a - \hbar\omega) \tag{13}$$

The final (real) states $|b\rangle$ of the photoelectric absorption are the intermediate (virtual) states of resonant scattering. Refer to Chapter 9 for a practical description of Eq. (13) and state $|b\rangle$. For calculations, one may expand the term $e^{\pm i\mathbf{k}\cdot\mathbf{r}} \approx 1 \pm i\mathbf{k}\cdot\mathbf{r} - 1/2(\mathbf{k}\cdot\mathbf{r})^2 \mp \ldots$ (multipole expansion) in Eqs (10), (11), and (13). This leads to the tensorial expression of the atomic scattering factor [36,37]. Also, the relation $[\mathbf{r}, H_0] = i\hbar(\mathbf{p}/m)$ is used to replace the momentum operator $\mathbf{p}$ by the position operator $\mathbf{r}$.

2.1.2. The case of forward scattering

Let us take the simple and very important case of forward scattering ($\mathbf{Q} = \mathbf{k}' - \mathbf{k} = 0$). For simplicity, we consider an atom with one electron. We will assume that the dipole approximation ($e^{i\mathbf{k}\cdot\mathbf{r}_\alpha}$) holds and that there is no change of polarization state. In that case one can easily write $f(\omega)$ as

$$f(\omega) = f^0(Q=0) + \Delta f(\omega) \approx -1$$
$$+ \sum_b \frac{|\langle b|\mathbf{p}\cdot\vec{\varepsilon}_\mu|a\rangle|^2}{m_e \hbar} \left[\frac{2\omega_{ba}}{(\omega_{ba}^2 - \omega^2)} + i\frac{\Gamma_b/2}{(\omega_{ba} - \omega)^2 + \Gamma_b^2/4}\right] \tag{14}$$

or by using the Reiche–Thomas–Kuhn sum rule $\sum_b 2|\langle b|\mathbf{p}\cdot\vec{\varepsilon}_\mu|a\rangle|^2/m_e\hbar\omega_{ba} = 1$ (i.e., the sum over all intermediate state $|b\rangle$ of the oscillator strength is equal to 1):

$$f(\omega) \approx \sum_b \frac{|\langle b|\mathbf{p}\cdot\vec{\varepsilon}_\mu|a\rangle|^2}{m_e \hbar} \left[\frac{2\omega^2}{\omega_{ba}(\omega_{ba}^2 - \omega^2)} + \frac{i\Gamma_b/2}{(\omega_{ba} - \omega)^2 + \Gamma_b^2/4}\right] \tag{15}$$

Equation (14) shows that the imaginary part of $\Delta f(\omega)$ is a narrow Lorentzian of width $\Gamma_b(\approx \pi\delta(\omega_{ba} - \omega))$. At resonance $\omega \approx \omega_{ba}$, the scattering amplitude is proportional to $1/(\omega_{ba} - \omega)^2 + \Gamma_b^2/4$. Note that the Thomson scattering as defined in Section 2.1.1 (Eq. (9)) corresponds to $\omega \gg \omega_{ba}$.

The real and imaginary parts of $\Delta f(\omega)$ in Eq. (14) are not independent and satisfy the relation

$$\text{Re}[f(\omega)] = \frac{2}{\pi} P \int_0^\infty \frac{\omega' \text{Im}[f(\omega')]}{\omega'^2 - \omega^2} d\omega', \quad \text{Im}[f(\omega)] = -\frac{2\omega}{\pi} P \int_0^\infty \frac{\text{Re}[f(\omega')]}{\omega'^2 - \omega^2} d\omega' \tag{16}$$

as well as the real and imaginary parts of $f(\omega)$ in Eq. (15) satisfy the relation

$$\text{Re}[f(\omega)] = \frac{2\omega^2}{\pi} P \int_0^\infty \frac{\text{Im}[f(\omega')] d\omega'}{\omega'(\omega'^2 - \omega^2)} \tag{17}$$

where P represents the Cauchy principal part. These relations are known as the dispersion relations for scattering of light or *Kramers–Krönig* relationships. It is a consequence of a general principle referred to as the causality principle: "no outgoing disturbance shall start until the incoming disturbance hits the scatterer" [33,38].

Now let us take, for simplicity, a material composed of identical atoms of atomic number Z and derive the relationship between the imaginary part of $f(\omega, \mathbf{Q}=0) = f(\omega) = -Z + f'(\omega) + if''(\omega)$ and the total cross section. By classical electromagnetism the refractive index is written as

$$n(\omega) = \sqrt{\varepsilon_r} = \sqrt{1+\chi} \simeq 1 + 2\pi r_0 \left(\frac{c}{\omega}\right)^2 N f(\omega) = 1 - \delta + i\beta \qquad (18)$$

where N is the number of scatterers per cubic meters, ε_r is the relative permittivity, and χ the dielectric susceptibility. The transmitted wave intensity traveling in the $z > 0$ direction is attenuated by the factor $e^{-\mu z} = e^{-\sigma_{total} N z} e^{-2(\omega/c)\Im(n)z}$, where $\mu = \sigma_{total} N$ is the linear absorption coefficient. This leads to an important result, which is expressed by the *optical theorem*:

$$\text{Im}[r_0 f(\omega, \mathbf{Q}=0)] = r_0 f'' = \frac{\omega}{4\pi c}\sigma_{total} \qquad (19)$$

where $\sigma_{total} = \sigma_{el} + \sigma_{pe} + \sigma_{in}$ is the total atomic cross section. In the photon energy range of 1–30 keV, the photoelectric cross section σ_{pe} is several orders of magnitude larger than the elastic and inelastic (Compton) cross sections σ_{el} and σ_{in}.

2.1.3. Atomic scattering factor computation

For calculating the scattering amplitude of an assembly of atoms, i.e., the structure factor (see Section 2.2.1), one usually writes the atomic scattering factor as follows: $f(\mathbf{Q},\omega) = (\vec{\varepsilon}_\mu \cdot \vec{\varepsilon}_{\mu'}) f^0(\mathbf{Q}) + f'(\omega, \mathbf{k}, \mathbf{k}', \vec{\varepsilon}_\mu, \vec{\varepsilon}_{\mu'}) + if''(\omega, \mathbf{k}, \mathbf{k}', \vec{\varepsilon}_\mu, \vec{\varepsilon}_{\mu'}) = f^0 + \Delta f$, where $f'' > 0$. Theoretical values of f^0 ($Q/2\pi = 2\sin\theta/\lambda$) and bare atom $\Delta f_0 = f'_0 + i\,f''_0$ are tabulated or can be calculated (International tables of Crystallography [39], Sasaki tables [40], *Cromer–Libermann* calculations [41], and *Chantler* tables [42]).

At the edge and up to few hundreds of electron volts above, experimental f' and f'' depend on the resonant atom valence and local environment as well as on the wave vectors and polarization directions of the incoming and outgoing photons (anisotropy of the tensor of suceptibility [36,37]). In this case, it is possible to perform f' and f'' ab initio calculations with the FDMNES code [13].

Within the dipole approximation ($e^{i\mathbf{k}\cdot\mathbf{r}_\alpha} \approx 1$), Δf does not depend on $\mathbf{k}$ and $\mathbf{k}'$, and therefore it can be determined in the forward scattering limit. The imaginary part $f''(E)$ is obtained from an absorption spectrum (optical theorem, Eq. (19)), and by applying the Kramers–Krönig transform [15,43,44], one can calculate the real part $f'(E)$. Note that according to the sign convention for f' and f'' ($f' < 0$, $f'' > 0$), f' is obtained from f'' by changing sign to the Kramers–Krönig transforms written in Eq. (16). As an example, Fig. 2 shows the anomalous scattering factors f' and f'' of As in bulk InAs, at the As K-edge (11.867 keV).

2.2. Multiwavelength anomalous diffraction

2.2.1. The DAFS spectrum parametrization

In the following one considers an assembly of N atoms comprising N_A resonant atoms A. The structure factor is given by $F = \sum_{l=1}^{N} f_l e^{-M_l Q^2} e^{i\mathbf{Q}\cdot\mathbf{r}_l} = |F|e^{i\varphi}$, where $f_l(Q,E) = f_l^0 + f_l' + if_l''$

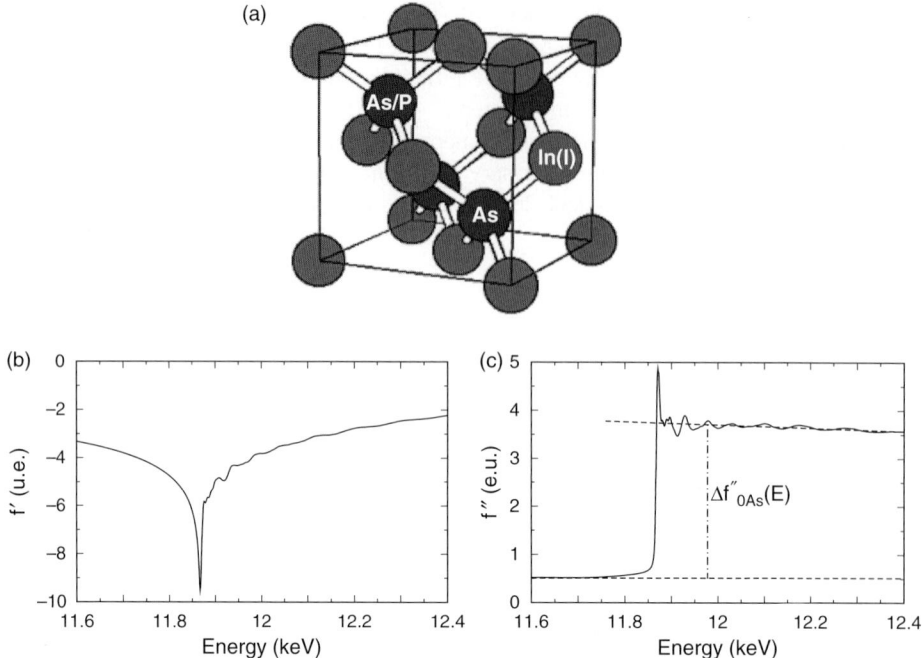

Fig. 2. (a) Sketch of the cubic InAs zinc-blende structure: fcc, In: (0, 0, 0); As: $(1/4, 1/4, 1/4)$. Experimental (b) f'_{As} and (c) f''_{As} anomalous scattering spectra of As in bulk InAs, at the As K-edge (11.867 keV). The f''_{As} spectrum was obtained from an absorption measurement, in transmission mode, of an InAs powder (optical theorem). The f'_{As} spectrum was calculated by applying the Kramers–Krönig transform to f''_{As}.

and $e^{-M_j Q^2}$ is the Debye–Waller factor. The diffracted intensity is proportional to $I = FF^*$. For simplicity, let us first deal with the smooth energy variations of the intensity, i.e., $f_{0l}(Q, E) = f_l^0 + f'_{0l} + if''_{0l}$, where $f'_{0l} + if''_{0l}$ represents the anomalous scattering of bare neutral atoms. According to the schematic representation in Fig. 3(a), the "smooth" structure factor F_0 is the sum of (a) the partial structure factor $F_A = \sum_j^{N_A} f_A^0(Q) e^{-M_j Q^2} e^{iQ \cdot r_j}$ of phase φ_A, which includes the Thomson scattering of all anomalous (A) atoms, (b) the partial structure factor F_N of phase φ_N that includes the scattering of all nonanomalous atoms, and (c) the anomalous scattering of all atoms A. One can write

$$F_0(Q, E) = F_T(Q) + \frac{F_A(Q)}{f_A^0}\left[f'_{0A}(E) + if''_{0A}(E)\right] \quad (20)$$

where $F_T = F_A + F_N$. From Fig. 3(a), one can readily write

$$|F_0(E)|^2 = \left(|F_T|\cos(\varphi_T - \varphi_A) + \frac{|F_A|}{f_A^0}f'_{0A}(E)\right)^2 + \left(|F_T|\sin(\varphi_T - \varphi_A) + \frac{|F_A|}{f_A^0}f''_{0A}\right)^2 \quad (21)$$

or

$$|F_0(E)|^2 = \left[|F_N|\cos(\varphi_N - \varphi_A) + |F_A|\left(1 + \frac{f'_{0A}}{f_A^0}\right)\right]^2 + \left(|F_N|\sin(\varphi_N - \varphi_A) + \frac{|F_A|}{f_A^0}f''_{0A}\right)^2 \quad (22)$$

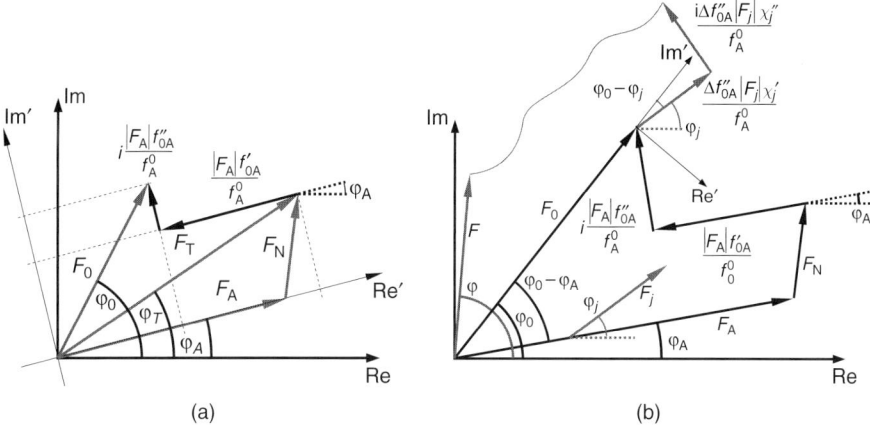

Fig. 3. (a) Schematic representation in the complex plane of the smooth structure factor factor F_0, and its relationship with F_T, F_A, and F_N (b) Schematic representation of the first-order contribution of χ'_j and χ''_j to the EDAFS oscillations. The wavy curve stands for all other atomic oscillatory contributions.

The modulus $|F_T|$, $|F_A|$, $|F_N|$ and phases $\varphi_T - \varphi_A$, $\varphi_N - \varphi_A$ can be considered as constant parameters as a function of the energy. Then by using tabulated or experimental f'_{0A} and f''_{0A} values and fitting the intensity variation of $I_0 = |F_0(E)|^2$, as a function of the energy, to the experimental DAFS spectrum, one can recover, without any structural model, either $\varphi_T - \varphi_A$ and $\beta_{TA} \doteq |F_A|/[f^0_A|F_T|]$ (Eq. (21)) or $\varphi_N - \varphi_A$ and $\beta_{NA} = |F_A|/|F_N|$ (Eq. (22)), at a given **Q** vector.

Considering now the structure factor F, Eq. (20) can be used with f'_j and f''_j (instead of $f'_{0j}(f''_{0j})$) provided that:

1. F_A is real ($\varphi_A = 0$), which is the case whenever the crystallographic structure is centrosymmetric,
2. the resonant atoms are located in equivalent crystallographic sites such that $\varphi_j = \varphi_A + n 2\pi$ (n integer),
3. the energy is such that all resonant atoms j have the same $f'_j(f''_j)$ values, i.e., below (few tens of electronvolts) and "far" above (few hundreds of electronvolts) the edge.

As an example, Fig. 4 shows diffracted intensity calculations as a function of energy (DAFS spectra) for bulk InAs at the As K-edge, using Eq. (21) and $f'_{As}(f''_{As})$ values as determined in Section 2.1.3. The InAs crystallographic structure has no center of symmetry but has only one crystallographic site for As atoms.

Note that regarding the sign convention for f' and f'', Eq (21) and (22) clearly show that computing the intensity with if'_A or $-if''_A$ gives exactly the same value provided that the phase $\varphi_{T(N)} - \varphi_A$ is changed to $-(\varphi_{T(N)} - \varphi_A)$.

2.2.2. What can we learn from the DAFS cusp shape?

In Chapter 10, the MAD method is presented and it is shown how to extract the partial structure factors F_A, F_N, and phase difference $\varphi_N - \varphi_A$ by measuring diffraction curves (or maps in the reciprocal space) at several energies in the vicinity of the absorption edge of one element (the resonant atom) that belongs to the nanostructures. The method provides a way to

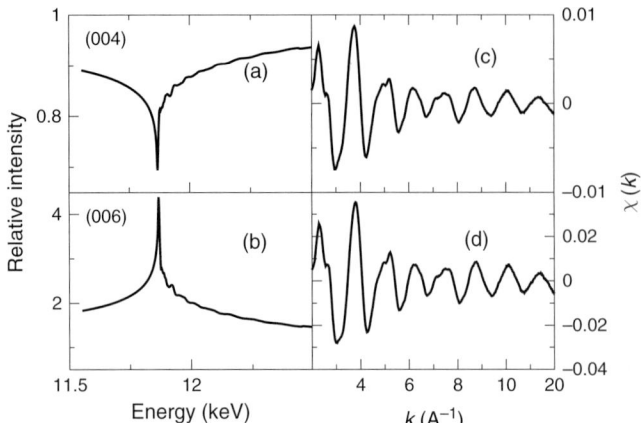

Fig. 4. InAs zinc-blende structure. If $h+k+l=4n$, $F=f_{In}+f_{As}$ (strong reflection); if $h+k+l=4n+2$, $F=f_{In}-f_{As}$ (weak reflection); $f_{As}=f^0_{As}+f'_{As}+if''_{As}$; (a) and (b) DAFS spectra calculated for (004) and (006) reflections, respectively, with bulk InAs f'_{As} and f''_{As}. For both reflections, the intensity was divided by $|F_T(E)|^2$; for reflections (004) and (006) the structure factor F_{As} is real. (c), (d) Normalized extended DAFS of the (004) and (006) InAs reflections, respectively (see Section 2.3).

disentangle, in the reciprocal space, the resonant atoms scattering amplitude from the substrate and/or matrix scattering amplitude. Once the extraction is done, one can recover model-free parameters as the average strain and size of the nanostructures [25,26,45].

Fitting of Eq. (22) to the DAFS spectrum cusp shape gives very precise values of $\beta_{NA}=|F_A|/|F_N|$ and $\varphi_N-\varphi_A$, which can be used to get information on composition. Notice that knowing β_{NA} and $\varphi_N-\varphi_A$ values is not sufficient to conclude about the actual nanostructure chemical composition in any case. Only in case of an isostrain *homogeneous binary alloy*, the $|F_A|/|F_N|$ ratio gives the exact intermixing amount ($\varphi_N-\varphi_A=0$), without any a priori crystallographic model. In such a case, it is convenient to combine MAD and "isostrain scattering" methods (see Chapter 10). The method can still be applied for *homogeneous alloys* provided that the crystallographic structure is known and can be used to simulate the DAFS spectrum and refine the occupation factor of the resonant atom [12,16,21,22].

Fitting of the DAFS spectrum cusp shape also gives the scale factor S_D and phase $\varphi_T-\varphi_A$ that are used to analyze the first-order EDAFS oscillations (see Section 2.3).

2.3. Extended diffraction anomalous fine structure

2.3.1. EDAFS to the first order

In the extended region above the edge, f'_j and f''_j can be split into "smooth" and oscillatory parts. $f'_j=f'_{0A}+\Delta f''_{0A}\chi'_j$ and $f''_j=f''_{0A}+\Delta f''_{0A}\chi''_j$, where $\Delta f''_{0A}$ represents the contribution of the resonant scattering to f''_{0A} (see Fig. 2(c)). The real and imaginary components of the complex EDAFS $\tilde{\chi}=\chi'_j+i\chi''_j$ are related by the Kramers–Krönig transforms, and the imaginary component χ''_j is related to EXAFS χ_j by the equation $\chi_j(E)=\text{Im}\tilde{\chi}(\mathbf{Q}=0,E)$.

One can write the structure factor as follows:

$$F(Q,E) = F_0(Q,E) + \frac{\Delta f''_{0A}(E)}{f^0_A(Q)} \sum_j F_j(Q)\left[\chi'_j(E) + i\chi''_j(E)\right] \quad (23)$$

To the first order, the intensity is given by

$$I = FF^* \approx I_0 + 2\frac{\Delta f''_{0A}|F_A||F_0|}{f^0_A}\chi_Q \quad (24)$$

where $F_0 = |F_0|e^{i\varphi_0}$ and χ_Q is the first-order EDAFS:

$$\chi_Q = S_D\left[\frac{I-I_0}{I_0}\right] = \cos(\varphi_0 - \varphi_A)\sum_{j=1}^{N_A} w'_j\chi'_j + \sin(\varphi_0 - \varphi_A)\sum_{j=1}^{N_A} w''_j\chi''_j \quad (25)$$

In the above equation, $S_D = \left[f^0_A|F_0|\right]/\left[2\Delta f''_{0A}|F_A|\right]$ plays a role of a normalization factor. By inspection of Eq. (21) and Fig. 3(a), one readily sees that determining β_{TA} and $\varphi_T - \varphi_A$ is sufficient to determine S_D and the phase difference $\varphi_0 - \varphi_A$, i.e., there is no need of a crystallographic model.

$$S_D = \frac{|F_0|}{|F_T|}\left[\frac{1}{2\beta_{TA}\Delta f''_{0A}}\right] \quad (26)$$

$$\tan(\varphi_0 - \varphi_A) = \frac{\cos(\varphi_T - \varphi_A) + \beta_{TA}f'_{0A}}{\sin(\varphi_T - \varphi_A) + \beta_{TA}f''_{0A}} \quad (27)$$

Here $|F_0|/|F_T| = \left[\cos(\varphi_T - \varphi_A) + \beta_{TA}f'_{0A}\right]^2 + \left[\sin(\varphi_T - \varphi_A) + \beta_{TA}f''_{0A}\right]^2$. Equation (27) shows that the phase φ_0 depends on the energy. The crystallographic weights $w'_j = |F_j|\cos(\varphi_0 - \varphi_j)/|F_A|\cos(\varphi_0 - \varphi_A)$ and $w''_j = |F_j|\sin(\varphi_0 - \varphi_j)/|F_A|\sin(\varphi_0 - \varphi_A)$ represent the orthogonal projection of χ'_j and χ''_j on the vector F_0 in the complex plane (see Fig. 3(b)), $\sum_j^{N_A} w'_j = \sum_j^{N_A} w''_j = 1$. They can be calculated provided that the average crystallographic structure is known, or determined if the individual $\chi'_j(\chi''_j)$ are known by fitting Eq. (25) to the experimental χ_Q. Note that $\sum_{j=1}^{N_A} w''_j\chi''_j$ is to be compared to the EXAFS oscillations given by $\chi_{EXAFS} = \sum_{j=1}^{N} \chi''_j$. The fundamental difference is the weights w''_j that give the DAFS site/spatial selectivity (see Section 5.2.2).

2.3.2. EDAFS path formalism

Using the paths formalism (see Chapter 9 and Ref. [44]), the *complex extended fine structure* that depends on the local atomic environment of a resonant atom j is given by [7,46]

$$\chi'_j + i\chi''_j = -\sum_{\gamma=1}^{\Gamma_j} A_{\gamma j}(k)e^{-i(2kR_{\gamma j} + \varphi_{\gamma j}(k))} \quad (28)$$

The sum runs over all scattering paths γ of the virtual photoelectron of state $|b\rangle$, $k = (1/\hbar)\sqrt{2m_e(E - E_{edge})}$ is the photoelectron wave number, $R_{\gamma j}$ is the effective length of path γj, and $\varphi_{\gamma j}(k)$ is the net scattering photoelectron phase shift.

Equations (25) and (28) leads to the following $\chi_Q(k)$ expression:

$$\chi_Q(k) = \sum_{j=1}^{N_A} \frac{|F_j|}{|F_A|} \sum_{\gamma=1}^{\Gamma_j} A_{\gamma j}(k) \sin\left(2kR_{\gamma j} + \varphi_{\gamma j}(k) + \varphi_0(k) - \varphi_j - \frac{\pi}{2}\right) \quad (29)$$

If the incoming and outgoing photons probe identical local atomic environment for all resonant atoms j, one can average over all identical photoelectron scattering paths. In that case $A_{\gamma j} = A_\gamma$, $\varphi_{\gamma j} = \varphi_\gamma$ and one can write Eq. (25) as

$$\chi_Q(k) = \sum_\gamma A_\gamma e^{-2k^2\sigma_\gamma^2}\left[-\cos(\varphi_0 - \varphi_A)\cos\left(2k\langle R\rangle_\gamma + \varphi_\gamma\right) \right. \\ \left. + \sin(\varphi_0 - \varphi_A)\sin\left(2k\langle R\rangle_\gamma + \varphi_\gamma\right)\right] \quad (30)$$

$$\chi_Q(k) = \sum_\gamma A_\gamma(k) e^{-2k^2\sigma_\gamma^2} \sin\left(2k\langle R\rangle_\gamma + \varphi_\gamma(k) + 2\delta_c(k) + \varphi_0(k) - \varphi_A - \frac{\pi}{2}\right) \quad (31)$$

where σ_γ represents the bond-length disorder (static and dynamic Debye–Waller factors) and $\langle R\rangle_\gamma$ is the average effective length of path γ. The weighting of individual $\chi'_j(\chi''_j)$ is implicit in Eqs (30) and (31). Equation (31) shows that χ_Q is written as in the case of one anomalous crystallographic site. The χ_Q formula is very similar to EXAFS one (see Chapter 9), the only difference is the crystallographic phase $\varphi_0(k) - \varphi_A - \pi/2$ in the sine argument.

3. Experimental setup

A DAFS experiment consists in measuring the elastic scattering intensity as a function of the incoming X-ray beam energy in regions spanning across an absorption edge, at fixed **Q** values. Experimental requirements have been discussed in a previous publication [15]; in the following, we give more details about GIDAFS that was developed at beamline BM2 at the European Synchrotron Radiation Facility (ESRF) in Grenoble. A DAFS experiment can be considered successful if EDAFS spectra are obtained with data quality that fits the standard criteria of the EXAFS community, i.e., with high signal-to-noise ratio (at least 1000) and without long-range distortions in the wide energy range measured.

To obtain high-quality DAFS spectra, beam *homogeneity* and *stability* at the sample position are required in a range of the order of 1 keV. All this makes the grazing incidence DAFS a quite demanding technique. For instance, at a grazing incident angle of 0.15°, a 4 mm × 4 mm sample surface intercepts about 10 μm of the incident beam. At beamline BM2, GIDAFS is performed with a beam size of 200 μm × 100 μm in the horizontal and vertical directions, respectively. The incident beam angular resolution is about 0.01°. Much effort must be paid to optics alignment, in particular regarding the tuning and the focalization of the second monochromator crystal, to maximize beam *homogeneity* and *stability* in the whole energy range. An accurate beam energy calibration is also needed.

Basically, the experimental diffraction setup consists of entrance slits, attenuators, and monitors. All these elements are mounted in a vacuum environment. Like for a XAFS experiment, the incoming beam must be carefully monitored. This is done by measuring the fluorescence signal emitted by a 4 μm thick, 99.6% pure Ti foil, also mounted in vacuum, at 45° with respect to the beam path. Homogeneity of the foil and high counting rate of the fluorescence signal are needed to ensure a high signal-to-noise ratio. To monitor the Ti fluorescence a PIN silicon photodiode (Canberra^{TM}) with an active area diameter of 19.5 mm is used. A 10μm thick, high-purity aluminum foil (99.999%) avoids visible light to reach the photodiode surface [15].

A NaI scintillator detector or a linear detector (Brucker^{TM}) is used to record the scattering intensity. The two detector slits are typically set at a distance of 260 and 650 mm from the sample, respectively. For in-plane reflections, the horizontal openings and offsets of the detector slits are adjusted to integrate the scattering intensity from $\alpha_f \simeq 0°$ to $\alpha_f \simeq 1°$, whereas the vertical gaps are adjusted to view the sample surface region at the center of the diffractometer [47] (see Fig. 6(c)). For an out-of-plane reflection, the slit gaps are adjusted in such a way that the exit angle be well defined and that the angular resolution minimizes the scattering overlap from different reciprocal space regions. Usually, for example, at the Ga and As K-edges, all detector slit gaps are set to 2 mm, corresponding to an angular resolution of 0.18°.

GIDAFS measurements are carried out with an eight-circle diffractometer equipment, using a Euler 4S + 2D geometry (ψ_c geometry [48]) that includes four circles for orienting the sample (μ, χ, φ, η) and two for moving the detector (ν, δ). The scattering intensity is measured either at near in-plane reflections, with grazing incidence angles α_i, and grazing exit angles α_f (Fig. 5), or in the vicinity of out-of-plane reflections, with grazing incidence angles and large exit angles. The plane of incidence that contains the incident and diffracted beams is vertical. In the case of in-plane reflections, the sample surface is perpendicular to the horizontal polarization vector of the incoming beam (a property of the light emitted at a bending magnet in the electron orbit plane). Figure 6(a) and (b) shows the grazing incidence setup for both the $(30\bar{3}0)$ and $(30\bar{3}2)$ reflections of the wurtzite structure (AIN, GaN, SiC). Figure 6(c) shows the incident and exit α_i and α_f angles, as well as the scattering angle 2θ, for an in-plane reflection.

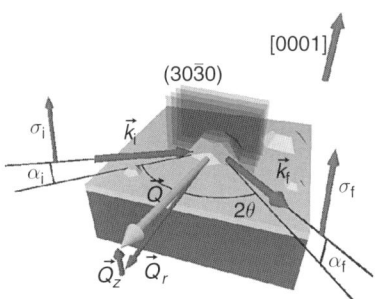

Fig. 5. Grazing incidence diffraction geometry for the GaN/AlN QDs$(30\bar{3}0)$ reflection. The incident and exit angles, α_i and α_f, are few tenths of a degree, yielding a nearly in-plane **Q** scattering vector. The σ_i polarization of the X-ray beam is perpendicular to the sample surface, along the [0001] direction.

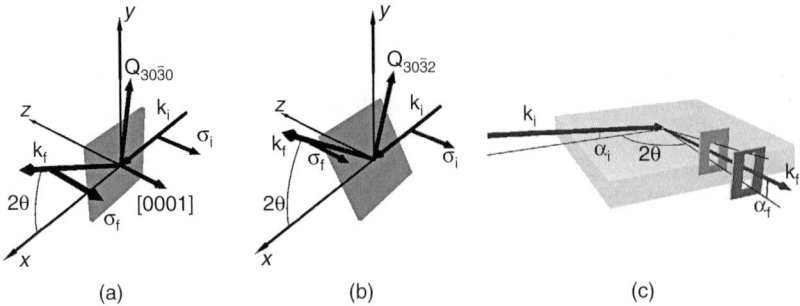

Fig. 6. (a) Grazing incidence diffraction geometry for the GaN/AlN QDs in-plane $(30\bar{3}0)$ reflection. (b) Grazing incidence geometry for the $(30\bar{3}2)$ reflection. The two reflections are denoted by their scattering **Q** $(30\bar{3}0)$ vectors, and **Q** $(30\bar{3}2)$. The scattering plane is vertical. The X-ray polarization vectors of the incident and exit beam, σ_i and σ_f, are perpendicular to the scattering plane. (c) Scattering angle (2θ), incident (α_i) and exit (α_f) angles, and detector slits setup for recording the scattering contributions along α_f, for an in-plane reflection.

4. Data analysis

4.1. X-ray multiple scattering in grazing incidence anomalous diffraction

In the past few years, the distorted wave born approximation (DWBA) has provided a better understanding of the scattering of nano-objects assemblies by taking into account X-ray MS related to the grazing incidence geometry [28]. Nevertheless, MS effects have not been included so far in the anomalous diffraction treatment (MAD and EDAFS), which obviously deserves a justification [27].

In this section the discussion focuses on free-standing nano-objects grown in a Stranski–Krastanow mode, either islands or wires, i.e., small size nano-objects that stand on top of a thin 2 ML thick 2D layer (wetting layer) matched to the substrate. In the framework of the DWBA, one defines an unperturbed system, M_O, which comprises the substrate and the wetting layer. The nanometer scale islands or wires are considered as a perturbation to M_O.

The scattered amplitude includes at least four main scattering paths as shown in Fig. 7. The first path (a) corresponds to the Born approximation (BA) (Fig. 7(a)), while paths (b)–(d) involve reflection(s) at M_O (Fig. 7(b)–(d)). We have neglected dynamical diffraction by M_O, and the strain field created by the nano-objects inside M_O. Both phenomena would add further scattering paths, increasing the scattering interference complexity [49]. In the following we assume a random distribution of the QDs on the substrate surface.

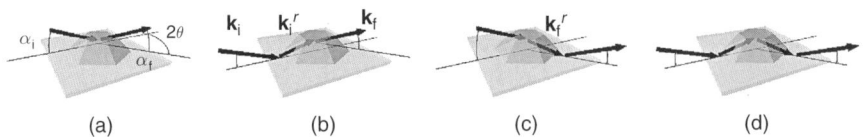

Fig. 7. Multiple scattering paths in the DWBA. α_i and α_f are the incident and exit angles, respectively. 2θ is the scattering angle. The scattering vectors of the four paths (a), (b), (c) and (d) are given by $\mathbf{k}_f - \mathbf{k}_i, \mathbf{k}_f - \mathbf{k}_i^r, \mathbf{k}_f^r - \mathbf{k}_i, \mathbf{k}_f^r - \mathbf{k}_i^r$, respectively.

In grazing incidence geometry, the diffraction anomalous effect in the vicinity of an absorption edge originates not only from the resonant atoms belonging to the nano-objects, but also from the same kind of atoms belonging to M_O, via the reflection coefficients. As for the sample, the wetting layer is about 2 ML thick. It represents a small amount of resonant atoms, giving negligible variations of M_O reflection coefficients as a function of energy.

4.1.1. Subcritical regime

For incidence angles (α_i) lower than the critical angle for total external reflection (α_c), the reflection coefficients of M_O experience a little dependence on energy across an absorption edge. As an example, for M_O made of a 2 ML GaN wetting layer on a 10 nm AlN buffer on a SiC substrate, and for $\alpha_i = 0.15°$ and an exit angle $\alpha_f = 0.15°$ ($\alpha_c \simeq 0.21°$ at 10.3 keV), the reflection coefficients introduce a few 1% additional anomalous effects on the total scattered intensity, at the Ga K-edge. The amplitude of the last two paths (Fig. 7(c) and (d)), which involve a reflection after scattering from the nano-object, can be even more reduced by integrating the scattered intensity as a function of α_f, over typically 1°. This is achieved experimentally by opening the two detector slits in the direction perpendicular to the sample surface.

For such an experimental setup and large scattering angles (2), one may approximate the scattering vectors to $k_f - k_i$ to compute the structure factor. Then the scattering amplitude is a sum of two terms:

$$F \simeq (1+r(\alpha_i)) \sum_{j \in \text{nano}} e^{iQ \cdot r_j} \qquad (32)$$

where r_j is the position of the atom (j) in the nano-object and $r(\alpha_i)$ is the reflection coefficient. Note that in the subcritical regime, $r(\alpha_i)$ experiences negligible variations as a function of energy, meaning that the usual treatment of anomalous diffraction, in the frame of the BA, remains justified [27].

4.1.2. Supercritical regime

For incident angles α_i higher than α_c, the variations of the M_O reflection coefficients are larger, but they are compensated by smaller absolute values of the reflection coefficients. As an example, for M_O made of a 2 ML GaN wetting layer on a 10 nm thick AlN buffer on a SiC substrate, for $\alpha_i = 0.3°$ and $\alpha_f = 0.3°$, the reflectivity exhibits 10–20% variations as a function of energy. However, owing to the low reflectivity (typically $1-2 \times 10^{-2}$), those variations have little effect on the total scattered intensity, typically 1–2% at the Ga K-edge. In other words, the importance of the three nonkinematic scattering paths (Fig. 7(b), (c), and (d)) is again negligible [27]. Moreover, the use of an experimental setup that integrates the scattered intensity as a function of α_f reduces even more the amplitudes of the last two paths in Fig. 7. It is therefore fully justified to write the total scattering amplitude as

$$F \simeq 1 \times \sum_{j \in \text{nano}} e^{iQ \cdot r_j} \qquad (33)$$

i.e., a kinematic formulation of the scattered amplitude, which obviously justifies a mere BA treatment.

The experimental conditions given above are representative of the situations encountered and reported in the section on applications (Section 5). We performed GIDAFS with a supercritical incidence angle, and integrated the scattered intensity as a function of α_f over 1°, both settings minimize the effect of MS paths. It is worth noting that the supercritical regime is definitely less demanding than the subcritical one, and therefore often preferred if possible. Indeed, if the surface sensitivity remains sufficient, the counting times may be shortened owing to a larger cutoff of the incident X-ray beam by the sample surface. In addition, the diffracted intensity as a function of the incident angle experiences slower variations in the supercritical than in the subcritical regime, and it is therefore more stable.

4.2. Data reduction

In grazing incidence an experimental DAFS spectrum is related to the square modulus of the structure factor according to the following formula:

$$I = r_0^2 S(Q) D(E) \frac{1}{\sin \alpha_i} L(Q, E) |F(Q, E)|^2 \qquad (34)$$

where I is the intensity corrected for fluorescence, S is a scale factor, D the detector efficiency, L the Lorentz and polarization factors for the Thomson scattering, and $1/\sin \alpha_i$ takes into account the variation of the beam footprint on the sample surface when changing the incidence angle α_i. The D factor takes into account the whole detection setup, comprising detector efficiency and absorption all the way from the monitor to the diffraction detector. It is not always possible to perform a direct measurement of the baseline, without inserting attenuators. However, the energy dependence of this factor is linear inside the energy range of interest, so that D may be fitted to the DAFS data with a straight line ($D = m(\Delta E + 1)$), where m is the only adjustable parameter, $\Delta E = E - E_0$, and E_0 is the edge energy. In this case, care should be taken to measure the DAFS spectrum far enough from the absorption edge up to the point where anomalous effect is negligible, otherwise the m parameter will be correlated to crystallographic phase $\Delta \varphi$. For a rotation scan, i.e., with the rotation axis perpendicular to the plane of incidence (containing **k** and **k'**), $L = \lambda^3/\sin 2\theta$. The polarization correction for the Thomson scattering is given by the dot product $\vec{\varepsilon}_\mu \cdot \vec{\varepsilon}_{\mu'}^*$. At a bending magnet, the polarization of the beam that propagates in the electron orbit plane is linear and perpendicular to the vertical plane of incidence (containing **k** and **Q**) that was used for all experiments reported hereafter; then $\vec{\varepsilon}_\mu \cdot \vec{\varepsilon}_\mu^* = 1$.

The very small fluorescence yield, which has no phase coherence with the anomalous scattering process, is systematically subtracted from the overall scattered intensity. For that purpose, one measures the background intensity variations as a function of the energy, in the vicinity of the reflections of interest. For instance, Fig. 8 shows the background intensity variations (mainly fluorescence yield) close to the GaN QDs (30$\bar{3}$0) reflection, measured at the Ga K-edge for one layer of GaN/AlN QDs capped by 11 ML of AlN. The signal was recorded with diffraction detectors slits opened as for the measurement of the GaN/AlN QDs (30$\bar{3}$0) reflection (see Fig. 6(c)). Compared to the GaN QDs GIDAFS spectrum (shown in Fig. 8), the fluorescence is weak.

Normalization of the EDAFS oscillations can readily be obtained by multiplying the extracted signal $(I - I_0)/I_0$ (where I_0 is the diffracted intensity without oscillations) by the normalization factor S_D, as defined in Section 2.3 (Eq. (26)).

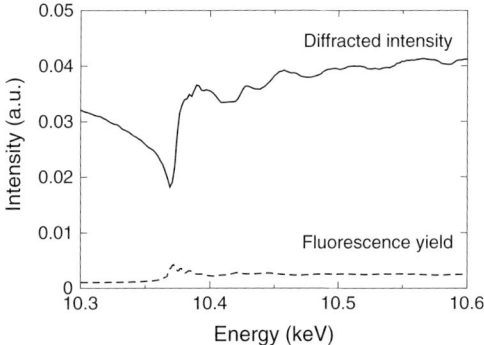

Fig. 8. Background intensity variations (mainly fluorescence yield) measured close to the GaN/AlN QDs $(30\bar{3}0)$ reflection, at $h = 3.03$, $k = 0$, and $\alpha_i = 0.3°$ (Dashed line), compared with the GaN/AlN QDs $(30\bar{3}0)$ GIDAFS spectrum (solid line).

In Eq. (34) the structure factor F is given by Eq. (32) or (33) in the case of uncapped nanostructures. In the case of embedded nanostructures with a flat sample surface one can often stay with the kinematic $F(\mathbf{Q}', E)$ structure factor that corresponds to the scattering of the dots and the surrounding matrix. However, the structure factor must be multiplied by the transmission function $T_i(\alpha_i, E)$ and $T_f(\alpha_f, E)$, and the vector $\mathbf{Q}' = \mathbf{k}'_f - \mathbf{k}'_i$ is the scattering vector in the average medium (see, for instance, Ref. [28] and Chapter 10).

5. Applications to semiconductor nanostructures
5.1. Introduction

In this section we report two recent examples of the application of EDAFS in grazing incidence (GIDAFS) to nanostructured systems, GaN QDs and InAs QWrs [25–27]. These two systems are quite emblematic as they have a straightforward application to nanotechnology and they have been studied as prepared for optical characterization and device integration. The nanostructure have a small size and they are encapsulated by a capping layer which makes part of the device as a barrier material and alters the strain state and shape of the nanoislands, contributing to their overall physical properties.

An important part of this section is dedicated to the comparison of the EDAFS results for some of the samples studied, with absorption spectroscopy, i.e., EXAFS. This also makes the selected examples particularly interesting as they illustrate in a very clear way the link between EDAFS and EXAFS. This is a new aspect of the absorption–diffraction complementarity, which is already well known, in a more general way, from the EXAFS early days. It is a routine approach to the structural problem comparing the short-range probe provided by EXAFS with the long-range order probe provided by diffraction, to get an overall picture of the system. Nevertheless, in this case this comparison acquires a new and more subtle meaning for one observes exactly the same phenomenon, the fine structure oscilllations showing up both in absorption and anomalous diffraction. We exploit the same physical probe, an outgoing photoelectron, real or virtual, traveling away and being scattered by the medium surrounding the resonant atom (absorber). We described in Section 2 a method to analyze the two phenomena in the same way, to have a direct comparison of the results.

5.2. GIDAFS study of GaN QDs

GaN/AlN QDs are a recent member of the self-assembled nanostructures family. The technological relevance of III–nitride compounds relies on their wide band gap, which can be adjusted to span the whole infrared to ultraviolet range, as well on their high radiative efficiency [50]. For actual 2D systems (high radiative efficiency at room temperature in InN-based 2D nanostructures is now suspected to arise from 3D confinement [51]), the main limitation to the thermal stability of the optical properties is the occurrence of high densities of structural defects (typically $10^{10}\,\text{cm}^{-2}$) due to the lack of substrates suitable for epitaxial growth. A solution to this problem has been the 3D Stransky–Krastanow growth that allows carriers confinement into nanometric, defect-free islands [52]. Obviously, the nature and morphology of the substrate can change the density, morphology, and strain of the QDs, modifying their optical properties. In the examples we have selected, the GaN QDs have been grown following a modified Stransky–Krastanow mode [53] by plasma-assisted molecular beam epitaxy (MBE) and two substrates have been compared: 6H–SiC(0001) with a thin (5 nm) AlN buffer layer and 2 μm thick AlN (0001) layers deposited by metalorganic vapor phase epitaxy (MOCVD) onto sapphire [54]. The in-plane misfit strain in the GaN epilayer deposited on top of the substrate, which is defined as $\varepsilon_{xx} = (a_{subs} - a_{GaN,\,bulk})/a_{GaN,\,bulk}$, varies from -2.4% for "bulk" AlN substrate to -3.4% for SiC. The strain state of the top of AlN thin buffers on SiC is unknown. Another important aspect is the effect of AlN capping. Depending on the kinematic parameters fixed by growth conditions and interface thermodynamics, this step can imply morphology and strain changes as well as intermixing, as observed, for instance, in the case of InAs/GaAs [55,56] and Ge/Si [57,58]. Capping by AlN of GaN QDs is known to result in an aspect ratio modification, with a flattening of the islands [59], and supposedly in a variation of strain, which is particularly important regarding the strong piezoelectric effects governing the optical properties in strained GaN(0001). We show in Fig. 9 an AFM image of typical GaN QDs grown on AlN buffer and SiC substrate.

The X-rays results we show hereafter have been collected at the ESRF beamline BM2 (see Section 3). The EDAFS experiment, as explained in the previous sections, is the last step of the following general MAD and DAFS experiment scheme:

1. Anomalous h scans measured at several energies close to the $(30\bar{3}0)$ AlN Bragg reflection.
2. MAD extraction of the partial structure factor F_{Ga} of Ga atoms selected by diffraction, i.e., that belong to the QDs.
3. EDAFS measurements at the Ga K-edge, i.e., measuring the diffracted intensity at a fixed $\mathbf{Q}$ value chosen according to the F_{Ga} profile (usually at the maximum of F_{Ga}).
4. Refinement of the DAFS lineshape to determine the phase difference $\varphi_0 - \varphi_{Ga}$ and scale factor SD, which are used to scale the experimental EDAFS and correct the virtual photoelectron phase shift.
5. Extraction of the experimental EDAFS and quantitative analysis according to EXAFS formalism.
6. Comparison with EXAFS results on the same sample, when possible.

Points 1 and 2 are detailed in Chapter 10, which deals with anomalous diffraction and MAD. In particular, we remind that a h scan in reciprocal space is achieved by keeping the Miller indices k and l fixed while varying h. h scans, in the range of 2.94–3.04, i.e., close to the $(30\bar{3}0)$ reflection, were performed at several energy values (10 or 12) close to the Ga K-edge. Typical results are shown in Fig. 10(a) and (b) for QDs grown on AlN/sapphire substrate with 10 ML AlN capping. The $h = h_{Ga} = 2.973$ position of the broad F_{Ga} peak maximum (Fig. 10(b)) is

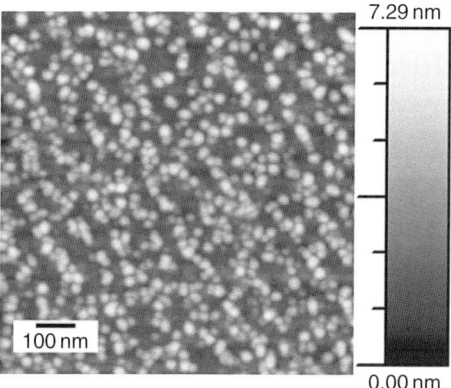

Fig. 9. 1 μm² atomic force microscope (AFM) images of uncapped GaN QDs grown on AlN/SiC.

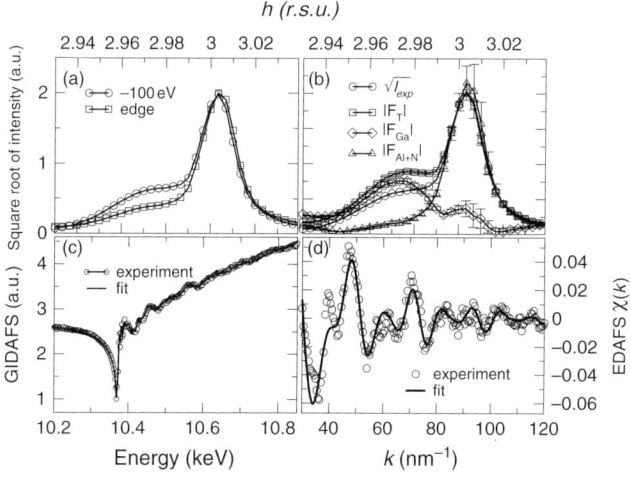

Fig. 10. Data corresponding to QDs grown on AlN/sapphire with 10 ML AlN coverage. (a) Experimental square root intensities $\sqrt{I_{exp}}$ measured below (-100 eV) and at the Ga K-edge (b) $\sqrt{I_{exp}}$ measured at 10.317 keV (50 eV below Ga K-edge), $|F_{Ga}|$ and $|F_{Al+N}|$ (c) GIDAFS spectra measured at the maximum of F_{GA} ($h = 2.973$) together with the crystallographic best-fit performed with experimental f'_{Ga} and f''_{Ga} of a GaN thin film. (d) experimental EDAFS compared with the best-fit result.

inversely proportional to the in-plane average lattice parameter a_{GaN}, as the distance between GaN $(30\bar{3}0)$ planes is $d_{30\bar{3}0} = a_{GaN}/\sqrt{3^2 \times 4/3} = a_{AlN}/\sqrt{h_{Ga}^2 \times 4/3}$ with $a_{AlN} \simeq 0.3081$ nm, for the AlN substrate peak used as a reference. Figure 11 shows the evolution of the in-plane lattice parameter as a function of the AlN cap thickness for the two different series of samples. The uncapped QDs are partially in-plane relaxed, with an average strain relative to bulk GaN, $\varepsilon_{xx,MAD} = (a_{GaN,MAD} - a_{GaN,bulk})/a_{GaN,bulk} \simeq -1\%$, with $a_{GaN,bulk} = 0.3189$ nm. The QDs are then progressively in-plane compressed by the AlN capping, but keep being relaxed, $\varepsilon_{xx,MAD} \simeq -1.3\%$, compared to pseudomorphic GaN ($\varepsilon_{xx,GIXRD} \simeq -2.4\%$).

Both in-plane and out-of-plane average size and strain of the nanostructures can be in principle recovered by using GIMAD. However, in that case achieving absolute out-of-plane

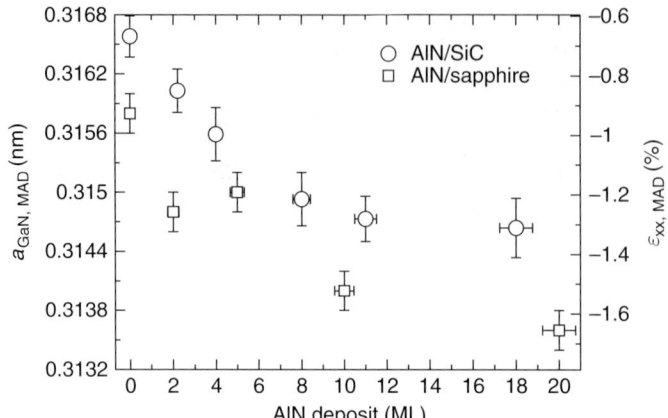

Fig. 11. In-plane lattice parameter a and strain (relative to bulk GaN) in the GaN QDs grown on AlN/SiC (open circles) and AlN/sapphire (open squares), deduced from the position of the F_{Ga} peak maximum. Bulk GaN gives $\varepsilon_{xx} = 0\%$ with $a_{GaN,bulk} = 0.3189$ nm while pseudomorphic GaN gives $\varepsilon_{xx} = -2.4\%$ with $a_{AlN,bulk} = 0.3112$ nm. Note that the uncertainties refer to the a_{GaN} values.

values is difficult. The main reason is that the out-of-plane diffraction data are affected by refraction. This effect is not the same for the substrate and the nanostructures, which means that the relative position of the substrate and the nanostructure peaks (in the reciprocal space direction perpendicular to the sample surface) is not known in a direct and absolute way. Also, in grazing incidence and exit geometry MAD cannot discriminate atomic intermixing and a mixing of in-plane isostrained phases (the pseudomorphic materials on top and/or beneath the nanostructure). This problem can be overcome, once partial "in-plane" structure factors have been extracted by MAD, by means of GIDAFS.

Typical GIDAFS and the corresponding EDAFS spectra, measured in grazing incidence and exit are shown in Fig. 10(c) and (d), for QDs grown on AlN/sapphire with 10 ML AlN capping. The solid line in Fig. 10(c); illustrates step 4, i.e., the cusp shape refinement according to formula 21. Oscillations also show up in the simulated GIDAFS spectrum because the f'_{Ga} and f''_{Ga} values for Ga have been obtained from EXAFS data of a GaN thin film. The spectrum was recorded at the maximum of F_{Ga}.

5.2.1. EDAFS results

Typical $\chi_Q(k)$ EDAFS oscillations are shown in Fig. 12 for two samples of the SiC series, with an AlN capping thickness of 4 and 11 ML. These were extracted and normalized to atomic background according to expression (25).

The EDAFS analysis was performed by the IFEFFIT code [60] implemented by Artemis interface [61]. The FEFF8 code [62] was used to generate theoretical phases and amplitudes, taking into account beam polarization, for a 0.6 nm radius GaN cluster with the wurtzite structure. Figure 13 shows the scattering paths that have been taken into account. All theoretical phases were corrected for $\varphi_0 - \varphi_A - \pi/2$. The possible presence of Al atoms in the QDs or at the substrate and capping interfaces, as well as Ga–Al and Ga–N–Al scattering paths were considered by calculating an AlN cluster with the Ga central atom as absorber. The "in-plane" statement refers to the plane of growth, and all the scattering paths, except the first one, were expressed in terms of a and c cell parameters, according to the hexagonal cell symmetry.

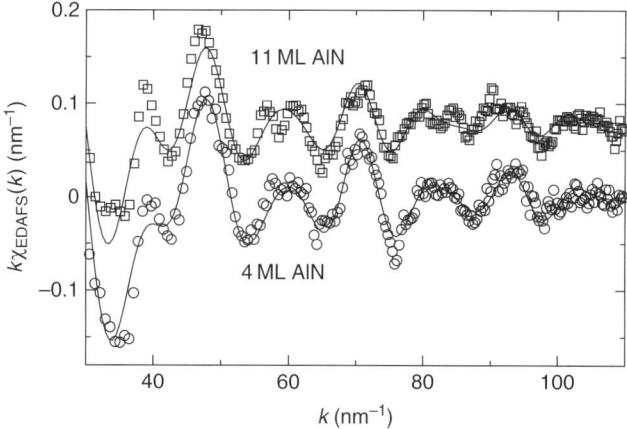

Fig. 12. Experimental EDAFS spectra for the GaN/AlN QDs grown on SiC with 4 (open circle) and 11 (open square) ML AlN capping, compared with the best-fit results (solid lines). The data corresponding to 4 ML AlN capping were shifted vertically for clarity.

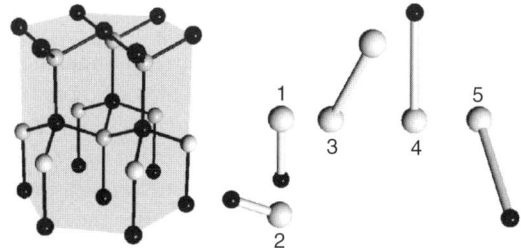

Fig. 13. Scheme of GaN wurzite structure, the most relevant virtual phoelectron scattering paths used for the EDAFS simulation are represented: (1) out-of-plane I shell Ga–N, (2) nearly in-plane I shell Ga–N, (3) out-of-plane II shell Ga–Ga (Al), (4) III shell Ga–N along [0001], (5) IV shell Ga–N. Ga atoms are represented by white spheres, N by black ones.

Fitting was performed with Eq. (31) by fixing the a parameter (in-plane) to the values found by diffraction, $a_{GaN,MAD}$ (Fig. 11), and letting the c parameter to vary. The Ga–N first shell distances were let free to vary independently of a and c since, as it is well known, the Vegard's law is not valid for semiconductor alloys, in which the bond-bending mechanism is dominant compared to bond stretching [63]. The best-fit curves are shown in Fig. 12 compared to the experimental raw data. The general finding are (a) the Ga–N first shell distances (Fig. 13, paths 1 and 2) show to be very close to each other, within the fit errors (0.01 Å), in agreement with previous studies [64], and (b) the Al content is always very low, showing that no intermixing takes place in the QDs as expected for the Al/Ga species [65].

From the fit results, the in-plane and out-of-plane strains are calculated, with respect to relaxed (bulk) GaN, as $\varepsilon_{xx} = (a_{GaN,MAD} - a_{GaN,bulk})/a_{GaN,bulk}$ and $\varepsilon_{zz} = (c_{GaN,EDAFS} - c_{GaN,bulk})/c_{GaN,bulk}$, ε_{xx} vs ε_{zz} is shown in Fig. 14 for the two series. These values are compared to the biaxial elastic deformation of GaN (straight line), which corresponds to $\varepsilon_{xx} = -2\varepsilon_{zz}c_{13}/c_{33}$. Figure 14 shows that the QDs are relaxed in the growth plane with respect to the pseudomorphic situation, i.e., $\varepsilon_{xx} = -3.4\%$ or -2.4% corresponding to

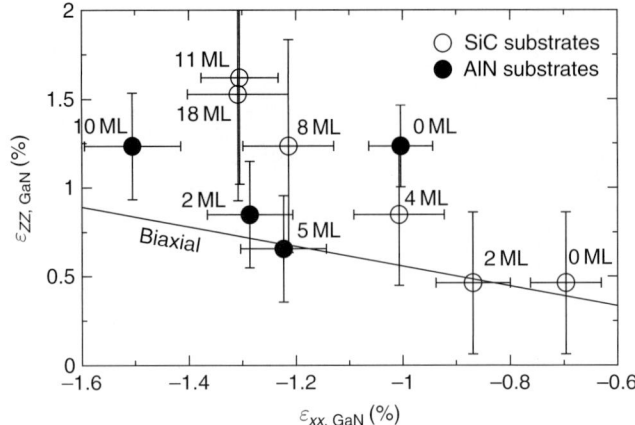

Fig. 14. GaN QDs in-plane strain ($\varepsilon_{xx} = (a_{GaN,MAD} - a_{GaN,bulk})/a_{GaN,bulk}$) versus out-of-plane strain ($\varepsilon_{zz} = (c_{GaN,EDAFS} - c_{GaN,bulk})/c_{GaN,bulk}$), as a function of the AlN capping thickness, for QDs grown on SiC/AlN (open circles) and on AlN/sapphire (open squares). The straight line represents the biaxial elastic deformation.

pseudomorphic GaN grown on bulk SiC or AlN, respectively. Simulations using elasticity model have shown that the main contribution to diffraction is the QDs core (see Section 5.2.2), which has a lower strain content than the QDs base due to a progressive relaxation along z (see Fig. 14). Nevertheless, even taking into account this aspect, strain is not well reproduced by finite element method (FEM) simulations and further strain relaxation mechanisms must be taken into account, as dislocation propagation through the AlN epilayer or buffer [66,67], misfit dislocations at the GaN/AlN interface [68] or large in-plane strain relaxation in AlN buffer layers arising from the platelet-like growth of AlN [69].

Also, the effect of capping on strain evolution is very interesting. For both series, an increasing deviation from the elastic biaxial deformation, with increasing AlN cap thickness, is observed. The progressive capping gives place to a remarkable increase to in-plane and out-of plane strain with a saturation at about 11 ML thickness. The influence of the substrate on the QDs morphology is also put in evidence, the SiC series is more relaxed, and in particular the uncapped sample, shows to be almost biaxial. The uncapped sample of the AlN series instead, is far from showing biaxial strain, in agreement with the different aspect ratio that is smaller in former case. Figure 14 also shows that the combination of MAD and DAFS provides a sensitivity to the cap thickness as low as 2 ML.

As mentioned above, the Al content found by EDAFS is practically negligible, excluding Al/Ga intermixing phenomena inside the QDs. One can compare Al content deduced by EDAFS, i.e., negligible, with the Al content obtained by fitting of the DAFS spectrum cusp shape [26]. The presence of Al is needed to reproduce correctly the shape of the DAFS spectrum at the Ga K-edge. The values obtained in Ref. [26] are shown in Fig. 15. We note, first, that the Al fraction detected by diffraction varies from a few percent to about 60%, i.e., not at all negligible. Due to the previous EDAFS results, one can state that the Al atoms contributing to diffraction belong to the selected isostrain region but not to the dots. They belong, indeed, to the strained AlN close to the QDs/capping interface. The further finding is that the Al content, as a function of capping thickness, shows a clear slope change. This can

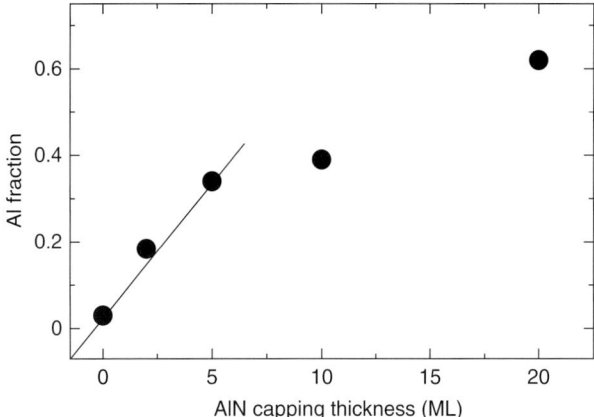

Fig. 15. Al fraction ($Al_{1-x}Ga_xN$) in the isostrain region selected by diffraction, as deduced from the fit of the cusp shape of the DAFS spectra measured at the maximum of F_{Ga} for the QDs samples grown on AlN pseudosubstrates. For low AlN coverage, a linear increase is observed.

be explained by a two-step capping mechanism: up to a thickness of 4–5 ML, AlN wets the QDs so that the AlN fraction contributing to diffraction increases linearly (Fig. 15). Above 5 ML, AlN start to fill the interdots valleys. This gives a lower slope increase as observed in Fig. 15. This capping growth mechanism is supported by plane-view AFM and cross section transmission electron microscopy (TEM) observations [70].

5.2.2. EDAFS oscillations and strain field

In this section we would like to illustrate the DAFS spectroscopy selectivity to isostrain regions of the nanostructures. In the previous section, for the EDAFS oscillations analysis, a single kind of local environment of the resonant atom in the QDs was considered (Section 2.3.2). The strain field in the QDs was taken into account by static Debye–Waller factors. Now we consider that all Ga atoms occupy unequivalent crystallographic sites as a consequence of the strain field in the QDs. Regarding the number of Ga atoms in a QDs, typically $1 \times 10^4 - 1 \times 10^5$ for 15 nm length by 3 nm height dots, it is obviously not possible to refine all Ga coordinates in the dots. To know the positions of all Ga_j atoms in the dots, one can calculate the strain field in the frame of an elasticity model with the FEM.

In the following, we will consider 15 nm length by 3 nm height pure GaN(0001) QDs grown on a AlN(0001) substrate, i.e., the average dimensions of the QDs as deduced from AFM quantitative analysis. FEM simulations are performed with a 2 ML GaN wetting layer pseudomorphic to the AlN substrate and for QDs capped with 5 nm of AlN [71]. Figure 16(a) and (b) show, the corresponding in- and out-of-plane strain maps along a central cross section of the QDs.

Atomic contributions to the EDAFS signal The atomic weights to the EDAFS signal of each Ga atom in the QDs, w'_j and w''_j, were defined in Section 2.3.1. For a given (h, k, l) position in reciprocal space, w'_j and w''_j depend on the actual positions of Ga_j atoms, (x_j, y_j, z_j), obtained by FEM simulations, through their crystallographic phase $\varphi_j = 2\pi (hx_j + ky_j + lz_j)$, *all other crystallographic factors being constant.*

Figure 16(c),(e),(g) and (d),(f),(h) shows kinematic calculations of w'_j and w''_j maps along a cross section of the GaN QD that goes through the middle, for $k = l = 0$, and three h values

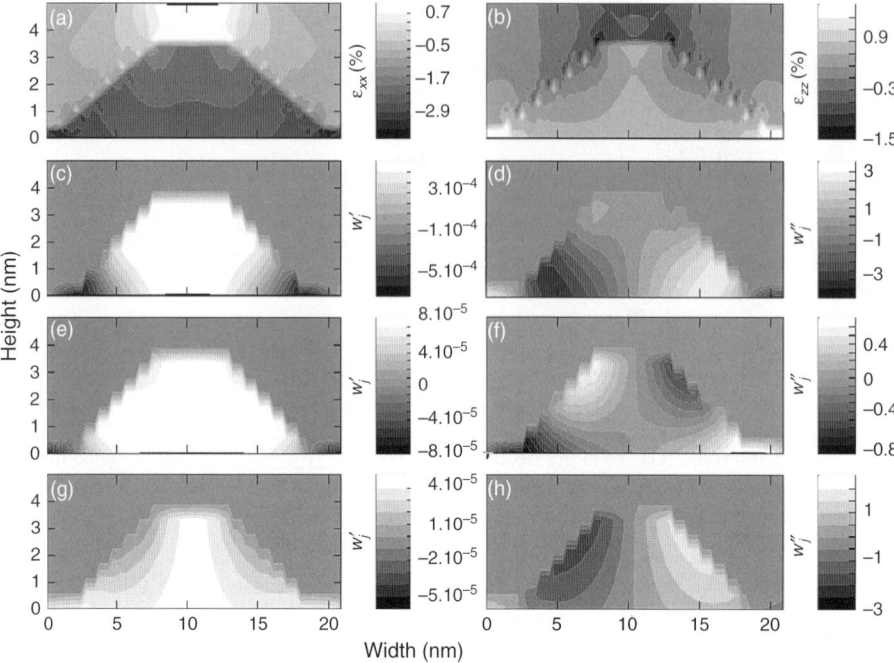

Fig. 16. GaN/AlN QDs. In-plane ε_{xx} (a) and out-of-plane ε_{xx} (b) strains simulated within an elastic model, using the FEM method, for 15 nm length by 3 nm height QDs on AlN(0001) substrate, capped with a 5 nm AlN layer. Weights w'_j (c,e,g) and w''_j (d,f,h) of the Ga atoms in the QDs, calculated with the atomic displacements obtained by FEM simulations. The weights are calculated at different Q values: $h = 2.978$, $(k = l = 0)$ (c,d), $h = 2.988$ (e,f) corresponding to the maximum of F_{Ga} determined from simulation of h-scan of the diffracted intensity, and $h = 2.998$ (g,h).

close to the h_{Ga} value of the maximum of $|F_{Ga}|$. The h_{Ga} value was determined by simulation of $|F_{Ga}|$ as a function of h, close to the $(30\bar{3}0)$ reflection of the AlN substrate ($h = 3$). The observed symmetry of w'_j and w''_j originates from that of the atomic displacements in the QDs, as well as from that of the sine and cosine functions occurring in w'_j and w''_j.

The w''_j individual atomic weights are larger than the corresponding w'_j values, for the (h, k, l) positions considered. However, careful inspection of Fig. 16(d),(f), and (h) puts in evidence that all w''_j contributions do cancel out as a whole. Such an observation does not apply for w'_j (Fig. 16(c),(e),(f) and (g)). This means that close to the $(30\bar{3}0)$ reflection for the QDs, the EDAFS signal is mainly x'_j contribution. As can be seen in Fig. 16(c),(e), and (g), w'_j weights cancel out in some regions of the QDs. Note that in-plane alternating black and white regions correspond to negative and positive weights. For $h = 2.978 < h_{Ga}$, w'_j do not cancel at the top of the dots only, for $h = 2.988 = h_{Ga}$, only Ga atoms in the core of the dots do contribute, and for $h = 2.998 > h_{Ga}$, the dominant contribution comes from a central vertical region of the dot. This figure also shows that, in this case, the isostrain approach in which the smaller the h value, the smaller the in-plane strain and the higher in the dots the selected region, is no longer valid. One cannot think of horizontal slices having approximately the same strain. This is due to the presence of capping that exerts stress at the QDs lateral surfaces and to the small dot size. Note that for $h = h_{Ga}$, which corresponds to the experimental condition used to measure the EDAFS oscillations, an enhanced sensitivity to the core of the dots is achieved. This point will be discussed in the following subsection.

5.2.3. Comparison of EDAFS with polarization-dependent EXAFS

In this section we show the comparison between EXAFS and EDAFS measurements of two samples, belonging to the series grown on AlN substrate: the sample with free-standing GaN QDs and the sample with GaN QDs covered by a 10 ML capping layer. As explained previously, the two techniques have different spatial selectivities. Therefore, one can change appreciably the information content of the oscillatory signal by switching from a technique to the other. Fluorescence EXAFS measurements have been performed at the French Collaborative Research Group (CRG) beamline BM30 at the ESRF, at the Ga K-edge with the polarization vector nearly parallel and perpendicular to the growth plane (the angle of incidence was in the 3–5° range). The EXAFS data analysis was performed, as for EDAFS, by using IFEFFIT code [60] implemented by Artemis interface [61]. Ab initio phase shifts and amplitudes were calculated by FEFF8.0 [62] that allows the choice of polarization of the incoming beam. Two set of theoretical signals were generated with polarization direction along [0001] and [11$\bar{2}$0] crystallographic directions. Calculation of theoretical phases and amplitudes was carried out by taking into account the same kind of scattering paths and the same R range, as for the EDAFS spectra analysis, and maintaining the hexagonal wurtzite symmetry of the cluster. The reason for recording parallel and perpendicular spectra is to exploit the anisotropy of the EXAFS probe to determine both cell parameters a and c. Indeed, for perpendicular polarization, the in-plane Ga next nearest neighbors (NNN) do not contribute to the signal, whereas for parallel polarization, the NNN signal is a combination of in-plane and out-of-plane Ga atoms contribution. In the case of EDAFS, due to the grazing incidence diffraction geometry, only spectra with perpendicular polarization were available, and the a lattice parameter was provided by independent MAD measurements. Simultaneous fitting of both EXAFS spectra corresponding to the two polarizations was performed to determine both a and c cell parameters. The fit and experimental curves are shown, for the sample with 10 ML AlN capping, in Fig. 17.

Table 1 provides the most relevant fit parameters, the interatomic Ga–N and Ga–Ga distances together with the Debye–Waller factors, the c values, and the Al occupation factor.

The most evident and important finding is that for the uncapped sample, the EXAFS and EDAFS results are quite close to each other. In particular the Al concentration is found to be small in both cases. Instead, for the capped sample EDAFS and EXAFS give quite different best-fit values. As explained in Section 2.3, a way to compare directly EDAFS and EXAFS oscillations is to reconstruct EDAFS from EXAFS according to Eq. (25) where $\sum_{j=1}^{N_A} w_j'' \chi_j''$ and $\sum_{j=1}^{N_A} w_j' \chi_j'$ are replaced by χ_{EXAFS}'' and χ_{EXAFS}', respectively; χ_{EXAFS}'' is the experimental EXAFS signal measured with the incoming X-ray polarization perpendicular to the growth plane and χ_{EXAFS}' is the Kramers–Krönig transform of χ_{EXAFS}''. One can clearly see in Fig. 18 the difference between the EXAFS and EDAFS for the capped sample. In particular, the Al concentration is found to be negligible for EDAFS but no longer negligible for EXAFS. The presence of Al is in principle expected due to the Al interface atoms that belong to the buffer and capping layer. From the QDs shape and size, given by TEM and AFM, one can make an estimation of the interface contribution considering sharp interface. For perpendicular polarization, it would give approximately 10% of Al as out-of-plane NNN (scattering path number 3 in Fig. 13). This is a lower limit estimation, as the interface could span over 1 or 2 ML. The presence of Al as NNN

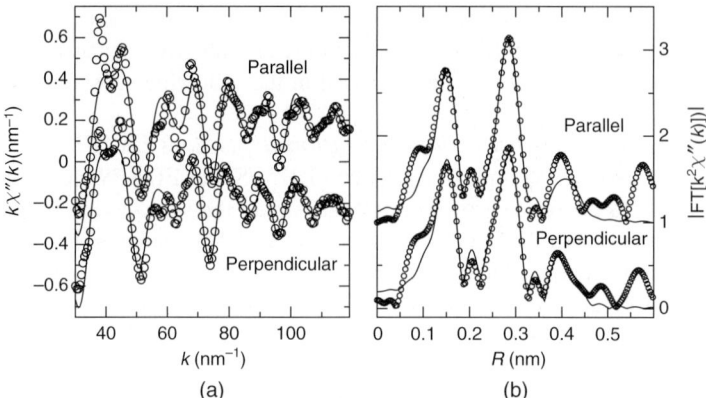

Fig. 17. EXAFS spectra (left panel) and the corresponding Fourier transforms (right panel) for GaN QDs grown on AlN substrate with 10 ML AlN capping. The labels parallel and perpendicular refer to light polarization of X-rays with respect to the sample surface. FTs were calculated in about the same range as EDAFS (2.7–11 Å^{-1}) with a Hanning window.

Table 1
Effective paths length (r_1, r_3 r_4), Debye–Waller coefficients (σ_1^2, σ_4^2), and Al proportions (x_3, x_4) around Ga atoms, deduced from best fits of EXAFS and EDAFS oscillations for two QDs planes, grown on AlN/sapphire. One is capped by AlN (10 ML) the other is uncapped (r_4 corresponds to scattering path number 3 in Fig. 13).

	Relaxed	Biaxial	0 ML EDAFS	0 ML EXAFS	10 ML EDAFS	10 ML EXAFS
r_1(GaN) (nm)	–	–	0.193 ± 0.001	0.194 ± 0.001	0.194 ± 0.001	0.195 ± 0.001
σ_1^2 (nm^2)	–	–	$2.10^{-5} \pm 2.10^{-5}$	$4.10^{-5} \pm 2.10^{-5}$	$4.10^{-5} \pm 2.10^{-5}$	$2.10^{-5} \pm 2.10^{-5}$
r_3(GaGa)$_\parallel$ (nm)	0.3189	0.3110	0.3156*	0.3150 ± 0.001	0.3140*	0.3150 ± 0.001
r_4(GaGa) (nm)	0.318	–	0.319 ± 0.001	0.318 ± 0.001	0.318 ± 0.002	0.316 ± 0.002
σ_1^2 (nm^2)	–	–	6.10^{-5}	8.10^{-5}	4.10^{-5}	7.10^{-5}
r_3–r_4(GaAl) (nm)	0.307	–	0.311 ± 0.003	0.313 ± 0.003	–	0.315 ± 0.003
c_{GaN} (nm)	0.5185	0.526	0.525 ± 0.02	0.522 ± 0.03	0.523 ± 0.03	0.519 ± 0.03
c_{GaN}/a_{GaN}	1.626	1.69	1.66	1.66	1.67	1.64
x_3	–	–	–	0.14 ± 0.07	–	0.18 ± 0.07
x_4	–	–	0.1 ± 0.1	0.12 ± 0.07	0.0 ± 0.1	0.26 ± 0.07

*Obtained from MAD measurements.

atom can be observed qualitatively by comparing the FT curves of the capped and unncapped sample. For the capped sample the intensity of the Ga–Ga(Al) peak is strongly reduced with respect to the uncapped one, due to the destructive interference between Ga–Ga and Ga–Al NNN contributions. As a matter of fact, the experimental Al concentration determined by EXAFS is 26% for the out-of-plane NNN (Table 1). A larger difference, with respect to the uncapped sample, is detected for the out-of-plane Ga–Ga distance, and therefore for the c value, i.e., 5.25 Å in EDAFS and 5.19 Å in EXAFS. The lower EXAFS values is the result of an average of different Ga environments. We now have a stronger interface contribution due to the presence of the capping with a GaAlN interface region in which the increasing of c due to strain is limited by alloying (Vegard's law for the cation sublattice). In any case, owing to the independent EDAFS results we

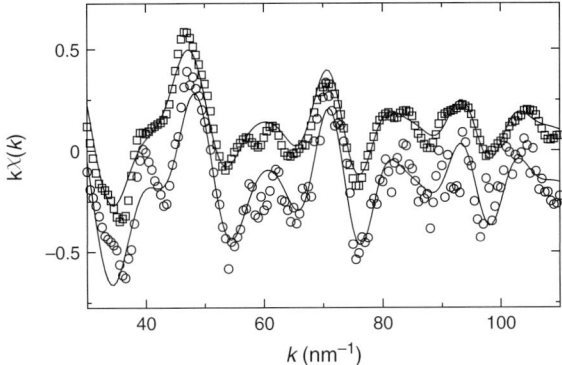

Fig. 18. Direct comparison between EDAFS (open circle) and reconstructed EDAFS (open square) for GaN QDs grown on AlN substrate covered with a 10 ML AlN capping. Best-fit curves are also shown (solid lines).

can exclude Al intermixing inside the dots. This experiment highlights the importance of comparing EDAFS and EXAFS. EDAFS provides the composition and strain inside the dot, EXAFS reveals a interface phenomenon of intermixing skipped by diffraction but also relevant for the structural and optical properties of the materials since it can change either the strain relaxation and the electronic wave function confinement.

5.3. GIDAFS study of InAs quantum wires

We show here a further example about the application of GIDAFS to another self-assembled nanostructures system. We report a study on the determination of strain and composition of buried InAs QWrs [25]. InAs QWrs and QDs can be employed as active region materials of lasers and light-emitting diodes as their typical emission wavelengths, 1.3–1.6 μm, fall into one of the optical fibers transmission windows, suitable for telecommunications technology. To be suitable for devices, the nanostructures are encapsulated with InP or embedded into a superlattice. They must be homogeneous in size, shape, and composition, to provide well-defined emission wavelengths. The knowledge of strain field, chemical gradients, and chemical mixing at the interface is again of major importance. In this study, two different samples are compared. For the first one (sample 1), the MBE growth parameters were tuned to minimize the As/P exchange and reduce the height dispersion of the InAs islands [72]. The growth temperature was about 520°C, the As pressure ranged from 2 to 4×10^{-6} Torr, giving a growth rate of 0.25 μm/h. The second one (sample 2) was also grown by MBE, but at a lower substrate temperature, with a higher substrate roughness. The growth temperature was about 400°C followed by an annealing at 480°C to favor QWrs formation. The As pressure ranged from 1.5 to 2×10^{-6} Torr, giving a growth rate of 0.5 μm/h. [73]. Both samples were grown on InP and encapsulated by an InP cap layer. The experiment scheme is the same as described previously for the GaN QDs. The MAD results are described in Chapter 10, where the modulus of the partial structure factors F_T and F_{As} are shown along the [001] direction in the vicinity of the (442) Bragg diffraction peak of the InP substrate.

5.3.1. EDAFS and polarized EXAFS results
Grazing incidence anomalous diffraction at the As K-edge (11.867 keV) was performed at the French CRG beamlines BM32 and BM2. The GIDAFS measurements were taken at a **Q** value corresponding to the maximum of $|F_{As}|$ profile. With respect to the previous experiment on

GaN QDs, in this case, the (442) Bragg reflection is asymmetric, so that the diffraction vector **Q** has a Q_z component (perpendicular to the sample surface), and both ε_{xx} and ε_{zz} are probed. The incidence angle was slightly above the InP critical angle ($\alpha_i \simeq 0.2°$), and the exit angle was $\alpha_f \simeq 20°$. Figure 19 shows the GIDAFS spectrum for sample 2. The normalized EDAFS oscillations are shown in Fig. 20 for both samples.

They have been analyzed by fitting the theoretical signal to the experiment with IFEFFIT code. Theoretical phase and amplitudes have been calculated for a bulk InAs cluster taking into account the [001] polarization of the incident beam. All phases were corrected for $\varphi_0 - \varphi_{As} - /\pi 2$. Theoretical phases and amplitudes were also calculated for an InP cluster in which one of the P atoms was substituted by an As absorber to investigate the presence of P atoms, as NNN, in the As local environment.

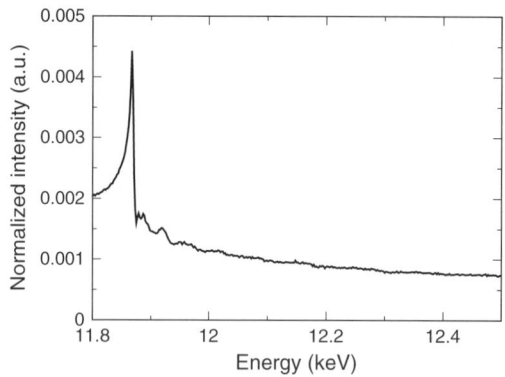

Fig. 19. GIDAFS spectrum of sample 2, recorded at the As K-edge at the maximum of F_{As} profile ($h = k = 3.98$, $l = 1.9$).

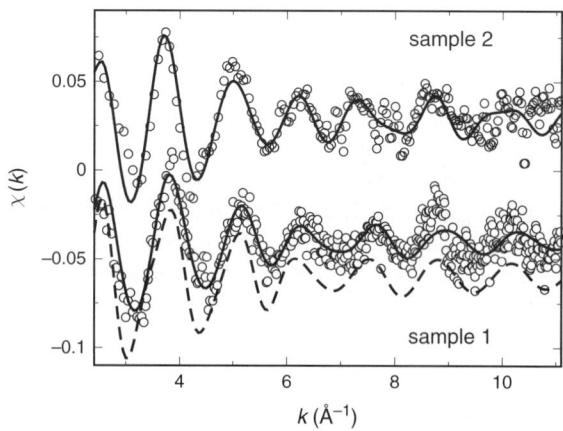

Fig. 20. EDAFS oscilllations for sample 1 (lower curve) and sample 2 (upper curve), the two spectra are shifted along y-axis for clarity. The best-fit curves are also shown (solid lines). For sample 1, the EDAFS calculated from EXAFS best fit is also shown for comparison (dashed line).

Table 2

EDAFS and EXAFS best fit values for interatomic distances and P concentration for sample 1 and sample 2. Labels I, II, and III refer to the first, second, and third coordination shells composed of In, As or P, In atoms, respectively (see Fig. 2(a)). The symbol $_\parallel$ and $_{out}$ stand for in-plane (the plane of growth) and out-of-plane atoms, respectively

Sample/path (Å)	Bulk InAs	InAs/InP (pseudmorphic)	Sample 1 (EXAFS)	Sample 1 (DAFS)	Sample 2 (EXAFS)	Sample 2 (DAFS)
As–In$_I$	2.632	2.60	2.593 ± 0.003	2.57 ± 0.02	2.593 ± 0.003	2.63 ± 0.02
(As–As$_{II}$)$_\parallel$	4.284	4.15	4.16 ± 0.06	–	4.15 ± 0.06	–
(As–As$_{II}$)$_{out}$	4.284	4.29	4.23 ± 0.04	4.30 ± 0.04	4.25 ± 0.04	4.22 ± 0.04
(As–P$_{II}$)$_\parallel$	–	–	4.19 ± 0.07	–	4.15 ± 0.07	–
As–P$_{II}$)$_{out}$	–	–	4.17 ± 0.03	4.20 ± 0.06	4.18 ± 0.03	4.19 ± 0.06
As–In$_I$–As$_{II}$	4.765	5.26	4.71	4.72	4.72	4.73
As–In$_I$–P$_{II}$	–	–	4.66	4.66	4.66	4.68
(As–In$_{III}$)$_\parallel$	5.023	4.87	4.88 ± 0.03	–	4.87 ± 0.03	–
(As–In$_{III}$)$_{out}$	5.023	5.16	4.93 ± 0.06	–	4.94 ± 0.06	–
$X_\parallel$ (As$_{1-x}$–P$_x$)$_\parallel$	–	–	0.3 ± 0.1	–	0.4 ± 0.1	–
X_{out} (As$_{1-x}$–P$_x$)$_{out}$	–	–	0.5 ± 0.1	0.4 ± 0.2	0.6 ± 0.1	0.4 ± 0.3

The fit results for interatomic distances are reported in Table 2. The scattering paths found to be relevant were (see Fig. 2(a) and also Ref [24]): the nearest neighbor (NN) As–In path, the NNN paths As–As and As–P, and the three legs MS paths As–In–As and As–In–P. Figure 20 shows the comparison of the DAFS spectra with the best-fit theoretical curves.

EXAFS measurements have been performed, on the same samples, at the As K-edge with the polarization vector perpendicular and parallel to the growth plane at the French CRG FAME beamline BM30. The perpendicular and parallel EXAFS spectra were fitted simultaneously with the same codes as EDAFS spectra. The interatomic distances iterated in the fit procedure were first (As–In), second (As–As and As–P), and third (As–In) coordination shells. In-plane and out-of-plane scattering paths have their own P population, x and y, respectively, as fit parameters. The multifit results are shown in Table 2 and Fig. 21, where the EXAFS results for both samples are compared.

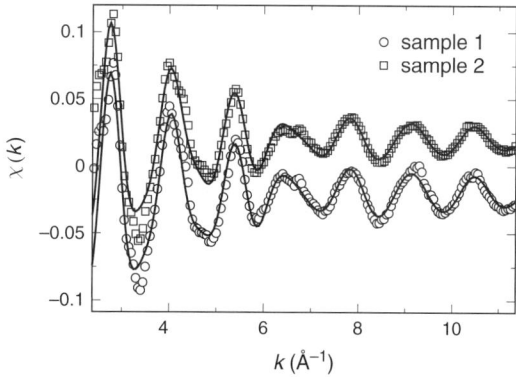

Fig. 21. EXAFS oscilllations for sample 1 (lower curve) and sample 2 (upper curve), measured with the incoming X-ray beam polarization perpendicular to the sample surface, the two spectra are shifted along y-axis for clarity. The best-fit curves are also shown as solid curves.

5.3.2. Comparison of DAFS and EXAFS results

A graphic comparison between EDAFS and EXAFS is given in Figs 20 and 21. We see that the two EDAFS spectra show differences beyond the noise level, whereas EXAFS measurements look very similar to each other. Regarding EXAFS results, one can observe the following findings:

1. The EXAFS spectra and the corresponding best-fit parameters are very similar for the two samples studied.
2. The As–In distance is slightly contracted in agreement with the presence of mismatch strain [74].
3. For the As–As distance two values, 4.15–4.16 Å (in-plane) and 4.23–4.25 Å (out-of-plane), are found. The first one is close to the InP distance in bulk InP, i.e., is the value expected for pseudomorphic InAs on InP. The latter is an intermediate value in between of pseudomorphic InAs (4.29 Å) and InAsP alloy [74].
4. The in-plane and out-of-plane As–P distances are found in the range of 4.15–4.19 Å, which are values typical of bulk InP and InAsP alloy, respectively.
5. The As–In third coordination shell distance shows two values: 4.88–4.87 and 4.93–4.94 Å. The first one is close to the in-plane value expected for pseudomorphic InAs, i.e., the same as the InP substrate (4.87 Å), whereas the second one is in between of the values observed for a strained and relaxed alloy with $x \sim 0.4$. The pseudomorphic elongated distance (5.13 Å) which should be associated to a pure tetragonal deformation is not observed.
6. The P concentration is found to be quite high. We can observe a difference between in-plane and out-of-plane P coordination, the latter being the highest.

We have to note that the equivalent thickness of InAs is of about 2.2 ML, so one must expect quite an important effect on the distances and coordination due to the InAs/InP interface. A rough estimation of P concentration seen by the As absorbers inside the wire can be obtained by calculating the ratio between the number of As atoms close to the wire interfaces with the capping layer and the substrate, and the number of inner As atoms. This gives a few P percent for in-plane polarization and about 15% for out-of-plane polarization, i.e., a strong anisotropic effect. The experimental values reported are higher but this estimation holds only for a sharp interface. This is not a realistic view, at least for group V elements as As and P, which usually show atomic intermixing. The GIDAFS results show instead that the intermixing is limited to at most a couple of MLs [25]. If the model is reconsidered making the hypothesis of 2 InAsP MLs at the capping interface with a 30% P concentration, one gets an in-plane $x_\| \simeq 0.35$ and $x_{out} \simeq 0.45$, which is close to the experimental values reported. This does not exclude an As gradient spreading over few MLs in the capping layer. EXAFS indeed probes simultaneously all the different As environments: the As atoms inside the wires and the As atoms that spread over the interface. This explains on the one hand, the presence of a short in-plane As–As distance as in a biaxially strained InAs lattice, and on the other, the intermediate As–As and As–In out-of-plane distances due to the formation of an InAsP strained alloy.

EDAFS results provide us with a different image of the QWrs. This is due to the different selectivities of the two techniques. EXAFS spectroscopy is a chemically selective probe; EDAFS, in addition, is a spatially selective probe. Different sample regions are probed depending on the kind of spectroscopy used: the As atoms that belong to the wires in case of DAFS and, in case of EXAFS, all the As atoms, i.e., As in the wires, interface, capping, and wetting layer. For the InAs QWrs, the main difference is found for the interatomic distances of the II coordination shell, i.e., As–As(P) pairs. The two samples are no longer similar as found by EXAFS: the As–As and As–P distances for sample 1 are 4.30 and 4.20 Å, respectively, and 4.22 and 4.19 Å for sample 2. They are much more apart from each other for sample 1 than for sample 2. Sample 1 shows an (As–As)$_{out}$ distance close to the bulk (4.28 Å) or the biaxial deformation (4.29 Å) values, whereas for sample 2 the value found is in favour of strained InAsP alloys. An increase of the third (III) shell As–In distance is also found by EDAFS for sample 1, together with a lower P concentration, in agreement with the increased As–As distance.

The difference between the EXAFS and DAFS data is also well illustrated in Fig. 20 where the DAFS best fit (solid line) is compared with the EDAFS calculated according to Eq. (25), where $\sum_{j=1}^{N_A} w_j'' \chi_j''$ and $\sum_{j=1}^{N_A} w_j' \chi_j'$ have been replaced by the corresponding χ_{EXAFS}'' and χ_{EXAFS}', respectively, obtained with the EXAFS best-fit parameters. Note that one can directly compare data obtained with the beam polarization perpendicular to the sample surface, as, due to the scattering geometry, in-plane polarization was not allowed for EDAFS.

The EXAFS versus EDAFS comparison shows that

1. In both samples 1 and 2 an intermixing As/P takes place.
2. The QWrs core is pure InAs in both samples.
3. The interatomic distances are closer to the values found for InAsP for sample 2, whereas for sample 1 tetragonal deformations still shows up.

The interpretation of the experimental findings is that the II and III shells interatomic distances result from an average of different environments; biaxially strained pure InAs at the QWrs core and InAsP strained regions at interface QWrs/capping. Switching from EXAFS to EDAFS a different contrast is obtained due to the different selectivities of the two techniques. This allows to establish that a stronger As/P intermixing mechanism takes place for sample 2 compared to sample 1. For sample 1 it looks to be sharp whereas for sample 2 a wider interface region with an As gradient in the InP capping layer is detected.

This kind of structural information is very relevant to interpret the optical properties of the wires, which strongly depends on composition for emission wavelengths and electron wave function confinement, and also, to get a deeper knowledge of the growth mechanism.

6. Summary and perspectives

In this chapter we give a detailed view of DAFS spectroscopy, a technique based on anomalous diffraction, which has recently shown to be a powerful tool in the study of the structural properties of nanostructures. We give a brief quantum mechanics description of the atomic elastic resonant scattering and review the major experimental and data analysis issues in connection with grazing incidence geometry. We focus on a few applications that recently appeared, in particular the structural study of InAs QWrs and GaN QDs. This chapter, coming after the two previous review chapters (chapters 9 and 10) on X-ray absorption and X-ray anomalous diffraction, respectively, has also the aim of showing a special combination of these two physical phenomena. To this end, we report the comparison of EXAFS and EDAFS results on the same samples, which gives a new insight to the classic complementarity existing between X-ray diffraction and absorption. This chapter is in itself more a perspective than a review contribution to this book. Indeed, despite its great potentialities, DAFS is not as widely used as it could be and the groups involved in DAFS experiments all around the world are still a few. Newer perspectives will be given by experimental and technical improvements of diffraction. In the nanostructures domain, X-ray diffraction with submicron size X-ray beam obtained with achromatic optics like the Kirkpatrick–Baez mirrors, represents a promising tool to study individual nano-object. For instance, DAFS could provide both strain and composition of QDs built into nanowires, which are the subject of intense research due to strong technological impact (see Ref. [75] for instance).

Regarding synchrotron sources, new perspectives to DAFS and GIDAFS will be given in the intermediate energy range (2–7 keV) with dedicated beamlines. For instance, at the SIRIUS beamline at the French synchrotron SOLEIL, it is possible to perform DAFS (GIDAFS) at the In L-edge. This will allow a deeper insight into the study of In-based nanostructures, which are a very important class of the semiconductor nanostructures and devices that are at present under development.

Acknowledgments

We acknowledge the French CRG for granting beam time and supporting the development of the DAFS spectroscopy at the beamlines BM2. We are very grateful to J.F. Bérar, S. Arnaud, B. Caillot, and N. Boudet for their constant help in the development of the DAFS spectroscopy at BM2. We are also very grateful to S. Grenier and A. Letoublon, who were the first PhD student and postdoctoral fellow, respectively, to cope with the GIDAFS spectroscopy and apply it to nanostructures. The authors are indebted to Vincent Favre-Nicolin for the MAD measurements and data analysis and C. Priester for FEM simulations. A special thanks to L. González and Y. González, M. Gendry, and B. Daudin, who have manifested interest in DAFS and shared their deep knowlegde on the materials.

References

[1] J. García, G. Subias, M. G. Proietti, et al., *Resonant "forbidden" reflections in Magnetite*, Phys. Rev. Lett., **85** (2000), 578–581.
[2] Y. Cauchois, *Distribution spectrale observée dans une région d'absorption propre de divers cristaux*, C. R. Acad. Sci, **242** (1956), 100.
[3] Y. Cauchois and C. Bonnelle, *Distribution spectrale observée dans une région d'absorption propre du quartz*, C. R. Acad. Sci, **242** (1956), 1596.
[4] T. Fukamachi, S. Hosoya, T. Kawamura, and J. Hastings, *Measurements of integrated intensity near the absorption edge with synchrotron radiation*, J. Appl. Crystallogr., **10** (1977), 321.
[5] U. W. Arndt, T. J. Greenhough, J. R. Helliwell, et al., *Optimized anomalous dispersion in crystallography: A synchrotron X-ray polychromatic simultaneous profile method*, Nature, **298** (1982), 835.
[6] I. Arcon, A. Kodre, D. Glavic, and M. Hribar, *Extended fine structure of Bragg reflectivity of copper sulphate in the vicinity of copper K-edge*, J. Phys – Coll, **9** (1987), 1105.
[7] H. Stragier, J. O. Cross, J. J. Rehr, et al., *Diffraction anomalous fine structure: A new X-ray structural technique*, Phys. Rev. Lett., **69** (1992), 3064.
[8] I. J. Pickering, M. Sansone, J. Marsch, and G. N. George, *Site-specific X-ray absorption spectroscopy using DIFFRAXAFS*, Jpn. J. Appl. Phys., **32** (1993), 206.
[9] I. J. Pickering, M. Sansone, J. Marsch, and G. N. George, *Diffraction Anomalous Fine Structure: A new technique for probing local atomic environment*, J. Amer. Chem. Soc., **115** (1993), 6302.

[10] L. B. Sorensen, J. O. Cross, M. Newville, et al., in *Diffraction anomalous fine structure: unifying X-ray diffraction and X-ray absorption with DAFS*, Resonant Anomalous X-Ray Scattering: Theory and Applications (G. Materlik, C. J. Sparks, and K. Fischer, eds), Elsevier Science, North-Holland, Amsterdam, 1994.

[11] J. -L. Hodeau, V. Favre-Nicolin, S. Bos, et al., *Resonant diffraction*, Chemi. Rev., **101** (2001), 1843.

[12] M. G. Proietti, H. Renevier, J. -L. Hodeau, et al., *Diffraction Anomalous Fine Structure spectroscopy applied to the study of III-V strained semiconductors*, Phys. Rev. B, **59** (1999), 5479.

[13] Y. Joly, *X-ray absorption near-edge structure calculations beyond the muffin-tin approximation*, Phys. Rev. B, **63** (2001), 125120.

[14] D. C. Meyer, K. Richter, A. Seidel, et al., *Absorption and extinction correction in quantitative DAFS analysis*, J. Synchrotron Rad., **10** (2003), 144.

[15] H. Renevier, S. Grenier, S. Arnaud, et al., *Diffraction anomalous fine structure spectroscopy at the beam line BM2-D2AM at the European Synchrotron Radiation Facility*, J. Synchrotron Rad., **10** (2003), 435.

[16] H. Renevier, J. L. Hodeau, P. Wolfers, et al., *Selective study of Fe atoms at the interfaces of an Fe/Ir(100) superlattice by means of diffraction anomalous fine structure*, Phys. Rev. Lett., **78** (1997), 2775.

[17] J. C. Woicik, J. O. Cross, C. E. Bouldin, et al., *Diffraction anomalous fine structure study of strained $Ga_{1-x}In_xAs$ on GaAs(001)*, Phys. Rev. B, **58** (1998), R4215.

[18] D. C. Meyer, K. Richter, P. Paufler, and G. Wagner, *X-ray analysis of the short-range order in the ordered-alloy domains of epitaxial (Ga,In)P layers by diffraction anomalous fine structure of superlattice reflections*, Phys. Rev. B, **59** (1999), 15253–15260.

[19] T. Bigault, F. Bocquet, S. Labat, et al., *Interfacial structure in (111) Au/Ni multilayers investigated by anomalous X-ray diffraction*, Phys. Rev. B, **64** (2001), 125414.

[20] G. M. Luo, Z. H. Mai, T. P. A. Hase, et al., *Variable wavelength grazing incidence X-ray reflectivity measurements of structural changes on annealing Cu/NiFe multilayers*, Phys. Rev. B, **64** (2001), 245404.

[21] H. H. Lee, M. S. Yi, H. W. Jang, et al., *Determination of absolute indium content in InGaN/GaN multiple quantum wells using anomalous x-ray scattering*, Appl. Phys. Lett., **81** (2002), 5120.

[22] O. Ersen, V. Pierron-Bohnes, M. H. Tuilier, et al., *Short- and long-range order in iron and cobalt disilicides thin films investigated by the diffraction anomalous fine structure technique*, Phys. Rev. B, **67** (2003), 094116.

[23] S. Grenier, M. G. Proietti, H. Renevier, et al., *Glancing-angle diffraction anomalous fine structure of InAs quantum dots and quantum wires*, J. Synchrotron Rad., **8** (2001), 536.

[24] S. Grenier, M. G. Proietti, H. Renevier, et al., *Grazing-incidence diffraction anomalous fine structure of InAs on InP(001) self-assembled quantum wires*, Europhys. Lett., **57** (2002), 499.

[25] A. Letoublon, V. Favre-Nicolin, H. Renevier, et al., *Strain, size, and composition of InAs quantum sticks embedded in InP determined via grazing incidence X-ray anomalous diffraction*, Phys. Rev. Lett., **92** (2004), 186101.

[26] J. Coraux, V. Favre-Nicolin, M. G. Proietti, et al., *Grazing incidence diffraction anomalous fine structure study of GaN/AlN quantum dots*, Phys. Rev. B, **73** (2006), 205343.

[27] J. Coraux, V. Favre-Nicolin, M. G. Proietti, et al., *Grazing incidence diffraction anomalous fine structure: Application to the structural investigation of III-nitride quantum dots*, Phys. Rev. B, **75** (2007), 1.

[28] U. Pietsch, V. Holy, and T. Baumbach, *High-resolution X-ray scattering: From thin films to lateral nanostructures*, Springer Verlag, New York, 2004.

[29] J. Stangl, V. Holy, and G. Bauer, *Structural properties of self-organized semiconductor nanostructures*, Rev. Mod. Phys., **76** (2004), 725.

[30] G. J. Williams, M. A. Pfeifer, I. A. Vartanyants, and I. K. Robinson, *Internal structure in small Au crystals resolved by three-dimensional inversion of coherent X-ray diffraction*, Phys. Rev. B, **73** (2006), 094112.

[31] S. Grenier, *Spectroscopie de diffraction résonante. Etudes de nanostructures de semiconducteurs III-V et de l'ordre de charge dans α'-NaV$_2$O$_5$*, Ph.D. thesis, Université Joseph Fourier, Grenoble, 2001.
[32] J. Coraux, *Etude par spectroscopie X en condition de diffraction de la croissance et de l'encapsulation de boîtes quantiques GaN/AlN*, Ph.D. thesis, Université Joseph Fourier, Grenoble, 2006.
[33] J. J. Sakurai, *Advanced Quantum Mechanic*, Addison-Wesley Publishing Company, New York, 1984.
[34] M. Blume, *Magnetic scattering of X-rays*, J. Appl. Phys., **57** (1985), 3615.
[35] C. Cohen-Tannoudji, J. Dupont-Roc, and G. Grynberg, *Processus d'interaction entre photons et atomes*, InterEditions/Editions du CNRS, Paris, 1988.
[36] D. H. Templeton, L. K. Templeton, J. C. Phillips, and K. O. Hodgson, *Anomalous scattering of X-rays by Cesium and Cobalt measured with synchrotron radiation*, Acta Crystallogr., **A36** (1980), 436.
[37] V. E. Dimitrienko, *Forbidden reflections due to anisotropic X-ray susceptibility of crystals*, Acta Crystallogr., **39** (1983), 29.
[38] J. S. Toll, *Causality and the dispersion relation: Logical foundations*, Phys. Rev., **104** (1956), 1760.
[39] E. Prince, *International tables for crystallography volume C: Mathematical, physical and chemical tables*, Springer, New York, 2004.
[40] S. Sasaki, *Numerical tables of anomalous scattering factors calculated by the cromer and liberman method*, KEK Report, **88** (1989), 1.
[41] D. T. Cromer and D. Libermann, *Relativistic calculation of anomalous scattering factors for X-rays*, J. Chem. Phys., **53** (1970), 1891.
[42] C. T. Chantler, *Theoretical form factor, attenuation, and scattering tabulation for Z = 1−92 from E = 1−10eV to E = 0.4−1.MeV*, J. Phys. Chem. Ref. Data, **24** (1995), 71.
[43] J. S. Cross, *Analysis of diffraction anomalous fine structure*, Ph.D. thesis, University of Washington, 1996.
[44] J. O. Cross, M. Newville, J. J. Rehr, et al., *Inclusion of local structure effects in theoretical X-ray resonant scattering amplitudes using ab initio X-ray absorption spectra calculations*, Phys. Rev. B, **58** (1998), 11215.
[45] J. Coraux, H. Renevier, V. Favre-Nicolin, et al., *In situ resonant X-ray study of vertical correlation and capping effects during GaN/AlN quantum dots growth*, Appl. Phys. Lett., **88** (2006), 153125.
[46] H. J. Stragier, *DAFS: A new X-ray structural technique*, Ph.D. thesis, University of Washington, 1993.
[47] H. Dosch, *Critical Phenomena at Surfaces and Interfaces*, Springer Verlag, Berlin-Heidelberg, 1992.
[48] H. You, *Angle calculations for a '4S + 2D' six-circle diffractometer*, J. Appl. Crystallogr., **32** (1999), 614.
[49] M. Schmidbauer, D. Grigoriev, M. Hanke, et al., *Effects of grazing incidence conditions on the X-ray diffuse scattering from self-assembled nanoscale islands*, Phys. Rev. B, **71** (2005), 115324.
[50] F. A. Ponce and D. P. Bour, *Nitride-based semiconductors for blue and green light-emitting devices*, Nature, **386** (1997), 351.
[51] S. Nakamura, *The roles of structural imperfections in InGaN-based blue light emitting diodes and laser diodes*, Science, **281** (1998), 956.
[52] Y. Arakawa and H. Sakaki, *Multidimensional quantum well laser and temperature dependence of its threshold current*, Appl. Phys. Lett., **40** (1982), 939.
[53] N. Gogneau, D. Jalabert, E. Monroy, et al., *Structure of GaN quantum dots grown under "modified stranski-krastanow" conditions on AlN*, J. Appl. Phys., **94** (2003), 2254.
[54] T. Shibata, K. Asai, S. Sumiya, et al., *High-quality AlN epitaxial films on (0001)-faced sapphire and 6H-SiC substrate*, Phys. Stat. Sol. C, **0** (2003), 2023.
[55] J. M. Garcia, G. Medeiros-Ribeiro, K. Schmidt, et al., *Intermixing and shape changes during the formation of InAs self-assembled quantum dots*, Appl. Phys. Lett., **71** (1997), 2014.

[56] P. B. Joyce, T. J. Krzyzewski, G. R. Bell, and T. S. Jones, *Surface morphology evolution during the overgrowth of large InAs/GaAs quantum dots*, Appl. Phys. Lett., **79** (2001), 3615.

[57] P. Sutter and M. G. Lagally, *Embedding of nanoscale 3D SiGe islands in a Si matrix*, Phys. Rev. Lett., **81** (1998), 3471.

[58] A. Hesse, J. Stangl, V. Holy, et al., *Effect of overgrowth on shape, composition, and strain of SiGe islands on Si(001)*, Phys. Rev. B, **66** (2002), 085321.

[59] N. Gogneau, E. Monroy, D. Jalabert, et al., *Influence of AlN overgrowth on structural properties of GaN quantum wells and quantum dots grown by plasma-assisted molecular beam epitaxy*, J. Appl. Phys., **96** (2004), 1104.

[60] M. Newville, *IFEFFIT: Interactive XAFS analysis and FEFF fitting*, J. Synchrotron Rad., **8** (2001), 322.

[61] B. Ravel and M. Newville, *ATHENA, ARTEMIS, HEPHAESTUS: Data analysis for X-ray absorption spectroscopy using IFEFFIT*, J. Synchrotron Rad., **12** (2005), 537.

[62] A. L. Ankudinov, B. Ravel, J. J. Rehr, and S. D. Conradson, *Real-space multiple-scattering calculation and interpretation of X-ray-absorption near-edge structure*, Phys. Rev. B, **58** (1998), 7565.

[63] F. Romanato, D. De Salvador, M. Berti, et al., *Bond-length variation in $In_xGa_{1-x}As/InP$ strained epitaxial layers*, Phys. Rev. B, **57** (1998), 14619.

[64] F. d'Acapito, F. Boscherini, S. Mobilio, et al., *Epitaxy and strain in the growth of GaN on AlN: A polarized X-ray absorption spectroscopy study*, Phys. Rev. B, **66** (2002), 205411.

[65] M. Arlery, J. L. Rouvière, F. Widmann, et al., *Quantitative characterization of GaN quantum-dot structures in AlN by high-resolution transmission electron microscopy*, Appl. Phys. Lett., **74** (1999), 3287.

[66] X. H. Wu, L. M. Brown, D. Kapolnek, et al., *Defect structure of metal-organic chemical vapor deposition-grown epitaxial (0001) GaN/Al_2O_3*, J. Appl. Phys., **80** (1996), 3228.

[67] J.-L. Rouvière, J. Simon, N. Pelekanos, et al., *Preferential nucleation of GaN quantum dots at the edge of AlN threading dislocations*, Appl. Phys. Lett., **75** (1999), 2632.

[68] A. Bourret, C. Adelmann, B. Daudin, et al., *Strain relaxation in (0001) AlN/GaN heterostructures*, Phys. Rev. B, **63** (2001), 245307.

[69] S. Founta, J. Coraux, D. Jalabert, et al., *Anisotropic strain relaxation in a-plane GaN quantum dots*, J. Appl. Phys., **101** (2007), 063541.

[70] J. Coraux, B. Amstatt, J. A. Budagoski, et al., *Mechanism of GaN quantum dots capped with AlN: An AFM, electron microscopy, and X-ray anomalous diffraction study*, Phys. Rev. B, **74** (2006), 195302.

[71] V. Ranjan, G. Allan, C. Priester, and C. Delerue, *Self-consistent calculations of the optical properties of GaN quantum dots*, Phys. Rev. B, **68** (2003), 115305.

[72] M. Gendry, C. Monat, J. Brault, et al., *From high to low height dispersion for InAs quantum sticks emitting at 1.55µm on InP(001)*, J. Appl. Phys., **95** (2004), 4761.

[73] L. González, J. M. García, R. García, et al., *Influence of buffer-layer surface morphology on the self-organized growth of InAs on InP(001) nanostructures*, Appl. Phys. Lett., **76** (2000), 1104.

[74] S. Pascarelli, F. Boscherini, C. Lamberti, and S. Mobilio, *Tetragonal-strain-induced local structural modifications in $InAs_xP_{1-x}/InP$ superlattices: A detailed X-ray-absorption investigation.*, Phys. Rev. B, **56** (1997), 1936.

[75] J. H. Na, R. A. Taylor, J. H. Rice, et al., *Time-resolved and time-integrated photoluminescence studies of coupled asymmetric GaN quantum discs embedded in AlGaN barriers*, Appl. Phys. Lett., **86** (2005), 083109.

12

The role of photoemission spectroscopies in heterojunction research

Giorgio Margaritondo

VPAA, Ecole Polytechnique Fédérale de Lausanne (EPFL), CH-1015 Lausanne, Switzerland

Abstract Experimental techniques based on the photoelectric effect are very powerful probes of the electronic structure of solids. Because of their surface sensitivity, they are particularly useful in the study of semiconductor interfaces such as those of semiconductor–semiconductor heterojunctions. We begin our treatment of this topic by a discussion of the use of photoemission to measure and study heterojunction band discontinuities with different levels of accuracy and sophistication. We then deal with a broader range of interface properties and their analysis by angle-resolved photoemission and photoemission spectromicroscopy. This background discussion is followed by the presentation of examples and case studies, in particular the issues of interface parameter control and of the lateral fluctuations of interface barriers. Finally, we briefly treat discontinuity measurements by internal photoemission.

Keywords photoemission, photoelectron, synchrotron radiation, spectromicroscopy, heterojunction, semiconductor, band discontinuities

1. Introduction

The electronic structure of semiconductor heterojunction interfaces is one of the most interesting subjects in condensed-matter research [1,2]. As such, it strongly profited from the development of increasingly sophisticated experimental probes based on the photoelectric effect, starting from the pioneering work of P. Perfetti et al. in 1978 [3]. These probes, collectively known as "photoemission spectroscopies," actually consist [4–7] in a collection of diverse techniques that can explore fine and diverse details of the electronic properties of condensed-matter systems.

In memory of Neville V. Smith, a quiet scientific giant and father of angle-resolved photoemission: "I trust you are having a great time with Mort, as we did all together thirty years ago"

1.1. Photoemission measurements of band discontinuities: a simple description

In order to understand how the photoelectric effect can be used to study heterojunction properties [1,2], we can start from the most important among the heterojunction features: the band discontinuities. This concept is discussed in detail in other parts of this book (Chapters 1–3 and 6), so that here we can simply summarize its main aspects. As shown in Fig. 1, when two semiconductors A and B are combined together in a heterojunction, the difference between the two forbidden gaps E_{gA} and E_{gB} creates interface discontinuities in the valence and conduction band edges, E_v and E_c. These discontinuities ΔE_v and ΔE_c determine many of the important properties of the heterojunction [1] and dominate the behavior and performances of the corresponding devices.

Many different techniques were developed to study band discontinuities [1]. Among them, photoemission probes are the most direct: Fig. 2 shows why. First (Fig. 2(a)), we see the scheme of a photoemission experiment in which an electron absorbs the energy of an incoming photon, $h\nu$, is emitted into the vacuum, and is then captured and analyzed – measuring in particular its energy.

Figure 2(b) illustrates a photoemission experiment for the semiconductor A only. During a photoelectron emission process, the energy of the involved electron is first augmented by an amount equal to the energy of the absorbed photon, $h\nu$. Some of the excited electrons are then able to reach the vacuum without losing any energy. Their energy distribution (photoemission intensity spectrum) is, in first approximation, a fingerprint of the density of occupied states $N(E)$ inside the solid – shifted upwards by $h\nu$. Note, in particular, that after correcting for the $h\nu$-shift the upper edge of the spectrum corresponds to E_v.

Figure 2(c) shows the same scheme for the two semiconductors forming the heterojunction. Both in the density of states and in the photoemission spectrum, we see a double edge reflecting the valence-band discontinuity ΔE_v. From the analysis of this double edge, the magnitude and other fundamental properties of the discontinuity can be extracted. Note that, once ΔE_v is known, the conduction band discontinuity is simply given by:

$$\Delta E_c = \left| E_{gA} - E_{gB} \right| - \Delta E_v \tag{1}$$

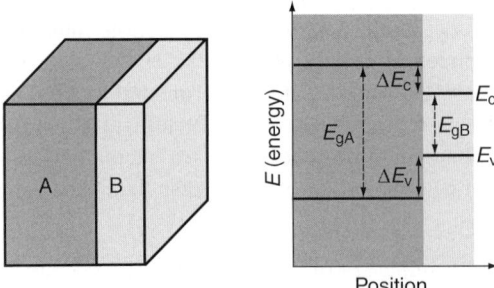

Fig. 1. Schematic explanation of the origin of heterojunction band discontinuities. Left: a heterojunction consists of the combination of two different semiconductors, A and B. Right: the band structures of the two materials include two different forbidden energy gaps, E_{gA} and E_{gB}. The difference between the gaps gives rise at the A–B interface to two "jumps" or discontinuities in the conduction and valence-band edges, ΔE_v and ΔE_c.

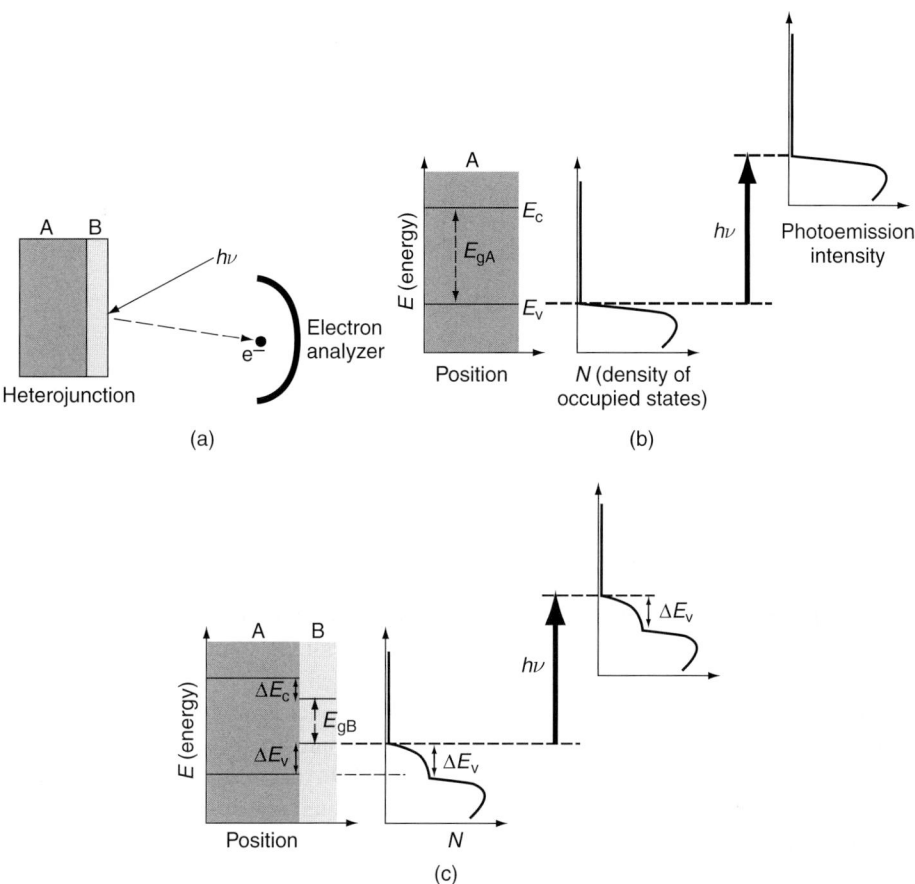

Fig. 2. Photoemission measurements of valence-band discontinuities. (a) Scheme of a photoemission experiment for a heterojunction; a photon of energy $h\nu$ is absorbed by an electron in the heterojunction; the electron thus reaches a sufficiently high energy to leave the solid system. It is then captured by an analyzer that measures its energy. (b) Energy diagram of a photoemission experiment involving the semiconductor A alone. The highest-energy photoelectrons that are emitted correspond to the highest occupied states in the semiconductor, i.e., to the top of the valence-band. The maximum photoelectron energy thus equals the valence-band edge E_v plus the photon energy $h\nu$. (b) When the experiment is performed for the A–B heterojunction, two edges are visible and their distance in energy can be used to evaluate the valence band discontinuity E_v.

Figure 3 shows a nice example [8] of this approach: the photoelectron energy distribution of the CdS–Si heterojunction interface. Without any data processing, the double edge in the spectrum reveals the valence-band discontinuity that can be thus measured and analyzed. Early experiments of this type [1,3,8] practically transformed band discontinuities from merely theoretical notions only indirectly measurable to very concrete and tangible entities, relatively easy to visualize and study.

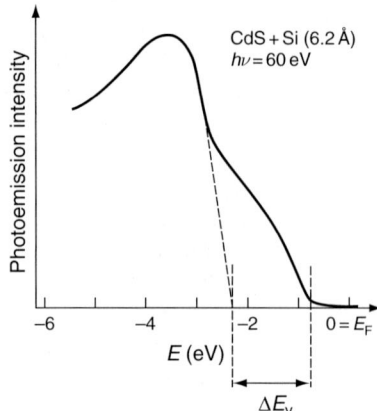

Fig. 3. A practical example of the approach illustrated by Fig. 2: the CdS–Si valence-band discontinuity is reflected in the double edge of the photoemission spectrum of a CdS substrate covered by a thin (6.2 Å) overlayer of silicon (data extracted from Ref. [8]).

1.2. A more refined approach

The above analysis does not take into account a number of theoretical and practical complications [1] that affect this experimental approach. First of all, the photoemission process is a complicated combination of different phenomena entangled together that affect the relation between the electronic structure of the system and the photoemission spectra. The direct relation suggested by Fig. 2 is an oversimplified first approximation.

In the spirit of William E. Spicer's "three-step model" [9], the photoemission process can be described as the combination of three different (but interacting) phenomena: photon absorption and excitation of the electron, travel of the excited electron to the surface, and emission into vacuum. Each of these phenomena affects the photoemission spectra [4–7]. For example, as the excited electron moves towards the surface there can be inelastic scattering effects that decrease its energy: as a result, in the photoemission spectra the "primary" contribution from lossless electrons is superimposed to a "secondary" photoelectron background [3]. Fortunately, such a background does not significantly affect the study of heterojunction band discontinuities and can be neglected in our present discussion.

The initial excitation process itself is more complicated than a simple upwards shift in energy by $h\nu$. It specifically depends on the initial and final states as well as on the optical transition probability for such states [4–7]. As a consequence, spectra taken at different photon energies can significantly differ from each other because a given initial state corresponds to different final states. It is, therefore, desirable to take spectra at several photon energies and compare them: this makes it possible to reliably identify the spectral features that directly reflect the electronic structure of the interface – as opposed to spurious features.

The need to work with different $h\nu$-values is one of the reasons that make synchrotron light sources very important [4,6] for photoemission studies – in particular, photoemission experiments on heterojunctions [2]. Considering its importance, we must elaborate a bit on the role of synchrotron sources.

In order to produce photoelectrons, the absorbed photons must excite the initial-state electrons to energies higher than the minimum level required to overcome the surface barrier and exit into vacuum (the so-called "vacuum level"). Only ultraviolet and X-ray photons provide the required energy. In the early years of photoemission spectroscopy, such photons were obtained from conventional sources such as discharge lamps or electron-bombarded anodes. Conventional source, however, offer very limited performances; in particular, they cannot produce intense photon beams with tunable photon energies.

This problematic situation changed radically in the 1970s with the advent of synchrotron sources [4] that emit very intense and tunable ultraviolet and X-ray beams – thanks to the action of magnets on electrons circulating under vacuum in a "storage ring" at nearly the speed of light. In addition to intensity and tunability, synchrotron sources possess other very valuable properties for photoemission spectroscopy. They are linearly polarized (and in some cases elliptically polarized), a feature that can be exploited to study symmetry effects in the excitation processes. They operate under ultrahigh vacuum, facilitating the study of interfaces without risking contamination. Moreover, their emitted beams have small cross sections and are strongly collimated, simplifying the task of focusing them into small areas. Nor, surprisingly, the overwhelming majority of photoemissions studies of heterojunctions are performed with synchrotron sources [1,2].

Surface sensitivity is another critical issue in photoemission heterojunction studies. The excited electrons can only travel for a very short distance before losing energy. Therefore, "primary" photoelectrons originate from a very thin slab at the surface. The thickness of this slab (called the photoelectron "escape depth") depends on the excited-electron energy and therefore on $h\nu$, and ranges in most cases from a few angstroms to $100\,\text{Å}$. Photoemission experiments, therefore, probe a surface region whose thickness does not exceed a few atomic planes.

Photoemission studies are thus very sensitive to surface contamination and must be performed on ultraclean surfaces under ultrahigh vacuum. In addition to this requirement, the surface sensitivity has additional consequences that depend on the specific system under investigation.

The specific consequences for heterojunction studies are illustrated by Figs 4–6. In the case of Fig. 4(a), the thickness of the semiconductor side B is larger than the escape depth: the photoelectrons cannot originate from the interface region and photoemission spectroscopy cannot probe the valence-band discontinuity. On the contrary, in Fig. 4(b), the thickness of the B layer is smaller than the escape depth: photoemission does probe the valence-band discontinuity.

Photoemission, therefore, cannot probe "deep" heterojunction interfaces. The experiments must be conducted on semiconductor substrates covered by thin semiconductor overlayers. This raises a question: Is the overlayer thickness large enough for the interface to be representative of the "bulk" interfaces properties for two thick semiconductors? The answer is empirical: experiments must be performed for increasing overlayer thicknesses to follow the evolution of the interface properties – including the valence-band discontinuity – and verify the saturation of their thickness dependence. Figure 5 shows an example [10] of this approach: it is quite evident that the evolution of ΔE_v saturates very rapidly and that an overlayer of a few angstroms already produces, in practice, a bulk interface.

Band bending is another phenomenon that must be taken into account because of the surface sensitivity of photoemission. For simplicity, in Figs 1 and 4 we showed only flat bands with no bending at all. This picture, however, is not realistic: at most semiconductor interfaces the energy bands are bent because of the local charge distribution.

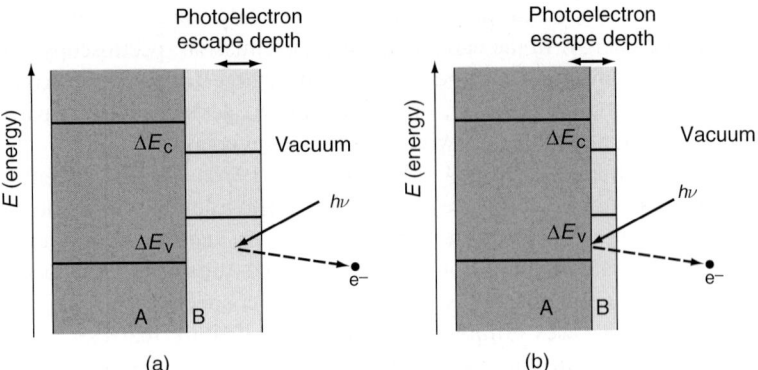

Fig. 4. Schematic explanation of the impact of the surface sensitivity of photoemission on heterojunction studies. (a) Here the overlayer thickness is larger than the photoelectron escape depth: photoelectrons cannot originate from the interface region and the experiment cannot probe the valence-band discontinuity. (b) The contrary is true when the overlayer thickness is smaller than the escape depth.

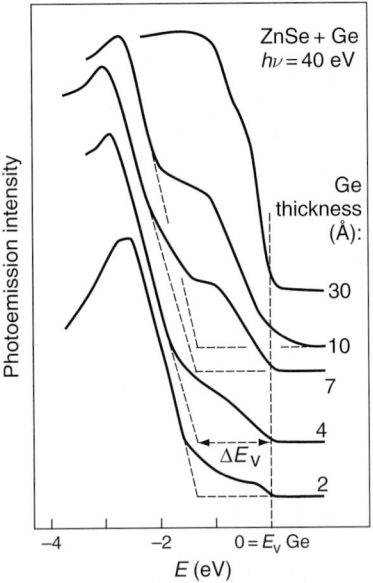

Fig. 5. The evolution of the photoemission results for a heterojunction interface as a function of the overlayer thickness. Tests of this kind are required to verify that the measured interface properties are representative of those for bulk interfaces. In this case, the evolution of the measured ZnSe–Ge valence-band discontinuity vs the Ge overlayer thickness saturates very rapidly: bulk-like interface properties are already present when the overlayer is only a few angstrom thick (data derived from Ref. [10]).

The top part of Fig. 6 shows band-bending effects [1] for (a) a metal–semiconductor interface (Schottky barrier), (b) a p–n semiconductor homojunction, and (c) a heterojunction. The bottom part of Fig. 6 illustrates the specific case of the formation of a heterojuction

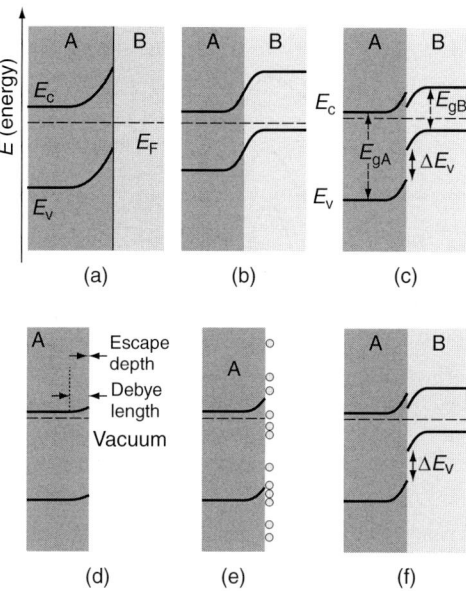

Fig. 6. For most semiconductor interfaces, the bands are not flat as in the previous figures but "bent" because of the effects of the local charge distribution. The top part of the figure shows band-bending effects for different types of interfaces: (a) a metal–semiconductor B–A junction, (b) a p–n homojunction, and (c) a heterojunction. The bottom part illustrates band-bending effects during the formation of a heterojunction interface. We start (d) from a clean substrate with its own band bending. Note that the photoelectron escape depth is shorter than the "Debye length," over which the band bending occurs: therefore, the photoemission data reflect the band energies at the surface rather than in the bulk. (e) With a thin overlayer, the band bending changes. This change continues until (f) the final situation of a bulk interfaces is reached.

interface by deposition of a semiconductor overlayer B on the semiconductor substrate A. The first step (a) corresponds to the clean surface of the semiconductor A. Even for the clean surface, there is a band bending due to the microscopic charge distribution near the surface. The band bending occurs over a distance determined by the so-called Debye length of the semiconductor [1].

As shown in Fig. 6(d), the Debye length is typically much larger than the photoelectron escape depth. Therefore, photoemission spectra of clean surfaces reflect the band structure at the surface rather than in the bulk. In Fig. 6(e), the first overlayer particles are deposited, the local charge distribution changes, and so does the band bending: the photoemission spectra change accordingly. This evolution continues as the overlayer becomes thicker, until (Fig. 6(f)) the final "bulk" interface situation is reached.

The photoemission spectra during the formation of semiconductor interfaces do reflect these effects. Figure 7 shows spectra (referred to the Fermi level E_F) that are modified due to the band-bending evolution: specifically, the CdS substrate valence-band edge shifts as the Ge overlayer thickness increases.

One of the main effects of band bending is the spectral broadening. For an infinitely small photoelectron escape depth, a photoemission experiment would only detect the energy positions at the surface. Because the escape depth is small but finite, the experiments detect

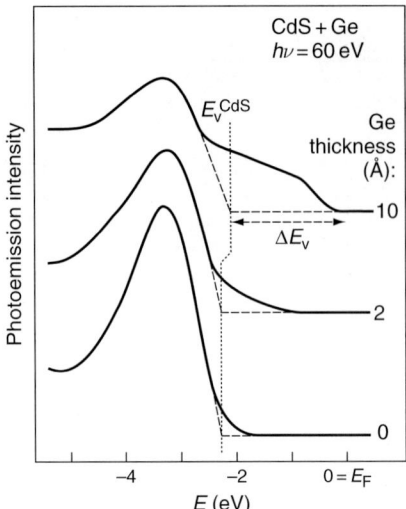

Fig. 7. A practical case of band-bending effects during the formation of a heterojunction interface. As the thickness of the Ge overlayer increases during the interface formation process, the valence-band edge of the CdS substrate shifts. This reflects the band-bending evolution on the CdS side (data derived from Ref. [11]).

energy positions in a thin slab – that change across the slab because of band bending. In practice, this effect broadens the spectral features including the valence-band edges and negatively impacts the accuracy of heterojunction discontinuity measurements [1,2].

1.3. An alternate general method to measure band discontinuities

Other factors besides the band bending affect the accuracy of photoemission experiments on heterojunctions, including instrumental resolution and intrinsic broadening [1,2]. When the valence-band discontinuity is too small with respect to the combined broadening effects, a double edge like those of Figs 3, 5, and 7 is no longer visible: this is the case of Fig. 8 [11]. The value of ΔE_v can no longer be immediately extracted from the spectra and an alternate method must be devised and applied.

By analyzing Fig. 8 in a superficial way, one could imagine that the valence-band edge position of the substrate corresponds to the high-energy edge of the bottom spectrum (no coverage), whereas E_v for the overlayer corresponds to the spectral edge for large coverage. If this was true, then the value of ΔE_v could be simply derived from the difference between the two edges. This, however, is not correct in most cases because the substrate band bending changes during the interface formation and so does the substrate valence-band edge position.

However, if the band-bending changes can be measured, then the result can be used [12] to correct the edge difference and obtain the right value of ΔE_v. Unfortunately, the band-bending changes cannot be easily followed by monitoring the substrate valence-band spectral features because they are progressively concealed by the overlayer features.

There exist, however, another possibility: the substrate band-bending changes can be monitored using substrate core-level spectral peaks. Such peaks are isolated in the spectrum

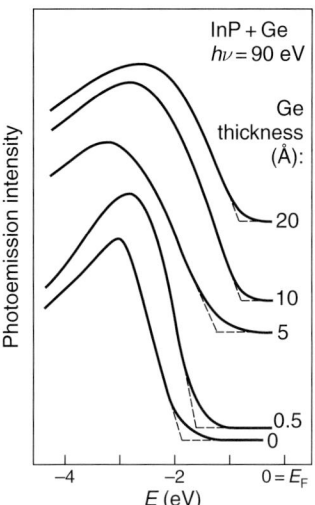

Fig. 8. Results illustrating the need for an alternate method to derive band discontinuities from photoemission spectra. In this case, the valence-band discontinuity is too small with respect to the spectral broadening to observe a double edge as in Fig. 3, 5, and 7 and to extract from it the discontinuity (data derived again from Ref. [11]).

and not heavily affected by overlayer features. Nonetheless, they are not entirely immune from problems. The interface formation process can deform them; such deformations reflect the creation of local chemical bonds. Careful spectral analysis is therefore necessary to find the correct band-bending corrections.

Figures 9 and 10 illustrate the basis of this method and an example of its practical use. In Fig. 9(a), we see the electronic structure of the clean A-substrate with its band bending and the corresponding photoemission spectrum. Note that the structure includes not only the valence band but also a core level CL. From the spectrum, one can measure the energy distance X between the core level and the valence-band edge of the A-semiconductor.

Figure 9(b) shows the situation after deposition of a thin overlayer of the B-semiconductor. The band bending changes for the A-side (for simplicity, we assume no band bending on the B-side). The overall photoemission spectrum is the combination of the contributions from the two sides (dashed lines). In this total spectrum, it is easy to measure the distance Y between the core level CL and the valence-band edge – that now corresponds to E_v for the B-side.

Although the overall picture might seem rather complex, the derivation of the valence-band discontinuity is quite easy. It is indeed evident from Fig. 9 that

$$\Delta E_v = |Y - X| \qquad (2)$$

Thus, by simply measuring the two distances X and Y in the clean-substrate and thin-overlayer spectra, the discontinuity can be derived using Eq. (2) that implements an automatic correction for the band-bending effects.

Figure 10 shows a practical example of this approach – applied to the same interface as in Fig. 3 – illustrating some additional complications with respect to above discussion. The plots show the evolution of several features: the top of the CdS substrate valence band (before the

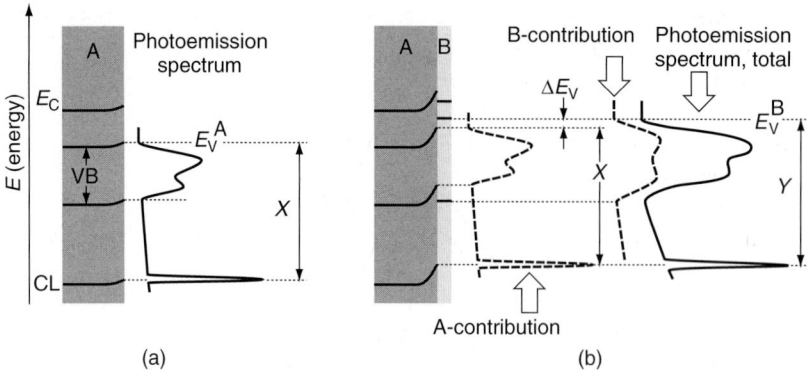

Fig. 9. Schematic explanation of the alternate photoemission method to measure valence-band discontinuities. (a) On the left side, we see the band structure of the clean substrate A with the conduction-and valence-band edges; below the valence band, we also see a core level CL. In the proximity of the surface, all these features are affected by the band bending. The photoemission spectrum on the right-hand side thus reflects the energy position of the above features *at the surface* rather than in the bulk. (b) After deposition of a thin B-overlayer, the photoemission spectrum is the combination of the contributions from both the substrate and the overlayer. The substrate spectral features are shifted in energy because of the overlayer-induced band-bending changes. However, the distance X between CL and the substrate valence-band edge remains unchanged. By subtracting it from the corresponding distance Y for the overlayer valence-band edge-one can derive the valence band discontinuity [Eq. (2)].

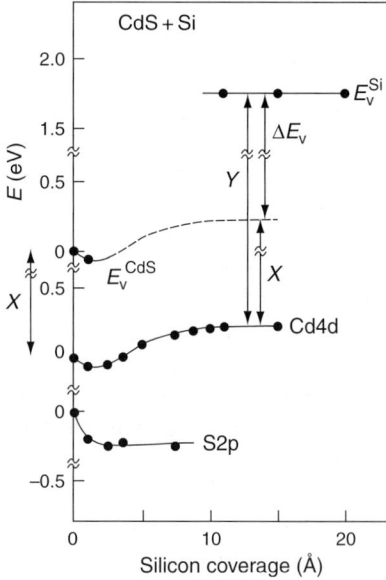

Fig. 10. Practical implementation of the method of Eq. (2) and Fig. 9 in the case of CdS–Si (see the discussion in the text; data derived from Ref. [8]). The plots show the evolution of the photoemission spectral features with respect to the overlayer thickness, reflecting the changes in the substrate band bending as well as local chemical phenomena that affect to the position of the S2p peak – making it unsuitable to track band-bending changes.

overlayer becomes too thick and makes it difficult to detect), the top of the Si overlayer (after the thickness is large enough to allow its detection), and the core-level peaks Cd3d and S2p.

First of all, it is clear that the overlayer does change the band bending: this is revealed by the shifts of the substrate spectral features for increasing Si coverage. Second, the band-bending-induced shift of the CdS valence-band edge makes it impossible to evaluate ΔE_v by simply taking the distance in energy between the CdS and Si edges. Third, the initial shifts of the Cd3d peak and of the CdS edge are similar, suggesting that they are both due to the band-bending changes; on the contrary, the S2p shifts does not track the other two, indicating additional effects besides the band bending.

This is quite reasonable, because the creation of interface chemical bonds does affect the core levels of the involved elements [4]. In this case, the experimental results suggest that S is involved in the interface chemical bonds that affect the S2p peak; this peak, therefore, cannot be used to track the band-bending changes.

Such changes can be derived instead from the shift of the Cd3d peak: the dashed line in the figure shows the corresponding band-bending-induced evolution of the CdS valence-band edge. As also shown in the figure, this makes it possible to evaluate the valence-band discontinuity; this procedure is equivalent to simply using Eq. (2), which would automatically take into account the band-bending effects.

This discussion reveals the practical difficulties encountered in the use of photoemission to measure heterojunction band discontinuities and provides some hints about the corresponding accuracy limits. Great efforts were dedicated [13] to improving this accuracy, for example by sophisticated modeling of the edge and core-level lineshapes. Realistically speaking, however, we believe that the accuracy of the above methods cannot be systematically improved below the 100 meV level, even if better accuracies were reported for specific systems [13].

2. Angle-resolved photoemission

Our previous discussion illustrated how photoemission techniques can be exploited to study heterojunction discontinuities. Photoemission, however, can deliver a much broader variety of information [4–7]. In particular, experiments detecting not only the energy of the photoelectrons but also the direction along which they propagate can be quite helpful in the study of crystalline heterojunction interfaces.

So far, we discussed photoemission techniques under the implicit assumption that the investigated properties are averages over all the directions. Specifically, the state of a photoelectron in vacuum is characterized by both the magnitude and the direction of its k-vector: our previous discussion assumed the averaging over all k-directions. This is partially true when photoelectrons are collected by a large-area detector, although the averaging is not over all directions but limited to those captured by the detector. And it is definitely true when the sample is polycrystalline because the mixture of different grain orientations automatically "scrambles" the k-directions.

Direction averaging makes it specifically possible to consider the energy distribution of the photoelectrons as an image of the energy distribution of the electrons inside the sample. This was the implicit foundation of our previous discussion of photoemission measurements of valence-band discontinuities. Direction averaging does simplify the interpretation of photoemission data. It also corresponds, however, to a loss of the potential information carried by the photoelectrons. This loss can be avoided by using a small-area detector that collects photoelectrons only along a specific direction.

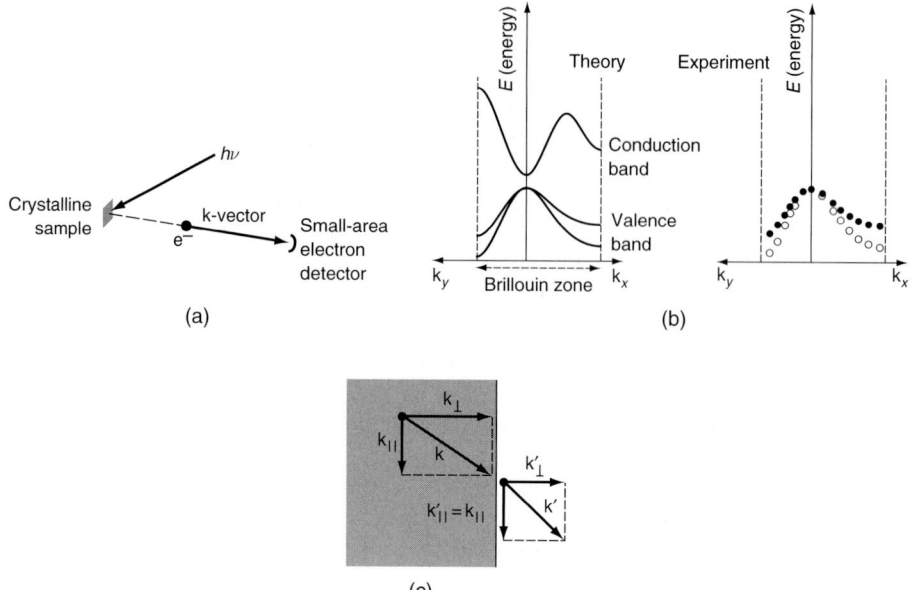

Fig. 11. Angle-resolved photoemission and band-structure mapping. (a) With a small-area detector, one can measure both the energy and the direction of the photoelectrons; in turn, these parameters entirely define the k-vector and therefore the state of each photoelectron in vacuum – and can be used to evaluate the corresponding parameters for the ground-state electron in the sample. (b) The energy vs **k** plots so obtained provide an experimental map of the band structure (right) and a test of the theoretical band structure for occupied states. (c) However, the extraction of the ground-state k-vector in the solid from measurements on the photoelectrons in vacuum is not trivial: the surface barrier changes indeed the perpendicular component of the k-vector, whereas the parallel component is not modified.

Figure 11 shows a practical example of the use of angle-resolved photoemission. In Fig. 11(a), we see the capture of a photoelectron along a specific direction. By measuring its energy, one determines the magnitude of the k-vector. Combined with the information about the direction, this gives a complete knowledge of the k-vector **k** and therefore of the free-electron state of the photoelectron.

By combining measurements of energies and k-vectors, one can obtain an experimental map [14–17] of the "band structure" of crystalline systems: indeed, the band structure corresponds to the three-dimensional function $E(\mathbf{k})$. The left-hand side of Fig. 11(b) shows plots of a calculated (theoretical) band structure along two different k-vector directions. On the right-hand side, we see the corresponding experimental "map" consisting of measured $E(\mathbf{k})$ points. Note that only occupied states can be mapped: this technique cannot be used to map the conduction band.

The band-mapping technique can be used in many different ways. For example, we can see on the left-hand side of Fig. 11(b) the presence of two branches in the valence band, corresponding to two different hole effective masses. The experimental maps on the right-hand side of the same figure can then be used to evaluate such effective masses.

Figure 11(c) illustrates the main difficulty encountered in practical band mapping. The experimental data provide the energy and k-vector of the photoelectrons in vacuum. The band structure corresponds instead to the ground-state energy and k-vector of the electrons inside

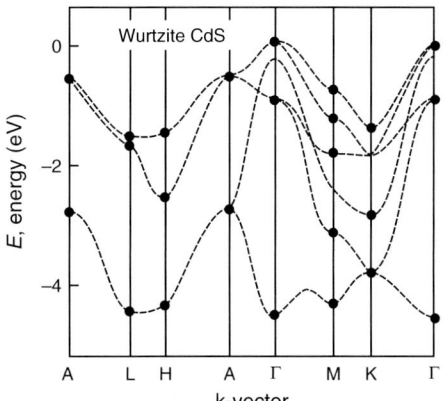

Fig. 12. An example of band structure mapping in three dimensions: results by N. G. Stoffel for wurtzite CdS (extracted from Ref. [16]).

the sample (more specifically, the k-vector corresponds here to the so-called "crystal momentum"). Band mapping requires then a retrieval of such parameters from those measured for the free photoelectron in vacuum.

The discussion of Fig. 2(b) demonstrated that the photoelectron energy (for primary photoelectrons) can be converted into the ground-state energy by simply subtracting the photon energy $h\nu$. The correction is more complicated for the k-vector: as shown in Fig. 11(c), the k-vector component perpendicular to the sample surface is modified by the surface energy barrier as the excited electron crosses it to become a photoelectron. On the contrary, the parallel component remains unchanged.

Band mapping is thus easy [14,15] for strongly two-dimensional crystalline samples such as layer compounds – for which the parallel k-vector component is in practice the only important one. For three-dimensional system [16,17], it requires instead rather delicate and in many cases complicated procedures to take into account the change in the perpendicular component. With such procedures, however, band structure mapping can be applied to a wide variety of crystalline samples [4]. Figure 12 shows one of the early examples: the mapping of the three-dimensional band structure of wurtzite CdS [16].

Band mapping is quite relevant to the studies of crystalline heterojunction. One could imagine, in principle, to use accurate maps of the band structures on the two sides to determine with high accuracy the positions in energy of the two valence bands' tops – and therefore the valence-band discontinuity. This approach is practical unfeasible, however, because of the overlap of the valence-band spectral features of the two sides.

More realistically and without fully using band mapping, angle-resolved photoemission can be used to detect the valence-band edge where it occurs in k-space, such as for example at the center of the Brillouin zone. This can increase the practical accuracy of discontinuity measurements [18]; such a method is limited of course to interfaces with good crystal quality on both sides.

3. Local chemical analysis

Photoelectrons carry valuable information on the local chemical properties that can be harvested to achieve a better understanding of the heterojunction properties – in particular, of the link between interface parameters and local chemistry. Photoemission was in fact originally developed as a probe of chemical properties – hence the name ESCA or "electron spectroscopy for chemical analysis" that is often used for X-ray photoemission [4]. ESCA probes indeed the energy distribution of electrons inside condensed-matter systems that is strongly influenced by the formation of chemical bonds. In order to understand this point, we should distinguish between valence electrons – the electrons directly involved in the formation of bonds – and core electrons.

The energies of valence electrons are directly modified by the bond formation process, for example by the transfer of electrons between atoms for ionic bonds or by quantum-state modifications for covalent bonds. A detailed photoemission analysis of the electron energies can thus yield a great deal of information on the chemical bonding mechanism. This analysis, however, is not simple and normally requires a sophisticated theoretical treatment of the underlying phenomena.

The study of core electrons provides instead more limited but more readily extractable information. Although the core electrons do not directly contribute to the bond formation, their energies are modified by the corresponding changes in the valence charge distribution – and can thus indirectly probe the bond properties.

For example, during the formation of an ionic bond, valence electrons are transferred from the cations to the anions. Compared with the neutral atom, the cation has less valence electron (negative) charge. This means that the core electrons interact with an overall local charge distribution that is more positive than for an isolated atom – therefore, they are more bound and their energy decreases. The opposite is true for core electrons in anions, whose energies increase with respect to neutral isolated atoms. Such energy changes can be easily measured with photoemission spectroscopy.

This discussion explains how photoemission experiments can yield information on the local chemical reactions during the formation of heterojunction interfaces. This is a very important point because the discontinuities as well as many other heterojunction properties are influenced or even determined by the local chemistry [1].

Consider for example the extreme case of an interface formation process involving no interface chemical reactions at all. The electron energy structure would then be merely the result of electrostatic processes. This is the basis of the "Schottky model" of metal–semiconductor interfaces [19] that justifies the Schottky barrier only in terms of the metal work function and of the semiconductor electron affinity.

The corresponding model for heterojunctions is called the "Anderson model" [20]. The electron affinity in a semiconductor is defined as the energy difference between the bottom of the conduction band and the "vacuum level" (the minimum energy of an electron in vacuum). Thus, in the absence of defects or chemical interactions, the conduction band discontinuity would simply correspond to the difference between the electron affinities of the two semiconductors (and the valence-band discontinuity can be derived using Eq. (1)).

Local chemical interactions would invalidate the basis for the Anderson model: their absence or presence is thus a key question for heterojunctions – and photoemission is a powerful instrument to study the issue. Whenever interface chemical interactions do occur, photoemission spectra can effectively monitor them. We already saw in Section 1.3 and Fig. 10 how chemical effects can influence photoemission features during a heterojunction formation process. Figure 13 shows another good example: the In4d photoemission spectra of an InP substrate progressively covered by a Ge overlayer [21].

We observe that the In4d peak is both shifted in energy and deformed during the interface formation. The shift is justified at least in part by the bend-bending changes, as discussed in

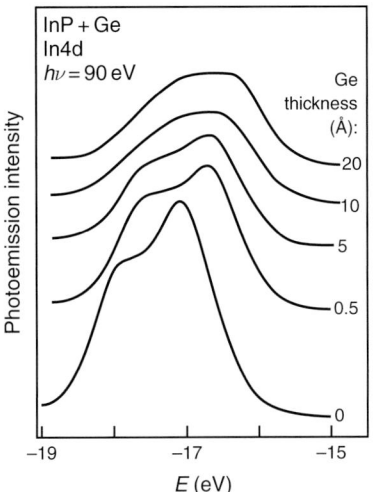

Fig. 13. During the InP–Ge interface formation process, the In4d photoemission peak is both shifted in energy and deformed. These changes reflect in part the previously discussed band-bending modifications; in addition, they are also influenced by local chemical interactions accompanying the interface formation (data derived from Ref. [21]).

Sections 1.2 and 1.3. The other peak modifications reflect the formation of local chemical bonds involving indium atoms. The interface is thus chemically reactive and its properties are influenced by the local chemical bonds: this requires a theoretical analysis more sophisticated than the simple Anderson model [20].

4. Photoemission spectromicroscopy

Surface sensitivity, as we saw in Section 1.2, enables photoemission spectroscopy to probe electronic properties localized within a few atomic planes of the sample surface. On the contrary, photoemission does not automatically provide spatial resolution along the surface plane. Until the late 1980s, photoemission experiments were forced to average the surface and interface properties over areas with a typical size of a few millimeters.

This was a significant disadvantage because many important properties fluctuate on a much shorter scale. The issue is specifically important for semiconductor interfaces: Are their properties constant for the entire interface plane or do they fluctuate? Without lateral resolution, photoemission was unable to address this crucial question.

The situation changed in the past decade [4,7,22–25], thanks to a series of technological breakthroughs. First, the brightness (or brilliance) of the new synchrotron radiation sources [4,7] provided the necessary conditions to effectively focus the photon beam thus reducing the size of the photoelectron-emitting area. Second, the technical means to effectively focus X-rays dramatically improved. Important progress was also made in the sample positioning and lateral scanning, in the data taking and processing, and in other key domains [22–27]. As a result, the combination of photoemission spectroscopy and lateral resolution – conventionally called "photoemission spectromicroscopy" - is now an established technique [22–28].

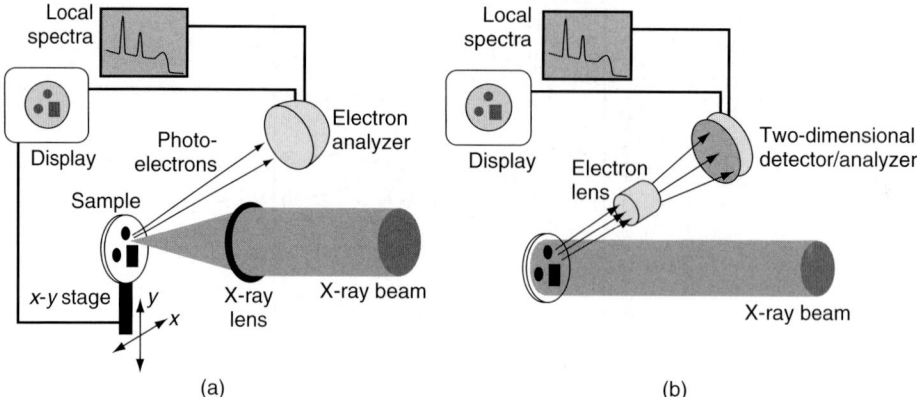

Fig. 14. The two modes of photoemission spectromicroscopy: in (a), high lateral resolution is achieved by focusing the photon beam; in (b), it is produced by an electron optics system similar to an electron microscope.

As shown in Fig. 14, lateral resolution can be achieved in photoemission with two different approaches [4,7,22–28]. In Fig. 14(a), we see the focusing–scanning approach: the X-ray beam is processed by one of the suitable focusing lenses – zone plates, Schwarzschild objectives, compound refractive lenses (CRLs), and others [4,7] – reducing its transverse size. Photoelectrons are thus emitted only from a small sample area and the corresponding photoelectron spectra yield local information.

As also shown in Fig. 14(a), the sample is mounted on a lateral (x–y) scanning stage. The lateral scanning makes it possible to obtain maps of the lateral distribution of the photoelectron intensity. Suppose for example that the x–y scanning is performed while collecting photoelectrons that originate from the atoms of a given chemical element: the resulting image in the display shows the lateral distribution of the same element. With sufficient energy resolution, this approach can even detect the lateral distribution of the element in a specific chemical configuration.

The "photon-focusing and scanning" approach was realized by a series of practical experimental systems, from the precursors MAXIMUM and SuperMAXIMUM to the more recent Elettra Spectromicroscopy beamline in Trieste [4,7,22–28]. The required brightness or brilliance is provided by synchrotron X-ray sources of the "undulator" type [7].

Figure 14(b) shows the alternate spectromicroscopy approach based on electron optics. The X-ray beam is no longer focused (or only weakly focused). The photoelectrons are processed by the equivalent of an electron microscope, with the photon-bombarded sample acting as the electron emitter. The basic instruments of this type – like the photoelectron emission microscope (PEEM) [27] – do not perform any electron energy filtering and yield images that are primarily created by the secondary electrons. More advanced instruments can now combine spectral analysis with excellent lateral resolution [4,7,25].

Figure 15 shows an early example of the applications of photoemission spectromicroscopy to semiconductor interface research [26]. The images were taken with the "photon-focusing and scanning" approach of Fig. 14(a) using the Schwarzschild-objective spectromicroscopy facility developed by F. Barbo et al. on Elettra, Trieste [26]. They show cross-sectional pictures of a sequence of GaAs p–n homojunctions. Figure 15(a) was obtained with photoelectrons corresponding to the Ga3d core-level energy in n-type GaAs and Fig. 15(b)

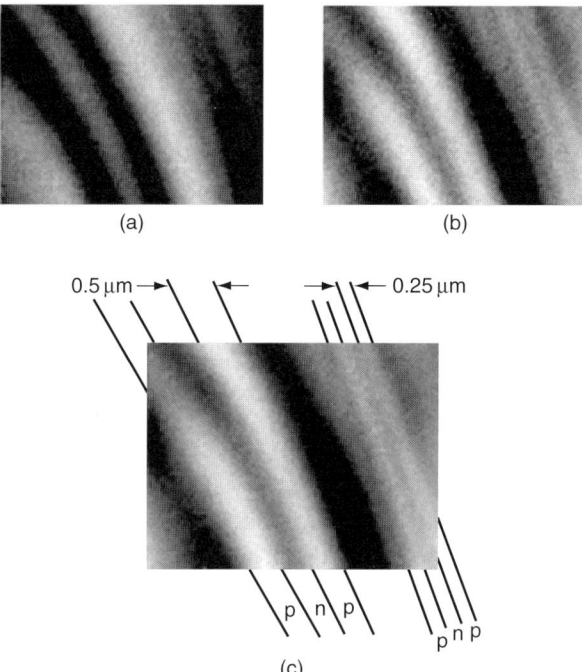

Fig. 15. Example of spectromicroscopy study of a semiconductor interface: photoemission intensity maps for the Ga3d peak in a sequence of p–n GaAs homojunctions. (a) map for the Ga3d peak in n–GaAs; (b) map for the Ga3d peak in p–GaAs; (c) digit-by-digit subtraction image of Fig. 15(b) from Fig. 15(a). Note the reversed contrast between (a) and (b), further enhanced in the subtraction image (data extracted from Ref. [26]).

for the same core level in p-type GaAs. Therefore, bright areas correspond to n-type zones in Fig. 15(a) and vice-versa in Fig. 15(b). The n–p contrast is further enhanced in Fig. 15(c), which corresponds to the digit-by-digit subtraction of Fig. 15(b) from Fig. 14(a).

5. Case studies

The previous sections established the fact that photoemission provides a powerful arsenal for studying band discontinuities and many other interesting issues concerning heterojunction interfaces. We now discuss some of the corresponding practical applications. The objective is not to present a complete review of results: we discuss significant cases with the goal to illustrate not only the concrete use of photoemission techniques but also their impact.

5.1. Empirical approaches and their limitations

One of the major consequences of the use of photoemission techniques in heterojunction research was the possibility to analyze general trends. Such techniques, in fact, can be applied to many different interfaces and therefore identify properties that span beyond the boundaries of a specific system or class of systems. On the contrary, before the advent of photoemission techniques it was difficult to test theoretical models because the experiments were typically limited to one interface or to a small group of similar interfaces.

Systematic photoemission data on heterojunction band discontinuities made it specifically possible to address general conceptual questions concerning these important parameters. Many different theoretical models were developed since the 1960s [1,2], but practically all of them can be grouped in two broad categories: (1) "linear" models that relate each discontinuity to the difference of two parameters, each one related to one of the two component semiconductors, and (2) models that attribute the discontinuities to more complex properties of the interface and of its formation process that cannot be modeled with the simple difference of two parameters. The truth probably resides in a combination of the two approaches, and both of them are important for a complete understanding of heterojunctions. We will discuss here some practical examples to better comprehend their foundations and the differences between them.

Consider for example the case of Fig. 16(a): here we assume that – for whatever reason – the distance in energy between the Fermi level and the top of the valence band is fixed at the surface of each one of the two component semiconductors, and equal to E_{mA} for semiconductor A and to E_{mB} for semiconductor B. This determines the band bending near the surface because in the bulk of each semiconductor the distance between E_F and E_v is fixed by the doping.

Assume now (Fig. 16(b)) that the situation of each surface is not changed when the interface is formed – except that thermodynamics and the continuous electrical contact require the Fermi level to be the same throughout the heterojunction. This immediately leads to the following expression for the valence-band discontinuity:

$$\Delta E_v = E_{mA} - E_{mB} \tag{3}$$

This equation is one specific example of the above general class (1) (linear models) of discontinuity models: in fact, it expresses the valence-band discontinuity as the difference between the two parameters E_{mA} and E_{mB}, each determined only by one of the two component semiconductors.

There exists a variety of theoretical models [1,2] that, although different from each other, all belong to the linear model class (1) and lead to results with the same mathematical structure as Eq. (3) – although the nature of the terms E_{mA} and E_{mB} changes from model to model. For example, the already mentioned Anderson model [20] expresses the conduction

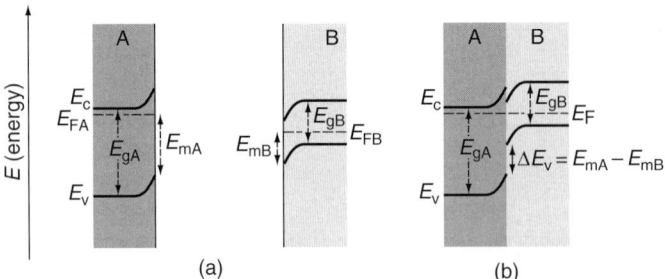

Fig. 16. Schematic explanation of one type of the "linear" models corresponding to Eq. (3). (a) Two different clean semiconductor surfaces A and B, with the band bending accommodating the difference between the bulk and surface positions of the Fermi level in the gap. (b) If we assume that the same band bending is present after the AB interface is formed (and the Fermi levels for the two semiconductors E_{FA} and E_{FB} converge to the same position E_F), then the valence-band discontinuity can be expressed by an equation with the mathematical structure of Eq. (3).

band discontinuity as the difference of the electron affinities of the two semiconductors: $\Delta E_c = \chi_A - \chi_B$; combined with Eq. (1) this gives: $\Delta E_v = |E_{gA} - E_{gB}| - (\chi_A - \chi_B)$ and therefore, assuming $E_{gA} > E_{gB}$:

$$\Delta E_v = (E_{gA} - \chi_A) - (E_{gB} - \chi_B), \tag{4}$$

an expression with the same mathematical structure as Eq. (3), i.e. the difference of two terms each related to only one of the two semiconductors.

Another interesting example of linear model class (1) is the midgap-energy-point approach derived independently by Tersoff [29] and by Flores and Tejedor [30]. In both cases, the interface formation process is assumed to change significantly the local electronic structure and therefore the "pinning" position of the Fermi level inside each gap. However, in first approximation the final distance between the valence-band edge and the Fermi level is fixed for each semiconductor. This gives once again the situation of Fig. 16(b) starting from a more complex interface formation process – and leads to an expression like Eq. (3).

Katnani and Margaritondo [1,2,31] addressed the general issue of the validity of "linear" models by adopting an empirical approach. Suppose that a rule like that of Eq. (3) is valid (without making any assumption about the reasons for which this is true): then, it is possible to empirically derive the values of the "interface parameters" E_{mA} and E_{mB} from those of measured valence-band discontinuities.

Consider in fact three different semiconductors X, Y, and Z and their interfaces XZ and YZ: Eq. (3) gives, with obvious meaning of the terms:

$$\begin{aligned} \Delta E_{vXZ} &= E_{mX} - E_{mZ} \\ \Delta E_{vYZ} &= E_{mY} - E_{mZ} \end{aligned} \tag{5}$$

and therefore:

$$\begin{aligned} E_{mX} &= \Delta E_{vXZ} + E_{mZ} \\ E_{mY} &= \Delta E_{vYZ} + E_{mZ} \end{aligned} \tag{6}$$

Equations like Eq. (6) can be extended to any number of semiconductors, with an obvious implication: by measuring the discontinuities of a set of semiconductors with respect to a reference semiconductor Z, one practically measures the interface parameter of the semiconductor in the group – except for a constant equal to the interface parameter E_{mZ} of the semiconductor. Because the interface parameters are always used in differences [Eq. (3)], the constant is irrelevant.

This approach was implemented by Katnani and Margaritondo [1,2,31] using as the reference semiconductor first Si and then Ge. The corresponding two sets of data were then averaged to obtain the empirical interface parameters of Table 1; the constant term is conventionally fixed here by assuming that the interface parameter for Ge is zero.

The values listed in Table 1 can be used in several different ways. The most obvious is to predict the valence-band discontinuity of any of the interfaces that involve pairs of materials present in the table. Consider for example the pair AlAs–GaAs. The difference of the corresponding terms of Table 1 would predict this heterojunction $\Delta E_v = 0.43$ eV (with the valence-band edge of GaAs above that of AlAs). The experimental values reported by Waldrop et al. [32] range between 0.42 and 0.55 eV depending on the interface orientation and growth sequence; the agreement is reasonable considering the accuracy limitations of the experimental values and of the theoretical approach, which are discussed later.

Table 1

Empirical interface parameters (in eV) derived by Katnani and Margaritondo [1,2,31] from systematic measurements of valence-band discontinuities of different semiconductors with respect to Si and Ge. In the first approximation [i.e., if Eq. (3) is valid], the valence-band discontinuity between two semiconductors can be evaluated by taking the difference of the corresponding terms from this table

Material	Interface Parameter	Material	Interface Parameter
Ge	0 (reference)	CdS	−1.74
Si	−0.16	CdSe	−1.33
α-Sn	0.22	CdTe	−0.88
AlAs	−0.78	ZnSe	−1.40
AlSb	−0.61	ZnTe	−1.00
GaAs	−0.35	PbTe	−0.35
GaP	−0.89	HgTe	−0.75
GaSb	−0.21	CuBr	−0.87
InAs	−0.28	GaSe	−0.95
InP	−0.69	$CuInSe_2$	−0.33
InSb	−0.09	$CuGaSe_2$	−0.62
		$ZnSnP_2$	−0.48

The second use of Table 1 is to assess the general limitations of "linear" models and of equations with the structure of Eq. (3). This equation implies two important properties of the discontinuity: reversibility and transitivity. Reversibility means that if the XY interface is fabricated by depositing the semiconductor overlayer Y on the semiconductor X, and the YX interface is fabricated with the opposite deposition sequence, then the two valence-band discontinuities are equal in magnitude and opposite in sign (the opposite sign means that the same semiconductor has the highest valence-band edge for both interfaces).

Transitivity can be understood by taking again three semiconductors X, Y, and Z. With obvious meaning of the terms, Eq. (3) implies that:

$$\Delta E_{vXY} + \Delta E_{vYZ} + \Delta E_{vZX} = 0 \qquad (7)$$

and therefore, for example:

$$\Delta E_{vXY} = -(\Delta E_{vYZ} + \Delta E_{vZX}) \qquad (8)$$

Similar transitivity rules can of course be found for groups of four or more semiconductors.

The reversibility and transitivity rules were extensively tested with experiments on a variety of interfaces [1,2,31]. Together with other tests, the results agree in demonstrating that no "linear" model can exceed an *average* accuracy of 0.1–0.2 eV in predicting heterojunction band discontinuity.

Note the subtle points underlying this conclusion: (1) the accuracy limit of 0.1–0.2 eV applies to all "linear" model as a class, but it does not guarantee that a given model will reach this limit and does not support the validity of any given model against the others; (2) this being an average limit, in specific cases the accuracy can be better or worse.

The consequences of the above conclusion are quite important. An accuracy of 0.1–0.2 eV is not sufficient for technological applications. Therefore, no "linear" model, no matter how sophisticated, can cope with the needs of heterojunction technology. There are factors in the physics of heterojunction interfaces that cannot be accounted for by linear models.

On the contrary, the implication of all linear models is that the band discontinuities are fixed once the two interface components are selected. If this was absolutely true, it would be impossible to modulate the discontinuities by suitable interface manipulations. Therefore, the accuracy limit for "linear" model opens the door to the exploitation of effects beyond their scope to the modification and control of such parameters. This is a pillar of the so-called "bandgap engineering" proposed by F. Capasso [1].

5.2. Controlling the heterojunction interface parameters

The identification of practical methods to modify heterojunction band discontinuities and other interface parameters has been, for a long time, a rather active research subfield [1,2]. Within the conceptual framework discussed in the previous section, several approaches were tested to change the discontinuities of a given heterojunction interface AB between two semiconductors. These include [32], in particular, manipulations of the crystalline orientation of the interface and the growth sequence ("A over B" or "B over A").

For example, Ref. [32] reported that in the case of the GaAs–AlAs interface the valence-band discontinuity changes with the growth sequence by 0.10–0.13 eV. Furthermore, differences of 0.06–0.09 eV were observed between the (100) and (110) interface orientations.

Such results are conceptually very interesting but they cannot be easily utilized to control interface parameters. A different, interesting approach consists in adding a very thin intralayer of a third material between the two sides of the interface. This approach produced significant modifications in a controlled way for a variety of interfaces [1,2,33].

Figure 17 shows [34] one of the most spectacular cases: thin intralayers of cesium or hydrogen inserted at the interface between Si and SiO_2 produce clearly visible changes in the valence-band discontinuity, in opposite directions. Figure 18 shows another example [35] with more limited but still clearly visible intralayer-induced changes in the valence-band discontinuity. The heterojunction in this case is CdS–Ge and the intralayer is made of aluminum.

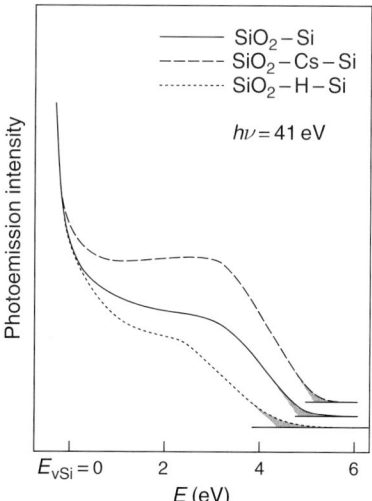

Fig. 17. Very large modifications in the valence-band discontinuity of the Si–SiO_2 interfaces are induced by thin cesium or hydrogen intralayers. The photoemission spectra were derived from Ref. [34].

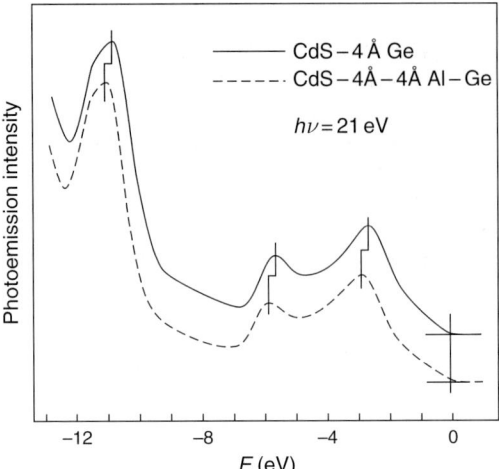

Fig. 18. Valence-band discontinuity changes caused by a thin Al intralayer at the CdS–Ge interface. The changes are revealed by the fact that the uppermost edges of the two spectra (the Ge valence-band edge) coincide with each other whereas the rest of the spectral features (related to CdS) are rigidly shifted (the spectra were derived from Ref. [35]).

The potential applications of intralayer-induced interface modifications stimulated remarkable theoretical efforts to understand and predict them. Such theories targeted in particular the interface dipoles expected to produce an electrostatic potential jump across the interface and therefore to modify the band positions. This kind of modeling also predicted [36] the possibility to induce discontinuities at semiconductor homojunctions by inserting a suitable dipole-creating intralayer between the two sides.

Figure 19 shows [36] an experimental verification of this intriguing possibility. By analyzing the Ge3d core-level lineshapes for Ge–Ge homojunctions with thin double intralayers made of Ga and As, we can see that the band structure on one side is shifted with respect to that on the other side – see the top-right panel. When the intralayer sequence is changed from As–Ga to Ga–As, the shift is consistently reversed. Thus, the double intralayer and its interface dipole create discontinuities at homojunction interfaces, where normally they do not exist.

5.3. Lateral fluctuations of the interface parameters

This is one of the most crucial unresolved issues in semiconductor interface science [1]. The background hypothesis of almost all interface models is that the interface parameters – such as the Schottky barriers and the heterojunction band discontinuities – are global properties with the same value for all parts of the interface. The implicit assumption is that even if the parameters do fluctuate from point to point the models can use average values.

These assumptions can be quite far from reality. Specifically, if interface barriers change from point to point, the transport properties are not likely to be determined by an average value but to be dominated by the weakest barrier values. Therefore, the possibility of interface parameter fluctuations constitutes a very important albeit often neglected issue.

Theoretical considerations do not rule out such a possibility. In fact, the effects of fluctuations can only be smoothed on a scale much smaller than the Debye length. It should

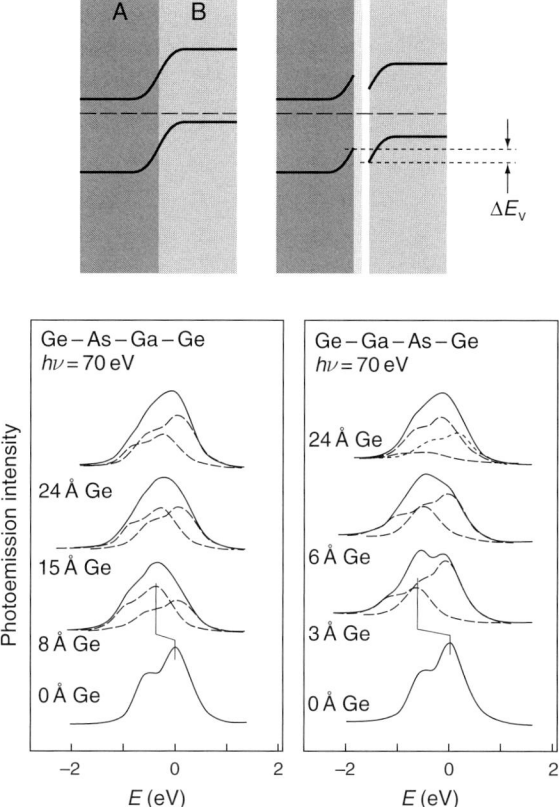

Fig. 19. Top, left: normally, a semiconductor n–p homojunction has no band discontinuities. Top, right: theory [36], however, predicts that discontinuities can be induced by thin double intralayers. Bottom: a detailed analysis of the Ge3d photoemission peaks verifies that discontinuities are indeed created when a Ga–As intralayer is inserted between the two sides of a Ge–Ge homojunction. The direction of the discontinuities changes as expected with the Ga and As deposition sequence (spectra derived from Ref. [36]).

be noted that the Debye length relevant for the interface does not necessarily coincide with the bulk value determined by the bulk doping – and in certain cases can be smaller.

The experimental evidence reveals [37–40] that fluctuations do exist and can be quite significant; Fig. 20 shows a rather straightforward case: the interface between GaSe and Ge [37]. The micrograph (obtained with a photoelectron spectromicroscope) reveals inhomogeneities in the plane of the interface. The spectra show core-level peaks from the Ge overlayer (Ge3d) and for the substrate, taken in two different points of the interface.

The Ge3d peak position is essentially the same for the two points. On the contrary, the two substrate peaks exhibit a rigid shift between them. This immediately implies that the relative positions of the substrate and overlayer band structure – including the valence-band edges – change between the two points – and that the valence-band discontinuity changes by the same amount.

Such a discovery was rather surprising in the case of GaSe–Ge. Due to the chemically unreactive character of the GaSe surface, interfaces involving GaSe were considered for many years [41,42] as prototype cases of "Schottky-like" semiconductor interfaces. In fact, the original Schottky model [19] explains the interface properties and the transport rectification effect without

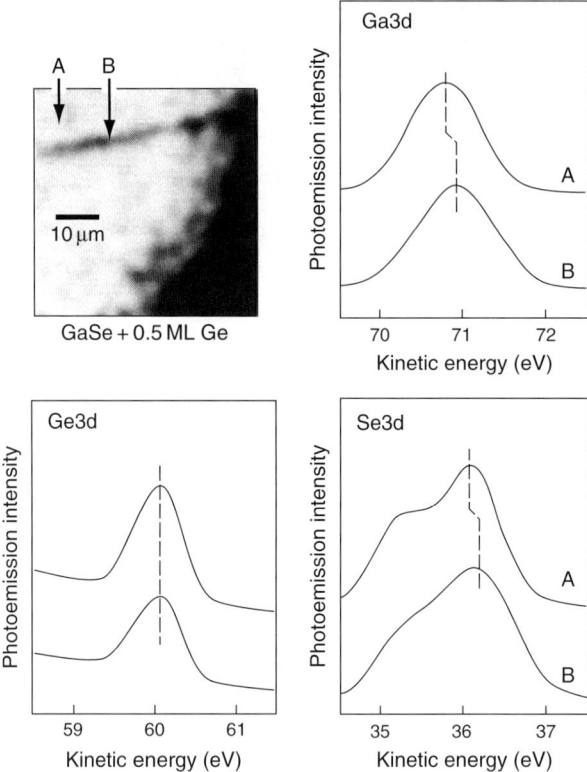

Fig. 20. Top, left: photoemission micrographs reveal lateral inhomogeneities at the GaSe–Ge heterojunction. The spectra in the remaining parts of the figure show that the valence-band discontinuity is also inhomogeneous (data derived from Ref. [37]).

any role of local chemical reactions. The experimental data [41,42] prior to photoemission spectromicroscopy with high lateral resolution seemed to corroborate the idea that GaSe interfaces (and III–VI interfaces in general) are formed without local chemical interactions.

Results like those of Fig. 20 destroyed this notion. A local chemical analysis by photoemission spectromicroscopy [37–40] was able to link the interface parameter fluctuations to the occurrence of local chemical reactions on a microscopic scale. As a consequence, no case is known today of a semiconductor interface with properties close to the Schottky model [19].

This conclusion – and in general the existence of lateral fluctuations of the interface parameters revealed by experiments like those of Fig. 20 – demonstrates that real semiconductor interfaces are quite different from the standard "textbook" pictures. Unfortunately, a realistic textbook description has not yet been developed – and this causes a regrettable discrepancy between what is taught at the elementary level and the real situation.

5.4. How this field is developing – internal photoemission

Photoemission studies of semiconductor heterojunctions are a mature field: after the early evolution, the corresponding techniques are now well established and routinely used to analyze new systems – see for example Ref. [43]. The limitations of other approaches and the rather straightforward interpretation of the photoemission measurements of band discontinuities were important factors in this success.

In parallel, new techniques were developed with different degrees of success. We would like to discuss here the one that was crowned by full success becoming part of the standard arsenal for heterojunction studies: internal photoemission [44–46].

The approach is schematically explained by the top part of Fig. 21. A photon beam with tunable photon energy $h\nu$ is sent into a heterojunction from the side of the larger-gap semiconductor. An external circuit can detect the induced photocurrent. Figure 21 illustrates three kinds of optical transitions that can be stimulated by photon absorption: α is a transition across the gap of the smallest-gap semiconductor, whereas in β a larger photon energy brings the electron above the conduction-band discontinuity. Finally, γ is a transition across the gap of the largest-gap semiconductor.

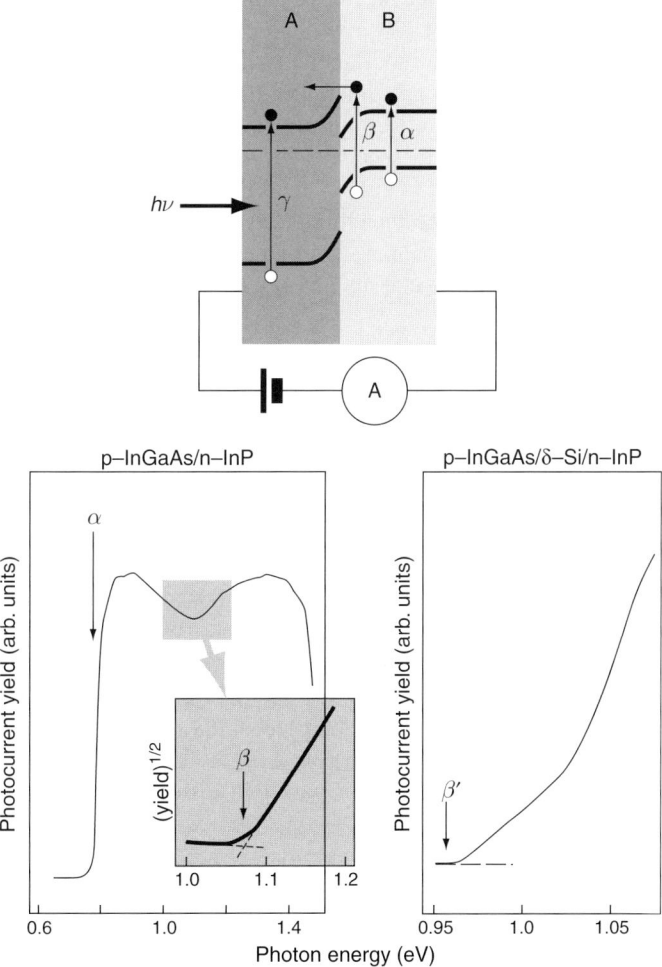

Fig. 21. Use of internal photoemission to measure heterojunction conduction band discontinuities. Top: scheme of the experiment with the different optical transitions that create threshold in the plots of the photocurrent (detected by the external circuit) vs $h\nu$ (discussion in the text). Bottom: two practical applications of the technique (data derived from Ref. [44]).

Each type of transition corresponds to a threshold in the photocurrent vs $h\nu$ plots. Specifically, threshold α reveals the small-gap width and threshold γ the large-gap width, whereas threshold β corresponds to the sum of the small gap plus the conduction band discontinuity. It is clear, therefore, that from an analysis of the photocurrent photon-energy thresholds one can derive ΔE_c.

The lower part of Fig. 21 shows a practical application of this approach [44]. On the lower-left plot, we see a photocurrent spectrum for a p–InGaAs/n–InP heterojunction with the two thresholds α and β. The inset shows a Fowler plot of the β-threshold based on the theoretical functional dependence of the yield on the square of the photon energy above threshold. From the α and β thresholds, the estimated value of ΔE_c is 0.29 eV.

On the bottom-right plot of Fig. 21, we see the results for a p–InGaAs/n–InP heterojunction with an inserted layer (δ-doping) of silicon. Similar to the cases discussed in Section 5.2, the Si doping modifies the band discontinuities. This is put in evidence by the presence of a new threshold β'; the shift with respect to the previous β-threshold reveals a ΔE_c change of 0.11 eV.

This rather powerful method is not, however, entirely immune from problems. As we have seen, the key threshold β (or β') does not directly give the valence-band discontinuity but a combination of ΔE_c with the smallest gap. Interface effects can locally modify this gap so that the measured "bulk" value may not reflect the gap value that contributes to the β-threshold.

It would be desirable, therefore, to spectrally extend the measurements to lower photon energies and detect the threshold that is directly related to ΔE_c – and is caused by transitions starting from states at the bottom of the "notch" created in the conduction band by the discontinuity. This is not easy with conventional infrared source. However, direct measurements of this kind become possible by using a high-intensity tunable infrared free-electron laser (FEL) [6,45].

6. Summary and conclusions

Experimental techniques related to the photoelectric effect constitute at present a powerful arsenal for the study of semiconductor heterojunctions. Reliable measurements of band discontinuities by photoemission and internal photoemission are now routine and can be systematically applied to a wide variety of systems.

Photoemission technique, however, can also explore more advanced issues that must still be entirely clarified from the theoretical point of view. Among those, the most relevant are the microscopic control of interface parameters (including heterojunction discontinuities) and the very critical issue of the lateral fluctuations of the same parameters. We believe that in these problems resides the frontier of this exciting domain at the interface between science and technology.

Additional refinements of the existing photoemission techniques are quite likely. We would like to mention, in particular, the recent breakthrough in time-resolved photoemission that opened the way to studies of dynamic properties on the femtosecond range [47]. The impact of new photon sources such as the X-ray FELs under development must still be assessed and it could be substantial [6].

References

[1] F. Capasso and G. Margaritondo, Heterojunction Band Discontinuities: Physics and Device Applications, North-Holland, Amsterdam, 1987; G. Margaritondo, Electronic Structure of Semiconductor Heterojunctions, Jaca, Milan, and Kluwer, North-Holland, Dordrecht, 1988.
[2] C. Lamberti, Surf. Sci. Rep. 53 (2004) 1.

[3] P. Perfetti, D. Denley, K. A. Mills, and D. A. Shirley, Appl. Phys. Lett. 33 (1978) 66.
[4] G. Margaritondo, Introduction to Synchrotron Radiation, Oxford, New York, 1988.
[5] J. H. Weaver and G. Margaritondo, Science 206 (1979) 151.
[6] G. Margaritondo, Elements of Synchrotron Light for Biology, Chemistry, and Medical Research, Oxford, New York, 2002.
[7] W. Schattke and M. A. Van Hove, Solid-State Photoemission and Related Methods: Theory and Experiment, Wiley, Berlin, 2003.
[8] A. D. Katnani, N. G. Stoffel, R. R. Daniels et al., J. Vac. Sci. Technol. A 1 (1983) 692.
[9] W. E. Spicer, Phys. Rev. 112 (1958) 114.
[10] G. Margaritondo, A. D. Katnani, N. G. Stoffel et al., Sol. State Commun. 43 (1982) 163.
[11] A. D. Katnani, R. R. Daniels, T.-X. Zhao, and G. Margaritondo, J. Vac. Sci. Technol. 20 (1982) 662.
[12] J. R. Waldrop and R. W. Grant, Appl. Phys. Lett. 68 (1996) 2879.
[13] E. A. Kraut, R. W. Grant, J. R. Waldrop, and S. P. Kowalczyk, Phys. Rev. Lett. 44 (1980) 1620.
[14] P. K. Larsen, G. Margaritondo, M. Schluter et al., Phys. Lett. A 58 (1976) 623.
[15] M. M. Traum and N. V. Smith, Phys. Rev. B 9 (1974) 1353.
[16] N. G. Stoffel, Phys. Rev. B 28 (1983) 3306.
[17] T.-C. Chiang, J. A. Knapp, M. Aono, and D. E. Eastman, Phys. Rev. B 21 (1980) 3513.
[18] D. Niles and H. Hochst, Phys. Rev. B 39 (1989) 7769.
[19] W. Schottky, Z. Phys. 41 (1940) 570.
[20] R. L. Anderson, Solid-State Electronics 5 (1962) 341.
[21] A. D. Katnani, R. Daniels, T.-X. Zhao, and G. Margaritondo, J. Vac. Sci. Technol. 20 (1982) 662.
[22] G. Margaritondo and F. Cerrina, Nucl. Instr. Meth. A 291 (1990) 26.
[23] G. De Stasio, G. F. Lorusso, T. Droubay et al., Rev. Sci. Instrum. 67 (1996) 737.
[24] G. Margaritondo, Prog. Surf. Sci. 56 (1997) 311.
[25] G. De Stasio, B. Gilbert, L. Perfetti et al., Rev. Sci. Instrum. 69 (1998) 3106.
[26] F. Barbo, M. Bertolo, A. Bianco et al., Rev. Sci. Instrum. 71 (2000) 5.
[27] B. P. Tonner and G. R. Harp, Rev. Sci. Instrum. 63 (1992) 564.
[28] M. Marsi, L. Casalis, L. Gregoratti et al., J. Electron. Spectrosc. Relat. Phenom. 84 (1997) 73.
[29] J. Tersoff, Phys. Rev. B 30 (1984) 4874.
[30] F. Flores and C. Tejedor, J. Phys. C 12 (1979) 731.
[31] A. D. Katnani and G. Margaritondo, Phys. Rev. B 28 (1983) 1984.
[32] J. R. Waldrop, R. W. Grant, and E. A. Kraut, J. Vac. Sci. Technol. B 5 (1987) 1209.
[33] D. W. Niles, G. Margaritondo, E. Colavita et al., J. Vac. Sci. Technol. A 4 (1986) 962.
[34] P. Perfetti, C. Quaresima, C. Coluzza et al., Phys. Rev. Lett. 57 (1986) 2065.
[35] D. W. Niles, G. Margaritondo, P. Perfetti et al., Appl. Phys. Lett. 47 (1985) 1092.
[36] M. Marsi, S. La Rosa, Y. Hwu et al., J. Appl. Phys. 71 (1992) 2048.
[37] F. Gozzo, H. Berger, I. R. Collins et al., Phys. Rev. B 51 (1995) 5024.
[38] M. Zacchigna, L. Sirigu, J. Almeida, H. Berger et al., Appl. Phys. Lett. 73 (1998) 1859.
[39] F. Gozzo, M. Marsi, H. Berger et al., Phys. Rev. B 48 (1993) 17163.
[40] M. Zacchigna, H. Berger, L. Sirigu et al., J. Electron Spectr. 101–103 (1999) 671.
[41] R. R. Daniels, G. Margaritondo, C. Quaresima et al., Sol. State Commun. 51 (1984) 495.
[42] R. R. Daniels, G. Margaritondo, P. Perfetti et al., J. Vac. Sci. Technol. A 3 (1985) 979.
[43] J. J. Chen, B. P. Gila, M. Hlad et al., Appl. Phys. Lett. 88 (2006) 142225.
[44] J. Almeida, T. dell'Orto, C. Coluzza et al., J. Appl. Phys. 78 (1995) 3258.
[45] J. T. McKinley, R. G. Albridge, A. V. Barnes et al., J. Vac. Sci. Technol. B 11 (1993) 1614.
[46] T. dell'Orto, J. Almeida, C. Coluzza et al., Appl. Phys. Lett. 64 (1994) 2111.
[47] L. Perfetti, P. A. Loukakos, M. Lisowski et al., Phys. Rev. Lett. 97 (2006) 067402.

13

ESR of interfaces and nanolayers in semiconductor heterostructures

Andre Stesmans and Valery V. Afanas'ev

Semiconductor Physics Laboratory, Department of Physics and Astronomy, University of Leuven, Belgium

Abstract This work addresses the application of conventional electron-spin resonance (ESR) spectrometry, as routinely operated in the 8–35 GHz microwave range, to semiconductor heterostructures (SHs), in particular semiconductor/insulator entities with focused interest on interfaces and thin dielectric (inter) layers. This predominantly entails probing of point defects where, by nature, the ESR method is only directly applicable to the subclass of paramagnetic defects. Of particular interest is the family of intrinsic point defects, inherently or inevitably occurring at interfaces and in interlayers, which play a crucial role in device-integrated semiconductor/insulator entities, often in the negative sense through operating as traps detrimentally affecting transport properties. The study implies various aspects: Obviously, a major one concerns characterization, desirably atomic identification, of occurring imperfections for which ESR appears the technique of choice, in fact eponymous for point defects. But once characterized, to the benefit, the defects offer great potential when exploited as true atomic probes of utmost sensitivity to their atomic environment. Through a selective compilation of representative experimental results, the usefulness of ESR is illustrated in assessing various aspects as diverse as atomic identification of defects, nature of interfaces, local strain, interface stability, structural/chemical evolution of interfaces and interlayers as a function of varying conditions (e.g., thermal treatment, stress), chemical activity of defects, and their anneal kinetics. While not a routine method in terms of experimental accessibility, sensitivity, and type of probing, i.e., magnetic rather than electric, ESR will continue to offer great potential in the study of basic SHs, in particular when turned into a multiple resonance spectroscopy through combination with other methods.

Keywords semiconductor/heterostructures, interfaces, point defects, electron-spin resonance, dielectric layers, thermal silica, hydrogen

1. Introduction

There is little doubt that with the invention of the (Ge) transistor in 1947 by J. Bardeen [1] and W. Brattain, the interest in the study of semiconductor heterostructures (SHs) got a main impetus from the technological side. The SHs form the basis of semiconductor-based microelectronic devices. The ensuing research has been that successful with such an unprecedented amount of technological realizations that it has taken us to the current semiconductor-based information age, perhaps better designated as the *Si age* – for Si, rather than the initially groundbreaking Ge, has been and still is technologically the most successful semiconductor. Except for very specific (niche) applications such as high speed, high frequency, and high temperature (T), Si has met little competition. This success story of the Si age owes much to the plain fact that Si, as a gift from nature, has a native insulator, SiO_2, of superb physical and electrical quality, which has culminated in this epic in an all-dominant role of the basic Si/SiO_2 entity – without doubt the most intensely and widely studied SH in history. More precisely, in practice, it concerns the thermal c-Si/ a-SiO_2 unit, where an amorphous (a) SiO_2 film is grown by thermal oxidation on a device grade crystalline (c) Si substrate of utmost purity and quality.

Since its inception, the microelectronic industry has experienced about four decades of explosive growth generally based on two pillars: (a) invention [2] of the planar integrated circuit (IC) later combined with the complementary metal oxide silicon (CMOS) planar transistor concept; (b) dimensional scaling (shrinking) approach of solid-state devices [3]. The latter continued scaling, rather than diversification, has appeared and still is the basic motor in this triumphal story. Shrinking transistors not only enables packing more devices per unit area, but also shortens the distance between the source and drain, i.e., the gate length, which can improve the switching speed. The continued progress through unabated scaling has witnessed an exponential decrease in minimum feature size in a transistor with time, resulting in a doubling of the number of active components in an IC about every 2–3 years (denoted as one component of Moore's law [4], an empirical (economical) observation rather than a physical law). And industry is taking great pains in combining efforts, however complicated and costly, to stay on track.

In this saga where the continuous development of devices floats amid the turmoil of the downscaling race, the basic guide has persistently been the realization of top grade electrical performance, on the work floor meaning undaunted elimination or suppression of ever emerging electrical deficiencies. Obviously, the electrical performance of a CMOS transistor (e.g., involving items such as carrier mobility, switching time, charge trapping, etc.) is, per definition, primarily an issue addressed using electrical techniques, such as capacitance–voltage (C–V) and current–voltage (I–V) analysis. Optimization implies continuous control of electrical deficiencies, showing up under different disguises in ever more complexity as crucial dimensions are scaled to the sub 100 nm range. Currently, the 65 nm logic technology node, featuring transistors of ~30 nm gate length, evolves in production [5].

In this strive for performance, *point defects* play a key role electrically. As amply evidenced over time for the Si/SiO_2 entity, point defects were demonstrated to be at the origin, or at least involved with, virtually all detrimental electrical aspects encountered [6], such as adverse interface traps [7], oxide-fixed charge [8], irradiation-induced degradation [9,10], and stress-induced leakage current [11], resulting in degradation phenomena such as thermally induced oxide breakdown [12,13], mobility degradation [14], threshold (V_t) shift, and negative bias temperature instability (NBTI) [15]. Perhaps the most crucial issue here concerns the quality of the c-Si/insulator interface, particularly as regards the presence of (inherent) electrically active defects at the origin of harmful charge traps [7]. Mercilessly, performance dictates the formation of these should be ad hoc prevented, or when unavoidably introduced, post hoc efficiently inactivated electrically, e.g., through pairing off dangling orbits by binding to hydrogen. It constitutes a permanent matter of concern on the route of progress as the defect situation changes unavoidably with any modification of the devising process, which alterations may be as diverse as introduction of new processing techniques (lithography, thermal steps), changing device architecture to introduction of new materials.

Obviously, when occurring, knowing the nature of traps would be of invaluable help. But superb as they are, the routinely invoked electrical methods inherently fall short in assessing the physico-chemical nature of traps, leaving the longed-for ultimate characterization unsatisfied. There is a lack of atomic discriminative power. It is here, when applicable, where the electron-spin resonance (ESR) technique (or enhanced ESR-based methods) can come to the rescue. Also denoted as electron paramagnetic resonance (EPR), it is the technique of choice for atomic identification and quantification of point defects – implicitly the tool eponymous for such defects with atomic level physico-chemical sensitivity. As yet, there seems to be no other experimental technique to rival ESR on this respect, although in fairness it should be added that it is not the only technique with atomic level discriminative power. (In varying detail, there are some other methods that may provide information on defect structure and chemical composition. Notably, we refer to high-resolution infrared spectroscopy and UV-visible absorption and photo-luminescence spectroscopy, probing the vibronic modes and electron levels of defects, respectively. These optical methods can yield the symmetry of a defect, information on involved impurities, its electronic energy levels, and the vibrational modes. However, provided it can be resolved, conclusive information is generally inferred from analysis of the hyperfine (hf) interaction structure in ESR observations. Unlike ESR though, the above spectroscopies do not require unpaired spins, which would render them of much wider potential use.)

So, one would hope the ESR technique to be applied as a standard when dealing with semiconductor/insulator structures, yet it does not figure as a key electrical diagnostic tool in the device world. The application faces obstacles: A first one is that the defects envisioned need to be in a spin-active (paramagnetic) state (suitable charge state), which appears often not the case. This, however, can in many cases be overcome by suitable sample treatment inducing charge transfer. Second, the sensitivity is limited: Current top performance continuous wave (CW) ESR spectrometers may detect $\sim 1 \times 10^9$ centers (spin $S = 1/2$) of 1 G line width at low T within acceptable averaging time. Many resonance signals, however, appear much broadened, which strongly impairs detection, thus generally rendering the conventional ESR technique less sensitive than typical state-of-the-art electrical observations, such as C–V or I–V. Third, there is the nonconventional probing approach, i.e., sensing the *magnetic moment* instead of the *charge* of the electrically active species, making the technique less popular in the electrical device world.

Concisely stated, when ESR is applied to SHs in technology, the main goal is *to achieve, in close combination with electrical measurements, (atomic) identification of traps*, i.e., point defects. In passing, the notion of point defects of course does not mean that these are null in volume, but rather encompass a limited region (say, 5–20 atoms) of irregularity within the solid-state matrix (be it crystalline, polycrystalline, or amorphous). Upon observation, the aim is to correlate the ESR spectrum with the geometrical and chemical properties of the tentatively conceived site giving rise to it. In combination with sophisticated quantum mechanical modeling, the detailed local structure of magnetic point defects may ultimately be attained.

The physical probing principle of ESR can be stated in simple terms: As to point defects, the ESR technique generally senses the total magnetic moment (spin) of unpaired electrons residing in atomic or molecular orbitals at these defects via a (weak) electromagnetic excitation (photon absorption). Through the unique relationship of the magnetic moment with the angular momentum, in turn serving as orbital quantizing quantity, the local structure of a defect site may thus be probed through the orbital distribution. Among experimental techniques, ESR may thus appear as the one that may offer the most detailed information on defect structure as it probes the interaction of the unpaired electron wave function with its surrounding. The characterization – desirably identification – is attained through assessing the various parameters (often in bountiful detail) provided by the ESR spectrum including g value (resonance field), symmetry, g matrix (symmetry), hf structure (interaction with spin-bearing isotopes), and fine structure (inter spin interaction). Some more details are given in Section 2. When resolved, most powerful though appears the observed hf structure, which in combination with quantum mechanical interpretation and modeling provides the most reliable atomic identification of a defect [16,17]. Other parameters fall short on this matter. However, over the recent years, it has surged that modeling, mostly based on density functional theory, applied to extended model clusters of atoms has progressed to such a level that it becomes within range of calculating g matrices of projected atomic defect models to sufficient reliable accuracy to stand comparison with experimental data [18]. So, in principle, it would enable defect identification to a large extent solely on the basis of this quantity. This field is in fast progress [19].

The ESR technique has been quite successfully applied to the Si/SiO_2-based entities (vide infra). The considerable arsenal of acquired knowledge serves as backdrop for exploration of any other semiconductor/insulator structure. But the usefulness and applicability of ESR extends far beyond, carried by the basic universality of the method laid down in sensing unpaired electron magnetic moments in whatever environment. Accordingly, the aim of this chapter is to provide, through compilation of experimental results, some synthesizing flavor of the scientifically and technologically useful application of ESR in research on SHs. However, trying to reach completeness in overviewing, if only for one sub-aspect of the matter dealt with, is considered beyond the scope of the current chapter, so, it will not be attempted. Instead, we will choose the "illustrative" approach, aiming to highlight the various powerful attributes of the ESR technique through presenting of some pertinent examples, hopefully excelling in relevance and significance for the rich field of the SHs, thus advocating the usefulness and potential of ESR in the field of SHs.

To put things correctly from the onset, we should clearly state that this chapter will only deal with the application of the conventional CW ESR method usually driven in the adiabatic mode, with first or second derivative absorption detection and using spectrometers operating in the $\sim$8–40 GHz range (encompassing the X, K (Ku, Ka), and Q microwave bands). It figures as the classic among the ESR techniques, as distinct from the more elaborate and focused techniques, such as electron nuclear double resonance (ENDOR) or pulsed ESR [20–22], often in combination with other spectroscopic techniques using other detection schemes (e.g., optically detected magnetic resonance [23]). By no means does this restriction entail any slightest lack of appreciation for these superb methods, indispensable when conventional ESR just fails to make progress and beautifully enriching and extending the ESR field. On the contrary, these techniques deserve all attention. Besides lack of space, the main reason for the current limitation is that these advanced techniques have been most competently addressed elsewhere [22,23]. Nevertheless, in doing so, it may enable us to highlight for the field of SHs some well-appreciated advantages of the ESR mother technique, such as versatility, universality, straightforwardness in interpretation, and reliability.

2. ESR technique
2.1. Principles and inferred information

CW ESR experiments generally consist of measuring the (derivative) absorption of fixed-frequency (v) microwaves (of power P_μ) directed into a high-quality resonating cell (usually, a metallic *cavity* of eigenfrequency v_c) containing the sample with the system of magnetic moments to be studied as a function of the magnitude B of an externally applied magnetic induction field $\bar{B}$. Usually, to improve the signal-to-noise ratio, a small-amplitude sinusoidal modulation is superimposed on $\bar{B}$ to enable phase-sensitive (lock-in) detection so that generally the first-derivative of the absorption signal (dP_μ/dB) is recorded. In a *nondestructive* way, ESR probes an interaction with magnetic moments, so only unpaired electrons ("spins" in usual jargon) contribute. The magnetic-field values ("resonance fields") of the observed signals (resonances) are theoretically described in terms of the so-called *effective spin* Hamiltonian, which contains all of the pertinent interactions of the unpaired spin with its environment. The theory and practice of ESR has been expounded into great detail in numerous textbooks [16,20,22–24]. In the following sections, we will briefly only consider those ESR elements most commonly encountered in semiconductor ESR research.

2.1.1. Basics of theory

In the classical point of view, a magnetic moment $\bar{\mu}$ placed in a magnetic field $\bar{B}$ generally precesses around the direction of the field at natural frequency v_L, called Larmor frequency. This precession can be driven in phase by an AC electromagnetic field of frequency v (injected irradiation, usually in the microwave range for electron spins) when $v = v_L$, at which instance microwave energy (photon) is resonantly absorbed by the precessing spin. This is the principle of resonant absorption of energy by a (magnetic) spin, which can be either an electron or nuclear spin, leading to the ESR and nuclear magnetic resonance (NMR) spectroscopies, respectively. Translating this into the language of quantum mechanics, it basically boils down to the detection of direct transitions between (adjacent) Zeeman levels, which result from splitting of spin-degenerate energy levels by an externally applied field $\bar{B}$, called the Zeeman effect. We now focus on ESR.

Assuming that the local magnetic field $\bar{B}_{loc}$ sensed by the electron spin equals the laboratory applied field $\bar{B}$, and neglecting any electron orbital contributions, the electron Zeeman levels are given by

$$E_{ez} = -\bar{\mu}_e \cdot \bar{B} = g\beta_e \bar{S} \cdot \bar{B} = -\gamma \hbar \bar{S} \cdot \bar{B} \tag{1}$$

where $\bar{\mu}_e = -g\beta_e \bar{S}$ is the magnetic moment of the electron. Here, g is the *spectroscopic splitting factor*, a dimensionless number often simply referred to as the electron g *value*; for a free electron it is isotropic with $g = g_e = 2.002319$ (by convention, g is positive for electrons, while the similar quantity g_N for nuclei takes the algebraic sign of the gyromagnetic ratio of the nucleus concerned); $\bar{S}$ is the dimensionless electron-spin angular momentum operator in units of $\hbar = h/2\pi$, and $\beta_e = |e|h/2m_e$ the Bohr magneton, with e the charge and m_e the rest mass of the electron, and h Planck's constant; $\gamma = -g\beta_e/\hbar$ is the gyromagnetic ratio, the proportionality constant between magnetic moment and angular momentum. The projections of the electron spin vector $\bar{S}$ on the direction of $\bar{B}$ may take only the values $M_S = S, S-1, \ldots, -S$, where M_S is the magnetic quantum number, leading to an equidistant ladder of Zeeman levels. Thus, for $\bar{B}\|\bar{z}$ direction of the laboratory reference system, Eq. (1) becomes

$$E_{ez} = g\beta_e S_z B \qquad (2)$$

with a separation of $g\beta_e B$ between adjacent levels. Appropriate electromagnetic stimulation (coupling by an AC magnetic field) can cause magnetic dipole transitions, obeying the selection rule $\Delta M_S = \pm 1$ in the case of ESR (and analogously, $\Delta M_I = \pm 1$ for NMR, where $\hbar I$ is the nuclear angular momentum). For the case of a single unpaired electron ($S = 1/2$), solving Schrödinger's equation leads to the resonance condition specifying the magnetic field B_{res} at which applied photons of fixed frequency v can stimulate direct transitions (*magnetic resonance*) between adjacent Zeeman levels given by

$$hv = \Delta E_{ez} = g\beta_e B_{res} \qquad (3)$$

The g value is the measured quantity from Eq. 3, serving as a quite unique ID for each type of spin (defect). For customary laboratory magnetic fields (~ 1 T $= 10{,}000$ G), v situates in the microwave region in the case of ESR. In Fig. 1, for a single electron with $S = 1/2$, the Zeeman energies of Eq. (1) are diagrammed as a function of the magnitude of $\bar{B}$ together with an illustration of a measured curve of resonant absorption of radiation (hv) occurring at $B = B_{res}$, as given by Eq. (3).

It appears that each spin system can be efficiently described by a (unique) spin Hamiltonian H_{In} consisting of all measurable useful interaction terms only comprising spin operators and parameters involving quantitative information about the nature of defects under study. In the general Hamiltonian picture, this spin part Hamiltonian, considered as the perturbation Hamiltonian, is distinct from the spin-unperturbed part H_0 comprising all the kinetic energy and Coulomb interaction terms, including the crystal field contributions. Here, we will elaborate on a few terms most encountered in semiconductor research.

In its most simple case, H_{In} only contains the interaction between the electron spin and $\bar{B}$, the Zeeman interaction (Eq. (1)). In general, however, there is admixture of electron orbital angular momentum $\bar{L}$ into the spin ground state, resulting in a total electron magnetic moment.

$$\bar{\mu} = -\beta_e(\bar{L} + g_e \bar{S}) \qquad (4)$$

This makes that g, rather than being a simple isotropic scalar, will usually become anisotropic.

If $\bar{L}$ is degenerate ($L \neq 0$), the resulting Hamiltonian H_{In} is the sum of the Zeeman interaction, $-\mu \cdot \bar{B}$, and the spin–orbit interaction, $\lambda \bar{L} \cdot \bar{S}$, giving

$$H_{In} = \beta_e(\bar{L} + g_e \bar{S}) \cdot \bar{B} + \lambda \bar{L} \cdot \bar{S} \qquad (5)$$

where λ is the spin–orbit coupling constant and $\bar{B}$ the local magnetic field, usually understood to be the externally applied field. But for most defects embedded in a solid, $\bar{L}$ is quenched, meaning that due to the action of local "crystal" electric fields only a small residual of the orbital moment $\bar{\mu}_L = -\beta_e \bar{L}$ is left. Consequently, the contributions of $\bar{L}$ to the electron spin can be treated by

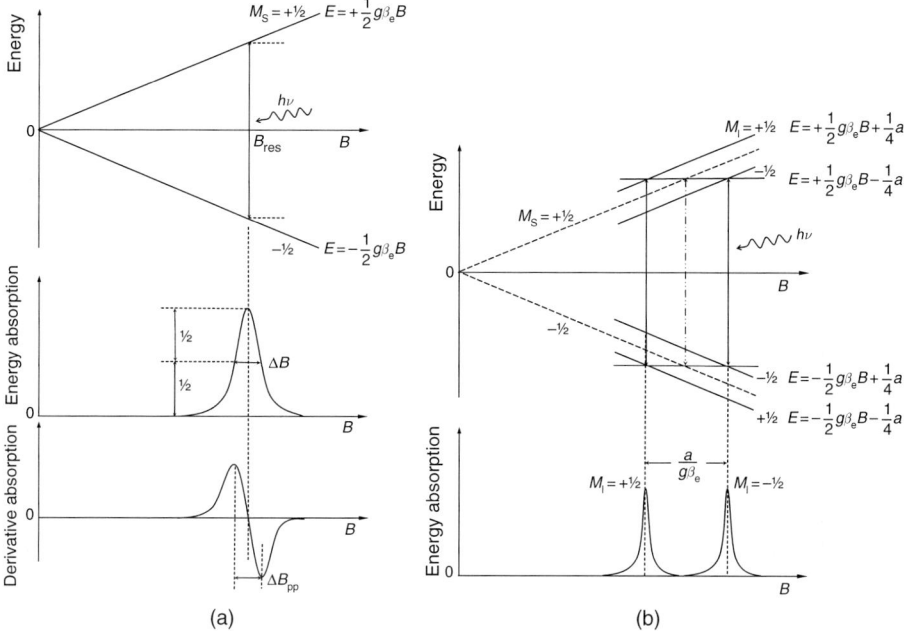

Fig. 1. (a) Electron energy level scheme for the Zeeman interaction of a single electron (S = 1/2) as a function of applied magnetic field B and illustration of the (ESR) resonant absorption spectrum occurring at $B = B_{\text{res}}$ in response to absorption of incident (microwave) photons of $h\nu = \gamma\beta_e B_{\text{res}}$, as detected in the direct absorption and first derivative mode. The width of the spectrum (ΔB) indicates that the levels, rather than (as drawn) sharply defined, have a finite spread. The constants g and β_e are defined in Section 2.1.1. (b) Level scheme for additional interaction of the unpaired electron (S = 1/2) with a magnetic nucleus with $I = 1/2$ (positive g_N) with positive isotropic hf interaction a (e.g., the free hydrogen atom). The corresponding allowed ESR transitions induced by (microwave) quanta $h\nu$ are indicated together with the fixed frequency hf absorption signals split by $a/(g\beta_e)$.

quantum mechanical (second order) perturbation theory. From Eq. (5), the perturbation treatment results in the *effective spin* Hamiltonian H_s

$$H_s = \beta_e \bar{B} \cdot \hat{g} \cdot \bar{S} \tag{6}$$

in which picture an effective spin $\mathbf{g} \cdot \mathbf{S}/g_e$ can be considered as being quantized along $\bar{B}$.

The $\hat{g}$ matrix is given as

$$\hat{g} = g_e \hat{1} + 2\lambda \hat{\Lambda}$$

where $\hat{1}$ is the unit matrix. For an unpaired spin delocalized on a single atom, the matrix $\hat{\Lambda}$ is given by

$$\Lambda_{ij} = \sum_{n \neq 0} \frac{\langle \Psi_0 | L_i | \Psi_n \rangle \langle \Psi_n | L_j | \Psi_0 \rangle}{\varepsilon_0 - \varepsilon_n} \tag{7}$$

a symmetric matrix where Ψ_0 is the ground-state spacial wave function of the unpaired electron, Ψ_n the excited-state wave function, and ε_0 and ε_n the corresponding energies. The principal values of the effective g matrix will thus deviate from g_e as a result of the admixing of the excited states into the ground state of the electron wave function through the spin–orbit coupling, providing us with a unique fingerprint of the defect center.

For $S > 1/2$ systems, the electron spin–spin interaction comes into play adding the term $\bar{S} \cdot \hat{D} \cdot \bar{S}$ to H_s. This term is variably referred to as the fine-structure or zero-field splitting. Part of its origin is the magnetic dipolar interaction between unpaired electrons, leading to a splitting of the ESR lines in zero magnetic field. In its principal axis system x, y, z, it is most frequently expressed in the handy alternative form

$$\bar{S} \cdot \hat{D} \cdot \bar{S} = D\left[S_z^2 - \frac{1}{3}S(S+1)\right] + E\left(S_x^2 - S_y^2\right) \tag{8}$$

with

$$D = D_{zz} - \frac{1}{2}\left(D_{xx} + D_{yy}\right) \tag{9a}$$

and

$$E = \frac{1}{2}\left(D_{xx} + D_{yy}\right) \tag{9b}$$

where D is the axially symmetric part and E is the asymmetry parameter.

As final term we discuss the hf interaction, which describes the interaction of the unpaired electron with nearby nuclear magnetic moments. For an $S = 1/2$ unpaired electron interacting with one central magnetic nucleus, the effective spin Hamiltonian becomes

$$H_s = \beta_e \bar{B} \cdot \hat{g} \cdot \bar{S} + \bar{S} \cdot \hat{A} \cdot \bar{I} - g_N \beta_N \bar{B} \cdot \bar{I} \tag{10}$$

where $\bar{I}$ is the nuclear spin angular-momentum operator in units of $\hbar$, g_N the nuclear g factor, $\beta_N = |e|\hbar/2m_p$ (m_p is the mass of the proton) the nuclear magneton, and $\hat{A}$ the hf tensor. The term $H_{hf} = \bar{S} \cdot \hat{A} \cdot \bar{I}$, representing the hf interaction, is replaced by a sum of similar terms when the electron spin interacts with several magnetic nuclei at different neighboring sites or with different $\bar{I}$. The last term in Eq. (10) represents the *nuclear* Zeeman term, which in usual practical cases can in first order be neglected as its value is small compared to the electronic Zeeman interaction. In the event that the electronic Zeeman term dominates energetically, and with $|B|$ much larger than the hf field of the electron imparted at the nucleus (which is not always the case, see Refs [16,20]), both $\bar{S}$ and $\bar{I}$ can be quantified along $\bar{B}$.

The hf tensor $\hat{A}$ is conveniently written as a sum of an isotropic and an anisotropic part

$$(\hat{A} = a\hat{1} + \hat{B}) \tag{11}$$

where a is the isotropic scalar. In its principal axes system, the anisotropic traceless diagonal tensor $\hat{B}$ can be described by two anisotropic hf constants b and b',

$$\hat{B} = \begin{pmatrix} B_{xx} & & \\ & B_{yy} & \\ & & B_{zz} \end{pmatrix} = \begin{pmatrix} -b+b' & & \\ & -b-b' & \\ & & 2b \end{pmatrix} \quad (12)$$

so that $b = B_{zz}/2$ and $b' = (B_{xx} - B_{yy})/2$, the latter describing the deviation from axial symmetry (rhombicity parameter). In the simple case that the unpaired electron (or hole) can be described by a one-particle wave function $\Psi(\bar{r})$ at site $\bar{r}$ (e.g., a simple atomic orbital or an expansion in orbitals all belonging to the same spin, such as carried out in the linear combination of atomic orbitals (LCAO) approach (vide infra)), the isotropic hf contribution in the coordinate system with the nucleus situated at the origin is given by the Fermi-contact Hamiltonian $H_F = a_{iso} \bar{S} \cdot \bar{I}$, with

$$a_{iso} = \frac{2}{3} \mu_0 g_e \beta_e g_N \beta_N |\Psi(0)|^2 \quad (13)$$

where $\mu_0 = 4\pi \times 10^{-7} \, \text{NA}^{-2}$ is the permeability of free space and a_{iso} the isotropic coupling constant, representing a measure for the unpaired electron probability density at the spin-active nucleus. It is non-vanishing only for the unpaired electron occupying an s orbital, while for all other orbitals, having a node at the origin, the term vanishes. Figure 1(b) depicts the energy level scheme including electron Zeeman and hf interaction for a simple $S = 1/2$, $I = 1/2$ system, corresponding to the energy levels $E = g\beta_e B M_s + aM_s M_I - g_N \beta_N B M_I$ as obtained from solving to first order Eq. (10) for isotropic g value and isotropic hf interaction, with the latter small compared to the electronic Zeeman interaction.

The anisotropic tensor elements result, in a classical view, from the magnetic point dipole–dipole interaction, with conventional expression of interaction energy

$$E_{dd} = \frac{\mu_0}{4\pi} \left(\frac{\bar{\mu}_e \cdot \bar{\mu}_N}{r^3} - 3 \frac{(\bar{\mu}_e \cdot \bar{r})(\bar{\mu}_N \cdot \bar{r})}{r^5} \right) \quad (14)$$

where $\bar{r}$ is the average distance vector between the electron and the nucleus. Inserting into this equation the correct quantum mechanical operators gives the dipolar Hamiltonian

$$H_{dd} = \frac{\mu_0}{4\pi} g_e \beta_e g_N \beta_N \left[\frac{3(\bar{S} \cdot \bar{r})(\bar{I} \cdot \bar{r})}{r^5} - \frac{\bar{S} \cdot \bar{I}}{r^3} \right] = \bar{S} \cdot \hat{T} \cdot \bar{I} \quad (15)$$

which by integrating over the electron distribution, gives

$$B_{ij} = \frac{\mu_0}{4\pi} g_e \beta_e g_N \beta_N \int \left(\frac{3 x_i x_j}{r^5} - \frac{\delta_{ij}}{r^3} \right) |\Psi(r)|^2 \, dV \quad (16)$$

If $\Psi(r)$ becomes a δ-function, then the classical point dipole–dipole interaction is obtained. Thus, the anisotropic hf constants b and b' (cf. Eqs (12, 16)) are essentially an average of $1/r^3$ over the wave function of the defect, reflecting its radial fall off.

With respect to the aimed determination of defect structure, it should be noted that all of the hf constants are proportional to the nuclear g factor. Thus, the ratio of the hf constants pertaining to different isotopes of the same element should reflect the *respective* g_N values, which is a powerful tool for chemical identification from ESR observations of the elements (impurities) involved in a defect.

In semiconductors and glasses, such as SiO_2, a wide range of fundamental defect centers exhibit an isotropic hf tensor or one with axial symmetry. In the latter case, we get from Eqs (10–12)

$$H_{hf} = A_\| I_z S_z + A_\perp (I_x S_x + I_y S_y) \\ = a_{iso} \bar{I} \cdot \bar{S} + b(3 I_z S_z - \bar{I} \cdot \bar{S}) \tag{17}$$

giving the relationships for the measured $A_\|$ and $A_\perp$ hf components of

$$\hat{A} = \begin{pmatrix} A_\perp & & \\ & A_\perp & \\ & & A_\| \end{pmatrix}$$

$$A_\| = a + B_\| = a + 2b \\ A_\perp = a + B_\perp = a - b \tag{18}$$

In the most simple, though often occurring case, the axially symmetric g and hf matrices are collinear, but this is not general (Eqs (16, 20)). Usually, the ratio $b/a \ll 1$.

In further deepening in this simple one-particle wave function approximation, the inferred information from hf interaction parameters can be spun out into more detail by focusing on the adduced nature of the possible wave function pertaining to the unpaired spin. Indeed, many point defects in semiconductors (e.g., Si, Ge) and glasses (e.g., SiO_2), the ground-state wave function $\Psi_0(\mathbf{r})$ of an unpaired spin on a defect can be described by a (highly) localized sp-type molecular hybrid, so that, in a LCAO approach, it can be expanded in terms of sp hybrids Ψ_i on atoms i near the defect where the wave function has an appreciable probability density, i.e.,

$$\Psi_0(\bar{r}) = \sum_i \eta_i \Psi_i = \sum_i \eta_i [\alpha_i (\Psi_{s,i}(\bar{r})) + \beta_i (\Psi_{p,i}(\bar{r}))] \tag{19}$$

where normalization requires $\sum_i \eta_i^2 = 1$ and $\alpha_i^2 + \beta_i^2 = 1$. It is generally a good approximation for point defects in solids with top valence band and conduction band formed by sp^3 hybrids. Due to its short-range character, the hf interaction of $\Psi_0(\mathbf{r})$ with nucleus i will be dominated by the (principal) component of $\Psi_0(\mathbf{r})$ at that site (neglect of overlap), i.e., Ψ_i, and we obtain the hf interaction as the sum of components at each site. Hence, to first order, the hf interaction will be axially symmetric along the p-orbital lobe. Also, for a single unpaired sp-hybrid, the symmetry of g and A matrices will be similar. Based on Eqs (13, 16, 18, 19), the wave function parameters α_i^2, β_i^2, and η_i^2 can be determined from the experimental hf coupling constants $(A_\parallel, A_\perp)$ according to

$$\alpha_i^2 \eta_i^2 = \frac{a_{\text{iso}}}{a_0} = \frac{A_\parallel^i + 2A_\perp^i}{3a_0} \tag{20a}$$

$$\beta_i^2 \eta_i^2 = \frac{b}{b_0} = \frac{A_\parallel^i - 2A_\perp^i}{3b_0} \tag{20b}$$

where a_0 and b_0 are the *atomic* s-state and p-state coupling constants, i.e., expressions Eqs (13 and 16) for $\Psi(\mathbf{r}) = \Psi_s(\mathbf{r})$ and $\Psi_p(\mathbf{r})$, respectively. Theoretical estimates for a_0 and b_0 can be found tabulated in several sources [17,20,25–27], exhibiting a substantial spread ($\sim 30\%$) according to the theoretical approach applied. Yet, the accuracy currently attained may be within a few percent. As evident from Eq. (20), the inferred values for α_i^2, β_i^2, and η_i^2 will directly depend on the used set of a_0 and b_0, so the accuracy of the free atom hf constants is important. This appears to be more so for η_i^2, which depends on the magnitude of a_0 and b_0 (wave function averages), than for α_i^2, β_i^2 as these are functions of the ratio a_0/b_0 for which the various theoretical estimates are much closer. In comparing various sources of hybridization and localization parameters, one should therefore ensure that the inferred data are referenced to the same set of free atom hf coupling constants a_0, b_0.

Thus the LCAO description permits to differentiate between the s (α_i^2) and p (β_i^2) character of the orbital composition of the unpaired spin wave function and to assess its localization η_i^2 at the ith site; obviously, a pure sp^3 hybrid would entail $\alpha^2/\beta^2 = 0.25$.

Normally, the algebraic sign of the hf coupling constants is not explicitly provided by the ESR data. (Generally, in multielectron systems, the sign of the hf coupling constants is determined by the net electron-spin polarization around the nucleus, which can be positive or negative. In the simple case of one-electron atoms, the sign of the hf constants is determined by that of g_N.) However, the sign may be reliably obtained from dipole–dipole interaction calculations [20].

Values of g_N and μ_N for specific nuclei can be found in tables in the literature [28]. The quantities a and b have units of energy, but often these are expressed in units of frequency (MHz) using $E = h\nu$ or as a wave number (in $10^{-4}\,\text{cm}^{-1}$) using $E = hc/\lambda$ (to convert from MHz units to $10^4\,\text{cm}^{-1}$ ones, divide by 3) or, often also, in units of magnetic field by dividing the coupling constants a and b (in energy units) by $g_e \beta_e$, where the quantities $a/g_e \beta_e$ and $b/g_e \beta_e$ are now commonly referred to as *hf splitting constants*. For conversion, B (mT) $= 0.0357 \times \nu$ (MHz). However, as to the latter, it should be noted that this entails an (usually negligible) approximation, because the correct values inferred from experiment are rather $\left(A_\parallel/g_e\beta_e\right)g_e/g_\parallel$ and $(A_\perp/g_e\beta_e)g_e/g_\perp$. Often, however, g_e is sufficiently close to $g_\parallel, g_\perp$ to allow this approximation. Finally, we add that in many texts the convention is adopted to use capital symbols to refer to hf *coupling constants*, while lower case ones are used for hf *splitting constants*, thus bearing out the difference in units applied explicitly [20].

Revealing the occurrence of hf structures in ESR spectra, without doubt, provides the most powerful information obtained from ESR. From the hf spectra, the nuclear spin of the interacting nuclei and their (natural) abundance can be determined together with $\hat{A}$. In this way, it is possible to identify the atoms involved, their arrangement, and, in favorable cases, even information on the position of the various nuclei involved in the center structure.

But extraction of the latter information requires theoretical backing, i.e., comparison with results from theoretical modeling of envisioned possible structures. So, in conjunction with theoretical modeling, detection of the hf structure can lead to a full atomic identification of the defect, the attained degree of success and reliability depending on adequate quantum mechanical modeling of the center.

2.1.2. Practice

The main parameters one attempts to measure are the three principal-axes components and directions of the $\hat{g}$ matrix and of the $\hat{A}$ tensor. In general $\hat{g}$ and $\hat{A}$ are not isotropic, and the observed g value and hf splitting depend on the orientation of $\hat{g}$ and $\hat{A}$, respectively, relative to the direction of $\bar{B}$. The orientation and symmetry of $\hat{g}$ and $\hat{A}$ reflect those of the involved electronic wave function which is co-determined by the defect's local surrounding. Especially in a crystalline environment, where defects occur in discrete orientations determined by the lattice symmetry, useful information concerning defect and lattice orientation may be obtained.

The g matrix in Eq. (6) can be expressed in terms of its principal axes. In this principal-axis coordinate system dictated by the *local symmetry of the defect*, the g matrix takes a diagonal form with principal values g_1, g_2, and g_3. For an orthorhombic g matrix ($g_x \neq g_y \neq g_z$), g is given by

$$g = \left(g_x^2 \cos^2 \theta_x + g_y^2 \cos^2 \theta_y + g_z^2 \cos^2 \theta_z \right)^{1/2} \tag{21}$$

where θ_x, θ_y, and θ_z represent the angles $\bar{B}$ makes with the respective principal axes. For many cases, g is axially symmetric (axial spin Hamiltonian) and Eq. (21) reduces to

$$g = (g_\perp^2 \sin^2\theta + g_\parallel^2 \cos^2\theta)^{1/2} \tag{22}$$

where θ is the angle between $\bar{B}$ and the symmetry axis. This unique symmetry axis of $\hat{g}$ may or may not lie parallel to a crystallographic axis of the crystal the defect is embedded in. The directions of the principal axes may be obtained from ESR goniometry.

However, in the case of a powdered crystal, or disordered material in general, this crystallographic information would be lost. In nonoriented media, such as polycrystalline or amorphous materials, however, anisotropic spectral properties are continuously averaged over all orientations of the defect, leading to the observation of less informative, smeared-out spectra. If all orientations of the local coordinate system with respect to the magnetic field are present in the sample, the ESR spectrum will then be a so-called *powder pattern*, i.e., a summation over all contributions, each multiplied with its transition probability. However, very favorably, the principal axis g values can be recovered from the observed ESR powder spectra. Many approximations exist for the calculation of such powder spectra [16,17,20]. Calculated theoretical powder patterns for an axial and orthorhombic system are illustrated in Fig. 2.

Other salient spectral features are line width and line shape. The first derivative absorption signal is characterized by the peak-to-peak width ΔB_{pp}, the added result of various line-broadening mechanisms. Studying the line width variations as a function of some varied parameters (e.g., temperature, spin density, pressure) can provide a wealth of information on the mechanisms of interaction of the unpaired spin with its environment (see Refs [16,24] for more details). The most frequently encountered basic absorption shapes are the Gaussian,

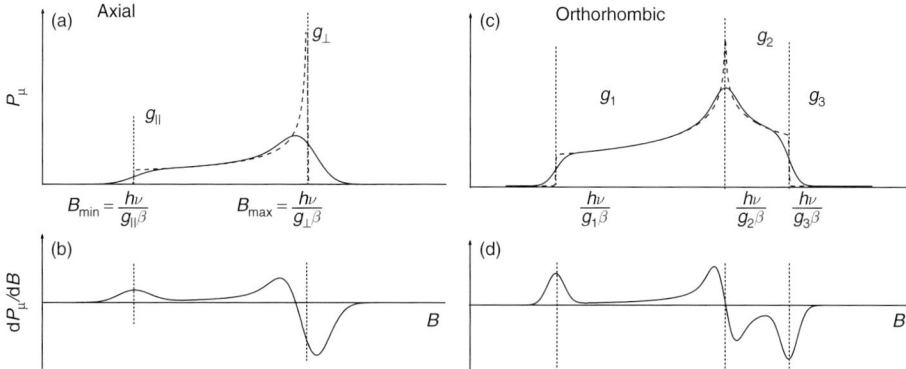

Fig. 2. Illustration of first-derivative powder pattern spectra for a $S = 1/2$ system with no hf interaction present for (a, b) axial ($g_1 = g_\parallel$, $g_2 = g_3 = g_\perp$) and (c, d) orthorhombic (g_1, g_2, g_3) g matrix symmetry. The dashed curves in (a) and (c) are computer-simulated powder patterns, resulting in the full curves (representing observed absorption spectra) after subsequent convolution with a Gaussian broadening function. The routinely detected first derivatives of the convoluted spectra in (a) and (c) are shown in (b) and (d), respectively. A well-known example of a defect in the Si/SiO$_2$ system exhibiting nearly axial symmetry is the E$'_\gamma$ center in the thermal SiO$_2$ layer, with $g_1 = g_\parallel = 2.0018$, $g_2 = 2.0006$, $g_3 = 2.0003$. The orthorhombic symmetry is typical for oxygen-related hole centers in SiO$_2$ [17].

Lorentzian, and Voigt line, where the latter is the convolution of a Gaussian and Lorentzian line. The line shape can be typified by the line shape factor $\kappa \equiv I_a/(A\Delta B_{pp}^2)$, where A is the signal amplitude, equal to half the signal's peak-to-peak height A_{pp}, and I_a the signal intensity (area under the absorption curve). The Lorentzian and Gaussian shapes are characterized by $\kappa = 1.033$ and 3.628, respectively.

Canonically, the signal intensity is properly determined by double numerical integration of the ESR spectra. In comparing relative signal intensities, one is often tempted to use A rather than I_a as the comparing parameter because of ease of measurement and "enhanced" accuracy. But obviously, this will be generally incorrect and misleading, unless line width and shape would not alter over the measurement sequence. As I_a is linearly proportional to the number of spins, the spin density can very accurately be determined by comparison of the intensity of an unknown signal with that of a calibrated marker signal, preferentially recorded in one trace. Such a marker is generally calibrated against a primary absolute spin standard, of which the number of defects has been determined reliably by one or more independent methods. Another reason favoring the use of a marker signal is that it provides, when accurately calibrated, a very accurate and easy way to read off g values [29].

2.2. Application to semiconductor heterostructures

2.2.1. Specific modalities

In the application of ESR to SHs, the inference of the g map (and correlated, the hf map), i.e., angular dependence (*anisotropy*) of the resonance field on the orientation of the applied magnetic field $\bar{B}$ with respect to the crystalline substrate, takes a special place for analyzing defects, more specifically with respect to interface defects. The presence of interfaces between dissimilar materials is perhaps the salient aspect per se in the concept of an SH. And it goes without saying how important it is to put the interface under tight control, i.e., establish high electrical quality, for attaining functional devices. The quality of the interface plays such a crucial role that its influence can hardly be overestimated.

A key item here is that generally, when two dissimilar materials are connected through chemical bonding, unless perhaps in a perfect pseudo-epitaxial growth, defects are incorporated as a result of mismatch, which may devastate the electrical quality (vide infra). An archetype example is the P_b defect of C_{3v} point symmetry at the c-(111)Si/a-SiO$_2$ interface [30–32]. If located in the top interfacial layer(s) of the semiconductor, these defects will be in registry with the crystallinity of the underlying substrate. If such defect would be distributed throughout the bulk of the substrate, in principle, a separate branch would be observed in the g map for each of the defect-site orientations equivalent through the symmetry properties of the crystal. But when restricted to an interface (or surface, for that matter) plane, only those sites interrelated through the symmetry of the plane will occur, so generally, a limited number of g branches will be observed. It provides the authentication of an interface (surface) constrained point defect per se.

In the case of the P_b center in standard (111)Si/SiO$_2$, only one g branch is observed, corresponding to the variant with the unpaired sp^3-type hybrid oriented along the [111] interface normal $\bar{n}$ [30]. The other variants that would be expected for defects distributed in a bulk sample, with the sp^3 hybrid oriented along the crystallographic equivalent [11 $\bar{1}$], [1 $\bar{1}$1], and [$\bar{1}$ 11] orientations, are not observed in standard thermal (111)Si/SiO$_2$. For completeness, however, we should add that these are observed in (111)Si/SiO$_2$ entities of inferior quality deliberately grown under reduced O$_2$ pressure (<6 mPa) or low T [33]; it refers to a dimensional (perpendicular) extension, i.e., roughening, of the interface region. Another well-known example of an interface constrained defect is the P_{b1} center in (100)Si/SiO$_2$ [34,35]. The P_{b1} g map is illustrated in Fig. 3, with indication of the experimentally "missing" branches.

A few more remarks concern sample handling. In dealing, e.g., with Si substrates, extreme care has to be exercised about left sample cutting damage, i.e., cracks at lateral edges, unavoidably giving rise to an intense, often prohibitively interfering, D signal (Si DBs in disordered Si environment) [36] on top of investigated weaker signals. Related to this, when studying one side grown Si/insulator structures, as often encountered, care should also be taken to exclude interference from unwanted possible signals originating from the less-controlled back surface/interface. The threat of the occurrence of perturbing irrelevant sample "background" signals is not equal for all substrates, e.g., it appears much less a concern for Ge [37].

2.2.2. Sensitivity

As to conventional ESR, a first general limitation in applicability concerns the comparatively low-sensitivity in general of magnetic resonance experiments resulting from the low energy involved in spin transitions (~tens of μeV for ESR). Current top quality conventional absorption-mode ESR spectrometers may detect $\sim 1 \times 10^9$ centers (spin $S=1/2$) of 1 G line width at cryogenic temperatures within acceptable signal averaging times. This state-of-the-art sensitivity may be put in light of the usefulness in applying conventional ESR, i.e., potentiality, to the study of SHs. To this, it may be instructive to recall the general expression for the average absorbed microwave power P_a by a sample (spin system) during magnetic resonance, given for Curie paramagnetism, as $P_a \propto \nu H_1^2 S(S+1) N/T$; H_1 is the microwave magnetic field amplitude at the sample site and N the number of spins in the sample. So, as compared with the more comfortable study of bulk samples, when focusing on interface and

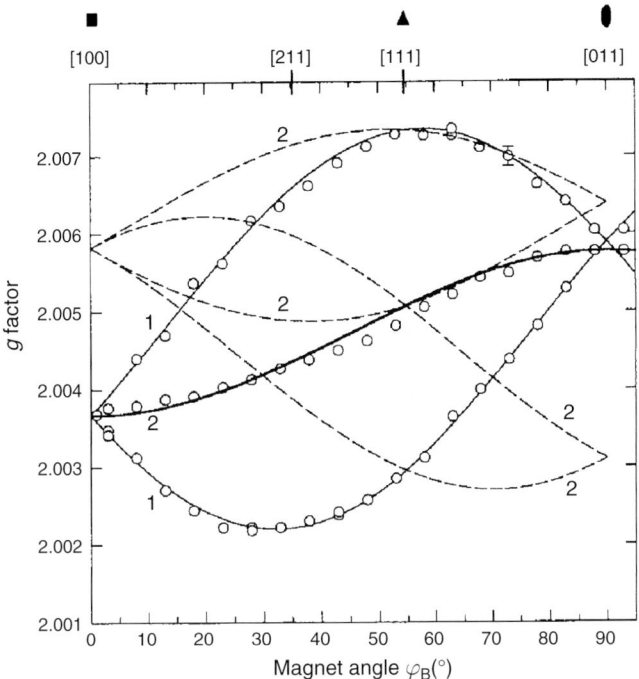

Fig. 3. Angular dependence of the P_{b1} g value observed in (100)Si/SiO$_2$ thermally grown at 970°C (1.1 atm O$_2$; oxide thickness $d_{ox} \sim 130$ nm) for $\bar{B}$ rotating in the Si substrate $(0\,\bar{1}1)$ plane; φ_B is the angle $\bar{B}$ makes with the [100] normal. The curves represent the optimized theoretical fit for monoclinic-I point group symmetry, from which the principle g values $g_1 = 2.00577$, $g_2 = 2.00735$, $g_3 = 2.0022$ are inferred. The branches in dashed curves are not observed experimentally; only the four defect orientations equivalent through the four-fold symmetry of the (100)Si face are observed, clearly exposing the interface-restricted nature of the defect. The g_2 axis of this defect in the $(0\,\bar{1}1)$ plane is at an angle of $3 \pm 1°$ with $[1\,\bar{1}\,\bar{1}]$ towards $[0\,\bar{1}\,\bar{1}]$. The added numbers indicate the relative intensities of the branches.

thin dielectric film properties, there appears an inherent main hurdle – i.e., the limited number of defects one can physically pack into one ESR sample fitting into the microwave cavity, inherent to studying interfaces and thin layers. As the ESR signal is proportional to the number of spin centers in the cavity, this may put severe limits on sensitivity and detectability. A routine way to improve on this consists in maximized stacking of (thinned down) sample slices. It seems the best one can do. Of course, one could as well resort to defect-enhancing additional (damage) treatments, such as, γ-irradiation. But, obviously, as this will divert the sample from its pristine as-grown state, the benefit of such procedures has to be balanced against the ultimate goal of study.

Another way to boost sensitivity may entail measuring at reduced temperature (cf. $P_a \propto T^{-1}$). For some type of defects (and spectrometers, as the success will also depend on the specific predominant noise characteristics), this may result in higher sensitivity through enhancement of paramagnetism by the Boltzmann factor. Moreover, this may advantageously be accompanied by a higher loaded Q due to freezing out of any substrate carriers. True, microwave saturability of signals generally increases rapidly with decreasing T because of increasing spin-lattice relaxation time T_1 [38]; but, with the commonly envisioned semiconductor/insulator-based entities, with necessarily ultrathin dielectric layers, saturability of defects, such as P_b-type interface defects, appears naturally reduced, similarly as observed for Si/SiO$_2$ with ultrathin SiO$_2$ layers [35,39]. This may be related to significant changes in the density of phonon states in the dielectric layer, or more generally, in the local environment of the P_b-type defects. Then higher microwave power $P_\mu (\propto H_1^2)$ can be applied, with attendant gain in sensitivity.

2.2.3. ESR and devices

As mentioned, the current compilation exclusively addresses the application of conventional CW ESR to the SH field. And as discussed in Section 2.2.2, however universal it may be, when it comes to useful applicability in the SH field, overall sensitivity is a major concern. It is disappointingly exposed in various cases when ESR analysis is directed to actual (individual) state-of-the-art devices, with intricate architecture, ultrathin layers (sub-nanometer), and lateral size scaled down to the tens of nanometers range. Detection of any detrimental imperfections by standard ESR just becomes elusive.

One then has to resort to next-level techniques combining standard ESR with other spectroscopies, leading to drastically enhanced sensitivity in specific applications. The common denominator of these techniques is that instead of observing the spin system resonance directly through its microwave absorption, it is detected indirectly through it triggering another electrically accessible sample parameter, which as it appears can be realized in many ways [40]. A basic phenomenon here is the spin-dependent nature of carrier recombination in semiconductors, denoted as spin-dependent recombination (SDR) [41, 42]. The underlying reason for the generally impressive gain in sensitivity is transformation of the stimulating energy quantum (microwave photon of $h\nu \sim$ tens of μeV) for transitions between ESR Zeeman levels in usual laboratory fields to energies of the order of eV pertaining to transitions between electronic states governed by spin selection rules. According to the parameter monitored, such as electrical conductivity, capacitance, noise, and optical response, this gives rise to the combined spectroscopic techniques denoted as electrically (EDMR), capacitively (CDMR), noise (NDMR), and optically (ODMR) detected magnetic resonance. These techniques have been successfully applied to study defects in the Si/insulator entities [43–47].

Obviously providing much detailed information on spin-dependent effects with impressive sensitivity, these combined ESR-based spectroscopies are the route of progress beyond the stage where conventional ESR falls short, in particular for single-device analysis. The techniques enjoy increasing attention, from which much is still to be expected, e.g., in developing fields such as spintronics and photonics. While of huge interest, the subject cannot be dealt with here. For more details, the reader may refer to many excellent reviews [40] and reports [43,45,47] in the literature. Yet, while outstanding in sensitivity and specificity, the combined methods may still not outperform conventional ESR in aspects such as versatility, universality, quantification, and identification of defects, so standard ESR will likely keep playing a basic (complementary) role.

2.2.4. Defect activation

As well outlined for the case of P_b-type interface centers, paramagnetic point defects are prone to ESR inactivation (at moderate temperatures) through binding with H [39,48]. In this respect, see also E′-type [49,50] and EX [51] defects in Si/SiO$_2$ physics. Hence, with the application of some particular fabrication techniques (H-rich ambient) for insulator film deposition, such as employed in high-κ layer deposition using, e.g., atomic layer deposition from H-rich precursors (e.g., Al(CH$_3$)$_3$, H$_2$O; see Section 3.2.2), it may be anticipated that in such cases, most point defects will be left effectively passivated (invisible for ESR). In the case of P$_b$-type defects, the usual method [39,52,53] to efficiently detach H is thermal treatment at elevated T (typically $\sim$600°C; $\sim$1 h). Clearly, however, such step might irreversibly alter the initial physico-chemical structure of interfaces and dielectric layers, with embedded defects, thus obfuscating the study of the initial state.

As an alternative method, one might prefer to subject samples, prior to ESR observations, to appropriate photon irradiation in air to photo-dissociate H-terminated dangling bonds (DBs) [54] and/or possibly unveil nonideal (strained, weak) bonding. Often used here is room ambient vacuum UV (VUV) irradiation obtained from a Xe-resonant (8.48 eV) or Kr-resonant discharge lamp (10.02 eV; typical flux $\sim 5 \times 10^{14}$ cm^{-2}) provided with MgF$_2$ windows. The method has been demonstrated to be a most efficient for both oxide and interface defects in Si/SiO$_2$. The front side interfaces and dielectric layers generally are found not affected by the VUV treatment, except for the ESR defect activation through H photodesorption. Also, high-dose UV irradiation, obtained from an Hg lamp, may be applied for this purpose. An additional bonus of the photon irradiation is that it may turn diamagnetic defects into the paramagnetic ESR-active state through charge transfer. In this perspective, every defect site is potentially available for ESR observation because electrons, if not already unpaired, can be made so by application of sufficient suitable irradiation.

3. Probing semiconductor/insulator interfaces through inherent point defects

3.1. Prototype interface: Si/SiO$_2$

3.1.1. The P$_b$-type interface defects and atomic identification

As mentioned, the first pioneering reports on the ESR observation of interface-associated defects in c-Si/SiO$_2$ entities date back to 1971, where the symbol P$_b$ was assigned to a particular anisotropic ESR signal [55]. Since then, more types have been revealed, which as detected by ESR are now generally termed P$_b$-type centers, the class of paramagnetic trivalent Si defects – Si DB centers – containing one unpaired spin ($S=1/2$) and correlating with Si substrate interface orientation. So, "P$_b$ center" is used as a generic term for the different variants of Si DB defects at the Si/SiO$_2$ interface. Mainly under technological impetus, these have been intensely studied [30,32,34,55–57] and various works have overviewed the salient properties [7,31,57]. The defects are inherently introduced in response to network-lattice mismatch between a-SiO$_2$ and c-Si, correlating with stress in the oxide layer [58]. Three types are common [31,34]: At the (111)Si/SiO$_2$ interface, only one type is generally observed, specifically termed P$_b$, identified as [32] prototype interfacial trivalent Si, denoted Si$_3 \equiv$Si$^\bullet$. From electric-field controlled ESR in combination with electrical measurements (C–V), the center was shown to be an amphoteric trap of effective correlation energy $U_e \sim 0.5$ eV with the electronic +/0 and 0/– transition levels deep in the Si bandgap at $\sim$0.3 and $\sim$0.8 eV above the valence band (E$_v$), respectively [59]. The latter is illustrated in Fig. 4 from quasi static C–V measurements, showing the electrical density of states D$_t$ across the bandgap. The technologically foremost (100)Si/SiO$_2$ interface usually exhibits two prominent types, called P$_{b0}$ and

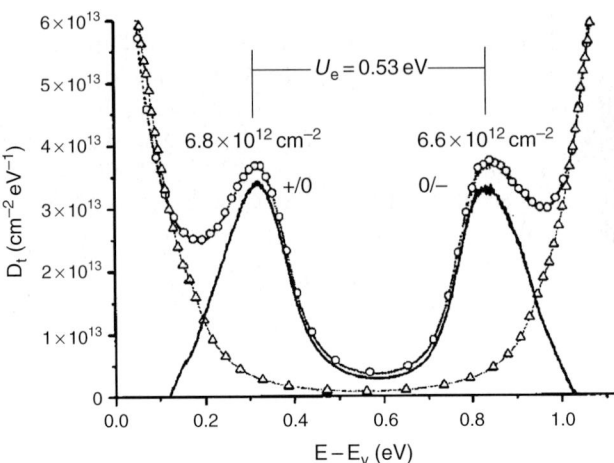

Fig. 4. Density of states (D_t) across the Si bandgap calculated from quasi-static C–V measurements on thermal (111)Si/SiO$_2$(11.5 nm) thermally grown at 850°C in pure O$_2$ after (o) rapid thermal annealing (RTA) at 1040°C (20 s; N$_2$), and subsequent (Δ) anneal in forming gas (450°C; 30 min; 15% H$_2$ in N$_2$). The latter P_b-passivation treatment reveals the U-shaped background density of states. The solid curve, exposing the P_b peaks, is obtained by subtracting the former two data curves. The two peak features are ascribed to the $+/0$ and $0/-$ electron transitions of the P_b interface defect, approximately centered at 0.31 and 0.84 eV above the valence band edge E_v, respectively, illustrating the amphoteric nature of the defect. The ~0.53 eV separation of the peaks is the positive effective correlation energy (U_e) for the P_b defect, which is the difference between the energies needed to add one electron to a singly occupied (neutral) Si sp$_3$-like orbital and to an unoccupied (positively charged) site (after P. Hurley et al. [61]).

P_{b1} [34]. All three variants were shown to be trivalent Si centers, where the experimental evidence indicates that P_{b0} is chemically identical to P_b, but now resides at imperfections (e.g., steps, kinks, ledges, (111) microfacets) at the macroscopic (100)Si/SiO$_2$ interface – effectively, it would indicate that some fabrication process-bound <111> faceting may be endemic for the (100)Si/SiO$_2$ interface, as the appearance of P_{b0} defects would thus point to nonideal (planar) Si/SiO$_2$ transition. In compliance, both were established as adverse electrical interface traps [59–61]. The electrical activity of the P_{b1} centers as interface traps, initially accepted, is still in dispute [52]. The P_{b1} is assigned to a distorted defected interfacial dimer ($\equiv$S–Si$^{\bullet}$=Si$_2$ defect, where the long hyphen symbolizes a strained bond, with an approximately <211> oriented unpaired Si sp^3 hybrid [62,63]). The current unrelaxed structure models of the P_b-type defects are illustrated in Fig. 5, while Fig. 6 displays a typical K-band P_b ESR spectrum observed on thermal (111)Si/SiO$_2$ (with oxide thickness $d_{ox} = 4$ nm). Like (111)Si/SiO$_2$, at the (110)Si/SiO$_2$ interface as well, only one type – the P_b variant – is observed [64]. For standard oxidation temperatures T_{ox} (800–960°C), the physical defect sites (including both ESR active and inactive ones) occur in *natural* densities of [P_b] ~ 5×10^{12} cm^{-2} and [P_{b0}], [P_{b1}] ~ 1×10^{12} cm^{-2} [35,57,65].

The successful atomic identification of these harmful interface defects probably figures as the first, both in time and achievement, principal triumph of the application of ESR technique to the research of Si/insulator structures. However, it took quite some time to reach that stage, without doubt, as outlined before mainly due to the limited ESR sensitivity inherent to the study of

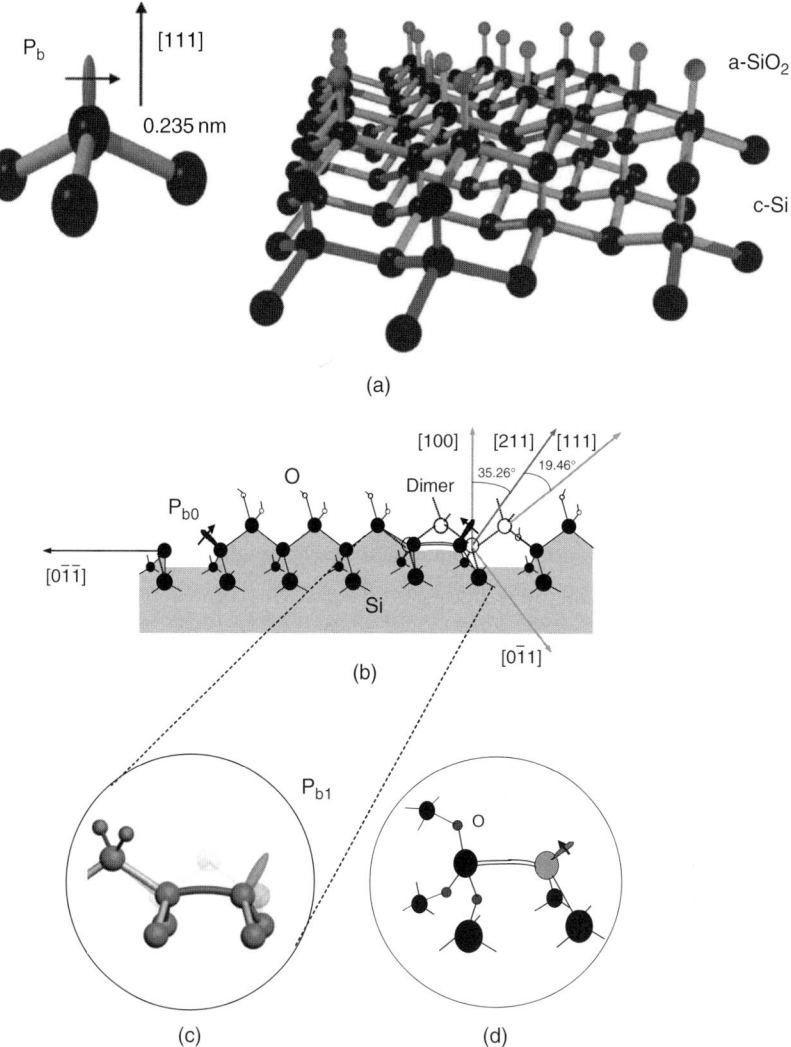

Fig. 5. Steric pictures of the model structures for the P_b (panel a) and P_{b0}, P_{b1} (panel b) interface defects at the (111)Si/SiO$_2$ and (100)Si/SiO$_2$ interfaces, respectively. Inset (c) gives a close up of the P_{b1} defect model first proposed from experimental hf interaction results (Ref. [62]); inset (d) represents the model favored by theory (Ref. [63]). (see Colour Plate 4)

interfaces. Since the first observation of P_b centers (1971) [55], breakthrough in the atomic identification of P_b in (111)Si/SiO$_2$ was provided in 1983 through detection of the ^{29}Si primary hf ESR structure originating from the central defected ^{29}Si atom [32], while breakthrough for P_{b1} had to await such observation in 1998 [62].

3.1.2. Thermal impact

The appearance of the P_b-type defects is susceptible to post oxidation annealing (POA). Their thermochemical properties appear dominated by interaction with hydrogen [39,48,53,66–68],

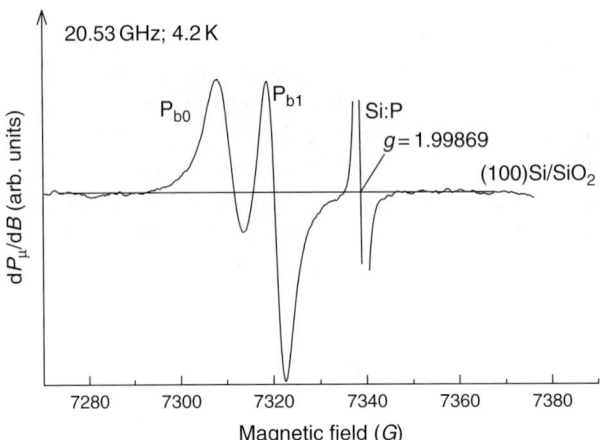

Fig. 6. Typical K-band (~20.5 GHz) ESR spectrum observed at 4.2 K for $\bar{B}\|\bar{n}$ on thermal (100)Si/SiO$_2$ ($d_{ox} \sim 4$ nm). The signal at ($g = 1.99869 \pm 0.00002$), not belonging to the P$_b$-type spectrum, stems from a co-mounted Si:P sample serving as g and intensity reference. The applied microwave power and modulation field amplitude was ~0.3 nW and ~1 G, respectively.

as will be dealt with later (Section 3.3). When heating in H$_2$ ambient, defect inactivation, i.e., passivation, is pictured as chemical saturation of the failing Si bond by hydrogen, symbolized as SiH formation. Passivation of these centers is part of the goal of the technologically routinely applied post-oxidation anneal in forming gas (FG; ~10% H$_2$ in N$_2$). Similar behavior was observed for all three P$_b$ variants; more details can be found elsewhere [48]. In later work, it has been found that appropriate POA in vacuum may lead to irreversible creation of substantial *additional* amounts of P$_b$s (e.g., up to 7.5×10^{12} cm^{-2} at ~950°C) [69]. This creation mechanism is strongly enhanced (~6 times for $T \geq 800$°C) when performed in H$_2$, allowing to achieve densities up to 3.1×10^{13} cm^{-2} [70].

3.2. Semiconductor/high-κ interface

3.2.1. The high-κ issue

As mentioned before, scaling down of the Si/SiO$_2$-based MOS transistor has been one of the main pillars at the origin of the unprecedented explosive growth of microelectronic device industry. Lateral scaling will enable to pack ever more devices in a given area, where the attendant shortening of the crucial channel length (distance between source and drain) may boost switching speed, resulting in faster devices. Yet according to the constant-field scaling theory [3], the vertical dimensions must be scaled down together with the lateral dimensions to keep control by the gate terminal on the charge carriers in the channel. Thus, compensation for the lost areal capacitance is mandatory, which in pace with relentlessly progressing scaling was pragmatically realized by gradual reduction in the thickness of the conventional thermal SiO$_2$ (SiO$_x$N$_y$) gate insulator, i.e., not giving in on the gate capacitance $C = \kappa\varepsilon_0 A/d$, where ε_0 is the vacuum permittivity, κ the dielectric constant, A the area, and d the insulator thickness. For device scaling beyond the 65 nm node with ~30 nm gate length transistors, currently evolving in production [5], the thickness of SiO$_2$-based gate oxides needs to be reduced to <1 nm. However, at such extreme gate insulator thicknesses, the application of SiO$_2$ faces fundamental limits [71]; serious problems arise, such as excessive leakage currents caused by direct electron tunneling, mobility degradation due to the increased electric field, dopant diffusion, and reliability problems [72–74].

An obvious solution would be the use of another insulating material with a dielectric constant higher than that of SiO$_2$ ($\kappa = 3.9$), the so-called high-κ materials, enabling us to use physically thicker films of equivalent (SiO$_2$ electrical) oxide thickness (EOT), defined as

$$\text{EOT} = d_{\text{high}-\kappa} \frac{\varepsilon_{\text{SiO}_2}}{\varepsilon_{\text{high}-\kappa}} \qquad (23)$$

to obtain the same specific capacitance, in that way reducing direct tunneling [72–74]. The subject has been overviewed in several excellent and complementary works [71,73–77].

Though addressed before in connection with IC technology, this has propelled research in appropriate alternative high-κ materials for replacing SiO$_2$ gates in future deep sub-micrometer MOS generations. As a result, the interest in high-κ dielectrics was renewed in the mid-1990s. Indeed, there has been interest previously in high-κ materials for reasons as, for example, their application in storage capacitors (e.g., Ta$_2$O$_3$, TiO$_2$, and BaSrTiO$_3$) [78] or, very initially in the IC era, just as a potential alternative for SiO$_2$ gates because of comparable attractive properties, such as Al$_2$O$_3$ ($\kappa_{\text{Al}_2\text{O}_3} \sim 8.5 > 2\varepsilon_{\text{SiO}_2}$; bandgap $= 8.8$ eV, similar to SiO$_2$) [79].

Intensively investigated candidate materials include [80] elemental metal oxides such as Al$_2$O$_3$, ZrO$_2$, HfO$_2$, Ta$_2$O$_3$, La$_2$O$_3$, Y$_2$O$_3$, Lu$_2$O$_3$, Sc$_2$O$_3$, and the binary oxides SrTiO$_2$ and LaAlO$_3$ as well as many of their composites such silicates and aluminates, often with N added to improve performance. Many have already been disfavored, often because of stability reasons (see Refs [73–76,80] for up-to-date details). The renewed high-κ research quickly converged on the set Al$_2$O$_3$, HfO$_2$, and ZrO$_2$, with most attention addressed to the latter two. Over the past few years, particular interest has been extended [81–86] to the Hf-based materials, e.g., HfO$_2$, HfSi$_x$O$_y$, HfAl$_x$O$_y$. Currently, the leading contenders are the family of Hf-based insulators, with the main focus on nitrided Hf silicate. Reasons for this are distinguishing characteristics including high-κ value ($\sim$3.9–20, depending on Hf content), good thermal and chemical stability in contact with Si [81,85,87–91], and favorable Si/HfO$_2$ conduction and valence band offsets (2.0 $\pm$ 0.1 and 2.5 $\pm$ 0.1 eV, respectively) [92], which limit carrier tunneling.

High-κ films may be deposited by various methods, such as atomic layer chemical vapor deposition (ALCVD) [74], metal-organic CVD (MOCVD) [93], and molecular beam deposition (MBD) [94], at temperatures in the range 100–400°C. Generally, amorphous films are deposited – preferred in terms of thermal stability – but the pseudo-epitaxial growth has also received a great deal of attention [95].

However, a suitable replacement dielectric must have not only a high κ. In order to be successful, these alternative interfaces should have comparable (or, should it be possible at all, even better) electrical properties than the standard Si/SiO$_2$ interface and should be able to endure the various thermal steps required for device manufacturing. To withstand extreme technological demands, when incorporated in the MOS entity, it must additionally combine superb electrical qualities with excellent structural stability, preferably not inferior to the conventional M/SiO$_2$/Si building block model. This includes a range of prerequisites, including good thermodynamic stability in direct contact with Si, preferably residing in the amorphous state to be maintained during necessary device fabrication thermal steps, negligible dopant penetration, subcritical metal penetration into Si, and an overall minimal amount of contamination. As for the Si/dielectric interface, the challenge concerns realization of smooth, uncontaminated interfaces, with well-controllable and stable inter(face) layers as thin as possible.

Obviously, with the replacement of the gate dielectric, "new" semiconductor/insulator entities are conceived de novo, which when introduced in MOS-structures should result in functional microelectronic devices of top electrical quality. This means that, in principle, the whole of the electrical characterization, implying quality of the dielectric and interfaces, would have to be redone for each of the newly conceived heterostructures as done before for the standard Si/SiO$_2$ case. Quite naturally, in view of the technological importance, the major top sensitivity topographic/imaging techniques are applied in conjunction to explore microstructural, compositional, and bonding chemistry aspects of SHs with ultrathin (sub 5 nm) high-κ films (see Refs [80,96] for some overview on these techniques as applied to high-κ dielectrics). Clearly, a major issue here is the occurrence of traps, in the oxide layer as well at crucial interfaces. Evidently then, as crucial next step, the electrical performance is analyzed using the standard electrical techniques C–V, I–V, and conductance–voltage (G–V), indicating that besides interface traps, there appear substantial densities of bulk traps in, for example, Al$_2$O$_3$, ZrO$_2$, and HfO$_2$ layers, the density and type (charge) of the traps depending on the kind of film deposition technique and experimental parameters (e.g., chemicals) used [97–99]. So, as with the Si/SiO$_2$ case, with the view to improve on the quality of the composed high-κ SHs, one would hope to understand what these traps are (atomic nature) – desired information for which inference, as outlined, the above techniques appear inadequate. Here again ESR appears as the truly complementary technique to assess the physico-chemical atomic nature of the point defects at the origin of the traps. There are two main parts in identification: defects located in (a) the high-κ layer and (b) at interfaces. The tool has been applied quite successfully to the study of SHs with high-κ layers in combination with Si [46, 99–111] and Ge [37]. The number of ESR studies here is increasing steadily.

3.2.2. ESR study of Si/high-κ insulator structures

Here, we will illustrate by two examples of how probing of ESR-active point defects, in combination with post deposition heat treatment, can reveal atomic scale structural information on the Si/high-κ interface and interlayer. More specifically, through the occurrence/absence of interface-specific (P$_{b0}$, P$_{b1}$) and/or interlayer-associated (EX, E') point defects, we will in an illustrative comparison demonstrate the appearance, additional growth, and modification of a SiO$_{2(x)}$ interlayer for (100)Si/SiO$_x$/HfO$_2$ and (100)Si/LaAlO$_3$ entities.

Initial K-band ESR works [101,102] on high-κ SHs reported on stacks of (100)Si with nanometer-thick ($\sim$0.5–20 nm) ALCVD layers of ZrO$_2$, Al$_2$O$_3$, and SiO$_{2(x)}$, and simultaneously addressed the role of SiO$_x$ and Al$_2$O$_3$ interlayers. The precursors used (pressure $\sim$1 Torr) were Al(CH$_3$)$_3$ and H$_2$O, and ZrCl$_4$ and H$_2$O for the deposition of the Al$_2$O$_3$ ($\sim$0.5–3 nm) and ZrO$_2$ (5–20 nm) layers, respectively, while the SiO$_x$ films ($\sim$0.5 nm) were grown in H$_2$O at 300°C (20 min).

Upon first application on samples in the as-received state, only disappointingly feeble ESR signals could be observed. But from previous experience with the Si/SiO$_2$ system, this was kind of anticipated given the potential passivating role of occurring H during layer deposition (cf. H-rich ambient used in ALCVD). This suspicion was confirmed as upon

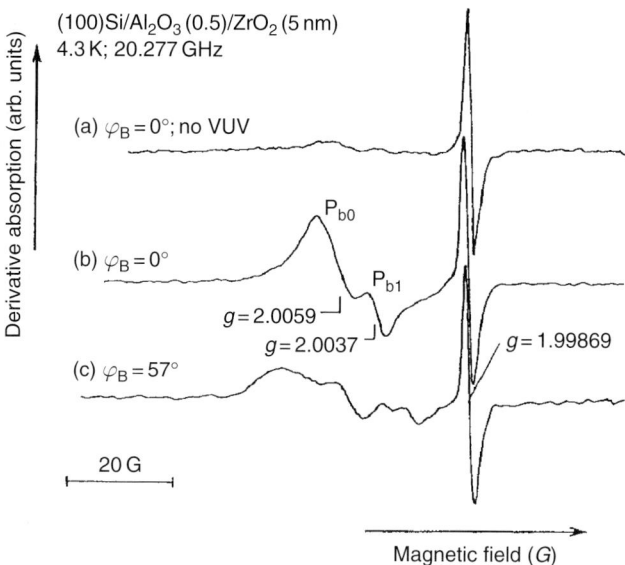

Fig. 7. Derivative-absorption ESR spectra observed at 4.3 K for two directions of $\bar{B}$ in the $(0\bar{1}1)$ plane on a (100)Si/Al$_2$O$_3$ (0.5)/ZrO$_2$(5 nm) stack (ALCVD; 300°C): (a) as grown; (b, c) after RT VUV irradiation (8.48 eV; ~5 min) to photodissociate H from passivated defects, clearly revealing the presence of P_{b0}, P_{b1} defects at the Si/Al$_2$O$_3$ interface. The signal at $g = 1.99869$ stems from a comounted Si:P marker sample. Applied spectrometer settings implied a modulation field amplitude of 0.4 G and incident $P_\mu \sim 2$ nW.

additional VUV irradiation, distinct signals did appear. As main observation, prominent P_{b0} and P_{b1} signals are observed (the sole defects observed), the archetypical defects for the (100)Si/SiO$_2$(SiO$_x$N$_y$) interfaces. This is illustrated in Fig. 7 for the (100)Si/Al$_2$O$_3$(0.5)/ZrO$_2$(5 nm) system for the $\bar{B}\|\bar{n}$ case (magnet angle $\varphi_B = 0°$), showing the appearance of P_{b0} and P_{b1} signals at zero-crossing g values of 2.0059 and 2.0037, respectively (curves a and b). The signals have been identified based on observed g values and specific anisotropy (g mapping; cf. Fig. 3) observed for $\bar{B}$ rotating in the $(0\bar{1}1)$ substrate plane. Curve (c) in Fig. 5 illustrates the spectral branching (anisotropy) for $\varphi_B = 57°$ ($\bar{B} \sim \|[111]$), with the P_{b0} and P_{b1} signals split in two and three components, respectively. The strict crystallographic correlation with the Si substrate implies that the defects pertain to the nominal (100)Si/Al$_2$O$_3$ interface, a conclusion confirmed by the study of plain (100)Si/Al$_2$O$_3$(3 nm) ALCVD structures.

With respect to the application of ESR, as a main finding, through the observation of the P_{b0}/P_{b1} interface defect attribute (system), the results reveal the primary interface to be Si/SiO$_2$ like. The defects are observed in enhanced densities, i.e., $P_{b0} \sim 5.9$ $(\pm 0.3) \times 10^{12}$ cm^{-2}, $P_{b1} \sim 1.5$ $(\pm 0.2) \times 10^{12}$ cm^{-2}, compared with standard Si/SiO$_2$, where both $[P_{b0}]$, $[P_{b1}] \sim 1 \times 10^{12}$ cm^{-2}. Moreover, additionally, line shape and width analysis reveals enhanced strain at this interface. Thus, while basically Si/SiO$_{2(x)}$ like, the interface appears of inferior technological quality. Similar results were obtained for the (100)Si/SiO$_x$(0.5)/ZrO$_2$(5 nm) system, although here it should be added that, given the incorporation of an SiO$_x$ interlayer at the onset, the observation of the P_{b0}/P_{b1} defect attribute might rather have been expected. The finding about the P_bs and the SiO_x interlayer was confirmed by subsequent X-band ESR works [102,104] on ALCVD (100)Si/Al_2O_3 and an electrically detected magnetic resonance investigation on ALCVD (100)Si/Al_2O_3 and (100)ZrO_2 entities [103]. The latter ESR works also reported an additional signal, the D center (generically a $Si_3\equiv Si^{\bullet}$ defect), which is generally ascribed to unpaired Si bonds in disordered/amorphous Si environment and has likely originated from damage in the Si substrate [100,103].

But, as a second major finding of ESR here, this situation may be drastically improved through additional post deposition annealing (PDA). Using the P_{b0}, P_{b1} densities as pertinent criterion (recall that the P_{b0}s constitute the dominant interface trap system) [59,60], this is illustrated in Fig. 8(a) for the (100)Si/Al_2O_3(0.5)/ZrO_2(5 nm) system. It is seen that (close to)

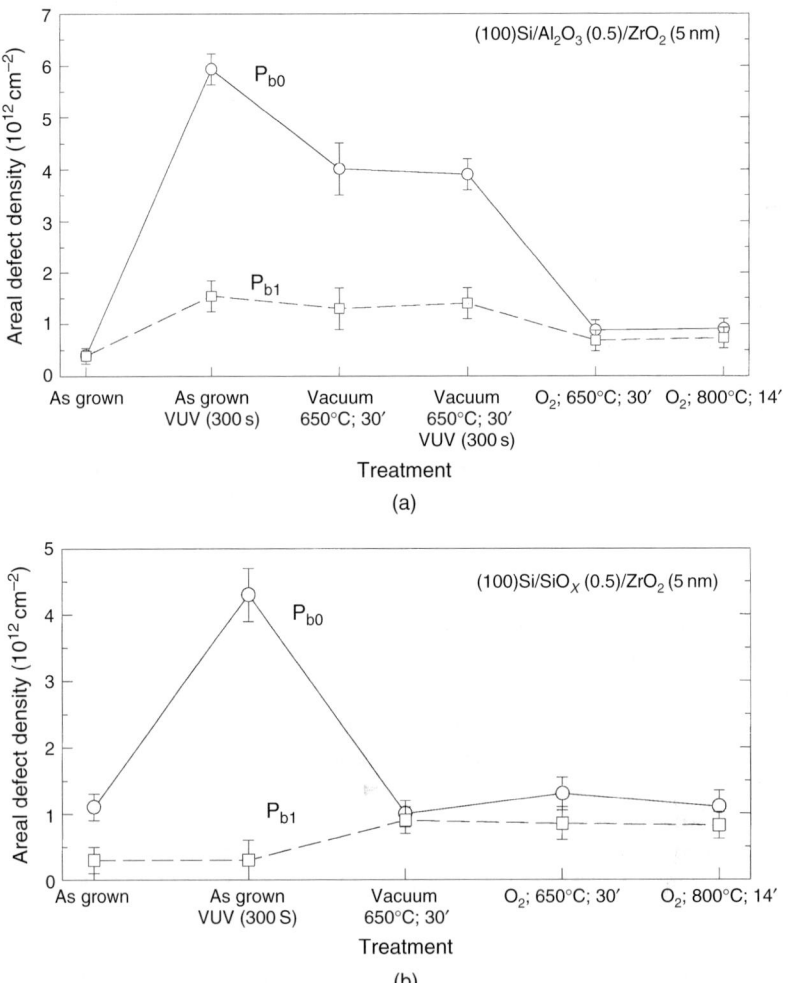

Fig. 8. Evolution of ESR-active P_{b0}, P_{b1} defects at the (100)Si/SiO_x interface in ALCVD grown (100)Si/SiO_x(0.5)/ZrO_2(5 nm) (panel a) and (100)Si/Al_2O_3 (0.5)/ZrO_2(5 nm) (panel b) stacks under various post deposition treatments.

standard quality Si/SiO$_2$ interface may be attained by PDA (30 min) at 650°C. Yet, to be noticed is that this has to be performed in O$_2$ ambient (1 atm), as vacuum ($<4\times 10^{-6}$ Torr) appears inadequate. As depicted in Fig. 8(b), this contrasts with the (100) Si/SiO$_x$(0.5)/ZrO$_2$(5 nm) case, where annealing in vacuum suffices. An additional supply of oxygen is needed in the (100)Si/Al$_2$O$_3$ case, which is understood in microscopic context [101,112], i.e., the initial abruptness of the interface prevents the required thermal adaptation to occur without additional SiO$_x$ interlayer growth. As one more result, a minimal SiO$_x$ interlayer thickness (≥ 0.5 nm) appears requisite. It comes as one more merit of ESR in that the method also succeeds to bring out the subtle difference between the two kinds of interfaces studied also in this respect.

This illustrates another valuable potential of ESR, namely, tracking the interface (quality) as a function of some post formation treatments (e.g., annealing, electrical stressing, diffusion of reactive species) through monitoring the properties (in first place densities, but also strain-sensitive ESR parameters such as line width and g spread [58,101,113]) of (unavoidably) inherently incorporated interface point defects. Put somewhat picturesquely, the defects serve as kind of horse of Troy with respect to the interface, of extreme (atomic) local sensitivity by nature.

Pertinently, we should add that model calculations based on constraint theory also bear out the necessity for interposing an ultrathin SiO$_2$ layer between the Si substrate and high-κ oxides (Al$_2$O$_3$, Ta$_2$O$_3$, TiO$_2$) to attain device-grade interfaces [114].

The finding of the occurrence of an SiO$_{2(x)}$ interlayer with attendant P$_b$-type interface defects was soon also reported by ESR for the Si/HfO$_2$ system in an ESR study [105] on (100)Si/HfO$_2$ HSs grown by three variants of CVD, i.e., ALCVD at 300°C using the HfCl$_4$ and H$_2$O precursors, MOCVD at 485°C from tetrakisdiethylaminohafnium ([(C$_2$H$_5$)$_2$N]$_4$Hf) and O$_2$, and nitrato CVD (NCVD) at 350°C using the nitrato precursor Hf(NO$_3$)$_4$, that is, nominally in H and C-free conditions. Also here, in the as-grown state, the P$_b$-type defects were reported as predominant centers, with no defects observed from the HfO$_2$ layers, as illustrated in Fig. 9. For the first two types, signals were observed only after 8.48 eV H photo-dissociation treatment. For the third type, no such treatment was needed, indicating indeed a negligible presence of H during the NCVD. As also clear from Fig. 9, the detected density of P$_{b0}$ defects is large, referring to a much inferior Si/SiO$_{2(x)}$ interface quality. It might have arisen from the incorporation of N in the near-interfacial oxide. In compliance, a room temperature X-band ESR work [106] also reported the observation of P$_b$ defects at the interface of nominally (111)Si/HfO$_2$(145 nm) entities grown by NCVD.

So, it appears that deposition of metal oxides directly on Si typically results in the formation of an Si-oxide-like interfacial layer. This was also revealed by numerous topographic/imaging techniques, such as medium-energy ion scattering (MEIS), high-resolution transmission electron microscopy (HRTEM), and X-ray photoelectron spectroscopy (XPS) [73,80,81,115–117]. Thus, these results would indicate that the formation of an SiO$_{2(x)}$ interlayer appears endemic for Si/high-κ metal oxide systems. However, both by independent electrical measurements [97,118–120] and by ESR observations [100,101,106,110], it has been evidenced that in this way an interface of technological quality can be realized in terms of interface-state density and passivation behavior in hydrogen, much alike the standard thermal Si/SiO$_2$ system. While highly beneficial in this respect, it obviously conflicts with the EOT restriction, the more so with increasing relative thickness of the "low-κ" SiO$_2$ ($\kappa=3.9$) interlayer. Yet, the insertion of a well-controlled (in thickness) SiO$_2$ layer of minimal thickness is currently adopted as modus vivendi route of technological progress [81,121].

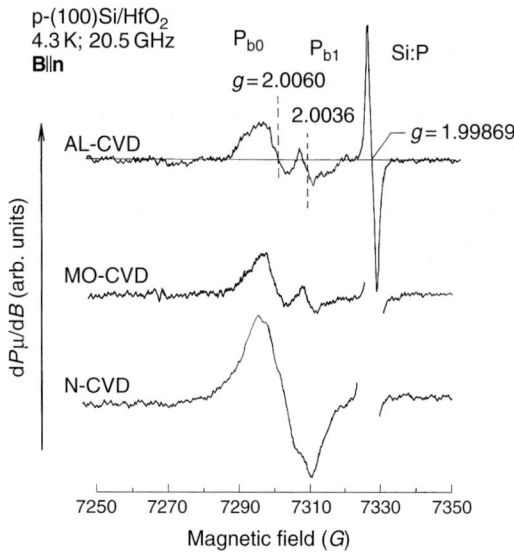

Fig. 9. Derivative-absorption ESR spectra observed at 4.3 K with $\bar{B}\|\bar{n}$ on (100)Si/HfO$_2$ samples grown by three variants of chemical vapor deposition (CVD), i.e., from top to bottom, atomic layer CVD at 300°C, metal-organic CVD at 485°C, and nitrato CVD at 350°C. All samples were subjected to hydrogen photo-dissociation prior to the measurement. Signals observed at $g = 2.0060$, $g = 2.0036$, and $g = 1.99869 \pm 0.00002$ stem from the P$_{b0}$, P$_{b1}$ centers and from the comounted Si:P marker, respectively. Spectrometer settings include an applied modulation field amplitude of 0.4 G and incident microwave power 0.8 nW.

Facing these facts, there are two possible modus operandi: The presence of an interlayer can be accepted as something unavoidable, making it more useful to focus on ways to control and optimize the thin SiO$_2$ interlayer rather than trying to eliminate it [77,110]. A second possibility is concentrating on "new" high-κ oxides such as LaAlO$_3$ ($\kappa = 20$–27), which keeps promises to form a truly abrupt interface with Si [122], with no perceivable SiO$_x$ interlayer. In this respect, much is also expected from pseudo-epitaxial growth of insulating layers [95].

3.3. Affecting point defects

3.3.1. Monitoring specific influences on point defects

The unique potentiality of ESR to isolate defects species individually enables one to assess particular properties of defects even without any kind of atomic identification having been reached. Having a defect type well isolated in terms of its ESR signature (signal) serves as a unique identity mark. From then on, through monitoring of the very particular ESR resonance parameters such as signal intensity (number of spins), width, resonance field (g factor), it enables the very focused appraisal of internally and externally induced influences on the defects.

These, of course, may be numerous encompassing virtually all agents that may affect solid-state structure, both atomic/configurationally and electrically. By way of example, we may mention: (a) applying external *stress* to the specimen under study, e.g., hydrostatic (negative) pressure [123], leading to densification linked with interatomic rearrangements potentially resulting in crystalline phase changes and structural reordering in the case of amorphous solids. As to the latter, in silica glasses, it may be linked with alteration of the distribution in *n*-member (SiO_4 tetrahedra) ring distribution. (b) *Stress in thin layers* and interfaces; this may be present inherently as a result of solid–solid structural mismatch (e.g., at the interface of a crystalline solid with an amorphous layer – as outlined above, a prototype here is the c-Si/a-SiO_2 interface – or a pseudo-epitaxial grown structure with some degree of lattice mismatch [95]) or mechanical stress may be applied through, for example, mechanical bending of the substrates to which the layers adhere [124]. (c) Injection of *electrical charge*, which may affect defects directly (ESR activity) through trapping of charge directly in an electronic level of the defect site [125,126], or via Coulombic interaction with charges trapped more remotely. While the effect of charge on defects is of course a long-known and widely studied fact in general, it takes a specific point of interest in the electronic device world based on semiconductor/insulator hetero junctions: In crucial electronic components, the tight control of charge trapping/detrapping by point defects is crucial for device performance, or even, for attaining any functionality of a device at all. (d) *Chemical interaction* – assessing the chemical activity of defects, i.e., chemical interaction with supplied elements (impurities), leading to chemical bonding, is a most notable attribute of ESR, well known of course in chemistry. As an immense field in applied chemistry, numerous such interactions have been studied compiled in numerous authoritative overviews [127].

Chemical activity takes a special position for the SH structures. Without doubt a most notorious example here concerns the interaction of embedded defects with *hydrogen* [49,128], supplied either from an external source or internally through stimulated release (e.g. irradiation by energetic particles (X and γ photons, electrons, ions) [49], hot carrier impact [129], charge trapping [130–132], etc.). The reasons for this may appear obvious: (1) if extreme precautions are not taken, hydrogenic species (H, H_2, H_2O, OH, H_3O, ...) abound in semiconductor/insulator entities (as in nature in general); (2) these species, particularly the very swiftly diffusing atomic H, may readily inactivate unpaired orbitals through bonding, often considered a beneficial attribute. In particular with respect to the thermal Si/SiO_2 system, it appears hydrogen has a hand in almost every essential structural effect. The ubiquity here of hydrogenic species makes the incorporation of H in the system virtually unavoidable, to such extent that hydrogen, next to Si and O, is considered a third intrinsic element of the Si/SiO_2 entity. Its effect may be benign (e.g., passivation of interface defects [6,30,31]), as well as detrimental (e.g., H-induced interface state generation [7,70,133–135]).

A first well-known example is the effect of H on the carrier-supplying dopants in semiconductors – at the heart of the enabling of semiconductors per se – e.g., passivation of acceptors (B) [136] and donors (P) [137] in Si, simply pictured as P-H and B-H bond formation, respectively.

Second, zooming in on the workhorse and still dominant semiconductor, Si, the interaction of the Si DB defect has taken a particular place in ESR research, often crucially related to aimed applications. Here two main cases come to the fore: (1) In the well-known case of a-Si, the level of attained success in passivation of the electrically and nefarious Si DBs (pictured as $Si_3 \equiv Si^\bullet$), inevitably introduced during deposition of the amorphous layer, is primordial to its solar energy applicability without which no such layer would ever result in useful functionality [138]. (2) A most notable case concerns the interfacial Si DB – P_b-type defect in ESR jargon, as outlined before – in the basic c-Si/SiO_2 entity underlying the success of the current CMOS-based IC technology. As well-known now, these defects, which operate as detrimental fast interface states, need to be carefully passivated or eliminated to sub-critical levels. The interaction with H has been long hinted at before, yet, inherently, without atomic-specific proving power. Technological progress here has mainly occurred empirically. As a merit, the kinetics of the P_b-type defects interaction with hydrogen (both atomic and molecularly) has been almost exclusively quantified through minute ESR analysis enabled through the fact that each such P_b-type of defect can be well isolated by its characteristic ESR spectrum. A predominant attribute of ESR here is its capability of monitoring the absolute density of defects, most accurately attained when referenced to the method of a comounted well-calibrated marker sample. Although the method is laborious, it has enabled the so far most accurate and reliable inference of interaction kinetics of such defects, with no doubt left as to the specific individuality of the defect(s) sampled. For example, with respect to P_b at the (111) Si/SiO_2 interface, it has led to the full H-interaction (passivation/depassivation) picture, spanning a broad temperature range (RT-1000°C). Among others, it puts on an insightful sound scientific basis items such as selecting the optimum temperature for applying forming gas (10% H_2 in N_2) anneal – a standard crucial step in the involved device-manufacturing process flow. Furthermore, through deeper analysis it may provide access to interaction modalities on atomic level, e.g., how thermally activated dissociation of an interfacial Si–H bond may proceed in an initial stadium; see next paragraph for more details.

Albeit generally far less extensive and convincingly, ESR has also been applied to the study of the interaction of H for other types of point defects in the Si/SiO_2 entity, with as main aim the inference of the activation energy for defect passivation by H. Examples here are the EX center in thermal SiO_2 layers [51], the E'_δ [139] defect in buried oxide layers of separation by implantation of oxygen (SIMOX) structures, and the P_{bN} center in SiC/SiO_2 entities [140].

Passivation by H of traps (point defects) is also a concern in the currently intensely studied semiconductor/high-κ insulator heterostructures, in which, generally, the occurrence of an enhanced density of traps in near-interfacial layers and in the oxide is a major concern [141]. By way of illustration of the adequacy of standard ESR, here we summarize below the results for the P_b center in (111)Si/SiO_2.

3.3.2. Chemical interaction kinetics: P_b–H case

With respect to the basic Si/SiO_2 structure, the interaction kinetics with hydrogen (in molecular as well as in atomic form) has been studied by ESR in more or less extent for all three kinds of P_b-type interface defects. Though it is well known that, rather than (111)Si, the (100)Si face is the preferred one in device technology, it has most completely be done so for the P_b center in (111)Si/SiO_2. Clearly, when referring to the limited sensitivity of ESR, in particular when applied to interface analysis, a main reason here is the most favorable spectroscopic situation for applying ESR: (a) there appears only one type of interface defect, so no marring interference with overlapping signals; (b) it exhibits favorable signal anisotropy (axial g matrix: $g_\parallel = 2.00136 \pm 0.00003$; $g_\perp = 2.0088 \pm 0.0001$) making the various g map branches corresponding to the chemically identical but geometrically differently oriented defect sites to coincide for the $\bar{B}\|\bar{n}$ situation, resulting in a singular narrow signal ($\Delta B_{pp} \sim 1.5–2\,G$); (c) among the three P_b-type defects known, its naturally occurring intrinsic density (N_i) is highest ($\sim 5 \times 10^{12}\,cm^{-2}$ for standard oxidation in the range $T_{ox} \sim 800–950°C$); (d) through combination of POA treatments at appropriate temperatures in vacuum and hydrogen ambient the density of *permanent* P_b centers (including the preexisting intrinsic ones and newly created ones) may be drastically enhanced (up to $\sim 3.2 \times 10^{13}\,cm^{-2}$) – the *POA-induced interface degradation effect* [52,69,70]. Obviously, when it comes to experimental sensitivity, the latter effect may be well applied to the benefit in studying the hydrogen-defect interaction kinetics.

Using ESR on thermal (111)Si/SiO$_2$ entities [39,66], Brower pioneered the first full quantification of the interaction kinetics of unpaired interface Si bonds, i.e., P$_b$s, with *molecular* H$_2$. His work resulted in an elegant model, henceforth called the simple thermal (ST) scheme, which was soon accepted as definitive upon disclosure. In fact, it concerns a least-complications scheme, the red wire being that the key interactions are simply rate limited by the availability of "reactive" P$_b$ sites, i.e., [P$_b$] and [P$_b$H] for passivation and dissociation, respectively. The P$_b$ passivation in *molecular* H$_2$ and dissociation in vacuum (studied isothermally in the temperature ranges 230–260 and 500–590°C, respectively) are modeled by the simple chemical reactions.

$$P_b + H_2 \xrightarrow{k_f} P_bH + H \qquad (24)$$

$$P_bH \xrightarrow{k_d} P_b + H \qquad (25)$$

leading to the respective first-order differential rate equations $d[P_b]/dt = -k_f [H_2][P_b]$ and $d[P_b]/dt = k_d(N_0 - [P_b])$, with solutions

$$[P_b]/N_0 = \exp(-k_f[H_2]t) \qquad (26)$$

$$[P_b]/N_0 = 1 - \exp(-k_d t) \qquad (27)$$

where the rate constants are given by the Arrhenius equations $k_f = k_{f0}\exp(-E_f/kT)$ and $k_d = k_{d0}\exp(-E_d/kT)$. Here, $N_0 = [P_b] + [HP_b]$ (maximum number of recoverable ESR-active P$_b$ centers), [H$_2$] the volume concentration of H$_2$ at the interface, and k Boltzmann's constant. The inferred values of the activation energies (E_f and E_d) and preexponential factors (k_{f0} and k_{d0}) are listed in Table 1. According to Eq. (24), the P$_b$ defect passivation is thus pictured as simple bonding to H – the basic hypothesis.

Regarding passivation, basic ingredients of the ST model include: (a) H$_2$ is physically absorbed into the SiO$_2$ layer and, as verified experimentally [39], diffuses molecularly among

Table 1
Kinetic parameters inferred within the GST model (see Eqs (28, 29)) for thermal passivation in molecular hydrogen H$_2$ and dissociation in vacuum of P$_b$ (P$_b$H) defects at the interface of standard (111)Si/SiO$_2$ samples grown at $T_{ox} \sim 870$°C (passivation) and 970 °C (dissociation)

	Dissociation[a]	
E_d (eV)	σE_d (eV)	$k_{d0}(10^{13}\,\mathrm{s^{-1}})$
2.83 ± 0.03	0.09 ± 0.03	1.6 ± 0.5
	Passivation[b]	
E_f	σE_f (eV)	k_{f0} ($10^{-8}\,\mathrm{cm^{-3}\,s^{-1}}$)
1.51 ± 0.04	0.060 ± 0.004	9.8 (+8/−5)

[a] Ref. [54].
[b] Ref. [49].

the accessible interstices of the SiO_2 network, including the reaction site at the P_b center; (b) rather than diffusion, the passivation is limited by direct P_b–H_2 reaction at the P_b reaction site; (c) there is no preliminary cracking at internal sites in the SiO_2 on their way toward the interface; (d) the interfacial concentration of H_2 is *assumed* equal to the physical solubility of H_2 in bulk vitreous silica, and the supply of H_2 is considered unlimited; (e) for both the passivation and dissociation steps, *single-valued* activation energies (E_f and E_d) are presumed; (f) it was argued that the deduced activation energies for Eqs (24) and (25) are equal or only slightly larger than the energy difference between the respective initial and final constituents. Hence, the reverse passivation and dissociation reactions would proceed essentially spontaneously (little or no barrier) in the presence of *atomic* hydrogen.

Here, we may recall some remarkable properties of the P_b system, some serving as basic pillars for the passivation/dissociation study. (a) The P_bs appear thermally stable for annealing at temperatures up to 1000°C; they may be repeatedly passivated in H_2 and dissociated in vacuum without removing the physical defect site [53,69]. (b) The high thermal stability indicates that the defects do not migrate over the interface. Once created, the defect appears persistently stabilized by the surface. Further oxidation is a way to affect them. (c) Very remarkably, the P_b cage exhibits a structural integrity resistant to intrusion of a range of impurities other than H. (d) In thermal (111)Si/SiO_2 ($T_{ox} \sim$ 300–950°C), a P_b density $N_0 = \sim 5 \times 10^{12}$ cm^{-2} is invariably incorporated at the interface [58].

On the theoretical side, the ST model found support [143] from semiempirical spin-unrestricted molecular orbital cluster calculations applied to a 22 Si/6 O (111)Si/SiO_2 cluster hosting a P_b defect, complemented with terminating H atoms. The values $E_f = 1.32$ eV and $E_d = 2.7$ eV were reported, in fair agreement with Brower's data [39,66] (1.66 and 2.56 eV, respectively). The theory purely bases on individual P_b–H_2 interaction, taking into account attendant configurational changes.

However, although the simple ST model was readily accepted as definitive upon disclosure, over time, there appeared deficiencies in the ST building. On the experimental side, some incorrect ESR practice was revealed from later experimental work focusing on the signal shape analysis. Also, there appeared problems with the consistency of inferred physical parameters with underlying physical insight on which the ST model was based. In hindsight, the accumulated deficiencies basically appear to have arisen from the fact that the ST model inference was based on inadequate ESR data [48] of insufficient extent.

The deficiencies were of various types [48,53]. In hindsight, a main one appeared to be that the ST model was based on the assumption of single-valued activation energies (E_f and E_d) for the passivation and dissociation steps, which appeared unfounded.

In later ESR work, these inadequacies were addressed separately, first for the passivation of P_b in H_2 [48] and later for the dissociation kinetics [53]. Some typical ESR spectra illustrating the effect of stepwise passivation of the P_bs are shown in Fig. 10. This resulted in physically consistent schemes for both interactions, referred to as the generalized simple model (GST). It matches underlying physical insight, and the inferred values of the pertinent physical parameters are collected in Table 1. As to this improvement, a major finding was the demonstration of the existence of significant spreads, σE_f and σE_d, in the activation energies E_f and E_d for passivation and dissociation, respectively. Without this recognition, no consistent description could be attained. The existence of such spread just traces back to the stress-induced variations in the P_b atomic morphology over the various defect sites [48,58,113]. There exists a statistical spread in interfacial stress – quite natural in a physical situation – resulting in a configurational distribution for the P_b system. In a simple description, the spreads emerge as

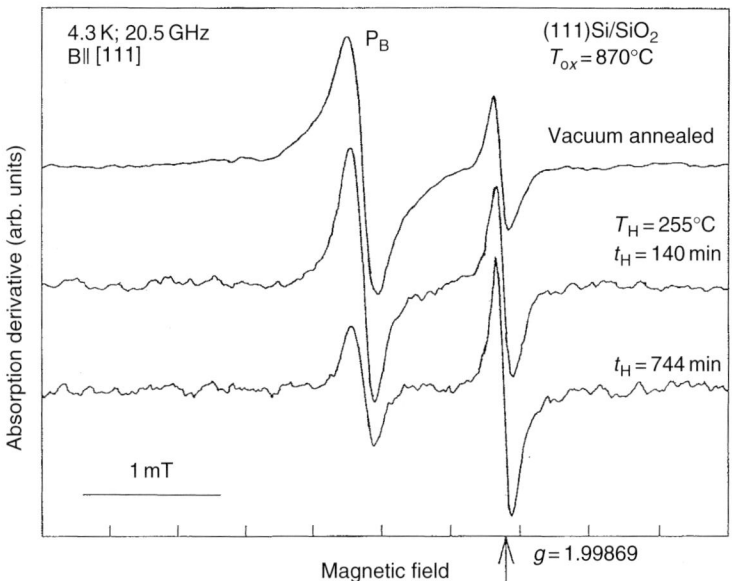

Fig. 10. ESR spectra illustrating the effect of stepwise passivation of P_b s in 1-atm H_2 carried out at 255°C for indicated times [48]. The top trace (pre-passivation state) corresponds to $[P_b] \sim 7 \times 10^{12}$ cm^{-2}. As evident from the SiP marker signal at $g = 1.99869$, the spectrometer gain was progressively enhanced with increasing passivation level. Correspondingly, the line shape and line width are seen to vary as a result of the gradually receding impact of interdefect dipolar interaction.

a measure of the interface structural quality. In a previous work, it could, for example, be quantified in a spread σh in the mean distance of the defect apex Si atom to the basal plane formed by the three back-bonded Si atoms. It is likely that the ensuing weak orbital rearrangements will cause this spread to be reflected in virtually any fundamental physical parameter of the point defect, such as g value and line width and, which is important for the current discussion, in the P_b–H bond strengths and, a fortiori, in the activation energies E_f and E_d. The known spreads in the $+/0$ and $0/-$ electronic P_b levels is an electronic manifestation related to this fact. Unless dealing with a point defect system residing in a *perfect* crystal, the presence of such spreads is, based on the quantum mechanical insight, a natural fact to be expected, and in this perspective it may appear surprising that the existence of such spreads in physical parameters is not taken into account from the onset in realistic modelling. It appears that such unrealistic "perfect crystal" approach has thwarted many interpretations before.

In later work, it has been exposed that the magnitudes of these spreads are not unique, i.e., these depend on the thermal history (e.g., T_{ox}) of the Si/SiO$_2$ structure studied [53, 68]. Hence, as the interfacial strain varies with T_{ox} (a significant relaxation was observed to initiate at $\sim$750°C upwards), so will σE_f and σE_d. Thus, it is important to emphasize, that the inferred value $\sigma E_f \sim 0.06$ eV pertains to standard thermal Si/SiO$_2$ grown in dry O$_2$ (1 atm) at 870°C. A decrease in σE_f from $\sim$0.11 eV to 0.042 eV has been found for T_{ox} increasing from 250 to $\sim$1100°C [68].

Including the spreads and basing on extended sets of ESR data, a generalized consistent simple thermal (GST) model was attained, also rooted on the chemical pathway laid down in

Eqs (24) and (25). The kinetics of passivation and dissociation are now found accurately described, respectively, by the generalized expressions

$$\frac{[P_b]}{N_0} = \frac{1}{\sqrt{2\pi}\sigma_{E_f}} \int_0^\infty e^{-\left[\left((E_{fi} - E_f)^2 / 2\sigma_{E_f}^2\right) + tk_{f0}[H_2]\exp\left(-E_{fi}/kT\right)\right]} dE_{fi} \qquad (28)$$

and

$$\frac{[P_b]}{N_0} = 1 - \frac{1}{\sqrt{2\pi}\sigma_{E_d}} \int_0^\infty e^{-\left[\left((E_{di} - E_d)^2 / 2\sigma_{E_d}^2\right) + tk_{d0}\exp\left(-E_{di}/kT\right)\right]} dE_{di} \qquad (29)$$

where now E_f and E_d represent *mean* activation energies. The values of the pertinent P_b kinetic parameters inferred for (111)Si/SiO$_2$ grown at 870–970°C are listed in Table 1. With this GST model, a consistent picture has been attained of the P_b–H interaction kinetics, accounting for all experimental data, with the inferred kinetic parameters matching underlying physical insight. In conjunction with theoretical work, a pathway for P_b–H dissociation has been inferred [53], and, in checking the model, separate supportive evidence was provided for the first-order character (i.e., no second-order kinetics) of the (P_bH) dissociation reaction.

A noteworthy aspect of the basic interaction scheme embraced by Eqs (24) and (25) is that its net effect is the dissociation of the H$_2$ molecule (both steps resulting in emission of a H atom) with a net apparent activation energy ~4.34 eV, close to the value of 4.52 eV for dissociation in vacuo. It effectively unveils P_b as a catalyst for H$_2$ molecule cracking.

Finally, in a synthesizing next step, with the consistent separate descriptions at hand of both the dissociation and passivation steps, the combination of these led to the GST model embodying the *full interaction case*, that is, facing the more realistic situation that both the reactions (24) and (25) continuously proceed in parallel (simultaneous dynamic processes). For example, during a passivation treatment in H$_2$, there will be a simultaneous countering probability for dissociation, which may not be negligible in general (for $T \geq 360$°C). This counteracting effect, as outlined, is intensified by the existence of spreads in activation energies.

In short, the essence of the complete interaction scheme is embedded in the total rate equation, obtained by combining the separate rate equations pertaining to the model chemical pathway represented by the chemical reactions (24) and (25) for passivation in H$_2$ and dissociation, respectively

$$\frac{d[P_b]}{dt} = (-k_d + k_f[H_2])[P_b] + k_d N_0 \qquad (30)$$

Assuming that the concentrations of molecular H$_2$ at the interface and in the ambient are continuously in equilibrium, Eq. (30) can be solved explicitly, giving the general expression

$$[P_b](t) = [P_b](\infty) + \{[P_b](0) - [P_b](\infty)\}\exp\{-(k_d + k_f[H_2])t\} \qquad (31)$$

with $[P_b](\infty) = k_d N_0/(k_d + k_f[H_2])$.

For the case $[P_b](0) = N_0$, i.e., starting passivation of an exhaustively dissociated P_b system, one gets from Eq. (31)

$$[P_b] = \frac{N_0}{(k_d + k_f[H_2])} \{k_d + k_f[H_2] \exp -(k_d + k_f[H_2])t\} \quad (32)$$

while for $[P_b](0) = 0$, i.e., starting from the exhaustively passivated case, expression (31) becomes

$$[P_b] = \frac{k_d N_0}{(k_d + k_f[H_2])} \{1 - \exp -(k_d + k_f[H_2])t\} \quad (33)$$

With inclusion of the spreads σE_f, σE_d, we obtain from Eq. (32)

$$\frac{[P_b]}{N_0} = \frac{1}{2\pi\sigma_{E_f}\sigma_{E_d}} \iint \exp\left\{-\frac{(E_{di} - E_d)^2}{2\sigma_{E_d}^2} - \frac{(E_{fj} - E_f)^2}{2\sigma_{E_f}^2}\right\} F(E_{di}, E_{fj}, k_{f0}, k_{d0}, t) dE_{di} dE_{fj} \quad (34)$$

where

$$F(E_{di}, E_{fj}, k_{f0}, k_{d0}, t) \equiv \frac{1}{(k_d + k_f[H_2])} \{k_d + k_f[H_2] \exp -(k_d + k_f[H_2])t\} \quad (35)$$

and $k_d = k_{d0}\exp(-E_{di}/kT)$ and $k_f = k_{f0}\exp(-E_{fj}/kT)$. The integration ranges span all occurring E_{di} and E_{fj} values. This perhaps represents the most interesting case as attaining optimum passivation is highly desired technologically.

Results of expression (34) have been calculated numerically. In a semilogarithmic plot, Fig. 11 shows the isochronal behavior of $[P_b]/N_0$ for four anneal times over the range 150–700°C (full curves) using the GST parameters tabulated in Table 1. The values inserted for $[H_2]$ correspond to the physical solubility of H_2 in vitreous silica as given in Ref. [142] [expression (5) there]. Each point on a curve in Fig. 11 gives the calculated $[P_b]/N_0$ value for annealing at the corresponding temperature for the indicated time, each time starting from the situation $[P_b]/N_0 = 1$. For reasons of comparison, also shown in Fig. 11 (dotted curves) are the results for the fictitious case of single-valued E_f and E_d, i.e., $\sigma E_f = \sigma E_d = 0$, which mirrors the initial ST model [39,66]. It illustrates the effect of the existence of the spreads σE_f, σE_d.

Starting from an exhaustively depassivated P_b bath ($[P_b] = N_0$) and for heating times in the range 10–120 min, the isochronal curves reach a minimum – i.e., optimum passivation – at anneal temperature $T_{an} \sim 390$–$430°C$. The $[P_b]/N_0$ values attained at these minima are $\sim 5 \times 10^{-7}$, i.e., about 0.5 ppm of P_bs remain unpassivated. As anticipated, it demonstrates that the P_b system, albeit rather efficient, cannot be fully passivated in this time window – obviously due to the counteracting dissociation during the essentially passivation treatment. Note that, when neglecting parallel dissociation, the GST passivation picture would predict $[P_b] \rightarrow 0$ for $T_{an} \rightarrow 410°C$ (cf. Fig. 11; dashed curve). More details about the significance of this effect with respect to technology can be found elsewhere [67].

Various interesting insights result from the developed GST picture. For one, from Fig. 11, it is seen that somewhat depending on the heating time (10–120 min), the temperature range for

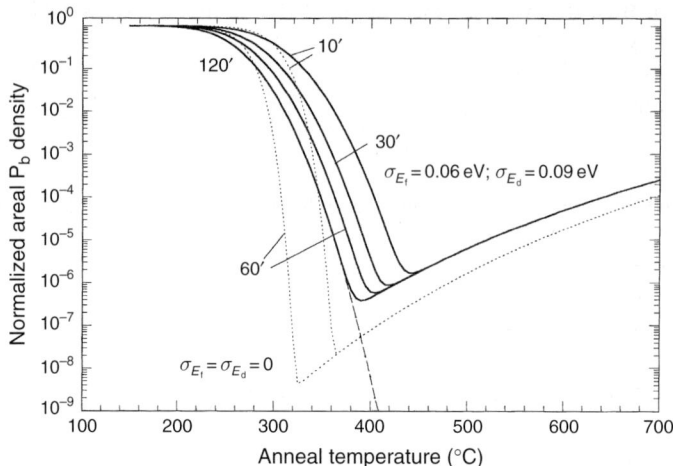

Fig. 11. Calculated isochronal behavior (solid curves) of the normalized P_b density in standard thermal (111)Si/SiO$_2$ for annealing in 1 atm H$_2$ starting form $[P_b]/N_0 = 1$, taking into account simultaneous passivation and dissociation [Eq. (34)]. The solid curves, picturing reality, are obtained using the GST parameters listed in Table 1. The dotted curves represent the fictitious case with $\sigma E_f = \sigma E_d = 0$, otherwise using the same parameters. The dashed curve represents the GST passivation description only ($t = 2$ h; simultaneous dissociation neglected) (after Ref. [67]).

optimum passivation is situated in the range $T_{an} \sim 390–430°C$ (for the $\sigma E_f = 0.06$ eV; $\sigma E_d = 0.09$ eV case). It provides insight and understanding of the procedure of the post-metallization anneal (PMA) in forming gas (10–30% H$_2$ in N$_2$), which as long before empirically established, is routinely applied at $\sim 400°C$ as standard step in device manufacture. As another example, it may be inferred that the P_b system serves as important mill of atomic hydrogen: for the case of $\sigma E_f = 0.06$ eV; $\sigma E_d = 0.09$ eV, it may be inferred that at $\sim 800°C$, an average P_b defect gets passivated and dissociated very second, meaning that two hydrogen atoms are liberated per defect per second (cf. elementary chemical reactions (24) and (25)). For an intrinsic P_b density of 5×10^{12} cm^{-2}, it implies that $\sim 10^{13}$ H atoms are generated per second at each cm^2 of interface. The atomic H production increases drastically with T_{an}. See Ref. [67] for more details.

3.4. Tracking interfaces

As outlined, the occurrence of point defects in SHs, particularly in (near) interfacial layers and dielectrics, is highly undesired because of the generally detrimental influence on device operation. However, when atomically well known (identified), we have also noticed these can be used as true atomic probes of the environment, i.e., the layers these are built in. So, conversely, the occurrence, or perhaps more precisely, sensing embedded point defects might be used beneficially to track the evolution of an interface or layer properties in general as a function of some varying (technological) parameters, such as annealing sequence, passivation treatments, gate contact material, electrical stressing, etc. One such example has already been presented in Section 3.2.2 where, in addressing the Si/high-κ structures, the evolution of the Si/SiO$_2$-type interface quality was tracked through monitoring the properties of the embedded P_b-type interface traps as a function of selected PDA treatments (cf. Fig. 8).

Here, we illustrate the ESR tracking power for another semiconductor/insulator system with a different type of interface. As addressed previously, for many (high-κ) metal oxides such as, e.g., Al_2O_3, HfO_2, and ZrO_2, deposition of the high-κ oxides directly on Si typically results in the formation of an $SiO_{2(x)}$-type interlayer [73,80,81,100,101,105,115–117]. As also remarked, this endemic feature could be beneficially exploited to establish a (close to) standard Si/SiO_2 interface quality, if it were not for the fact it conflicts with the low EOT requirements, as the SiO_x layer has a low κ value ($\sim$3.9). Facing these facts, there seem two modus operandi: First, the presence of an $SiO_{2(x)}$ interlayer can be accepted as something unavoidable, making it more sensible to focus on ways to control and optimize the thin interlayer rather than trying to eliminate it [110] – the route, as already mentioned, currently adopted for the application of the $HfSiO_xN_y$ layer as gate insulator. The second, opposite option, is concentrating on alternative high-κ dielectric that keep promises to form a truly abrupt interface with Si [122], i.e., with no perceivable SiO_x interlayer, at the same time hopefully of high electrical quality. One such candidate could be $LaAlO_3$ ($\kappa = 20$–27) as hinted [122] from first depositing $LaAlO_3$ thin films on (100)Si using MBD. Through combining various morphological/imaging and composition sensing techniques (including transmission infrared spectroscopy, MEIS, and XPS), Edge et al. [144] concluded that as-grown MBD $LaAlO_3$ films on (100)Si have less than 0.2 Å of SiO_2 at the interface, found to be stable up to temperatures of $\sim$935°C. But superb and outstanding as these are, by the nature of the methods employed, this inference of course does not track down to the single atomic level – so, checking the veracity of the conclusion would constitute a challenge for ESR, i.e., verification on true atomic level in terms of embedded occurring point defects.

Accordingly, an ESR study has been carried out on MBD (100)$Si/LaAlO_3$ structures [145], where the assessment of the (100)$Si/LaAlO_3$ interface and its evolution as a function of PDA provides interesting insight as exposed by the main results compiled in Fig. 12, showing the density of observed point defects as a function of isochronal PDA (5% O_2 in N_2 ambient; 1 atm) temperature. Several aspects are worth noting:

The as-grown state As it comes out, no ESR signals, in particular no P_b-type defects, could be observed in the as-deposited $Si/LaAlO_3$ structure, not even after applying VUV irradiation. So, indeed, on the basis of the "P_b criterion" it evidences, with atomic level sensitivity, that there is no SiO_x-type interlayer present or that this interlayer is at least substantially thinner than 3 Å [146]. It indicates the interface to be abrupt, in good agreement with previous results. In passing, we note that this is in sharp contrast with other stacks of (100)Si with high-κ layers, such as Al_2O_3, HfO_2, and ZrO_2, where the presence of such an interlayer appeared inevitable, even in the as-deposited state, which was revealed by ESR.

Post deposition heat treatment In terms of occurring defects, upon subsequent PDA treatments, the abrupt interface remains unaltered and proves to be thermally stable, even after annealing up to $T_{an} \leq 800°C$ and additional VUV treatment. But as can be seen, after annealing at $T_{an} \sim 860°C$, P_{b0} defects start becoming observed, compellingly indicating that an SiO_x-type interlayer has formed. Upon annealing at a slightly higher temperature $T_{an} \sim 888°C$, the SiO_2-associated EX defect appears, so, delayed over $\sim$30°C in terms of T_{an}. As EX constitutes an SiO_2-related defect, this observation corroborates the presence of an SiO_x-type interlayer and also indicates an additional growth of the interlayer. The defect was not observed in the sample annealed at $T_{an} \sim 860°C$ even though the presence of an SiO_x-type interlayer in this sample is signaled by the observation of P_{b0} defects. It suggests that a minimal thickness

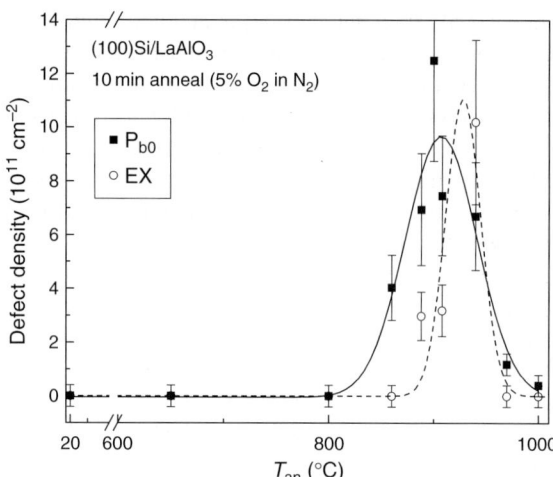

Fig. 12. Compilation of defect densities in (100)Si/LaAlO$_3$ entities inferred from K-band ESR observations as a function of post deposition isochronal annealing in N$_2$ + 5% O$_2$ (10 min) for P$_{b0}$ and EX centers, represented by closed and open symbols respectively. The solid and dashed lines are Gaussian curves, merely meant to guide the eye in exposing the peaking in defect generation and the somewhat lagging behind (~30°C) of EX production vis à vis P$_{b0}$. Data points at zero defect density symbolize that no signal could be detected.

of the SiO$_x$ interlayer is needed for (ESR) detectable formation of EX defects, at least more substantial than required for effective Si/SiO$_x$-interface formation. Thus the retardation in EX appearance vis-à-vis P$_{b0}$ would indicate an additional growth (or modification) of the interlayer. The interlayer thickness, however, is unknown. In broader context, it is interesting to note that the generation of EX centers on annealing at elevated temperatures in oxygen-containing ambients appears symptomatic for stacks of high-κ metal oxide layers on Si [110].

With increasing T_{an}, evolution of the defect densities show a peaked behavior, highlighted by the simulated Gaussian curves (cf. Fig. 12). The maximum P$_{b0}$ density [P$_{b0}$] ~ (1.3 ± 0.3) × 10^{12} cm^{-2} is obtained at T_{an} = 900°C, close to that found in standard thermal Si/SiO$_2$ entities. So, for T_{an} increasing above 900°C, the decreasing P$_{b0}$ density would indicate that the interface starts to break up. The EX density displays a similar peaked behavior as a function of T_{an} as the P$_{b0}$ density, but reaches its maximum [EX] ~ (1.0 ± 0.3) × 10^{12} cm^{-2} at a slightly higher T_{an} ~ 940°C. Interestingly, the EX density still increases in the range T_{an} ~ 900–940°C while the P$_{b0}$ density already starts to decrease, indicating that for annealing from T_{an} > 900°C onward the interface starts to break up first.

So, the overall picture emerging from the curves in Fig. 12 is that, as compared to P$_{b0}$, the manifestation of the EX peak is delayed in terms of T_{an}. The true character of the Si/SiO$_x$-type interface is disrupted first (elimination of interfacial Si DBs), but for T_{an} ≥ 940 °C, the defects rapidly disappear altogether, pointing to drastic disintegration of the interfacial region, i.e., elimination of the "pure" SiO$_x$ component. For clarity, this disappearance of the ESR-active centers is not due to inadvertent H-passivation, as verified by applying additional VUV irradiation after some PDA steps.

The atomic level information currently acquired from ESR could be well fit into the main previous results obtained from morphological/compositional studies, in particular with those of a recent study on Si/a-LaAlO$_3$ structures as a function of isochronal RTA steps (1 atm N$_2$; 20 s). A change in the structure of LaAlO$_3$ from amorphous to crystalline was revealed after a 935°C RTA together with La and Al penetration into the Si after RTA at $T_{an} \geq 950°$C. The ESR data may well fit within this morphological picture: in terms of T_{an}, the initiation of the formation of a Si/SiO$_x$-type interface may be linked with the very early onset of LaAlO$_3$ film crystallization, followed by some more substantial SiO$_x$-type interlayer growth with increasing T_{an}. (For clarity, though, the appearance of the P$_{b0}$ centers cannot be directly linked to the grain boundary regions per se, as this would conflict with the observed ESR spectral anisotropy in registry with the crystalline (100)Si surface.) Then, for T_{an} further increasing above ~940°C, the elimination of the pristine nature of the SiO$_x$ component and concomitantly the Si/SiO$_{2(x)}$ nature of the interface, could be linked to the progressing diffusion of La and Al into the Si substrate and possibly the formation of a silicate interlayer. It is possible that the onset temperature for significant La and Al out diffusion in the current case is somewhat lower considering the applied longer PDA treatment times (10 min) as compared to previous RTA work (10–20 s).

In this comparison, a noteworthy observation is that it appears that ESR detects the initiation of crystallization of the a-LaAlO$_3$ film somewhat at lower T_{an} than in previous work. In one interpretation, it may bear out the fact that ESR is prone to detect interfacial reshaping in a very embryotic stadium, ahead of standard morphological/compositional methods.

Thus, sensitive ESR analysis enables us to unveil elemental aspects (quality, nature, and stability) of evolving semiconductor/insulator interfaces (structures) on the atomic level. The ESR tool may thus come as a truly atomic level technique complementary to the state-of-the-art morphological/imaging methods and electrical techniques on true atomic level. In technological perspective, an outstanding bonus of ESR is that this is directly attained via sensing of potentially electrically detrimental traps.

3.5. Stress analysis

As mentioned previously, the general consensus is that the P$_b$-type interface defects are naturally incorporated at the Si/SiO$_2$ interface to account for the stress resulting from the structural mismatch between the dissimilar c-Si and a-SiO$_2$ solids [58]. When nature attempts complete bond formation at the interface, the origin of the stress is attributed to a component of the volume expansion parallel to the oxidation interface (there is a 120% expansion in molar volume when oxidizing Si to SiO$_2$). In other words, the difference in molar volume of both solids confronted at the Si/SiO$_2$ interface, i.e., the difference in bond length (the Si–Si interatomic spacing in c-Si and c-SiO$_2$ is ~2.352 and 3.06 Å, respectively), angles, and bond density gives rise to intrinsic interface strain. The main stress in the Si is compressive (negative) and lateral. Nature builds in these interface defects as a way to handle this stress.

The stress–defect density relationship has been explicitly established by ESR in previous work on (111)Si/SiO$_2$, revealing a close linear correlation between [P$_b$] and the average stress in the oxide film [58]. One might then speculate that tensile or compressive stressing of the Si substrate during oxidation may affect P$_b$-type defect formation and, in correlation, the interface quality. This conjecture has been explored in an ESR experiment on (111)Si, studying the effect on the inherent density of P$_b$ defects of in situ externally-applied in-plane compressive or tensile stress applied through wafer bending during thermal oxidation of (111)Si substrates. Although (100)Si/SiO$_2$ is the technologically favored structure, the (111)Si surface has been chosen for experimental reasons: The (111)Si/SiO$_2$ interface exhibits only one type of P$_b$ center, thus maximizing the chance on successful detection of changes in the Si/SiO$_2$ interface properties induced by the externally applied deformation force. The (100)Si/SiO$_2$ interface exhibits two types of interface defects, P$_{b0}$ and P$_{b1}$, and here the anticipated rather limited changes in interface properties induced by the external applied mechanical stress might be hard to quantify reliably due to the reduced inherent defect density and overlap of the two ESR spectra.

For this study, strips of 10×28 mm^2 were cut from two side polished B-doped (111)Si wafers ($\sim$75 μm thick). These samples were loaded in an in situ arranged apparatus, made of fused quartz, designed to bend the strips as beams in a four-point bending set-up during oxidation. Over the central portion of the sample ($\sim$12 mm in length), the bending moment is constant with a magnitude of $M = aP$, with $a = 8$ mm the spacing between the two points of load application on either side of the sample and where P is the applied load. Six samples were studied, with estimated imposed oxide film stresses of ± 25, ± 33.3, ± 35 MPa (see Ref. [147] for more details on the four-point bending method). An unloaded control sample was included in every test (labeled as 0 MPa). After oxidation at 800°C (dry O$_2$; 1 h) to an oxide thickness of $\sim$128 Å, the central portions of the samples, that were subjected to constant bending, were studied by conventional K-band ESR after being subjected to a 1-h 600°C anneal in vacuum to maximally ESR activate all P$_b$ defects through H-dissociation [53,67]. As a result of the externally applied moment during oxidation, the upper Si surface is under compression, while the lower one is under tension of equal magnitude. According to selection, one of the two Si/SiO$_2$ interfaces was used for ESR analysis and was protectively covered while the oxide on the other side was removed in a CP4-type etch.

As probed by the P$_b$ centers, the in situ applied mechanical stress was indeed found to affect the grown interface, as perhaps most impressively demonstrated by the measured [P$_b$]-applied stress relationship shown in Fig. 13. Clearly, the interface defect density has changed notably due to the externally applied stress. For compressive stress, [P$_b$]

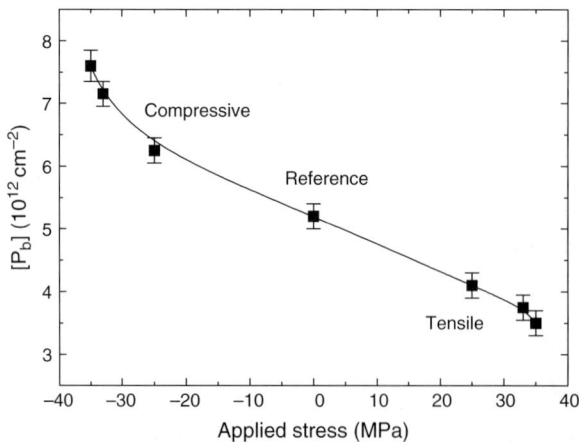

Fig. 13. Inherent P$_b$ defect density in thermal (111)Si/SiO$_2$ structures as a function of in situ applied mechanical stress during thermal oxidation. The curve guides the eye.

increases significantly as indicated by the solid line, while for tensile stress $[P_b]$ decreases, although somewhat smaller, relative to the measured expected value of $\sim 5\times 10^{12}$ cm^{-2} for the reference sample [58,69]. Concomitant changes due to the externally applied stress were also observed in other ESR properties of the P_b defect such as line width, line shape, and g-value. We refer the interested reader to Ref. [148] for more information and concentrate here on the observed change in defect density.

Here, we may suffice by analyzing somewhat deeper the effect induced by the applied stress on the P_b density. It was pointed out by Mihalyi et al. [147] and Fitch et al. [149] that the average strain (of the order of 10^{-7}) in the Si substrate is negligibly small. The weak strain is largest in the top near-interfacial Si layers, decaying further into the substrate. In these interfacial Si layers, the externally applied stress has its most significant impact and changes the stress from virtually zero from the unloaded condition to ± 56 MPa for an applied oxide stress of ± 35 MPa resulting in a lateral strain of 4×10^{-4} near the interfacial Si layers. Thus by applying tensile (compressive) bending stress, the Si–Si interatomic distance is enlarged (decreased). Comparing this with the drastic volume expansion accompanying SiO$_2$ growth, it would point to the interfacial atomic mismatch as the determining factor for P_b defect generation. This deduction is very much in keeping with the general notion that P_b centers are mismatch induced defects.

The above study addressed axial stress applied mechanically (wafer bending), which may appear a somewhat exotic approach. There are of course different means to apply strain to the Si substrate top layers. A well-adopted approach is starting from strained Si thin films, which may be obtained through pseudomorphically growing Si onto a lattice of larger unit cell dimensions, such as strain relaxed Si$_{1-x}$Ge$_x$ ($x=0.1$–0.4) layers used as virtual substrates. It results in an in-plane biaxial strain, rather than the axial strain realized in the above mechanical case. Being a well-developed method, the application of strained-Si (sSi) layers and stress spacers in Si is currently widely applied in CMOS integrated circuitry because of impacting benefits, including observed enhancement of device channel carrier mobility with increasing Si substrate tensile strain [150,151] as well as reduction in $1/f$ noise [152]. Actually, sSi is preferred above other high mobility substrates such as Ge and GaAs because of its easy integration in the present production lines. The "quality" of the Si/SiO$_2$ interfaces thermally grown on such sSi layers has recently also been investigated by ESR [37], more precisely on (100)sSi/SiO$_2$ entities thermally grown at 800°C, with results largely in compliance with those from the above described ESR study of P_b defects in thermal (111)Si/SiO$_2$ structures that were in situ mechanically stressed during oxidation [148]. As compared to co-processed standard (100)Si/SiO$_2$, a significant reduction ($>50\%$) is observed in the inherent density of the trivalent P_b-type interface defects (P_{b0}, P_{b1}). With the P_{b0}s established as detrimental fast interface traps, this result may adduce one more reason for the observed enhancement of device channel carrier mobility with increasing Si substrate tensile strain as well as reduction in $1/f$ noise. More details can be found elsewhere [37].

4. Paramagnetic defects in dielectric layers

4.1. General concern

The SiO$_2$ layer thermally grown on Si (technologically still the dominant semiconductor) has played and still is playing a crucial role in current IC technology based on the MOSFET concept, where it serves as workhorse gate insulator [64]. This SiO$_2$, an insulator native to Si, appears to be of unprecedented quality, which holds an almost perfect set of electrical and manufacturability properties that has enabled device scaling to proceed at steady pace for about four decades. It has a major contribution to the success story of the Si-based IC technology. Generally, electrical requirements impose that a gate insulator should be of superb quality in many respects, such as high thermodynamic stability, wide bandgap to provide large barriers for Schottky emission, low impurity diffusion coefficients, and inertness [75]. Very crucially, in its electrical role of gate insulator, the dielectric should form high-quality, spatially smooth, and well-passivated interfaces with the semiconductor, i.e., with as low as possible a density of interface states (defects) and low charge trapping in the oxide, reduced to subcritical limits. Indeed, the requirements to be met by a device performing gate insulator are very demanding. The reason for as low as possible a density of interface states and trapped charges is that these will be detrimental to the current transport in adjacent crucial current carrying channels in the MOSFET and detune its switching response and cause device aging (drifting). Eliminating charge traps, including precursors that could over time be turned into such traps (e.g., by some form of damaging irradiation), is a basic strive in device development. Needless to say that identification of these traps to the atomic level has received much attention. This has been the case for the Si/SiO_2 system, but is not different for the numerous semiconductor/insulator structures newly conceived with the intention to expand capabilities and explore new horizons in the device world. It constitutes a large field of research in the current high-κ issue (see Refs [73,74] and Section 3.2.1).

Here, we shall restrict ourselves to briefly discussing, by way of illustration, one specific case, i.e., the origin of positive charge (trapping) in SiO_2 layers on Si. It may serve as a typical example illustrating what it takes and how nontrivial it may appear to solve such at-first-sight-simple problem to some acceptable degree of scientific certainty, and how the road to the solution may be ridden with various obstacles and pitfalls.

4.2. Positive charge in SiO_2: combined ESR-electrical analysis

As amply addressed above, point defects are at the origin of virtually any detrimental electrical aspect of Si/SiO_2-based devices. Besides the interface states, which have been addressed in detail above, this very critically also concerns oxide charge trapping. It is a major concern in the semiconductor/insulator-based device world. As to the SiO_2 dielectric, indeed, oxide charging may pose problems, in particular positive charging. It may be present in the as-grown state, as the *oxide fixed charge* residing in near-interfacial layers [153]. Or there may occur positive charging of thermal SiO_2 on Si upon irradiation which has been known for decades for its detrimental effect on Si-based electron devices [154]. Identification and localization of the positive charge traps has generated intense and sophisticated research efforts [155].

Initial intense research had concluded the fundamental positive charging mechanism to be simple hole trapping by some oxide defects, i.e., a purely *electronic* process. But inherently, however sensitive these are, the electrical techniques fail when it comes to providing information on atomic scale of the trapping sites. For this, help of ESR was involved, and from initial electrical measurements [10,156] in conjunction with ESR probing, a correlation was inferred between induced positive charge and the generation of a paramagnetic center, the E′ defect, essentially the O vacancy in SiO_2 (vide infra).

The E′ center has been first observed in silica about half a century ago by R. Weeks [157] in a search for the origin of the coloration of shielding windows when exposed to γ-irradiation. Since then, numerous variants of the defect have been delineated in crystalline

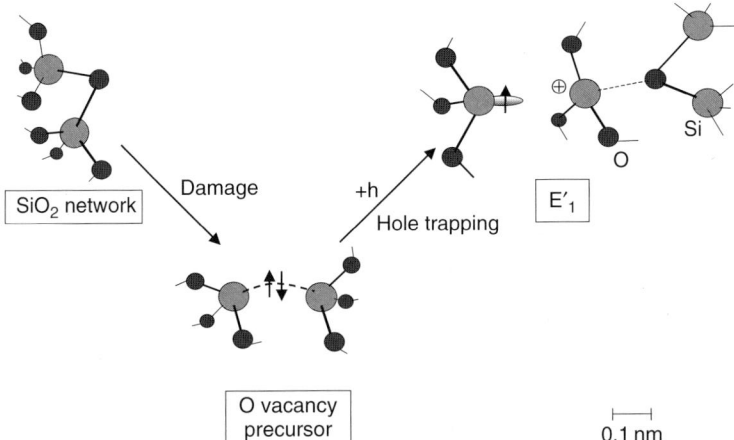

Fig. 14. Ball-and-stick picture of a formation scheme of the E'_1 (E'_γ) center and its current model as inferred from theory (after Refs [161,162]). As shown, the presumed precursor site pictures the defect as a hole trapped at a O-vacancy site.

SiO$_2$ (quartz) [158], fused silica, and all kind of glasses [17,159]; their properties have been compiled in various insightful overviews [17,158,159]. It is the most abundant and, without doubt, the most widely studied defect in SiO$_2$. It is a fundamental radiation-induced defect in amorphous silica (a-SiO$_2$). Among the various variants, we may mention: E'_1, and the H-associated E'_2 and E'_4 in α-quartz; E'_γ and E'_δ in a-SiO$_2$, where E'_γ is believed to be essentially identical to E'_1 [17]. Intense ESR studies supported by in-depth theoretical calculations and modeling have ascertained the basic generic part of the defect to be the paramagnetic O$_3\equiv$Si$^\bullet$ (the dot representing the unpaired electron) unit at the site of an O vacancy. More specifically, from theory, the E'_1 defect in α-quartz is ascribed to [17] the positively charged O$_3\equiv$Si$^\bullet$...$^+$Si$\equiv$O$_3$ unit (Fig. 14). [160]. In more detail, the current model pictures the É$_1$ center as involving a $\equiv$Si$^\bullet$...$^+$Si$\equiv$ pair around a missing O atom site in the SiO$_2$ matrix, where the unpaired electron containing Si is tetrahedral, while the positive charge bearing Si is removed from its tetrahedral counterpart, in fact puckered backwards through the plane of its three basal O atoms, indicated as back-bonded to a tree-fold coordinated O atom in the oxide matrix [161–164]. From this picture evolves the straightforwardly made inference as describing the E'_1 center as corresponding to a hole trapped at an oxygen vacancy, $V_0 + h^+ \rightarrow V_0^+$ (see Fig. 14). While the theoretical conclusion is strict for α-quartz, it should be added, in a correct objective scientific attitude, that so far the positive charge state of the E'_1 center has not been established by *direct* experimental measurement.

We now return to the origin of the positive charge in the a-SiO$_2$/c-Si entity: admittedly, the defect observed in thermal (amorphous) SiO$_2$ on Si is the E'_γ variant common for fused silica, suggesting the latter to be the hole trap with the unpaired electron facing a $^+$Si$\equiv$O$_3$ moiety. That assignment did match logically with the previously theoretically calculated E'_1 model [160–162,164] depicting the defect in α-quartz (crystalline) as the O$_3\equiv$Si$^\bullet$...$^+$Si$\equiv$O$_3$ unit when in the paramagnetic state. In apparent support, from later work it was concluded the E'_γ and E'_1 to be essentially identical [17], so it seemed a conclusive model was attained for the positive charge. It is important though to remark that the (puckered) O$_3\equiv$Si$^\bullet$...$^+$Si$\equiv$O$_3$ model was calculated for a crystalline form of SiO$_2$

(quartz), whereas thermal SiO_2 is in the amorphous state, and the crystalline model was tacitly carried over for the amorphous state. This need not necessarily to be valid as also hinted from more recent theoretical work [165], and may thus appear an unjustified link.

Indeed, that insight became beclouded as later on an increasing number of experiments concluded the origin of positive charge to be of different nature [126, 166–169], i.e., related to H. This ultimately found affirmation from a close correlative study [170] combining ESR with (electrical) measurement of VUV irradiation (10 eV) induced electrical charging under bias and probing motion of liberated atomic H through detection of changes in active dopant (B) density in the Si substrate (i.e., H-passivation [136]). The latter was done by monitoring the inversion capacitance C_{inv} in the studied p-type Si MOS structures. A principle set of results underlying this conclusion is shown in Fig. 15, comparing the variations of trapped charge density Q/q (panel a; q is the elemental charge) and E' density with relative changes in inversion capacitance δC_{inv} (panel b) during hole injection and later neutralization by electron injection of MOS structures prepared by evaporation of Al electrodes ($\sim$15 nm thick) on thermal p-type $(100)Si/SiO_2(60\,nm)$ grown at 970°C (type A; O) and after subsequent post-oxidation treatment at 970°C in Ar +10% H_2 or D_2 (type B; □); atomic H is detected through monitoring changes in the active dopant concentration in the B-doped Si substrate of the MOS capacitors [136]. This led to the conclusion that the origin of the positive charge is *protonic* in nature in which picture the E' centers still play a crucial role: for the main part, the E' precursor is seen as a

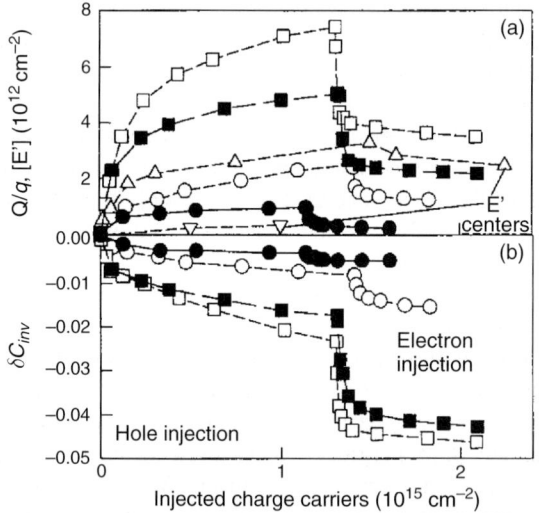

Fig. 15. Observed density of positively charged centers (a) and relative variation of inversion capacitance (b) as a function of injected electron/hole density in two types of MOS structures prepared by evaporation of Al electrodes ($\sim$15 nm thick) on thermal B-doped p-type $(100)Si/SiO_2(60$ nm) grown at 970°C (type A; O) and after subsequent post-oxidation treatment at 970°C in Ar +10% H_2 (type B, □). The filled symbols represent data obtained after subjecting the MOS samples to a post-metallization anneal in H_2 at 400°C (30 min). The charge densities Q/q (q is the elemental charge) were obtained from the shift of the flat band voltage (VFB) calculated from 1-MHz capacitance–voltage curves and the centroid of the trapped charge distribution. The E' center density $[E']$, inferred from K-band ESR observation at 4.3, is shown for samples A ($\triangledown$) and B ($\triangle$) in panel (a) (after Ref. [170]).

neutral diamagnetic $O_3{\equiv}Si$–H center which upon hole trapping releases a mobile proton (H^+; positive charge) leading to the formation of a paramagnetic $O_3{\equiv}Si^\bullet$ E' defect, as detected by ESR. (This is not in conflict with the previously attained conclusion that E'_γ and E'_1 are essentially identical; indeed, theoretical analysis [171] indicates that insofar the structural unit $O_3{\equiv}Si^\bullet$ is preserved as core of the defect, the main ESR quantities, such as hf parameters, appear practically unaffected by the charge state (positive or neutral) of the center.) A significant part of the protons become trapped in the SiO_2 network, resulting in trapped positive charge. Interestingly, the trapping of protons is found to occur with little sensitivity to the kind of SiO_2 growth technology [172], suggesting direct interaction (not mediated by any other defect) of the protons with the oxide network. So, the $O_3{\equiv}Si$–H precursors (O-deficiency centers; bonded H) would operate as crucial entities for subsequent potential positive charging as a result of some external influence (irradiation, current, etc.). The E' centers can be seen as constituting a H-storage system in the thermal amorphous silica layer.

Naturally, the presence of E' centers (O-vacancies) can be taken as a measure of oxide quality. Typically, in standard thermal thin SiO_2 films ($\leq$ few nm) grown on Si at $T \geq 800°C$, a density of $\sim 3 \times 10^{17}\,cm^{-3}$ is observed after VUV irradiation [110].

5. Summary and future perspectives

The work has addressed the application of conventional ESR spectroscopy to the study of SHs of point defects at the basis of detrimental charge traps and carrier recombination centers. While it has been applied most intensely quite successfully to the technologically far dominant Si/SiO_2-based entities, different SHs have been studied as well.

The method though does not figure among the arsenal of exquisite electrical charge sensing techniques routinely invoked in device characterization. Besides items such as experimental accessibility, this has two main reasons: (a) sensitivity and (b) the nonconventional sensing approach of ESR, that is, probing spin rather than charge. Thus, only paramagnetic species can be addressed, which need not necessarily be the common state of pertinent traps. If diamagnetic, this means that ESR has to be combined with charge modifying treatments (e.g., irradiation, electrical biasing) to turn defects into a paramagnetic state, which may not readily be attained. One more concern here may entail that such additional treatments might unacceptably affect the nature of a trap under study other than changing its charge state.

Nevertheless, there is little doubt the ESR tool will enjoy intensifying application to the study of SHs, if only for the fact the technique excels in providing the most direct identification of a point defect (trap) in a solid – an utmost aim in boosting technological progress based on fundamental insight. For one, in the current high-κ research, there are a lot of phenomena observed in high-κ materials, such as HfO_2 and ZrO_2, in which the occurrence of O vacancies is set down to play a key role [75]. For the sake of solid scientific progress, the direct convincing demonstration of the (numerous) presence of these and quantification by ESR would be a crucial step for insightful progress.

Without doubt, in this kind of ESR research a growing role of importance is ahead for the *next level* ESR-based techniques, combining conventional ESR with other spectroscopic techniques. Due to the drastically enhanced sensitivity and specificity, these techniques may enable one to probe deeper in properties of newly devised SH components. The techniques are crucial in the development of spin-based information processing (spintronics; quantum computing) and storage [173].

Typically, the major part of the probed defects turns out to be intrinsic, i.e., structural, not impurity-related imperfections, often inherently (naturally) incorporated in the formed SHs. Some may additionally be generated or ESR-activated through applying extra irradiation or charging treatment. Often, occurring defects are the main focus of interest in itself, i.e., tracing the atomic nature of defects at the origin of some detrimental electrical aspect of a MOSFET. In other cases, once isolated and/or identified, some types of defects may "beneficially" be employed as local probes to assess properties such as quality and stability of interfaces and thin (inter) layers. Thus, sensitive ESR analysis enables us to unveil elemental aspects (quality, nature, and stability) of evolving semiconductor/insulator interfaces (structures) on the atomic level. The tool complements the state-of-the-art morphological/imaging methods and electrical techniques on true atomic level.

References

[1] J. Bardeen and W. H. Brattain, Phys. Rev. 75 (1949) 1208.
[2] J. S. Kilby and E. Keonjian, Tech. Dig. IEDM 5 (1959) 76; J. S. Kilby, IEEE Trans. Electron Devices 23 (1976) 648.
[3] R. H. Dennard, F. H. Gaenssle, H. N. Yu et al., IEEE J. Solid State Circuits 9 (1974) 256.
[4] G. Moore, Electronics 38 (1965) 114; Tech. Dig. IEDM 21 (1975) 11.
[5] A. Chatterjee et al., Tech. Dig. IEDM (2004) 657.
[6] D. M. Brown and P. V. Gray, J. Electrochem. Soc. 115 (1968) 760; C.-T. Sah, J. Y.-C. Sun, and J. J.-T. Tsou, J. Appl. Phys. 55 (1984) 1525.
[7] See, e.g., H. Poindexter and P. J. Caplan, Prog. Surf. Sci. 14 (1983) 201, and references therein.
[8] B. E. Deal, B. Sklar, A. S. Grove, and E. H. Snow, J. Electrochem. Soc. 1214 (1967) 266.
[9] T. R. Oldham, F. B. McLean, H. E. Boesch, Jr, and J. M. McGarrity, Semicond. Sci. Technol. 4 (1989) 986.
[10] P. M. Lenahan and P. V. Dressendorfer, J. Appl. Phys. 55 (1984) 3495.
[11] B. E. Deal, J. Electrochem. Soc. 121 (1974) 198C; ibid. 127 (1980) 979; T. W. Hickmott, J. Appl. Phys. 48 (1977) 723.
[12] R. Degraeve, G. Groeseneken, R. Bellens et al., IEEE Trans. Electron Devices 45 (1998) 904.
[13] A. Stesmans and V. V. Afanas'ev, Appl. Phys. Lett. 69 (1996) 2056.
[14] G. J. Gerardi, E. H. Poindexter, P. J. Caplan, and N. M. Johnson, Appl. Phys. Lett. 49 (1986) 384.
[15] D. K. Schroder and J. A. Babcock, J. Appl. Phys. 94 (2003) 1.
[16] W. Gordy, Theory and Applications of Electron Spin Resonance, Wiley, New York, 1980.
[17] D. L. Griscom, in: D. R. Uhlmann and N. J. Kreidl (Eds), Glass: Science and Technology Vol 4B, Academic Press, New York, 1990, p. 151.
[18] C. J. Pickard and F. Mauri, Phys. Rev. Lett. 88 (2002) 086403-1.
[19] L. Giordano, P. V. Sushko, G. Pacchioni, and A. L. Shluger, Phys. Rev. B 75 (2007) 024109.
[20] J. A. Weil, J. R. Bolton, and J. E. Wertz, Electron Paramagnetic Resonance, Wiley, New York, 1994.
[21] A. Schweiger and G. Jeschke, Principles of Pulsed Electron Paramagnetic Resonance, Oxford University Press, Oxford, 2001.
[22] G. R. Eaton, S. S. Eaton, and K. M. Salikov (Eds), Foundation of Modern EPR, World Scientific, Singapore, 1998.
[23] J.-M. Spaeth, R. Niklas, and R. H. Batram, Structural Analysis of Point Defects in Solids, Springer-Verlag, Berlin, 1992; J.-M. Spaeth, H. Overhof, and H.-J. Queisser, Point Defects in Semiconductor and Insulators, Springer-Verlag, Berlin, 2003.
[24] C. P. Slichter, Principles of Magnetic Resonance, Springer-Verlag, Berlin, 1990.
[25] A. K. Koh and D. J. Miller, At. Data Nucl. Data Tables 33 (1985) 235.
[26] J. R. Morton and K. R. Preston, J. Magn. Res. 30 (1978) 577.

[27] C. G. Van de Walle and P. E. Blöchl, Phys. Rev. B 47 (1993) 4244.
[28] D. L. Lide (Ed.), CRC Handbook of Chemistry and Physics, 83rd edn, CRC Press, Boca Raton, FL, 2002, pp. 11–50.
[29] B. G. Segal, M. Kaplan, and G. K. Fraenkel, J. Chem. Phys. 43 (1965) 4191.
[30] P. J. Caplan, E. H. Poindexter, B. E. Deal, and R. R. Razouk, J. Appl. Phys. 50 (1979) 5847.
[31] R. Helms and E. Poindexter, Rep. Prog. Phys. 57 (1994) 791.
[32] K. L. Brower, Appl. Phys. Lett. 43 (1983) 1111.
[33] A. Stesmans, Appl. Phys. Lett. 48 (1986) 972.
[34] E. H. Poindexter, P. Caplan, B. Deal, and R. Razouk, J. Appl. Phys. 52 (1981) 879.
[35] A. Stesmans and V. V. Afanas'ev, J. Appl. Phys. 83 (1998) 2449.
[36] P. C. Taylor, in: J. I. Pankove (Ed.), Hydrogenated Amorphous Silicon, Semiconductors and Semimetals Vol 21C (academic, New York, 1984), p. 99, and references therein.
[37] A. Stesmans, P. Somers, V. V. Afanas'ev et al., Appl. Phys. Lett. 89 (2006) 152103-1.
[38] R. Orbach and M. J. Stapleton, in: S. Geschwind (Ed.), Electron Spin-Lattice Relaxation in Electron Spin Resonance, Plenum, New York, 1972.
[39] K. L. Brower, Phys. Rev. B 38 (1988) 9757.
[40] M. Stutzmann, M. S. Brandt, and M. W. Bayerl, J. Non-Cryst. Solids 266–269 (2000) 1.
[41] D. J. Lepine, Phys. Rev. B 6 (1972) 436.
[42] D. Kaplan, I. Solomon, and N. F. Mott, J. Phys. Lett. 39 (1978) L51.
[43] B. Henderson, M. Pepper, and R. V. Vranch, Semicond. Sci. Technol. 4 (1989) 1045.
[44] J. T. Krick, P. M. Lenahan, and G. J. Dunn, Appl. Phys. Lett. 59 (1991) 3437; J. P. Campbell, P. M. Lenahan, A. T. Krishnan, and S. Krishnan, IEEE 45th Intern. Reliability Physics Symp., Phoenix (2007) 503.
[45] B. Stich, S. Greulich-Weber, and J.-M. Spaeth, J. Appl. Phys. 77 (1995) 1546.
[46] M. Fanciulli, O. Costa, S. Baldovino et al., in: E. Gusev (Ed.), Defects in High-κ Gate Dielectric Stacks, Springer, Dordrecht, 2006, p. 263.
[47] M. S. Brandt, R. Neuberger, and M. Stutzmann, Appl. Phys. Lett. 76 (2000) 1467.
[48] A. Stesmans, Appl. Phys. Lett. 68 (1996) 2076; ibid. 68 (1996) 2723.
[49] D. L. Griscom, J. Appl. Phys. 58 (1985) 2524.
[50] A. H. Edwards, J. Non-Cryst. Solids 187 (1995) 232.
[51] A. Stesmans, J. Non-Cryst. Solids 179 (1994) 10.
[52] A. Stesmans and V. V. Afanas'ev, Phys. Rev. B 57 (1998) 10030.
[53] A. Stesmans, Phys. Rev. B 61 (2000) 8393.
[54] A. Pusel, U. Wetterauer, and P. Hess, Phys. Rev. Lett. 81 (1998) 645.
[55] Y. Nishi, Jpn. J. Appl. Phys. 10 (1971) 52.
[56] Y. Nishi, K. Tanaka, and A. Ahwada, Jpn. J. Appl. Phys. 11 (1972) 85.
[57] E. H. Poindexter, Semicond. Sci. Technol. 4 (1989) 961.
[58] A. Stesmans, Phys. Rev. B 48 (1993) 2418; Phys. Rev. Lett. 70 (1993) 1723.
[59] E. H. Poindexter, G. J. Gerardi, M.-E. Rueckel et al., J. Appl. Phys. 56 (1984) 2844.
[60] A. Stesmans and V. V. Afanas'ev, J. Phys. Condens. Matter 10 (1998) L19.
[61] P. K. Hurley, A. Stesmans, V. V. Afanas'ev et al., J. Appl. Phys. 93 (2003) 3971.
[62] A. Stesmans, B. Nouwen, and V. Afanas'ev, Phys. Rev. B 58 (1998) 15801.
[63] A. Stirling, A. Pasquarello, J.-C.Charlier, and R. Car, Phys. Rev. Lett 85 (2000) 2773.
[64] E. H. Poindexter, P. J. Caplan, and G. J. Gerardi, in: C. R. Helms and B. E. Deal (Eds), Physics and Chemistry of SiO_2 and the Si/SiO_2 Interface, Plenum, New York, 1988, p. 279.
[65] W. Futako, M. Nishikawa, T. Yasusa et al., in: S. Ohmi, F. Fujita, and H. S. Momose (Eds), International Workshop on Gate Insulator 2001 Jap. Soc. of Appl. Phys., Tokyo, 2001 p.130; W. Futako, N. Mizuochi, and S. Yamasaki, Phys. Rev. Lett. 92 (2004) 105505-1.
[66] K. L. Brower, Phys. Rev. B 42 (1990) 3444.
[67] A. Stesmans, J. Appl. Phys. 88 (2000) 489.
[68] A. Stesmans, J. Appl. Phys. 92 (2002) 1317.

[69] A. Stesmans and V. V. Afanas'ev, Phys. Rev. B 54 (1996) R11129.
[70] A. Stesmans and V. V. Afanas'ev, Appl. Phys. Lett. 72 (1998) 2271.
[71] M. L. Green, E. P. Gusev, R. Degraeve, and E. Garfunkel, J. Appl. Phys. 90 (2001) 2057.
[72] D. A. Buchanan and S. H. Lo, Microelectron. Eng. 36 (1997) 13.
[73] G. D. Wilk, R. M. Wallace, and J. M. Anthony, J. Appl. Phys. 89 (2001) 5243.
[74] M. Houssa and M. M. Heyns, in: M. Houssa (Ed.), High-κ Gate Dielectrics, Institute of Physics, Bristol, 2004, p. 3.
[75] J. Robertson, Eur. Phys. J. Appl. Phys. 28 (2004) 265.
[76] H. R. Huff and D. C. Gilmer (Eds), High Dielectric Constant Materials: VLSI MOSFET Applications, Springer Series in Advanced Microelectronics, (Springer-Verlag, Berlin, 2004).
[77] M. Houssa, L. Pantisano, L.-A. Ragnarsson et al., Mat. Sci. Eng. R 51 (2006) 37.
[78] H. Fukumoto, M. Morita, and Y. Osaka, J. Appl. Phys. 65, (1989) 5210; S. A. Campbell, D. C. Gilmer, X. C. Wang et al., IEEE Trans. Electron Devices 44 (1997) 104.
[79] M. T. Duffy and A. G. Revesz, J. Electrochem. Soc. 117, (1970) 372.
[80] E. P. Gusev, in: G. Pacchioni, L. Skuja, and D. L. Griscom (Eds), Defects in SiO_2 and Related Dielectrics: Science and Technology, NATO Science Series, Kluwer, Dordrecht, 2000, p. 557.
[81] M. A. Quevedo-Lopez, M. El-Bouanani, B. E. Gnade et al., J. Appl. Phys. 92 (2002) 3540, and references therein.
[82] H. Lee, L. Kang, R. Nieh et al., Appl. Phys. Lett. 76, (2000) 1926.
[83] M.-H. Cho, Y. S. Roh, C. N. Whang et al., Appl. Phys. Lett. 81, (2002) 472.
[84] M. Crivelli, S. Alessandri, F. Alberici et al., Mat. Res. Soc. Symp. Proc. Vol 745, (2003) 149.
[85] S. Van Elshocht, M. Caymax, S. De Gendt et al., Mat. Res. Soc. Symp. Proc. Vol 745, (Electrochem. Soc., Pennington, 2003) 197.
[86] M. Copel and M. C. Reuter, Appl. Phys. Lett. 83, (2003) 3398.
[87] G. D. Wilk and R. M. Wallace, Appl. Phys. Lett. 76, (2000) 11.
[88] H. Lee, L. Kang, R. Nieh et al., Appl. Phys. Lett. 76, (2000) 1926.
[89] M. Gutowski, J. E. Jaffe, C.-L. Liu et al., Appl. Phys. Lett. 80, (2002) 1897.
[90] M. R. Visokay, J. J. Chambers, A. L. Rotondaro et al., Appl. Phys. Lett. 80, (2002) 3183.
[91] Y. Morisaki, Y. Sugita, K. Irino, and T. Aoyama, in: S. Ohmi, F. Fujita, and H. S. Momose (Eds), International Workshop on Gate Insulators, Jap. Soc. of Appl. Phys., Tokyo, 2001, p. 184.
[92] V. V. Afanas'ev, A. Stesmans, F. Chen et al., Appl. Phys. Lett. 81, (2002), 1053.
[93] J. P. Senateur, C. Dubourdieu, F. Weiss et al., Adv. Mater. Opt. Electron. 10 (2000) 155.
[94] J. Lettieri, J. H. Haeni, and D. G. Schlom, J. Vac. Sci. Technol. A 20 (2002) 1332.
[95] S. C. Choi, M. H. Cho, S. W. Whanbo et al., Appl. Phys. Lett. 71 (1997) 903; R. A. McKee, F. J. Walker, and M. F. Chisholm, Phys. Rev. Lett. 81 (1998) 3014; H. J. Osten, J. P. Liu, H.-J. Müssig, and P. Zaumseil, Microelectron. Reliability 41 (2001) 991.
[96] E. G. Gusev, E. Cartier, D. A. Buchanan et al., Microelectron. Eng. 59 (2001) 341.
[97] M. Houssa, A. Stesmans, M. Naili, and M. M. Heyns, Appl. Phys. Lett. 77 (2000) 1381.
[98] J. R. Chavez, R. A. B. Devine, and L. Koltunski, J. Appl. Phys. 90 (2001) 4284.
[99] V. V. Afanas'ev and A. Stesmans, Appl. Phys. Lett. 80 (2002) 1261.
[100] A. Stesmans and V. V. Afanas'ev, J. Phys.: Condens. Matter 13 (2001) L673.
[101] A. Stesmans and V. V. Afanas'ev, Appl. Phys. Lett. 80 (2002) 1957.
[102] J. L. Cantin and H. J. von Bardeleben, J. Non-Cryst. Solids 303 (2002) 175.
[103] S. Baldovino, S. Nokrin, G. Scarel et al., J. Non-Cryst. Solids 322 (2003) 168.
[104] J. Jones and R. C. Barklie, Microelectron. Eng. 80 (2005) 74.
[105] A. Stesmans and V. V. Afanas'ev, Appl. Phys. Lett. 82 (2003) 4074.
[106] Y. Kang, P. M. Lenahan, J. F. Conley, Jr, and R. Solanski, Appl. Phys. Lett. 81 (2002) 1128.
[107] Y. Kang, P. M. Lenahan, and J. F. Conley, Jr, Appl. Phys. Lett. 83 (2003) 3407.
[108] A. Stesmans, V. Afanas'ev, F. Chen, and S. A. Campbell, Appl. Phys. Lett. 84 (2004) 4574.
[109] V. Lowalekar and S. Raghavan, J. Non-Cryst. Solids 351 (2005) 1559.
[110] A. Stesmans and V. V. Afanas'ev, Appl. Phys. Lett. 85 (2004) 3792; J. Appl. Phys. 97 (2004) 033510.

[111] H. J. von Bardeleben, J. L. Cantin, J. J. Ganem, and I. Trimaille, in: E. Gusev (Ed.), Defects in High-κ Gate Dielectric Stacks, Springer, Dordrecht, 2006, p. 249.
[112] E. P. Gusev, M. Copel, E. Cartier et al., Appl. Phys. Lett. 76 (2000) 176.
[113] K. L. Brower, Phys. Rev. 33 (4471) 1986.
[114] G. Lucovsky, Y. Wu, N. Niimi et al., App. Phys. Lett. 74 (1999) 2005.
[115] M. Houssa, N. Naili, C. Zhao et al., Semicond. Sci. Technol. 16 (2001) 31.
[116] A. J. Craven, M. MacKenzie, D. W. McComb, and F. T. Docherty, Microelectron. Eng. 80 (2005) 90.
[117] B. Crivelli, M. Alessandri, S. Alberici et al., Mat. Res. Soc. Symp. Proc. Vol. 745 (Electrochem. Soc., Pennington, 2003) 149.
[118] J. R. Chavez, R. A. B. Devine, and L. Koltunski, J. Appl. Phys. 90 (2001) 4284.
[119] V. V. Afanas'ev and A. Stesmans, Appl. Phys. Lett. 80 (2002) 1261.
[120] R. J. Carter, E. Carier, M. Caymax et al. in: S. Ohmi, F. Fujita, and H. S. Momose (Eds), International Workshop on Gate Insulators, Jap. Soc. of Appl. Phys., Tokyo, 2001, p. 24.
[121] W. Tsai, R. J. Carter, H. Nohira et al., Microelectron. Eng. 65 (2003) 259.
[122] B. E. Park and H. Ishiwara, Appl. Phys. Lett. 79 (2001) 806; 82 (2003) 1197.
[123] R. A. B. Devine and J. Arndt, Phys. Rev. 35 (1987) 9376.
[124] A. Stesmans, D. Pierreux, R. J. Jaccodine et al., Appl. Phys. Lett. 82 (2003) 3038.
[125] W. L. Warren, D. M. Fleetwood, M. R. Shaneyfelt et al., Appl. Phys. Lett. 63 (1993) 3330.
[126] D. Hervé, J. L. Leray, and R. A. B. Devine, J. Appl. Phys. 72 (1992) 3634.
[127] F. Gerson and W. Huber, Electron Spin Resonance Spectroscopy of Organic Radicals, Wiley, Weinheim, 2003.
[128] J. Chadi, Phys. Rev. B 64 (2001) 195403.
[129] J. DiMaria, E. Cartier, and D. Arnold, J. Appl. Phys. 73 (1993) 3367.
[130] G. Revesz, IEEE. Trans. Nucl. Sci. 24 (1977) 2102; F. B. McLean, IEEE Trans. Nucl. Sci. 27 (1980) 1651.
[131] N. Saks and R. W. Rendell, IEEE Trans. Nucl. Sci. 39 (1992) 2220.
[132] V. V. Afanas'ev, J. M. M. de Nijs, and P. Balk, Appl. Phys. Lett. 66 (1995) 1738.
[133] A. Rivera, A. van Veen, H. Shut et al., Solid State Electron. 46 (2002) 1775.
[134] J. H. Stathis and E. Cartier, Phys. Rev. Lett. 72 (1994) 2745; E. Cartier and J. H. Stathis, Appl. Phys. Lett. 69 (1996) 103.
[135] G. Lucovsky, H. Y. Yang, Z. Jing, and J. L. Witten, Phys. Stat. Sol. (a) 159 (1997) 5.
[136] J. I. Pankove, D. E. Carlson, J. E. Berkeyheiser, and R. O. Wance, Phys. Rev. Lett. 51 (1983) 2224.
[137] K. Murakami, H. Suhara, A. Fujita, and K. Masuda, Phys. Rev. B 44 (1991) 3409.
[138] R. A. Street, Physica B 170 (1991) 69.
[139] W. L. Warren, J. R. Schwank, M. R. Shaneyfelt et al., Appl. Phys. Lett. 62 (1993) 1661.
[140] J. L. Cantin, H. J. von Bardeleben, Y. Ke et al., Appl. Phys. Lett. 88 (2006) 092108.
[141] B. J. O'Sullivan, P. K. Hurley, E. O'Connor et al., J. Electrochem. Soc. 151 (2004) G493.
[142] J. E. Shelby, J. Appl. Phys. 48 (1977) 3387.
[143] A. H. Edwards, Phys. Rev. B 44 (1991) 1832.
[144] L. F. Edge, D. G. Schlom, R. T. Brewer et al., Appl. Phys. Lett. 84 (2004) 4629.
[145] A. Stesmans, K. Clémer, V. V. Afanas'ev et al., Appl. Phys. Lett. 89 (2006) 112121.
[146] W. Futako, T. Umeda, M. Nishizawa et al., J. Non-Cryst. Solids, 299–302 (2002) 575.
[147] A. Mihalyi, R. J. Jaccodine, and T. J. Delph, Appl. Phys. Lett. 74 (1999) 1981.
[148] A. Stesmans, D. Pierreux, R. J. Jaccodine et al., Appl. Phys. Lett. 82 (2003) 3038.
[149] J. T. Fitch, C. H. Bjorkman, G. Lucovsky et al., J. Vac. Sci. Technol. B 7 (1989) 775.
[150] J. Welser, J. L. Hoyt, S. Takagi, and J. F. Gibbons, International Electron Device Meeting Digest 1994, IEEE, New York, 1994, p. 373.
[151] C. W. Leitz, M. T. Curie, M. L. Lee et al., J. Appl. Phys. 92 (2002) 3745.
[152] E. Simoen, G. Eneman, P. Verheyen et al., IEEE Trans. Electron Devices 53 (2006) 1039.
[153] Y. C. Cheng, Prog. Surf. Sci. 8 (1977) 181; V. V. Afanas'ev and A. Stesmans, Phys. Rev. Lett. 80 (1998) 5176.

[154] C. T. Sah, Fundamentals of Solid State Electronics, World Scientific, Singapore, 1991.
[155] D. M. Fleetwood, P. S. Winokur, J. Reber, and T. L. Meisenheimer, J. Appl. Phys. 73 (1993) 10; D. J. DiMaria, D. A. Buchanan, J. H. Stathis, and R. Stahlbush, J. Appl. Phys. 77 (1995) 2032; D. Bauza, S. Bayon, and O. Ghobar, ECS Transactions 6 (2007) 3.
[156] Y. Y. Kim and P. M. Lenahan, J. Appl. Phys. 64 (1988) 3551; H. S. Witham and P. M. Lenahan, IEEE Trans. Nucl. Sci. 54 (1987) 1147.
[157] R. A. Weeks, J. Appl. Phys. 27 (1956) 1376.
[158] J. A. Weil, Phys. Chem. Minerals 10 (1984) 149.
[159] R. A. Weeks, J. Non-Cryst. Solids 179 (1994) 1.
[160] F. J. Feigl, W. B. Fowler, and K. L. Yip, Solid State Commun. 14 (1974) 225.
[161] K. L. Yip and W. B. Fowler, Phys. Rev. B 11 (1975) 2327.
[162] J. K. Rudra and W. B. Fowler, Phys. Rev. B 35 (1987) 8223.
[163] K. S. Snyder and W. B. Fowler, Phys. Rev. B 48 (1993) 13238; M. Boero, A. Pasquarello, J. Sarnthein, and R. Car, Phys. Rev. Lett. 78 (1997) 887.
[164] For a recent theoretical review, see G. Pacchioni, in: G. Pacchioni, L. Skuja, and D. L. Griscom (Eds), Defects in SiO_2 and Related Dielectrics: Science and Technology, NATO Science Series, Kluwer, Dordrecht, 2000, p.161.
[165] T. Uchino, M. Takahashi, and T. Yoko, Phys. Rev. Lett. 86 (2001) 5522.
[166] L. P. Trombetta, G. J. Gerardi, D. J. DiMaria, and E. Tierney, J. Appl. Phys. 64 (1988) 2434.
[167] B. J. Mrstik, V. V. Afanas'ev, A. Stesmans et al., J. Appl. Phys. 85 (1999) 6577.
[168] W. L. Warren, E. H. Poindexter, M. Offenberg, and W. Müller-Warmuth, J. Electrochem. Soc. 139 (1992) 872.
[169] J. F. Conley, Jr, P. M. Lenahan, H. L. Evans et al., J. Appl. Phys. 85 (1999) 6577.
[170] V. V. Afanas'ev and A. Stesmans, Europhys. Lett. 53 (2001) 233; J. Phys, Condens. Matter 12 (2000) 2285.
[171] A. Stirling and A. Pasquarello, Phys. Rev. B 66 (2002) 245201.
[172] V.V Afanas'ev, F. Ciobaru, G. Pensl, and A. Stesmans, Solid State Electron. 46 (2002) 1815.
[173] M. Xiao, I. Martin, E. Yablonovitch, and H. W. Jiang, Nature 430 (2004) 435.

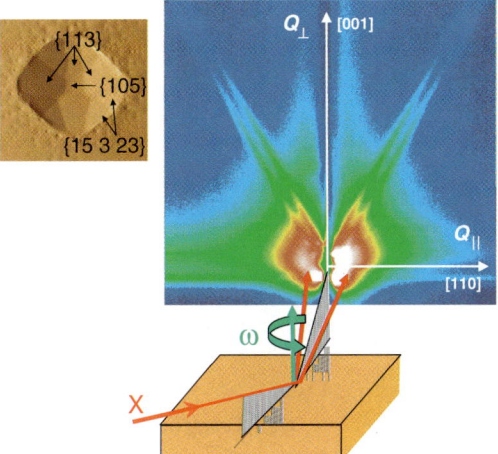

Plate 1. Two-dimensional GISAXS pattern performed on an assembly of Ge domes (see AFM image at the upper left), with {113}, {15 3 23} and {105} facets. These facets yield intensity rods of scattering perpendicular to them. (see Chapter 10, Fig. 5 on page 346)

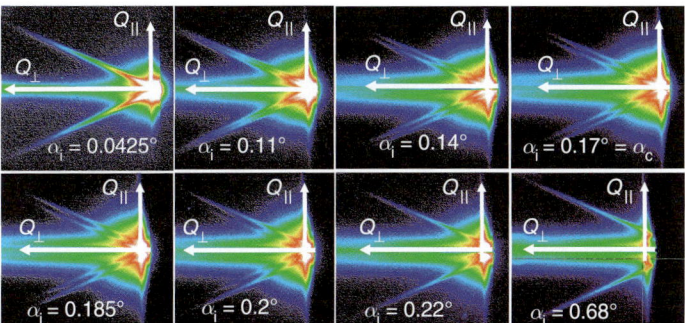

Plate 2. Series of GISAXS pattern, recorded on an assembly of large Ge domes grown on a Si(001) surface. The azimuth setting in <110> allows for the rods from {113} facets to be recorded. Note the splitting of the rods for incident angles close to the critical angle for total external reflection, $\alpha_c=0.17°$. (see Chapter 10, Fig. 7 on page 348)

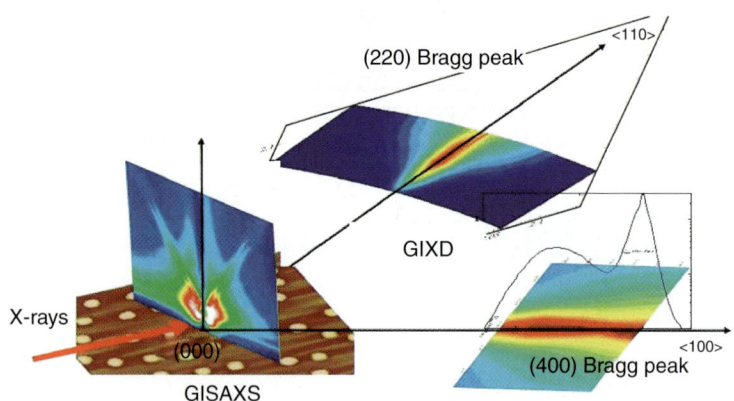

Plate 3. Schematic representation of the reciprocal space intensity distribution of Ge islands coherently grown on Si(001), with corresponding GISAXS and GID measurements, respectively around the origin (000) and around Bragg peaks (220) and (400). (see Chapter 10, Fig. 8 on page 349)

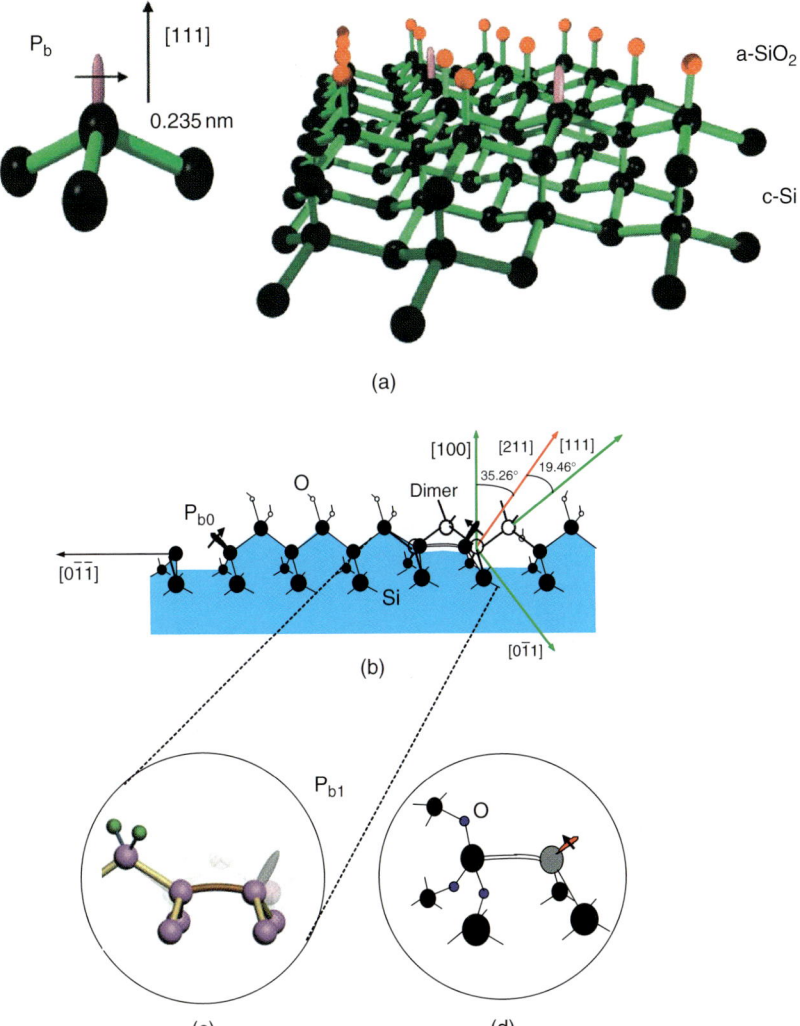

Plate 4. Steric pictures of the model structures for the P_b (panel a) and P_{b0}, P_{b1} (panel b) interface defects at the (111)Si/SiO$_2$ and (100)Si/SiO$_2$ interfaces, respectively. Inset (c) gives a close up of the P_{b1} defect model first proposed from experimental hf interaction results (Ref. [62]); inset (d) represents the model favored by theory (Ref. [63]). (see Chapter 13, Fig. 5 on page 453)

Subject Index

A

ab initio calculations 8, 17–51, 271, 285, 293–4, 377, 395
activation energy for defect passivation 462, 466
AlGaAs/GaAs 190, 196, 201–202, 275–7
AlGaAsN/GaAs 266
AlGaN 264
AlGaN/GaN 283–4
alloy 18, 23, 31, 37–42, 48–50, 97–8, 101, 108–110, 111, 129, 133, 138, 148–9, 155–6, 159, 163–6, 181, 190, 195, 197–9, 225–8, 249, 263–6, 268, 279–80, 289, 298–9, 307–309, 318, 319–24, 343, 353–4, 356, 358, 366, 380, 391, 396, 400–401
amorphous solids 176, 461
angle-resolved photoemission 417–19
angular momentum 294, 298, 304, 373, 438–42
annihilation operator 252, 373–5
anomalous scattering (or Anomalous X-ray diffraction) 37, 310, 331–67, 372–3, 375–82, 384–6, 401–402
anti-Stokes lines 250–1, 253–4, 261
Arrhenius plot (or equation) 68–9, 83, 203–204, 463
atomic absorption cross sections (of X-rays) 290–2, 294–5, 298–9, 302, 305, 318
atomic force microscopy (AFM) 127, 210, 279, 303, 333, 346, 354, 357–9, 388–9, 393, 395
atomic scattering factors (for electrons) 147–8, 158, 166–7
atomic scattering factors (for X-rays) 122, 127, 336, 338–9, 342–3, 352–6, 358, 361–2, 372, 375–7
axial spin Hamiltonian 446

B

band discontinuities (or band offset or band alignment) 17, 18–20, 29–30, 33–5, 37–9, 40, 42–3, 48–9, 176, 180–1, 220, 408–410, 414–17, 419, 420, 423–7, 428–30, 431–2
band edges 17–51, 63, 78, 84–5, 87–8, 111, 180, 183–5, 218, 223, 226, 276, 408–409, 414–17, 425–6, 428–30, 452
band-gap modulation 209–45
band mapping 418–19
Beer's law 191
Bloch functions (or Bloch states or Bloch waves) 44, 169–72, 183–4
Born–Oppenheimer approximation 26
bowing parameter for the energy gap 23, 40, 181
bowing parameter for the lattice parameter 109–110
Bragg diffraction 98–9, 135, 140–7, 337–42, 349–67, 388
Bragg's law 140
bright field (BF) image 135, 145–6
Brillouin zone (BZ) 22, 25, 44–5, 48, 111, 183–5, 250–1, 263, 272–3, 418–19

C

Car-Parrinello approach 23
carbon nanotubes 282–3
carrier dynamics 177, 179, 200–203, 226, 228, 240, 244–5
cathodoluminescence 83, 176, 209–45
CdS 419
CdS/Si 409–410, 415–17
charge neutrality levels 19–20
chemical-sensitive imaging 147–9, 160
CMOS (complementary metal oxide semiconductor) 436, 462, 473
coherent diffraction imaging (CDI) 331, 367
commutativity rule 35
composition evaluation 93–129, 133–72, 180–1, 190, 209, 225–6, 228–9, 260, 263–9, 307–308, 331–3, 341–4, 353–8, 371–3, 380, 397, 437, 469
composition graded heterostructures 113–14, 116, 117–20
composition measurement 108–110, 113–14, 116–20
core levels 20, 290, 292, 294, 414–17, 422–3, 428–9
creation operator 252, 373–5
critical angle (foe total external reflection of X-rays) 304, 336, 347–8, 354, 385, 398
critical diameter for GaN nanowires 80, 82, 87
critical thickness for strain relaxation 104, 111–12, 116–17, 122, 127, 307–308, 364–5
crystal truncation rods (CTR) 338–41, 349–50
C–V (capacitance voltage) measurements 62–5, 236, 436–7, 451–2, 456

D

DAFS (diffraction anomalous fine structure) 8, 299, 332, 342–4, 371–402
dark field (DF) image 135, 147–9, 161–2, 165–6, 168–9
Debye–Waller (DW) factor 296, 339, 343, 377–8, 382, 393, 395–6
deep-level activation energy 67
deep-level capture cross section 57, 66–7, 69, 78, 83
deep levels 56–7, 60, 64, 65–6, 67–9, 70, 71, 75, 78, 83, 87, 159, 223, 227, 303
defects 11, 18, 37–40, 43, 46, 51, 56
density functional theory (DFT) 18, 22–3, 26, 29–30, 183, 285, 321–2, 325, 438
density of states (DOS) 10, 28–30, 44–5, 49–50, 182, 186, 191, 219–20, 253, 262, 263–4, 276, 375, 408, 451–2
depletion region 60–4, 73, 81, 85–7
dielectric layers 450–1, 473–7
differential scattering cross section 215, 373–6
diffraction contrast (in TEM) 138–49, 157–8, 165–9
dipolar Hamiltonian 443
dipole approximation 291–4, 376–7
dipole–dipole interaction 443, 445
DLTS (deep level transient spectroscopy) 56–7, 65, 67–8, 69–71, 79, 83
dynamical electron diffraction theory 142–3, 149
dynamical X-ray diffraction theory 102, 104, 384

E

edge dislocation 145–7
effective charge 34
effective mass 86, 111, 183–8, 221, 276, 418
effective mass approximation (EMA) 183–8
electrical characterization 55–89, 176, 456

electrical measurements 55–89, 268, 437, 451–2, 459, 474, 476
electron beam writing 233, 234, 236, 240
electron energy loss spectroscopy (EELS, combined with TEM) 157–60
electron paramagnetic resonance (EPR), *see* Electron-spin resonance (ESR)
electron spectroscopy for chemical analysis (ESCA) 419–20
electron spin 17–51, 249–54, 276–9, 435–78
electron-spin resonance (ESR) 267, 276–9, 435–78
electron yield detection of XAFS 301–302
energy dispersive X-ray spectroscopy (EDX, combined with TEM) 157–60
energy gap 9–10, 22, 41, 44, 67, 70–5, 84–6, 111, 121, 127, 180–2, 188–203, 210, 217, 225–8, 232, 408
energy gap vs lattice parameter 181
escape depth (of a photoelectron) 411–13
ETS-10 quantum wire 8–11, 282
EXAFS (extended X-ray absorption fine structure) 8, 9, 11, 292, 295–8, 305, 311, 315–18, 320, 322–4, 372, 380–2, 388, 395–401
EXAFS function 292
excitons 10, 87, 163, 176, 187–91, 196, 210–11, 233, 243, 252, 275, 278, 282

F

Fermi-contact Hamiltonian 443
Fermi golden rule 187, 293
Fermi level 30, 39, 48–50, 60–1, 64–6, 71, 75, 80–1, 84–5, 219, 322–3
Feynman diagrams 252, 373
first-principle molecular dynamics simulation (FPMD) 23, 39–40, 110
first principles calculations 50, 267
fluorescence yield detection of XAFS 300–301
Franz–Keldysh effect 84–7

G

GaAlSb/GaAs 116–17
GaAlSb/GaSb 116–17
GaAs 65, 292
GaAs/AlAs 18, 25, 32, 40–3, 196–7, 273–5
GaAs homojunction 422–3
GaAsN 267, 319–21
GaAsN/GaAs 231–6
GaAsN:H/GaAs 231–6
GaMnAs 268–9, 321–4
GaMnN 264
GaN 78, 79–88, 147, 163–5, 215, 229, 240, 254–61, 265–8, 270–2, 280–1, 306, 312, 314, 317–19, 322–4, 333, 350–2, 359–61, 363–6, 364–5, 373, 383–6, 387, 388–97, 398
GaN/Al$_2$O$_3$ 147, 317–19
GaN/GaAlN 271–2
GaN nanowires 56, 79–88, 281
GaN/Si 317–19
GaSe/Ge 429–30
Ge/CdS 413–14, 428
Ge/GaAs 34–5
Ge/As/Ga/Ge and Ge/Ga/As/Ge (double intralayer homojunctions) 428–9

grazing incidence DAFS 371–402
grazing incidence diffraction (GID) 127, 331–67
grazing incidence (GI) of X-rays 127, 304–305, 331–67, 371–402
grazing incidence small angle X-ray scattering (GISAXS) 127, 331–67

H

Hamiltonian 19, 31, 154, 170, 184, 276, 374, 439–43, 446
heterojunction 17–51, 407–32
heterostructures 4–7, 8, 17–18, 19, 21, 23, 24–6, 27, 28–31, 32, 34, 35–6, 37–9, 40–3, 47–50, 55, 70, 93–129, 133–72, 175–205, 209–45, 249, 263–9, 272–82, 289–326, 332, 333–4, 352, 366, 372, 435–78
heterovalent heterostructure (or heterojunction or interface) 21, 24, 31, 34–5, 37, 38–9, 41–2, 43–5
Heusler compound 48
high pressure Raman experiments 260–3
high-resolution electron microscopy 133–72
homojunction 412–13, 422–3, 428–9
Howie–Whelan equations 142–3
hydrogen 436–7, 441, 453–4, 459, 460–4, 468

I

InAs 377–80
InAs/InP 309
InAs quantum wires in InP 359, 362–4, 397–401
in-depth resolution 222, 239, 283
index of refraction of X-rays 304, 332, 335
InGaAs 110
InGaAs/GaAs 96, 107–108, 117–20, 160–2
InGaAs/InAlAs 40–1
InGaAs/InP 309, 431–2
InGaAsN/GaAs 165–9
InGaN 265
InGaN/GaN 163–5, 228–31, 236–9
InGaP/GaAs 148
injection power 223, 227, 230, 239
InP 270
InP/Ge 414–15, 420–1
integrated circuit (IC) 436, 455, 462, 473–4
interface dipole 19–20, 26–7, 33, 35, 38, 42, 428
interface parameter (in semiconductor heterojunction) 407, 425–30
interface (between semiconductors) 17–51, 56, 59–63, 88, 95–123, 161–9, 175–205, 220–44, 240–3, 256–64, 271–5, 388–401, 407–32, 435–74
interface states 18, 43–6, 228, 271, 459–60, 461–2, 474
interlayers 41–3, 47, 237, 365, 456–7, 459, 460, 469–71
internal fields screening 209–45
intrinsic reference levels 19
isovalent heterostructure (or heterojunction or interface) 21, 26–7, 30–1, 32–5, 37–8, 40–1, 43, 45–6, 49
I–V (current voltage) measurements 57–62, 236, 436–7, 456

K

kinematical approximation of the electron diffraction theory 138–43, 148–9
kinematical approximation of the X-ray diffraction theory 122–4, 337, 385, 387–8, 393

Subject Index

L

lattice match/mismatch 20–1, 26–7, 31, 32–7, 38, 40, 43, 48, 93, 94–8, 103, 116, 122, 129, 155, 176, 225, 308–309, 318, 342, 350, 362, 384, 400, 451–2, 461
lattice parameter 27, 93–129, 163, 166, 181, 225
linear combination of atomic orbitals (LCAO) 443–5
linear response theory (LRT) 21, 31, 33–5, 38–9, 41–3, 44
local atomic structure 307, 318, 319, 325
local density approximation (LDA) 22, 29–30, 31, 32–4, 36, 40–1, 42–3, 48
local density of states (LDOS) 30, 44–5
longitudinal acoustic (LA) phonons 194, 200, 249–85
longitudinal optic (LO) phonons 194, 249–85

M

macroscopic average 25–6, 29, 32–3, 44
MAD (multiwavelength anomalous diffraction) 332, 342–3, 344, 359–66, 371–2, 377–80, 384, 388–9, 392, 395, 397
magnetic dipole transitions 440
magneto-Raman scattering 279–82
majority carriers 55–89
microscopy 133–72, 259, 265, 283, 285
micro-XAFS 306
minority carriers 55–89
models 9–11, 19–21, 30, 39–40, 42, 47, 57, 58, 80–1, 83, 86–7, 110, 111–12, 113, 114–16, 117–19, 121, 123–4, 138, 150, 160, 195, 198–200, 203, 220–1, 222, 224, 229, 231, 233, 240, 250, 264, 275–6, 296, 298, 308–309, 316–17, 334, 342–3, 351, 356, 358, 362–3, 366–7, 379–81, 392–4, 400, 410, 420–1, 424–7, 430, 438, 453, 456, 459, 463–7, 475–6
model-solid theory (or Van de Walle and Martin theory) 20–1, 39
momentum transfer (between incident and scattered radiation) or momentum conservation 99–100, 251–2, 255, 260, 272, 274–5, 310, 334–7, 341–2, 345, 347, 350, 353–4, 356, 373, 417–19
Monte Carlo simulation (MCS) 215–16, 240–5
MOS (metal oxide semiconductor) 454–6, 476
MOSFET (metal oxide semiconductor field effect transistor) 473–4, 478
Mott cross section 215
multiple scattering (MS) processes in photoelectron path 295, 297, 381–2, 399
multiple scattering (MS) theory of EXAFS 293–5
multiple scattering (MS) of X-rays 384–6

N

nanoscience 2–3, 4, 6, 332
nanostructures 4–6, 7–8, 24–6, 55–89, 122, 123, 127, 129, 175–205, 210, 223, 225, 227, 230, 231, 280, 282, 289–326, 331–67, 371–402
nanotechnology 1–11, 285, 332, 387
nanowires 82, 83–6, 87–8, 222, 281
   see also Quantum wire (QWr)
near-band-edge (NBE) transitions 212, 227, 232, 239
NEXAFS (near edge X-ray absorption fine structure).
   see XANES
NiMnSb/GaAs 48–50
nitrides 79, 98, 110, 209–45, 265, 283, 298–9, 317–21
nonradiative recombination 203–204

O

optical capture cross section 78

P

phase contrast (in TEM) 134, 141, 149–57
phonon 62, 192, 194, 198, 200, 218, 221, 224, 243, 249–85, 450
photoelectric absorption cross section, see Atomic absorption cross sections (of X-rays)
photoelectron 8, 180, 291–2, 294, 296–7, 298, 306, 310, 381–2, 387–8, 408–409, 410–12, 413, 417–19, 420–2, 429
photoelectron emission microscope (PEEM) 421–3
photoemission 18, 43, 362, 363, 365, 407–32
photoemission spectromicroscopy 421–3, 430
photoinduced spectroscopy 69–73, 407–32
photoluminescence (PL) 8–9, 110, 175–205, 210, 262, 311, 316–17
planar average 25–6
point defects 213, 219, 221, 227, 232, 236–7, 240, 267, 319, 333, 436–8, 444–5, 448–9, 450–73, 474, 477
Poisson ratio 96–7, 102, 111, 114, 163, 166
porous Si 316–17
post oxidation annealing (POA) 453–4, 462
Power-dependent cathodoluminescence 209–45
pseudopotentials 18, 20, 22–3, 29, 31, 40–1, 183
pump and probe Raman experiments 260–1
PWscf (plane-wave self-consistent field) package 23, 285

Q

quantum confined Stark effect (QCSE) 194, 223, 226, 228
quantum dot (QD) 4, 9, 121, 126, 147, 156, 176, 182, 220–1, 260, 273, 279–82, 310–12, 314, 334, 350–2, 359, 361, 363–6, 373
quantum dots: CdS dots in SiO_2 312–13
quantum dots: CdSe dots in BeTe 271
quantum dots: CdZnSe dots in ZnSe 280
quantum dots: GaAs dots in AlAs 280–1
quantum dots: GaN dots in AlN 284–5, 350–2, 359–66, 386–7, 388–97
quantum dots: GaN dots in SiO_2 312–13
quantum dots: GaSb dots in GaAs 279
quantum dots: Ge dots in Si 272–3, 310–11, 346–50, 354–9
quantum dots: Ge dots in SiO_2 312–13
quantum dots: InAs dots in GaAs 127–9, 147
quantum dots: InAs dots in InGaAs 120–2
quantum dots: SiGe dots in Si 124–5, 354–9
quantum dots: Sn dots in SiO_2 314–15
Quantum-ESPRESSO package 23
quantum theory of Raman scattering 252–4
quantum wells (QWs) 4, 7, 42–3, 70, 103, 134, 160–2, 163, 187, 193–204, 219, 249, 253, 263, 272–6, 277, 279
quantum wire (QWr) 4, 8–11, 182, 220–1, 279–82, 311, 359, 362–3, 373, 397–401
   see also Nanowires

R

Raman selection rules 254–8
Raman spectroscopy 8, 110, 121, 249–85, 311

random-walk modeling 240, 241–5, 351
rapid thermal annealing (RTA) 452, 471
reciprocal lattice maps (or reciprocal space mapping, RSM) 106–108, 120–1, 125–6, 128, 342–4, 361–4
reciprocal lattice vector 99, 139–40, 339–40
reciprocal space (RS) 22, 24, 27, 99–100, 139–40, 292, 337, 372, 379–80
refractive index of X-rays 231, 235, 335–6, 377
 see also Index of refraction
resonant Raman scattering (RRS) 253, 265, 273
resonant scattering, see Anomalous scattering (or Anomalous X-ray diffraction)
Rutherford cross section 215

S

scanning near-field optical microscopy (SNOM or NSOM) 259–60
scanning tunneling microscopy (STM) 6, 18, 28–30, 46–7, 124, 127, 260, 333
SCF (self-consistent field) 22–43
Schottky (barrier, contact, junction, diode) 19–20, 49, 57–8, 60–2, 67, 70, 412, 420, 428–30
semiconductor nanostructures 8, 175–205, 282, 285, 310, 367, 371, 372, 387–401
$Si/Al_2O_3/ZrO_2$ 455–9
SiC 268–70
Si/GaAs 46–7
Si:H 267–8
Si/HfO_2 455–6, 459–60, 469, 477
$Si/LaAlO_3$ 455–6, 460, 469–71, 469
single scattering (SS) processes in photoelectron path 292, 295–7, 381–2, 399
SiO_2 435–77
$SiO_2/Cs/Si$ 427
SiO_2/Si 427, 436–8, 447–8, 451–78
small angle X-ray scattering (SAXS) 127, 331–67
space charge region (SCR), see Depletion region
SPC (spectral photoconductivity) 56, 71–2, 77–9, 84–8
spectral photocurrent 71
spin Hamiltonian 276, 439–42
spin–orbit interaction (or spin–orbit factor) 33, 36–7, 40, 187, 276, 374, 440–2
spintronics 18, 47–50, 263, 321–2, 450, 477
spontaneous polarization (SP) 225
SPS (surface photovoltage spectroscopy) 71–2, 73, 75–6, 84–7
Stokes lines 250–1, 253–4
Stokes shift in (PL spectra) 197–200
strain 21, 27, 34, 35–7, 38–9, 40–1, 93–129, 149, 155–7, 160–2, 164, 167, 176, 179, 225–7, 231, 263–70, 276–7, 282–3, 307–310, 311, 313, 318, 332–4, 338, 341–4, 350–2, 354–9, 360–7, 372–3, 380, 384, 387–95, 396–7, 400–402, 451–2, 457, 459, 465, 471, 473
strain measurement 94–129
strain release models 111–13
stress 22, 27, 33, 37–8, 44, 48, 94–5, 96, 97, 101, 111–12, 124, 161, 168–9, 229, 236–7, 238–40, 263, 284, 293, 303, 342, 362, 372–3, 374–5, 436–7, 451, 459, 461, 464–5, 468, 471–3

structure factor 338–40, 343–4, 360–6, 377–80, 385–8, 390, 397
supercell 20, 25, 27, 29–30, 31–5, 39, 41–3, 44, 47, 48–50
surface enhanced Raman scattering (SERS) 259, 285
surface EXAFS (SEXAFS) 305
surface photovoltage 56, 71–2, 73–8, 79, 84
surface-sensitive techniques 331–67, 407–32
synchrotron radiation 290, 293, 299, 305–306, 325, 331–67, 410, 421

T

TEM simulation 160–72
Tersoff model 114–16
tetragonal distortion (or deformation) of the unit cell 27, 94–5, 101, 106, 155, 165–9, 307–310, 400–401
theory of elasticity 27, 95–8, 109, 124, 145, 155–6, 318, 392–3
thermal annealing 111, 113, 118, 236, 267–8, 312, 315, 320, 323, 325, 397, 452–4, 458–9, 462, 464, 467–71
thermal silica 455, 461, 464, 467, 471, 474–5, 477
thermal transient spectroscopy 65–71
tight-binding hamiltonian 19
TiO_2 quantum wires in SiO_2 8–11, 282
tip-enhanced Raman scattering (TERS) 259, 285
total external reflection of X-rays 304, 332, 336–7, 348
total scattering cross section 215, 377
transition dipole moment 189
transitivity rule 35, 426
transmission detection of XAFS 299–300
transmission electron microscopy (TEM) 8–9, 11, 18, 30, 46–8, 104, 117–18, 124, 127, 133–72, 190, 194, 210, 214, 228–9, 280–1, 314–15, 333, 360, 363, 365, 393, 395
transport mechanisms 56–7, 58–9, 60–2, 65–6, 76, 79, 83
transverse acoustic (TA) phonons 249–85
transverse optical (TO) phonons 249–85

V

Vegard's law 23, 108–110, 156, 307, 342, 391, 396
vibrational spectroscopy 249–50, 251–2, 254–5, 264, 266–7, 269, 271–2, 282, 285, 296, 310, 437
virtual crystal approximation (VCA) 23, 40, 42, 109

X

X-ray absorption fine structure (XAFS) 8, 289–326
X-ray absorption near-edge structure (XANES) 292–3, 298–9, 315–25
X-ray atomic absorption edges 159, 290–2, 315, 319, 321, 333, 342–3, 360–1, 371, 378, 380, 387, 389, 398
X-ray characterization of heterostructures 94, 98–108
X-ray characterization of quantum dot Heterostructures 94, 121, 122–4
X-ray reflectivity (XRR) 127, 304–305, 334–5, 348, 385

Z

Zeeman effect (or Zeeman splitting) 279–82, 439–43, 450
Zeeman splitting 276–7, 439–43
zeolites 5
ZnSe 278–9
ZnSe/Ge 44–6, 411–12